Trigonometric Functions

1. $\sin \theta = \dfrac{y}{r}$

2. $\cos \theta = \dfrac{x}{r}$

3. $\tan \theta = \dfrac{y}{x}$

4. $\cot \theta = \dfrac{x}{y}$

5. $\sec \theta = \dfrac{r}{x}$

6. $\csc \theta = \dfrac{r}{y}$

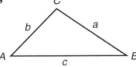

$r = \sqrt{x^2 + y^2}$

Geometric Formulas

Area

Triangle: $A = \dfrac{1}{2}bh$

Rectangle: $A = lw$

Parallelogram: $A = bh$

Trapezoid: $A = \dfrac{1}{2}h(a + b)$

Circle: $A = \pi r^2$; (Circumference: $C = \pi d = 2\pi r$)

Volume

Prism: $V = Bh$

Cylinder: $V = \pi r^2 h$

Pyramid: $V = \dfrac{1}{3} Bh$

Cone: $V = \dfrac{1}{3} \pi r^2 h$

Sphere: $V = \dfrac{4}{3} \pi r^3$

Oblique Triangles

Law of Sines: $\dfrac{a}{\sin A} = \dfrac{b}{\sin B} = \dfrac{c}{\sin C}$

Law of Cosines: $a^2 = b^2 + c^2 - 2bc \cos A$

$b^2 = a^2 + c^2 - 2ac \cos B$

$c^2 = a^2 + b^2 - 2ab \cos C$

Complex Numbers

$j = \sqrt{-1},\, j^2 = -1,\, j^3 = -j,\, j^4 = 1,\, j^5 = j, \ldots$

Rectangular form *Trigonometric form* *Exponential form*

$a + bj \quad = r(\cos \theta + j \sin \theta) = re^{j\theta}$

$r_1(\cos \theta_1 + j \sin \theta_1) \cdot r_2(\cos \theta_2 + j \sin \theta_2) = r_1 r_2 [\cos (\theta_1 + \theta_2) + j \sin (\theta_1 + \theta_2)]$

$\dfrac{r_1(\cos \theta_1 + j \sin \theta_1)}{r_2(\cos \theta_2 + j \sin \theta_2)} = \dfrac{r_1}{r_2}[\cos (\theta_1 - \theta_2) + j \sin (\theta_1 - \theta_2)]$

DeMoivre's Theorem

$[r(\cos \theta + j \sin \theta)]^n = r^n(\cos n\theta + j \sin n\theta)$

Technical Calculus

Technical Calculus

FOURTH EDITION

Dale Ewen
Parkland Community College

Joan S. Gary
Parkland Community College

James E. Trefzger
Parkland Community College

Prentice
Hall

Upper Saddle River, New Jersey
Columbus, Ohio

Library of Congress Cataloging-in-Publication Data

Ewen, Dale
 Technical calculus.—4th ed. / Dale Ewen, Joan S. Gary, James E. Trefzger.
   p. cm.
 Rev. ed. of: Technical calculus / Dale Ewin . . . [et al.]. 3rd ed. c1998.
 ISBN 0-13-093004-0
 1. Calculus. I. Gary, Joan S. II. Trefzger, James E. III. Technical calculus. IV. Title.

QA303 .E95 2002
515—dc21

2001032332

Editor in Chief: Stephen Helba
Executive Editor: Frank I. Mortimer, Jr.
Editorial Assistant: Barbara Rosenberg
Media Development Editor: Michelle Churma
Production Editor: Louise N. Sette
Production Supervision: **TECH**BOOKS
Design Coordinator: Robin G. Chukes
Cover Designer: Mark Shumaker
Cover photo: © FPG
Production Manager: Brian Fox
Marketing Manager: Tim Peyton

This book was set in Times Roman by **TECH**BOOKS. It was printed and bound by Courier
Westford, Inc. The cover was printed by Phoenix Color Corp.

Pearson Education Ltd., *London*
Pearson Education Australia Pty. Limited, *Sydney*
Pearson Education Singapore Pte. Ltd.
Pearson Education North Asia Ltd., *Hong Kong*
Pearson Education Canada, Ltd., *Toronto*
Pearson Educación de Mexico, S.A. de C.V.
Pearson Education—Japan, *Tokyo*
Pearson Education Malaysia Pte. Ltd.
Pearson Education, *Upper Saddle River, New Jersey*

10 9 8 7 6 5 4 3 2 1
ISBN 0-13-093004-0

Preface

Technical Calculus, Fourth Edition, provides the calculus skills for students in an engineering technology program that requires a development of practical calculus. This edition has been carefully reviewed, and special efforts have been taken to emphasize clarity and accuracy of presentation.

The text presents the following major areas: analytic geometry, differential calculus, integral calculus, partial derivatives, double integrals, series, and differential equations.

Key Features

- Numerous detailed, illustrated examples.
- Chapter review summaries.
- Chapter review exercises.
- Important formulas and principles are highlighted.
- Abundant two-color illustrations.
- Two-color format that effectively highlights and illustrates important principles.
- Instruction using a basic graphing calculator (Appendix C) and an advanced graphing calculator (Appendix D) is developed in the appendices with calculator examples integrated throughout the text. A graphing calculator may be used as a faculty option.
- Chapter introduction and chapter objectives.
- More than 3400 exercises.
- An Instructor's Manual with solutions for odd-numbered exercises, answers to even-numbered exercises, and answers to chapter tests.

Illustration of Some Key Features

Examples   Since many students learn by example, a large number of detailed and well-illustrated examples are used throughout the text. Page 261 illustrates this feature.

Page 342 illustrates the use of an advanced graphing calculator to evaluate a definite integral as an alternative to using a trigonometric substitution to integrate. Page 477 illustrates the use of an advanced graphing calculator to solve a nonhomogeneous differential equation. Each graphing calculator feature can easily be omitted without loss of continuity.

Illustrations and Boxes   Page 293 is an example of the abundant and effective use of illustrations and boxes to highlight important principles.

Chapter End Matter   A chapter summary and a chapter review are provided at the end of each chapter to review concept understanding and to help students review for quizzes and examinations.

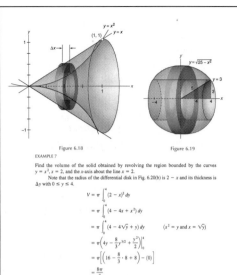

Figure 6.18   Figure 6.19

EXAMPLE 7

Find the volume of the solid obtained by revolving the region bounded by the curves $y = x^2$, $x = 2$, and the x-axis about the line $x = 2$.

Note that the radius of the differential disk in Fig. 6.20(b) is $2 - x$ and its thickness is Δy with $0 \le y \le 4$.

$$V = \pi \int_0^4 (2 - x)^2 \, dy$$

$$= \pi \int_0^4 (4 - 4x + x^2) \, dy$$

$$= \pi \int_0^4 (4 - 4\sqrt{y} + y) \, dy \qquad (x^2 = y \text{ and } x = \sqrt{y})$$

$$= \pi \left(4y - \frac{8}{3} y^{3/2} + \frac{y^2}{2} \right) \Big|_0^4$$

$$= \pi \left[\left(16 - \frac{8}{3} \cdot 8 + 8 \right) - (0) \right]$$

$$= \frac{8\pi}{3}$$

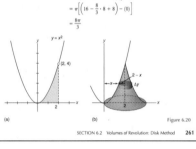

(a)   (b)   Figure 6.20

$$= \frac{1}{50} \left(-\frac{1}{\sin \theta} \right) + C$$

$$= -\frac{1}{50} \csc \theta + C \qquad \left(\csc \theta = \frac{2x}{\sqrt{4x^2 - 25}} \text{ from Fig. 7.5} \right)$$

$$= \frac{-x}{25\sqrt{4x^2 - 25}} + C$$

EXAMPLE 4

Evaluate $\displaystyle\int_0^4 \frac{dx}{(\sqrt{x^2 + 9})^3}$ (see Fig. 7.6).

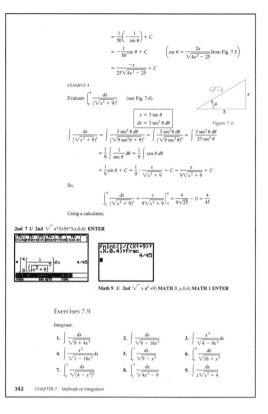

Figure 7.6

$$x = 3 \tan \theta$$
$$dx = 3 \sec^2 \theta \, d\theta$$

$$\int \frac{dx}{(\sqrt{x^2 + 9})^3} = \int \frac{3 \sec^2 \theta \, d\theta}{(\sqrt{9 \tan^2 \theta + 9})^3} = \int \frac{3 \sec^2 \theta \, d\theta}{(\sqrt{9 \sec^2 \theta})^3} = \int \frac{3 \sec^2 \theta \, d\theta}{27 \sec^3 \theta}$$

$$= \frac{1}{9} \int \frac{1}{\sec \theta} \, d\theta = \frac{1}{9} \int \cos \theta \, d\theta$$

$$= \frac{1}{9} \sin \theta + C = \frac{1}{9} \cdot \frac{x}{\sqrt{x^2 + 9}} + C = \frac{x}{9\sqrt{x^2 + 9}} + C$$

So,

$$\int_0^4 \frac{dx}{(\sqrt{x^2 + 9})^3} = \frac{x}{9\sqrt{x^2 + 9}} \Big|_0^4 = \frac{4}{9\sqrt{25}} - 0 = \frac{4}{45}$$

Using a calculator,

2nd 7 1/ 2nd $\sqrt{\ }$ x^2+9)^3,x,0,4) ENTER

Math 9 1/ 2nd $\sqrt{\ }$ x x² +9) MATH 3 ,x,0,4) MATH 1 ENTER

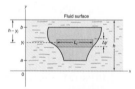

Exercises 7.9

Integrate.

1. $\displaystyle\int \frac{dx}{\sqrt{9 + 4x^2}}$

2. $\displaystyle\int \frac{dx}{\sqrt{9 - 16x^2}}$

3. $\displaystyle\int \frac{x^2}{\sqrt{4 - 9x^2}} \, dx$

4. $\displaystyle\int \frac{x^2}{\sqrt{1 - 16x^2}} \, dx$

5. $\displaystyle\int_0^2 \frac{dx}{\sqrt{9 - x^2}}$

6. $\displaystyle\int_0^2 \frac{dx}{\sqrt{16 + x^2}}$

7. $\displaystyle\int_0^1 \frac{dx}{\sqrt{(4 - x^2)^3}}$

8. $\displaystyle\int_2^3 \frac{dx}{\sqrt{4x^2 - 9}}$

9. $\displaystyle\int \frac{dx}{x\sqrt{x^2 + 4}}$

$$2C - 2B - 3A = 0 \qquad \text{(constants)} \qquad \textbf{(1)}$$
$$-4C - 3B \qquad = 0 \qquad \text{(coefficients of } x\text{)} \qquad \textbf{(2)}$$
$$-3C \qquad\qquad = 1 \qquad \text{(coefficients of } x^2\text{)} \qquad \textbf{(3)}$$
$$5D \qquad\qquad = 1 \qquad \text{(coefficients of } e^{-2x}\text{)} \qquad \textbf{(4)}$$

Solving this system of four equations, we find that

$$D = \frac{1}{5} \qquad C = -\frac{1}{3} \qquad B = \frac{4}{9} \qquad \text{and} \qquad A = -\frac{14}{27}$$

The particular solution is then

$$y_p = -\frac{14}{27} + \frac{4}{9} x - \frac{1}{3} x^2 + \frac{1}{5} e^{-2x}$$

GENERAL SOLUTION OF NONHOMOGENEOUS DIFFERENTIAL EQUATION

To find the general solution of the nonhomogeneous equation

$$a_0 D^2 y + a_1 Dy + a_2 y = b:$$

1. Find the complementary solution y_c by solving the homogeneous equation $a_0 D^2 y + a_1 Dy + a_2 y = 0$ using the methods developed in Sections 12.1 and 12.2.
2. Find a particular solution y_p by using the method of undetermined coefficients described in Examples 2 and 3.
3. Find the general solution y by adding the complementary solution y_c from step 1 and the particular solution y_p from Step 2:

$$y = y_c + y_p$$

EXAMPLE 4

Find the general solution of

$$y'' - 2y' - 3y = e^x$$

Step 1: Find the complementary solution y_c, which is the solution of the homogeneous equation $y'' - 2y' - 3y = 0$. We solved this equation in Example 2, Section 12.1:

$$y_c = k_1 e^{3x} + k_2 e^{-x}$$

Step 2: Find a particular solution y_p, which we obtained in Example 2 of this section:

$$y_p = -\frac{1}{4} e^x$$

Step 3: The desired general solution is

$$y = y_c + y_p$$

$$y = k_1 e^{3x} + k_2 e^{-x} - \frac{1}{4} e^x$$

Using a calculator,

F3 alpha C y 2nd = 2nd = -2y 2nd = -3y= green diamond e^x x),x,y) ENTER

For example, the total force on the bottom of a rectangular swimming pool 12 ft × 30 ft when the water is 8 ft deep is

$$F = (62.4 \text{ lb/ft}^3)(8 \text{ ft})(12 \text{ ft} \times 30 \text{ ft})$$

$$F = 180,000 \text{ lb (approx.)}$$

The more difficult problem is finding the total force against the *vertical sides* of a container because the pressure is not constant. The pressure increases as the depth increases.

Let a vertical plane region be submerged into a fluid of constant density ρ as shown in Fig. 6.51. We need to find the total force against this region from depth $h - a$ to $h - b$. First, divide the interval $a \le y \le b$ into n rectangles each of width Δy. The ith rectangle has length L_i, area $L_i \, \Delta y$, and depth $h - y_i$. The force on the ith rectangle is

$$\Delta F_i = \rho g (h - y_i) L_i \, \Delta y$$

Figure 6.51   Finding the force exerted by a fluid against a submerged vertical plane.

Summing the forces on all such rectangles gives the following integral.

FORCE EXERTED BY A FLUID

The force F exerted by a fluid of constant mass density ρ against a submerged vertical plane region from $y = a$ to $y = b$ is given by

$$F = \rho g \int_a^b (h - y) L \, dy$$

where h is the total depth of the fluid and L is the horizontal length of the region at y.

Note: In the metric system, the mass density, ρ, must be known and $g = 9.80 \text{ m/s}^2$. In the English system, the weight density, ρg, must be known.

EXAMPLE 7

A vertical gate in a dam is in the shape of an isosceles trapezoid 12 ft across the top and 8 ft across the bottom, with a height of 10 ft. Find the total force against the gate if the water surface is at the top of the gate.

The solution can be simplified if we position the trapezoid in the plane as shown in Fig. 6.52. The equation of the line through $(4, 0)$ and $(6, 10)$ is

$$y - 0 = 5(x - 4)$$

$$x = \frac{y + 20}{5}$$

To the Faculty

The topics have been arranged with the assistance of faculty who teach in a variety of technical programs. However, we have also allowed for many other compatible arrangements. The topics are presented in an intuitive manner with technical applications integrated throughout whenever possible. The large number of detailed examples and exercises are features that students and faculty alike find essential.

The students are assumed to have a mathematics background of algebra and trigonometry. This text is intended for use in Associate Degree programs as well as ABET (Accrediting Board for Engineering Technology) programs and BIT (Bachelor of Industrial Technology) programs. The companion text, *Mathematics for Technical Education,* serves as a smooth transition to this book, although other equivalent texts are also feasible.

Chapter 1 provides the basic analytic geometry needed for a study of a practical calculus. Chapters 2 through 4 present intuitive discussions about the limit and develop basic techniques and applications of differentiation. Chapters 5 through 7 develop basic integration concepts, some appropriate applications, and more complicated methods of integration. Chapter 8 presents partial derivatives and double integrals. Chapters 9 and 10 provide a basic understanding of progressions and series. Chapters 11 and 12 provide an introduction to differential equations with technical applications.

To the Student

Mathematics provides the essential framework for and is the basic language of all the technologies. With this basic understanding of mathematics, you will be able to quickly understand your chosen field of study and then be able to independently pursue your own life-long education. Without this basic understanding, you will likely struggle and often feel frustrated not only in your mathematics and support sciences courses but also in your technical courses.

Technology and the world of work will continue to change rapidly. Your own working career will likely change several times during your working lifetime. Mathematical, problem-solving, and critical-thinking skills will be crucial as opportunities develop in your own career path in a rapidly changing world.

Acknowledgments

The authors especially thank the many faculty and students who have used the previous editions and those who have offered suggestions. If anyone wishes to correspond with us regarding suggestions, criticisms, questions, or errors, please contact Dale Ewen directly through Prentice Hall or e-mail the authors at MathComments@aol.com.

We extend our sincere and special thanks to our reviewers: Joe Jordan, John Tyler Community College (VA); Maureen Kelly, North Essex Community College (MA); Carol A. McVey, Florence-Darlington Technical College (SC); John D. Meese, DeVry Institute of Technology (OH); Kenneth G. Merkel, Ph.D., PE, University of Nebraska-Lincoln; Susan L. Miertschin, University of Houston; and Pat Velicky, Florence-Darlington Technical College (SC). We would also like to extend thanks to our Prentice Hall editor—Stephen Helba, to our media development editor—Michelle Churma, to our production editor—Louise Sette, Wendy Druck at **TECH**BOOKS, and to Joyce Ewen for her superb proofing assistance.

Dale Ewen
Joan S. Gary
James E. Trefzger

Contents

1
Analytic Geometry

INTRODUCTION

Analytic geometry is the study of the relationships between algebra and geometry. We will study equations with a dependent and an independent variable and sketch a corresponding figure in two dimensions. We begin with linear equations (straight lines) and then study the conic sections, i.e., parabolas, circles, ellipses, and hyperbolas. These curves occur often in nature and play an important role in applied mathematics.

For example, the existence of the focus of a parabola is what makes flashlights, microphones, and satellite dishes work. Planets orbit the sun in elliptical paths, and circular wheels and gears help to keep us mobile. Analytic geometry, which we use to study these curves, evolved from the work of René Descartes, a French mathematician in the seventeenth century.

Objectives

- Determine if a relation is a function.
- Use functional notation.
- Graph equations.
- Find the slope of a line.
- Write the equation of a line given defining properties.
- Determine if lines are parallel, perpendicular, or neither.
- Use the distance formula.
- Use the midpoint formula.
- Graph circles, ellipses, parabolas, and hyperbolas.
- Find the center and radius of a circle.
- Find the vertex, directrix, and focus of a parabola.
- Find the vertices, foci, and lengths of the major and minor axes of an ellipse.
- Find the vertices, foci, and lengths of the transverse and conjugate axes of a hyperbola.
- Use translation of axes in sketching graphs and identifying key features.
- Solve systems of quadratic equations.
- Graph using polar coordinates.
- Convert between polar and rectangular coordinates.

1.1 FUNCTIONS

The **positive integers** are the counting numbers; that is, 1, 2, 3, ... Note that the positive integers form an infinite set. The **negative integers** may be defined as the set of opposites of the positive integers; that is, -1, -2, -3, ... **Zero** is the dividing point between the positive integers and the negative integers and is neither positive nor negative. The set of **integers** consists of the positive integers, the negative integers, and zero. The set may be represented on a number line as in Fig. 1.1.

Figure 1.1   Number line.

The **rational numbers** are those numbers that can be represented as the ratio of two integers, such as $\dfrac{3}{4}$, $\dfrac{-7}{5}$, and $\dfrac{5}{1}$. The **irrational numbers** are those numbers that cannot be represented as the ratio of two integers, such as $\sqrt{3}$, $\sqrt[3]{16}$, and π.

The set of **real numbers** is the set consisting of the rational numbers and the irrational numbers. With respect to the number line, we say there is a one-to-one correspondence between the real numbers and the points on the number line; that is, for each real number there is a corresponding point on the number line, and for each point on the number line there is a corresponding real number. As a result, we say the number line is dense, or "filled."

Inequalities are statements involving less than or greater than and may be used to describe various intervals on the number line, as follows:

Type of interval	Symbols	Meaning	Number line graph
Open	$x > a$	x is greater than a	
	$x < b$	x is less than b	
	$a < x < b$	x is between a and b	
Half-open	$x \geq a$	x is greater than or equal to a	
	$x \leq b$	x is less than or equal to b	
	$a < x \leq b$	x is between a and b, including b but excluding a	
	$a \leq x < b$	x is between a and b, including a but excluding b	
Closed	$a \leq x \leq b$	x is between a and b, including both a and b	

Analytic geometry is the study of the relationships between algebra and geometry. The concepts of analytic geometry provide us with ways of algebraically analyzing a geometrical problem. Likewise, with these concepts we can often solve an algebraic problem by viewing it geometrically. We will develop several basic relations between equations and their graphs.

In common usage, a relation means that two or more things have something in common. We say that a brother and a sister are related because they have the same parents or that a person's career potential is related to his or her education and work experience. In mathematics a **relation** is defined as a set of ordered pairs of numbers in the form (x, y). Sometimes an equation, a rule, a data chart, or some other type of description is given that states the relationship between x and y. In an ordered pair the first element or variable, called the **independent variable,** may be represented by any letter, but x is normally used. The second element or variable is normally represented by the letter y and is called the **dependent variable** because its value depends on the particular choice of the independent variable.

All of the numbers that can be used as the first element of an ordered pair or as replacements for the independent variable of a given relation form a set of numbers called the **domain.** The domain is often referred to as the set of all x's. We can think of these x-values as "inputs." The **range** of a relation is the set of numbers that can be used as the second element of an ordered pair or as replacements for the dependent variable. The range is often referred to as the set of all y's. We can think of these y-values as "outputs."

EXAMPLE 1

Given the relation described in ordered pair form $A = \{(1, 2), (3, 5), (7, 9), (6, 3)\}$, find its domain and its range.

The domain is the set of first elements: $\{1, 3, 6, 7\}$. The range is the set of second elements: $\{2, 3, 5, 9\}$.

Note: Braces $\{\ \}$ are normally used to group elements of sets.

EXAMPLE 2

Given the relation in equation form $y = x^2$, find its domain and its range.

The domain is the set of possible replacements for the independent variable x. Note that there are no restrictions on the numbers that you may substitute for x. That is, we may replace x by any real number. We say that the domain is the set of real numbers.

After each replacement of x, there is no possible way that we can obtain a negative value for y because the square of any real number is always positive or zero. Thus, the range is the set of nonnegative real numbers, or $y \geq 0$.

EXAMPLE 3

Find the domain and the range of the relation $y = \sqrt{x - 4}$.

Note that no value of x less than 4 may be used because the square root of any negative number is not a real number. Thus, the domain is the set of real numbers greater than or equal to 4, or $x \geq 4$.

After each possible x-replacement, the square root of the resulting value is never negative, so the range is $y \geq 0$.

FUNCTION

A **function** is a special relation: a set of ordered pairs in which no two distinct ordered pairs have the same first element.

In equation form, a relation is a function when for each possible value of the first or independent variable, there is only one corresponding value of the second or dependent variable. In brief, for a relation to be a function, each value of x must correspond to one, and only one, value of y.

EXAMPLE 4

Is the relation $B = \{(3, 2), (6, 7), (5, 3), (1, 1), (3, 7)\}$ a function? Find its domain and its range.

B is not a function because it contains two different ordered pairs that have the same first element: $(3, 2)$ and $(3, 7)$. In other words, the fact that both 2 and 7 correspond to 3 causes the relation B not to be a function. The domain of B is $\{1, 3, 5, 6\}$. The range of B is $\{1, 2, 3, 7\}$.

Does the set $A = \{(1, 2), (3, 5), (7, 9), (6, 3)\}$ from Example 1 describe a function? Yes, because no two ordered pairs have the same first element.

EXAMPLE 5

Is the relation $x = y^2$ a function? Find its domain and its range.

Can we find two ordered pairs that have the same first element? Yes, for example, $[(9, 3)$ and $(9, -3)$ as well as $(16, 4)$ and $(16, -4)]$ and many others. Therefore, $x = y^2$ is not a function because for at least one x-value, there corresponds more than one y-value.

To find the domain, note that each x-value is the square of a real number and can never be negative. Thus, the domain is $x \geq 0$.

There are no restrictions on replacements for y; therefore, the range is the set of all real numbers.

Consider the relations in Examples 2 and 3. Are they functions? Note that in the relation $y = x^2$, for each value of x there is only one corresponding value of y. For example, $(2, 4)$, $(-2, 4)$, $(3, 9)$, $(-3, 9)$, $(4, 16)$, $(-4, 16)$, and so forth. Therefore, $y = x^2$ is a function.

Example 3 was the relation $y = \sqrt{x - 4}$. Here we find that for each x-value there corresponds only one y-value [for example, $(5, 1)$, $(8, 2)$, $(10, \sqrt{6})$, . . .]. Therefore, $y = \sqrt{x - 4}$ is a function.

In summary, a function is a relationship between two sets of numbers, the domain and the range, that relates each number, x, in the domain to one and only one number, y, in the range.

Next let's consider the following, more intuitive function: On a summer vacation trip, you are driving 65 mi/h using cruise control on an interstate highway. You want to relate the distance and the time you are traveling. First, we know that distance equals rate times time, or $d = rt$. Also, since $r = 65$, we have the relation $d = 65t$. As you drive along, you begin to think how far you can drive in one hour:

$$d = 65t = 65(1) = 65 \text{ mi}$$

How far can you drive in three hours?

$$d = 65t = 65(3) = 195 \text{ mi}$$

Is this relation a function? Yes, because for each value of t, there is one and only one value of d; that is, during each driving time period, there is one and only one distance traveled. What are the domain and the range? First, note that $t \geq 0$ and $d \geq 0$. While there are no theoretical upper limits on t and d, the practical limits depend on the amount of time and the distance that you want to travel.

Functional Notation

To say that y is a function of x means that for each value of x from the domain of the function, we can find exactly one value of y from the range. This statement is said so often that we have developed the following notation, called **functional notation,** to write that y is a function of x:

$$y = f(x)$$

with $f(x)$ read "f of x." *Note: $f(x)$ does **not** mean f times x.*

In each of the following equations, y can be replaced by $f(x)$, and the resulting equation is written in functional notation.

Equation	Functional notation form
$y = 3x - 4$	$f(x) = 3x - 4$
$y = 5x^2 - 8x + 7$	$f(x) = 5x^2 - 8x + 7$
$y = \sqrt{6 - 2x}$	$f(x) = \sqrt{6 - 2x}$

Functional notation can be used to simplify statements. For example, find the value of $y = 3x^2 + 5x - 6$ for $x = 2$. Using substitution, we replace x with 2 as follows:

$$y = 3x^2 + 5x - 6$$
$$y = 3(2)^2 + 5(2) - 6 = 16$$

The statement "Find the value of $y = 3x^2 + 5x - 6$ for $x = 2$" may be abbreviated using functional notation as follows:

$$\text{Given } f(x) = 3x^2 + 5x - 6, \text{ find } f(2).$$

EXAMPLE 6

Given the function $f(x) = 5x - 4$, find each of the following:
(a) $f(0)$

Replace x with 0 as follows:

$$f(0) = 5(0) - 4 = 0 - 4 = -4$$

(b) $f(7)$

Replace x with 7 as follows:

$$f(7) = 5(7) - 4 = 35 - 4 = 31$$

A function is usually named by a specific letter, such as $f(x)$, where f names the function. Other letters, such as g in $g(x)$ and h in $h(x)$, are often used to represent or name functions.

EXAMPLE 7

Given the function $g(x) = \sqrt{x + 4} + 3x^2$, find each of the following:
(a) $g(5)$

Replace x with 5 as follows:

$$g(5) = \sqrt{5 + 4} + 3(5)^2 = 3 + 75 = 78$$

(b) $g(-3)$

Replace x with -3 as follows:

$$g(-3) = \sqrt{-3 + 4} + 3(-3)^2 = 1 + 27 = 28$$

(c) $g(-10)$

Replace x with -10 as follows:

$$g(-10) = \sqrt{-10 + 4} + 3(-10)^2 = \sqrt{-6} + 300$$

which is not a real number because $\sqrt{-6}$ is not a real number. Another way of responding to Part (c) is to say, "Since -10 is not in the domain of $g(x)$, $g(-10)$ has no real value."

Letters may also be used with functional notation as illustrated by the following example.

EXAMPLE 8

Given the function $f(x) = x^2 - 4x$, find each of the following:

(a) $f(a)$

Replace x with a as follows:

$$f(a) = a^2 - 4a$$

(b) $f(3c^2)$

Replace x with $3c^2$ as follows:

$$f(3c^2) = (3c^2)^2 - 4(3c^2) = 9c^4 - 12c^2$$

(c) $f(a + 5)$

Replace x with $a + 5$ as follows:

$$f(a + 5) = (a + 5)^2 - 4(a + 5)$$
$$= a^2 + 10a + 25 - 4a - 20$$
$$= a^2 + 6a + 5$$

Using a calculator,

2nd CUSTOM   F2   6   **x^2-4x   ENTER**   (**2nd CUSTOM** restores standard menus.)

  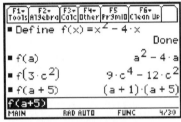

Note that the last answer is factored.
Use **expand** to multiply if you wish.

Other letters, such as t in $f(t)$ and r in $f(r)$, are used in applications to name independent variables.

EXAMPLE 9

Given the function $f(t) = 0.50t + 5.4$, find each of the following:

(a) $f(3.2)$

Replace t with 3.2 as follows:

$$f(t) = 0.50t + 5.4$$
$$f(3.2) = 0.50(3.2) + 5.4$$
$$= 1.6 + 5.4$$
$$= 7.0$$

(b) $f(t_0)$

Replace t with t_0 as follows:

$$f(t) = 0.50t + 5.4$$
$$f(t_0) = 0.50t_0 + 5.4$$

Exercises 1.1

Determine whether or not each relation is a function. Write its domain and its range.

1. $A = \{(2, 4), (3, 7), (9, 2)\}$
2. $B = \{(5, 2), (3, 3), (1, 2)\}$
3. $C = \{(2, 5), (7, 3), (2, 1), (1, 3)\}$
4. $D = \{(0, 2), (5, -1), (2, 7), (5, 1)\}$
5. $E = \{(3, 2), (5, 2), (2, 2), (-2, 2)\}$
6. $F = \{(3, 4), (3, -4), (-3, -4), (-3, 4)\}$
7. $y = 2x + 5$
8. $y = -3x$
9. $y = x^2 + 1$
10. $y = 2x^2 - 3$
11. $x = y^2 - 2$
12. $x = 3y^2 + 4$
13. $y = \sqrt{x + 3}$
14. $y = \sqrt{3 - 6x}$
15. $y = 6 + \sqrt{2x - 8}$
16. $y = 16 - \sqrt{x + 5}$

17. Given the function $f(x) = 8x - 12$, find
    (a) $f(4)$ (b) $f(0)$ (c) $f(-2)$

18. Given the function $g(x) = 20 - 4x$, find
    (a) $g(6)$ (b) $g(0)$ (c) $g(-3)$

19. Given $g(x) = 10x + 15$, find
    (a) $g(2)$ (b) $g(0)$ (c) $g(-4)$

20. Given $f(x) = x^2 - 4$, find
    (a) $f(6)$ (b) $f(0)$ (c) $f(-6)$

21. Given $h(x) = 3x^2 + 4x$, find
    (a) $h(5)$ (b) $h(0)$ (c) $h(-2)$

22. Given $f(x) = -2x^2 + 6x - 7$, find
    (a) $f(3)$ (b) $f(0)$ (c) $f(-1)$

23. Given $f(t) = \dfrac{5 - t^2}{2t}$, find
    (a) $f(1)$ (b) $f(-3)$ (c) $f(0)$

24. Given $g(t) = \sqrt{21 - 5t}$, find
    (a) $g(1)$ (b) $g(-3)$ (c) $g(2)$ (d) $g(8)$

25. Given $f(x) = 6x + 8$, find
    (a) $f(a)$ (b) $f(4a)$ (c) $f(c^2)$

26. Given $g(x) = 8x^2 - 7x$, find
    (a) $g(z)$ (b) $g(2y)$ (c) $g(3t^2)$

27. Given $h(x) = 4x^2 - 12x$, find
    (a) $h(x + 2)$ (b) $h(x - 3)$ (c) $h(2x + 1)$

28. Given $f(y) = y^2 - 3y + 6$, find
    (a) $f(y - 1)$ (b) $f(y^2 + 1)$ (c) $f(1 - 4y)$

29. Given $f(x) = 3x - 1$ and $g(x) = x^2 - 6x + 1$, find
    (a) $f(x) + g(x)$ (b) $f(x) - g(x)$ (c) $[f(x)][g(x)]$ (d) $f(x + h)$

30. Given $f(t) = 5 - 2t + t^2$ and $g(t) = t^2 - 4t + 4$, find
    (a) $f(t) + g(t)$ (b) $g(t) - f(t)$ (c) $[f(t)][g(t)]$ (d) $g(t + h)$

Find the domain of each function.

31. $f(x) = \dfrac{3x + 4}{x - 2}$
32. $f(t) = \dfrac{8}{6t + 3}$
33. $g(t) = \dfrac{2t + 4t^2}{(t - 6)(t + 3)}$
34. $g(x) = \dfrac{3x - 10}{x^2 + 4}$
35. $f(x) = \dfrac{12}{\sqrt{15 - 3x}}$
36. $g(t) = \dfrac{9}{\sqrt{5t + 20}}$

1.2 GRAPHING EQUATIONS

Consider a plane in which two number lines intersect at right angles. Let the point of intersection be the zero point of each line and call it the **origin.** Each line is called an **axis.** The horizontal number line is usually called the **x-axis,** and the vertical line is usually called the **y-axis.** On each axis the same scale (unit length) is preferred but not always possible in all applications. Such a system is called the **rectangular coordinate system,** or the **Cartesian coordinate system.** (The name *Cartesian* is after Descartes, a seventeenth-century French mathematician. He first conceived this idea of combining algebra and geometry together in such a way that each could aid the study of the other.) The plane is divided by the axes into four regions called **quadrants.** The quadrants are numbered as shown in Fig. 1.2.

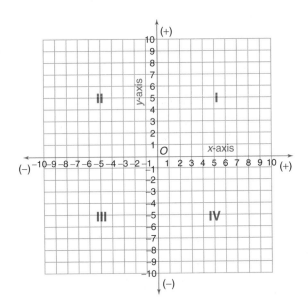

Figure 1.2   The rectangular coordinate system.

In the plane there is a point that corresponds to each ordered pair of real numbers (x, y). Likewise, there is an ordered pair (x, y) that corresponds to each point in the plane. Together x and y are called the **coordinates** of the point; x is called the **abscissa,** and y is called the **ordinate.** This relationship is called a **one-to-one correspondence.** The location, or position, of a point in the plane corresponding to a given ordered pair is found by first counting right or left from O (origin) the number of units along the x-axis indicated by the first number of the ordered pair (right if positive, left if negative). Then from this point reached on the x-axis, count up or down the number of units indicated by the second number of the ordered pair (up if positive, down if negative).

EXAMPLE 1

Plot the point corresponding to each ordered pair in the number plane:

$A(3, 1)$ $B(2, -3)$ $C(-4, -2)$ $D(-3, 0)$ $E(-6, 2)$ $F(0, 2)$ (See Fig. 1.3.)

To graph equations we plot a sample of ordered pairs and connect them with a smooth curve. To obtain the sample, we need to generate ordered pairs from a given equation. One way to generate these ordered pairs is by randomly choosing a value for x, replacing this value for x in the equation, and solving for y.

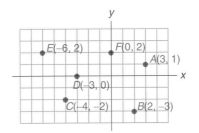

Figure 1.3

EXAMPLE 2

Graph $y = 2x - 3$.

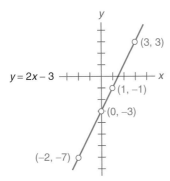

x	y	$y = 2x - 3$ or $f(x) = 2x - 3$
1	-1	$y = 2(1) - 3 = -1$
3	3	$y = 2(3) - 3 = 3$
-2	-7	$y = 2(-2) - 3 = -7$
0	-3	$y = 2(0) - 3 = -3$

Figure 1.4

Plot the ordered pairs and connect them with a smooth line as in Fig. 1.4.

A **linear equation** with two unknowns is an equation of degree one in the form $ax + by = c$ with a and b not both 0. Its graph is always a straight line. Therefore, two ordered pairs are sufficient to graph a linear function, since two points determine a straight line. However, finding a third point provides good insurance against a careless error.

EXAMPLE 3

Graph $y = -3x + 5$.

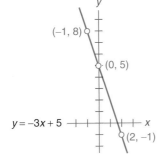

x	y	$y = -3x + 5$ or $g(x) = -3x + 5$
0	5	$y = -3(0) + 5 = 5$
2	-1	$y = -3(2) + 5 = -1$
-1	8	$y = -3(-1) + 5 = 8$

See Fig. 1.5.

Figure 1.5

The graph of an equation that is not linear is usually a curve of some kind and hence requires several points to sketch a smooth curve.

EXAMPLE 4

Graph $y = x^2 - 4$.

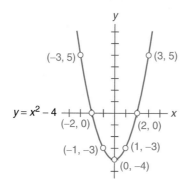

Figure 1.6

x	y	$y = x^2 - 4$
0	-4	$y = (0)^2 - 4 = -4$
1	-3	$y = (1)^2 - 4 = -3$
2	0	$y = (2)^2 - 4 = 0$
3	5	$y = (3)^2 - 4 = 5$
-1	-3	$y = (-1)^2 - 4 = -3$
-2	0	$y = (-2)^2 - 4 = 0$
-3	5	$y = (-3)^2 - 4 = 5$

See Fig. 1.6.

EXAMPLE 5

Graph $y = 2x^2 + x - 5$.

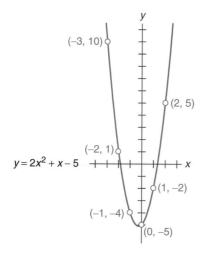

Figure 1.7

x	y	$y = 2x^2 + x - 5$
0	-5	$y = 2(0)^2 + (0) - 5 = -5$
1	-2	$y = 2(1)^2 + (1) - 5 = -2$
2	5	$y = 2(2)^2 + (2) - 5 = 5$
-1	-4	$y = 2(-1)^2 + (-1) - 5 = -4$
-2	1	$y = 2(-2)^2 + (-2) - 5 = 1$
-3	10	$y = 2(-3)^2 + (-3) - 5 = 10$

See Fig. 1.7.

For a more complicated function, more ordered pairs are usually required to obtain a smooth curve. It may also be necessary to change the scale of the graph in order to plot enough ordered pairs to obtain a smooth curve. To change the scale means to enlarge or reduce the unit length on the axes according to a specified ratio. This ratio is chosen on the basis of fitting the necessary values in a given space allowed for the graph.

EXAMPLE 6

Graph $y = x^3 + 4x^2 - x - 4$.

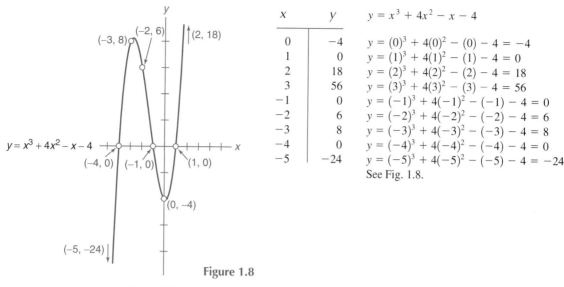

x	y
0	-4
1	0
2	18
3	56
-1	0
-2	6
-3	8
-4	0
-5	-24

$y = x^3 + 4x^2 - x - 4$

$y = (0)^3 + 4(0)^2 - (0) - 4 = -4$
$y = (1)^3 + 4(1)^2 - (1) - 4 = 0$
$y = (2)^3 + 4(2)^2 - (2) - 4 = 18$
$y = (3)^3 + 4(3)^2 - (3) - 4 = 56$
$y = (-1)^3 + 4(-1)^2 - (-1) - 4 = 0$
$y = (-2)^3 + 4(-2)^2 - (-2) - 4 = 6$
$y = (-3)^3 + 4(-3)^2 - (-3) - 4 = 8$
$y = (-4)^3 + 4(-4)^2 - (-4) - 4 = 0$
$y = (-5)^3 + 4(-5)^2 - (-5) - 4 = -24$
See Fig. 1.8.

Figure 1.8

EXAMPLE 7

Graph $y = \sqrt{2x - 6}$.

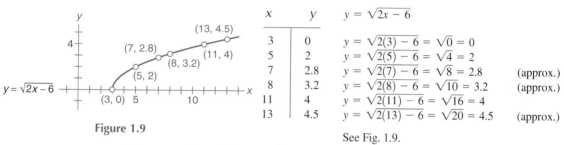

x	y
3	0
5	2
7	2.8
8	3.2
11	4
13	4.5

$y = \sqrt{2x - 6}$

$y = \sqrt{2(3) - 6} = \sqrt{0} = 0$
$y = \sqrt{2(5) - 6} = \sqrt{4} = 2$
$y = \sqrt{2(7) - 6} = \sqrt{8} = 2.8$ (approx.)
$y = \sqrt{2(8) - 6} = \sqrt{10} = 3.2$ (approx.)
$y = \sqrt{2(11) - 6} = \sqrt{16} = 4$
$y = \sqrt{2(13) - 6} = \sqrt{20} = 4.5$ (approx.)

Figure 1.9

See Fig. 1.9.

Using a graphing calculator, we have

green diamond **Y =**   green diamond **TblSet**   **3** down arrow **2 ENTER ENTER**   green diamond **TABLE**

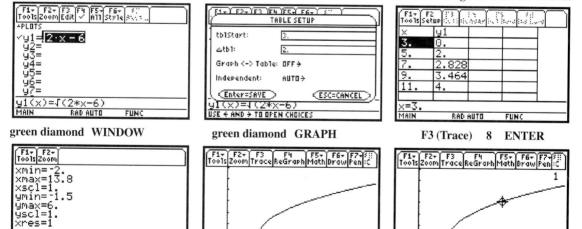

green diamond **WINDOW**   green diamond **GRAPH**   **F3 (Trace)**   **8**   **ENTER**

Specific function values can be calculated on the **Trace** screen.

Solving Equations by Graphing

Equations may be solved graphically. This method is particularly useful when an algebraic method is very cumbersome, cannot be recalled, or does not exist; it is especially useful in technical applications.

Solving for y = 0   To solve the equation $y = x^2 - x - 6$ for $y = 0$ graphically means to find the point or points, if any, where the graph crosses the line $y = 0$ (the x-axis).

EXAMPLE 8

Solve $y = x^2 - x - 6$ for $y = 0$ graphically.
First, graph the equation $y = x^2 - x - 6$ (see Fig. 1.10).

x	y
1	−6
2	−4
3	0
4	6
0	−6
−1	−4
−2	0
−3	6

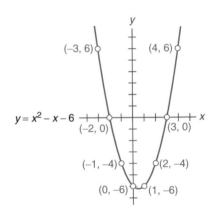

Figure 1.10

Then note the values of x where the curve crosses the x-axis: $x = -2$ and $x = 3$. Therefore, from the graph the solutions of $y = x^2 - x - 6$ for $y = 0$ are $x = -2$ and $x = 3$.

Sometimes the curve crosses the x-axis between the unit marks on the x-axis. In this case we must estimate as closely as possible the point of intersection of the curve and the x-axis. If a particular problem requires greater accuracy, we can scale the graph to allow a more accurate estimation.

EXAMPLE 9

Solve $y = x^2 + 2x - 4$ for $y = 0$ graphically.
First, graph the equation $y = x^2 + 2x - 4$ (see Fig. 1.11).

x	y
0	−4
1	−1
2	4
−1	−5
−2	−4
−3	−1
−4	4

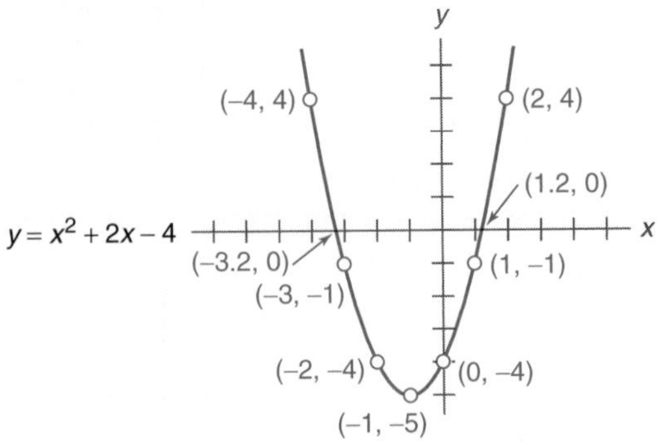

$y = x^2 + 2x - 4$
(−4, 4) (2, 4)
(1.2, 0)
(−3.2, 0) (1, −1)
(−3, −1)
(−2, −4) (0, −4)
(−1, −5)

Figure 1.11

The values of x where the curve crosses the x-axis are approximately 1.2 and -3.2. Therefore, the approximate solutions of $y = x^2 + 2x - 4$ for $y = 0$ are $x = 1.2$ and $x = -3.2$.

Solving for $y = k$

EXAMPLE 10

Solve the equation from Example 9, which was $y = x^2 + 2x - 4$, for $y = 4$ and $y = -3$.

First find the values of x where the curve crosses the line $y = 4$. From the graph in Fig. 1.12, the x-values are 2 and -4. Therefore, solving for $y = 4$, we find that the solutions of $y = x^2 + 2x - 4$ are $x = 2$ and $x = -4$.

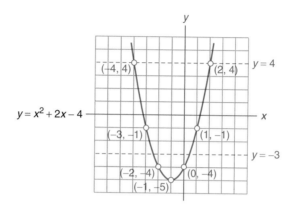

Figure 1.12

Next find the values of x where the curve crosses the line $y = -3$. From the graph in Fig. 1.12, the approximate x-values are 0.4 and -2.4. That is, for $y = -3$, the solutions of $y = x^2 + 2x - 4$ are approximately $x = 0.4$ and $x = -2.4$.

Note that $y = x^2 + 2x - 4$ has no solutions for $y = -7$.

EXAMPLE 11

The voltage, V in volts, in a given circuit varies with time, t in ms, according to the equation $V = 6t^2 + t$. Solve for t when $V = 0$, 70, and 120.

t	V
0	0
1	7
2	26
3	57
4	100
5	155

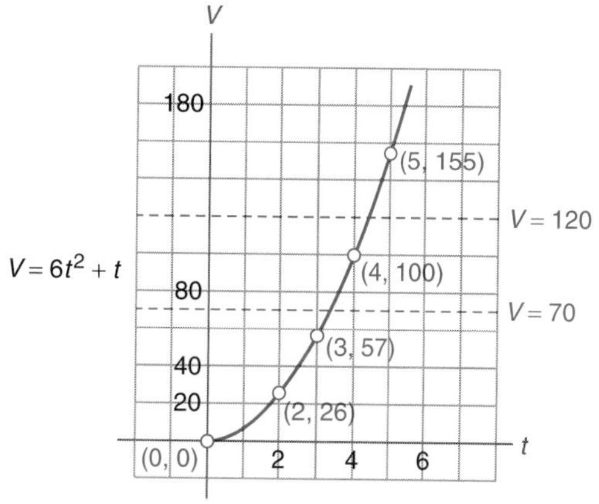

Figure 1.13

In Fig. 1.13, we used the following scales.

$$t: \quad 1 \text{ square} = 1 \text{ ms}$$
$$V: \quad 1 \text{ square} = 20 \text{ V}$$

From the graph,

$$\text{at} \quad V = 0 \text{ V}, \qquad t = 0 \text{ ms}$$
$$\text{at} \quad V = 70 \text{ V}, \qquad t = 3.3 \text{ ms}$$
$$\text{at} \quad V = 120 \text{ V}, \qquad t = 4.4 \text{ ms}$$

Negative values of time, t, are not meaningful in this example.

EXAMPLE 12

The work, w, done in a circuit varies with time, t, according to the equation $w = 8t^2 + 4t$. Solve for t when $w = 60$, 120, and 250.

t	w
0	0
1	12
2	40
3	84
4	144
5	220
6	312

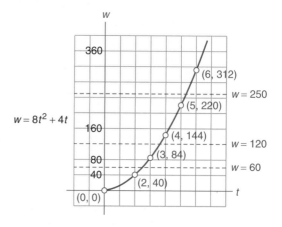

Figure 1.14

In Fig. 1.14, we used these scales

$$t: \quad 1 \text{ square} = 1 \text{ unit}$$
$$w: \quad 1 \text{ square} = 40 \text{ units}$$

From the graph,

$$\text{at} \quad w = 60, \qquad t = 2.5$$
$$\text{at} \quad w = 120, \qquad t = 3.6$$
$$\text{at} \quad w = 250, \qquad t = 5.4$$

Exercises 1.2

Graph each equation.

1. $y = 2x + 1$

2. $y = 3x - 4$

3. $-2x - 3y = 6$

4. $2y = -4x - 3$

5. $y = x^2 - 9$

6. $y = x^2 + x - 6$

7. $y = x^2 - 5x + 4$

8. $y = x^2 + 3$

9. $y = 2x^2 + 3x - 2$

10. $y = -x^2 + 2x + 4$

11. $y = x^2 + 2x$

12. $y = x^2 - 4x$

13. $y = -2x^2 + 4x$

14. $y = -\frac{1}{4}x^2 - \frac{3}{2}x + 2$

15. $y = x^3 - x^2 - 10x + 8$

16. $y = x^3 - 4x^2 + x + 6$

17. $y = x^3 + 2x^2 - 7x + 4$

18. $y = x^3 - 8x - 3$

19. $y = \sqrt{x + 4}$

20. $y = \sqrt{3x - 12}$

21. $y = \sqrt{12 - 6x}$

22. $y = \sqrt{3 - x}$

Solve each equation graphically for the given values.

23. Exercise 5 for $y = 0, -5$, and 2.   **24.** Exercise 6 for $y = 0, 6$, and -3.

25. Exercise 7 for $y = 0, 2$, and -4.   **26.** Exercise 8 for $y = 0, 4$, and 6.

27. Exercise 9 for $y = 0, 3$, and 5.   **28.** Exercise 10 for $y = 0, 4$, and -2.

29. Exercise 11 for $y = 0, 3$, and 6.   **30.** Exercise 12 for $y = 0, -2$, and 3.

31. Exercise 13 for $y = 0, 5, -4$, and $-1\frac{1}{2}$.   **32.** Exercise 14 for $y = 0, -1$, and 1.5.

33. Exercise 15 for $y = 0, 2$, and -2.   **34.** Exercise 16 for $y = 0, 2$, and 8.

35. Exercise 17 for $y = 0, 4$, and 8.   **36.** Exercise 18 for $y = 0, 2$, and -3.

Solve each equation graphically for the given values.

37. $y = x^2 + 3x - 4$ for $y = 0, 6$, and -2.

38. $y = 2x^2 - 5x - 3$ for $y = 2, 0$, and -3.

39. $y = -\frac{1}{2}x^2 + 2$ for $y = 0, 4$, and -4.

40. $y = -\frac{1}{4}x^2 + x$ for $y = 0, \frac{1}{2}$, and -4.

41. $y = x^3 - 3x^2 + 1$ for $y = 0, -2$, and -0.5.

42. $y = -x^3 + 3x + 2$ for $y = 2, 0$, and 3.

43. The resistance, r, of a resistor in a circuit of constant current varies with time, t in ms, according to the equation
$$r = 10t^2 + 20$$
Solve for t when $r = 90\ \Omega$, $180\ \Omega$, and $320\ \Omega$.

44. An object dropped from an airplane 2500 m above the ground falls according to the equation $h = 2500 - 4.95t^2$, where h is the height in metres above the ground and t is the time in seconds. Find the times for the object to fall to a height of 2000 m, 1200 m, and 600 m above the ground. Also find the time it takes to hit the ground.

45. The energy dissipated (work lost), w, by a resistor varies with the time, t in ms, according to the equation $w = 5t^2 + 6t$. Solve for t when $w = 2, 4$, and 10.

46. The resistance r in ohms, in a given circuit varies with time, t in ms, according to the equation $r = 10 + \sqrt{t}$. Find t when $r = 14.1\ \Omega$, $14.3\ \Omega$, and $14.7\ \Omega$. (*Hint:* Choose a suitable scale for the graph and graph only the part you need.)

47. A given inductor carries a current expressed by the equation $i = t^3 - 15$, where i is the current in amperes and t is the time in seconds. Find t when i is 5 A and 15 A.

48. A charge, q in coulombs, flowing in a given circuit varies with the time, t in μs, according to the equation $q = t^2 - \dfrac{t^3}{3}$. Find t when q is 5 coulombs (C) and 10 C.

49. A machinist needs to drill four holes 2.00 in. apart in a straight line in a metal plate as shown in Fig. 1.15. The first hole is placed at the origin, and the line forms an angle of 36.0° with the vertical axis. Find the coordinates of the other three holes.

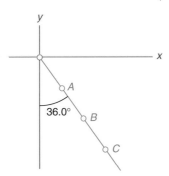

Figure 1.15

50. A machinist often uses a coordinate system to drill holes by placing the origin at the most convenient location. A bolt circle is the circle formed by completing an arc through the centers of the bolt holes in a piece of metal. Find the coordinates of the centers of eight equally spaced, $\frac{1}{4}$-in. holes on a bolt circle of radius 4.00 in. as shown in Fig. 1.16.

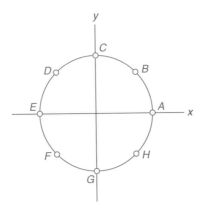

Figure 1.16

1.3 THE STRAIGHT LINE

The **slope** of a nonvertical line is the ratio of the difference of the y-coordinates of any two points on the line to the difference of their x-coordinates when the differences are taken in the same order (see Fig. 1.17).

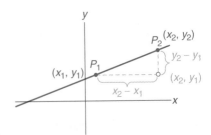

Figure 1.17   Slope of line through P_1 and P_2.

SLOPE OF A LINE

If $P_1(x_1, y_1)$ and $P_2(x_2, y_2)$ represent any two points on a straight line, then the slope m of the line is

$$m = \frac{y_2 - y_1}{x_2 - x_1}$$

EXAMPLE 1

Find the slope of the line passing through $(-2, 1)$ and $(3, 5)$.

If we let $x_1 = -2, y_1 = 1, x_2 = 3$, and $y_2 = 5$ as in Fig. 1.18, then

$$m = \frac{y_2 - y_1}{x_2 - x_1} = \frac{5 - 1}{3 - (-2)} = \frac{4}{5}$$

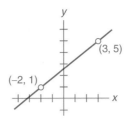

Figure 1.18

Note that if we reverse the order of taking the differences of the coordinates, the result is the same.

$$\frac{y_1 - y_2}{x_1 - x_2} = \frac{1 - 5}{-2 - 3} = \frac{-4}{-5} = \frac{4}{5} = m$$

EXAMPLE 2

Find the slope of the line passing through $(-2, 4)$ and $(6, -6)$.

If we let $x_1 = -2$, $y_1 = 4$, $x_2 = 6$, and $y_2 = -6$ as in Fig. 1.19, then

$$m = \frac{y_2 - y_1}{x_2 - x_1} = \frac{-6 - 4}{6 - (-2)} = \frac{-10}{8} = -\frac{5}{4}$$

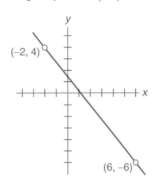

Figure 1.19

Note that in Example 1, the line slopes upward from left to right, while in Example 2 the line slopes downward. In general, we have the following:

1. If a line has positive slope, then the line slopes upward from left to right ("rises").
2. If a line has negative slope, then the line slopes downward from left to right ("falls").
3. If the line has zero slope, then the line is horizontal ("flat").
4. If the line is vertical, then the line has undefined slope, because $x_1 = x_2$, or $x_2 - x_1 = 0$. In this case, the ratio $\dfrac{y_2 - y_1}{x_2 - x_1}$ is undefined because division by zero is undefined.

We can use these facts to assist us in graphing a line if we know the slope of the line and one point P on the line. The line can be sketched by drawing a line through the given point P and a point Q which is plotted by moving one unit to the right of P, then moving vertically m units. That is, a point moving along a line will move vertically an amount equal to m, the slope, for every unit move to the right as in Fig. 1.20.

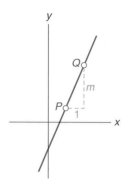

Figure 1.20 The slope *m* corresponds to the vertical change for each horizontal change of 1.

EXAMPLE 3

Graph a line with slope -2 that passes through the point $(1, 3)$.

Since the slope is -2, points on the line drop 2 units for every unit move to the right. The line passes through $(1, 3)$ and $(2, 1)$ as in Fig. 1.21.

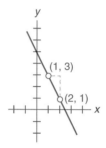

Figure 1.21

Knowing the slope and one point on the line will also determine the equation of the straight line. Let *m* be the slope of a given nonvertical straight line, and let (x_1, y_1) be the coordinates of a point on this line. If (x, y) is any other point on the line as in Fig. 1.22, then we have

$$\frac{y - y_1}{x - x_1} = m$$

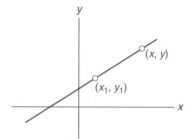

Figure 1.22

By multiplying each side of the equation by $(x - x_1)$, we obtain the following.

POINT-SLOPE FORM OF A STRAIGHT LINE

If *m* is the slope and (x_1, y_1) is any point on a nonvertical straight line, its equation is

$$y - y_1 = m(x - x_1)$$

EXAMPLE 4

Find the equation of the line with slope 3 that passes through the point $(-1, 2)$.

Here $m = 3$, $x_1 = -1$, and $y_1 = 2$. Using the point-slope form, we have the equation

$$y - y_1 = m(x - x_1)$$
$$y - 2 = 3[x - (-1)]$$

Simplifying, we have

$$y - 2 = 3x + 3$$
$$y = 3x + 5$$

The point-slope form can also be used to find the equation of a straight line that passes through two points.

EXAMPLE 5

Find the equation of the line passing through the points $(2, -3)$ and $(-2, 5)$.

First, find the slope.

$$m = \frac{y_2 - y_1}{x_2 - x_1} = \frac{5 - (-3)}{-2 - 2} = \frac{5 + 3}{-4} = \frac{8}{-4} = -2$$

Substitute $m = -2$ and the point $(2, -3)$ in the point-slope form.

$$y - y_1 = m(x - x_1)$$
$$y - (-3) = -2(x - 2)$$
$$y + 3 = -2x + 4$$
$$2x + y - 1 = 0$$

Note: We could have used the other point $(-2, 5)$ in the point-slope form to obtain the equation

$$y - 5 = -2[x - (-2)]$$

which also simplifies to

$$2x + y - 1 = 0$$

A nonvertical line will intersect the y-axis at some point in the form $(0, b)$ as in Fig. 1.23. This point $(0, b)$ is called the **y-intercept** of the line. If the slope of the line is m, then

$$y - y_1 = m(x - x_1)$$
$$y - b = m(x - 0)$$
$$y - b = mx$$
$$y = mx + b$$

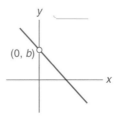

Figure 1.23 b is the y-coordinate of the point where the line crosses the y-axis.

EXAMPLE 6

Find the equation of the line with slope $\frac{1}{2}$ that crosses the y-axis at $b = -3$.

Using the slope-intercept form, we have

$$y = mx + b$$

$$y = \frac{1}{2}x + (-3)$$

$$y = \frac{1}{2}x - 3$$

or

$$x - 2y - 6 = 0$$

A line parallel to the x-axis has slope $m = 0$ (see Fig. 1.24). Its equation is

$$y = mx + b$$
$$y = (0)x + b$$
$$y = b$$

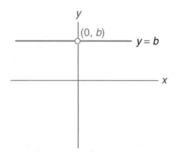

Figure 1.24 Horizontal line

EXAMPLE 7

Find the equation of the line parallel to and 3 units above the x-axis.

The equation is $y = 3$.

By writing the equation of a nonvertical straight line in the slope-intercept form, we can quickly determine the line's slope and a point on the line (the point where it crosses the y-axis).

EXAMPLE 8

Find the slope and the y-intercept of $3y - x + 6 = 0$. Graph the line.

Write the equation in slope-intercept form; that is, solve for y.

$$3y - x + 6 = 0$$

$$3y = x - 6$$

$$y = \frac{x}{3} - 2$$

$$y = \left(\frac{1}{3}\right)x + (-2)$$

So $m = \frac{1}{3}$ and $b = -2$ (see Fig. 1.25).

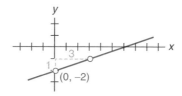

Figure 1.25

EXAMPLE 9

Describe and graph the line whose equation is $y = -5$.

This is a line parallel to and 5 units below the x-axis (see Fig. 1.26).

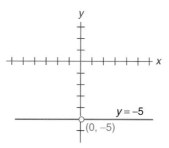

Figure 1.26

If a line is vertical, then we cannot use any of these equations since the line has undefined slope. However, note that in this case, as shown in Fig. 1.27, the line crosses the x-axis at some point in the form $(a, 0)$. All points on the line have the same abscissa as the point $(a, 0)$. This characterizes the line, giving us the following equation.

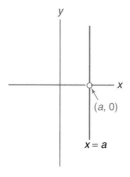

Figure 1.27 Vertical Line

If a vertical line passes through the point (a, b), its equation is

$$x = a$$

EXAMPLE 10

Describe and graph the line whose equation is $x = 2$.

This is a line perpendicular to the x-axis that crosses the x-axis at the point $(2, 0)$ (see Fig. 1.28).

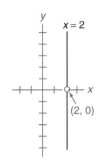

Figure 1.28

EXAMPLE 11

Write the equation of the line perpendicular to the x-axis that crosses the x-axis at the point $(-3, 0)$.

The equation is $x = -3$.

Note: All the equations presented in this section can be put in the form

$$Ax + By + C = 0 \quad \text{with } A \text{ and } B \text{ not both } 0.$$

This is known as the **general form** of the equation of the line and agrees with our definition of a linear equation.

Exercises 1.3

Find the slope of each line passing through the given points.

1. $(4, 2), (3, 1)$ **2.** $(-3, 2), (-1, -2)$ **3.** $(4, -5), (2, 3)$

4. $(-6, -4), (5, -3)$ **5.** $(-3, 2), (6, 2)$ **6.** $(4, -7), (4, 3)$

7. $(5, 7), (-3, 2)$ **8.** $(-3, 6), (-1, 3)$

Graph each line passing through the given point with the given slope.

9. $(2, -1), m = 2$ **10.** $(0, 1), m = -3$ **11.** $(-3, -2), m = \frac{1}{2}$

12. $(4, 4), m = -\frac{1}{3}$ **13.** $(4, 0), m = -2$ **14.** $(-3, 1), m = 4$

15. $(0, -3), m = -\frac{3}{4}$ **16.** $(5, -2), m = \frac{3}{2}$

Find the equation of the line with the given properties.

17. Passes through $(-2, 8)$ with slope -3.

18. Passes through $(3, -5)$ with slope 2.

19. Passes through $(-3, -4)$ with slope $\frac{1}{2}$.

20. Passes through $(6, -7)$ with slope $-\frac{3}{4}$.

21. Passes through $(-2, 7)$ and $(1, 4)$.

22. Passes through $(1, 6)$ and $(4, -3)$.

23. Passes through $(6, -8)$ and $(-4, -3)$.

24. Passes through $(-2, 2)$ and $(7, -1)$.

25. Crosses the y-axis at -2 with slope -5.

26. Crosses the y-axis at 8 with slope $\frac{1}{3}$.

27. Has y-intercept 7 and slope 2.

28. Has y-intercept -4 and slope $-\frac{3}{4}$.

29. Parallel to and 5 units above the x-axis.

30. Parallel to and 2 units below the x-axis.

31. Perpendicular to the x-axis and crosses the x-axis at $(-2, 0)$.

32. Perpendicular to the x-axis and crosses the x-axis at $(5, 0)$.

33. Parallel to the x-axis containing the point $(2, -3)$.

34. Parallel to the y-axis containing the point $(-5, -4)$.

35. Perpendicular to the x-axis containing the point $(-7, 9)$.

36. Perpendicular to the y-axis containing the point $(4, 6)$.

Find the slope and the y-intercept of each straight line.

37. $x + 4y = 12$ **38.** $-2x + 3y + 9 = 0$ **39.** $4x - 2y + 14 = 0$

40. $3x - 6y = 0$ **41.** $y = 6$ **42.** $x = -4$

Graph each equation.

43. $y = 3x - 2$ **44.** $y = -2x + 5$ **45.** $5x - 2y + 4 = 0$

46. $4x + 3y + 6 = 0$ **47.** $x = 7$ **48.** $x = -2$

49. $y = -3$ **50.** $y = 2$ **51.** $6x + 8y = 24$

52. $3x - 5y = 30$ **53.** $x - 3y = -12$ **54.** $x + 6y = 8$

55. A certain metal rod with temperature $-15.0°C$ is 43.0 cm long and at $55.0°C$ is 43.2 cm long. These data can be listed in (x, y) form as $(-15.0, 43.0)$ and $(55.0, 43.2)$. Find the slope (as a simplified fraction) of the straight line passing through these two points.

1.4 PARALLEL AND PERPENDICULAR LINES

PARALLEL LINES

Two lines are parallel if either one of the following conditions holds:

1. They are both perpendicular to the x-axis [see Fig. 1.29(a)].

2. They both have the same slope [see Fig. 1.29(b)]. That is, if the equations of the lines are

$$L_1: \quad y = m_1 x + b_1 \quad \text{and} \quad L_2: \quad y = m_2 x + b_2$$

then

$$m_1 = m_2$$

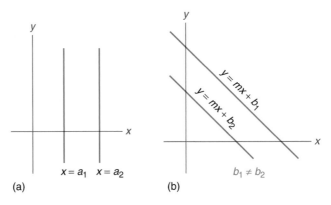

Figure 1.29

PERPENDICULAR LINES

Two lines are perpendicular if either one of the following conditions holds:

1. One line is vertical with equation $x = a$, and the other is horizontal with equation $y = b$.

2. Neither is vertical and the slope of one line is the negative reciprocal of the other. That is, if the equations of the lines are

$$L_1: \quad y = m_1 x + b_1 \quad \text{and} \quad L_2: \quad y = m_2 x + b_2$$

then

$$m_1 = -\frac{1}{m_2}$$

To show this last relationship, consider the triangle in Fig. 1.30, where L_1 is perpendicular to L_2. Let

$(c, 0)$ represent the point P

$(d, 0)$ represent the point R

(e, f) represent the point Q

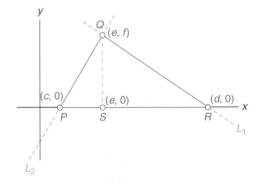

Figure 1.30

Draw QS perpendicular to the x-axis. Then S must be represented by $(e, 0)$.

Triangles PSQ and QSR in Fig. 1.30 are similar. (Note that angle PQS equals angle QRS.) From geometry we know that

$$\frac{PS}{QS} = \frac{QS}{SR} \tag{1}$$

In this case,

$$PS = e - c \quad \text{(the distance from } c \text{ to } e \text{ on the } x\text{-axis)}$$
$$QS = f \quad \text{(the distance from 0 to } f \text{ on the } y\text{-axis)}$$
$$SR = d - e \quad \text{(the distance from } e \text{ to } d \text{ on the } x\text{-axis)}$$

Substituting these values in Equation (1), we have

$$\frac{e - c}{f} = \frac{f}{d - e}$$

Multiplying each side of the equation by $(d - e)f$ gives

$$f^2 = (d - e)(e - c) \qquad\qquad (2)$$

Compute slopes m_1 and m_2 as follows:

$$m_1 = \frac{f - 0}{e - d} = \frac{f}{e - d}$$

$$m_2 = \frac{f - 0}{e - c} = \frac{f}{e - c}$$

$$(m_1)(m_2) = \frac{f}{e - d} \cdot \frac{f}{e - c} = \frac{f^2}{(e - d)(e - c)}$$

Substituting from Equation (2) we have

$$(m_1)(m_2) = \frac{(d - e)(e - c)}{(e - d)(e - c)} = \frac{d - e}{e - d} = -\left(\frac{e - d}{e - d}\right) = -1$$

or

$$(m_1)(m_2) = -1$$

Dividing each side of this equation by m_2 we have

$$m_1 = \frac{-1}{m_2}$$

EXAMPLE 1

Determine whether the lines given by the equations $3y + 6x - 5 = 0$ and $2y - x + 7 = 0$ are perpendicular.

Change each equation into slope-intercept form; that is, solve for y.

$$y = -2x + \frac{5}{3} \qquad \text{(Slope is } -2.\text{)}$$

and

$$y = \frac{1}{2}x - \frac{7}{2} \qquad \left(\text{Slope is } \frac{1}{2}.\right)$$

Since

$$-2 = \frac{-1}{\frac{1}{2}} \qquad \left(-2 \text{ is the negative reciprocal of } \frac{1}{2}.\right)$$

the lines are perpendicular.

EXAMPLE 2

Find the equation of the line through $(-3, 2)$ and perpendicular to $2y - 3x + 5 = 0$.

We can find the slope of the desired line by finding the negative reciprocal of the slope of the given line. First find the slope of the line $2y - 3x + 5 = 0$. Writing this equation in slope-intercept form, we have

$$y = \frac{3}{2}x - \frac{5}{2}$$

The slope of this line is $m = \frac{3}{2}$. The slope of the line perpendicular to this line is then equal to $-\frac{2}{3}$, the negative reciprocal of $\frac{3}{2}$. Now using the point-slope form, we have

$$y - y_1 = m(x - x_1)$$

$$y - 2 = -\frac{2}{3}[x - (-3)]$$

$$y - 2 = -\frac{2}{3}(x + 3)$$

or

$$2x + 3y = 0$$

EXAMPLE 3

Find the equation of the line through $(2, -5)$ and parallel to $3x + y = 7$.

First, find the slope of the given line by solving its equation for y.

$$y = -3x + 7$$

Its slope is -3. The slope of any line parallel to this line has the same slope. Now, write the equation of the line with slope -3 passing through $(2, -5)$.

$$y - y_1 = m(x - x_1)$$
$$y - (-5) = -3(x - 2)$$
$$y + 5 = -3x + 6$$
$$y = -3x + 1 \quad \text{or} \quad 3x + y = 1$$

EXAMPLE 4

Judging the slopes of lines using a graphing calculator can be misleading unless a "square" viewing window is chosen. In particular, perpendicular lines don't look like they are intersecting at a 90° angle unless incremental changes along the x- and y-axes have the same meaning. The **ZoomSqr** feature will square up the current viewing window by choosing the *smaller* unit and using it on both axes (the effect is to zoom *out* to square the viewing window). As the sixth frame shows, **ZoomDec** is a square viewing window to begin with (each pixel on the x- and y-axes represents 0.1 units).

Graph $y = 3x - 4$ and $y = -\frac{1}{3}x - 2$.

green diamond Y=   **F2   6   (ZoomStandard)**   (not a square viewing window)

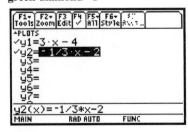   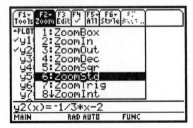   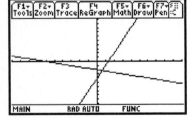

   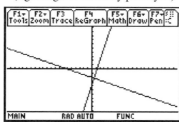   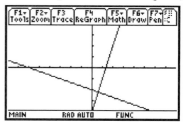

Exercises 1.4

Determine whether each given pair of equations represents lines that are parallel, perpendicular, or neither.

1. $x + 3y - 7 = 0; -3x + y + 2 = 0$ **2.** $x + 2y - 11 = 0; x + 2y + 4 = 0$

3. $-x + 4y + 7 = 0; x + 4y - 5 = 0$ **4.** $2x + 7y + 4 = 0; 7x - 2y - 5 = 0$

5. $y - 5x + 13 = 0; y - 5x + 9 = 0$ **6.** $-3x + 9y + 22 = 0; x + 3y - 17 = 0$

Find the equation of the line that satisfies each set of conditions.

7. Passes through $(-1, 5)$ and is parallel to $-2x + y + 13 = 0$.

8. Passes through $(2, -2)$ and is perpendicular to $3x - 2y - 14 = 0$.

9. Passes through $(-7, 4)$ and is perpendicular to $5y = x$.

10. Passes through $(2, -10)$ and is parallel to $2x + 3y - 7 = 0$.

11. Passes through the origin and is parallel to $3x - 4y = 12$.

12. Passes through the origin and is perpendicular to $4x + 5y = 17$.

13. Has x-intercept 6 and is perpendicular to $4x + 6y = 9$.

14. Has y-intercept -2 and is parallel to $6x - 4y = 11$.

15. Has y-intercept 8 and is parallel to $y = 2$.

16. Has x-intercept -4 and is perpendicular to $y = 6$.

17. Has x-intercept 7 and is parallel to $x = -4$.

18. Has y-intercept -9 and is perpendicular to $x = 5$.

19. The vertices of a quadrilateral are $A(-2, 3)$, $B(2, 2)$, $C(9, 6)$, and $D(5, 7)$.
   (a) Is the quadrilateral a parallelogram? Why or why not?
   (b) Is the quadrilateral a rectangle? Why or why not?

20. The vertices of a quadrilateral are $A(-4, 1)$, $B(0, -2)$, $C(6, 6)$, and $D(2, 9)$.
   (a) Is the quadrilateral a parallelogram? Why or why not?
   (b) Is the quadrilateral a rectangle? Why or why not?

1.5   THE DISTANCE AND MIDPOINT FORMULAS

We now wish to find the distance between two points on a straight line. Suppose P has the coordinates (x_1, y_1) and Q has the coordinates (x_2, y_2). Then a triangle similar to that in Fig. 1.31 can be constructed. Note that R must have the coordinates (x_2, y_1). (Point R has the same x-coordinate as Q and the same y-coordinate as P.)

Using the Pythagorean theorem, we have

$$PQ^2 = PR^2 + QR^2 \qquad\qquad \textbf{(3)}$$

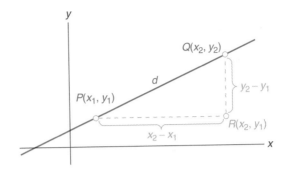

Figure 1.31   *d* is the distance between points *P* and *Q*.

Observe that

$$PR = x_2 - x_1 \qquad \text{(the horizontal distance between } x_1 \text{ and } x_2 \text{ on the } x\text{-axis)}$$

$$QR = y_2 - y_1 \qquad \text{(the vertical distance between } y_1 \text{ and } y_2 \text{ on the } y\text{-axis)}$$

Substituting these values for *PR* and *QR* in Equation (3) gives

$$PQ^2 = (x_2 - x_1)^2 + (y_2 - y_1)^2$$

DISTANCE FORMULA

The distance between two points $P(x_1, y_1)$ and $Q(x_2, y_2)$ is given by the formula

$$d = PQ = \sqrt{(x_2 - x_1)^2 + (y_2 - y_1)^2}$$

EXAMPLE 1

Find the distance, *d*, between $(3, 4)$ and $(-2, 7)$.

$$d = \sqrt{(x_2 - x_1)^2 + (y_2 - y_1)^2}$$
$$d = \sqrt{(-2 - 3)^2 + (7 - 4)^2}$$
$$= \sqrt{(-5)^2 + (3)^2} = \sqrt{25 + 9} = \sqrt{34}$$

Note that we can reverse the order of $(3, 4)$ and $(-2, 7)$ in the formula for computing *d* without affecting the result.

$$d = \sqrt{(3 - (-2))^2 + (4 - 7)^2}$$
$$= \sqrt{(5)^2 + (-3)^2} = \sqrt{25 + 9} = \sqrt{34}$$

MIDPOINT FORMULA

The coordinates of point $Q(x_m, y_m)$ which is midway between two points $P(x_1, y_1)$ and $R(x_2, y_2)$ are given by

$$x_m = \frac{x_1 + x_2}{2} \qquad y_m = \frac{y_1 + y_2}{2}$$

Figure 1.32 illustrates the midpoint formula. First look at points *P*, *Q*, and *R*. Triangles *PSQ* and *QTR* are congruent. This means that

$$PS = QT$$

Since

$$PS = x_m - x_1$$

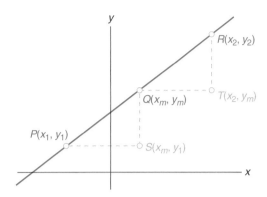

Figure 1.32 Point Q is the midpoint between points P and R.

and

$$QT = x_2 - x_m$$

then

$$x_m - x_1 = x_2 - x_m$$

or

$$2x_m - x_1 = x_2$$
$$2x_m = x_1 + x_2$$
$$x_m = \frac{x_1 + x_2}{2}$$

The formula for y_m is found in the same manner.

EXAMPLE 2

Find the point midway between $(2, -3)$ and $(-4, 6)$.

$$x_m = \frac{x_1 + x_2}{2} = \frac{2 + (-4)}{2} = \frac{-2}{2} = -1$$

$$y_m = \frac{y_1 + y_2}{2} = \frac{-3 + 6}{2} = \frac{3}{2}$$

The midpoint is $\left(-1, \frac{3}{2}\right)$.

Exercises 1.5

Find the distance between each pair of points.

1. $(4, -7); (-5, 5)$ **2.** $(4, 3); (-2, -1)$ **3.** $(3, -2); (10, -2)$

4. $(6, -2); (6, 4)$ **5.** $(5, -2); (1, 2)$ **6.** $(2, -3); (-1, 1)$

7. $(3, -5); (3, 2)$ **8.** $(2, -4); (6, -4)$

Find the coordinates of the point midway between each pair of points.

9. $(2, 3); (5, 7)$ **10.** $(0, 5); (2, -4)$ **11.** $(3, -2); (0, 0)$

12. $(2, -3); (4, -3)$ **13.** $(11, 4); (-11, -9)$ **14.** $(4, 10); (-6, -8)$

The vertices of each △ABC are given below. For each triangle, find (a) the perimeter, (b) whether it is a right triangle, (c) whether it is isosceles, and (d) its area if it is a right triangle.

15. $A(2, 8)$; $B(10, 2)$; $C(10, 8)$ **16.** $A(0, 0)$; $B(3, 3)$; $C(3, -3)$

17. $A(-3, 6)$; $B(5, 0)$; $C(4, 9)$ **18.** $A(-6, 3)$; $B(-3, 7)$; $C(1, 4)$

19. Given △ABC with vertices $A(7, -1)$, $B(9, 1)$, and $C(-3, 5)$, find the distance from A to the midpoint of side BC.

20. Find the distance from B to the midpoint of side AC in Exercise 19.

21. Find the equation of the line parallel to the line $3x - 6y = 10$ and through the midpoint of AB, where $A(4, 2)$ and $B(8, -6)$.

22. Find the equation of the line perpendicular to the line $2x + 5y = 12$ and through the midpoint of AB, where $A(-3, -4)$ and $B(7, -8)$.

23. Find the equation of the line perpendicular to the line $4x + 8y = 16$ and through the midpoint of AB, where $A(-8, 12)$ and $B(6, 10)$.

24. Find the equation of the line parallel to the line $5x - 6y = 30$ and through the midpoint of AB, where $A(3, 11)$ and $B(7, -5)$.

In Exercises 25–28, start with a graph and then use the distance formula and the slopes to confirm the given geometric figure.

25. Show that $ABCD$ is a rectangle, where $A(-2, 2)$; $B(1, 3)$; $C(2, 0)$; and $D(-1, -1)$.

26. Show that $ABCD$ is a parallelogram, where $A(2, 6)$; $B(7, 2)$; $C(8, 5)$; and $D(3, 9)$.

27. Show that $ABCD$ is a trapezoid with one right angle, where $A(-12, 8)$; $B(3, 2)$; $C(5, 7)$; $D(-5, 11)$.

28. Show that if the coordinates of the vertices of a triangle are (a, b), $(a + c, b)$, and $(a + c, b + c)$, the triangle is a right triangle.

1.6 THE CIRCLE

Equations in two variables of second degree in the form

$$Ax^2 + Bxy + Cy^2 + Dx + Ey + F = 0$$

are called **conics.** We begin a systematic study of conics with the circle.

    The **circle** consists of the set of points located the same distance from a given point, called the **center.** The distance at which all points are located from the center is called the **radius.** A circle may thus be graphed in the plane given its center and radius.

EXAMPLE 1

Graph the circle with center at $(1, -2)$ and radius $r = 3$.

    Plot all points in the plane located 3 units away from the point $(1, -2)$ as in Fig. 1.33. (You may wish to use a compass.)

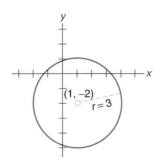

Figure 1.33

From the definition of a circle we can determine the equation of a circle. Let (h, k) be the coordinates of the center, and let r represent the radius. If any point (x, y) is located on the circle, it must be a distance r from the center (h, k) as in Fig. 1.34.

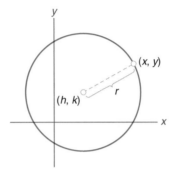

Figure 1.34 The set of points (x, y) located at the same distance r from the given point (h, k).

Using the distance formula, we have

$$\sqrt{(x_2 - x_1)^2 + (y_2 - y_1)^2} = d$$
$$\sqrt{(x - h)^2 + (y - k)^2} = r$$

Squaring each side, we have

STANDARD FORM OF A CIRCLE

$$(x - h)^2 + (y - k)^2 = r^2$$

where r is the radius and (h, k) is the center.

Any point (x, y) satisfying this equation lies on the circle.

EXAMPLE 2

Find the equation of the circle with radius 3 and center $(1, -2)$. (See Example 1.)
Using the standard form of the equation of a circle, we have

$$(x - h)^2 + (y - k)^2 = r^2$$
$$(x - 1)^2 + [y - (-2)]^2 = (3)^2$$
$$(x - 1)^2 + (y + 2)^2 = 9$$

EXAMPLE 3

Find the equation of the circle with center at $(3, -2)$ and passing through $(-1, 1)$.

To write the equation, we need to know the radius r of the circle. Although r has not been stated, we do know that every point on the circle is a distance r from the center, $(3, -2)$. In particular, the point $(-1, 1)$ is a distance r from $(3, -2)$ (see Fig. 1.35). Using the distance formula,

$$d = \sqrt{(x_2 - x_1)^2 + (y_2 - y_1)^2}$$
$$d = r = \sqrt{(x - h)^2 + (y - k)^2}$$
$$r = \sqrt{(-1 - 3)^2 + [1 - (-2)]^2}$$
$$= \sqrt{(-4)^2 + 3^2} = \sqrt{16 + 9} = \sqrt{25} = 5$$

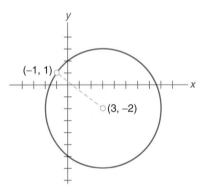

Figure 1.35

Now write the equation of the circle.

$$(x - h)^2 + (y - k)^2 = r^2$$
$$(x - 3)^2 + [y - (-2)]^2 = 5^2$$
$$(x - 3)^2 + (y + 2)^2 = 25$$

If we remove parentheses in the equation

$$(x - h)^2 + (y - k)^2 = r^2$$

we have

$$x^2 - 2xh + h^2 + y^2 - 2yk + k^2 = r^2$$

Rearranging terms, we have

$$x^2 + y^2 - 2hx - 2ky + h^2 + k^2 - r^2 = 0$$

If we let $D = -2h$, $E = -2k$, and $F = h^2 + k^2 - r^2$, we obtain the equation

> **GENERAL FORM OF A CIRCLE**
>
> $$x^2 + y^2 + Dx + Ey + F = 0$$

Any equation in this form represents a circle.

EXAMPLE 4

Write the equation $(x - 3)^2 + (y + 2)^2 = 25$ obtained in Example 3 in general form.

$$(x - 3)^2 + (y + 2)^2 = 25$$
$$x^2 - 6x + 9 + y^2 + 4y + 4 = 25$$
$$x^2 + y^2 - 6x + 4y - 12 = 0$$

EXAMPLE 5

Find the center and radius of the circle given by the equation

$$x^2 + y^2 - 4x + 2y - 11 = 0$$

Looking back at how we arrived at the general equation of a circle, we see that if we rearrange the terms of the equation

$$(x^2 - 4x \quad) + (y^2 + 2y \quad) = 11$$

then $(x^2 - 4x \quad)$ represents the first two terms of

$$(x - h)^2 = x^2 - 2hx + h^2$$

and $(y^2 + 2y \quad)$ represents the first two terms of

$$(y - k)^2 = y^2 - 2ky + k^2$$

This means that

$$-4 = -2h \quad \text{and} \quad 2 = -2k$$
$$h = 2 \qquad\qquad k = -1$$

To complete the squares $(x - h)^2$ and $(y - k)^2$, we must add $h^2 = 2^2 = 4$ and $k^2 = (-1)^2 = 1$ to each side of the equation.

$$(x^2 - 4x \quad) + (y^2 + 2y \quad) = 11$$
$$(x^2 - 4x + 4) + (y^2 + 2y + 1) = 11 + 4 + 1$$
$$(x - 2)^2 + (y + 1)^2 = 16$$
$$(x - 2)^2 + [y - (-1)]^2 = 16 = 4^2$$

From this we see that we have the standard form of the equation of a circle with radius 4 and center at the point $(2, -1)$.

This process is called **completing the square** of the x- and y-terms. In general, if the coefficients of x^2 and y^2 are both equal to 1, then these values can be found as follows: Add h^2 and k^2 to each side of the equation, where

$$h^2 = (\tfrac{1}{2} \text{ the coefficient of } x)^2 = (\tfrac{1}{2}D)^2$$
$$k^2 = (\tfrac{1}{2} \text{ the coefficient of } y)^2 = (\tfrac{1}{2}E)^2$$

EXAMPLE 6

Find the center and radius of the circle given by the equation

$$x^2 + y^2 + 6x - 4y - 12 = 0$$

Sketch the graph of the circle.

$$h^2 = [(\tfrac{1}{2})(6)]^2 = 3^2 = 9$$
$$k^2 = [\tfrac{1}{2}(-4)]^2 = (-2)^2 = 4$$

Rewrite the equation and add 9 and 4 to each side.

$$(x^2 + 6x \quad) + (y^2 - 4y \quad) = 12$$
$$(x^2 + 6x + 9) + (y^2 - 4y + 4) = 12 + 9 + 4$$
$$(x + 3)^2 + (y - 2)^2 = 25 = 5^2$$

The center is at $(-3, 2)$ and the radius is 5. Plot all points that are at a distance of 5 from the point $(-3, 2)$ as in Fig. 1.36.

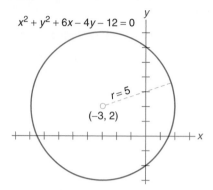

$$x^2 + y^2 + 6x - 4y - 12 = 0$$

$r = 5$

$(-3, 2)$

Figure 1.36

If the center of a circle is at the origin, then $h = 0$ and $k = 0$, and its standard equation becomes

$$x^2 + y^2 = r^2$$

where r is the radius and the center is at the origin.

EXAMPLE 7

Find the equation of the circle with radius 3 and center at the origin. Also, graph the circle.

$$x^2 + y^2 = r^2$$
$$x^2 + y^2 = (3)^2$$
$$x^2 + y^2 = 9 \qquad \text{(See Fig. 1.37.)}$$

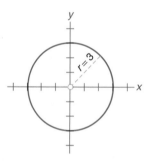

$r = 3$

Figure 1.37

Exercises 1.6

Graph the circle with the given center and radius.

1. Center at $(2, -1)$, $r = 3$.   **2.** Center at $(3, 3)$, $r = 2$.

3. Center at $(0, 2)$, $r = 4$.   **4.** Center at $(-4, -5)$, $r = 3$.

Find the equation of the circle (in standard form) with the given properties.

5. Center at $(1, -1)$, radius 4.   **6.** Center at $(-2, 3)$, radius $\sqrt{5}$.

7. Center at $(-2, -4)$, passing through $(1, -9)$.

8. Center at $(5, 2)$, passing through $(-2, -6)$.

9. Center at $(0, 0)$, radius 6.

10. Center at $(0, 0)$, passing through $(3, -4)$.

Find the center and radius of the given circle.

11. $x^2 + y^2 = 16$

12. $x^2 + y^2 - 4x - 5 = 0$

13. $x^2 + y^2 + 6x - 8y - 39 = 0$

14. $x^2 + y^2 - 6x + 14y + 42 = 0$

15. $x^2 + y^2 - 8x + 12y - 8 = 0$

16. $x^2 + y^2 + 10x + 2y - 14 = 0$

17. $x^2 + y^2 - 12x - 2y - 12 = 0$

18. $x^2 + y^2 + 4x - 9y + 4 = 0$

19. $x^2 + y^2 + 7x + 3y - 9 = 0$

20. $x^2 + y^2 - 5x - 8y = 0$

21. Find the equation of the circle or circles whose center is on the y-axis and that contain the points $(1, 4)$ and $(-3, 2)$. Give the center and radius.

22. Find the equation of the circle with center in the first quadrant on the line $y = 2x$, tangent to the x-axis, and radius 6. Give its center.

23. Find the equation of the circle containing the points $(3, 1)$, $(0, 0)$, and $(8, 4)$. Give its center and radius.

24. Find the equation of the circle containing the points $(1, -4)$, $(-3, 4)$, and $(4, 5)$. Give its center and radius.

1.7 THE PARABOLA

While the parabola may not be as familiar a geometric curve as the circle, examples of the parabola are found in many technical applications. A **parabola** consists of all points that are the same distance from a given fixed point and a given fixed line. The fixed point is called the **focus.** The fixed line is called the **directrix.** This relationship is shown in Fig. 1.38 for the points P, Q, and V, which lie on a parabola with focus F and directrix D.

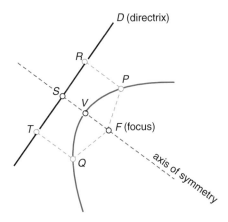

Figure 1.38 A parabola is a set of points that are the same distance from a given fixed point (focus) and a given fixed line (directrix).

Note:

$$RP = PF$$
$$SV = VF$$
$$TQ = QF$$

The point V midway between the directrix and the focus is called the **vertex.** The vertex and the focus lie on a line perpendicular to the directrix, which is called the **axis of symmetry.**

There are two standard forms for the equation of a parabola. The form depends on the position of the parabola in the plane. We first discuss the parabola with focus on the x-axis at $(p, 0)$ and directrix the line $x = -p$ as in Fig. 1.39. Let $P(x, y)$ represent any point on this parabola. Vertex V is then at the origin, and the axis of symmetry is the x-axis.

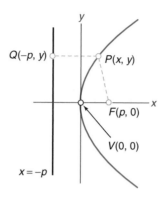

Figure 1.39

By the way we have described the parabola, the distance between P and F must equal the distance between P and Q. Using the distance formula, we have

$$PF = PQ$$
$$\sqrt{(x - p)^2 + (y - 0)^2} = \sqrt{[x - (-p)]^2 + (y - y)^2}$$

Squaring each side,

$$(x - p)^2 + y^2 = (x + p)^2$$
$$x^2 - 2xp + p^2 + y^2 = x^2 + 2xp + p^2$$
$$y^2 = 4px$$

STANDARD FORM OF PARABOLA

$$y^2 = 4px$$

with focus at $(p, 0)$ and with the line $x = -p$ as the directrix.

Note that in Fig. 1.39, $p > 0$.

EXAMPLE 1

Find the equation of the parabola with focus at $(3, 0)$ and directrix $x = -3$.
   In this case $p = 3$, so we have

$$y^2 = 4(3)x$$
$$y^2 = 12x$$

EXAMPLE 2

Find the focus and equation of the directrix of the parabola $y^2 = 24x$.

$$y^2 = 24x$$
$$y^2 = 4(6)x$$
$$y^2 = 4px$$

Since p must be 6, the focus is $(6, 0)$ and the directrix is the line $x = -6$.

EXAMPLE 3

Find the equation of the parabola with focus at $(-2, 0)$ and with directrix $x = 2$. Sketch the graph.

Here $p = -2$. The equation becomes

$$y^2 = 4(-2)x$$
$$y^2 = -8x \qquad \text{(See Fig. 1.40.)}$$

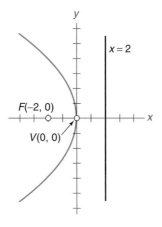

Figure 1.40

Observe the following,

1. If $p > 0$, the coefficient of x in the equation $y^2 = 4px$ is *positive* and the parabola opens to the *right* (see Fig. 1.41a.)

2. If $p < 0$, the coefficient of x in the equation $y^2 = 4px$ is *negative* and the parabola opens to the *left* (see Fig. 1.41b.)

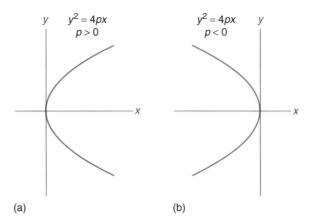

(a)   (b)

Figure 1.41   Standard form of the parabola $y^2 = 4px$.

We obtain the other standard form of the parabola when the focus lies on the y-axis and the directrix is parallel to the x-axis. Let $(0, p)$ be the focus F and $y = -p$ be the directrix. The vertex is still at the origin, but the axis of symmetry is now the y-axis (see Fig. 1.42).

> **STANDARD FORM OF PARABOLA**
>
> $$x^2 = 4py$$
>
> with focus at $(0, p)$ and with the line $y = -p$ as the directrix. (see Fig. 1.42).

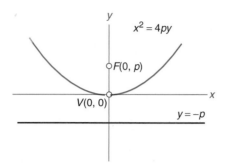

Figure 1.42

EXAMPLE 4

Find the equation of the parabola with focus at $(0, 3)$ and with directrix $y = -3$. Sketch the graph.

Since the focus lies on the y-axis and the directrix is parallel to the x-axis, we use the equation $x^2 = 4py$ with $p = 3$ (see Fig. 1.43).

$$x^2 = 4(3)y$$
$$x^2 = 12y$$

EXAMPLE 5

Find the equation of the parabola with focus at $(0, -1)$ and with directrix $y = 1$. Sketch the graph.

Again the focus lies on the y-axis with directrix parallel to the x-axis, so we use the equation $x^2 = 4py$ with $p = -1$ (see Fig. 1.44).

$$x^2 = 4(-1)y$$
$$x^2 = -4y$$

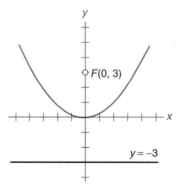

Figure 1.43

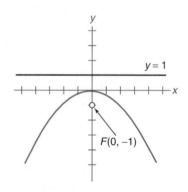

Figure 1.44

Observe the following.

1. If $p > 0$, the coefficient of y in the equation $x^2 = 4py$ is *positive* and the parabola opens *upward* (see Fig. 1.45a).

2. If $p < 0$, the coefficient of y in the equation $x^2 = 4py$ is *negative* and the parabola opens *downward* (see Fig. 1.45b).

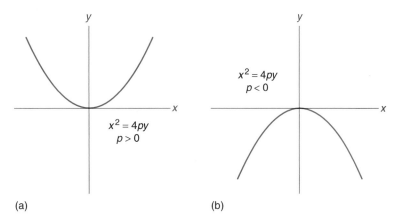

(a)   (b)

Figure 1.45   Standard form of the parabola $x^2 = 4py$.

We are now able to describe the graph of a parabola by inspection of its equation in standard form. We can also find the focus and directrix.

EXAMPLE 6

Describe the graph of the equation $y^2 = 20x$.

This is an equation of a parabola in the form $y^2 = 4px$. Since $p = 5$, this parabola has its focus at $(5, 0)$ and its directrix is the line $x = -5$. The parabola opens to the right (since $p > 0$).

EXAMPLE 7

Describe the graph of the equation $x^2 = -2y$.

This is an equation of a parabola in the form $x^2 = 4py$, where $p = -\frac{1}{2}$ [as $4(-\frac{1}{2}) = -2$]. The focus is at $(0, -\frac{1}{2})$ and the directrix is the line $y = \frac{1}{2}$. The parabola opens downward (since $p < 0$).

Of course, not all parabolas are given in standard position.

EXAMPLE 8

Find the equation of the parabola with focus at $(1, 3)$ and with the line $y = -1$ as directrix. We must use the definition of the parabola (see Fig. 1.46).

$$PF = PQ$$
$$\sqrt{(x - 1)^2 + (y - 3)^2} = \sqrt{(x - x)^2 + [y - (-1)]^2}$$
$$x^2 - 2x + 1 + y^2 - 6y + 9 = y^2 + 2y + 1 \qquad \text{(Square each side and}$$
$$x^2 - 2x - 8y + 9 = 0 \qquad \qquad \text{remove parentheses.)}$$

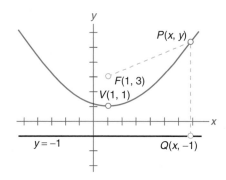

Figure 1.46

In fact, any equation of the form

$$Ax^2 + Dx + Ey + F = 0$$

or

$$Cy^2 + Dx + Ey + F = 0$$

represents a parabola.

In graphing a parabola in the form $y = f(x) = ax^2 + bx + c, a \neq 0$, it is most help-ful to graph the x-intercepts, if any, and the vertex. To find the x-intercepts, let $y = 0$ and solve $ax^2 + bx + c = 0$ for x. The solutions for this equation are given by the quadratic formula:

$$x = \frac{-b \pm \sqrt{b^2 - 4ac}}{2a}$$

Recall that the solutions are real numbers only if the *discriminant*, $b^2 - 4ac$, is nonnega-tive and that the solutions are imaginary if $b^2 - 4ac < 0$. Thus, the graph of the parabola $y = f(x) = ax^2 + bx + c, a \neq 0$, has

1. two different x-intercepts if $b^2 - 4ac > 0$,
2. only one x-intercept if $b^2 - 4ac = 0$ (the graph is tangent to the x-axis),
3. no x-intercepts if $b^2 - 4ac < 0$.

The **axis of symmetry** of the parabola in the form $y = f(x) = ax^2 + bx + c$ is a vertical line halfway between the x-intercepts. The equation of the axis is the vertical line passing through the midpoint of the line segment joining the two x-intercepts (see Fig. 1.47). This midpoint is

$$\frac{x_1 + x_2}{2} = \frac{\dfrac{-b - \sqrt{b^2 - 4ac}}{2a} + \dfrac{-b + \sqrt{b^2 - 4ac}}{2a}}{2} = \frac{\dfrac{-2b}{2a}}{2} = -\frac{b}{2a}$$

Thus, the equation of the axis is $x = -\dfrac{b}{2a}$.

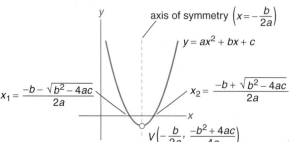

Figure 1.47 Axis of symmetry of the parabola $y = ax^2 + bx + c$.

Since the parabola contains the vertex, its x-coordinate is $-\dfrac{b}{2a}$. Its y-coordinate is then $f\left(-\dfrac{b}{2a}\right)$. To find the y-coordinate, evaluate

$$f(x) = ax^2 + bx + c$$

$$f\left(-\frac{b}{2a}\right) = a\left(-\frac{b}{2a}\right)^2 + b\left(-\frac{b}{2a}\right) + c$$

$$= \frac{b^2}{4a} - \frac{b^2}{2a} + c$$

$$= \frac{b^2}{4a} - \frac{2b^2}{4a} + \frac{4ac}{4a}$$

$$= \frac{-b^2 + 4ac}{4a}$$

AXIS AND VERTEX OF A PARABOLA

Given the parabola $y = f(x) = ax^2 + bx + c$, its axis is the vertical line $x = -\dfrac{b}{2a}$ and its vertex is the point

$$\left(-\frac{b}{2a}, f\left(-\frac{b}{2a}\right)\right) = \left(-\frac{b}{2a}, \frac{-b^2 + 4ac}{4a}\right)$$

The vertex is a maximum point if $a < 0$ and a minimum point if $a > 0$.

EXAMPLE 9

Graph $y = f(x) = 2x^2 - 8x + 11$. Find its vertex and the equation of the axis.

First, note that $b^2 - 4ac = (-8)^2 - 4(2)(11) = -24 < 0$, which means that the graph has no x-intercepts. The equation of the axis is

$$x = -\frac{b}{2a} = -\frac{-8}{2(2)} = 2$$

The vertex is the point $(2, f(2))$ or $(2, 2 \cdot 2^2 - 8 \cdot 2 + 11) = (2, 3)$.

This y-coordinate may also be found using the formula

$$\frac{-b^2 + 4ac}{4a} = \frac{-(-8)^2 + 4 \cdot 2 \cdot 11}{4 \cdot 2} = \frac{24}{8} = 3$$

Since $a > 0$, the vertex $(2, 3)$ is a minimum point and the graph opens upward. You may also find some additional ordered pairs to graph this equation depending on whether you need only a rough sketch or a fairly accurate graph. The graph is shown in Fig. 1.48.

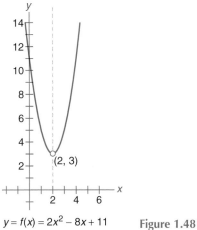

$y = f(x) = 2x^2 - 8x + 11$

Figure 1.48

Since the vertex of a parabola in the form $y = f(x) = ax^2 + bx + c$ is the highest point or the lowest point on the graph, we can use this fact to find a maximum or a minimum value of a quadratic function.

EXAMPLE 10

An object is thrown upward with an initial velocity of 48 ft/s. Its height after t seconds is given by $h = f(t) = 48t - 16t^2$. Find its maximum height and the time it takes the object to hit the ground.

First, find the vertex.

$$-\frac{b}{2a} = -\frac{48}{2(-16)} = \frac{3}{2}$$

Then the vertex is

$$\left(\frac{3}{2}, f\left(\frac{3}{2}\right)\right) = \left(\frac{3}{2}, 48\left(\frac{3}{2}\right) - 16\left(\frac{3}{2}\right)^2\right) = \left(\frac{3}{2}, 36\right)$$

Since $a = -16 < 0$, the vertex is a maximum point, and the maximum height is 36 ft.

The first coordinate of the vertex gives the amount of time it takes for the object to reach its maximum height. The time it takes for such a projectile to reach its maximum height is the same as the time it takes to drop back to the ground. Thus, the object hits the ground $2 \cdot \frac{3}{2}$ or 3 s after it is thrown.

Exercises 1.7

Find the focus and the directrix of each parabola. Sketch each graph.

1. $x^2 = 4y$ **2.** $x^2 = -8y$ **3.** $y^2 = -16x$ **4.** $x^2 = -6y$

5. $y^2 = x$ **6.** $y^2 = -4x$ **7.** $x^2 = 16y$ **8.** $y^2 = -12x$

9. $y^2 = 8x$ **10.** $x^2 = -y$

Find the equation of the parabola with given focus and directrix.

11. $(2, 0), x = -2$ **12.** $(0, -3), y = 3$ **13.** $(-8, 0), x = 8$

14. $(5, 0), x = -5$ **15.** $(0, 6), y = -6$ **16.** $(0, -1), y = 1$

17. Find the equation of the parabola with focus at $(-4, 0)$ and vertex at $(0, 0)$.

18. Find the equation of the parabola with vertex at $(0, 0)$ and directrix $y = -2$.

19. Find the equation of the parabola with focus $(-1, 3)$ and directrix $x = 3$.

20. Find the equation of the parabola with focus $(2, -5)$ and directrix $y = -1$.

21. The surface of a roadway over a bridge follows a parabolic curve with vertex at the middle of the bridge. The span of the bridge is 400 m. The roadway is 16 m higher in the middle than at the end supports. How far above the end supports is a point 50 m from the middle? 150 m from the middle?

22. The shape of a wire hanging between two poles closely approximates a parabola. Find the equation of a wire that is suspended between two poles 40 m apart and whose lowest point is 10 m below the level of the insulators. (Choose the lowest point as the origin of your coordinate system.)

23. A suspension bridge is supported by two cables that hang between two supports. The curve of these cables is approximately parabolic. Find the equation of this curve if the focus lies 8 m above the lowest point of the cable. (Set up the xy-coordinate system so that the vertex is at the origin.)

24. A culvert is shaped like a parabola, 120 cm across the top and 80 cm deep. How wide is the culvert 50 cm from the top?

Graph each parabola. Find its vertex and the equation of the axis.

25. $y = 2x^2 + 7x - 15$ **26.** $y = -x^2 - 6x - 8$

27. $f(x) = -2x^2 + 4x + 16$ **28.** $f(x) = 3x^2 + 6x + 10$

29. Starting at $(0, 0)$, a projectile travels along the path $y = f(x) = -\frac{1}{256}x^2 + 4x$, where x is in metres. Find **(a)** the maximum height and **(b)** the range of the projectile.

30. The height of a bullet fired vertically upward is given by $h = f(t) = 1200t - 16t^2$ (initial velocity is 1200 ft/s). Find **(a)** its maximum height and **(b)** the time it takes to hit the ground.

31. Enclose a rectangular area with 240 m of fencing. Find the largest possible area that can be enclosed.

32. A 36-in.-wide sheet of metal is bent into a rectangular trough with a cross section as shown in Fig. 1.49. What dimensions will maximize the flow of water? That is, what dimensions will maximize the cross-sectional area?

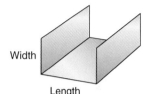

Figure 1.49

1.8   THE ELLIPSE

An **ellipse** consists of the set of points in a plane, the *sum* of whose distances from two fixed points is a positive constant. These two fixed points are called **foci.** As in Fig. 1.50, let the foci lie on the x-axis at $(-c, 0)$ and $(c, 0)$. Then any point $P(x, y)$ lies on the ellipse if its distance d_1 from P to the point $(-c, 0)$ plus its distance d_2 from P to the point $(c, 0)$ is equal to a given constant k. Let the constant be written as $k = 2a$; then

$$d_1 + d_2 = 2a$$

Again using the formula for computing the distance between two points, we have

$$\sqrt{[x - (-c)]^2 + (y - 0)^2} + \sqrt{(x - c)^2 + (y - 0)^2} = 2a$$

Rewrite the previous equation as follows.

$$\sqrt{(x + c)^2 + y^2} = 2a - \sqrt{(x - c)^2 + y^2}$$
$$(x + c)^2 + y^2 = 4a^2 - 4a\sqrt{(x - c)^2 + y^2} + (x - c)^2 + y^2 \quad \text{(Square each side.)}$$
$$x^2 + 2cx + c^2 + y^2 = 4a^2 - 4a\sqrt{(x - c)^2 + y^2} + x^2 - 2cx + c^2 + y^2$$
$$4cx - 4a^2 = -4a\sqrt{(x - c)^2 + y^2}$$
$$a^2 - cx = a\sqrt{(x - c)^2 + y^2} \quad \text{(Divide each side by } -4.\text{)}$$
$$(a^2 - cx)^2 = a^2[(x - c)^2 + y^2] \quad \text{(Square each side.)}$$
$$a^4 - 2a^2cx + c^2x^2 = a^2(x^2 - 2cx + c^2 + y^2)$$
$$a^4 - 2a^2cx + c^2x^2 = a^2x^2 - 2a^2cx + a^2c^2 + a^2y^2$$
$$a^4 - a^2c^2 = a^2x^2 - c^2x^2 + a^2y^2$$
$$a^2(a^2 - c^2) = (a^2 - c^2)x^2 + a^2y^2 \quad \text{(Factor.)}$$
$$1 = \frac{x^2}{a^2} + \frac{y^2}{a^2 - c^2} \quad [\text{Divide each side by } a^2(a^2 - c^2).]$$

If we now let $y = 0$ in this equation, we find that $x^2 = a^2$. The points $(-a, 0)$ and $(a, 0)$, which lie on the graph, are called **vertices** of the ellipse. Observe that $a > c$.

If we let $b^2 = a^2 - c^2$, the proceeding equation then becomes

$$\frac{x^2}{a^2} + \frac{y^2}{b^2} = 1$$

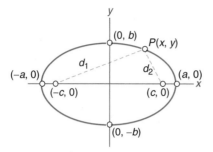

Figure 1.50 An ellipse is a set of points in a plane, the sum of whose distances from two fixed points (called foci) is a positive constant.

The line segment connecting the vertices $(a, 0)$ and $(-a, 0)$ is called the **major axis.** The point midway between the vertices is called the **center** of the ellipse. In this case the major axis lies on the x-axis and the center is at the origin. If we let $x = 0$ in the above equation, we find $y^2 = b^2$. The line connecting $(0, b)$ and $(0, -b)$ is perpendicular to the major axis and passes through the center (see Fig. 1.50). This line is called the **minor axis** of the ellipse. In this case the minor axis lies on the y-axis. Note that $2a$ is the length of the major axis and $2b$ is the length of the minor axis.

> **STANDARD FORM OF ELLIPSE**
>
> $$\frac{x^2}{a^2} + \frac{y^2}{b^2} = 1$$
>
> with center at the origin and with the major axis lying on the x-axis.
> *Note: $a > b$.*

One easy way to approximate the curve of an ellipse is to fix a string at two points (foci) on a piece of paper as in Fig. 1.51(a). Then using a pencil to keep the string taut, trace out the curve as illustrated. Note that $d_1 + d_2$ is always constant—namely, the length of the string. Detach the string and compare the length of the string with the length of the major axis; note that the lengths are the same, $2a$.

The relationship $b^2 = a^2 - c^2$ or $a^2 = b^2 + c^2$ can also be seen from this string demonstration as in Fig. 1.51(b). Put a pencil inside the taut string and on an end of the minor axis; this bisects the length of string and sets up a right triangle with a as its hypotenuse and b and c as the legs.

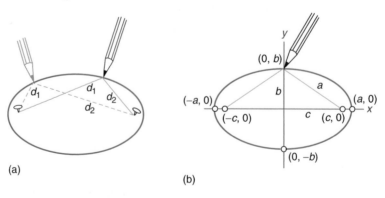

(a)

(b)

Figure 1.51 Drawing an ellipse with a string.

EXAMPLE 1

Find the vertices, the foci, and the lengths of the major and minor axes of the ellipse

$$\frac{x^2}{25} + \frac{y^2}{9} = 1$$

Sketch the graph.

Since $a^2 = 25$, the vertices are at $(5, 0)$ and $(-5, 0)$. The length of the major axis is $2a = 2(5) = 10$. Since $b^2 = 9$, then $b = 3$, and the length of the minor axis is $2b = 2(3) = 6$. We need the value of c to determine the foci. Since $b^2 = a^2 - c^2$, we can write

$$c^2 = a^2 - b^2 = 25 - 9 = 16$$
$$c = 4$$

The foci are thus $(4, 0)$ and $(-4, 0)$. The graph of the ellipse is shown in Fig. 1.52.

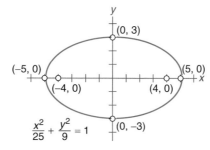

Figure 1.52

You will want to remember the equation relating a, b, and c for the ellipse.

$$c^2 = a^2 - b^2$$

When the major axis lies on the y-axis with center at the origin as in Fig. 1.53, the **standard form** of the equation of the ellipse becomes

STANDARD FORM OF ELLIPSE

$$\frac{y^2}{a^2} + \frac{x^2}{b^2} = 1$$

with center at the origin and with the major axis lying on the y-axis.
Note: $a > b$

This result may be shown similarly as the derivation of the first standard form. Notice that the larger denominator now lies below y^2 instead of below x^2 as in the first case. The vertices are now $(0, a)$ and $(0, -a)$.

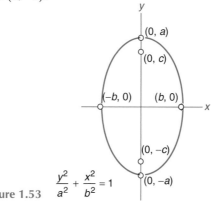

Figure 1.53

EXAMPLE 2

Given the ellipse $25x^2 + 9y^2 = 225$, find the foci, vertices, and lengths of the major and minor axes. Sketch the graph.

First divide each side of the equation by 225 to put the equation in standard form.

$$\frac{x^2}{9} + \frac{y^2}{25} = 1$$

Since the larger denominator belongs to the y^2 term, this ellipse has its major axis on the y-axis and a^2 must then be 25. So $a = 5$ and $b = 3$. The vertices are $(0, 5)$ and $(0, -5)$. The length of the major axis is $2a = 10$, and the length of the minor axis is $2b = 6$.

$$c^2 = a^2 - b^2 = 25 - 9 = 16$$
$$c = 4$$

Thus the foci are $(0, 4)$ and $(0, -4)$ (see Fig. 1.54).

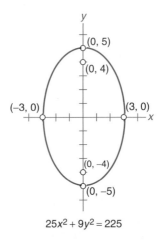

$$25x^2 + 9y^2 = 225$$

Figure 1.54

In general, a is always greater than b for an ellipse. The following are also true:

1. If the larger denominator belongs to the x^2-term, then this denominator is a^2, the major axis lies on the x-axis, and the vertices are $(a, 0)$ and $(-a, 0)$.

2. If the larger denominator belongs to the y^2-term, then this denominator is a^2, the major axis lies on the y-axis, and the vertices are $(0, a)$ and $(0, -a)$.

EXAMPLE 3

Find the equation of the ellipse with vertices at $(6, 0)$ and $(-6, 0)$ and foci at $(4, 0)$ and $(-4, 0)$.
Since $a = 6$ and $c = 4$, we have $a^2 = 36$ and $c^2 = 16$. Thus

$$b^2 = a^2 - c^2 = 36 - 16 = 20$$

Since the major axis lies on the x-axis, the equation in standard form is

$$\frac{x^2}{a^2} + \frac{y^2}{b^2} = 1$$
$$\frac{x^2}{36} + \frac{y^2}{20} = 1$$

Ellipses with centers not located at the origin will be presented in Section 1.10. If we were to determine the equation of the ellipse where the sum of the distances of all points from the foci $(-2, 3)$ and $(6, 3)$ is always 10, we would have

$$9x^2 + 25y^2 - 36x - 150y + 36 = 0$$

In general, an equation of the form

$$Ax^2 + Cy^2 + Dx + Ey + F = 0$$

represents an ellipse with axes parallel to the coordinate axes, where A and C are both positive (or both negative) and, unlike a circle, $A \neq C$.

Exercises 1.8

Find the vertices, foci, and lengths of the major and minor axes of each ellipse. Sketch each graph.

1. $\dfrac{x^2}{25} + \dfrac{y^2}{16} = 1$ **2.** $\dfrac{x^2}{36} + \dfrac{y^2}{64} = 1$ **3.** $9x^2 + 16y^2 = 144$

4. $25x^2 + 16y^2 = 400$ **5.** $36x^2 + y^2 = 36$ **6.** $4x^2 + 3y^2 = 12$

7. $16x^2 + 9y^2 = 144$ **8.** $x^2 + 4y^2 = 16$

Find the equation of each ellipse satisfying the given conditions.

9. Vertices at $(4, 0)$ and $(-4, 0)$; foci at $(2, 0)$ and $(-2, 0)$.

10. Vertices at $(0, 7)$ and $(0, -7)$; foci at $(0, 5)$ and $(0, -5)$.

11. Vertices at $(0, 9)$ and $(0, -9)$; foci at $(0, 6)$ and $(0, -6)$.

12. Vertices at $(12, 0)$ and $(-12, 0)$; foci at $(10, 0)$ and $(-10, 0)$.

13. Vertices at $(6, 0)$ and $(-6, 0)$; length of minor axis is 10.

14. Vertices at $(0, 10)$ and $(0, -10)$; length of minor axis is 18.

15. Foci at $(0, 5)$ and $(0, -5)$; length of major axis is 16.

16. Foci at $(3, 0)$ and $(-3, 0)$; length of major axis is 8.

17. A weather satellite with an orbit about the earth reaches a minimum altitude of 1000 mi and a maximum altitude of 1600 mi. The path of its orbit is approximately an ellipse with the center of the earth at one focus. Find the equation of this curve. Assume the radius of the earth is 4000 mi and the x-axis is the major axis.

18. An arch is in the shape of the upper half of an ellipse with a horizontal major axis supporting a foot bridge 40 m long over a stream in a park. The center of the arch is 8 m above the bridge supports. Find an equation of the ellipse. (Choose the point midway between the bridge supports as the origin.)

1.9 THE HYPERBOLA

A **hyperbola** consists of the set of points in a plane, the *difference* of whose distances from two fixed points is a positive constant. The two fixed points are called the **foci.**

Assume now as in Fig. 1.55 that the foci lie on the x-axis at $(-c, 0)$ and $(c, 0)$. Then a point $P(x, y)$ lies on the hyperbola if the difference between its distances to the foci is

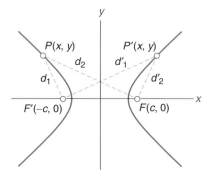

Figure 1.55 A hyperbola is the set of points in a plane, the difference of whose distances from two fixed points (called foci) is a positive constant.

equal to a given constant k. That is, $d_1 - d_2 = k$ or $d_2 - d_1 = k$. Again, this constant k equals $2a$; that is,

$$d_2 - d_1 = 2a$$

To obtain the equation of the hyperbola, use the distance formula.

$$d_2 - d_1 = 2a$$

$$\sqrt{(x - c)^2 + (y - 0)^2} - \sqrt{[x - (-c)]^2 + (y - 0)^2} = 2a$$

Rewrite the equation above as follows:

$$\sqrt{(x - c)^2 + y^2} = 2a + \sqrt{(x + c)^2 + y^2}$$
$$(x - c)^2 + y^2 = 4a^2 + 4a\sqrt{(x + c)^2 + y^2} + (x + c)^2 + y^2 \quad \text{(Square each side.)}$$
$$x^2 - 2cx + c^2 + y^2 = 4a^2 + 4a\sqrt{(x + c)^2 + y^2} + x^2 + 2cx + c^2 + y^2$$
$$-4a^2 - 4cx = 4a\sqrt{(x + c)^2 + y^2}$$
$$-a^2 - cx = a\sqrt{(x + c)^2 + y^2} \quad \text{(Divide each side by 4.)}$$
$$a^4 + 2a^2cx + c^2x^2 = a^2[(x + c)^2 + y^2] \quad \text{(Square each side.)}$$
$$a^4 + 2a^2cx + c^2x^2 = a^2(x^2 + 2cx + c^2 + y^2)$$
$$a^4 + 2a^2cx + c^2x^2 = a^2x^2 + 2a^2cx + a^2c^2 + a^2y^2$$
$$a^4 - a^2c^2 = a^2x^2 - c^2x^2 + a^2y^2$$
$$a^2(a^2 - c^2) = (a^2 - c^2)x^2 + a^2y^2 \quad \text{(Factor.)}$$
$$1 = \frac{x^2}{a^2} + \frac{y^2}{a^2 - c^2} \quad \text{[Divide each side by } a^2(a^2 - c^2).\text{]}$$
$$1 = \frac{x^2}{a^2} - \frac{y^2}{c^2 - a^2}$$

In triangle $F'PF$

$$PF' < PF + FF' \quad \text{(The sum of any two sides of a triangle is greater than the third side.)}$$

$$PF' - PF < FF'$$
$$2a < 2c \quad (PF' - PF = 2a \text{ by the definition of a hyperbola and } FF' = 2c.)$$
$$a < c$$
$$a^2 < c^2 \quad \text{(Since } a > 0 \text{ and } c > 0.)$$
$$0 < c^2 - a^2$$

Since $c^2 - a^2$ is positive, we may replace it by the positive number, b^2, as follows:

$$1 = \frac{x^2}{a^2} - \frac{y^2}{b^2}$$

where $b^2 = c^2 - a^2$.

The equation of the hyperbola with foci on the x-axis at $(c, 0)$ and $(-c, 0)$ is

$$\frac{x^2}{a^2} - \frac{y^2}{b^2} = 1$$

The points $(a, 0)$ and $(-a, 0)$ are called the **vertices.** The line segment connecting the vertices is called the **transverse axis.** The vertices and transverse axis in this case lie on the

x-axis. The length of the transverse axis is $2a$. The line segment connecting the points $(0, b)$ and $(0, -b)$ is called the **conjugate axis** and in this case lies on the y-axis. The length of the conjugate axis is $2b$. The **center** lies at the intersection of the conjugate and transverse axes.

STANDARD FORM OF HYPERBOLA

$$\frac{x^2}{a^2} - \frac{y^2}{b^2} = 1$$

with center at the origin and with the transverse axis lying on the x-axis.

If we draw the central rectangle as in Fig. 1.56 and draw lines passing through opposite vertices of the rectangle, we obtain lines called the **asymptotes** of the hyperbola. In this case the equations of these lines are

$$y = \frac{b}{a}x$$

$$y = -\frac{b}{a}x$$

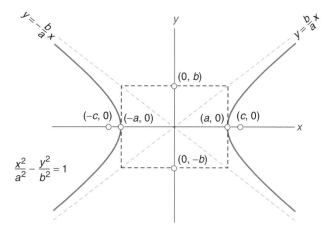

Figure 1.56

Asymptotes serve as guidelines to the branches of the hyperbola. That is, as the distance from the center of the hyperbola increases, the points on the branches get closer and closer to the asymptotes but never cross or touch them.

To sketch the graph of the hyperbola:

1. Locate the vertices $(a, 0)$ and $(-a, 0)$.
2. Locate the points $(0, b)$ and $(0, -b)$.
3. Sketch the central rectangle as in Fig. 1.56. [The coordinates of the vertices are $(a, b), (a, -b), (-a, b)$, and $(-a, -b)$.]
4. Sketch the two asymptotes (the lines passing through the pairs of opposite vertices of the rectangle).
5. Sketch the branches of the hyperbola.

EXAMPLE 1

Find the vertices, foci, and lengths of the transverse and conjugate axes of the hyperbola

$$\frac{x^2}{9} - \frac{y^2}{16} = 1$$

Sketch the graph. Find the equations of the asymptotes.

Since 9 is the denominator of the x^2-term, $a^2 = 9$ and $a = 3$. The vertices are therefore $(3, 0)$ and $(-3, 0)$, and the length of the transverse axis is $2a = 2(3) = 6$. Since 16 is the denominator of the y^2-term, $b^2 = 16$ and $b = 4$. So the length of the conjugate axis is $2b = 2(4) = 8$.

To find the foci we need to know c^2. Since $b^2 = c^2 - a^2$, we have

$$c^2 = a^2 + b^2 = (3)^2 + (4)^2 = 25$$
$$c = 5$$

The foci are $(5, 0)$ and $(-5, 0)$. The asymptotes are $y = \frac{4}{3}x$ and $y = -\frac{4}{3}x$ (see Fig. 1.57).

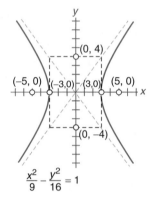

Figure 1.57

You will want to remember the equation relating a, b, and c for the hyperbola.

$$c^2 = a^2 + b^2$$

EXAMPLE 2

Write the equation of the hyperbola with foci at $(5, 0)$ and $(-5, 0)$ and whose transverse axis is 8 units in length.

Here we have $c = 5$. Since $2a = 8$, $a = 4$,

$$c^2 = a^2 + b^2$$
$$25 = 16 + b^2$$
$$b^2 = 9$$

The equation is then

$$\frac{x^2}{16} - \frac{y^2}{9} = 1$$

STANDARD FORM OF HYPERBOLA

$$\frac{y^2}{a^2} - \frac{x^2}{b^2} = 1$$

with center at the origin and with the transverse axis lying on the y-axis.

We obtain a graph as shown in Fig. 1.58.

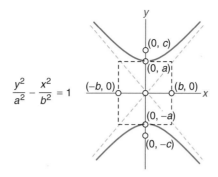

$$\frac{y^2}{a^2} - \frac{x^2}{b^2} = 1$$

Figure 1.58

Note that the difference between this equation and the first equation is that a^2 is now the denominator of the y^2-term, which is the positive term. This means that the vertices (and transverse axis) now lie on the y-axis.

The equations of the asymptotes are

$$y = \frac{a}{b}x$$

$$y = -\frac{a}{b}x$$

In general, the positive term indicates on which axis the vertices, foci, and transverse axis lie.

1. If the x^2 term is positive, then the denominator of x^2 is a^2, and the denominator of y^2 is b^2. The transverse axis lies along the x-axis, and the vertices are $(a, 0)$ and $(-a, 0)$.

2. If the y^2 term is positive, then the denominator of y^2 is a^2, and the denominator of x^2 is b^2. The transverse axis lies along the y-axis, and the vertices are $(0, a)$ and $(0, -a)$.

EXAMPLE 3

Sketch the graph of the hyperbola

$$\frac{y^2}{36} - \frac{x^2}{49} = 1$$

Since the y^2-term is positive, the vertices lie on the y-axis and $a^2 = 36$. Then $b^2 = 49$, $a = 6$, and $b = 7$. The graph is sketched in Fig. 1.59.

EXAMPLE 4

Write the equation of the hyperbola with foci at $(0, 8)$ and $(0, -8)$, and vertices at $(0, 6)$ and $(0, -6)$.

In this case $a = 6$ and $c = 8$, so $b^2 = c^2 - a^2 = 64 - 36 = 28$. Since the vertices and foci lie on the y-axis, the y^2-term is positive with denominator a^2. The equation is

$$\frac{y^2}{36} - \frac{x^2}{28} = 1$$

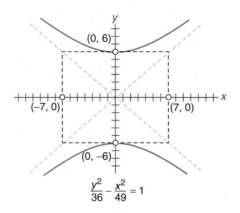

$$\frac{y^2}{36} - \frac{x^2}{49} = 1$$

Figure 1.59

As with the ellipse, not all hyperbolas are located with their centers at the origin. We have seen the standard forms of the equation of the hyperbola with center at the origin and whose transverse and conjugate axes lie on the x-axis and y-axis. In general, however, the equation of a hyperbola is of the form

$$Ax^2 + Bxy + Cy^2 + Dx + Ey + F = 0$$

where either (1) $B = 0$ and A and C differ in sign or (2) $A = 0$, $C = 0$, and $B \neq 0$.

A simple example of this last case is the equation $xy = k$. The foci and vertices lie on the line $y = x$ if $k > 0$ or on the line $y = -x$ if $k < 0$.

EXAMPLE 5

Sketch the graph of the hyperbola $xy = -6$.

Since there are no easy clues for sketching this equation (unlike hyperbolas in standard position), set up a table of values for x and y. Then plot the corresponding points in the plane as in Fig. 1.60.

x	y
6	-1
3	-2
2	-3
1	-6
-1	6
-2	3
-3	2
-6	1

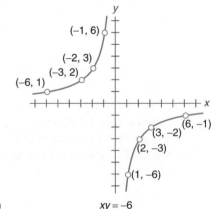

$xy = -6$

Figure 1.60

Exercises 1.9

Find the vertices, foci, and lengths of the transverse and conjugate axes of each hyperbola. Find the equations of the asymptotes and sketch each graph.

1. $\dfrac{x^2}{25} - \dfrac{y^2}{144} = 1$ **2.** $\dfrac{x^2}{144} - \dfrac{y^2}{25} = 1$ **3.** $\dfrac{y^2}{9} - \dfrac{x^2}{16} = 1$

4. $\dfrac{y^2}{16} - \dfrac{x^2}{9} = 1$ **5.** $5x^2 - 2y^2 = 10$ **6.** $3y^2 - 2x^2 = 6$

7. $4y^2 - x^2 = 4$ **8.** $4x^2 - y^2 = 4$

Find the equation of the hyperbola satisfying each of the given conditions.

9. Vertices at $(4, 0)$ and $(-4, 0)$; foci at $(6, 0)$ and $(-6, 0)$.

10. Vertices at $(0, 5)$ and $(0, -5)$; foci at $(0, 7)$ and $(0, -7)$.

11. Vertices at $(0, 6)$ and $(0, -6)$; foci at $(0, 8)$ and $(0, -8)$.

12. Vertices at $(2, 0)$ and $(-2, 0)$; foci at $(5, 0)$ and $(-5, 0)$.

13. Vertices at $(3, 0)$ and $(-3, 0)$; length of conjugate axis is 10.

14. Vertices at $(0, 6)$ and $(0, -6)$; length of conjugate axis is 8.

15. Foci at $(6, 0)$ and $(-6, 0)$; length of transverse axis is 10.

16. Foci at $(0, 8)$ and $(0, -8)$; length of transverse axis is 12.

17. Sketch the graph of the hyperbola given by $xy = 8$.

18. Sketch the graph of the hyperbola given by $xy = -4$.

1.10   TRANSLATION OF AXES

We have seen the difficulty in determining the equations of the parabola, ellipse, and hyperbola when these are not in standard position in the plane. It is still possible to find the equation of these curves fairly easily if the axes of these curves lie on lines parallel to the coordinate axes. This is accomplished by the translation of axes. We shall demonstrate this method with four examples.

EXAMPLE 1

Find the equation of the ellipse with foci at $(-2, 3)$ and $(6, 3)$, and vertices at $(-3, 3)$ and $(7, 3)$.

The center of the ellipse is at $(2, 3)$, which is midway between the foci or the vertices. The distance between the foci $(-2, 3)$ and $(6, 3)$ is 8. So $c = 4$. The distance between $(-3, 3)$ and $(7, 3)$ is 10. So $a = 5$.

$$c^2 = a^2 - b^2$$
$$16 = 25 - b^2$$
$$b^2 = 9$$
$$b = 3$$

Sketch the graph as in Fig. 1.61(a). Next, plot the same ellipse in another coordinate system with center at the origin as in Fig. 1.61(b). Label the coordinate axes of this new system x' and y'. We know that in this $x'y'$-coordinate system, the equation for this ellipse is

$$\frac{(x')^2}{a^2} + \frac{(y')^2}{b^2} = 1$$

And, since $a = 5$ and $b = 3$, we have

$$\frac{(x')^2}{25} + \frac{(y')^2}{9} = 1$$

Each point on the ellipse can now be seen as having coordinates (x, y) in the xy-plane and coordinates (x', y') in the $x'y'$-plane. If we compare coordinates in the two coordinate systems, we see, for example, that the right-hand vertex has coordinates $(7, 3)$ in the xy-plane, but the

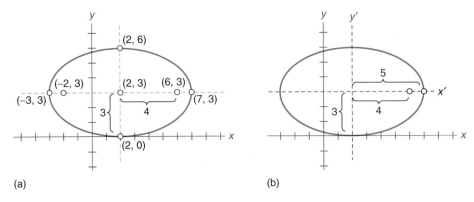

(a)　　　　　　　　　　　　　(b)

Figure 1.61

same point has coordinates $(5, 0)$ in the $x'y'$-plane. Likewise, the point at the upper end of the minor axis has coordinates $(2, 6)$ in the xy-plane, but the same point has coordinates $(0, 3)$ in the $x'y'$-plane.

In general, the x- and x'-coordinates are related as follows.

$$x = x' + 2$$

That is, the original x-coordinates are 2 larger than the new x'-coordinates. Note that this is the distance that the new origin was moved along the x-axis: the x-coordinate of the center of the ellipse. (See Fig. 1.61b.)

Similarly, the y- and y'-coordinates are related as follows.

$$y = y' + 3$$

Note that 3 is the distance that the new origin was moved along the y-axis: the y-coordinate of the center of the ellipse. (See Fig. 1.61b.) We now rearrange terms and have

$$x' = x - 2$$
$$y' = y - 3$$

Now replace x' by $x - 2$ and y' by $y - 3$ in the equation

$$\frac{(x')^2}{25} + \frac{(y')^2}{9} = 1$$

$$\frac{(x - 2)^2}{25} + \frac{(y - 3)^2}{9} = 1$$

This is the equation of the ellipse with center at $(2, 3)$ in the xy-plane.

To write an equation for a parabola, ellipse, or hyperbola whose axes are parallel to the x-axis and y-axis,

1. For a parabola, identify (h, k) as the vertex; for an ellipse or hyperbola identify (h, k) as the center.

2. Translate xy-coordinates to a new $x'y'$-coordinate system by using the translation equations

$$x' = x - h$$
$$y' = y - k$$

where (h, k) has been identified as in Step 1.

3. Write the equation of the conic, which is now in standard position in the $x'y'$-coordinate system.

4. Translate the equation derived in Step 3 back into the original coordinate system by making the following substitutions for x and y into the dervied equation.

$$x' = x - h$$
$$y' = y - k$$

The resulting equation is an equation for the conic in the xy-coordinate system.

EXAMPLE 2

Find the equation of the parabola with focus $(-1, 2)$ and directrix $x = -7$.

Step 1: The vertex of this parabola is halfway along the line $y = 2$ between the focus $(-1, 2)$ and the directrix $x = -7$ (see Fig. 1.62). Thus the vertex has coordinates $(-4, 2)$. This becomes the origin of the new coordinate system, so $h = -4$ and $k = 2$.

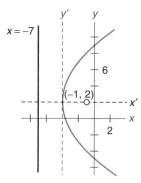

Figure 1.62

Step 2: $x' = x - h = x - (-4) = x + 4$

$y' = y - k = y - 2$

The new $x'y'$-coordinates of the focus become

$$x' = x - h = -1 - (-4) = 3$$
$$y' = y - k = 2 - 2 = 0$$

and the equation of the directrix $x = -7$ becomes

$$x' = x - h = -7 - (-4)$$
$$x' = -3$$

Step 3: Since the parabola is now in standard position in the new coordinate system with focus $(3, 0)$, we have $p = 3$. The equation in this system becomes

$$(y')^2 = 4px'$$
$$(y')^2 = 4(3)x'$$
$$(y')^2 = 12x'$$

Step 4: Replace x' with $x + 4$ and y' with $y - 2$.

$$(y')^2 = 12x'$$
$$(y - 2)^2 = 12(x + 4)$$
$$y^2 - 4y - 12x - 44 = 0$$

We sometimes know the equation of a curve and need to identify the curve and sketch its graph, as in the following example.

EXAMPLE 3

Describe and sketch the graph of the equation

$$\frac{(y-4)^2}{9} - \frac{(x+2)^2}{16} = 1$$

If we let

$$x' = x - h = x + 2 = x - (-2)$$
$$y' = y - k = y - 4$$

we have

$$\frac{(y')^2}{9} - \frac{(x')^2}{16} = 1$$

This is the equation of a hyperbola with center at $(-2, 4)$. Since $a^2 = 9$ and $b^2 = 16$, we have

$$c^2 = a^2 + b^2 = 9 + 16 = 25$$

so

$$a = 3, \quad b = 4, \quad \text{and} \quad c = 5$$

In terms of the $x'y'$-coordinates, the foci are at $(0, 5)$ and $(0, -5)$, the vertices are at $(0, 3)$ and $(0, -3)$, the length of the transverse axis is 6, and the length of the conjugate axis is 8.

To translate the $x'y'$-coordinates to xy-coordinates, we use the equations

$$x = x' + h \qquad y = y' + k$$

In this case

$$x = x' + (-2) \qquad y = y' + 4$$

So in the xy-plane the foci are at $(-2, 9)$ and $(-2, -1)$; the vertices are at $(-2, 7)$ and $(-2, 1)$; the length of the transverse axis is 6; and the length of the conjugate axis is 8 (see Fig. 1.63).

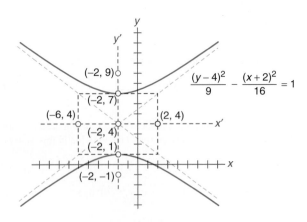

Figure 1.63

EXAMPLE 4

Name the equation $16x^2 + 9y^2 + 64x + 54y + 1 = 0$. Locate the vertex if it is a parabola or the center if it is an ellipse or a hyperbola.

First complete the square for x and y. (See Section 1.6.)

$(16x^2 + 64x \quad) + (9y^2 + 54y \quad) = -1$

$16(x^2 + 4x \quad) + 9(y^2 + 6y \quad) = -1$ (Factor out the coefficients of x^2 and y^2 before completing the square. The coefficients of x^2 and y^2 must be **one**.)

$16(x^2 + 4x + 4) + 9(y^2 + 6y + 9) = -1 + 16(4) + 9(9)$

$16(x + 2)^2 + 9(y + 3)^2 = 144$

$\dfrac{(x + 2)^2}{9} + \dfrac{(y + 3)^2}{16} = 1$ (Divide each side by 144.)

This is an equation of an ellipse. Noting that

$$x' = x - h = x + 2 = x - (-2)$$
$$y' = y - k = y + 3 = y - (-3)$$

we see that the center is at $(-2, -3)$.

GENERAL FORMS OF CONICS WITH AXES PARALLEL TO THE COORDINATE AXES

1. $(y - k)^2 = 4p(x - h)$
   is a parabola with vertex at (h, k) and axis parallel to the x-axis.

2. $(x - h)^2 = 4p(y - k)$
   is a parabola with vertex at (h, k) and axis parallel to the y-axis.

3. $\dfrac{(x - h)^2}{a^2} + \dfrac{(y - k)^2}{b^2} = 1 \quad (a > b)$
   is an ellipse with center at (h, k) and major axis parallel to the x-axis.

4. $\dfrac{(y - k)^2}{a^2} + \dfrac{(x - h)^2}{b^2} = 1 \quad (a > b)$
   is an ellipse with center at (h, k) and major axis parallel to the y-axis.

5. $\dfrac{(x - h)^2}{a^2} - \dfrac{(y - k)^2}{b^2} = 1$
   is a hyperbola with center at (h, k) and transverse axis parallel to the x-axis.

6. $\dfrac{(y - k)^2}{a^2} - \dfrac{(x - h)^2}{b^2} = 1$
   is a hyperbola with center at (h, k) and transverse axis parallel to the y-axis.

Exercises 1.10

Find the equation of each curve from the given information.

1. Ellipse with center at $(1, -1)$, vertices at $(5, -1)$ and $(-3, -1)$, and foci at $(3, -1)$ and $(-1, -1)$.

2. Parabola with vertex at $(-1, 3)$, focus at $(-1, 4)$, and directrix $y = 2$.

3. Hyperbola with center at $(1, 1)$, vertices at $(1, 7)$ and $(1, -5)$, and foci at $(1, 9)$ and $(1, -7)$.

4. Ellipse with center at $(-2, -3)$, vertices at $(4, -3)$ and $(-8, -3)$, and minor axis length 10.

5. Parabola with vertex at $(3, -1)$, focus at $(5, -1)$, and directrix $x = 1$.

6. Hyperbola with center at $(-2, -2)$, vertices at $(1, -2)$ and $(-5, -2)$, and conjugate axis length 10.

Name and graph each equation.

7. $(x - 2)^2 = 4(y + 3)$

8. $\dfrac{(x + 1)^2}{36} + \dfrac{(y - 2)^2}{64} = 1$

9. $\dfrac{y^2}{9} - \dfrac{(x + 2)^2}{16} = 1$

10. $y^2 = 8(x + 1)$

11. $9(x - 2)^2 + 16y^2 = 144$

12. $\dfrac{(x + 1)^2}{9} - \dfrac{(y + 3)^2}{16} = 1$

13. $\dfrac{(x - 3)^2}{36} + \dfrac{(y - 1)^2}{16} = 1$

14. $\dfrac{(x - 3)^2}{36} - \dfrac{(y - 1)^2}{16} = 1$

15. $(y + 3)^2 = 8(x - 1)$

16. $(x - 5)^2 = 12(y + 2)$

17. $\dfrac{(y + 1)^2}{9} - \dfrac{(x + 1)^2}{9} = 1$

18. $\dfrac{(y + 4)^2}{4} + \dfrac{(x - 2)^2}{9} = 1$

Name and sketch the graph of each equation. Locate the vertex if it is a parabola or the center if it is an ellipse or a hyperbola.

19. $x^2 - 4x + 2y + 6 = 0$

20. $9x^2 + 4y^2 - 18x + 24y + 9 = 0$

21. $x^2 + 4y^2 + 4x - 8y - 8 = 0$

22. $-2x^2 + 3y^2 + 8x - 14 = 0$

23. $4x^2 - y^2 - 8x + 2y + 3 = 0$

24. $y^2 + 6y - x + 12 = 0$

25. $25y^2 - 4x^2 - 24x - 150y + 89 = 0$

26. $25x^2 + 9y^2 - 100x - 54y - 44 = 0$

27. $x^2 + 16x - 12y + 40 = 0$

28. $9x^2 - 4y^2 + 54x + 40y - 55 = 0$

29. $4x^2 + y^2 + 48x + 4y + 84 = 0$

30. $y^2 - 10x - 6y + 39 = 0$

1.11   THE GENERAL SECOND-DEGREE EQUATION

The circle, parabola, ellipse, and hyperbola are all special cases of the second-degree equation

$$Ax^2 + Bxy + Cy^2 + Dx + Ey + F = 0$$

When $B = 0$ and at least one of the coefficients A or C is not zero, the following summarizes the conditions for each curve.

1. If $A = C$, we have a *circle*.
   In special cases, the graph of the equation may be a point, or there may be no graph. (The equation may have only one or no solution.)

2. If $A = 0$ and $C \neq 0$, or $C = 0$ and $A \neq 0$, then we have a *parabola*.

3. If $A \neq C$, and A and C are either both positive or both negative, then we have an *ellipse*.
   In special cases, the graph of the equation may be a point, or there may be no graph. (The equation may have only one or no solution.)

4. If A is positive and C is negative, or A is negative and C is positive, then we have a *hyperbola*.
   In some special cases the graph may be a pair of intersecting lines.

If $D \neq 0$ or $E \neq 0$ or both are not equal to zero, the curve does not have its center (or vertex in the case of the parabola) at the origin (see Section 1.10). If $B \neq 0$, then the axis of the curve does not lie along the x-axis or y-axis. The hyperbola $xy = k$ is the only such example we have studied (see Section 1.9).

EXAMPLE

Identify the curve

$$x^2 + 3y^2 - 2x + 4y - 7 = 0$$

Since $A \neq C$, A and C are both positive, and $B = 0$, the curve is an ellipse. (The center is not the origin since $D \neq 0$ and $E \neq 0$.)

The curves represented by the second-degree equation

$$Ax^2 + Bxy + Cy^2 + Dx + Ey + F = 0$$

are called **conic sections** because they can be obtained by cutting the cones with a plane, as in Fig. 1.64.

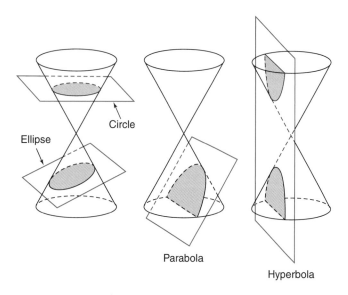

Figure 1.64 Conic sections

Exercises 1.11

Determine whether each equation represents a circle, a parabola, an ellipse, or a hyperbola.

1. $x^2 + 3y^2 + 4x - 5y - 40 = 0$ **2.** $x^2 + y^2 + 4x - 6y - 12 = 0$

3. $4y^2 - 8y + 3x - 2 = 0$ **4.** $9x^2 + 4y^2 + 36x - 8y + 4 = 0$

5. $4x^2 - 5y^2 - 16x + 10y + 20 = 0$ **6.** $x^2 + y^2 + 3x - 2y - 14 = 0$

7. $3x^2 + 3y^2 + x - y - 6 = 0$ **8.** $x^2 + 4x - 3y - 52 = 0$

9. $x^2 + y^2 + 2x - 3y - 21 = 0$ **10.** $x^2 - y^2 - 6x + 3y - 100 = 0$

11. $9x^2 + 4y^2 - 18x + 8y + 4 = 0$ **12.** $3x^2 - 2y^2 + 6x - 8y - 17 = 0$

13. $3x^2 - 3y^2 - 2x - 4y - 13 = 0$ **14.** $4x^2 + 4y^2 - 16x - 4y - 5 = 0$

15. $x^2 - 6x - 6y + 3 = 0$ **16.** $4x^2 - 4x - 4y - 5 = 0$

1.12 SYSTEMS OF QUADRATIC EQUATIONS

To solve systems of equations involving conics algebraically, try to eliminate a variable.

Substitution Method

EXAMPLE 1

Solve the system of equations using the substitution method.

$$y^2 = x$$
$$y = x - 2$$

Since $y^2 = x$, we can substitute y^2 for x in the second equation.

$$y = x - 2$$
$$y = (y^2) - 2$$
$$y^2 - y - 2 = 0$$
$$(y - 2)(y + 1) = 0$$

So

$$y = 2 \quad \text{or} \quad y = -1$$

Substituting these values for y in the second equation, $y = x - 2$, we have $x = 1$ when $y = -1$ and $x = 4$ when $y = 2$.

The solutions of the system are then $(1, -1)$ and $(4, 2)$. Check by substituting the solutions in each original equation.

	$y^2 = x$		$y = x - 2$
$(1, -1)$	$(-1)^2 = (1)$		$(-1) = (1) - 2$
	$1 = 1$		$-1 = -1$
$(4, 2)$	$(2)^2 = (4)$		$(2) = (4) - 2$
	$4 = 4$		$2 = 2$

(See Fig. 1.65 for a graphical solution.)

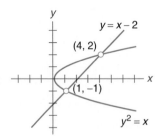

Figure 1.65

EXAMPLE 2

Solve the system of equations using the substitution method.

$$4x^2 - 3y^2 = 4$$
$$x^2 - 4x + y^2 = 0$$

If we solve the second equation for y^2, we have $y^2 = 4x - x^2$. We can now substitute $4x - x^2$ for y^2 in the first equation.

$$4x^2 - 3(4x - x^2) = 4$$
$$4x^2 - 12x + 3x^2 = 4$$
$$7x^2 - 12x - 4 = 0$$
$$(x - 2)(7x + 2) = 0$$
$$x = 2 \quad \text{or} \quad x = -\frac{2}{7}$$

To find y we substitute each of these values for x in one of the original equations. We use the second equation.

For $x = 2$:

$$(2)^2 - 4(2) + y^2 = 0$$
$$4 - 8 + y^2 = 0$$
$$y^2 = 4$$

Thus $y = 2$ or -2 when $x = 2$.

For $x = -\frac{2}{7}$:

$$\left(-\frac{2}{7}\right)^2 - 4\left(-\frac{2}{7}\right) + y^2 = 0$$

$$\frac{4}{49} + \frac{8}{7} + y^2 = 0$$

$$y^2 = -\frac{60}{49}$$

Since y^2 can never be negative, we conclude that there are no real solutions when $x = -\frac{2}{7}$. (This is what we call an *extraneous root.*) The solutions of the system are $(2, 2)$ and $(2, -2)$. The solutions should be checked in each original equation (see Fig. 1.66 for a graphical solution).

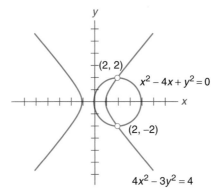

Figure 1.66

Addition-Subtraction Method

EXAMPLE 3

Solve the system of equations using the addition-subtraction method.

$$3x^2 + y^2 = 14$$
$$x^2 - y^2 = 2$$

By adding the two equations we can eliminate the y^2-term.

$$3x^2 + y^2 = 14$$
$$\underline{x^2 - y^2 = 2}$$
$$4x^2 = 16$$

Solving for x,

$$x^2 = 4$$
$$x = 2 \quad \text{or} \quad x = -2$$

Using the second equation, $x^2 - y^2 = 2$, we find the corresponding values for y. For $x = 2$:

$$(2)^2 - y^2 = 2$$
$$4 - y^2 = 2$$
$$y^2 = 2$$
$$y = \sqrt{2} \quad \text{or} \quad y = -\sqrt{2}$$

For $x = -2$:

$$(-2)^2 - y^2 = 2$$
$$4 - y^2 = 2$$
$$y^2 = 2$$
$$y = \sqrt{2} \quad \text{or} \quad y = -\sqrt{2}$$

The solutions are $(2, \sqrt{2})$, $(2 - \sqrt{2})$, $(-2, \sqrt{2})$, and $(-2, -\sqrt{2})$. These should be checked in each original equation.

When solving a system of two equations where one represents a conic and the other a line, the substitution method is usually preferred. Example 4 shows how the addition-subtraction method may be helpful in some cases.

EXAMPLE 4

Solve the system of equations.

$$y + 6x = 2$$
$$y^2 = 6x$$

We use the addition-subtraction method.

$$y + 6x = 2$$
$$\underline{y^2 - 6x = 0}$$
$$y^2 + y = 2 \qquad \text{(Add.)}$$
$$y^2 + y - 2 = 0$$
$$(y + 2)(y - 1) = 0$$
$$y = -2 \quad \text{or} \quad y = 1$$

Using the first equation we find that when $y = -2$, $x = \frac{2}{3}$ and when $y = 1$, $x = \frac{1}{6}$, the solutions are $(\frac{2}{3}, -2)$ and $(\frac{1}{6}, 1)$. These solutions should be checked in each original equation.

Using a calculator,

2nd CUSTOM F3 4 *twelve* left arrows y+6x=2 *five* right arrows y^2=6x **ENTER** up arrow **2nd** right arrow

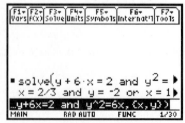
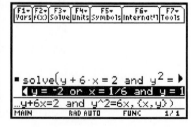

Exercises 1.12

Solve each system of equations.

1. $x^2 = 3y$
 $y = 2x - 3$

2. $x^2 - 2y^2 = 1$
 $3x^2 + 2y^2 = 3$

3. $x^2 + 4x + y^2 - 8 = 0$
 $x^2 + y^2 = 4$

4. $x^2 + 2y^2 = 12$
 $y = -x$

5. $y^2 - x^2 = 12$
 $x^2 = 4y$

6. $x^2 + y^2 = 9$
 $y = 4$

7. $x^2 + y^2 = 4$
 $x^2 - y^2 = 4$

8. $x^2 + y^2 - 6y = 0$
 $y = x$

9. $x^2 = 6y$
 $y = 6$

10. $\dfrac{y^2}{16} + \dfrac{x^2}{9} = 1$
 $4x + 3y = 12$

11. $y^2 = 4x + 12$
 $y^2 = -4x - 4$

12. $x^2 - y^2 = 2$
 $y^2 = x$

13. $x^2 + y^2 = 36$
 $y = x^2$

14. $y = x^2 - 3x - 10$
 $2x + y + 4 = 0$

15. $x^2 - y^2 = 9$
 $x^2 + 9y^2 = 169$

16. $x^2 + 4y^2 = 36$
 $x^2 + y^2 = 16$

17. $x^2 + y^2 = 17$
 $xy = 4$

18. $3x^2 + 4y^2 = 48$
 $xy = 6$

1.13 POLAR COORDINATES

Each point in the number plane has been associated with an ordered pair of real numbers (x, y), which are called rectangular or Cartesian coordinates. Point $P(x, y)$ is shown in Fig. 1.67. Point P can also be located by specifying an angle θ from the positive x-axis and a directed distance r from the origin, and described by the ordered pair (r, θ) called *polar coordinates*. The polar coordinate system has a fixed point in the number plane called the *pole* or *origin*. From the pole draw a horizontal ray directed to the right, which is called the *polar axis* (see Fig. 1.68).

Angle θ is a directed angle: $\theta > 0$ is measured counterclockwise; $\theta < 0$ is measured clockwise. Angle θ is commonly expressed in either degrees or radians. Distance r is a directed distance: $r > 0$ is measured in the direction of the ray (terminal side of θ); $r < 0$ is measured in the direction opposite the direction of the ray.

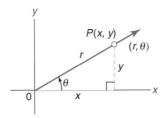

Figure 1.67 Polar coordinates

Figure 1.68 Polar axis

EXAMPLE 1

Graph each point whose polar coordinates are given: (a) $(2, 120°)$, (b) $(4, 4\pi/3)$, (c) $(4, -2\pi/3)$, (d) $(-5, 135°)$, (e) $(-2, -60°)$, (f) $(-3, 570°)$ (see Fig. 1.69).

From the results of Example 1, you can see that there is a major difference between the rectangular coordinate system and the polar coordinate system. In the rectangular system there is a one-to-one correspondence between points in the plane and ordered pairs

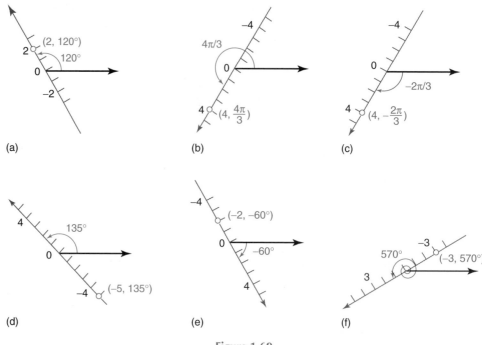

(a) (b) (c)

(d) (e) (f)

Figure 1.69

of real numbers. That is, each point is named by exactly one ordered pair, and each ordered pair corresponds to exactly one point. This one-to-one correspondence is not a property of the polar coordinate system. In Example 1, parts (a) and (e) describe the same point in the plane; and parts (b) and (c) describe the same point. In fact, each point may be named by infinitely many polar coordinates. In general, the point $P(r, \theta)$ may be represented by

$$(r, \theta + k \cdot 360°) \quad \text{or} \quad (r, \theta + k \cdot 2\pi)$$

where k is any integer. $P(r, \theta)$ may also be represented by

$$(-r, \theta + k \cdot 180°) \quad \text{or} \quad (-r, \theta + k\pi)$$

where k is any odd integer.

EXAMPLE 2

Name an ordered pair of polar coordinates that corresponds to the pole or origin.

Any set of coordinates in the form $(0, \theta)$, where θ is any angle, corresponds to the pole. For example, $(0, 64°)$, $(0, 2\pi/3)$, and $(0, -\pi/6)$ name the pole.

Polar graph paper is available for working with polar coordinates. Fig. 1.70 shows graph paper in both degrees and radians.

EXAMPLE 3

Plot each point whose polar coordinates are given. Use polar graph paper in degrees (see Fig. 1.71).

$$A(6, 60°), \quad B(4, 270°), \quad C(3, -210°), \quad D(-6, 45°), \quad E(-2, -150°), \quad F(8, 480°)$$

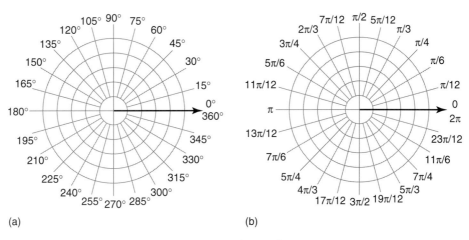

Figure 1.70   Polar graph paper

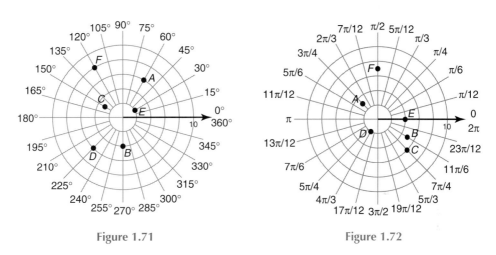

Figure 1.71   Figure 1.72

EXAMPLE 4

Plot each point whose polar coordinates are given. Use polar graph paper in radians (see Fig. 1.72).

$$A\left(3, \frac{3\pi}{4}\right), \quad B\left(5, \frac{11\pi}{6}\right), \quad C\left(6, -\frac{\pi}{4}\right), \quad D\left(-2, \frac{\pi}{3}\right), \quad E(-4, -\pi), \quad F\left(7, \frac{13\pi}{2}\right)$$

EXAMPLE 5

Given the point $P(4, 150°)$, name three other sets of polar coordinates for P such that $-360° \leq \theta \leq 360°$.

For $r > 0$ and $\theta < 0$: $(4, -210°)$

For $r < 0$ and $\theta > 0$: $(-4, 330°)$

For $r < 0$ and $\theta < 0$: $(-4, -30°)$

EXAMPLE 6

Graph $r = 10 \cos \theta$ by plotting points. Assign θ values of $0°$, $30°$, $45°$, $60°$, and so on until you have a smooth curve.

Make a table for the ordered pairs as follows. *Note:* Although r and θ are given in the same order as the ordered pair (r, θ), θ is actually the independent variable.

r	θ	$r = 10 \cos \theta$
10	0°	$r = 10 \cos 0° = 10$
8.7	30°	$r = 10 \cos 30° = 8.7$
7.1	45°	$r = 10 \cos 45° = 7.1$
5	60°	$r = 10 \cos 60° = 5$
0	90°	$r = 10 \cos 90° = 0$
-5	120°	$r = 10 \cos 120° = -5$
-7.1	135°	$r = 10 \cos 135° = -7.1$
-8.7	150°	$r = 10 \cos 150° = -8.7$
-10	180°	$r = 10 \cos 180° = -10$

Then plot the points as shown in Fig. 1.73.

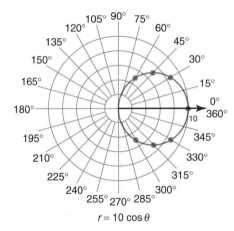

$r = 10 \cos \theta$

Figure 1.73

Note: You should plot values of θ from 0° to 360°, since the period of the cosine function is 360°. In this case choosing values of θ between 180° and 360° will give ordered pairs that duplicate those in Fig. 1.73.

Let point $P(x, y)$ be any point in the rectangular plane. Let the polar plane coincide with the rectangular plane so that $P(x, y)$ and $P(r, \theta)$ represent the same point, as shown in Fig. 1.74. Note the following relationships:

POLAR-RECTANGULAR RELATIONSHIPS

1. $\cos \theta = \dfrac{x}{r}$ or $x = r \cos \theta$

2. $\sin \theta = \dfrac{y}{r}$ or $y = r \sin \theta$

3. $\tan \theta = \dfrac{y}{x}$ or $\theta = \arctan \dfrac{y}{x}$

4. $x^2 + y^2 = r^2$

5. $\cos \theta = \dfrac{x}{\sqrt{x^2 + y^2}}$

6. $\sin \theta = \dfrac{y}{\sqrt{x^2 + y^2}}$

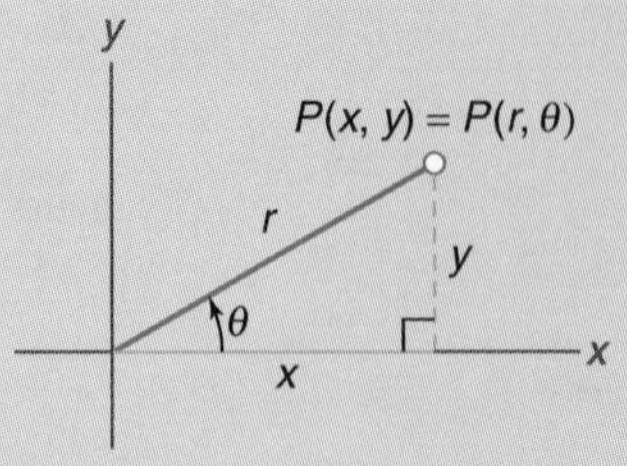

Figure 1.74

Suppose that we wish to change coordinates from one system to the other.

EXAMPLE 7

Change $A(4, 60°)$ and $B(8, 7\pi/6)$ to rectangular coordinates.

For point A:

$$x = r\cos\theta \qquad y = r\sin\theta$$

$$x = 4\cos 60° \qquad y = 4\sin 60°$$

$$= 4\left(\frac{1}{2}\right) \qquad = 4\left(\frac{\sqrt{3}}{2}\right)$$

$$= 2 \qquad = 2\sqrt{3}$$

Thus $A(4, 60°) = (2, 2\sqrt{3})$.

For point B:

$$x = r\cos\theta \qquad y = r\sin\theta$$

$$x = 8\cos\frac{7\pi}{6} \qquad y = 8\sin\frac{7\pi}{6}$$

$$= 8\left(-\frac{\sqrt{3}}{2}\right) \qquad = 8\left(-\frac{1}{2}\right)$$

$$= -4\sqrt{3} \qquad = -4$$

Thus $B(8, 7\pi/6) = (-4\sqrt{3}, -4)$.

EXAMPLE 8

Find polar coordinates for each point: $C(2\sqrt{3}, 2)$ in degrees, $0° \leq \theta < 360°$, and $D(6, -6)$ in radians, $0 \leq \theta < 2\pi$.

Note: The signs of x and y determine the quadrant for θ. That is, the signs of x and y determine in which quadrant the point lies and hence the quadrant in which θ must lie.

For point C:

$$r^2 = x^2 + y^2 \qquad\qquad \theta = \arctan\frac{y}{x}$$

$$r^2 = (2\sqrt{3})^2 + 2^2 = 16 \qquad \theta = \arctan\frac{2}{2\sqrt{3}}$$

$$r = 4 \qquad\qquad \theta = 30°$$

Thus $C(2\sqrt{3}, 2) = (4, 30°)$.

For point D:

$$r^2 = x^2 + y^2 \qquad\qquad \theta = \arctan\frac{y}{x}$$

$$r^2 = 6^2 + (-6)^2 = 72 \qquad \theta = \arctan\left(\frac{-6}{6}\right) = \arctan(-1)$$

$$r = 6\sqrt{2} \qquad\qquad \theta = \frac{7\pi}{4} \qquad \left(Note: \alpha^* = \frac{\pi}{4}.\right)$$

Thus $D(6, -6) = (6\sqrt{2}, 7\pi/4)$.

Some curves are most simply expressed and easiest to work with in rectangular coordinates; others are most simply expressed and easiest to work with in polar coordinates. As a result you must be able to change a polar equation to a rectangular equation and to change a rectangular equation to a polar equation.

*α is the reference angle.

EXAMPLE 9

Change $x^2 + y^2 - 4x = 0$ to polar form.

    Substituting $x^2 + y^2 = r^2$ and $x = r \cos \theta$, we have

$$r^2 - 4r \cos \theta = 0$$

$$r(r - 4 \cos \theta) = 0 \qquad \text{(Factor.)}$$

So

$$r = 0 \quad \text{or} \quad r - 4 \cos \theta = 0$$

But $r = 0$ (the pole) is a point that is included in the graph of the equation $r - 4 \cos \theta = 0$. Note that $(0, \pi/2)$ is an ordered pair that satisfies the second equation and names the pole. Thus the simplest polar equation is

$$r = 4 \cos \theta$$

EXAMPLE 10

Change $r = 4 \sin \theta$ to rectangular form.

    Multiply both sides of the equation by r:

$$r^2 = 4r \sin \theta$$

Substituting $r^2 = x^2 + y^2$ and $r \sin \theta = y$, we have

$$x^2 + y^2 = 4y$$

Note that by multiplying both sides of the given equation by r, we added the root $r = 0$. But the point represented by that root is already included in the original equation. So no new points are added to those represented by the original equation.

EXAMPLE 11

Change $r \cos^2 \theta = 6 \sin \theta$ to rectangular form.

    First multiply both sides by r:

$$r^2 \cos^2 \theta = 6r \sin \theta$$

$$(r \cos \theta)^2 = 6r \sin \theta$$

Substituting $r \cos \theta = x$ and $r \sin \theta = y$, we have

$$x^2 = 6y$$

EXAMPLE 12

Change $r = \dfrac{2}{1 - \cos \theta}$ to rectangular form.

$$r = \frac{2}{1 - \cos \theta} \qquad \qquad \textbf{(1)}$$

First multiply both sides by $1 - \cos \theta$:

$$r(1 - \cos \theta) = 2$$

$$r - r \cos \theta = 2$$

$$r = 2 + r \cos \theta \qquad \qquad \textbf{(2)}$$

Substituting $r = \pm \sqrt{x^2 + y^2}$ and $r \cos \theta = x$, we have

$$\pm \sqrt{x^2 + y^2} = 2 + x$$

Squaring both sides, we have

$$x^2 + y^2 = 4 + 4x + x^2$$
$$y^2 = 4x + 4$$

Note that squaring both sides was a risky operation because we introduced the possible extraneous solutions

$$r = -(2 + r \cos \theta) \qquad \qquad \textbf{(3)}$$

However, in this case both Equations (2) and (3) have the same graph. To show this, solve Equation (3) for r:

$$r = \frac{-2}{1 + \cos \theta} \qquad \qquad \textbf{(4)}$$

Recall that the ordered pairs (r, θ) and $(-r, \theta + \pi)$ represent the same point. Let us replace (r, θ) by $(-r, \theta + \pi)$ in Equation (4):

$$-r = \frac{-2}{1 + \cos (\theta + \pi)}$$

$$r = \frac{2}{1 - \cos \theta} \qquad [\text{Recall } \cos (\theta + \pi) = -\cos \theta.]$$

Equations (2) and (3) and thus Equations (1) and (4) have the same graph, and no extraneous solutions were introduced when we squared both sides. So our result $y^2 = 4x + 4$ is correct.

Exercises 1.13

Plot each point whose polar coordinates are given.

1. $A(3, 150°)$, $B(7, -45°)$, $C(2, -120°)$, $D(-4, 225°)$

2. $A(5, -90°)$, $B(2, -210°)$, $C(6, -270°)$, $D(-5, 30°)$

3. $A\left(4, \dfrac{\pi}{3}\right)$, $B\left(5, -\dfrac{\pi}{4}\right)$, $C\left(3, -\dfrac{7\pi}{6}\right)$, $D\left(-6, \dfrac{11\pi}{6}\right)$

4. $A\left(4, \dfrac{5\pi}{3}\right)$, $B\left(5, -\dfrac{3\pi}{2}\right)$, $C\left(3, -\dfrac{19\pi}{12}\right)$, $D\left(-6, -\dfrac{2\pi}{3}\right)$

For each point, name three other sets of polar coordinates such that $-360° \le \theta \le 360°$.

5. $(3, 60°)$ **6.** $(2, 240°)$ **7.** $(-5, 315°)$

8. $(-6, 90°)$ **9.** $(4, -135°)$ **10.** $(-1, -180°)$

For each point, name three other sets of polar coordinates such that $-2\pi \le \theta \le 2\pi$.

11. $\left(3, \dfrac{\pi}{6}\right)$ **12.** $\left(-7, \dfrac{\pi}{2}\right)$ **13.** $\left(-9, \dfrac{2\pi}{3}\right)$

14. $\left(-2, -\dfrac{5\pi}{6}\right)$ **15.** $\left(-4, -\dfrac{7\pi}{4}\right)$ **16.** $\left(5, -\dfrac{5\pi}{3}\right)$

Graph each equation by plotting points. Assign θ values of $0°$, $30°$, $45°$, $60°$, and so on, until you have a smooth curve.

17. $r = 10 \sin \theta$ **18.** $r = -10 \sin \theta$ **19.** $r = 4 + 4 \cos \theta$

20. $r = 4 + 4 \sin \theta$ **21.** $r \cos \theta = 4$ **22.** $r \sin \theta = -4$

Graph each equation by plotting points. Assign θ values of 0, π/6, π/4, π/3, and so on, until you have a smooth curve.

23. $r = -10 \cos \theta$ **24.** $r = 6 \sin \theta$ **25.** $r = 4 - 4 \sin \theta$

26. $r = 4 - 4 \cos \theta$ **27.** $r = \theta, 0 \le \theta \le 4\pi$ **28.** $r = 2\theta, 0 \le \theta \le 2\pi$

Change each set of polar coordinates to rectangular coordinates.

29. $(3, 30°)$ **30.** $(2, 180°)$ **31.** $\left(2, \dfrac{\pi}{3}\right)$

32. $\left(7, \dfrac{5\pi}{6}\right)$ **33.** $(-4, 150°)$ **34.** $(1, 420°)$

35. $\left(-6, \dfrac{3\pi}{2}\right)$ **36.** $(3, -\pi)$ **37.** $(-5, -240°)$

38. $(2, -120°)$ **39.** $\left(2, -\dfrac{7\pi}{4}\right)$ **40.** $\left(-1, -\dfrac{5\pi}{3}\right)$

Change each set of rectangular coordinates to polar coordinates in degrees, $0° \le \theta \le 360°$.

41. $(5, 5)$ **42.** $(-\sqrt{3}, 1)$ **43.** $(0, 4)$

44. $(-3, 0)$ **45.** $(-2, -2\sqrt{3})$ **46.** $(-1, 1)$

Change each set of rectangular coordinates to polar coordinates in radians, $0 \le \theta < 2\pi$.

47. $(-4, 4)$ **48.** $(-1, -\sqrt{3})$ **49.** $(-\sqrt{6}, \sqrt{2})$

50. $(5\sqrt{2}, -5\sqrt{2})$ **51.** $(0, -4)$ **52.** $(0, 0)$

Change each equation to polar form.

53. $x = 3$ **54.** $y = 5$ **55.** $x^2 + y^2 = 36$

56. $y^2 = 5x$ **57.** $x^2 + y^2 + 2x + 5y = 0$ **58.** $2x + 3y = 6$

59. $4x - 3y = 12$ **60.** $ax + by = c$ **61.** $9x^2 + 4y^2 = 36$

62. $4x^2 - 9y^2 = 36$ **63.** $x^3 = 4y^2$ **64.** $x^4 - 2x^2y^2 + y^4 = 0$

Change each equation to rectangular form.

65. $r \sin \theta = -3$ **66.** $r \cos \theta = 7$ **67.** $r = 5$

68. $r = 3 \sec \theta$ **69.** $\theta = \dfrac{\pi}{4}$ **70.** $\theta = -\dfrac{2\pi}{3}$

71. $r = 5 \cos \theta$ **72.** $r = 6 \sin \theta$ **73.** $r = 6 \cos\left(\theta + \dfrac{\pi}{3}\right)$

74. $r = 4 \sin\left(\theta - \dfrac{\pi}{4}\right)$ **75.** $r \sin^2 \theta = 3 \cos \theta$ **76.** $r^2 = \tan^2 \theta$

77. $r^2 \sin 2\theta = 2$ **78.** $r^2 \cos 2\theta = 6$ **79.** $r^2 = \sin 2\theta$

80. $r^2 = \cos 2\theta$ **81.** $r = \tan \theta$ **82.** $r = 4 \tan \theta \sec \theta$

83. $r = \dfrac{3}{1 + \sin \theta}$ **84.** $r = \dfrac{-4}{1 + \cos \theta}$ **85.** $r = 4 \sin 3\theta$

86. $r = 4 \cos 2\theta$ **87.** $r = 2 + 4 \sin \theta$ **88.** $r = 1 - \cos \theta$

89. Find the distance between the points whose polar coordinates are $(3, 60°)$ and $(2, 330°)$.

90. Find the distance between the points whose polar coordinates are $(5, \pi/2)$ and $(1, 7\pi/6)$.

91. Find a formula for the distance between two points whose polar coordinates are $P_1(r_1, \theta_1)$ and $P_2(r_2, \theta_2)$.

As you undoubtedly know, a graph of any equation may be made by finding and plotting "enough" ordered pairs that satisfy the equation and connecting them with a curve. As you also undoubtedly know, this is often tedious and time-consuming at best. We need a method for sketching the graph of a polar equation that minimizes the number of ordered pairs that must be found and plotted. One such method involves symmetry. We shall present tests for three kinds of symmetry:

SYMMETRY WITH RESPECT TO THE

1. *Horizontal axis:* Replace θ by $-\theta$ in the original equation. If the resulting equation is equivalent to the original equation, then the graph of the original equation is symmetric with respect to the *horizontal* axis.
2. *Vertical axis:* Replace θ by $\pi - \theta$ in the original equation. If the resulting equation is equivalent to the original equation, then the graph of the original equation is symmetric with respect to the *vertical* axis.
3. *Pole:*
   (a) Replace r by $-r$ in the original equation. If the resulting equation is equivalent to the original equation, then the graph of the original equation is symmetric with respect to the *pole*.
   (b) Replace θ by $\pi + \theta$ in the original equation. If the resulting equation is equivalent to the original equation, then the graph of the original equation is symmetric with respect to the *pole*.

You should note that these tests for symmetry are sufficient conditions for symmetry; that is, they are sufficient to assure symmetry. You should also note that these are not necessary conditions for symmetry; that is, symmetry may exist even though the test fails.

If either Test 3(a) or 3(b) is satisfied, then the graph is symmetric with respect to the pole. It is also true that if any two of the three kinds of symmetry hold, then the remaining third symmetry automatically holds. Can you explain why?

To help you quickly test for symmetry, the following identities are listed for your convenience.

POLAR COORDINATE IDENTITIES FOR TESTING SYMMETRY

$$\sin(-\theta) = -\sin\theta$$
$$\cos(-\theta) = \cos\theta$$
$$\tan(-\theta) = -\tan\theta$$
$$\sin(\pi - \theta) = \sin\theta$$
$$\cos(\pi - \theta) = -\cos\theta$$
$$\tan(\pi - \theta) = -\tan\theta$$
$$\sin(\pi + \theta) = -\sin\theta$$
$$\cos(\pi + \theta) = -\cos\theta$$
$$\tan(\pi + \theta) = \tan\theta$$

EXAMPLE 1

Graph $r = 4 + 2 \cos \theta$.

Replacing θ by $-\theta$, we see that the graph is symmetric with respect to the horizontal axis. The other tests fail. Thus we need to make a table as follows (note that because of symmetry with respect to the horizontal axis, we need only generate ordered pairs for $0° \le \theta \le 180°$):

r	θ	$r = 4 + 2 \cos \theta$
6	0°	$r = 4 + 2 \cos 0° = 6$
5.7	30°	$r = 4 + 2 \cos 30° = 5.7$
5	60°	$r = 4 + 2 \cos 60° = 5$
4	90°	$r = 4 + 2 \cos 90° = 4$
3	120°	$r = 4 + 2 \cos 120° = 3$
2.3	150°	$r = 4 + 2 \cos 150° = 2.3$
2	180°	$r = 4 + 2 \cos 180° = 2$

Plot the points as shown in Fig. 1.75(a). Because of the symmetry with respect to the horizontal axis, plot the corresponding mirror-image points below the horizontal axis (see Fig. 1.75b).

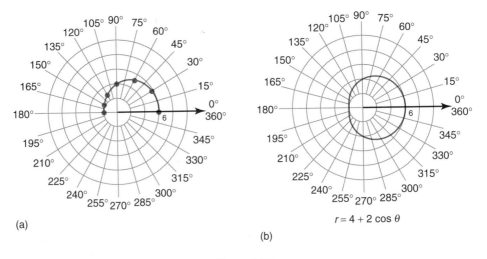

(a)

(b)

$r = 4 + 2 \cos \theta$

Figure 1.75

EXAMPLE 2

Graph $r = 4 + 4 \sin \theta$.

Replacing θ by $\pi - \theta$, we see that the graph is symmetric with respect to the vertical axis. The other tests fail. Thus, make a table as follows (note that because of symmetry with respect to the vertical axis, we need only generate ordered pairs for $-\pi/2 \le \theta \le \pi/2$).

r	θ	$r = 4 + 4 \sin \theta$
4	0	$r = 4 + 4 \sin 0 = 4$
6	$\pi/6$	$r = 4 + 4 \sin \pi/6 = 6$
7.5	$\pi/3$	$r = 4 + 4 \sin \pi/3 = 7.5$
8	$\pi/2$	$r = 4 + 4 \sin \pi/2 = 8$
2	$-\pi/6$	$r = 4 + 4 \sin (-\pi/6) = 2$
0.54	$-\pi/3$	$r = 4 + 4 \sin (-\pi/3) = 0.54$
0	$-\pi/2$	$r = 4 + 4 \sin (-\pi/2) = 0$

Plot the points as shown in Fig. 1.76 (a). Because of the symmetry with respect to the vertical axis, plot the corresponding mirror-image points to the left of the vertical axis (see Fig. 1.76b).

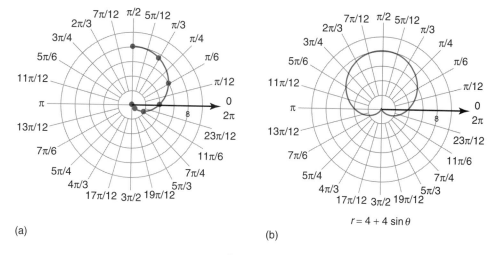

(a)

(b)

$$r = 4 + 4 \sin \theta$$

Figure 1.76

EXAMPLE 3

Graph $r^2 = 16 \sin \theta$.

Replacing r by $-r$, we see that the graph is symmetric with respect to the pole. Replacing θ by $\pi - \theta$, we see that the graph is also symmetric with respect to the vertical axis. Since two of the three kinds of symmetry hold, the graph is also symmetric with respect to the horizontal axis. (*Note:* Replacing θ by $-\theta$ gives the resulting equation $r^2 = -16 \sin \theta$, which is different from the original equation. However, its solutions when graphed give the same curve.)

r	θ	$r^2 = 16 \sin \theta$
0	0°	$r^2 = 16 \sin 0° = 0; r = 0$
2.8	30°	$r^2 = 16 \sin 30° = 8; r = 2.8$
3.7	60°	$r^2 = 16 \sin 60° = 13.9; r = 3.7$
4	90°	$r^2 = 16 \sin 90° = 16; r = 4$

Plot the points as shown in Fig. 1.77(a). Because of the symmetry with respect to the horizontal and vertical axes, plot the corresponding mirror-image points below the horizontal axis. Then plot the mirror image points of all resulting points to the left of the vertical axis (see Fig. 1.77b).

EXAMPLE 4

Graph $r^2 = 25 \cos 2\theta$.

By replacing θ by $-\theta$ and r by $-r$, we have symmetry with respect to the horizontal axis and the pole, respectively. Thus we also have symmetry with respect to the vertical axis. Working in the first quadrant, we have

r	θ	$r^2 = 25 \cos 2\theta$
5	0	$r^2 = 25 \cos 2(0) = 25; r = 5$
4.7	$\pi/12$	$r^2 = 25 \cos 2(\pi/12) = 21.7; r = 4.7$
3.5	$\pi/6$	$r^2 = 25 \cos 2(\pi/6) = 12.5; r = 3.5$
0	$\pi/4$	$r^2 = 25 \cos 2(\pi/4) = 0; r = 0$

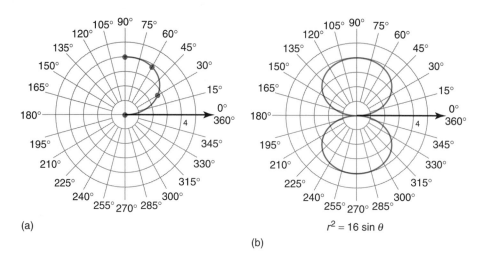

(a)

$r^2 = 16 \sin \theta$

(b)

Figure 1.77

Note: For the interval $\pi/4 < \theta \le \pi/2$, $r^2 < 0$ and r is undefined.

Plot the points as shown in Fig. 1.78(a). Because of the symmetry with respect to the horizontal and vertical axes, plot the corresponding mirror-image points below the horizontal axis and to the left of the vertical axis (see Fig. 1.78b).

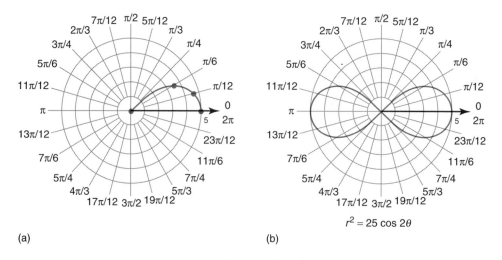

(a)

(b)

$r^2 = 25 \cos 2\theta$

Figure 1.78

There are various general polar equations whose graphs may be classified as shown in Fig. 1.79. What do you think the graphs of the various forms of $r = a + b \sin\theta$ are like?

Equations in the form

$$r = a \sin n\theta \quad \text{or} \quad r = a \cos n\theta$$

where n is a positive integer, are called *petal* or *rose curves*. The number of petals is equal to n if n is an *odd* integer, and is equal to $2n$ if n is an *even* integer. This is because the graph "retraces" itself as θ goes from 0° to 360° when n is odd, so there are only half as many distinct petals. (For $n = 1$ there is one circular petal. See Example 6, Section 1.13). The value of a corresponds to the length of each petal.

Limaçons ($r = a + b \cos \theta$)

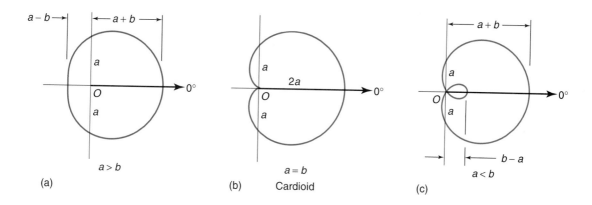

(a) $a > b$

(b) Cardioid $a = b$

(c) $a < b$

Lemniscates

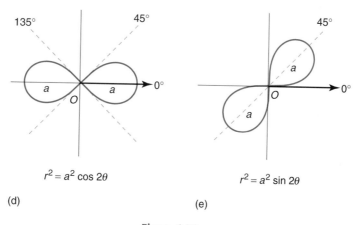

$r^2 = a^2 \cos 2\theta$

(d)

$r^2 = a^2 \sin 2\theta$

(e)

Figure 1.79

The tests for symmetry may be used to graph petal curves. However, we shall illustrate a somewhat different, as well as easier and quicker, method for graphing petal curves.

EXAMPLE 5

Graph $r = 6 \cos 2\theta$.

First, note that $n = 2$, which is even. Therefore, we have four petals. The petals are always uniform; each petal occupies $360°/4$, or $90°$, of the polar coordinate system. Next, find the tip of a petal; this occurs when r is maximum or when

$$\cos 2\theta = 1$$
$$2\theta = 0°$$
$$\theta = 0°$$

That is, $r = 6$ when $\theta = 0°$.

Finally, sketch four petals, each having a maximum length of six and occupying $90°$ (see Fig. 1.80). For more accuracy, you may graph the ordered pairs corresponding to a "half petal" ($0° \leq \theta \leq 45°$ in this case).

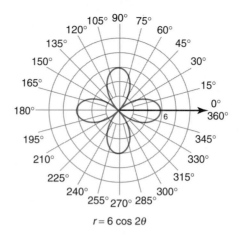

$$r = 6 \cos 2\theta$$

Figure 1.80

Polar coordinates are especially useful for the study and graphing of *spirals*. The *spiral of Archimedes* has an equation in the form

$$r = a\theta$$

Its graph is shown in Fig. 1.81. (The dashed portion of the graph corresponds to $\theta < 0$.)
The *logarithmic spiral* has an equation of the form

$$\log_b r = \log_b a + k\theta \quad \text{or} \quad r = a \cdot b^{k\theta}$$

Its graph is shown in Fig. 1.82.

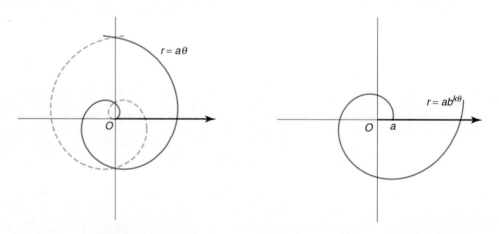

Figure 1.81   Spiral of Archimedes

Figure 1.82   Logarithmic spiral

For graphing calculator examples in polar coordinates, please see Appendix Section C.5 (TI-83 Plus) or D.7 (TI-89).

Exercises 1.14

Graph each equation.

1. $r = 6$

2. $r = 3$

3. $r = -2$

4. $r = -4$

5. $\theta = 30°$

6. $\theta = -120°$

7. $\theta = -\dfrac{\pi}{3}$

8. $\theta = \dfrac{7\pi}{6}$

9. $r = 5 \sin \theta$

10. $r = 8 \cos \theta$

11. $r = 6 \cos \left(\theta + \dfrac{\pi}{3} \right)$

12. $r = 4 \sin \left(\theta - \dfrac{\pi}{4} \right)$

13. $r = 4 + 2 \sin \theta$

14. $r = 8 + 2 \cos \theta$

15. $r = 4 - 2 \cos \theta$

16. $r = 4 - 2 \sin \theta$

17. $r = 3 + 3 \cos \theta$

18. $r = 5 + 5 \sin \theta$

19. $r = 2 + 4 \sin \theta$

20. $r = 2 + 8 \cos \theta$

21. $r = 2 - 4 \cos \theta$

22. $r = 2 - 4 \sin \theta$

23. $r = 3 - 3 \cos \theta$

24. $r = 5 - 5 \sin \theta$

25. $r \cos \theta = 6$

26. $r \sin \theta = -4$

27. $r^2 = 25 \cos \theta$

28. $r^2 = -9 \sin \theta$

29. $r^2 = 9 \sin 2\theta$

30. $r^2 = 16 \cos 2\theta$

31. $r^2 = -36 \cos 2\theta$

32. $r = -36 \sin 2\theta$

33. $r = 5 \sin 3\theta$

34. $r = 4 \cos 5\theta$

35. $r = 3 \cos 2\theta$

36. $r = 6 \sin 4\theta$

37. $r = 9 \sin^2 \theta$

38. $r = 16 \cos^2 \theta$

39. $r = 4 \cos \dfrac{\theta}{2}$

40. $r = 5 \sin^2 \dfrac{\theta}{2}$

41. $r = \tan \theta$

42. $r = 2 \csc \theta$

43. $r = 3\theta, \theta > 0$

44. $r = \dfrac{3}{\theta}, \theta > 0$

45. $r = 2^{3\theta}$

46. $r = 2 \cdot 3^{2\theta}$

47. $r = \dfrac{4}{\sin \theta + \cos \theta}$

48. $r = \dfrac{-2}{\sin \theta + \cos \theta}$

49. $r(1 + \cos \theta) = 4$

50. $r(1 + 2 \sin \theta) = -4$

CHAPTER SUMMARY

1. *Basic terms*
   (a) *Analytic geometry:* Study of relationships between algebra and geometry.
   (b) *Relation:* Set of ordered pairs, usually in the form (x, y).
   (c) *Independent variable:* First element of an ordered pair, usually x.
   (d) *Dependent variable:* Second element of an ordered pair, usually y.
   (e) *Domain:* Set of all first elements in a relation or set of all x's.
   (f) *Range:* Set of all second elements in a relation or set of all y's.
   (g) *Function:* A set of ordered pairs in which no two distinct ordered pairs have the same first element.
   (h) *Functional notation:* To write an equation in variables x and y in functional notation, solve for y, and replace y with $f(x)$.
   (i) *Linear equation with two unknowns:* An equation of degree one in the form $ax + by = c$ where a and b are not both 0.
   (j) *Positive integers:* $1, 2, 3, \ldots$.
   (k) *Negative integers:* $-1, -2, -3, \ldots$.
   (l) *Integers:* $\ldots, -3, -2, -1, 0, 1, 2, 3, \ldots$.
   (m) *Rational numbers:* Numbers that can be represented as the ratio of two integers $a/b, b \neq 0$.
   (n) *Irrational numbers:* Numbers that cannot be represented as the ratio of two integers.
   (o) *Real numbers:* Set of numbers consisting of the rational numbers and the irrational numbers.
   (p) *Inequalities:* Statements involving less than or greater than and may be used to describe various intervals on the number line as follows:

Type of interval	Symbols	Meaning	Number line graph
Open	$x > a$	x is greater than a	
	$x < b$	x is less than b	
	$a < x < b$	x is between a and b	
Half-open	$x \geq a$	x is greater than or equal to a	
	$x \leq b$	x is less than or equal to b	
	$a < x \leq b$	x is between a and b, including b but excluding a	
	$a \leq x < b$	x is between a and b, including a but excluding b	
Closed	$a \leq x \leq b$	x is between a and b, including both a and b	

2. *Slope of a line:* If $P_1(x_1, y_1)$ and $P_2(x_2, y_2)$ represent any two points on a line, then the slope m of the line is

$$m = \frac{y_2 - y_1}{x_2 - x_1}$$

  (a) If a line has positive slope, the line slopes upward from left to right.
  (b) If a line has negative slope, the line slopes downward from left to right.
  (c) If a line has zero slope, the line is horizontal.
  (d) If a line has undefined slope, the line is vertical.

3. *Point-slope form of a line:* If m is the slope and $P_1(x_1, y_1)$ is any point on a nonvertical line, its equation is

$$y - y_1 = m(x - x_1)$$

4. *Slope-intercept form of a line:* If m is the slope and $(0, b)$ is the y-intercept of a nonvertical line, its equation is

$$y = mx + b$$

5. *Equation of a horizontal line:* If a horizontal line passes through the point (a, b), its equation is

$$y = b$$

6. *Equation of a vertical line:* If a vertical line passes through the point (a, b), its equation is

$$x = a$$

7. *General form of the equation of a straight line:*

$$Ax + By + C = 0 \qquad \text{where } A \text{ and } B \text{ are not both } 0$$

8. *Parallel lines:* Two lines are parallel if either one of the following conditions holds:
   (a) They are both perpendicular to the x-axis.
   (b) They both have the same slope. That is, if the equations of the lines are

$$L_1: \quad y = m_1 x + b_1 \quad \text{and} \quad L_2: \quad y = m_2 x + b_2$$

then

$$m_1 = m_2$$

9. *Perpendicular lines:* Two lines are perpendicular if either one of the following conditions holds:
   (a) One line is vertical with equation $x = a$, and the other is horizontal with equation $y = b$.
   (b) Neither is vertical and the slope of one line is the negative reciprocal of the other. That is, if the equations of the lines are

$$L_1: \quad y = m_1 x + b_1 \quad \text{and} \quad L_2: \quad y = m_2 x + b_2$$

then

$$m_1 = -\frac{1}{m_2}$$

10. *Distance formula:* The distance between two points $P(x_1, y_1)$ and $Q(x_2, y_2)$ is given by the formula

$$d = PQ = \sqrt{(x_2 - x_1)^2 + (y_2 - y_1)^2}$$

11. *Midpoint formula:* The coordinates of the point $Q(x_m, y_m)$ that is midway between two points $P(x_1, y_1)$ and $R(x_2, y_2)$ are given by

$$x_m = \frac{x_1 + x_2}{2} \quad \text{and} \quad y_m = \frac{y_1 + y_2}{2}$$

12. *Conics:* Equations in the form $Ax^2 + Bxy + Cy^2 + Dx + Ey + F = 0$ are called conics.

13. *Circle*
   (a) *Standard form:* $(x - h)^2 + (y - k)^2 = r^2$, where r is the radius and (h, k) is the center.
   (b) *General form:* $x^2 + y^2 + Dx + Ey + F = 0$.
   (c) *Center at the origin:* $x^2 + y^2 = r^2$, where r is the radius.

14. *Parabola with vertex at the origin*
   (a) $y^2 = 4px$ with focus at $(p, 0)$ and $x = -p$ as the directrix.
      (i) When $p > 0$, the parabola opens to the right.
      (ii) When $p < 0$, the parabola opens to the left.
   (b) $x^2 = 4py$ with focus at $(0, p)$ and $y = -p$ as the directrix.
      (i) When $p > 0$, the parabola opens upward.
      (ii) When $p < 0$, the parabola opens downward.

15. *Ellipse with center at the origin*
   (a) $\dfrac{x^2}{a^2} + \dfrac{y^2}{b^2} = 1$ with the major axis on the x-axis and $a > b$.
   (b) $\dfrac{y^2}{a^2} + \dfrac{x^2}{b^2} = 1$ with the major axis on the y-axis and $a > b$.

16. *Hyperbola with center at the origin*
   (a) $\dfrac{x^2}{a^2} - \dfrac{y^2}{b^2} = 1$ with the transverse axis on the x-axis.
   (b) $\dfrac{y^2}{a^2} - \dfrac{x^2}{b^2} = 1$ with the transverse axis on the y-axis.

17. *Translation equations*

$$x' = x - h \quad \text{and} \quad y' = y - k$$

18. *General forms of conics with axes parallel to the coordinate axes*

(a) $(y - k)^2 = 4p(x - h)$

is a parabola with vertex at (h, k) and axis parallel to the x-axis.

(b) $(x - h)^2 = 4p(y - k)$

is a parabola with vertex at (h, k) and axis parallel to the y-axis.

(c) $\dfrac{(x - h)^2}{a^2} + \dfrac{(y - k)^2}{b^2} = 1, \quad (a > b)$

is an ellipse with center at (h, k) and major axis parallel to the x-axis.

(d) $\dfrac{(y - k)^2}{a^2} + \dfrac{(x - h)^2}{b^2} = 1, \quad (a > b)$

is an ellipse with center at (h, k) and major axis parallel to the y-axis.

(e) $\dfrac{(x - h)^2}{a^2} - \dfrac{(y - k)^2}{b^2} = 1$

is a hyperbola with center at (h, k) and transverse axis parallel to the x-axis.

(f) $\dfrac{(y - k)^2}{a^2} - \dfrac{(x - h)^2}{b^2} = 1$

is a hyperbola with center at (h, k) and transverse axis parallel to the y-axis.

19. *The general second-degree equation:* The circle, parabola, ellipse, and hyperbola are all special cases of the second-degree equation

$$Ax^2 + Bxy + Cy^2 + Dx + Ey + F = 0$$

When $B = 0$ and at least one of the coefficients A or C is not zero, the following summarizes the conditions for each curve:

(a) If $A = C$, we have a *circle*.

In special cases, the graph of the equation may be a point or there may be no graph. (The equation may have only one or no solution.)

(b) If $A = 0$ and $C \neq 0$ or $C = 0$ and $A \neq 0$, then we have a *parabola*.

(c) If $A \neq C$, and A and C are either both positive or both negative, then we have an *ellipse*.

In special cases, the graph of the equation may be a point or there may be no graph. (The equation may have only one or no solution.)

(d) If A and C differ in sign, then we have a *hyperbola*. In some special cases the graph may be a pair of intersecting lines.

If $D \neq 0$ or $E \neq 0$ or both are not equal to zero, the curve does not have its center (or vertex in the case of the parabola) at the origin. If $B \neq 0$, then the axis of the curve does not lie along the x-axis or y-axis.

20. Each point $P(x, y)$ in the rectangular coordinate system may be described by an ordered pair $P(r, \theta)$ in the polar coordinate system.

21. In the rectangular coordinate system, there is a one-to-one correspondence between points in the plane and ordered pairs of real numbers. This one-to-one correspondence is not a property of the polar coordinate system. In general the point $P(r, \theta)$ may be represented by

$$(r, \theta + k \cdot 360°) \quad \text{or} \quad (r, \theta + k \cdot 2\pi)$$

where k is any integer. $P(r, \theta)$ may also be represented by

$$(-r, \theta + k \cdot 180°) \quad \text{or} \quad (-r, \theta + k\pi)$$

where k is any odd integer.

22. The relationships between the rectangular and polar coordinate systems are

(a) $x = r\cos\theta$

(b) $y = r\sin\theta$

(c) $\tan\theta = \dfrac{y}{x}$ or $\theta = \arctan\dfrac{y}{x}$

(d) $x^2 + y^2 = r^2$

(e) $\cos\theta = \dfrac{x}{\sqrt{x^2 + y^2}}$

(f) $\sin\theta = \dfrac{y}{\sqrt{x^2 + y^2}}$

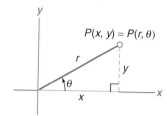

Figure 1.83

23. *Symmetry tests for graphing polar equations*

(a) *Horizontal axis:* Replace θ by $-\theta$ in the original equation. If the resulting equation is equivalent to the original equation, then the graph of the original equation is symmetric with respect to the *horizontal* axis.

(b) *Vertical axis:* Replace θ by $\pi - \theta$ in the original equation. If the resulting equation is equivalent to the original equation, then the graph of the original equation is symmetric with respect to the *vertical* axis.

(c) *Pole*

  (i) Replace r by $-r$ in the original equation. If the resulting equation is equivalent to the original equation, then the graph of the original equation is symmetric with respect to the *pole*.

  (ii) Replace θ by $\pi + \theta$ in the original equation. If the resulting equation is equivalent to the original equation, then the graph of the original equation is symmetric with respect to the *pole*.

CHAPTER 1 REVIEW

Determine whether or not each relation is a function. Write its domain and its range.

1. $A = \{(2, 3), (3, 4), (4, 5), (5, 6)\}$

2. $B = \{(2, 6), (6, 4), (2, 1), (4, 3)\}$

3. $y = -4x + 3$

4. $y = x^2 - 5$

5. $x = y^2 + 4$

6. $y = \sqrt{4 - 8x}$

7. Given $f(x) = 5x + 14$, find
  (a) $f(2)$ (b) $f(0)$ (c) $f(-4)$

8. Given $g(t) = 3t^2 + 5t - 12$, find
  (a) $g(2)$ (b) $g(0)$ (c) $g(-5)$

9. Given $h(x) = \dfrac{4x^2 - 3x}{2\sqrt{x - 1}}$, find
  (a) $h(2)$ (b) $h(5)$ (c) $h(-15)$ (d) $h(1)$

10. Given $g(x) = x^2 - 6x + 4$, find
  (a) $g(a)$ (b) $g(2x)$ (c) $g(z - 2)$

Graph each equation.

11. $y = 4x + 5$

12. $y = x^2 + 4$

13. $y = x^2 + 2x - 8$

14. $y = 2x^2 + x - 6$

15. $y = -x^2 - x + 4$

16. $y = \sqrt{2x}$

17. $y = \sqrt{-2 - 4x}$

18. $y = x^3 - 6x$

Solve each graphically.

19. Exercise 12 for $y = 5, 7,$ and 2.

20. Exercise 13 for $y = 0, -2,$ and 3.

21. Exercise 15 for $y = 2, 0,$ and -2.

22. Exercise 18 for $y = 0, 2,$ and -3.

23. The current, i, in a given circuit varies with the time, t, according to $i = 2t^2$. Find t when $i = 2, 6,$ and 8.

24. A given capacitor receives between its terminals a voltage, V, where $V = 4t^3 + t$ when t is in seconds. Find t when V is 40 and 60.

Use the points $(3, -4)$ and $(-6, -2)$ in Exercises 25–27.

25. Find the slope of the line through the two points.

26. Find the distance between the two points.

27. Find the coordinates of the point midway between the two points.

Find the equation of the line that satisfies each condition in Exercises 28–31.

28. Passes through $(4, 7)$ and $(6, -4)$.

29. Passes through $(-3, 1)$ with slope $\frac{2}{3}$.

30. Crosses the y-axis at -3 with slope $-\frac{1}{3}$.

31. Is parallel to and 3 units to the left of the y-axis.

32. Find the slope and the y-intercept of $3x - 2y - 6 = 0$.

33. Graph $3x - 4y = 12$.

Using the slope and the intercept of each line, determine whether each given pair of equations represents lines that are parallel, perpendicular, or neither.

34. $x - 2y + 3 = 0; 8x + 4y - 9 = 0$

35. $2x - 3y + 4 = 0; -8x + 12y = 16$

36. $3x - 2y + 5 = 0; 2x - 3y + 9 = 0$

37. $x = 2; y = -3$

38. $x = 4; y = 7$

39. Find the equation of the line parallel to the line $2x - y + 4 = 0$ that passes through the point $(5, 2)$.

40. Find the equation of the line perpendicular to the line $3x + 5y - 6 = 0$ that passes through the point $(-4, 0)$.

41. Write the equation of the circle with center at $(5, -7)$ and with radius 6.

42. Find the center and radius of the circle $x^2 + y^2 - 8x + 6y - 24 = 0$.

43. Find the focus and directrix of the parabola $x^2 = 6y$ and sketch its graph.

44. Write the equation of the parabola with focus at $(-4, 0)$ and directrix $x = 4$.

45. Write the equation of the parabola with focus at $(4, 3)$ and directrix $x = 0$.

46. Find the vertices and foci of the ellipse $4x^2 + 49y^2 = 196$ and sketch its graph.

47. Find the equation of the ellipse with vertices $(0, 4)$ and $(0, -4)$, and with foci at $(0, 2\sqrt{3})$ and $(0, -2\sqrt{3})$.

48. Find the vertices and foci of the hyperbola $4x^2 - 9y^2 = 144$ and sketch its graph.

49. Write the equation of the hyperbola with vertices at $(0, 5)$ and $(0, -5)$, and with foci at $(0, \sqrt{41})$ and $(0, -\sqrt{41})$.

50. Write the equation of the ellipse with center at $(3, -4)$, vertices at $(3, 1)$ and $(3, -9)$, and foci at $(3, 0)$ and $(3, -8)$.

51. Write the equation of the hyperbola with center at $(-7, 4)$ and vertices at $(2, 4)$ and $(-16, 4)$; the length of the conjugate axis is 6.

52. Name and sketch the graph of $16x^2 - 4y^2 - 64x - 24y + 12 = 0$.

Solve each system of equations.

53. $y^2 + 4y + x = 0$
$x = 2y$

54. $3x^2 - 4y^2 = 36$
$5x^2 - 8y^2 = 56$

Plot each point whose polar coordinates are given.

55. $A(6, 60°)$, $B(3, -210°)$, $C(-2, -270°)$, $D(-4, 750°)$

56. $A\left(5, \dfrac{\pi}{6}\right)$, $B\left(2, -\dfrac{5\pi}{4}\right)$, $C\left(-3, -\dfrac{\pi}{2}\right)$, $D\left(-5, \dfrac{19\pi}{2}\right)$

57. For point $A(5, 135°)$, name three other sets of polar coordinates for $-360° \le \theta < 360°$.

58. For point $B(-2, 7\pi/6)$, name three other sets of polar coordinates for $-2\pi \le \theta \le 2\pi$.

59. Change each set of polar coordinates to rectangular coordinates.

(a) $(3, 210°)$ (b) $(2, -120°)$ (c) $\left(-5, \dfrac{11\pi}{6}\right)$ (d) $\left(-6, -\dfrac{\pi}{2}\right)$

60. Change each set of rectangular coordinates to polar coordinates in degrees for $0° \le \theta < 360°$.
(a) $(-3, 3)$ (b) $(0, -6)$ (c) $(-1, \sqrt{3})$

61. Change each set of rectangular coordinates to polar coordinates in radians for $0 \le \theta < 2\pi$.
(a) $(-5, 0)$ (b) $(-6\sqrt{3}, 6)$ (c) $(1, -1)$

Change each equation to polar form.

62. $x^2 + y^2 = 49$ **63.** $y^2 = 9x$ **64.** $5x + 2y = 8$

65. $x^2 - 4y^2 = 12$ **66.** $y^3 = 6x^2$ **67.** $y(x^2 + y^2) = x^2$

Change each equation to rectangular form.

68. $r \cos \theta = 12$ **69.** $r = 9$ **70.** $\theta = \dfrac{2\pi}{3}$

71. $r = 8 \cos \theta$ **72.** $r \sin^2 \theta = 5 \cos \theta$ **73.** $r^2 \sin 2\theta = 8$

74. $r^2 = 4 \cos 2\theta$ **75.** $r = \csc \theta$ **76.** $r = 1 + \sin \theta$

77. $r = \dfrac{2}{1 - \sin \theta}$

Graph each equation.

78. $r = 7$ **79.** $\theta = -\dfrac{\pi}{4}$ **80.** $r = 5 \cos \theta$

81. $r = 6 + 3 \sin \theta$ **82.** $r = 6 - 3 \sin \theta$ **83.** $r = 4 + 4 \cos \theta$

84. $r = 3 - 6 \cos \theta$ **85.** $r \sin \theta = 5$ **86.** $r^2 = 36 \cos \theta$

87. $r = 6 \sin 5\theta$ **88.** $r^2 = 25 \sin 2\theta$ **89.** $r(1 - \sin \theta) = 6$

2

The Derivative

INTRODUCTION

Although algebra, trigonometry, and geometry are of fundamental importance to the mathematician and technician, a wide variety of technical problems cannot be solved using only these tools of mathematics. Many problems must be solved using the methods of the calculus. As early as the seventeenth century, scientists found the need for new techniques of mathematics to study the motion of projectiles, the motion of the moon and planets, and the motion of light. Scientists such as Isaac Newton began developing a new branch of mathematics to solve problems involving motion. This new branch of mathematics came to be known as the calculus. Today, the calculus remains a powerful development of mathematics. Even though the calculus began with the study of motion, it is useful in many varied technical areas today.

Objectives

- Find limits and instantaneous velocity.
- Find the slope and equation of a tangent line to a curve at a point.
- Use the definition of derivative to find derivatives.
- Differentiate polynomials, products, and quotients of polynomials.
- Use the chain rule to find derivatives.
- Use implicit differentiation.

2.1 MOTION

Motion is usually defined as a continual change in position. Linear motion is motion along a straight line. In this section, we limit our discussion to linear motion. You are familiar with finding the average speed of an object in motion. For example, if you drive 150 mi in 3 hours (h), then by dividing 150 mi by 3 h you find that you drove 50 mi/h on the average. This does not tell you how fast you were driving exactly 1 h and 32 minutes (min) after you began the trip. You may have been stopped at a traffic light or you may have been traveling 65 mi/h!

In attempting to solve this problem mathematically, we assume we can describe the distance traveled by an object as a function of time. That is, at each point in time t we can associate a number s representing the distance traveled by the object. For example, $s = 4t + 1$ is a function that describes the motion of an object as it moves along a straight line in terms of time, t. If t is measured in seconds (s) and s in metres (m), then after 2 s, the object is at $s = 4(2) + 1 = 9$ m along the line of motion. Three seconds later, $t = 2 + 3$, the object has moved to $s = 4(2 + 3) + 1 = 4(5) + 1 = 21$ m along the line of motion. (See Fig. 2.1.)

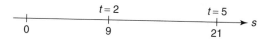

Figure 2.1 Distance traveled as a function of time.

The *average speed* $\bar{v}$ of an object in motion is the ratio of distance traveled by an object to the time taken to travel that distance. In our previous example, the distance traveled by the object is 21 m − 9 m = 12 m. It traveled this distance in 3 s. The average speed over this time period is then

$$\bar{v} = \frac{12 \text{ m}}{3 \text{ s}} = 4 \text{ m/s}$$

The mathematical symbol, the Greek letter Δ, indicates a change between two values of a variable. In this section Δt (read "delta t") represents the change in time t and Δs (read "delta s") represents the change in distance s. (The letter s is commonly used to represent distance. The letter d is used to represent diameter.) Above, the time interval $\Delta t = 3$ s. This is the change in time needed for the object to go from 9 m to 21 m along the line of motion. The change in distance for this time interval $\Delta t = 3$ s is $\Delta s = 21$ m − 9 m = 12 m. Using this notation we can write

$$\bar{v} = \frac{\Delta s}{\Delta t} = \frac{21 \text{ m} - 9 \text{ m}}{5 \text{ s} - 2 \text{ s}} = \frac{12 \text{ m}}{3 \text{ s}} = 4 \text{ m/s}$$

When both speed and direction are needed to describe motion, the term *velocity* is used. The **velocity** of an object is the time rate of change of its displacement, where displacement is the vector difference (net distance) between the initial and final positions of the object. Since both magnitude (speed) and direction are required to completely describe velocity, velocity is a vector quantity. The direction of velocity along a straight line is usually given as positive or negative.

Note: Since speed is the magnitude of velocity, it is common to use v for either speed or the magnitude of velocity. In addition, the terms *speed* and *velocity* are often used interchangeably because the direction of the motion is often understood. For example, velocity that results from motion in a straight line has the same direction as the direction of the motion, such as in a freely falling body.

Recall that a function is a set of ordered pairs, no two of which have the same first element. We find it helpful to use a special notation, called *functional notation,* to represent a functional relationship. For example, the function $y = x^2 + 3$ is written $f(x) = x^2 + 3$ using functional notation. The symbol $f(x)$, read "f of x," is used to represent the number y which corresponds to a number x under the given functional relationship. (See Section 1.1.) The following table gives $f(x)$ for various values of x.

x	$f(x) = x^2 + 3$
-3	$f(-3) = (-3)^2 + 3 = 12$
0	$f(0) = (0)^2 + 3 = 3$
1	$f(1) = (1)^2 + 3 = 4$
2	$f(2) = (2)^2 + 3 = 7$

x	$f(x) = x^2 + 3$
h	$f(h) = (h)^2 + 3 = h^2 + 3$
$3t$	$f(3t) = (3t)^2 + 3 = 9t^2 + 3$
$1 + \Delta x$	$f(1 + \Delta x) = (1 + \Delta x)^2 + 3 = 4 + 2\Delta x + (\Delta x)^2$

The use of the symbol $f(x)$ is helpful since we can use $f(x)$ to represent the number corresponding to x under the functional relationship without having to actually determine the number as done in the table above. For example, $f(3)$ represents the number corresponding to $x = 3$ under any given functional relationship. For this reason, $f(x)$ is often called the *value of the function at x*.

EXAMPLE 1

Find the value of the function $f(x) = x^3 - 2$ for (a) $x = -3$ and (b) $x = 2 + \Delta x$.

(a) $\qquad f(-3) = (-3)^3 - 2 = -27 - 2 = -29$

(b) $\ f(2 + \Delta x) = (2 + \Delta x)^3 - 2$
$$= 8 + 12(\Delta x) + 6(\Delta x)^2 + (\Delta x)^3 - 2$$
$$= 6 + 12(\Delta x) + 6(\Delta x)^2 + (\Delta x)^3$$

EXAMPLE 2

Evaluate the function $g(x) = \sqrt{2x + 3}$ at $x = 3$.
$$g(3) = \sqrt{2(3) + 3} = \sqrt{9} = 3$$

EXAMPLE 3

Evaluate the function $f(x) = x^2 + 3x - 5$ at $x = h + 2$.
$$f(h + 2) = (h + 2)^2 + 3(h + 2) - 5$$
$$= h^2 + 4h + 4 + 3h + 6 - 5$$
$$= h^2 + 7h + 5$$

On pages 84–85 we were considering an object that moved along a straight line according to the function $s = 4t + 1$. We can express this using functional notation: $s = f(t) = 4t + 1$.

Recall from this earlier discussion that Δs is the change in distance s, and Δt is the change in time after $t = 2$ s. Then using functional notation,

$$\Delta s = f(2 + \Delta t) - f(2)$$

represents, as we saw earlier, the change in distance traveled for a given change in time Δt. When $\Delta t = 3$ s, we have

$$\Delta s = f(2 + \Delta t) - f(2)$$
$$= f(2 + 3) - f(2) \qquad (\Delta t = 3)$$
$$= f(5) - f(2)$$
$$= [4(5) + 1] - [4(2) + 1]$$
$$= 21 - 9$$
$$= 12 \text{ m}$$

So the average speed over this time period is

$$\bar{v} = \frac{\Delta s}{\Delta t} = \frac{f(2 + \Delta t) - f(2)}{\Delta t}$$

$$= \frac{12 \text{ m}}{3 \text{ s}} = 4 \text{ m/s}$$

as we determined earlier.

In general, the distance traveled by an object from time t to time $t + \Delta t$ is given in functional notation by

$$\Delta s = f(t + \Delta t) - f(t)$$

The average speed of this object over the time interval from t to $t + \Delta t$ is then

> **AVERAGE SPEED**
>
> $$\bar{v} = \frac{\Delta s}{\Delta t} = \frac{f(t + \Delta t) - f(t)}{\Delta t}$$

EXAMPLE 4

Given that $s = f(t) = t^2 - 1$ describes the motion of an object moving along a straight line, where s is measured in feet (ft), (a) find Δs and $\bar{v}$; and (b) find the average velocity $\bar{v}$ over the time interval from 4 s to 7 s.

(a)
$$\Delta s = f(t + \Delta t) - f(t)$$
$$= [(t + \Delta t)^2 - 1] - (t^2 - 1)$$
$$= [t^2 + 2t(\Delta t) + (\Delta t)^2 - 1] - (t^2 - 1)$$
$$= 2t(\Delta t) + (\Delta t)^2$$

$$\bar{v} = \frac{\Delta s}{\Delta t} = \frac{2t(\Delta t) + (\Delta t)^2}{\Delta t}$$

$$= \frac{\Delta t(2t + \Delta t)}{\Delta t}$$

$$= 2t + \Delta t$$

(b) $\Delta t = 7 - 4 = 3$ s

From (a) we have

$$\bar{v} = 2t + \Delta t$$
$$= 2(4) + (3)$$
$$= 11 \text{ ft/s}$$

This same result can be obtained by calculating:

$$\bar{v} = \frac{f(4 + 3) - f(4)}{3} = \frac{\text{distance traveled}}{\text{time traveled}} = \frac{48 - 15}{3} = 11 \text{ ft/s}$$

From Example 4 we see that to calculate $\bar{v} = (\Delta s / \Delta t)$ we need to know the time t at which we begin measuring $\bar{v}$ as well as the change in time Δt. Note that both t and Δt can be negative. If $\Delta t = -1$, then $f(t + (-1))$ represents the position of the object 1 s before it reaches the position $f(t)$.

The use of functional notation, as well as the concept of function itself, will receive strong emphasis in the remaining material. The development of the calculus depends heavily on this concept.

To find instantaneous speeds, consider the motion of an object moving along a straight line described by $s = f(t) = 3t^2 + 1$ with s measured in feet. We will now find the "instantaneous" speed after exactly 2 s of travel.

The average speed over the time interval from 2 s to $(2 + \Delta t)$ s is given by

$$\bar{v} = \frac{\text{change in distance}}{\text{change in time}} = \frac{f(2 + \Delta t) - f(2)}{\Delta t}$$

$$= \frac{[3(2 + \Delta t)^2 + 1] - [3(2)^2 + 1]}{\Delta t}$$

$$= \frac{3[4 + 4(\Delta t) + (\Delta t)^2] + 1 - 13}{\Delta t}$$

$$= \frac{12(\Delta t) + 3(\Delta t)^2}{\Delta t}$$

$$= \frac{\Delta t[12 + 3(\Delta t)]}{\Delta t} = 12 + 3(\Delta t)$$

So, for example, with a change in time from $t = 2$ to $t = 6$, $\Delta t = 4$ s. The average speed is $12 + 3(4) = 24$ ft/s. We now tabulate $\bar{v}$ for different values of Δt:

Δt (s)	$\bar{v}$ (ft/s)
4	24
2	18
1	15
0.5	13.5
0.1	12.3
0.01	12.03
0.001	12.003
−0.001	11.997
−0.01	11.97
−0.1	11.7
−0.5	10.5
−2	6

From this table we can see that the closer Δt is to 0, the closer $\bar{v}$ is to 12 ft/s. As we consider the average speed over shorter and shorter time spans, we would expect that the average speed will better approximate the instantaneous speed of the object at 2 s. That is, $\bar{v} = 12.3$ ft/s after 0.1 s of travel (beyond the 2-s mark) is a better approximation than $\bar{v} = 24$ ft/s after 4 s of travel (beyond the 2-s mark). From this table, we see that the instantaneous speed at time $t = 2$ s appears to be 12 ft/s.

INSTANTANEOUS SPEED

To find the *instantaneous speed* of an object in motion at a given time t:

1. Find

$$\bar{v} = \frac{f(t + \Delta t) - f(t)}{\Delta t} = \frac{\Delta s}{\Delta t}$$

where $s = f(t)$ describes the motion of the object as a function of time.

2. Observe what number, if any, $\bar{v}$ approaches as values of Δt approach 0. If there is such a number, it is called the instantaneous speed v.

EXAMPLE 5

Find the instantaneous speed of an object moving according to $s = f(t) = 5t^2 - 4$ at $t = 3$ s.

Step 1: $\bar{v} = \dfrac{f(3 + \Delta t) - f(3)}{\Delta t}$

$= \dfrac{[5(3 + \Delta t)^2 - 4] - [5(3)^2 - 4]}{\Delta t}$

$= \dfrac{30(\Delta t) + 5(\Delta t)^2}{\Delta t}$

$= \dfrac{\Delta t[30 + 5(\Delta t)]}{\Delta t}$

$= 30 + 5(\Delta t)$

Step 2: As Δt approaches (gets close to) 0, $\bar{v}$ approaches 30. We conclude that

$$v = 30 \text{ ft/s}$$

Note: It is tempting to simply substitute $\Delta t = 0$ in the formula for $\bar{v}$. This would be an attempt to compute an average velocity over a time interval of length 0 s. This gives us a zero time interval over which to average! We would be attempting to divide by zero, which is undefined.

$$\frac{f(3 + 0) - f(3)}{0} = \frac{0}{0}!!!$$

As in Example 5, we must find a way to simplify the expression for $\bar{v}$ so that Δt does not remain in the denominator. Only then can we begin to see what number $\bar{v}$ approaches as Δt approaches 0.

EXAMPLE 6

Find v at $t = 2$ when $s = f(t) = \dfrac{1}{t}$.

Step 1: $\bar{v} = \dfrac{f(2 + \Delta t) - f(2)}{\Delta t}$

$= \dfrac{\dfrac{1}{2 + \Delta t} - \dfrac{1}{2}}{\Delta t}$

$= \dfrac{\dfrac{2 - (2 + \Delta t)}{2(2 + \Delta t)}}{\Delta t}$

$= \dfrac{-\Delta t}{2(2 + \Delta t)\Delta t}$

$= \dfrac{-1}{2(2 + \Delta t)}$

Step 2: As Δt approaches 0, $\bar{v}$ approaches $-\frac{1}{2(2)}$. So $v = -\frac{1}{4}$.

Exercises 2.1

Evaluate each function at the given value.

1. $f(x) = 2x^2 + 7, x = 1$

2. $g(x) = x^3 - x + 3, x = -1$

3. $h(x) = 3x^3 - 2x + 4, x = -2$

4. $k(x) = 2x^3 + x - 5, x = 2$

5. $f(x) = \dfrac{x^2 - 3}{x + 5}, x = 2$ **6.** $g(x) = \dfrac{(x^3 - 2x + 3)\sqrt{x - 2}}{x - 4}, x = 3$

7. $f(z) = \sqrt{z^2 + 3}, z = -5$ **8.** $f(t) = \sqrt{t^2 - 1}, t = 4$

9. $f(x) = 3x - 2, x = h + 3$ **10.** $g(x) = x^2 + x - 3, x = w + 4$

11. $f(t) = 3t^2 + 2t - 5, t = 2 + \Delta t$ **12.** $f(t) = t^2 - 3t + 4, t = -3 + \Delta t$

*Find (**a**) Δs and (**b**) $\bar{v}$ for each function expressing distance s in terms of time t. (Express results in terms of t and Δt.)*

13. $s = 3t - 4$ **14.** $s = 2t + 2$

15. $s = 2t^2 + 5$ **16.** $s = 3t^2 - 7$

17. $s = t^2 - 2t + 8$ **18.** $s = 5t^2 + 3t - 9$

Find $\bar{v}$ for each function $s = f(t)$ at the values of t and Δt (s is measured in metres and t in seconds).

19. $s = 5t^2 + 6, t = 3, \Delta t = 4$ **20.** $s = 2t^2 - 5, t = 2, \Delta t = 3$

21. $s = 3t^2 - t + 4, t = 2, \Delta t = 2$ **22.** $s = 6t^2 + 2t - 7, t = 5, \Delta t = 1$

23. A charged particle moves from 0.2 m to 0.5 m from a fixed reference point in 0.5 microseconds (μs). Find its average speed during that interval ($1 \ \mu s = 10^{-6} \ s$).

24. An electron moved a distance of 0.2 m in 1 μs. Find its average speed in m/s during that interval.

25. The average current in a capacitor over a time interval Δt is given by $i_{av} = C(\Delta V / \Delta t)$ amperes, where C is the capacitance in farads and V is the voltage across the capacitor in volts (V). Find the average current (in μA) in a 10-microfarad (μF) capacitor from 2 s to 5 s where the voltage is given by $V = t^2 + 3t + 160$.

26. Find the average current (in μA) in a 15-μF capacitor from 3 s to 7 s where the voltage is given by $V = 2t^2 - 4t + 200$ (see Exercise 25).

Find the instantaneous velocity v of an object moving along a straight line for each given expression for s (measured in metres) at the given value of t (measured in seconds).

27. $s = f(t) = 3t^2 - 6t + 1, t = 2$ **28.** $s = f(t) = -4t^2 + 8, t = 3$

29. $s = f(t) = 5t^2 - 7, t = 1$ **30.** $s = f(t) = 8t^2 + 3t - 11, t = 4$

31. $s = f(t) = \dfrac{1}{2t}, t = 3$ **32.** $s = f(t) = \dfrac{1}{t + 1}, t = 2$

33. $s = f(t) = \dfrac{1}{t - 2}, t = 4$ **34.** $s = f(t) = \dfrac{1}{3t}, t = 4$

35. A free-falling object (neglecting friction) falls from rest according to $s = 16t^2$ when time t is measured in seconds and distance s is measured in feet. Find the speed of an object after falling 2 s.

36. A circuit-breaker contact moves approximately $s = 200,000t^3$, where s is in centimetres (cm) and t is in seconds. Find the speed v of the contact when $t = 0.1$ s.

2.2 THE LIMIT

The process we developed in problems involving motion is very useful in other applications. The technique used is often called "the limit process." Given any function, we may find whether the functional values approach some number when the value of the variable approaches a specified number.

EXAMPLE 1

Let $f(x) = x^2 - 3x + 2$. What number, if any, does $f(x)$ approach as x approaches -1?

Since x^2 approaches $(-1)^2 = 1$ as x approaches -1 and $-3x$ approaches $(-3)(-1) = 3$ as x approaches -1, we conclude that $f(x) = x^2 - 3x + 2$ approaches $1 + 3 + 2 = 6$ as x approaches -1.

We use symbols to describe this limit process more compactly. The symbol " $\rightarrow$ " means "approaches," so x approaches -1 would be written $x \rightarrow -1$.

LIMIT

If $f(x) \rightarrow L$ as $x \rightarrow a$, then L is called "the limit of the function as $x \rightarrow a$." This process is written as

$$\lim_{x \to a} f(x) = L$$

and read "the limit of f of x as x approaches a equals L."

The expression in Example 1 would be written $\lim_{x \to -1} (x^2 - 3x + 2) = 6$.

Note in this case that the limit and the functional value are equal.

THEOREM

For any polynomial $f(x)$,

$$\lim_{x \to a} f(x) = f(a)$$

EXAMPLE 2

Given $f(x) = \dfrac{\tan x}{x}$, what number, if any, does $f(x)$ approach as x approaches 0?

Notice that $f(0)$ does not exist, but that does not mean that the limit does not exist because finding a limit is not the same as finding a functional value. That is, find $\lim_{x \to 0} \dfrac{\tan x}{x}$.

Let's use a calculator and make a table of values for x close to zero. (Be certain that your calculator is in the radian mode.)

x	$\dfrac{\tan x}{x}$	x	$\dfrac{\tan x}{x}$
0.5	1.092605	-0.5	1.092605
0.25	1.021368	-0.25	1.021368
0.15	1.007568	-0.15	1.007568
0.10	1.003347	-0.10	1.003347
0.05	1.000834	-0.05	1.000834
0.01	1.0000332	-0.01	1.0000332
0.001	1.0000005	-0.001	1.0000005

As you can see, as x gets closer and closer to 0, $\dfrac{\tan x}{x}$ approaches 1 (see Fig. 2.2). Thus

$$\lim_{x \to 0} \frac{\tan x}{x} = 1$$

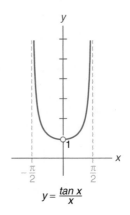

$$y = \frac{\tan x}{x}$$

Figure 2.2

Note: This example does not constitute a proof; it gives only an intuitive idea about limits.

This example leads to a fact that is of paramount importance in physics, namely that $\dfrac{\tan x}{x}$ approaches 1 if x approaches 0 and so $\tan x$ approaches x if x approaches 0.

EXAMPLE 3

Find $\displaystyle\lim_{x \to 3} \left(\frac{x^2 - 9}{x - 3} \right)$

As $x \to 3$, the denominator approaches 0. We cannot divide by zero. However,

$$\frac{x^2 - 9}{x - 3} = \frac{(x + 3)(x - 3)}{x - 3} = x + 3 \qquad (x \neq 3)$$

In the limit process we are not concerned about what happens at $x = 3$, but only what happens as $x \to 3$ (Fig. 2.3). As $x \to 3$, $x + 3 \to 6$. So

$$\lim_{x \to 3} \left(\frac{x^2 - 9}{x - 3} \right) = \lim_{x \to 3} (x + 3) = 6$$

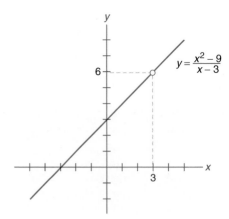

Figure 2.3

Note that in Example 3 we can find the limit of $f(x) = \dfrac{x^2 - 9}{x - 3}$ as $x \to 3$ even though the function is not defined at $x = 3$. However, we will now see that limits do not always exist.

EXAMPLE 4

Find $\lim\limits_{x\to 0} \sqrt{x - 5}$.

Since the square root of a negative number is not real, the function $f(x) = \sqrt{x - 5}$ cannot be evaluated for x less than 5. It is impossible then to observe the values of $\sqrt{x - 5}$ as x takes on values close to 0 (since the quantity $x - 5$ would be negative).

We conclude that $\lim\limits_{x\to 0} \sqrt{x - 5}$ does not exist.

Sometimes a function approaches a limiting number L as $x \to \infty$: that is, the function approaches L as x is allowed to get large without bound.

EXAMPLE 5

Find $\lim\limits_{x\to\infty} (1/x)$.

As the denominator $x \to \infty$, the fraction $1/x$ approaches 0. So

$$\lim_{x\to\infty} \frac{1}{x} = 0$$

Note in Fig. 2.4 that as x increases in value, the graph gets closer and closer to the x-axis. That is, y approaches 0. From the graph, also note that $\lim\limits_{x\to -\infty} \frac{1}{x} = 0$.

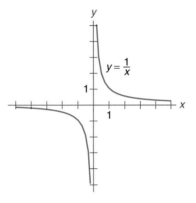

Figure 2.4

EXAMPLE 6

Find $\lim\limits_{x\to\infty} \dfrac{2x^2 + x}{7x^2 - 3}$.

As $x \to \infty$, both numerator and denominator approach ∞ separately. However, if we divide numerator and denominator by the highest power of x in the denominator, x^2, we have

$$\frac{2 + \dfrac{1}{x}}{7 - \dfrac{3}{x^2}}$$

$$\lim_{x\to\infty} \frac{2x^2 + x}{7x^2 - 3} = \lim_{x\to\infty} \frac{2 + \dfrac{1}{x}}{7 - \dfrac{3}{x^2}} = \frac{2 + 0}{7 - 0} = \frac{2}{7}$$

Note: As $x \to \infty$, $\dfrac{1}{x} \to 0$ and $\dfrac{3}{x^2} \to 0$.

Using a calculator,

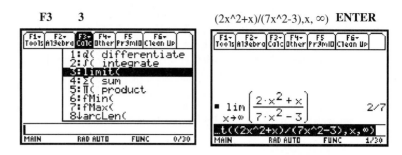

F3   3   (2x^2+x)/(7x^2-3),x, ∞) **ENTER**

Finding instantaneous velocity is an application of the limit process. As the time interval, Δt, decreases in the average velocity formula

$$\bar{v} = \frac{\Delta s}{\Delta t} = \frac{f(t + \Delta t) - f(t)}{\Delta t}$$

the average velocity approaches the instantaneous velocity, v–the velocity at a given instant. That is,

INSTANTANEOUS VELOCITY

$$v = \lim_{\Delta t \to 0} \frac{f(t + \Delta t) - f(t)}{\Delta t}$$

EXAMPLE 7

Find the instantaneous velocity v at $t = 3$ when $s = f(t) = t^2 - 7$.

$$v = \lim_{\Delta t \to 0} \frac{f(3 + \Delta t) - f(3)}{\Delta t}$$

$$= \lim_{\Delta t \to 0} \frac{[9 + 6(\Delta t) + (\Delta t)^2 - 7] - [9 - 7]}{\Delta t}$$

$$= \lim_{\Delta t \to 0} \frac{6(\Delta t) + (\Delta t)^2}{\Delta t}$$

$$= \lim_{\Delta t \to 0} \frac{\Delta t(6 + \Delta t)}{\Delta t}$$

$$= \lim_{\Delta t \to 0} (6 + \Delta t)$$

$$= 6$$

It can be shown that the limit process follows these formulas:

$$\lim_{x \to a} [f(x) \pm g(x)] = \lim_{x \to a} f(x) \pm \lim_{x \to a} g(x) \tag{1}$$

EXAMPLE 8

$$\lim_{x \to 3} (x^3 + x^2) = \lim_{x \to 3} x^3 + \lim_{x \to 3} x^2$$

$$36 = 27 + 9$$

$$\lim_{x \to a} [k \cdot f(x)] = k \cdot \lim_{x \to a} f(x), \text{ where } k \text{ is a constant.} \tag{2}$$

EXAMPLE 9

$$\lim_{x \to -2} 12x^2 = 12 \lim_{x \to -2} x^2$$

$$48 = 12(4)$$

$$\lim_{x \to a} k = k, \text{ where } k \text{ is a constant.} \tag{3}$$

EXAMPLE 10

$$\lim_{x \to 2} 8 = 8$$

Note: No matter what x approaches, $f(x) = 8$; so $f(x)$ not only approaches 8, but actually is 8.

$$\lim_{x \to a} [f(x) \cdot g(x)] = \lim_{x \to a} f(x) \cdot \lim_{x \to a} g(x) \tag{4}$$

EXAMPLE 11

$$\lim_{x \to 3} [x^2(x - 1)] = \lim_{x \to 3} x^2 \cdot \lim_{x \to 3} (x - 1)$$

$$18 = 9 \cdot 2$$

Using Formula (4), one could show that

$$\lim_{x \to a} x^n = a^n \qquad (\text{where } n \text{ is a positive integer})$$

$$\lim_{x \to a} \frac{f(x)}{g(x)} = \frac{\lim_{x \to a} f(x)}{\lim_{x \to a} g(x)} \qquad \text{where} \quad \lim_{x \to a} g(x) \neq 0 \tag{5}$$

Note: $\lim_{x \to a} f(x)$ and $\lim_{x \to a} g(x)$ must both exist in Formulas (1)–(5).

EXAMPLE 12

$$\lim_{x \to 1} \frac{x^2 - 4}{x + 2} = \frac{\lim_{x \to 1} (x^2 - 4)}{\lim_{x \to 1} (x + 2)}$$

$$-1 = \frac{-3}{3}$$

The idea of continuity is very closely related to the limit idea. A function is *continuous* if its graph is an unbroken curve. The function $f(x) = x^2$ is continuous, as you can see from its graph in Fig. 2.5.

The function

$$f(x) = \begin{cases} x & \text{if } x \geq 0 \\ 1 & \text{if } x < 0 \end{cases}$$

is not continuous because its graph in Fig. 2.6 is broken at $x = 0$.

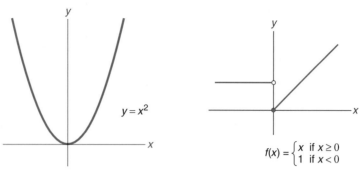

$$f(x) = \begin{cases} x & \text{if } x \geq 0 \\ 1 & \text{if } x < 0 \end{cases}$$

Figure 2.5   **Figure 2.6**

More formally,

CONTINUITY

A function is continuous at $x = a$ if and only if

1. $f(a)$ is defined,
2. $\lim_{x \to a} f(x)$ exists, and
3. $\lim_{x \to a} f(x) = f(a)$.

Figure 2.7 shows the graphs of three functions that are not continuous at $x = a$.

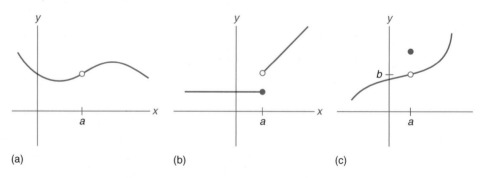

(a)   (b)   (c)

Figure 2.7   Functions not continuous at $x = a$.

Figure 2.7(a) is not continuous because $f(a)$ is not defined.

Figure 2.7(b) is not continuous because $\lim_{x \to a} f(x)$ does not exist.

Figure 2.7(c) is not continuous because $\lim_{x \to a} f(x) = b \neq f(a)$.

If a function is continuous at all points in a given interval, it is continuous on that interval.

Exercises 2.2

Find each limit using a calculator. (For trigonometric functions, be certain that your calculator is in the radian mode.)

1. $\lim\limits_{x \to 2} \dfrac{x^2 - 4}{x - 2}$

2. $\lim\limits_{x \to -\frac{1}{2}} \dfrac{4x^2 - 1}{2x + 1}$

3. $\lim\limits_{x \to \infty} \dfrac{3x + 2}{x}$

4. $\lim\limits_{x \to \infty} \dfrac{3x^2 + 4x}{2x^2 - 1}$

5. $\lim\limits_{x \to 0} \dfrac{\sin x}{x}$

6. $\lim\limits_{x \to 0} \dfrac{\cos x - 1}{x}$

Find each limit without using a calculator.

7. $\lim\limits_{x \to 2} (x^2 - 5x)$

8. $\lim\limits_{x \to -1} (3x^2 + 7x + 1)$

9. $\lim\limits_{x \to -1} (2x^3 + 5x^2 - 2)$

10. $\lim\limits_{x \to 2} (x^3 - 3x^2 + x + 4)$

11. $\lim\limits_{x \to 1} \dfrac{x^2 - 1}{x - 1}$

12. $\lim\limits_{x \to -3} \dfrac{x^2 - 9}{x + 3}$

13. $\lim\limits_{x \to -3/2} \dfrac{4x^2 - 9}{2x + 3}$

14. $\lim\limits_{x \to 4/3} \dfrac{9x^2 - 16}{3x - 4}$

15. $\lim\limits_{x \to -1} \sqrt{2x + 3}$

16. $\lim\limits_{x \to 4} \sqrt{3x - 3}$

17. $\lim\limits_{x \to 6} \sqrt{4 - x}$

18. $\lim\limits_{x \to -1} \sqrt{2x + 1}$

19. $\lim\limits_{x \to \infty} \dfrac{1}{2x}$

20. $\lim\limits_{x \to \infty} \dfrac{1}{x^2}$

21. $\lim\limits_{x \to \infty} \dfrac{3x^2 - 5x + 2}{4x^2 + 8x - 11}$

22. $\lim\limits_{x \to \infty} \dfrac{7x^3 + 2x - 13}{4x^3 + x^2}$.

Find the instantaneous velocity v for each expression of s and value of t.

23. $s = f(t) = 4t^2 - 3t,\ t = 2$

24. $s = f(t) = 3t^2 - 5t + 2,\ t = 3$

25. $s = f(t) = t^2 + 3t - 10,\ t = 4$

26. $s = f(t) = 5t^2 - 7t + 8,\ t = 2$

Find each limit using Formulas (1)–(5).

27. $\lim\limits_{x \to 2} (x^2 + x)$

28. $\lim\limits_{x \to 3} (x^3 + x^2)$

29. $\lim\limits_{x \to 1} (4x^2 + 100x - 2)$

30. $\lim\limits_{x \to -1} (3x^2 + 5x - 8)$

31. $\lim\limits_{x \to 1} (x + 3)(x - 4)$

32. $\lim\limits_{x \to 4} (2x + 1)(x - 3)$

33. $\lim\limits_{x \to -2} (x^2 + 3x + 1)(x^4 - 2x^2 + 3)$

34. $\lim\limits_{x \to 2} (x^2 + 5x - 10)(x^3 + 6x^2 - x)$

35. $\lim\limits_{x \to 2} \dfrac{x^2 + 3x + 2}{x^2 + 1}$

36. $\lim\limits_{x \to 3} \dfrac{x^2 - 4x + 5}{x^2 + 2x}$

37. $\lim\limits_{x \to -7} \dfrac{x^2 - 49}{x + 7}$

38. $\lim\limits_{x \to 2} \dfrac{x^2 - 4}{x - 2}$

39. $\lim\limits_{x \to 5/2} \dfrac{4x^2 - 25}{2x - 5}$

40. $\lim\limits_{x \to -4/3} \dfrac{9x^2 - 16}{3x + 4}$

41. $\lim\limits_{x \to 3} \dfrac{(x^2 + 3x + 1)(x + 5)}{x - 2}$

42. $\lim\limits_{x \to -2} \dfrac{(x^2 + x - 5)(x - 3)}{x + 3}$

43. $\lim\limits_{h \to 0} \dfrac{(x + h)^2 - x^2}{h}$

44. $\lim\limits_{x \to a} \dfrac{x^3 - a^3}{x - a}$

45. $\lim\limits_{h \to 0} \dfrac{\dfrac{1}{x + h} - \dfrac{1}{x}}{h}$

46. $\lim\limits_{x \to a} \dfrac{\dfrac{1}{x^2} - \dfrac{1}{a^2}}{x - a}$

47. $\lim\limits_{x \to a} \dfrac{\sqrt{x} - \sqrt{a}}{x - a}$ (*Hint:* Rationalize the numerator.)

48. $\lim\limits_{h \to 0} \dfrac{\sqrt{x + h} - \sqrt{x}}{h}$

In Exercises 49–56, find $\lim\limits_{x \to a} f(x)$ if it exists.

49.
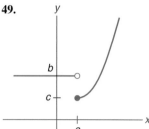

50.
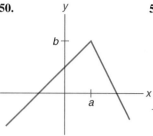

51.
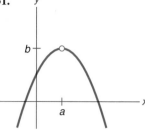

52.
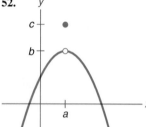

53.
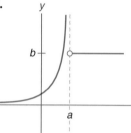

54.
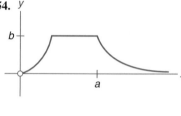

55.

56.
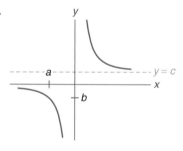

57–64. In Exercises 49–56, is the function continuous at $x = a$?

65. In Exercise 49, find $\lim\limits_{x \to +\infty} f(x)$ if it exists.

66. In Exercise 53, find $\lim\limits_{x \to +\infty} f(x)$ if it exists.

67. In Exercise 54, find $\lim\limits_{x \to +\infty} f(x)$ if it exists.

68. In Exercise 55, find $\lim\limits_{x \to +\infty} f(x)$ if it exists.

69. In Exercise 49, find $\lim\limits_{x \to -\infty} f(x)$ if it exists.

70. In Exercise 53, find $\lim\limits_{x \to -\infty} f(x)$ if it exists.

71. In Exercise 54, find $\lim\limits_{x \to -\infty} f(x)$ if it exists.

72. In Exercise 56, find $\lim\limits_{x \to +\infty} f(x)$ if it exists.

2.3 THE SLOPE OF A TANGENT LINE TO A CURVE

The limit process so far has been applied only to motion problems. We now look at its geometric application. We saw in Section 1.3 how to find the slope of a line. But how can we describe the slope of a tangent line to a nonlinear curve at a given point? As in Fig. 2.8, assume the curve is the graph of a given function $y = f(x)$. We wish to find the slope of the tangent line, m_{tan}, at the point P with coordinates $(x, f(x))$. We can determine the slope of a line passing through P and any other point Q on the curve (the secant line). Observe the slopes of these secant lines as we choose points Q closer and closer to the point P. As Q approaches P, the values of the slopes of these secant lines become closer and closer to the slope of the tangent line, m_{tan}. We can express this process in terms of the coordinates of P and Q as in Fig. 2.9. In this figure, $\Delta y = f(x + \Delta x) - f(x)$.

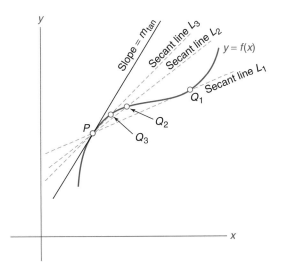

Figure 2.8 As Q approaches P, the slope of the secant line approaches the slope of the tangent line at point P.

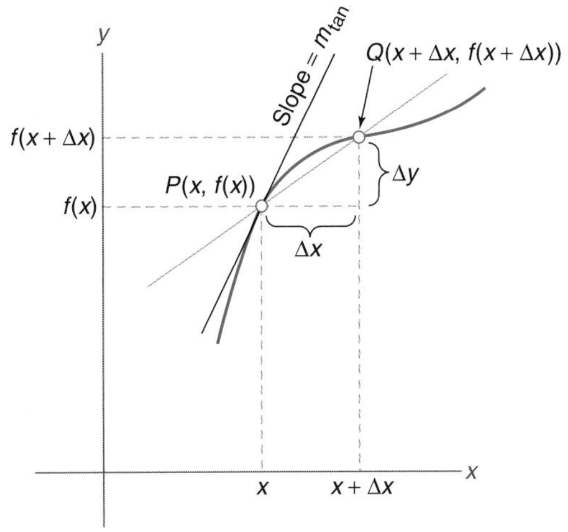

Figure 2.9

As we choose values of Δx closer to 0, point Q moves closer to P along the curve. Thus the slope of the secant line approaches m_{tan}, the slope of the tangent line. The slope of the secant line through P and Q is given by

$$\frac{f(x + \Delta x) - f(x)}{(x + \Delta x) - x} = \frac{f(x + \Delta x) - f(x)}{\Delta x} = \frac{\Delta y}{\Delta x}$$

so

SLOPE OF TANGENT LINE

$$m_{\text{tan}} = \lim_{\Delta x \to 0} \frac{\Delta y}{\Delta x} = \lim_{\Delta x \to 0} \frac{f(x + \Delta x) - f(x)}{\Delta x}$$

EXAMPLE 1

Find the slope of the tangent line to the curve $y = x^2 + 3$ at $(1, 4)$.

$$
\begin{aligned}
m_{\text{tan}} &= \lim_{\Delta x \to 0} \frac{\Delta y}{\Delta x} = \lim_{\Delta x \to 0} \frac{f(x + \Delta x) - f(x)}{\Delta x} \\
&= \lim_{\Delta x \to 0} \frac{[(1 + \Delta x)^2 + 3] - [(1)^2 + 3]}{\Delta x} \\
&= \lim_{\Delta x \to 0} \frac{2(\Delta x) + (\Delta x)^2}{\Delta x} \\
&= \lim_{\Delta x \to 0} \frac{\Delta x(2 + \Delta x)}{\Delta x} \\
&= \lim_{\Delta x \to 0} (2 + \Delta x) \\
&= 2
\end{aligned}
$$

The curve and the tangent line appear in Fig. 2.10.

The process used to solve the geometric problem is the same as used for the motion problem. This process, the limit, is the foundation of the calculus.

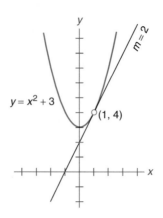

Figure 2.10

EXAMPLE 2

Find the equation of the tangent line to the curve $y = 2x^2 - 5$ at $(2, 3)$.

Step 1:   Find m_{tan}:

$$m_{\text{tan}} = \lim_{\Delta x \to 0} \frac{\Delta y}{\Delta x} = \lim_{\Delta x \to 0} \frac{f(x + \Delta x) - f(x)}{\Delta x}$$

$$= \lim_{\Delta x \to 0} \frac{[2(2 + \Delta x)^2 - 5] - [2(2)^2 - 5]}{\Delta x}$$

$$= \lim_{\Delta x \to 0} \frac{\Delta x(8 + 2\Delta x)}{\Delta x}$$

$$= \lim_{\Delta x \to 0} (8 + 2\Delta x)$$

$$= 8$$

Step 2:   Find the equation of the line:
Using the point-slope formula, we have

$$y - y_1 = m(x - x_1)$$
$$y - 3 = 8(x - 2)$$
$$y = 8x - 13$$

Exercises 2.3

Find the slope of the tangent line to each curve at the given point.

1. $y = x^2; (3, 9)$ **2.** $y = 3x^2; (1, 3)$

3. $y = 3x^2 - 4; (-1, -1)$ **4.** $y = 4x^2 + 3; (2, 19)$

5. $y = 2x^2 + x - 3; (2, 7)$ **6.** $y = 3x^2 + 8x - 10; (-2, -14)$

7. $y = -4x^2 + 3x - 2; (1, -3)$ **8.** $y = -5x^2 - 3x - 1; (2, -27)$

9. $y = x^3; (2, 8)$ **10.** $y = x^3 + 1; (-1, 0)$

Find the equation of the tangent line to each curve at the given point.

11. $y = x^2; (-2, 4)$ **12.** $y = -5x^2; (2, -20)$

13. $y = 2x^2 - 3; (-2, 5)$ **14.** $y = -4x^2 + 2; (-1, -2)$

15. $y = 5x^2 - 3x + 2; (-1, 10)$ **16.** $y = 4x^2 - 7x + 5; (3, 20)$

17. $y = -3x^2 + 5x + 2; x = 3$ **18.** $y = -2x^2 + 4x - 7; x = -2$

19. $y = x^3 + x - 1; x = 1$ **20.** $y = x^3 - 3x + 4; x = 2$

Find a point or points on the curve where the slope of the tangent line is the given value.

21. $y = x^2; m = -\frac{1}{3}$ **22.** $y = 3x^2 - 4; m = 6$

23. $y = x^3 + x; m = 4$ **24.** $y = x^3 - 3x^2 - 5x; m = 4$

2.4 THE DERIVATIVE

Instantaneous velocity and the slope of the tangent line to a curve are only two examples of using the concept of limit. A more general use of the limit is found in the derivative. When we developed instantaneous velocity and the slope of the tangent line, we applied the limit concept to related functions as follows:

$$v = \lim_{\Delta t \to 0} \frac{\Delta s}{\Delta t} = \lim_{\Delta t \to 0} \frac{f(t + \Delta t) - f(t)}{\Delta t}$$

and

$$m_{\text{tan}} = \lim_{\Delta x \to 0} \frac{\Delta y}{\Delta x} = \lim_{\Delta x \to 0} \frac{f(x + \Delta x) - f(x)}{\Delta x}$$

For any function $y = f(x)$,

$$\frac{\Delta y}{\Delta x} = \frac{f(x + \Delta x) - f(x)}{\Delta x}$$

is the *average rate of change* of the function f at x, and

DERIVATIVE

$$\lim_{\Delta x \to 0} \frac{\Delta y}{\Delta x} = \lim_{\Delta x \to 0} \frac{f(x + \Delta x) - f(x)}{\Delta x}$$

is the *derivative* of the function f at x. (The derivative could also be thought of as the instantaneous rate of change of f at x.) The process of finding this limit, the derivative, is called *differentiation*.

EXAMPLE 1

Find the derivative of $f(x) = x^2$ at $x = 3$.

$$\lim_{\Delta x \to 0} \frac{\Delta y}{\Delta x} = \lim_{\Delta x \to 0} \frac{f(x + \Delta x) - f(x)}{\Delta x}$$

$$= \lim_{\Delta x \to 0} \frac{(3 + \Delta x)^2 - (3)^2}{\Delta x}$$

$$= \lim_{\Delta x \to 0} \frac{9 + 6\Delta x + (\Delta x)^2 - 9}{\Delta x}$$

$$= \lim_{\Delta x \to 0} \frac{6\Delta x + (\Delta x)^2}{\Delta x}$$

$$= \lim_{\Delta x \to 0} \frac{\Delta x(6 + \Delta x)}{\Delta x}$$

$$= \lim_{\Delta x \to 0} (6 + \Delta x)$$

$$= 6$$

In Example 1 we found the derivative at $x = 3$. However, we can also find the derivative without specifying a value for x.

EXAMPLE 2

Find the derivative of $f(x) = x^2$.

$$\lim_{\Delta x \to 0} \frac{\Delta y}{\Delta x} = \lim_{\Delta x \to 0} \frac{f(x + \Delta x) - f(x)}{\Delta x}$$

$$= \lim_{\Delta x \to 0} \frac{(x + \Delta x)^2 - x^2}{\Delta x}$$

$$= \lim_{\Delta x \to 0} \frac{x^2 + 2x(\Delta x) + (\Delta x)^2 - x^2}{\Delta x}$$

$$= \lim_{\Delta x \to 0} \frac{\Delta x(2x + \Delta x)}{\Delta x}$$

$$= \lim_{\Delta x \to 0} (2x + \Delta x)$$

$$= 2x$$

Note that when finding the above limit, x is held constant. Only Δx is approaching 0. Note also that if we let $x = 3$ in Example 2, the derivative has the value $2x = 2(3) = 6$ as in Example 1.

The following notations for the derivative are often used in place of the symbol $\lim_{\Delta x \to 0} \dfrac{\Delta y}{\Delta x}$:

$$\frac{dy}{dx}, \quad y', \quad \frac{d}{dx}[f(x)], \quad f'(x), \quad \text{and} \quad Dy$$

So for Example 1 we could write

$$\frac{dy}{dx} = 6 \quad \text{at} \quad x = 3$$

$$y' = 6 \quad \text{at} \quad x = 3$$

$$\frac{d}{dx}[f(x)] = 6 \quad \text{at} \quad x = 3$$

$$f'(3) = 6$$

or

$$Dy = 6 \quad \text{at} \quad x = 3$$

For Example 2 we could write

$$\frac{dy}{dx} = 2x$$

$$y' = 2x$$

$$\frac{d}{dx}[f(x)] = 2x$$

$$f'(x) = 2x$$

or

$$Dy = 2x$$

Note that the expressions above for the derivative determined in Example 2 show that the derivative becomes a new function of x when the value for x is not specified. That is, $y' = 2x$ is a function that relates to each x the value $2x$.

EXAMPLE 3

Find the derivative of $y = x^3$.

$$\frac{dy}{dx} = \lim_{\Delta x \to 0} \frac{f(x + \Delta x) - f(x)}{\Delta x}$$

$$= \lim_{\Delta x \to 0} \frac{(x + \Delta x)^3 - x^3}{\Delta x}$$

$$= \lim_{\Delta x \to 0} \frac{x^3 + 3x^2(\Delta x) + 3x(\Delta x)^2 + (\Delta x)^3 - x^3}{\Delta x}$$

$$= \lim_{\Delta x \to 0} \frac{3x^2(\Delta x) + 3x(\Delta x)^2 + (\Delta x)^3}{\Delta x}$$

$$= \lim_{\Delta x \to 0} \frac{\Delta x[3x^2 + 3x(\Delta x) + (\Delta x)^2]}{\Delta x}$$

$$= \lim_{\Delta x \to 0} [3x^2 + 3x(\Delta x) + (\Delta x)^2]$$

$$= 3x^2$$

EXAMPLE 4

Find the derivative of $y = \dfrac{1}{x + 2}$.

$$\frac{dy}{dx} = \lim_{\Delta x \to 0} \frac{f(x + \Delta x) - f(x)}{\Delta x}$$

$$= \lim_{\Delta x \to 0} \frac{\dfrac{1}{x + \Delta x + 2} - \dfrac{1}{x + 2}}{\Delta x}$$

$$= \lim_{\Delta x \to 0} \frac{\dfrac{(x + 2) - (x + \Delta x + 2)}{(x + \Delta x + 2)(x + 2)}}{\Delta x}$$

$$= \lim_{\Delta x \to 0} \frac{\dfrac{-\Delta x}{(x + \Delta x + 2)(x + 2)}}{\Delta x}$$

$$= \lim_{\Delta x \to 0} \frac{-1}{(x + \Delta x + 2)(x + 2)}$$

$$= \frac{-1}{(x + 2)^2}$$

EXAMPLE 5

Find the derivative of $y = 3x^2 + 2x - 4$.

$$\frac{dy}{dx} = \lim_{\Delta x \to 0} \frac{f(x + \Delta x) - f(x)}{\Delta x}$$

$$= \lim_{\Delta x \to 0} \frac{[3(x + \Delta x)^2 + 2(x + \Delta x) - 4] - (3x^2 + 2x - 4)}{\Delta x}$$

$$= \lim_{\Delta x \to 0} \frac{3x^2 + 6x(\Delta x) + 3(\Delta x)^2 + 2x + 2(\Delta x) - 4 - 3x^2 - 2x + 4}{\Delta x}$$

$$= \lim_{\Delta x \to 0} \frac{6x(\Delta x) + 3(\Delta x)^2 + 2(\Delta x)}{\Delta x}$$

$$= \lim_{\Delta x \to 0} \frac{\Delta x[6x + 3(\Delta x) + 2]}{\Delta x}$$

$$= \lim_{\Delta x \to 0} [6x + 3(\Delta x) + 2]$$

$$= 6x + 2$$

EXAMPLE 6

Find the derivative of $f(x) = \sqrt{x}$.

$$\frac{dy}{dx} = \lim_{\Delta x \to 0} \frac{f(x + \Delta x) - f(x)}{\Delta x}$$

$$= \lim_{\Delta x \to 0} \frac{\sqrt{x + \Delta x} - \sqrt{x}}{\Delta x} \qquad (\textit{Note: } \text{To evaluate this limit, you must rationalize the numerator.})$$

$$= \lim_{\Delta x \to 0} \frac{\sqrt{x + \Delta x} - \sqrt{x}}{\Delta x} \cdot \frac{\sqrt{x + \Delta x} + \sqrt{x}}{\sqrt{x + \Delta x} + \sqrt{x}}$$

$$= \lim_{\Delta x \to 0} \frac{x + \Delta x - x}{\Delta x(\sqrt{x + \Delta x} + \sqrt{x})}$$

$$= \lim_{\Delta x \to 0} \frac{\Delta x}{\Delta x(\sqrt{x + \Delta x} + \sqrt{x})}$$

$$= \lim_{\Delta x \to 0} \frac{1}{\sqrt{x + \Delta x} + \sqrt{x}}$$

$$= \frac{1}{\sqrt{x} + \sqrt{x}} = \frac{1}{2\sqrt{x}}$$

Using a calculator,

 F3   3   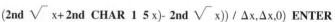 **(2nd** $\sqrt{}$ **x+ 2nd CHAR 1 5** x)- **2nd** $\sqrt{}$ x)) / Δx,Δx,0) **ENTER**

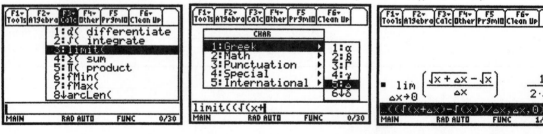

Note that the symbol Δ can be obtained on the TI-89 by using the keystrokes **2nd CHAR 1 5**.

Exercises 2.4

Find the derivative of each function.

1. $y = 3x + 4$ **2.** $y = 2 - 6x$ **3.** $y = 1 - 2x$ **4.** $y = 4x - 5$

5. $y = 3x^2$ **6.** $y = -4x^2$ **7.** $y = x^2 - 2x$ **8.** $y = x^2 + 1$

9. $y = 3x^2 - 4x + 1$ **10.** $y = 6x^2 - 8x + 2$ **11.** $y = 1 - 6x^2$ **12.** $y = 3 - 4x - 7x^2$

13. $y = x^3 + 4x$ **14.** $y = 1 - 2x^3$ **15.** $y = \dfrac{1}{x}$ **16.** $y = \dfrac{1}{x - 1}$

17. $y = \dfrac{2}{x - 3}$ **18.** $y = \dfrac{-1}{2x + 1}$ **19.** $y = \dfrac{1}{x^2}$ **20.** $y = \dfrac{1}{x^2 - 1}$

21. $y = \dfrac{1}{4 - x^2}$ **22.** $y = \dfrac{4}{x^2 + 1}$ **23.** $y = \sqrt{x + 1}$ **24.** $y = \sqrt{x - 2}$

25. $y = \sqrt{1 - 2x}$ **26.** $y = \sqrt{x^2 + 1}$ **27.** $y = \dfrac{1}{\sqrt{x - 1}}$ **28.** $y = \dfrac{1}{\sqrt{4 - x}}$

Find the slope of the tangent line to each curve at the given point.

29. $y = 3x^2 - 4$; $(1, -1)$ **30.** $y = \dfrac{1}{x - 4}$; $\left(1, -\dfrac{1}{3}\right)$ **31.** $y = \dfrac{1}{1 - 3x^2}$; $\left(1, -\dfrac{1}{2}\right)$

32. $y = \sqrt{x + 5}$; $(4, 3)$

Find the equation of the tangent line to each curve at the given point.

33. $y = x^2 - 3x$; $(2, -2)$ **34.** $y = \dfrac{6}{x}$; $(-2, -3)$ **35.** $y = \sqrt{x - 7}$; $(11, 2)$

36. $y = \dfrac{2}{x^2 + 1}$; $(-1, 1)$

Find a point or points on the curve where the slope of the tangent line is the given value.

37. $y = \dfrac{1}{x - 3}$; $m = -1$ **38.** $y = \dfrac{1}{x^2}$; $m = \dfrac{1}{4}$ **39.** $y = \sqrt{x + 4}$; $m = \dfrac{1}{2}$

40. $y = \dfrac{2}{\sqrt{x}}$; $m = -\dfrac{1}{8}$

2.5   DIFFERENTIATION OF POLYNOMIALS

Recall from algebra that a *polynomial* is defined as follows:

$$a_n x^n + a_{n-1} x^{n-1} + a_{n-2} x^{n-2} + \cdots + a_2 x^2 + a_1 x + a_0$$

where n is a positive integer and the coefficients $a_n, a_{n-1}, a_{n-2}, \ldots, a_2, a_1, a_0$ are real numbers. As we saw, the derivative definition in Section 2.4 can be used to find the derivative of a polynomial. However, the process is tedious. We now develop some formulas that will shorten the process to find such a derivative.

First, let's find the derivative of a constant function.

EXAMPLE 1

Find $\dfrac{dy}{dx}$ when $y = C$, where C is a constant.

$$\frac{dy}{dx} = \lim_{\Delta x \to 0} \frac{f(x + \Delta x) - f(x)}{\Delta x}$$

$$= \lim_{\Delta x \to 0} \frac{C - C}{\Delta x}$$

$$= \lim_{\Delta x \to 0} \frac{0}{\Delta x}$$

$$= \lim_{\Delta x \to 0} 0$$

$$= 0$$

So, if $f(x)$ is a constant function, i.e., $f(x) = C$ for all x, then $f'(x) = 0$ for all x.

EXAMPLE 2

Find the derivative of $f(x) = 10$.

Since this is a constant function, $f'(x) = 0$ for all x.

Suppose that $y = x^n$, where n is some positive integer. By using the binomial theorem (see Section 9.3), we have

$$(x + \Delta x)^n = x^n + nx^{n-1}(\Delta x) + \frac{n(n-1)}{2!}x^{n-2}(\Delta x)^2$$

$$+ \frac{n(n-1)(n-2)}{3!}x^{n-3}(\Delta x)^3 + \cdots$$

$$+ nx(\Delta x)^{n-1} + (\Delta x)^n$$

where the right-hand side consists of $(n+1)$ terms, each of which has a factor of Δx except for the first term. So if

$$\Delta y = (x + \Delta x)^n - x^n$$

$$= x^n + nx^{n-1}(\Delta x) + \frac{n(n-1)}{2!}x^{n-2}(\Delta x)^2$$

$$+ \frac{n(n-1)(n-2)}{3!}x^{n-3}(\Delta x)^3 + \cdots + nx(\Delta x)^{n-1} + (\Delta x)^n - x^n$$

then

$$\frac{\Delta y}{\Delta x} = nx^{n-1} + \frac{n(n-1)}{2!}x^{n-2}(\Delta x) + \frac{n(n-1)(n-2)}{3!}x^{n-3}(\Delta x)^2 + \cdots$$

$$+ nx(\Delta x)^{n-2} + (\Delta x)^{n-1}$$

We then have

$$\frac{dy}{dx} = \lim_{\Delta x \to 0} \frac{\Delta y}{\Delta x} = nx^{n-1}$$

Note that every term but the first term has a factor of Δx, which will make each of these other terms approach 0 as Δx approaches 0.

$$\frac{d}{dx}x^n = nx^{n-1}$$

EXAMPLE 3

Find the derivative of $y = x^{32}$. Since $n = 32$,

$$\frac{dy}{dx} = 32x^{32-1}$$

$$= 32x^{31}$$

We now make the following observation.

$$\frac{d}{dx}[c f(x)] = c\frac{d}{dx}[f(x)]$$

This formula is shown below:

$$\frac{d}{dx}[c\,f(x)] = \lim_{\Delta x \to 0} \frac{c[f(x + \Delta x)] - c[f(x)]}{\Delta x}$$

$$= \lim_{\Delta x \to 0} \frac{c[f(x + \Delta x) - f(x)]}{\Delta x}$$

$$= c \lim_{\Delta x \to 0} \frac{f(x + \Delta x) - f(x)}{\Delta x}$$

$$= c \frac{d}{dx}[f(x)]$$

EXAMPLE 4

Find the derivative of $y = -5x^{12}$.

$$\frac{dy}{dx} = \frac{d}{dx}(-5x^{12})$$

$$= -5\frac{d}{dx}(x^{12})$$

$$= -5(12x^{12-1})$$

$$= -60x^{11}$$

We can now differentiate functions of the form $y = cx^n$, where n is a positive integer, by use of a simple formula. We now develop formulas to differentiate other functions.

If $h(x) = f(x) + g(x)$, the sum of two functions, then

$$\frac{d}{dx}h(x) = \frac{d}{dx}[f(x) + g(x)]$$

$$= \lim_{\Delta x \to 0} \frac{[f(x + \Delta x) + g(x + \Delta x)] - [f(x) + g(x)]}{\Delta x}$$

$$= \lim_{\Delta x \to 0} \frac{[f(x + \Delta x) - f(x)] + [g(x + \Delta x) - g(x)]}{\Delta x}$$

$$= \lim_{\Delta x \to 0} \left\{ \frac{[f(x + \Delta x) - f(x)]}{\Delta x} + \frac{[g(x + \Delta x) - g(x)]}{\Delta x} \right\}$$

$$= \lim_{\Delta x \to 0} \frac{[f(x + \Delta x) - f(x)]}{\Delta x} + \lim_{\Delta x \to 0} \frac{[g(x + \Delta x) - g(x)]}{\Delta x}$$

$$= \frac{d}{dx}[f(x)] + \frac{d}{dx}[g(x)]$$

That is, the derivative of a sum of functions is the sum of the derivatives of the functions:

$$\frac{d}{dx}[f(x) + g(x)] = \frac{d}{dx}[f(x)] + \frac{d}{dx}[g(x)]$$

Similarly,

$$\frac{d}{dx}[f(x) - g(x)] = \frac{d}{dx}[f(x)] - \frac{d}{dx}[g(x)]$$

That is, the derivative of a difference of functions is the difference of the derivatives of the functions.

We now have the formulas we need to differentiate polynomials.

EXAMPLE 5

Differentiate $y = 7x^5 - 2x^3 + x^2 - 8$.

$$\frac{dy}{dx} = \frac{d}{dx}(7x^5 - 2x^3 + x^2 - 8)$$

$$= \frac{d}{dx}(7x^5) - \frac{d}{dx}(2x^3) + \frac{d}{dx}(x^2) - \frac{d}{dx}(8)$$

$$= 7\frac{d}{dx}(x^5) - 2\frac{d}{dx}(x^3) + \frac{d}{dx}(x^2) - \frac{d}{dx}(8)$$

$$= 7 \cdot 5x^{5-1} - 2 \cdot 3x^{3-1} + 2x^{2-1} - 0$$

$$= 35x^4 - 6x^2 + 2x$$

EXAMPLE 6

Find the equation of the tangent line to the curve $y = 5x^3 - 2x^2 + 3$ at the point $(1, 6)$.

The slope of the tangent line, $m_{\tan}$, is the derivative of $y = 5x^3 - 2x^2 + 3$ at $x = 1$.

$$\frac{dy}{dx} = \frac{d}{dx}(5x^3 - 2x^2 + 3)$$

$$= 5 \cdot 3x^{3-1} - 2 \cdot 2x^{2-1} + 0$$

$$= 15x^2 - 4x$$

Thus the derivative $\dfrac{dy}{dx}$ at $x = 1$ is $15(1)^2 - 4(1) = 11$. So, $m_{\tan} = 11$. The equation of the tangent line written in point-slope form is

$$y - y_1 = m(x - x_1)$$
$$y - 6 = 11(x - 1)$$
$$y = 11x - 5$$

In this example we needed to evaluate the derivative of a function after applying the differentiation formulas. To show that the derivative is to be evaluated at $x = a$, we usually write

$$\frac{dy}{dx}\bigg|_{x=a} \quad \text{or} \quad f'(a)$$

So in Example 6 we would write

$$m_{\tan} = \frac{dy}{dx}\bigg|_{x=1} = 15(1)^2 - 4(1) = 11 \quad \text{or} \quad m_{\tan} = f'(1) = 15(1)^2 - 4(1) = 11$$

Using a calculator,

F3   **1**   5x^3−2x^2+3,x)   **ENTER**   right arrow | x=1 **ENTER**

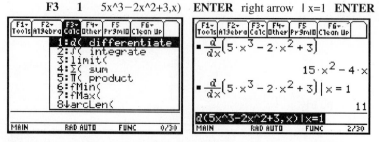

Note that the "with" key | is located directly to the left of **7** on the TI-89.

For examples of numerical derivatives on the TI-83 Plus, see Appendix section C.11.

EXAMPLE 7

Find the instantaneous velocity v of an object falling freely from a building 100 ft tall after 2 s. The position of this free-falling object is given by $s = -16t^2 + 100$ (measured in feet.)

v is the derivative of $s = -16t^2 + 100$ at $t = 2$. So

$$v = \frac{ds}{dt}\bigg|_{t=2} = \frac{d}{dt}(-16t^2 + 100)\bigg|_{t=2}$$
$$= (-32t + 0)|_{t=2}$$
$$= -32(2) = -64 \text{ ft/s}$$

Note: The negative sign indicates that the velocity of the object is in the negative direction (downward). A positive velocity would indicate motion in the positive direction (upward).

While we showed that $\dfrac{d}{dx}(x^n) = nx^{n-1}$ based on n being a positive integer, in fact n can be any real number.

EXAMPLE 8

Find the derivative of $y = \sqrt{x}$,

Since $y = \sqrt{x} = x^{1/2}$, then

$$\frac{dy}{dx} = \tfrac{1}{2}x^{(1/2)-1}$$
$$= \tfrac{1}{2}x^{-1/2}$$
$$= \frac{1}{2\sqrt{x}}$$

EXAMPLE 9

Find the derivative of $y = 1/x^3$.

$$y = \frac{1}{x^3} = x^{-3}$$

so

$$\frac{dy}{dx} = -3x^{-3-1}$$
$$= -3x^{-4}$$
$$= -\frac{3}{x^4}$$

EXAMPLE 10

Find the derivative of $y = 3/\sqrt{x}$.

$$y = \frac{3}{\sqrt{x}} = 3x^{-1/2}$$

so

$$\frac{dy}{dx} = \frac{d}{dx}(3x^{-1/2})$$

$$= 3\frac{d}{dx}(x^{-1/2})$$

$$= 3\left(-\tfrac{1}{2}x^{-(1/2)-1}\right)$$

$$= -\frac{3}{2}x^{-3/2}$$

Exercises 2.5

Find the derivative of each polynomial function.

1. $y = 32$
2. $y = -100$
3. $y = x^5$
4. $y = x^{16}$
5. $y = 4x + 1$
6. $y = 6x - 2$
7. $y = 1 - 3x$
8. $y = x$
9. $y = 5x^2$
10. $y = -3x^2$
11. $y = x^2 - 3x$
12. $y = x^2 - 4x + 2$
13. $y = 4x^2 - 3x - 2$
14. $y = 9x^2 + 8x - 7$
15. $y = 1 - 8x^2$
16. $y = 4 - 6x - 5x^2$
17. $y = 3x^3 + 2x^2 - 6x$
18. $y = 13x^4 - \dfrac{7x^2}{2} + 2$
19. $y = 4x^5 - 2x^3 + x + 3$
20. $y = 5x^6 - 3x^4 + 2x^3 - 8x^2 + 4x - 7$
21. $y = \tfrac{5}{2}x^8 - \tfrac{6}{5}x^5 + \tfrac{15}{2}x^4 - x^3 + \sqrt{2}$
22. $y = \tfrac{8}{3}x^6 - \tfrac{7}{3}x^5 + 4x^4 - 3x^3 - 7x + \pi$
23. $y = \sqrt{7}x^4 - \sqrt{5}x^3 - \sqrt{3}x + \sqrt{7}$
24. $y = \sqrt{10}x^{10} - \sqrt{7}x^7 + \sqrt{5}x^5 - \sqrt{3}$

Find $f'(a)$ for each function.

25. $y = 3x^2 + 2x - 1;\ a = -1$
26. $y = 4x^2 + 6x;\ a = 2$
27. $y = 2x^3 - 6x^2 + 2x + 9;\ a = -3$
28. $y = 6x^3 - 5x + 3;\ a = -2$
29. $y = 4x^5 + 3x^2 - 2;\ a = 1$
30. $y = 5x^4 - 2x^2 + 18;\ a = 2$
31. $y = 5x^4 + 8x^3 + 2x - 1;\ a = 0$
32. $y = 9x^6 + 6x^5 - 7x^4 + 2;\ a = 0$
33. $y = 1 - 8x^2 - 5x^3 + 5x^6;\ a = 3$
34. $y = 3 - 4x^2 - 6x^4 + 5x^9;\ a = -2$

35. Find the equation of the tangent line to the curve $y = x^3 + 4x^2 - x + 2$ at $(-2, 12)$.

36. Find the equation of the tangent line to the curve $y = x^{100} + 5$ at $(-1, 6)$.

37. Given that $p = Ri^2$, find the rate of change dp/di of the power p in a 30-ohm (Ω) resistor when $i = 2$ amperes (A).

38. An object is falling according to $h = 5000 - 3.28t^2$ where h is in metres and t is in seconds. Find the speed (*instantaneous velocity*) of the object after 5 s.

39. Given that $V = ir$, find the rate of change dV/dr of the voltage drop for a 0.4-A current when $r = 4\ \Omega$.

40. In a circuit the voltage V varies according to $V = t^4$ where t is time in seconds. Find dV/dt when $t = 3$ s.

Find the derivative of each function.

41. $y = x^{3/2}$
42. $y = \sqrt[3]{x}$
43. $y = \dfrac{1}{x^4}$
44. $y = \dfrac{4}{x^3}$
45. $y = 6x^{20}$
46. $y = 35x^{10}$
47. $y = \dfrac{14}{x^8}$
48. $y = \dfrac{52}{x^{13}}$

49. $y = \dfrac{5}{\sqrt[3]{x}}$ **50.** $y = \dfrac{2}{x\sqrt{x}}$

51. Given that $V = ir$, find the (instantaneous) rate of change dV/dr of the voltage drop V with respect to r (measured in ohms) for a 0.5-A current.

52. Given that $p = Ri^2$, find the rate of change dp/di of the power p in a 20-Ω resistor where p is in watts (W) and i is in amperes (A).

2.6 DERIVATIVES OF PRODUCTS AND QUOTIENTS

We need to differentiate products and quotients of functions. For example, the techniques developed so far would not apply to $y = (x^2 + 4x)\sqrt{x - 1}$ or $y = \dfrac{32x^2 - 9}{5x^5 - 7x^2}$. We would be forced to find the derivative $\dfrac{dy}{dx}$ by using the limit process method of Section 2.4. Fortunately, we can develop formulas to do such examples more efficiently.

DERIVATIVE OF A PRODUCT

If $y = f(x)$ can be written as the product of two functions, $u = g(x)$ and $v = h(x)$, then we can write

$$y = g(x) \cdot h(x)$$

or

$$y = u \cdot v$$

In Section 2.9 we show that

$$\frac{dy}{dx} = u\frac{dv}{dx} + v\frac{du}{dx}$$

That is, *the derivative of a product of two functions is the product of the first function and the derivative of the second function plus the product of the second function and the derivative of the first function.*

EXAMPLE 1

Find $\dfrac{d}{dx}[(4x + 3)(7x - 1)]$.

Let $u = 4x + 3$, $v = 7x - 1$, and $y = u \cdot v$.

$$\frac{dy}{dx} = u\frac{dv}{dx} + v\frac{du}{dx}$$

$$= (4x + 3)\frac{d}{dx}(7x - 1) + (7x - 1)\frac{d}{dx}(4x + 3)$$

$$= (4x + 3)(7) + (7x - 1)(4)$$

$$= 28x + 21 + 28x - 4$$

$$= 56x + 17$$

EXAMPLE 2

Find $\dfrac{d}{dx}[(3x^2 - 4)(5x^3 - 7)]$.

Let $u = 3x^2 - 4$, $v = 5x^3 - 7$, and $y = u \cdot v$.

$$\frac{dy}{dx} = u\frac{dv}{dx} + v\frac{du}{dx}$$

$$= (3x^2 - 4)\frac{d}{dx}(5x^3 - 7) + (5x^3 - 7)\frac{d}{dx}(3x^2 - 4)$$

$$= (3x^2 - 4)(15x^2) + (5x^3 - 7)(6x)$$

$$= 45x^4 - 60x^2 + 30x^4 - 42x$$

$$= 75x^4 - 60x^2 - 42x$$

DERIVATIVE OF A QUOTIENT

If $y = f(x)$ can be written as the quotient of two functions, $u = g(x)$ and $v = h(x)$, then we can write

$$y = \frac{g(x)}{h(x)}$$

or

$$y = \frac{u}{v}$$

In Section 2.9 we show that

$$\frac{d}{dx}\left(\frac{u}{v}\right) = \frac{v\dfrac{du}{dx} - u\dfrac{dv}{dx}}{v^2}$$

That is, *the derivative of a quotient of two functions is the denominator times the derivative of the numerator minus the numerator times the derivative of the denominator all divided by the square of the denominator.*

EXAMPLE 3

Find $\dfrac{d}{dx}\left(\dfrac{5x + 3}{4x^2 - 7}\right)$.

Here $u = 5x + 3$ and $v = 4x^2 - 7$. Then

$$\frac{d}{dx}\left(\frac{u}{v}\right) = \frac{v\dfrac{du}{dx} - u\dfrac{dv}{dx}}{v^2}$$

$$= \frac{(4x^2 - 7)\dfrac{d}{dx}(5x + 3) - (5x + 3)\dfrac{d}{dx}(4x^2 - 7)}{(4x^2 - 7)^2}$$

$$= \frac{(4x^2 - 7)(5) - (5x + 3)(8x)}{(4x^2 - 7)^2}$$

$$= \frac{20x^2 - 35 - 40x^2 - 24x}{(4x^2 - 7)^2}$$

$$= \frac{-20x^2 - 24x - 35}{(4x^2 - 7)^2}$$

EXAMPLE 4

Find $\dfrac{dy}{dx}$ if $y = \dfrac{32x^2 - 9}{5x^5 - 7x^2}$.

$$\frac{dy}{dx} = \frac{d}{dx}\left(\frac{32x^2 - 9}{5x^5 - 7x^2}\right)$$

$$= \frac{(5x^5 - 7x^2)(64x) - (32x^2 - 9)(25x^4 - 14x)}{(5x^5 - 7x^2)^2}$$

$$= \frac{320x^6 - 448x^3 - 800x^6 + 225x^4 + 448x^3 - 126x}{(5x^5 - 7x^2)^2}$$

$$= \frac{-480x^6 + 225x^4 - 126x}{(5x^5 - 7x^2)^2}$$

Using a calculator,

2nd 8 (32x^2-9)/(5x^5-7x^2),x) **ENTER**

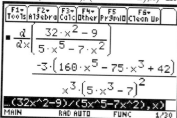

Show the two results are equivalent.

Exercises 2.6

*Find the derivative of each function (**a**) by finding the derivative of the product as given using the formula for the derivative of a product and (**b**) by first multiplying the given expression, then finding the derivative of the resulting polynomial.*

1. $y = x^2(2x + 1)$ **2.** $y = x^3(3x^2 + 2x - 1)$

3. $y = 2x(4x^2 + 3x - 5)$ **4.** $y = 4x^2(5 - 6x - 3x^2)$

5. $y = (2x + 3)(5x - 4)$ **6.** $y = (6x - 2)(4x - 3)$

7. $y = (4x + 7)(x^2 - 1)$ **8.** $y = (3x^3 - 1)(2x - 1)$

Find the derivative of each product.

9. $y = (x^2 + 3x + 4)(x^3 - 4x)$ **10.** $y = (x^3 + 2x - 7)(x^2 + 4x - 2)$

11. $y = (x^4 - 3x^2 - x)(2x^3 - 4x)$ **12.** $y = (x^2 - 3x + 8)(1 - 3x^4)$

Find the derivative of each quotient.

13. $y = \dfrac{x}{2x + 5}$ **14.** $y = \dfrac{3x}{x - 4}$ **15.** $y = \dfrac{1}{x^2 + x}$

16. $y = \dfrac{3}{x^2 - 1}$ **17.** $y = \dfrac{3x - 1}{2x + 4}$ **18.** $y = \dfrac{5x - 1}{3x + 2}$

19. $y = \dfrac{x^2}{2x + 1}$ **20.** $y = \dfrac{4x^2}{3x^2 - 1}$ **21.** $y = \dfrac{x - 1}{x^2 + x + 1}$

22. $y = \dfrac{3x - 7}{x^2 - 2}$ **23.** $y = \dfrac{4x^2 + 9}{3x^3 - 4x^2}$ **24.** $y = \dfrac{4x^3 - 1}{6x^2 + 3}$

25. Find $f'(2)$ when $f(x) = (x^2 - 4x + 3)(x^3 - 5x)$.

26. Find $f'(2)$ when $f(x) = (3x^2 - 1)(x^4 - 6x)$.

27. Find $f'(-1)$ when $f(x) = \dfrac{3x - 4}{x + 2}$.   **28.** Find $f'(3)$ when $f(x) = \dfrac{x - 2}{x^2 + 5}$.

29. Find the slope of the tangent line to the curve $y = \dfrac{x-3}{2-5x}$ at $\left(2, \dfrac{1}{8}\right)$.

30. Find the slope of the tangent line to the curve $y = \dfrac{2x^2 + 1}{x - 2}$ at $x = 1$.

31. Find the equation of the tangent line to the curve $y = \dfrac{x+3}{x-2}$ at $x = 3$.

32. Find the equation of the tangent line to the curve $y = \dfrac{1}{x+3}$ at $x = -1$.

33. Given $V = ir$, the voltage across a resistor, find $\dfrac{dV}{dt}$ at $t = 3$ s if the current $i = 6 + 0.02t^3$ A and the resistance $r = 20 - 0.05t\ \Omega$.

34. Given $i = V/r$, find di/dt at $t = 10$ s if $V = 40 + 0.1t^2$ V and $r = 60 - 0.01t\ \Omega$.

2.7 THE DERIVATIVE OF A POWER

If $y = (3x + 2)^{50}$, $\dfrac{dy}{dx}$ could be found by applying the product formula repeatedly. But this would be quite a long process. Also, there is a need to differentiate such a function as $y = \sqrt{3x + 2}$. We now use a formula for such powers.

We show in Section 2.9 that if $y = u^n$ where $u = f(x)$ then

> **POWER RULE**
>
> $$\frac{dy}{dx} = \frac{d}{dx}(u^n) = nu^{n-1}\frac{du}{dx}$$

The power rule should not be confused with the formula in Section 2.5.

$$\frac{d}{dx}(x^n) = nx^{n-1}$$

Note that the formula for $\dfrac{d}{dx}(u^n)$ involves an additional factor $\dfrac{du}{dx}$. This is because u itself is a function of x. Be sure to keep this distinction in mind. For example, if $y = u^3$ and $u = x$, then

$$y = x^3$$
$$\frac{dy}{dx} = 3x^{3-1} = 3x^2$$

But if $u = x^2 + 1$, then

$$y = (x^2 + 1)^3 = u^3$$
$$\frac{dy}{dx} = 3u^{3-1}\frac{du}{dx}$$
$$= 3(x^2 + 1)^2 \frac{d}{dx}(x^2 + 1)$$
$$= 3(x^2 + 1)^2(2x)$$
$$= 6x(x^2 + 1)^2$$

Thus, only when $u = x$ are the formulas the same, for then the additional factor $\dfrac{du}{dx} = 1$.

EXAMPLE 1

Find $\dfrac{dy}{dx}$ if $y = (3x + 2)^{50}$.

$$\frac{dy}{dx} = \frac{d}{dx}(3x + 2)^{50}$$

$$= 50(3x + 2)^{50-1}\frac{d}{dx}(3x + 2)$$

$$= 50(3x + 2)^{49}(3)$$

$$= 150(3x + 2)^{49}$$

Compare this result with the following example.

EXAMPLE 2

Find $\dfrac{dy}{dx}$ if $y = x^{50}$.

$$\frac{dy}{dx} = 50x^{50-1} = 50x^{49}$$

The formula $\dfrac{d}{dx}(u^n) = nu^{n-1}\dfrac{du}{dx}$ is also valid for n, any real number.

A **composite function** is a function of another function. For example,

$$y = (3x - 2)^4$$

may be written as

$$y = u^4 \quad \text{where} \quad u = 3x - 2$$

Here y is a function of u, and u is a function of x. So y is a function of a function of x.
More generally, let

$$y = f(g(x))$$

or

$$y = f(u) \quad \text{where} \quad u = g(x)$$

be any composite function. Then

$$\frac{dy}{dx} = \lim_{\Delta x \to 0} \frac{\Delta y}{\Delta x}$$

$$= \lim_{\Delta x \to 0}\left(\frac{\Delta y}{\Delta u} \cdot \frac{\Delta u}{\Delta x}\right) \qquad (\Delta u \neq 0)$$

$$= \lim_{\Delta x \to 0}\frac{\Delta y}{\Delta u} \cdot \lim_{\Delta x \to 0}\frac{\Delta u}{\Delta x}$$

$$= \lim_{\Delta u \to 0}\frac{\Delta y}{\Delta u} \cdot \lim_{\Delta x \to 0}\frac{\Delta u}{\Delta x} \qquad (\textit{Note: } \Delta u \to 0 \text{ as } \Delta x \to 0.)$$

$$= \frac{dy}{du} \cdot \frac{du}{dx}$$

Thus

> **CHAIN RULE**
>
> For a function $y = f(g(x))$, where $u = g(x)$,
>
> $$\frac{dy}{dx} = \frac{dy}{du} \cdot \frac{du}{dx}$$

The power rule is a special case of the *chain rule*. The chain rule is used to find derivatives of a function, which is a function of another function. The chain rule is used extensively beginning with the transcendental functions in Chapter 4.

Using the chain rule to find the derivative of $y = (3x - 2)^4$, let

$$y = u^4 \quad \text{and} \quad u = 3x - 2$$

$$\frac{dy}{du} = 4u^3 \qquad \frac{du}{dx} = 3$$

Then

$$\frac{dy}{dx} = \frac{dy}{du} \cdot \frac{du}{dx}$$
$$= (4u^3)(3)$$
$$= 12(3x - 2)^3$$

EXAMPLE 3

Find $\dfrac{dy}{dx}$ if $y = \sqrt{3x^2 + 5}$.

$$\frac{dy}{dx} = \frac{d}{dx}\left(\sqrt{3x^2 + 5}\right)$$
$$= \frac{d}{dx}\left[(3x^2 + 5)^{1/2}\right]$$
$$= \frac{1}{2}(3x^2 + 5)^{(1/2)-1}\frac{d}{dx}(3x^2 + 5) \qquad \left(n = \frac{1}{2}\right)$$
$$= \frac{1}{2}(3x^2 + 5)^{-1/2}(6x)$$
$$= \frac{3x}{\sqrt{3x^2 + 5}}$$

EXAMPLE 4

Find $\dfrac{dy}{dx}$ if $y = \dfrac{1}{(3x^2 - 2)^5}$

First, rewrite the expression as a power:

$$y = \frac{1}{(3x^2 - 2)^5} = (3x^2 - 2)^{-5}$$

Then

$$\frac{dy}{dx} = \frac{d}{dx}(3x^2 - 2)^{-5}$$

$$= -5(3x^2 - 2)^{-6}\frac{d}{dx}(3x^2 - 2)$$

$$= -5(3x^2 - 2)^{-6}(6x)$$

$$= \frac{-30x}{(3x^2 - 2)^6}$$

EXAMPLE 5

Find $\dfrac{dy}{dx}$ if $y = (8x + 1)^{1/4}(6x^2 + 2)$.

Using the derivative of a product form,

$$\frac{dy}{dx} = (8x + 1)^{1/4}\frac{d}{dx}(6x^2 + 2) + (6x^2 + 2)\frac{d}{dx}(8x + 1)^{1/4}$$

Note: The factor $(8x + 1)^{1/4}$ is in the form $u^{1/4}$ where $u = 8x + 1$. Using the chain rule on this factor, we have

$$\frac{1}{4}u^{-3/4}\frac{du}{dx} = \frac{1}{4}(8x + 1)^{-3/4}(8)$$

$$\frac{dy}{dx} = (8x + 1)^{1/4}(12x) + (6x^2 + 2)\left(\frac{1}{4}\right)(8x + 1)^{-3/4}(8)$$

$$= 12x(8x + 1)^{1/4} + (12x^2 + 4)(8x + 1)^{-3/4}$$

$$= (8x + 1)^{-3/4}[12x(8x + 1) + (12x^2 + 4)] \qquad \text{(Factor out the power}$$
$$\text{with a negative exponent.)}$$

$$= \frac{96x^2 + 12x + 12x^2 + 4}{(8x + 1)^{3/4}}$$

$$= \frac{108x^2 + 12x + 4}{(8x + 1)^{3/4}}$$

EXAMPLE 6

Find $\dfrac{dy}{dx}$ if $y = \dfrac{(2x - 3)^3}{(3x + 1)^4}$.

Using the derivative of a quotient form,

$$\frac{dy}{dx} = \frac{(3x + 1)^4\dfrac{d}{dx}(2x - 3)^3 - (2x - 3)^3\dfrac{d}{dx}(3x + 1)^4}{[(3x + 1)^4]^2}$$

$$= \frac{(3x + 1)^4(3)(2x - 3)^2(2) - (2x - 3)^3(4)(3x + 1)^3(3)}{[(3x + 1)^4]^2}$$

$$= \frac{6(3x + 1)^3(2x - 3)^2[(3x + 1) - 2(2x - 3)]}{(3x + 1)^8} \qquad \text{(Factor.)}$$

$$= \frac{6(2x - 3)^2(-x + 7)}{(3x + 1)^5}$$

EXAMPLE 7

Find $\dfrac{dy}{dx}$ if $y = \dfrac{(3x + 1)^{2/3}}{(2x - 3)^{1/2}}$.

$$\frac{dy}{dx} = \frac{(2x - 3)^{1/2}(\frac{2}{3})(3x + 1)^{-1/3}(3) - (3x + 1)^{2/3}(\frac{1}{2})(2x - 3)^{-1/2}(2)}{[(2x - 3)^{1/2}]^2}$$

$$= \frac{2(2x - 3)^{1/2}(3x + 1)^{-1/3} - (3x + 1)^{2/3}(2x - 3)^{-1/2}}{2x - 3}$$

Next, factor the negative exponent factors in the numerator as follows:

$$= \frac{(3x + 1)^{-1/3}(2x - 3)^{-1/2}[2(2x - 3) - (3x + 1)]}{(2x - 3)^1}$$

$$= \frac{4x - 6 - 3x - 1}{(3x + 1)^{1/3}(2x - 3)^{3/2}}$$

$$= \frac{x - 7}{(3x + 1)^{1/3}(2x - 3)^{3/2}}$$

EXAMPLE 8

Find $\dfrac{dy}{dx}$ if $y = (2x + 1)^{1/2}(4x - 5)^{3/4}$.

$$\frac{dy}{dx} = (2x + 1)^{1/2}(\tfrac{3}{4})(4x - 5)^{-1/4}(4) + (4x - 5)^{3/4}(\tfrac{1}{2})(2x + 1)^{-1/2}(2)$$

$$= 3(2x + 1)^{1/2}(4x - 5)^{-1/4} + (4x - 5)^{3/4}(2x + 1)^{-1/2}$$

Next, factor the negative exponent factors as follows:

$$= (4x - 5)^{-1/4}(2x + 1)^{-1/2}[3(2x + 1) + (4x - 5)]$$

$$= (4x - 5)^{-1/4}(2x + 1)^{-1/2}[6x + 3 + 4x - 5]$$

$$= \frac{10x - 2}{(4x - 5)^{1/4}(2x + 1)^{1/2}}$$

Exercises 2.7

Find the derivative of each function.

1. $y = (4x + 3)^{40}$

2. $y = 5(x^2 - 7)^{16}$

3. $y = (3x^2 - 7x + 4)^5$

4. $y = (x^4 - 3x^2 + 1)^6$

5. $y = \dfrac{1}{(x^3 + 3)^4}$

6. $y = \dfrac{1}{(3x^4 - 2x + 3)^3}$

7. $y = \sqrt{5x^2 - 7x + 2}$

8. $y = \sqrt{4x^3 + 2x^2 - x}$

9. $y = (8x^3 + 3x)^{2/3}$

10. $y = (6x^4 - 5x + 2)^{3/8}$

11. $y = (2x + 3)^{-3/4}$

12. $y = (4x^2 - 1)^{-1/2}$

13. $y = 3x(4x + 5)^4$

14. $y = 2x^2(3 - x^2)^5$

15. $y = x^3(x^3 - x)^3$

16. $y = 2x(x^2 - 4)^4$

17. $y = (2x + 1)^2(x^2 + 1)^2$

18. $y = (x^2 - 2)^3(3x - 4)^2$

19. $y = (x^2 + 1)\sqrt{9x^2 - 2x}$

20. $y = (x + 1)^{3/2}x^4$

21. $y = (3x + 4)^{3/4}(4x^2 + 8)$

22. $y = (3x - 1)^{2/3}\sqrt{4x + 5}$

23. $y = \dfrac{1}{x^4} - (2x + 1)^4$

24. $y = (5x - 2)^3 - (1 - x)^4$

25. $y = \dfrac{5x^2}{(3x - 1)^2}$

26. $y = \dfrac{10x^3}{(1 - 5x^2)^2}$

27. $y = \dfrac{(x^3 + 2)^4}{4x^2 - 3x}$

28. $y = \dfrac{8x^2 + 2}{\sqrt{1 - x}}$

29. $y = \dfrac{(3x + 2)^5}{(2x - 1)^3}$

30. $y = \dfrac{(4x^2 - 1)^3}{(3x + 4)^5}$

31. $y = \dfrac{(3x - 1)^{2/3}}{\sqrt{4x + 3}}$

32. $y = \dfrac{\sqrt[4]{8x^2 + 4x}}{(3x^2 + 1)^{5/3}}$

33. $y = \left(\dfrac{1 + x}{1 - x}\right)^4$

34. $y = \left(\dfrac{1 + x^2}{1 - x^2}\right)^5$

35. Find the velocity of an object after 3 s of travel where the distance s (in metres) traveled by the object is given by $s = \dfrac{t + 1}{\sqrt{t^2 - 1}}$.

2.8 IMPLICIT DIFFERENTIATION

Often a function $y = f(x)$ is defined *implicitly*. That is, y is not directly expressed in terms of x. Such a function is called an *implicit function* of x. For example, the relation $y^2 = x$ actually defines two implicit functions of x:

$$y = f(x) = \sqrt{x} \quad \text{and} \quad y = g(x) = -\sqrt{x}$$

since both $[f(x)]^2 = x$ and $[g(x)]^2 = x$ (see Fig. 2.11). In this example, we were able to determine the two functions of x defined by the relation $y^2 = x$. Often it is not possible or at least not easy, to determine directly the functions defined implicitly by a given relation, say, $2y^3 + 3x^2y - 8x^3 + 2x - y = 0$. That is, could you solve this equation for y in terms of x?

It is still possible, however, to obtain an expression for the derivative $\dfrac{dy}{dx}$. The general procedure is to differentiate both sides of the given equation with respect to the independent variable.

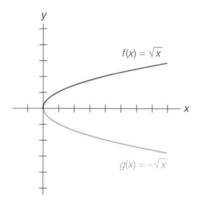

Figure 2.11

EXAMPLE 1

Find $\dfrac{dy}{dx}$ if $y^2 = x$.

$$\frac{d}{dx}(y^2) = \frac{d}{dx}(x)$$

$$2y^{2-1}\frac{d}{dx}(y) = 1 \qquad \text{(Use the power or chain rule.)}$$

$$2y\frac{dy}{dx} = 1 \qquad \left(\text{Solve for } \frac{dy}{dx}.\right)$$

$$\frac{dy}{dx} = \frac{1}{2y}$$

This process is called *implicit differentiation*.

EXAMPLE 2

Find $\dfrac{dy}{dx}$ if $2y^3 + 3x^2y - 8x^3 + 2x - y = 0$ defines y as an implicit function of x.

$$\frac{d}{dx}(2y^3 + 3x^2y - 8x^3 + 2x - y) = \frac{d}{dx}(0)$$ (Use product rule on second term.)

$$6y^2\frac{dy}{dx} + 3x^2\frac{dy}{dx} + 6xy - 24x^2 + 2 - \frac{dy}{dx} = 0$$ $\left(\text{Solve for }\frac{dy}{dx}.\right)$

$$6y^2\frac{dy}{dx} + 3x^2\frac{dy}{dx} - \frac{dy}{dx} = -6xy + 24x^2 - 2$$

$$(6y^2 + 3x^2 - 1)\frac{dy}{dx} = -6xy + 24x^2 - 2 \quad \text{(Factor.)}$$

$$\frac{dy}{dx} = \frac{-6xy + 24x^2 - 2}{6y^2 + 3x^2 - 1}$$

Note: When differentiating the term $3x^2y,$ we must consider this term as the product of two functions of x: $u = 3x^2$ and $v = y$. Then apply the product formula for differentiation.

EXAMPLE 3

Find $\frac{dy}{dx}$ if $(y + 3)^3 = (5x^2 - 4)^2$.

$$\frac{d}{dx}[(y + 3)^3] = \frac{d}{dx}[(5x^2 - 4)^2]$$

$$3(y + 3)^{3-1}\frac{d}{dx}(y + 3) = 2(5x^2 - 4)^{2-1}\frac{d}{dx}(5x^2 - 4)$$

$$3(y + 3)^2\left(\frac{dy}{dx} + 0\right) = 2(5x^2 - 4)(10x)$$

$$3(y + 3)^2\frac{dy}{dx} = 20x(5x^2 - 4) \qquad \left(\text{Solve for }\frac{dy}{dx}.\right)$$

$$\frac{dy}{dx} = \frac{20x(5x^2 - 4)}{3(y + 3)^2}$$

EXAMPLE 4

Find $\frac{dy}{dx}$ if $(2y^3 + 1)^4 = 8x^2 + 4y^2$.

Note that the left side of the equation is in the form u^4, where $u = 2y^3 + 1$. Using the chain rule on this term, we have

$$4u^3\frac{du}{dx} = 4(2y^3 + 1)^3\left(6y^2\frac{dy}{dx}\right)$$

So,

$$\frac{d}{dx}(2y^3 + 1)^4 = \frac{d}{dx}(8x^2 + 4y^2)$$

$$4(2y^3 + 1)^3\left(6y^2\frac{dy}{dx}\right) = 16x + 4\left(2y\frac{dy}{dx}\right)$$

$$24y^2(2y^3 + 1)^3\frac{dy}{dx} = 16x + 8y\frac{dy}{dx}$$

$$24y^2(2y^3 + 1)^3\frac{dy}{dx} - 8y\frac{dy}{dx} = 16x$$

$$[24y^2(2y^3 + 1)^3 - 8y]\frac{dy}{dx} = 16x$$

$$\frac{dy}{dx} = \frac{16x}{24y^2(2y^3 + 1)^3 - 8y} = \frac{2x}{3y^2(2y^3 + 1)^3 - y}$$

Exercises 2.8

Find the derivative of each expression using implicit differentiation.

1. $4x + 3y = 7$ **2.** $10y - 8x = 4$

3. $x^2 - y^2 = 9$ **4.** $3x^2 + 4y^2 = 12$

5. $x^2 + y^2 + 4y = 0$ **6.** $x^2 - y^2 + 8x = 0$

7. $3x^2 - y^3 - 3xy = 0$ **8.** $y^3 = 3x^2y - 4$

9. $y^4 - y^2x + x^2 = 0$ **10.** $y^2 - 2xy + x^3 = 0$

11. $y^4 - 2y^2x^2 + 3x^2 = 0$ **12.** $3x^2y^2 + 4y^5 + 8x^2y^3 + xy = 5$

13. $(y^2 + 2)^2 = (x^3 - 4x)^3$ **14.** $(y + 4)^4 = (2x^2 - 3x + 2)^5$

15. $(x + y)^3 = (x - y + 4)^2$ **16.** $(x^2 + y^2)^2 = 3x^2 + 4y^2$

17. $\dfrac{x + y}{x - y} = y^2$ **18.** $(x + y)^{2/3} = x^2$

Find the slope of the tangent line to each curve at the given point.

19. $4x^2 + 5y^2 = 36; (2, -2)$ **20.** $x^2 - y^2 = 16; (5, 3)$

21. $x^2 + y^2 - 6x - 2y = 0; (2, 4)$ **22.** $y^2 + 3x - 8y + 3 = 0; (4, 3)$

Find the equation of the tangent line to each curve at the given point.

23. $y = 3x^2 + 4x + 9; (0, 9)$ **24.** $y = \dfrac{(x + 2)^2}{\sqrt{3x^2 + 1}}; (0, 4)$

25. $y^2 + 3xy = 4; (1, -4)$ **26.** $xy + y^2 = 6; (1, 2)$

*Find $\dfrac{dy}{dx}$ at the given point by (**a**) solving the given equation for y and finding the derivative explicitly, and (**b**) using implicit differentiation.*

27. $x^2 + 2y = 7$ at $(1, 3)$ **28.** $4x^2 - 3y + 3 = 0$ at $(6, 49)$

29. $y^2 = x - 2$ at $(11, 3)$ **30.** $y^2 = 3x + 1$ at $(1, 2)$

2.9 PROOFS OF DERIVATIVE FORMULAS

Derivative of a Product

In Section 2.6 we claimed that if $y = f(x) = g(x) \cdot h(x)$ then

$$\frac{dy}{dx} = u\frac{dv}{dx} + v\frac{du}{dx}$$

where $u = g(x)$ and $v = h(x)$. We will now demonstrate this result.

Using the notation of increments we have that

$$\Delta u = g(x + \Delta x) - g(x)$$
$$\Delta v = h(x + \Delta x) - h(x)$$

and

$$\Delta y = g(x + \Delta x)h(x + \Delta x) - g(x)h(x)$$

$$\frac{dy}{dx} = \lim_{\Delta x \to 0} \frac{\Delta y}{\Delta x} = \lim_{\Delta x \to 0} \frac{g(x + \Delta x)h(x + \Delta x) - g(x)h(x)}{\Delta x}$$

$$= \lim_{\Delta x \to 0} \frac{g(x + \Delta x)h(x + \Delta x) - g(x + \Delta x)h(x) + g(x + \Delta x)h(x) - g(x)h(x)}{\Delta x}$$

Note: $-g(x + \Delta x)h(x) + g(x + \Delta x)h(x)$, which is zero, is added to the numerator.

$$= \lim_{\Delta x \to 0} \frac{g(x + \Delta x)[h(x + \Delta x) - h(x)] + h(x)[g(x + \Delta x) - g(x)]}{\Delta x}$$

$$= \lim_{\Delta x \to 0} \left[g(x + \Delta x) \frac{h(x + \Delta x) - h(x)}{\Delta x} + h(x) \frac{g(x + \Delta x) - g(x)}{\Delta x} \right]$$

$$= \lim_{\Delta x \to 0} \left[g(x + \Delta x) \frac{h(x + \Delta x) - h(x)}{\Delta x} \right]$$

$$+ \lim_{\Delta x \to 0} \left[h(x) \frac{g(x + \Delta x) - g(x)}{\Delta x} \right] \qquad \text{[By limit formula (1) page 94.]}$$

$$= \left[\lim_{\Delta x \to 0} g(x + \Delta x) \right]\left[\lim_{\Delta x \to 0} \frac{h(x + \Delta x) - h(x)}{\Delta x} \right]$$

$$+ \left[\lim_{\Delta x \to 0} h(x) \right]\left[\lim_{\Delta x \to 0} \frac{g(x + \Delta x) - g(x)}{\Delta x} \right] \qquad \text{[By limit formula (4) page 95.]}$$

$$= g(x) \frac{dv}{dx} + h(x) \frac{du}{dx}$$

Note: $h(x)$ is not affected by Δx and so is constant as we apply the limit as $\Delta x \to 0$.

$$= u \frac{dv}{dx} + v \frac{du}{dx}$$

or

$$\frac{d}{dx}[g(x)h(x)] = g(x) \frac{d}{dx}[h(x)] + h(x) \frac{d}{dx}[g(x)]$$

Derivative of a Quotient

The formula for $\dfrac{d}{dx}\left(\dfrac{u}{v}\right)$ from Section 2.6 can be found as follows:

Let $y = f(x) = \dfrac{g(x)}{h(x)}$ where $y = g(x)$ and $v = h(x)$. Using the same increment notation as before, we have

$$\Delta u = g(x + \Delta x) - g(x)$$
$$\Delta v = h(x + \Delta x) - h(x)$$

and

$$\Delta y = \frac{g(x + \Delta x)}{h(x + \Delta x)} - \frac{g(x)}{h(x)}$$

Then

$$\frac{dy}{dx} = \lim_{\Delta x \to 0} \frac{\Delta y}{\Delta x} = \lim_{\Delta x \to 0} \frac{\dfrac{g(x + \Delta x)}{h(x + \Delta x)} - \dfrac{g(x)}{h(x)}}{\Delta x}$$

$$= \lim_{\Delta x \to 0} \frac{h(x)g(x + \Delta x) - h(x + \Delta x)g(x)}{\Delta x\, h(x)h(x + \Delta x)}$$

$$= \lim_{\Delta x \to 0} \frac{\dfrac{h(x)g(x + \Delta x) - h(x + \Delta x)g(x)}{\Delta x}}{h(x)h(x + \Delta x)}$$

$$= \lim_{\Delta x \to 0} \frac{\dfrac{h(x)g(x + \Delta x) - h(x)g(x) + h(x)g(x) - h(x + \Delta x)g(x)}{\Delta x}}{h(x)h(x + \Delta x)}$$

Note: $-h(x)g(x) + h(x)g(x)$, which is zero, is added to the numerator.

$$= \lim_{\Delta x \to 0} \frac{h(x)\left[\dfrac{g(x + \Delta x) - g(x)}{\Delta x}\right] - g(x)\left[\dfrac{h(x + \Delta x) - h(x)}{\Delta x}\right]}{h(x)h(x + \Delta x)}$$

$$= \frac{\left[\displaystyle\lim_{\Delta x \to 0} h(x)\right]\left[\displaystyle\lim_{\Delta x \to 0}\dfrac{g(x + \Delta x) - g(x)}{\Delta x}\right] - \left[\displaystyle\lim_{\Delta x \to 0} g(x)\right]\left[\displaystyle\lim_{\Delta x \to 0}\dfrac{h(x + \Delta x) - h(x)}{\Delta x}\right]}{\left[\displaystyle\lim_{\Delta x \to 0} h(x)\right]\left[\displaystyle\lim_{\Delta x \to 0} h(x + \Delta x)\right]}$$

[By limit Formulas (1), (4), and (5), pages 94–95.]

$$= \frac{h(x)\dfrac{du}{dx} - g(x)\dfrac{dv}{dx}}{h(x)h(x)}$$

$$= \frac{v\dfrac{du}{dx} - u\dfrac{dv}{dx}}{v^2}$$

Derivative of a Power

In Section 2.7 we introduced the formula $\dfrac{d}{dx}u^n = nu^{n-1}\dfrac{du}{dx}$. To show this result, let $u = f(x)$ and $y = u^n$ where n is a positive integer, then

$$\Delta y = (u + \Delta u)^n - u^n$$

$$= u^n + nu^{n-1}(\Delta u) + \frac{n(n - 1)}{2}u^{n-2}(\Delta u)^2 + \cdots + (\Delta u)^n - u^n$$

$$= nu^{n-1}(\Delta u) + \frac{n(n - 1)}{2}u^{n-2}(\Delta u)^2 + \cdots + (\Delta u)^n$$

$$= nu^{n-1}(\Delta u)$$

$$+ \left[\frac{n(n - 1)}{2}u^{n-2} + (\text{other terms involving both } u \text{ and } \Delta u) + (\Delta u)^{n-2}\right](\Delta u)^2$$

So

$$\frac{\Delta y}{\Delta x} = nu^{n-1}\frac{\Delta u}{\Delta x}$$

$$+ \left[\frac{n(n - 1)}{2}u^{n-2} + (\text{other terms involving both } u \text{ and } \Delta u) + (\Delta u)^{n-2}\right]\left(\frac{\Delta u}{\Delta x}\right)(\Delta u)$$

Then, as $\Delta x \to 0$, all terms except $nu^{n-1}\dfrac{\Delta u}{\Delta x}$ approach 0 since

$$\lim_{\Delta x \to 0}\left(\frac{\Delta u}{\Delta x} \cdot \Delta u\right) = \lim_{\Delta x \to 0}\frac{\Delta u}{\Delta x} \cdot \lim_{\Delta x \to 0}\Delta u$$

$$= \frac{du}{dx} \cdot \lim_{\Delta x \to 0} [f(x + \Delta x) - f(x)]$$

$$= \frac{du}{dx} \cdot 0$$

$$= 0$$

and $\left(\dfrac{\Delta u}{\Delta x} \cdot \Delta u \right)$ is a factor of all the remaining terms. So

$$\frac{dy}{dx} = \lim_{\Delta x \to 0} \frac{\Delta y}{\Delta x}$$

$$= \lim_{\Delta x \to 0} nu^{n-1} \frac{\Delta u}{\Delta x}$$

$$= nu^{n-1} \cdot \lim_{\Delta x \to 0} \frac{\Delta u}{\Delta x}$$

$$= nu^{n-1} \frac{du}{dx}$$

2.10 HIGHER DERIVATIVES

The derivative of a function is also a function. For example, given

$$f(x) = 5x^4$$

then

$$f'(x) = 20x^3$$

Since $f'(x)$ is a function, its derivative (if it exists) is also a function. This result, called the second derivative of $f(x)$, is written $f''(x)$. Thus

$$f''(x) = 60x^2$$

Continuing, we have

$$f'''(x) = 120x$$
$$f^{(4)}(x) = 120$$
$$f^{(5)}(x) = 0$$
$$f^{(6)}(x) = 0$$

and so forth.

Some of the various notations used for derivatives are given below.

First derivative	Second derivative	Third derivative	Fourth derivative
$f'(x)$	$f''(x)$	$f'''(x)$	$f^{(4)}(x)$
y'	y''	y'''	$y^{(4)}$
$\dfrac{dy}{dx}$	$\dfrac{d^2y}{dx^2}$	$\dfrac{d^3y}{dx^3}$	$\dfrac{d^4y}{dx^4}$
$\dfrac{d}{dx}f(x)$	$\dfrac{d^2}{dx^2}f(x)$	$\dfrac{d^3}{dx^3}f(x)$	$\dfrac{d^4}{dx^4}f(x)$
Dy	D^2y	D^3y	D^4y
$Df(x)$	$D^2f(x)$	$D^3f(x)$	$D^4f(x)$

EXAMPLE 1

Find the first four derivatives for $y = x^6 - 8x^4 + 3x^3 - 2$.

$$\frac{dy}{dx} = 6x^5 - 32x^3 + 9x^2$$

$$\frac{d^2y}{dx^2} = 30x^4 - 96x^2 + 18x$$

$$\frac{d^3y}{dx^3} = 120x^3 - 192x + 18$$

$$\frac{d^4y}{dx^4} = 360x^2 - 192$$

EXAMPLE 2

Find the first three derivatives of $y = x^2 + x^{1/2}$.

$$y' = 2x + \frac{1}{2}x^{-1/2} = \frac{4x^{3/2} + 1}{2\sqrt{x}}$$

$$y'' = 2 - \frac{1}{4}x^{-3/2} = \frac{8x^{3/2} - 1}{4x^{3/2}}$$

$$y''' = \frac{3}{8}x^{-5/2} = \frac{3}{8x^{5/2}}$$

Using a calculator,

2nd 8 x^2+x^(1/2),x) **ENTER**   right arrow left arrow **, 2 ENTER**   right arrow left arrow ← **3 ENTER**

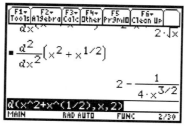   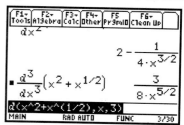

The second derivative of an implicit function may be found as shown in the following example.

EXAMPLE 3

Given $y^2 - xy = 5$, find $\dfrac{d^2y}{dx^2}$.

First, find y' as shown in Section 2.8.

$$2yy' - xy' - y = 0$$

Then solve for y'.

$$(2y - x)y' = y$$

$$y' = \frac{y}{2y - x}$$

Next, find y'' using the quotient formula.

$$y'' = \frac{(2y - x)y' - y(2y' - 1)}{(2y - x)^2}$$

$$= \frac{2yy' - xy' - 2yy' + y}{(2y - x)^2}$$

$$= \frac{-xy' + y}{(2y - x)^2} \qquad \text{(Next, substitute for } y' \text{ and simplify.)}$$

$$= \frac{-x\left(\dfrac{y}{2y - x}\right) + y}{(2y - x)^2} \cdot \left(\frac{2y - x}{2y - x}\right) \qquad \text{(Simplify the complex fraction.)}$$

$$= \frac{-xy + 2y^2 - xy}{(2y - x)^3}$$

$$= \frac{2y^2 - 2xy}{(2y - x)^3}$$

$$= \frac{2(y^2 - xy)}{(2y - x)^3}$$

$$= \frac{2(5)}{(2y - x)^3} \qquad \text{(\textit{Note}: From the original equation } y^2 - xy = 5.)$$

$$= \frac{10}{(2y - x)^3}$$

Recall that *velocity* is the instantaneous rate of change of displacement with respect to time. That is,

VELOCITY

$$v = \frac{ds}{dt}$$

Acceleration is the instantaneous rate of change of velocity with respect to time. So, acceleration is the first derivative of velocity with respect to time and the second derivative of displacement (distance) with respect to time. That is,

ACCELERATION

$$a = \frac{dv}{dt} = \frac{d^2s}{dt^2}$$

EXAMPLE 4

The displacement of an object (in metres) is given by $s = 5t^4 - 6t^3 + 8t + 3$. Find the equation that describes the acceleration of the object. Find its acceleration at $t = 2$ s.

First, find the velocity equation:

$$v = \frac{ds}{dt} = 20t^3 - 18t^2 + 8$$

Then find the acceleration equation:

$$a = \frac{dv}{dt} = 60t^2 - 36t$$

At $t = 2$ s,

$$a = 60(2)^2 - 36(2) = 168 \text{ m/s}^2$$

Exercises 2.10

Find the first four derivatives of each function.

1. $y = x^5 + 3x^2$

2. $y = 3x^6 - 8x^3 + 2$

3. $y = 5x^5 + 2x^3 - 8x$

4. $y = 3x^2 + 4x - 7$

Find the indicated derivative.

5. $y = \dfrac{1}{x}; \dfrac{d^3y}{dx^3}$

6. $y = \dfrac{3}{x^2}; y'''$

7. $y = (3x - 5)^3; y'''$

8. $y = (4x^2 - 2)^4; \dfrac{d^2y}{dx^2}$

9. $y = \sqrt{3x + 2}; \dfrac{d^4y}{dx^4}$

10. $y = (2x + 1)^{3/2}; \dfrac{d^4y}{dx^4}$

11. $y = \dfrac{1}{x^2 + 1}; y''$

12. $y = \dfrac{4}{(x^2 + 5)^2}; y''$

13. $y = \dfrac{x + 1}{x - 1}; \dfrac{d^2y}{dx^2}$

14. $y = \dfrac{3x + 1}{2x - 4}; \dfrac{d^2y}{dx^2}$

Find y' and y'' implicitly and express the result in terms of x and y.

15. $x^2 + y^2 = 1$

16. $(x - 3)^2 + y^2 = 9$

17. $x^2 - xy + y^2 = 1$

18. $y^2 + xy = 1$

19. $\sqrt{x} + \sqrt{y} = 1$

20. $x + \dfrac{x}{y} = 8$

21. $\dfrac{1}{x} - \dfrac{1}{y} = 1$

22. $x = \dfrac{y + 1}{y - 1}$

23. $x = (1 + y)^2$

24. $\sqrt{1 + y^2} = x$

Each equation describes the displacement of an object. Find the equation that describes the acceleration of the object.

25. $s = 0.5t^4 - 6t^3 - 4t^2 - 1$

26. $s = \dfrac{4}{2t + 1}$

27. $s = \sqrt{6t - 4}$

28. $s = \dfrac{8t + 3}{5 - 4t}$

29. The curve $x^2 - xy + y^2 = 7$ has two tangents at $x = 1$. Find their equations.

30. Show that the tangent to the parabola $y^2 = 4cx$ at (x_0, y_0) is $y_0 y = 2c(x + x_0)$.

31. Show that the tangent to the ellipse $\dfrac{x^2}{a^2} + \dfrac{y^2}{b^2} = 1$ at (x_0, y_0) is $\dfrac{x_0 x}{a^2} + \dfrac{y_0 y}{b^2} = 1$.

32. Show that the tangent to the hyperbola $\dfrac{x^2}{a^2} - \dfrac{y^2}{b^2} = 1$ at (x_0, y_0) is $\dfrac{x_0 x}{a^2} - \dfrac{y_0 y}{b^2} = 1$.

CHAPTER SUMMARY

1. *Average velocity:*

$$\bar{v} = \frac{\text{change in distance}}{\text{change in time}} = \frac{\Delta s}{\Delta t} = \frac{f(t + \Delta t) - f(t)}{\Delta t}$$

2. *Limit:* If $f(x) \to L$ as $x \to a$, then L is called the "limit of the function as $x \to a$." This is written

$$\lim_{x \to a} f(x) = L$$

and read "the limit of f of x as x approaches a equals L."

3. *Instantaneous velocity:*

$$v = \lim_{\Delta t \to 0} \frac{f(t + \Delta t) - f(t)}{\Delta t} \quad \text{or} \quad v = \frac{ds}{dt}$$

4. *Limit formulas* $\left(\lim_{x \to a} f(x) \text{ and } \lim_{x \to a} g(x) \text{ must both exist.}\right)$

(a) $\lim_{x \to a} [f(x) \pm g(x)] = \lim_{x \to a} f(x) \pm \lim_{x \to a} g(x)$

(b) $\lim_{x \to a} [k \cdot f(x)] = k \cdot \lim_{x \to a} f(x)$, where k is a constant

(c) $\lim_{x \to a} k = k$, where k is a constant

(d) $\lim_{x \to a} [f(x) \cdot g(x)] = \lim_{x \to a} f(x) \cdot \lim_{x \to a} g(x)$

(e) $\lim_{x \to a} \dfrac{f(x)}{g(x)} = \dfrac{\lim\limits_{x \to a} f(x)}{\lim\limits_{x \to a} g(x)}$, where $\lim_{x \to a} g(x) \neq 0$

5. *Continuity:*

(a) A function is continuous at $x = a$ if

   (i) $f(a)$ is defined,

   (ii) $\lim_{x \to a} f(x)$ exists, and

   (iii) $\lim_{x \to a} f(x) = f(a)$.

(b) If a function is continuous at all points in a given interval, it is continuous on that interval.

6. *Slope of a tangent line to* $y = f(x)$:

$$m_{\text{tan}} = \lim_{\Delta x \to 0} \frac{\Delta y}{\Delta x} = \lim_{\Delta x \to 0} \frac{f(x + \Delta x) - f(x)}{\Delta x}$$

7. *Derivative:* The derivative of the function $f(x)$ is

$$\frac{dy}{dx} = \lim_{\Delta x \to 0} \frac{\Delta y}{\Delta x} = \lim_{\Delta x \to 0} \frac{f(x + \Delta x) - f(x)}{\Delta x}$$

8. *Basic differentiation formulas:*

(a) $\dfrac{d}{dx}(c) = 0$

(b) $\dfrac{d}{dx}(x) = 1$

(c) $\dfrac{d}{dx}(x^n) = nx^{n-1}$

(d) $\dfrac{d}{dx}(cu) = c\dfrac{du}{dx}$

(e) $\dfrac{d}{dx}(u + v) = \dfrac{du}{dx} + \dfrac{dv}{dx}$

(f) $\dfrac{d}{dx}(u - v) = \dfrac{du}{dx} - \dfrac{dv}{dx}$

(g) $\dfrac{d}{dx}(uv) = u\dfrac{dv}{dx} + v\dfrac{du}{dx}$

(h) $\dfrac{d}{dx}\left(\dfrac{u}{v}\right) = \dfrac{v\dfrac{du}{dx} - u\dfrac{dv}{dx}}{v^2}, \quad (v \neq 0)$

(i) $\dfrac{d}{dx}(u^n) = nu^{n-1}\dfrac{du}{dx}$

9. *Acceleration:*

$$a = \frac{dv}{dt} = \frac{d^2s}{dt^2}$$

CHAPTER 2 REVIEW

Find (a) Δs and (b) $\bar{v}$ for each function expressing distance s in terms of t and Δt.

1. $s = 3t^2 + 4$ **2.** $s = 5t^2 - 6$ **3.** $s = t^2 - 3t + 5$

4. $s = 3t^2 - 6t + 8$

Find $\bar{v}$ for each function $s = f(t)$ at the values of t and Δt (s is measured in metres and t in seconds).

5. $s = 3t^2 - 7, t = 2, \Delta t = 5$ **6.** $s = 5t^2 - 3, t = 1, \Delta t = 2$

7. $s = 2t^2 - 4t + 7, t = 2, \Delta t = 3$ **8.** $s = 4t^2 - 7t + 2, t = 3, \Delta t = 4$

9–12. Find the instantaneous velocity v for the functions given in Exercises 5–8 at the values of t. Find each limit.

13. $\lim_{x \to 3} (2x^2 - 5x + 1)$ **14.** $\lim_{x \to -2} (x^2 + 4x - 7)$ **15.** $\lim_{x \to -2} \frac{x^2 - 4}{x + 2}$

16. $\lim_{x \to 5} \frac{25 - x^2}{5 - x}$ **17.** $\lim_{x \to 2} \sqrt{3 - 4x}$ **18.** $\lim_{x \to -3} \sqrt{6 + x^2}$

19. $\lim_{x \to 2} \frac{5x^2 + 2}{3x^2 - 2x + 1}$ **20.** $\lim_{x \to -3} \frac{2x^2 - 4x + 7}{x^3 - x}$

21. $\lim_{x \to -3} (x^2 - 4x + 3)(2x^2 + 5x + 4)$ **22.** $\lim_{x \to 2} (x^3 + x - 2)(x^3 + x^2 + x)$

23. $\lim_{h \to 0} \frac{\dfrac{1}{2 + h} - \dfrac{1}{2}}{h}$ **24.** $\lim_{h \to 0} \frac{\sqrt{1 + h} - 1}{h}$

25. $\lim_{x \to \infty} \frac{5x^2 - 2x + 3}{2x^2 - 4}$ **26.** $\lim_{x \to \infty} \frac{7x^3 - 4x + 2}{10x^3 - x^2 + 5}$

In Exercises 27–32, use Fig. 2.12.

27. Find $\lim_{x \to a} f(x)$ if it exists.

28. Find $\lim_{x \to d} f(x)$ if it exists.

29. Find $\lim_{x \to +\infty} f(x)$ if it exists.

30. Find $\lim_{x \to -\infty} f(x)$ if it exists.

31. Is the function continuous at $x = a$?

32. Is the function continuous at $x = d$?   **Figure 2.12**

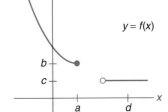

Find the slope and equation of the tangent line to each curve at the given point.

33. $y = 3x^2 - 4x + 5; (-1, 12)$ **34.** $y = x^2 - 5x - 12; (2, -18)$

35. $y = 2x^2 + 2x + 7; (3, 31)$ **36.** $y = 4x^2 - 8x + 3; (-2, 35)$

37. An object is moving along a straight-line path according to the equation $s = 3/t^2$, where s is measured in centimetres. Find its velocity after 5 s.

38. An object is falling freely according to the equation $s = 150 - 16t^2$, where s is measured in feet. Find its velocity after 2 s.

Find each derivative dy/dx:

39. $y = 5x^4 - 3x^3 + 2x^2 + 5x - 9$ **40.** $y = x^{100} + 80x^5 + 16$

41. $y = (x^3 + 4)(x^3 - x + 1)$ **42.** $y = (3x^2 - 5)(x^5 + x^2 - 4x)$

43. $y = \dfrac{x^2 + 1}{3x - 4}$

44. $y = \dfrac{2x - x^2}{3x^4 + 2}$

45. $y = (3x^2 - 8)^5$

46. $y = (x^4 + 2x^3 + 7)^{3/4}$

47. $y = \dfrac{1}{(3x + 5)^4}$

48. $y = \dfrac{\sqrt{7x^2 - 5}}{(x + 3)^2}$

49. $y = \dfrac{x\sqrt{2 - 3x}}{x + 5}$

50. $x^2 - 4xy^3 + y^2 = 0$

51. $y^4 - y^2 = 2xy$

52. $(y^2 + 1)^3 = 4x^2 + 3$

53. $(y + 2)^4 = (2x^3 - 3)^3$

Find $f'(a)$ for each.

54. $f(x) = (x^2 + 2)^3(x + 1)$; $a = -2$

55. $f(x) = \dfrac{x^2 - 8}{x + 3}$; $a = -4$

56. $f(x) = \dfrac{x^2 + 3x - 2}{x - 2}$; $a = 3$

57. $f(x) = \dfrac{\sqrt{3x^2 - 1}}{x + 5}$; $a = 1$

Find the equation of the tangent line to each curve at the given point.

58. $y = 3x^2 + x - 2$ at $(-2, 8)$

59. $y = x^3 - x + 8$ at $(-3, -16)$

60. $y = \dfrac{x^2 - 2}{\sqrt{x} - 3}$ at $(4, 14)$

61. $y = \dfrac{\sqrt{x^2 + 7}}{x - 2}$ at $(3, 4)$

Find the instantaneous velocity of an object moving along a straight line according to each function $s = f(t)$ at the given time t (s is measured in metres and t in seconds).

62. $s = 3t^2 - 8t + 4$; $t = 2$

63. $s = t^3 - 9t^2 + 3$; $t = -2$

64. $s = \dfrac{t^2 - 5}{\sqrt{t} + 1}$; $t = 3$

65. $s = \dfrac{\sqrt{t^2 - 3}}{t + 5}$; $t = 4$

66. Find the equation of the tangent line to the curve $y = \dfrac{x^2 - 6}{x + 3}$ at $(-4, -10)$.

67. The instantaneous current $i = dq/dt$ at any point in an electrical circuit where q is the charge (in coulombs, C) and t is the time (in seconds). Find i (in amperes) where $q = 1000t^3 + 50t$ when $t = 0.01$ s.

68. The specific heat c of a gas as a function of the temperature T is given by

$$c = 8.40 + 0.5T + 0.000006T^2$$

Find the rate of change of the specific heat with respect to temperature, dc/dT.

69. The resonant frequency of a series ac circuit is given by

$$f = \dfrac{1}{2\pi\sqrt{LC}}$$

where L and C are the inductance and capacitance in the circuit, respectively. Find the rate of change f with respect to C, assuming L remains constant.

70. Find the first four derivatives of $y = 4x^6 - 8x^4 + 9x^3 - 6x + 9$.

Find the indicated derivative.

71. $y = \sqrt{2x - 3}$; y''

72. $y = \dfrac{3}{2x^2 + 1}$; $\dfrac{d^2y}{dx^2}$

Find y' and y'' implicitly and express the result in terms of x and y.

73. $y^2 + 2xy = 4$

74. $\dfrac{1}{\sqrt{x}} - \dfrac{1}{\sqrt{y}} = 1$

75. The equation $s = (2t + 3)^{1/4}$ describes the displacement of an object. Find the equation that describes its acceleration.

3

Applications of the Derivative

INTRODUCTION

Many problems require the use of a derivative for their solutions. With derivatives, we can determine the shape of a container that will give maximum volume using a fixed amount of material, we can find the right combination of materials to produce a product that meets specifications but costs as little as possible, and we can calculate the rate at which the radius of a balloon is changing as it is being inflated. Problems such as these are discussed in this chapter.

Objectives

- Find relative maximum and minimum values of a function.
- Determine intervals in which a curve is concave upward and concave downward.
- Use derivatives to sketch curves.
- Solve maximum and minimum application problems.
- Solve related rate problems.

3.1 CURVE SKETCHING

In this course, as well as in your other technical courses, you will find it most helpful to sketch quickly the graph of an equation without plotting several points. In this section you will be given a series of graphing aids, each designed to give you valuable information about the graph, that will enable you to sketch quickly the graph of an equation.

Intercepts

The intercepts are those points where the graph crosses either the x-axis or the y-axis.

1. To find the x-intercepts, substitute $y = 0$ into the given equation and solve for x. Each resulting value of x is an x-intercept.
2. To find the y-intercepts, substitute $x = 0$ into the given equation and solve for y. Each resulting value of y is a y-intercept.

EXAMPLE 1

Find the intercepts for the graph of $y = (x - 1)(x + 3)^2$.

Step 1: To find the x-intercepts, substitute $y = 0$ and solve for x:

$$y = (x - 1)(x + 3)^2$$
$$0 = (x - 1)(x + 3)^2$$
$$x - 1 = 0 \quad \text{or} \quad (x + 3)^2 = 0$$
$$x = 1 \quad \text{and} \quad x = -3 \quad \text{are } x\text{-intercepts}$$

Step 2: To find the y-intercepts, substitute $x = 0$ and solve for y:

$$y = (x - 1)(x + 3)^2$$
$$y = (0 - 1)(0 + 3)^2$$
$$y = -9 \quad \text{is the } y\text{-intercept}$$

Location of the Curve above and below the *x*-Axis

Knowing in which regions the curve is above and below the x-axis can be most important information. If the expression can be factored, this information can be quickly found as follows:

1. Solve the given equation for y.
2. Factor the result from step 1 as much as possible. [If this is a rational (fractional) expression, factor both the numerator and the denominator as much as possible.]
3. Set each factor from step 2 equal to zero. These solutions are called *points of division* because they divide the number line (x-axis) into regions where the graph is above or below the x-axis. Plot each solution on the x-axis.
4. Find the sign of y in each region by choosing a test point. Substitute this test point value into each factor of the given equation and determine whether each factor is positive or negative in the region. Determine the sign of y by using the rules for multiplication and/or division of signed numbers. If the sign of y is *positive*, the curve is *above* the x-axis in this region. If the sign of y is *negative*, the curve is *below* the x-axis in the region.

EXAMPLE 2

Determine where the graph of $y = (x - 1)(x + 3)^2$ is above and below the x-axis.
From Example 1, the points of division are 1 and -3. Plot them on the x-axis.

Signs of factors $(x - 1)$ $\quad\quad -\quad\quad\quad\quad -\quad\quad\quad +$
in each interval: $(x + 3)^2$ $\quad +\quad\quad\quad\quad +\quad\quad\quad +$

$\xleftarrow{\quad\quad\quad}\overset{\qu\quad}{|}\overset{\quad}{\quad}\overset{}{|}\xrightarrow{\quad\quad} x$
$\quad\quad\quad\quad\quad\quad\quad -3\quad\quad\quad 1$

Sign of y in
each interval: $\quad\quad\quad\quad -\quad\quad\quad\quad -\quad\quad\quad +$
$\quad\quad\quad\quad\quad\quad\text{(below)}\quad\text{(below)}\quad\text{(above)}$

As you can see, we have three regions.

For $x > 1$, the product is *positive*, which means the curve is *above* the x-axis.

For $-3 < x < 1$, the product is *negative*, which means the curve is *below* the x-axis.

For $x < -3$, the product is *negative*, which means the curve is *below* the x-axis.

From the information gained from these first two graphing aids, we can now make a quick sketch of the graph of the equation in Fig. 3.1. We will be able to refine this procedure even more as we progress through the chapter.

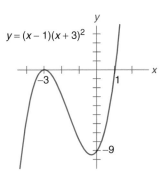

$$y = (x - 1)(x + 3)^2$$

Figure 3.1

Symmetry

Another graphing aid or characteristic that is helpful in sketching a curve is *symmetry*. There are two kinds of symmetry: symmetry about a line and symmetry about a point.

A curve is symmetric about a line if one half of the curve is the mirror image of the other half on opposite sides of the given line. That is, for every point P on the curve on one side of the given line, there is a point P' on the curve on the opposite side of the line so that PP' is perpendicular to the given line and bisected by it. For example, the graph of $y = 1/x^2$ is symmetric about the y-axis (see Fig. 3.2).

A curve is symmetric about a point, say O, if for every other point $P \neq O$ on the curve, there is a point P' on the curve so that PP' is bisected by point O. For example, the graph of $y = 1/x$ is symmetric about the origin (see Fig. 3.3).

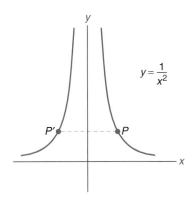

Figure 3.2 Symmetry about the y-axis.

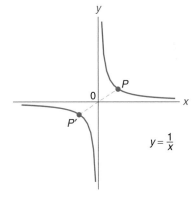

Figure 3.3 Symmetry about the origin.

Of the many possible lines and points of symmetry, we shall present the three most common: symmetry about the y-axis, symmetry about the x-axis, and symmetry about the origin.

1. A curve is symmetric about the y-axis if for every point $P(x, y)$ on the curve, there is a point $P'(-x, y)$ on the curve. This symmetry can be tested as follows:

> **SYMMETRY ABOUT THE Y-AXIS**
>
> The graph of an equation is symmetric about the y-axis if replacement of x by $-x$ in the original equation results in an equivalent equation.

Note that if we replace x by $-x$ in the earlier equation $y = 1/x^2$, we get

$$y = \frac{1}{(-x)^2}$$

which is equivalent to the original equation.

2. A curve is symmetric about the x-axis if for every point $P(x, y)$ on the curve, there is a point $P'(x, -y)$ on the curve. This symmetry can be tested as follows:

SYMMETRY ABOUT THE X-AXIS

The graph of an equation is symmetric about the x-axis if replacement of y by $-y$ in the original equation results in an equivalent equation.

Note that if we replace y by $-y$ in the equation $x = y^2$, we get

$$x = (-y)^2$$

which is equivalent to the original equation (see Fig. 3.4).

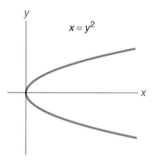

Figure 3.4   Symmetry about the x-axis.

3. A curve is symmetric about the origin if for every point $P(x, y)$ on the curve, there is a point $P'(-x, -y)$ on the curve. This symmetry can be tested as follows:

SYMMETRY ABOUT THE ORIGIN

The graph of an equation is symmetric about the origin if replacement of both x by $-x$ and y by $-y$ in the original equation results in an equivalent equation.

Note that if we replace x by $-x$ and y by $-y$ in the earlier equation

$$y = \frac{1}{x}$$

we get

$$-y = \frac{1}{-x}$$

which is equivalent to the original equation.

If any two of these three symmetry tests are satisfied, then the graph contains all three symmetries. Do you see why?

Asymptotes

An *asymptote* to a curve is a line that the curve approaches (gets closer to) as the distance from the origin increases without bound.

A *rational function* is an equation in the form

$$y = f(x) = \frac{g(x)}{h(x)}$$

where $g(x)$ and $h(x)$ are polynomials. Such rational functions may have asymptotes. We will present three possible asymptotes. In all cases, first solve the given equation for y in terms of x.

1. *Vertical asymptotes:* Set the denominator equal to zero and solve for x. These solutions give the points where the vertical asymptotes cross the x-axis. For example, if $x = a$ is a solution, one vertical asymptote crosses the x-axis at a and $x = a$ is the equation of the vertical asymptote.

2. *Horizontal asymptotes:* If $\lim\limits_{x \to +\infty} f(x) = b$ or $\lim\limits_{x \to -\infty} f(x) = b$, then $y = b$ is a horizontal asymptote. We have two cases:

   (a) If the degree of the numerator is less than the degree of the denominator, then $y = 0$ is a horizontal asymptote.

   Given the equation $y = 1/x^2$, do you see that as $x \to \infty$, $y \to 0$?

   (b) If the rational function is in the form

   $$f(x) = \frac{a_n x^n + \cdots + a_0}{b_n x^n + \cdots + b_0} \qquad (b_n \neq 0)$$

   then $y = \dfrac{a_n}{b_n}$ is a horizontal asymptote. That is, if the numerator and the denominator are of the same degree, the horizontal asymptote is the ratio of the coefficients of the terms of highest degree.

   Given the equation

   $$y = \frac{3x^2 - 10x - 8}{x^2 + 3x - 10}$$

   divide numerator and denominator by x^2, which is the largest power of x:

   $$\lim_{x \to \infty} \frac{3 - \dfrac{10}{x} - \dfrac{8}{x^2}}{1 + \dfrac{3}{x} - \dfrac{10}{x^2}} = 3$$

   So $y = 3$ is a horizontal asymptote.

3. *Slant asymptotes:* If the degree of the numerator is one greater than the degree of the denominator, there may be a slant asymptote. To find it, divide the numerator by the denominator using polynomial division and drop the remainder. Then $y = $ *the quotient* is the equation of the slant asymptote.

   Given the equation

   $$y = \frac{x^2 - 1}{x - 2}$$

divide the numerator by the denominator:

$$
\begin{array}{r}
x + 2 + \dfrac{3}{x-2} \\[4pt]
x - 2\overline{)x^2 \qquad\quad -\ 1} \\[2pt]
\underline{x^2 - 2x \qquad} \\[2pt]
2x - 1 \\[2pt]
\underline{2x - 4} \\[2pt]
3
\end{array}
$$

The slant asymptote is $y = x + 2$.

Do you see that as $x \to \infty$, the remainder, $\dfrac{3}{x-2}$, approaches zero?

Restricted Domains

Some equations have restricted domains by the nature of their definitions. For example, the domain of

$$y = \sqrt{3 - x}$$

is $x \le 3$ since the square root of a negative number is not real. The domain of

$$y = \log_{10}(x + 2)$$

is $x > -2$ since only the logarithm of a positive number is defined.

These graphing aids will now be illustrated with the following examples. You will find it most helpful to follow a consistent pattern of applying each aid.

EXAMPLE 3

Graph $y = x^2(x^2 - 9)$.
Factor: $\quad y = x^2(x + 3)(x - 3)$

Intercepts: If $y = 0$, then $x = 0, 3, -3$.

If $x = 0$, then $y = 0$.

Plot points of division on the x-axis:

Signs of factors in each interval:				
x^2	$+$	$+$	$+$	$+$
$(x + 3)$	$-$	$+$	$+$	$+$
$(x - 3)$	$-$	$-$	$-$	$+$

$$\xleftarrow{\qquad\underset{-3}{\mid}\qquad\underset{0}{\mid}\qquad\underset{3}{\mid}\qquad}\xrightarrow{}\; x$$

Sign of y in each interval:			
$+$	$-$	$-$	$+$
(above)	(below)	(below)	(above)

Symmetry: y-axis (Replacement of x by $-x$ results in an equivalent equation.)
Asymptotes: None
Restricted domain: None
(See Fig. 3.5.)

EXAMPLE 4

Graph $y = \dfrac{3}{(x + 1)^2}$.

Intercepts: Note that $y \ne 0$ because $\dfrac{3}{(x + 1)^2} \ne 0$; therefore, there is no x-intercept.

If $x = 0$, then $y = 3$.

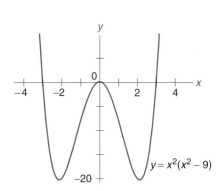

Figure 3.5

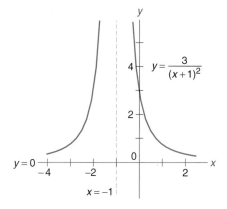

Figure 3.6

Points of division on the *x*-axis:
Signs of factor in
each interval: $(x + 1)^2$ +   +

$$\xleftarrow{\hspace{3cm}}\underset{-1}{\mid}\xrightarrow{\hspace{3cm}} x$$

Note: *y* is always positive; therefore, this curve is always above the
x-axis.

Symmetry: None
Asymptotes:
  Vertical: $x = -1$ $[$*Note:* $(x + 1)^2 = 0$ when $x = -1.]$
  Horizontal: $y = 0$ (Degree of numerator is less than degree of denominator.)
  Slant: None
Restricted domain: None except $x \neq -1$. (See Fig. 3.6.)

EXAMPLE 5

Graph $y = \dfrac{3x^2 - 10x - 8}{x^2 + 3x - 10}$.

  Factor: $y = \dfrac{(3x + 2)(x - 4)}{(x + 5)(x - 2)}$

Intercepts: If $y = 0$, then $x = -\frac{2}{3}, 4$.
            If $x = 0$, then $y = \frac{4}{5}$.

Plot points of division on the *x*-axis: (Set each factor equal to zero.)

Signs of factors	$(3x + 2)$	−	−	+	+	+
in each interval:	$(x - 4)$	−	−	−	−	+
	$(x + 5)$	−	+	+	+	+
	$(x - 2)$	−	−	−	+	+

$$\xleftarrow{\hspace{1cm}}\underset{-5}{\mid}\hspace{0.5cm}\underset{-2/3}{\mid}\hspace{0.5cm}\underset{2}{\mid}\hspace{0.5cm}\underset{4}{\mid}\xrightarrow{\hspace{1cm}} x$$

Sign of *y* in   +   −   +   −   +
each interval: (above)   (below)   (above)   (below)   (above)

Symmetry: None
Asymptotes:
  Vertical: $x = -5, x = 2$ (Set the denominator equal to zero.)
  Horizontal: $y = 3$.   (Degree of numerator equals degree of denominator;
                                  the ratio of coefficients of highest degree is 3/1 or 3.)
  Slant: None
Restricted domain: None except $x \neq -5, 2$. (See Fig. 3.7.)

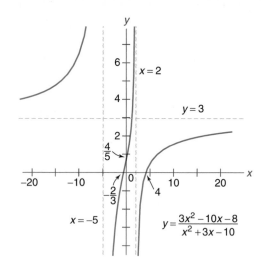

$$x = 2$$

$$y = 3$$

$$\frac{4}{5}$$

$$-\frac{2}{3}$$

$$x = -5$$

$$y = \frac{3x^2 - 10x - 8}{x^2 + 3x - 10}$$

Figure 3.7

EXAMPLE 6

Graph $y = \dfrac{x^2 - 1}{x - 2}$.

Factor: $y = \dfrac{(x + 1)(x - 1)}{x - 2}$

Intercepts: If $y = 0$, then $x = -1, 1$.
If $x = 0$, then $y = \frac{1}{2}$.

Plot points of division on the x-axis:

Signs of factors $(x + 1)$ $-$ $+$ $+$ $+$
in each interval: $(x - 1)$ $-$ $-$ $+$ $+$
                  $(x - 2)$ $-$ $-$ $-$ $+$

                            -1 1 2 $\longrightarrow x$

Sign of y in
each interval: $-$ $+$ $-$ $+$
                  (below)   (above)   (below)   (above)

Symmetry: None

Asymptotes:

   Vertical: $x = 2$

   Horizontal: None

   Slant: $y = x + 2$ (Degree of numerator is one greater than degree of denominator. Divide the numerator by the denominator as shown on page 136.)

Restricted domain: None except $x \neq 2$.

(See Fig. 3.8.)

EXAMPLE 7

Graph $y = \sqrt{\dfrac{x + 4}{1 - x}}$.

Intercepts: If $y = 0$, then $x = -4$.
If $x = 0$, then $y = 2$.

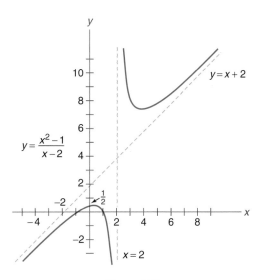

Figure 3.8

Plot points of division on the *x*-axis:

| Signs of factors | $(x + 4)$ | $-$ | $+$ | $+$ |
| in each interval: | $(1 - x)$ | $+$ | $+$ | $-$ |

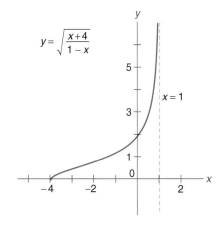

Sign under radical in each interval: $-$ $+$ $-$

Note: This function has a *restricted domain* of $-4 \le x < 1$ and the graph is never below the *x*-axis.

Symmetry: None

Asymptotes:
  Vertical: $x = 1$
(See Fig. 3.9.)

$$y = \sqrt{\frac{x+4}{1-x}}$$

Figure 3.9

EXAMPLE 8

Graph $y^2 = \dfrac{x^2}{x^2 - 4}$.

Solve for *y* and factor: $y = \pm \sqrt{\dfrac{x^2}{(x + 2)(x - 2)}}$

Intercepts: If $y = 0$, then $x = 0$.
            If $x = 0$, then $y = 0$.

Plot points of division on the x-axis:

Signs of factors x^2 $\quad+\quad\quad+\quad\quad+\quad\quad+$

in each interval: $(x + 2)$ $\quad-\quad\quad+\quad\quad+\quad\quad+$

$(x - 2)$ $\quad-\quad\quad-\quad\quad-\quad\quad+$

Sign under radical

in each interval: $\quad+\quad\quad\quad\quad-\quad\quad\quad-\quad\quad\quad+$

at $-2, 0, 2$ on the axis

Note: This equation has a *restricted domain* of $x < -2$,

$x > 2, x = 0$.

Symmetry: y-axis, x-axis, and origin.

Asymptotes:

Vertical: $x = 2, x = -2$.

Horizontal: $y = 1, y = -1$ (Divide numerator and denominator of the given

equation by x^2: $y^2 = \dfrac{1}{1 - \dfrac{4}{x^2}}$. Do you see that

$y^2 \to 1$ as $x \to \infty$?)

(See Fig. 3.10.)

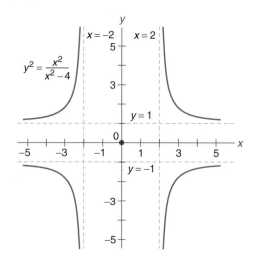

Figure 3.10

Exercises 3.1

Graph each equation.

1. $y = 2x(x + 1)(x - 4)$ **2.** $y = (x + 2)(x - 3)(x + 4)$

3. $y = (x - 1)(x - 3)(x + 5)$ **4.** $y = (3 - x)(x^2 - 4)$

5. $y = x^3 + 2x^2 - 15x$ **6.** $y = x^3 - 10x^2 + 24x$

7. $y = x^2(x + 1)(3 - 2x)$ **8.** $y = x(x + 4)(x - 1)^2$

9. $y = x^4 - 13x^2 + 36$ **10.** $y = (x + 1)^3(x - 7)^2(x + 2)$

11. $y = x^2(x - 2)^2(x + 4)^2$ **12.** $y = (x - 3)(2x + 1)(x - 2)^2$

13. $y = \dfrac{3}{2x + 1}$ **14.** $y = \dfrac{2}{(x - 4)^2}$ **15.** $y = \dfrac{2x}{(x + 1)(x - 3)}$

16. $y = \dfrac{2x - 1}{(x - 2)(x - 4)}$ **17.** $y = \dfrac{3}{x^2 + 4}$ **18.** $y = \dfrac{4x}{x^2 - 9}$

19. $y = \dfrac{4x}{x - 2}$ **20.** $y = \dfrac{3x}{1 - 2x}$ **21.** $y = \dfrac{3x^2}{x^2 - 4}$

22. $y = \dfrac{x - 5}{x + 2}$

23. $y = \dfrac{x^2 - 6}{x + 2}$

24. $y = \dfrac{2x^2 - 11x + 12}{x - 1}$

25. $y = \dfrac{2x^2 - x - 3}{x - 4}$

26. $y = \dfrac{x^2 - x}{x + 3}$

27. $y = \sqrt{x + 4}$

28. $y = \sqrt{1 - 2x}$

29. $y = \sqrt{\dfrac{x}{x - 3}}$

30. $y = -\sqrt{\dfrac{9}{x + 4}}$

31. $y^2 = x + 9$

32. $y^2 = (x + 1)(x - 3)$

33. $y^2 = \dfrac{x}{x + 4}$

34. $y^2 = \dfrac{x}{(x + 1)(x - 3)}$

35. $y^2 = \dfrac{x^2}{x^2 + 4}$

36. $y^2 = \dfrac{x^2}{(1 - x)(x + 2)}$

3.2 USING DERIVATIVES IN CURVE SKETCHING

The concept of a derivative is very useful in curve sketching. We can obtain information about the curve $y = f(x)$ that would be unavailable without the use of differentiation. From the derivative of a function $y = f(x)$, we can determine where a function is increasing and where it is decreasing.

> **INCREASING AND DECREASING FUNCTIONS**
>
> A function $y = f(x)$ is *increasing* in an interval if $f(x_2) > f(x_1)$ for any two points $x_2 > x_1$ in the interval.
> A function $y = f(x)$ is *decreasing* in an interval if $f(x_2) < f(x_1)$ for any two points $x_2 > x_1$ in the interval.

In Fig. 3.11 the function is increasing for $x < a$ and for $x > b$. The function is decreasing for $a < x < b$.

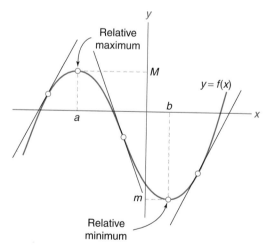

Figure 3.11

Observe that $f'(x)$, the slope of the tangent line at x, is positive when $f(x)$ is increasing and negative when $f(x)$ is decreasing. That is,

> $f(x)$ is increasing when $f'(x) > 0$ and $f(x)$ is decreasing when $f'(x) < 0$

Observe also, in Fig. 3.11, that at $x = a$ the function changes from increasing to decreasing. At such a point the value of the function, $f(a) = M$, is called a *relative maximum* (or *local maximum*). At $x = b$, the function changes from decreasing to increasing. The value of the function, $f(b) = m$, is called a *relative minimum* (or *local minimum*).

In Fig. 3.11 note that at both $x = a$ and $x = b$ the derivative of $y = f(x)$ is zero. That is, $f'(a) = 0$ and $f'(b) = 0$. Note that if the function $y = f(x)$ is to change smoothly from increasing to decreasing or from decreasing to increasing, then the derivative $f'(x)$ must be zero at that point.

A relative maximum or relative minimum can also occur at a point x where the curve has undefined slope (the derivative does not exist). In Fig. 3.12, $y = f(x)$ has a relative maximum at $x = a$, but there is no derivative.

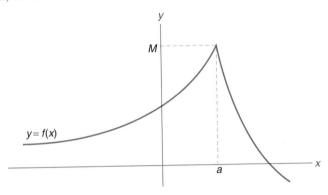

Figure 3.12 A critical point at $x = a$.

CRITICAL POINTS

Critical points are points of continuity $(x, f(x))$ on $y = f(x)$ where

(a) $f'(x) = 0$, or
(b) $f'(x)$ is undefined.

That is, where $f'(x)$ is neither positive nor negative.

The following test uses the first derivative to determine whether a given critical point is a relative maximum, a relative minimum, or neither.

FIRST DERIVATIVE TEST

Given $(c, f(c))$ is a critical point on $y = f(x)$.

(a) If $f'(x)$ changes from positive to negative at $x = c$, then $f(x)$ is changing from increasing to decreasing and the point $(c, f(c))$ is a relative maximum.
(b) If $f'(x)$ changes from negative to positive at $x = c$, then $f(x)$ is changing from decreasing to increasing and the point $(c, f(c))$ is a relative minimum.
(c) If $f'(x)$ does not change sign at $x = c$, then the point $(c, f(c))$ is neither a relative maximum nor a relative minimum.

EXAMPLE 1

Find any relative maximum or minimum values of the function $f(x) = x^3 - 3x$ and sketch the curve.

$$f'(x) = 3x^2 - 3 = 3(x^2 - 1)$$

so

$$f'(x) = 0 \quad \text{when} \quad 3(x^2 - 1) = 0$$
$$3(x + 1)(x - 1) = 0$$
$$x = -1 \quad \text{or} \quad x = 1$$

So $x = -1$ and $x = 1$ are critical points. Check to see if the function has relative maximums or minimums at these points. Observe that

$$\text{if } x < -1 \qquad f'(x) = 3(x + 1)(x - 1) > 0$$
$$\text{if } -1 < x < 1 \qquad f'(x) = 3(x + 1)(x - 1) < 0$$
$$\text{if } x > 1 \qquad f'(x) = 3(x + 1)(x - 1) > 0$$

Note: The following diagram can help determine the sign of the derivative. In this example, the expression for the derivative involves two factors. Determine the sign of each factor for each interval determined by the critical points. Applying the rule of signs for multiplication, determine the sign of the derivative:

| Sign of factor | $3(x + 1)$ | $-$ | $+$ | $+$ |
| in each interval: | $(x - 1)$ | $-$ | $-$ | $+$ |

$\xleftarrow{\hspace{6cm}} x$

$\qquad\qquad -1 \qquad\qquad 1$

| Sign of $f'(x)$: | $+$ | $-$ | $+$ |
| Function is: | (increasing) | (decreasing) | (increasing) |

Thus the function $f(x) = x^3 - 3x$ is

increasing for $x < -1$

decreasing for $-1 < x < 1$

increasing for $x > 1$

We conclude that $f(-1) = (-1)^3 - 3(-1) = 2$ or $(-1, 2)$ is a relative maximum and $f(1) = (1)^3 - 3(1) = -2$ or $(1, -2)$ is a relative minimum (see Fig. 3.13).

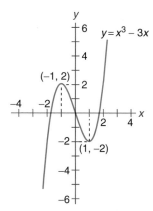

Figure 3.13

Note: When the *original equation factors easily,* we may use the discussion in Section 3.1 to determine where the curve is above and below the *x*-axis:

$$f(x) = x^3 - 3x = x(x^2 - 3) = x(x + \sqrt{3})(x - \sqrt{3})$$

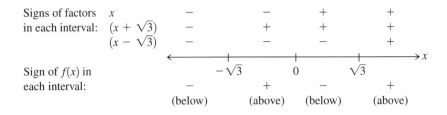

Signs of factors x $-$ $-$ $+$ $+$

in each interval: $(x + \sqrt{3})$ $-$ $+$ $+$ $+$

$(x - \sqrt{3})$ $-$ $-$ $-$ $+$

                                $-\sqrt{3}$ 0 $\sqrt{3}$ x

Sign of $f(x)$ in

each interval: $-$ $+$ $-$ $+$

                    (below)   (above)   (below)   (above)

EXAMPLE 2

Find any relative maximum or minimum values of $f(x) = x^3$ and sketch the curve.

$$f'(x) = 3x^2$$

so

$$f'(x) = 0 \quad \text{when} \quad 3x^2 = 0$$
$$x = 0$$

We find that $x = 0$ is the only critical point. But since $f'(x) = 3x^2$ is never negative, the function $f(x) = x^3$ is always increasing. Thus, there is no relative maximum or minimum. Observe, however, in Fig. 3.14 that while the curve is always increasing there is a difference in the shape of the curve for $x < 0$ and for $x > 0$.

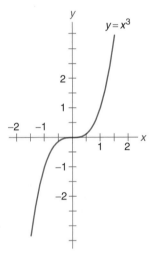

Figure 3.14

The number line diagram confirms this discussion.

Sign of $f'(x)$: $3x^2$ $+$ $+$

                                        0 x

Function is: (increasing)   (increasing)

EXAMPLE 3

Find any relative maximum or minimum values of the function $f(x) = x^4 + \frac{8}{3}x^3$ and sketch the curve.

$$f'(x) = 4x^3 + 8x^2 = 4x^2(x + 2) = 0$$

when

$$x = 0 \quad \text{or} \quad x = -2$$

Using the number line diagram, we have:

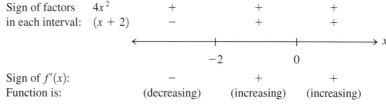

Sign of factors $4x^2$ $+$ $+$ $+$
in each interval: $(x + 2)$ $-$ $+$ $+$

               -2 0

Sign of $f'(x)$: $-$ $+$ $+$
Function is: (decreasing)   (increasing)   (increasing)

We conclude that $f(-2) = (-2)^4 + \frac{8}{3}(-2)^3 = 16 - \frac{64}{3} = -\frac{16}{3}$ or $(-2, -\frac{16}{3})$ is a relative minimum, and that $x = 0$ or $(0, 0)$ is neither a relative maximum nor a relative minimum (see Fig. 3.15).

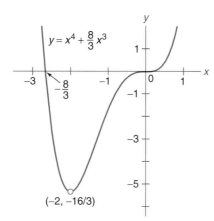

Figure 3.15

Exercises 3.2

Find any relative maximum or minimum values of each function and sketch the curve.

1. $y = x^2 + 6x - 16$ **2.** $y = 3x^2 + 12x + 9$ **3.** $y = 4 - 4x - 3x^2$

4. $y = 24 + 4x - 4x^2$ **5.** $y = x^3 + 3x^2 + 4$ **6.** $y = x^3 - 3x - 6$

7. $y = \frac{1}{3}x^3 - 9x - 4$ **8.** $y = \frac{5}{3}x^3 + 10x^2 + 8$ **9.** $y = 3x^5 - 5x^3$

10. $y = x^4 - 4x^2 + 3$ **11.** $y = (x - 2)^5$ **12.** $y = (x + 1)^4$

13. $y = (x^2 - 1)^4$ **14.** $y = (x^2 - 4)^3$ **15.** $y = \dfrac{x^2}{x - 4}$

16. $y = \dfrac{x^2}{x + 9}$ **17.** $y = \dfrac{x}{x + 1}$ **18.** $y = \dfrac{x}{x - 3}$

19. $y = x + (1/x)$ **20.** $y = 4x + (16/x)$ **21.** $y = \sqrt{x}$

22. $y = \sqrt[3]{x}$ **23.** $y = x^{2/3}$ **24.** $y = x^{-1/3}$

3.3 MORE ON CURVE SKETCHING

The derivative of a function $f(x)$ tells us where that function increases and where it decreases. So, $f''(x)$, the derivative of $f'(x)$, will tell us where $f'(x)$ increases or decreases (just as $f'(x)$ has told us where $f(x)$ increases or decreases).

As with any derivative, those values of x where $f''(x) = 0$ are possible points where the function $f'(x)$ can change from increasing to decreasing or from decreasing to increasing.

For example, if $y = x^3$, $f''(x) = 6x$.

1. For $x < 0$, $f''(x) = 6x < 0$. (This corresponds to our observation in Fig. 3.16 that the derivative is decreasing for $x < 0$.)

2. For $x > 0$, $f''(x) = 6x > 0$. (This tells us that the derivative is increasing as shown in Fig. 3.17.)

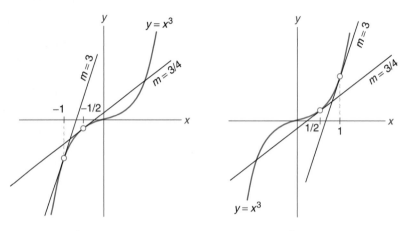

Figure 3.16   Figure 3.17

The curve $y = f(x)$ is *concave upward* (opens up) where the derivative $f'(x)$ is increasing, that is, where $f''(x) > 0$. The curve is *concave downward* (opens down) where the derivative $f'(x)$ is decreasing, that is, where $f''(x) < 0$. Any point $(x, f(x))$ where the curve changes from concave upward to concave downward or from concave downward to concave upward is called a *point of inflection.*

So $y = x^3$ is concave downward for $x < 0$ and concave upward for $x > 0$ and $(0, 0)$ is a point of inflection.

The curve in Fig. 3.18 is concave upward (opens up) for $a < x < b$. This curve is concave downward (opens down) for $b < x < c$.

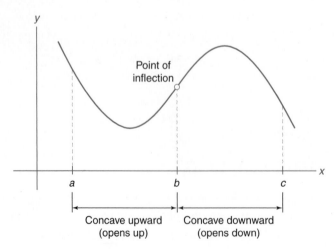

Figure 3.18

EXAMPLE 1

Find the concavity of the function $y = x^3 - 3x$.

$$f'(x) = 3x^2 - 3 \quad \text{so} \quad f''(x) = 6x$$

As in $y = x^3$, the function is concave downward for $x < 0$, concave upward for $x > 0$, and $(0, 0)$ is a point of inflection. The sketch of the curve appears in Fig. 3.13 in Section 3.2.

The following method is used to aid us in sketching a curve $y = f(x)$:

CURVE SKETCHING

1. Find any x- and y-intercepts. The curve crosses the x-axis at an x-intercept. That is, an x-intercept is a point where $y = 0$. The curve crosses the y-axis at a y-intercept. That is, a y-intercept is a point where $x = 0$. If the original equation factors easily, you may use the discussion in Section 3.1 to determine where the curve is above and below the x-axis.

2. Find all critical points for $y = f(x)$. That is, find all values of x where $f'(x) = 0$ or where $f'(x)$ is undefined.

3. Knowing the critical points, determine where the function is increasing and where it is decreasing.

$$y = f(x) \quad \text{is increasing where} \quad f'(x) > 0$$
$$y = f(x) \quad \text{is decreasing where} \quad f'(x) < 0$$

4. Find the relative maximum and minimum values. That is, find all values of x where $f'(x) = 0$ or is undefined. Check these critical points for possible relative maximums and relative minimums using the information obtained in Step 3.

5. Find the x-coordinate of each possible point of inflection. That is, find all values of x where $f''(x) = 0$ or is undefined.

6. From the possible points of inflection, determine the concavity of $y = f(x)$.

$$y = f(x) \quad \text{is concave upward where} \quad f''(x) > 0$$
$$y = f(x) \quad \text{is concave downward where} \quad f''(x) < 0$$

7. Find which points from Step 5 are points of inflection using the information obtained in Step 6. Note that it is possible for the second derivative to be zero at $x = a$ and $f''(a) = 0$, while $f(x)$ does not have a point of inflection at $x = a$. That is, the curve does not necessarily change concavity where $f''(x) = 0$. This must be checked in Step 6.

8. Sketch the curve.

EXAMPLE 2

Find the maximum and minimum values of the function $y = 3x^2 - 2x^3$. Determine the concavity and sketch the curve.

Step 1: Setting $y = 0$, we have

$$3x^2 - 2x^3 = 0$$
$$x^2(3 - 2x) = 0$$

so that $x = 0$ and $x = \frac{3}{2}$ are x-intercepts.
Setting $x = 0$, we have $y = 0$ as the only y-intercept.

To find where the curve is above and below the x-axis:

Signs of factors $\quad x^2$
in each interval: $\quad (3 - 2x)$

$$
\begin{array}{ccc}
+ & + & + \\
+ & + & -
\end{array}
$$

(number line with points at 0 and $\frac{3}{2}$, axis labeled x)

Sign of y in
each interval:
Function is:

$$
\begin{array}{ccc}
+ & + & - \\
\text{(above)} & \text{(above)} & \text{(below)}
\end{array}
$$

Step 2: Setting $f'(x) = 0$, we have

$$f'(x) = 6x - 6x^2 = 0$$
$$6x(1 - x) = 0$$

so that $x = 0$ and $x = 1$ are the critical points.

Step 3: $f'(x) = 6x(1 - x)$

To find where the curve is increasing and decreasing:

Signs of factors $\quad 6x$
in each interval: $\quad (1 - x)$

$$
\begin{array}{ccc}
- & + & + \\
+ & + & -
\end{array}
$$

(number line with points at 0 and 1, axis labeled x)

Sign of $f'(x)$:
Function is:

$$
\begin{array}{ccc}
- & + & - \\
\text{(decreasing)} & \text{(increasing)} & \text{(decreasing)}
\end{array}
$$

So $y = f(x)$ is increasing for $0 < x < 1$ and decreasing for $x < 0$ and for $x > 1$.

Step 4: Since $y = f(x)$ changes from decreasing to increasing at $x = 0$, $f(0) = 0$ is a relative minimum. Since $y = f(x)$ changes from increasing to decreasing at $x = 1$, $f(1) = 1$ is a relative maximum.

Step 5: Setting $f''(x) = 0$ we have

$$f''(x) = 6 - 12x = 0$$

so that $x = \frac{1}{2}$ will give us a possible point of inflection.

Step 6: $f''(x) = 6 - 12x$

To find where the curve is concave upward and concave downward:

Sign of $f''(x)$
in each interval: $\quad (6 - 12x)$

$$
\begin{array}{cc}
+ & -
\end{array}
$$

(number line with point at $\frac{1}{2}$, axis labeled x)

Function is: $\quad \left(\begin{array}{c}\text{concave} \\ \text{up}\end{array}\right) \quad \frac{1}{2} \quad \left(\begin{array}{c}\text{concave} \\ \text{down}\end{array}\right)$

So $y = f(x)$ is concave upward for $x < \frac{1}{2}$ and concave downward for $x > \frac{1}{2}$.

Step 7: Since $y = f(x)$ changes concavity at $x = \frac{1}{2}$, and $f(\frac{1}{2}) = 3(\frac{1}{2})^2 - 2(\frac{1}{2})^3 = \frac{1}{2}$, $(\frac{1}{2}, \frac{1}{2})$ is a point of inflection.

Step 8: The curve is sketched in Fig. 3.19.

From Fig. 3.19 note that the curve is concave downward at a relative maximum and concave upward at a relative minimum. We can use this information to distinguish relative minimums from relative maximums in place of the first derivative test. This procedure uses the second derivative to test critical points and is known as the *second derivative test*. That is,

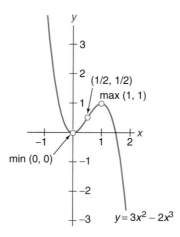

Figure 3.19

SECOND DERIVATIVE TEST

1. If $f'(a) = 0$ and $f''(a) < 0$, then $M = f(a)$ is a relative maximum.
2. If $f'(a) = 0$ and $f''(a) > 0$, then $m = f(a)$ is a relative minimum.

Note: If $f''(a) = 0$ or if $f''(a)$ is undefined, the test fails and the first derivative test should be used instead.

In Example 2, we saw that $f''(x) = 6 - 12x$. At $x = 0, f''(0) = 6 > 0$, so $0 = f(0)$ is a relative minimum (as already determined by first derivative test). At $x = 1, f''(1) = -6$, so that $1 = f(1)$ is a relative maximum.

EXAMPLE 3

Find any relative maximum and minimum values for $y = x^4 - 2x^2$.

$$f'(x) = 4x^3 - 4x = 4x(x + 1)(x - 1)$$

Setting $f'(x) = 0$, we have $4x(x + 1)(x - 1) = 0$ so that $x = 0, x = -1$, and $x = 1$ are critical points.

$$f''(x) = 12x^2 - 4$$

Since $f''(0) = -4 < 0, f(0) = 0$ is a relative maximum.

Since $f''(-1) = 8 > 0, f(-1) = -1$ is a relative minimum.

Since $f''(1) = 8 > 0, f(1) = -1$ is a relative minimum.

(See Fig. 3.20.)

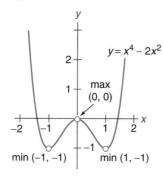

Figure 3.20

EXAMPLE 4

Find any relative maximum and minimum values for $y = \dfrac{2x + 1}{x - 3}$. Determine the concavity and sketch the curve.

Using techniques from Section 3.1:

Symmetry: None

Asymptotes:

Vertical: $x = 3$

Horizontal: $y = 2$ (Degree of numerator equals degree of denominator.)

Using the curve sketching outline in this section:

Step 1: Setting $y = 0$, we have

$$\frac{2x + 1}{x - 3} = 0$$

so that $x = -\frac{1}{2}$ is the x-intercept.

Setting $x = 0$, we have $y = -\frac{1}{3}$ as the y-intercept.
To find where the curve is above and below the x-axis:

Signs of factors $(2x + 1)$ $-$ $+$ $+$
in each interval: $(x - 3)$ $-$ $-$ $+$

Sign of y in
each interval: $+$ $-$ $+$
Function is: (above)   (below)   (above)

Step 2: $f'(x) = \dfrac{(x - 3)(2) - (2x + 1)(1)}{(x - 3)^2} = \dfrac{2x - 6 - 2x - 1}{(x - 3)^2} = \dfrac{-7}{(x - 3)^2}$

Note that $f'(x)$ cannot equal 0 and that $x = 3$ makes $f'(x)$ undefined because $f(3)$ is undefined.

Step 3: To find where the curve is increasing or decreasing:

Sign of $f'(x)$ in
each interval: $-$ $-$

Function is: (decreasing)   (decreasing)

So $y = f(x)$ is decreasing on the intervals $x < 3$ and $x > 3$.

Step 4: There is no relative maximum or minimum.

Step 5: $f''(x) = \dfrac{(x - 3)^2(0) - (-7)(2)(x - 3)(1)}{(x - 3)^4} = \dfrac{14(x - 3)}{(x - 3)^4} = \dfrac{14}{(x - 3)^3}$

Since the second derivative cannot equal 0, there are no points of inflection.

Step 6: Since the second derivative is undefined at $x = 3$, we need to check the concavity of $f(x)$ as follows:

Sign of $f''(x)$ in
each interval: $-$ $+$

Function is: (concave down)   (concave up)

The curve is sketched in Fig. 3.21.

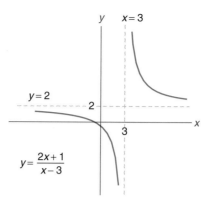

Figure 3.21

Figure 3.22 compares the relationship between the curve $y = f(x)$, its first derivative $y' = f'(x)$, and its second derivative $y'' = f''(x)$.

From A to B, $y = f(x)$ is decreasing and $y' = f'(x) < 0$.

From B to D, y is increasing and $y' > 0$.

From D to E, y is decreasing and $y' < 0$.

At B, $y' = 0$ and $y'' > 0$ (relative minimum).

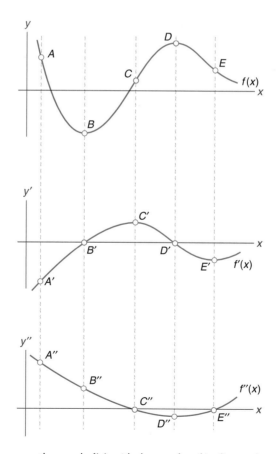

Figure 3.22   Compare the graph $f(x)$ with the graphs of its first and second derivatives.

At D, $y' = 0$ and $y'' < 0$ (relative maximum).

From A to C, y' is increasing, $y'' > 0$, and y is concave upward.

From C to E, y' is decreasing, $y'' < 0$, and y is concave downward.

At C and E, $y'' = 0$ (points of inflection).

The graphs of functions that describe most physical relationships are both continuous and "smooth"; that is, they contain neither breaks (missing points or displaced points) nor make sudden or abrupt changes. Physical changes are most often smooth and gradual. Functions whose graphs are both continuous and smooth are differentiable at each and every point. Polynomial functions are both continuous and smooth, and the derivative exists at every value of x. The function in Fig. 3.23 is not smooth at $x = a$, and the derivative does not exist at $x = a$. Why not?

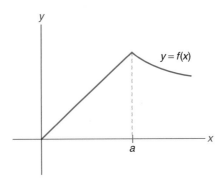

Figure 3.23

Exercises 3.3

*For each function find (**a**) intervals for which $f(x)$ is increasing and decreasing, (**b**) relative maximums and relative minimums, (**c**) intervals for which $f(x)$ is concave upward and concave downward, and (**d**) points of inflection, and (**e**) sketch the curve.*

1. $y = x^2 - 4x$ **2.** $y = x^2 - x - 2$ **3.** $y = x^4$

4. $y = 5 - x^2$ **5.** $y = 3x - x^3$ **6.** $y = 3x^4 - 4x^3$

7. $y = -x^2 + 4x - 1$ **8.** $y = x^4 - 18x^2 + 40$ **9.** $y = x^4 - 8x^2 + 5$

10. $y = x^7$ **11.** $y = \dfrac{1}{x^2}$ **12.** $y = \dfrac{1}{x^3}$

13. $y = \dfrac{x}{x + 2}$ **14.** $y = \dfrac{x}{x - 1}$ **15.** $y = \dfrac{-1}{(x + 1)^3}$

16. $y = \dfrac{1}{(x - 2)^2}$ **17.** $y = \dfrac{x}{x^2 + 1}$ **18.** $y = \dfrac{x^2 + 1}{x}$

19. $y = \dfrac{x - 2}{x + 3}$ **20.** $y = \dfrac{x + 3}{x - 2}$ **21.** $y = \dfrac{4}{x^2 + 4}$

22. $y = \dfrac{4}{x^2 - 4}$ **23.** $y = \dfrac{x - 1}{x^2}$ **24.** $y = \dfrac{x + 4}{x^2}$

3.4 MAXIMUM AND MINIMUM PROBLEMS

Many applications involve finding the maximum or minimum values of a variable quantity under given conditions. If we can express this quantity as a function of some other variable, then we can use the techniques of the last two sections to locate the possible relative maximum or relative minimum values.

In addition to the techniques developed in those sections, we must

1. Translate the information of a stated word problem into a mathematical function.

2. Determine that maximum (or minimum) value which is the largest (or smallest) value that the quantity can attain.

EXAMPLE 1

Using 120 m of fencing, find the largest possible rectangular area that can be enclosed.

The area of a rectangle is given by $A = xy$, where x is the length of one side and y is the length of the other side (see Fig. 3.24). The perimeter P of a rectangle is given by

$$P = 2x + 2y$$

which in this case must be 120 m.

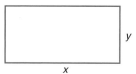

Figure 3.24

Possible choices for x and y are $x = 10$ and $y = 50$, where $A = xy = (10)(50) = 500$ m^2 or $x = 20$ and $y = 40$ where $A = xy = (20)(40) = 800$ m^2. But what is the maximum area A? To find the maximum area we will use the techniques of the previous sections. We must express A as a function of one variable only. To do this, solve for x in $2x + 2y = P$ where $P = 120$:

$$2x + 2y = 120$$
$$2x = 120 - 2y$$
$$x = 60 - y$$

and substitute for x in the formula for area,

$$A = xy$$
$$A = (60 - y)y = 60y - y^2$$

The area is now expressed as a function of one variable y. Using the techniques of the last sections, locate the maximum value:

$$\frac{dA}{dy} = \frac{d}{dy}(60y - y^2)$$
$$= 60 - 2y$$

Setting $\dfrac{dA}{dy} = 0$, we have

$$0 = 60 - 2y$$
$$2y = 60$$
$$y = 30$$

Since $\dfrac{d^2A}{dy^2} = -2$, the curve $A = 60y - y^2$ is always concave downward.

So $y = 30$ is a relative maximum and the maximum area is obtained when $y = 30$ m. To find the maximum area, find the corresponding value of x when $y = 30$ m:

$$x = 60 - y$$
$$= 60 - (30)$$
$$= 30 \text{ m}$$

The maximum area occurs when the rectangle is a square and its value is

$$A = xy$$
$$= (30 \text{ m})(30 \text{ m})$$
$$= 900 \text{ m}^2$$

When an expression that is to be maximized or minimized is given in more than one other variable, a formula, the Pythagorean theorem, or using similar triangles may be used to eliminate a variable. In Example 1 it was necessary to combine the formula for perimeter with the formula for area. Only then could the area be expressed in terms of one variable.

If there is only one critical point, the information contained in the problem will often make it clear whether the critical point corresponds to a maximum or a minimum (without using the standard tests).

EXAMPLE 2

Each side of a square piece of metal is 12 cm. A small square is to be cut from each corner and the metal folded to form a box without a top. Determine the dimensions of the box that will have the largest volume.

If a square with dimension s cm is cut out from each corner, the resulting box will have a square base with each side $12 - 2s$ cm and a height of s cm (see Fig. 3.25). The volume of the box will be

$$V = (12 - 2s)^2 s$$

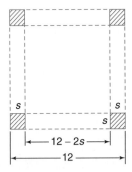
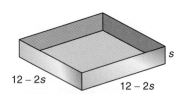

Figure 3.25

From the given conditions, the maximum value will occur when $0 \le s \le 6$ (since $2s \le 12$). Differentiating,

$$\frac{dV}{ds} = (12 - 2s)^2(1) + s(2)(12 - 2s)(-2)$$
$$= (12 - 2s)[(12 - 2s) - 4s]$$
$$= (12 - 2s)(12 - 6s)$$
$$= 12(6 - s)(2 - s)$$

Alternatively, since $V = (12 - 2s)^2 s = 144s - 48s^2 + 4s^3$,

$$\frac{dV}{ds} = 144 - 96s + 12s^2$$
$$= 12(12 - 8s + s^2)$$
$$= 12(6 - s)(2 - s)$$

Setting $\dfrac{dV}{ds} = 12(6 - s)(2 - s) = 0$, we have $s = 6$ or $s = 2$. When $s = 6$ cm, the volume is $V = 0$ cm^3 (an obvious minimum). When $s = 2$ cm, $\dfrac{d^2V}{ds^2} = -96 + 24s$,

$$\left.\frac{d^2V}{ds^2}\right|_{s=2} = -96 + 24(2) = -48$$

While our intuition indicates that there must be a maximum volume, the second derivative test guarantees a relative maximum at $s = 2$. So indeed, the maximum volume must occur when $s = 2$ cm. That is,

$$V = [12 - 2(2)]^2(2)$$
$$= 128 \text{ cm}^3$$

Note that the second derivative test cannot be used if the value of the second derivative is zero or undefined at a critical point. Should this occur, the first derivative test must be used.

SOLVING MAXIMUM AND MINIMUM PROBLEMS

1. Make a sketch of the conditions given in the problem whenever possible.
2. Label appropriate quantities and assign symbols for variable quantities.
3. Determine which variable quantity is to be maximized or minimized.
4. Write an equation that expresses the quantity identified in Step 3 as a function of one variable. You may need to combine different equations to obtain the desired function of one variable. These equations are obtained either from given conditions or from your understanding of the relationship between quantities involved in the problem.
5. Differentiate the function obtained in Step 4.
6. Set the derivative from Step 5 equal to zero. Solve for the values of the variable that make the derivative zero or undefined.
7. Determine which values of the variable from Step 6 provide the desired maximum or minimum.

EXAMPLE 3

The height of a bullet fired vertically upward is given by $s = 1200t - 16t^2$ (initial velocity is 1200 ft/s). Find the maximum height that the bullet will rise (neglecting air resistance).

Differentiate and set $\dfrac{ds}{dt} = 1200 - 32t = 0$.

$$t = 37.5$$

Since $\dfrac{d^2s}{dt^2} = -32$, there is a relative maximum at $t = 37.5$ s. The maximum height is then

$$s = 1200(37.5) - 16(37.5)^2$$
$$= 45,000 - 22,500 = 22,500 \text{ ft}$$

EXAMPLE 4

Make a cylindrical can with 24π cm^2 of metal. Find the dimensions of the can that give the maximum volume (see Fig. 3.26).

The volume of a cylinder is given by

$$V = \pi r^2 h$$

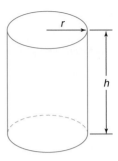

Figure 3.26

The lateral surface area of a cylinder is given by $A = 2\pi rh$. The sum of the areas of the two circular bases or ends is $2\pi r^2$. The total surface area is then

$$A = 2\pi rh + 2\pi r^2$$

We are to maximize the volume while the total surface area is held constant. We must express the volume as a function of only one variable. So substitute $A = 24\pi$ into

$$A = 2\pi rh + 2\pi r^2$$
$$24\pi = 2\pi rh + 2\pi r^2$$

and solve for h. First, divide both sides of this equation by 2π:

$$12 = rh + r^2$$
$$12 - r^2 = rh$$
$$\frac{12 - r^2}{r} = h$$

Then substitute for h into the volume formula:

$$V = \pi r^2 h$$
$$V = \pi r^2 \left(\frac{12 - r^2}{r}\right)$$
$$V = \pi r(12 - r^2)$$
$$V = \pi(12r - r^3)$$

Then differentiate:

$$\frac{dV}{dr} = \pi(12 - 3r^2)$$

and set $\dfrac{dV}{dr} = 0$.

$$\frac{dV}{dr} = \pi(12 - 3r^2) = 0$$
$$12 - 3r^2 = 0$$
$$12 = 3r^2$$
$$4 = r^2$$
$$\pm 2 = r$$

Since $r > 0$, $r = 2$. Then substitute into

$$h = \frac{12 - r^2}{r}$$
$$h = \frac{12 - 2^2}{2} = \frac{12 - 4}{2} = \frac{8}{2} = 4$$

Thus $r = 2$ cm and $h = 4$ cm give the maximum volume.

Using a calculator,

F2 1 F3 1 π(12r-r^3),r)=0,r) **ENTER F3 1** π(12r-r^3),r,2)|r=2 **ENTER**

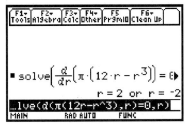

Finding the critical values.   The second derivative test indicates a maximum when $r = 2$.

Exercises 3.4

1. The sum of two positive numbers is 56. Find the two numbers if their product is to be a maximum.

2. Find a positive number such that the sum of this number and its reciprocal is a minimum.

3. An open box is to be made from a square piece of aluminum, 3 cm on a side, by cutting equal squares from each corner and then folding up the sides. Determine the dimensions of the box that will have the largest volume.

4. An open rectangular box is to be made from a rectangular piece of metal 16 in. by 10 in. by cutting equal squares from each corner and then folding up the sides. Determine the maximum volume of the box.

5. A man wishes to fence in a rectangular plot lying next to a river. No fencing is required along the river bank. If he has 800 m of fence and he wishes the maximum area to be fenced, find the dimensions of the desired enclosed plot.

6. A long rectangular sheet of metal, 12 in. wide, is to be made into a trough by turning up two sides forming right angles. Find the dimensions of the trough to give it maximum capacity.

7. Find the maximum possible area of a rectangle whose perimeter is 36 cm.

8. A rancher wants to fence in two rectangular corrals, equal in area, using an inner fence as shown in Fig. 3.27. If 300 ft of fencing is used, find the maximum area of the combined corrals.

9. A farmer wants to fence in 80,000 m^2 of land and then divide it into three plots of equal area as shown in Fig. 3.28. Find the minimum amount of fence needed.

10. Four neighbors want to fence in an 8100-ft^2 garden plot into equal plots as shown in Fig. 3.29. Find the minimum amount of fence needed.

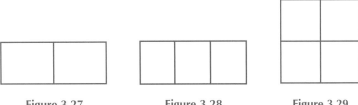

Figure 3.27   Figure 3.28   Figure 3.29

11. Find the maximum area of the rectangle with two vertices on the parabola $y = 36 - x^2$ and the remaining two vertices on the x-axis.

12. Find the dimensions of the largest rectangle that can be inscribed in a semicircle of radius 4.

13. Find the area of the largest rectangle that can be inscribed in the ellipse $x^2 + 9y^2 = 16$.

14. The charge transmitted through a circuit varies according to $q = \dfrac{4t}{t^2 + 1}$ coulombs (C). Find the maximum charge at time t in milliseconds (ms).

15. The power P in watts (W) in a circuit with resistance R in ohms (Ω) varies according to $P = \dfrac{36R}{(R + 4)^2}$. Find the resistance that gives the maximum power.

16. A rectangular yard 24,300 ft^2 in area is bounded on three sides by a fence that costs \$6/ft and in the front by a fence that costs \$10/ft. Find the most economical dimensions of the yard.

17. Show that the rectangle of largest area that can be inscribed in a given circle is a square.

18. Find the minimum slope of a tangent line to the curve $y = 4x^2 + 8x^3$.

19. Find the maximum slope of a tangent line to the curve $y = 3x^2 - 2x^3$.

20. The work done by a solenoid in moving an armature varies according to $w = 2t^2 - 3t^4$ joules (J). Find the greatest power in watts (W) developed in t seconds $\left(\text{power } p = \dfrac{dw}{dt}\right)$.

21. The charge transmitted through a circuit varies according to $q = t^4 - 4t^3$ coulombs (C). Find the time t in seconds when the current i (in amperes, A) $i = \dfrac{dq}{dt}$ reaches a minimum.

22. Find the greatest current in a capacitor with capacitance C equal to $\frac{4}{3} \times 10^{-6}$ farad (F) if the applied voltage is given by $V = 250t^2 - 200t^3$ volts (V) $\left(i = C\dfrac{dV}{dt}\right)$.

23. A rectangular box, open at the top, with a square base is to have a volume of 4000 cm^3. Find the dimensions if the box is to contain the least amount of material.

24. If the box in Exercise 23 has a closed top, find its dimensions.

25. The total cost C of making x units of a certain commodity is given by $C = 0.005x^3 + 0.45x^2 + 12.75x$. All units made are sold at \$36.75 per unit. The profit P is then given by $P = 36.75x - C$. Find the number of units to make to maximize profit.

26. Find the volume of the largest right circular cylinder that can be inscribed in a sphere of radius 3 in.

27. A cylindrical can with one end is to be made with 24π cm^2 of metal. Find the dimensions of the can that give the maximum volume.

• 28. The current I in a cell is given by $I = \dfrac{E}{R + r}$, where E is the electromotive force, R is the external resistance, and r is the internal resistance. In a given cell, E and r are fixed and determined by the cell. The power developed is $P = RI^2$. Show that P is a maximum when $R = r$.

29. The strength of a rectangular beam is proportional to the product of its width and the square of its depth; that is, $S = kwd^2$, where k is a constant. Find the dimensions of the strongest wooden beam that can be cut from a circular log of radius r in Fig. 3.30.

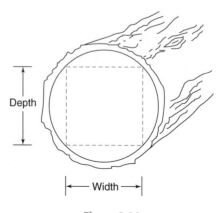

Figure 3.30

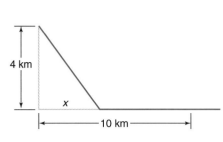

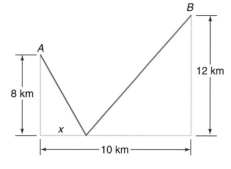

Figure 3.31   **Figure 3.32**

30. A person is in a boat 4 km off a straight coast. The person needs to reach a point 10 km down and along the coast in the least possible time. If the person can row at 6 km/h and run at 8 km/h, how far, x, down the coast should he land the boat? (See Fig. 3.31.)

31. Two ships are anchored off a straight coast as shown in Fig. 3.32. A boat from ship A is to land a passenger and then proceed to ship B in the shortest distance. Find x.

32. Show that any cylindrical can with two ends made with a given amount of material has a maximum volume when the height equals the diameter.

3.5   RELATED RATES

If an equation relates two or more variables, the rate at which one variable changes will affect the rates of change of the remaining variables. Problems involving variables whose rates of change are related by some equation are called *related rate* problems. In such problems, one or more rates are given and another rate is to be found.

EXAMPLE 1

A hot air balloon (spherical) is being inflated at the rate of 3 m³/s. Find the rate at which the radius is increasing when the radius is 10 m.

First determine the relationship between the volume of the balloon and its radius. The volume of a sphere is given by $V = \frac{4}{3}\pi r^3$. Differentiate each side of the equation with respect to time.

$$\frac{dV}{dt} = \frac{d}{dt}\left(\frac{4}{3}\pi r^3\right)$$

$$= \frac{4}{3}\pi \frac{d}{dt}(r^3)$$

$$= \frac{4}{3}\pi\left(3r^2\frac{dr}{dt}\right) \quad \text{(Note that } r \text{ is a function of time.)}$$

$$= 4\pi r^2\frac{dr}{dt}$$

$\dfrac{dV}{dt}$ is given as 3 m³/s, and we wish to find $\dfrac{dr}{dt}$ when $r = 10$ m. Substituting,

$$3 \text{ m}^3/\text{s} = 4\pi(10 \text{ m})^2\frac{dr}{dt}$$

$$3 \text{ m}^3/\text{s} = (400\pi \text{ m}^2)\frac{dr}{dt}$$

$$\frac{3 \text{ m}^3/\text{s}}{400\pi \text{ m}^2} = \frac{dr}{dt}$$

$$\frac{dr}{dt} = \frac{3}{400\pi} \text{ m/s} \quad \text{or} \quad 0.00239 \text{ m/s} \quad \text{or} \quad 2.39 \text{ mm/s}$$

SOLVING RELATED RATE PROBLEMS

1. Determine the equation for the variables whose rates are related. This equation may or may not be directly stated in the problem.

2. Differentiate each side of the resulting equation with respect to time (or other desired variable).

3. *After* differentiating in Step 2, substitute all given rates and given values of variables into the equation.

4. Solve for the unknown rate of change quantity.

EXAMPLE 2

Two airplanes pass each other in flight at 9:00 A.M. One is traveling east at 90 mi/h. The other is traveling south at 180 mi/h. How fast are they separating at 11:00 A.M.?

Step 1: Relate the distance that each plane has traveled with the distance traveled since separation at 9:00 A.M. From Fig. 3.33 note that the distances are related as the sides of a right triangle. The Pythagorean theorem provides the relationship:

$$x^2 + y^2 = z^2$$

where x is the distance traveled by A, y is the distance traveled by B, and z is the distance of separation.

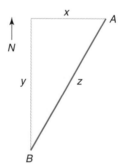

Figure 3.33

Step 2:
$$\frac{d}{dt}(x^2 + y^2) = \frac{d}{dt}(z^2)$$

$$2x\frac{dx}{dt} + 2y\frac{dy}{dt} = 2z\frac{dz}{dt}$$

$$x\frac{dx}{dt} + y\frac{dy}{dt} = z\frac{dz}{dt}$$

Step 3: We are given that $\frac{dx}{dt} = 90$ mi/h and $\frac{dy}{dt} = 180$ mi/h. Now after 2 h of travel,

$$x = (90 \text{ mi/h})(2 \text{ h}) = 180 \text{ mi} \qquad (d = rt)$$

$$y = (180 \text{ mi/h})(2 \text{ h}) = 360 \text{ mi}$$

So, again from the Pythagorean theorem,

$$z = \sqrt{x^2 + y^2} = \sqrt{(180 \text{ mi})^2 + (360 \text{ mi})^2} = 402 \text{ mi}$$

Step 4: $\dfrac{dz}{dt} = \dfrac{x\dfrac{dx}{dt} + y\dfrac{dy}{dt}}{z}$ (from step 2)

$$= \frac{(180 \text{ mi})(90 \text{ mi/h}) + (360 \text{ mi})(180 \text{ mi/h})}{402 \text{ mi}}$$

$$= 201 \text{ mi/h}$$

EXAMPLE 3

The electrical resistance of a resistor is given by $R = 6 + 0.002T^2$ (in ohms, Ω) where T is the temperature (in °C). If the temperature is increasing at the rate of 0.2°C/s, find how fast the resistance is changing when $T = 120$°C.

Step 1: The relationship is given:

$$R = 6 + 0.002T^2$$

Step 2: $\dfrac{dR}{dt} = 2(0.002T)\dfrac{dT}{dt}$

$$= 0.004T\frac{dT}{dt}$$

Step 3: $\dfrac{dR}{dt} = 0.004(120)(0.2)$

Step 4: $\dfrac{dR}{dt} = 0.096 \ \Omega/s$

Exercises 3.5

1. Given $y = 3x^2$ and $\dfrac{dx}{dt} = 2$ at $x = 3$. Find $\dfrac{dy}{dt}$.

2. Given $y = 4x^3 + 2x^2$ and $\dfrac{dx}{dt} = \dfrac{1}{2}$ at $x = 1$. Find $\dfrac{dy}{dt}$.

3. Given $x^2 + y^2 = 25$ and $\dfrac{dy}{dt} = \dfrac{9}{2}$ at $y = 4(x > 0)$. Find $\dfrac{dx}{dt}$.

4. Given $x^2 + y^2 = z^2$, $\dfrac{dx}{dt} = \dfrac{1}{2}$, and $\dfrac{dy}{dt} = 4$ at $x = 6$ and $y = 8$ $(z > 0)$. Find $\dfrac{dz}{dt}$.

5. The electrical resistance of a resistor is given by $R = 3 + 0.001T^2$ (in ohms, Ω) where T is the temperature (in °C). If the temperature is increasing at the rate of 0.3°C/s, find how fast the resistance is changing when $T = 100$°C.

6. A man begins walking due north at 3 mi/h. From the same place at the same time a woman begins walking east at 4 mi/h. Find the rate of change of the distance between them after 1 h of walking.

7. A square plate is being heated so that the length of a side increases at a rate of 0.08 cm/min. How fast is the area increasing when the length of a side is 12 cm?

8. A circular plate is being heated. Find the rate of increase in the area of the plate as it expands if the radius is increasing by 0.2 cm/h, when the radius is 5 cm.

9. The side of a cube due to heating is increasing at the rate of 0.1 cm/min. Find how fast the volume of the container is changing when a side measures 12 cm.

10. A stone is dropped into a lake and forms ripples of concentric circles of increasing radii. Find the rate at which the area of one of these circles is increasing when its radius is 3 ft and when its radius is increasing at the rate of 0.5 ft/s.

11. The sides of an equilateral triangular plate, being cooled, are decreasing at a rate of 0.04 cm/min. At what rate is the area changing when the sides are 8 cm in length? The area of an equilateral triangle is given by $A = \dfrac{\sqrt{3}}{4} s^2$.

12. The area of an equilateral triangular plate, being heated, is increasing at a rate of 150 mm²/min. At what rate is the length of a side changing when the sides are 250 mm long?

13. Mineral waste is falling into a conical pile at the rate of 5 m³/min. The height is equal to $\frac{3}{4}$ the radius of the base. Find how fast the radius is changing when the pile is 10 m high.

14. The relation between the voltage V that produces a current i in a wire of radius r is given by $V = 0.02\, i/r^2$. Find the rate at which the voltage is increasing when the current is increasing at the rate of 0.04 A/s in a wire of radius 0.02 in.

15. The power in a circuit is given by $P = I^2 R$. Find how fast the power changes in W/s when the current is 8 A and decreasing at a rate of 0.4 A/s, and the resistance is 75 Ω and increasing at a rate of 5 Ω/s.

16. Boyle's law states that the pressure of a gas varies inversely as its volume with constant temperature. We begin with 800 cm³ at a pressure of 250 kilopascals (kPa). Then the volume is decreased at a rate of 20 cm³/min. How is the pressure changing when the volume is 500 cm³?

17. A cylinder with an inside radius of 75 mm is sealed at one end, and a piston is at the other end. At what rate is the piston moving if fluid is being pumped into the cylinder at the rate of 90 cm³/s?

18. A man on a dock throws a rope to a woman in a boat, who fastens the rope to the boat. The man then pulls the boat in. If the man's hands are 8 ft above the water and if he hauls the rope in at 2 ft/s, find how fast the boat is approaching the base of the dock when there is still 10 ft of rope remaining to be pulled in.

19. A 5-m ladder is leaning against a wall. Its upper end is sliding down the wall at a rate of 1 m/s. Find how fast the bottom end of the ladder is moving at the point 3 m from the wall.

20. The work W done by an electromagnet varies according to $W = 36t^3 - t^2$. Find the power $\dfrac{dW}{dt}$ at $t = \dfrac{1}{4}$ s.

21. When a gas is compressed adiabatically (with no loss or gain of heat), its behavior is described by $PV^{1.4} = k$, where P is the pressure, V is the volume, and k is a constant. At a given instant, the pressure is 60 lb/in² and the volume is 56 in³ and decreasing at a rate of 8 in³/min. At what rate is the pressure changing?

22. In Exercise 21, find the rate at which the volume is changing when the volume is 5600 cm³, the pressure is 400 kilopascals (kPa), and the pressure is increasing at a rate of 50 kPa/min.

23. The equivalent resistance R of a parallel circuit with two resistances R_1 and R_2 is given by

$$R = \frac{R_1 R_2}{R_1 + R_2}$$

If R_1 is a constant 120 Ω and R_2 is decreasing at a rate of 15 Ω/s, find the rate at which R is changing when R_2 is 180 Ω.

24. In Exercise 23, assume that R_1 is 150 Ω and increasing at 15 Ω/s while R_2 is 300 Ω and decreasing at 25 Ω/s. Find the rate at which R is changing.

25. The power P, in watts, in a circuit varies according to $P = Ri^2$, where R is the resistance in ohms and i is the current in amperes. Find the rate of change, namely $\dfrac{dP}{di}$, in a 30-Ω resistor when the current $i = 4$ A.

26. Given $P = Ri^2$. If $R = 100\ \Omega$ and i varies according to $i = t^2 + 3t$, find the rate of change of the power with respect to time when $t = 2$ s.

27. A voltage $V = 0.02i/r^2$ produces a current i in a wire of radius r. Find the rate at which the voltage is increasing when the current is increasing at the rate of 0.04 A/s in a wire of radius 0.01 in.

28. A light inside a garage is 10 ft above the floor and 6 ft from the door opening. If the overhead garage door opener lets the door down at a rate of 1 ft/s, at what rate is the door's shadow approaching the garage when the bottom of the door is 2 ft above the floor? Assume that the bottom of the garage door stays in the same vertical plane as the door opening.

29. A tank is in the shape of an inverted cone with height 8 m and radius 2 m. Water is being pumped in at a rate of 2π m³/min. How fast is the depth of the water changing when the depth is **(a)** 2 m and **(b)** 6 m? (*Hint:* Use similar triangles to express the volume in terms of the depth only.)

30. The weight in pounds of an object varies according to

$$W = W_e\left(1 + \frac{r}{3960}\right)^{-2}$$

where W_e is the object's weight on the earth's surface and r is the distance above the earth's surface in miles. Find the rate at which a person's weight is decreasing in a space shuttle 500 miles above the earth when its altitude is increasing at 16 mi/s. The person's weight on earth is 175 lb.

3.6 DIFFERENTIALS

Up to now, we have defined and treated dy/dx as the limit:

$$\frac{dy}{dx} = \lim_{\Delta x \to 0} \frac{\Delta y}{\Delta x}$$

Now we need to discuss separate meanings of dy and dx.

First, solve the expression $\dfrac{dy}{dx} = f'(x)$ for dy.

> **DIFFERENTIAL**
>
> $dy = f'(x)\, dx$

The differential of y, denoted dy, is defined as $dy = f'(x)\, dx$, where $y = f(x)$ and dx is taken to represent a change in the value of x. Fig. 3.34 shows the geometric relationships between dy, dx, Δy, and Δx. Note that L_1 is the line tangent to the curve at point $P(x, y)$ and L_2 is a secant line through point P and a second point "close" to point P. Note that

1. $f'(x)$ is the slope of the tangent line L_1 to $y = f(x)$.

2. $\dfrac{dy}{dx}$ is the instantaneous rate of change of the function $y = f(x)$.

3. $\dfrac{\Delta y}{\Delta x}$ is the slope of the secant line and the rate of change of the function between the points (x, y) and $(x + \Delta x, y + \Delta y)$.

4. dx and Δx are each the amount of change in x.

5. Δy is the amount of change in y as measured along the secant line.

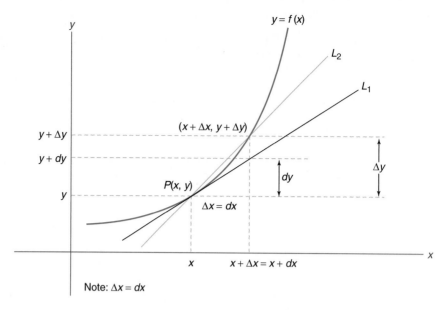

Note: $\Delta x = dx$

Figure 3.34 Geometric relationships between dy, dx, Δy, and Δx.

6. dy is the amount of change in y as measured along the tangent line.

7. If Δx is "small," dy is approximately equal to Δy; that is, $|dy - \Delta y|$ is approximately zero.

To find the differential, dy, of a function, simply find the derivative, $\dfrac{dy}{dx}$, and then multiply both sides of the equation by dx.

EXAMPLE 1

Given $y = 8x^3 - 4x + 2$, find dy.

$$y = 8x^3 - 4x + 2$$
$$\frac{dy}{dx} = 24x^2 - 4$$
$$dy = (24x^2 - 4)\,dx$$

EXAMPLE 2

Given $y = \dfrac{8x + 4}{2x - 1}$, find dy.

$$y = \frac{8x + 4}{2x - 1}$$
$$\frac{dy}{dx} = \frac{(2x - 1)(8) - (8x + 4)(2)}{(2x - 1)^2}$$
$$\frac{dy}{dx} = \frac{-16}{(2x - 1)^2}$$
$$dy = \frac{-16}{(2x - 1)^2}\,dx \quad \text{or} \quad \frac{-16\,dx}{(2x - 1)^2}$$

Approximations can be made using differentials.

EXAMPLE 3

The radius of a sphere measures 30.00 cm with a maximum possible error of 0.15 cm. Find the maximum possible error in the volume.

The volume of a sphere is found by

$$V = \frac{4}{3}\pi r^3$$

Find dV:

$$dV = \frac{4}{3}\pi(3r^2)\,dr$$

$$dV = 4\pi r^2\,dr$$

Here $r = 30.00$ cm and $dr = 0.15$ cm:

$$dV = 4\pi(30.00 \text{ cm})^2(0.15 \text{ cm})$$

$$= 1696.46 \text{ cm}^3$$

which is the differential approximation of the maximum possible error.

Note: $\Delta V = \frac{4}{3}\pi(30.15 \text{ cm})^3 - \frac{4}{3}\pi(30.00 \text{ cm})^3 = 1704.96 \text{ cm}^3$, which is the calculated maximum possible error by substitution into the volume formula. That is, when Δr or dr is small, $|dV - \Delta V|$ is small.

An expression that better describes some approximations is relative error or percentage error.

$$\text{relative error in } x = \frac{\text{error in } x}{x} = \frac{dx}{x}$$

$$\text{percentage error in } x = \frac{\text{error in } x}{x} \cdot 100\% = \frac{dx}{x} \cdot 100\%$$

EXAMPLE 4

Find the percentage error in Example 3.

$$\frac{dV}{V} \cdot 100\% = \frac{4\pi r^2\,dr}{\frac{4}{3}\pi r^3} \cdot 100\%$$

$$= \frac{3\,dr}{r} \cdot 100\%$$

$$= \frac{3(0.15 \text{ cm})}{30.00 \text{ cm}} \cdot 100\%$$

$$= 1.5\%$$

EXAMPLE 5

The current in a circuit changes according to $i = (3t + 2)^{1/3}$. Find the approximate change in current using differentials as t changes from 2.00 s to 2.05 s.

First, find the differential di:

$$i = (3t + 2)^{1/3}$$

$$di = \frac{1}{3}(3t + 2)^{-2/3}(3)\,dt$$

$$= (3t + 2)^{-2/3}\,dt$$

$$= [3(2.00) + 2]^{-2/3}(0.05) \qquad (\textit{Note: } dt = 0.05.)$$

$$= \left(\frac{1}{4}\right)(0.05) = 0.0125 \text{ A} \quad \text{or} \quad 12.5 \text{ mA}$$

Exercises 3.6

Find the differential for each expression.

1. $y = 5x^2 - 8x^3$

2. $y = 4t^5 - 6\sqrt{t} + 7$

3. $y = \dfrac{x + 3}{2x - 1}$

4. $s = \dfrac{2t + 1}{t + 6}$

5. $y = (2t^2 + 1)^4$

6. $i = 4t^2(t^2 + 1)^3$

7. $s = (t^4 - t^{-2})^{-2}$

8. $m = \left(3t - \dfrac{1}{3t}\right)^{-3}$

Find dy for each expression.

9. $x^2 + 4y^2 = 6$

10. $y^2 + 4xy = x$

11. $(x + y)^3 = \sqrt{x} + \sqrt{y}$

12. $(x^2 + y^2)^{1/2} = x + y$

Using a differential expression, find the change in each expression for the given change in the given independent variable.

13. $y = 8x^4$ from $x = 3.00$ to 3.05

14. $y = 4t^{2/3}$ from $t = 8.0$ to 8.1

15. $v = r^3 - 3r^2$ from $r = 4.00$ to 4.05

16. $y = (x - 2)^3$ from $x = 5.0$ to 5.1

17. $V = \frac{4}{3}\pi r^3$ from $r = 15.00$ to 15.10

18. $A = 4\pi r^2$ from $r = 21.0$ to 21.5

19. The side of a square measures 12.00 cm with a maximum possible error of 0.05 cm. **(a)** Find the maximum possible error in the area using differentials. **(b)** Find the maximum possible error by substituting into the formula for the area of a square. **(c)** Find the percentage error.

20. The side of a cube measures 12.00 cm with a maximum possible error of 0.05 cm. **(a)** Find the maximum possible error in the volume using differentials. **(b)** Find the maximum possible error by substituting into the formula for the volume of a cube. **(c)** Find the percentage error.

21. Suppose you want to build a spherical water tower with an inner diameter of 26.00 m and sides of thickness 4.00 cm. **(a)** Find the approximate volume of steel needed using differentials. **(b)** If the density of steel is 7800 kg/m^3, find the approximate amount of steel used.

22. The current in a resistor varies according to $i = 0.08t^6 - 0.04t^2$. Find the approximate change in current using differentials as t changes from 2.00 s to 2.10 s.

23. The horsepower of an internal combustion engine is given by $p = nd^2$, where n is the number of cylinders and d is the diameter of each bore. Find the approximate increase in horsepower using differentials for an engine with eight cylinders when the bore of each cylinder is increased from 3.750 in. to 3.755 in.

24. A freely falling body drops according to $s = \frac{1}{2}gt^2$, where s is the distance in metres, $g = 9.80$ m/s^2, and t is the time in seconds. Approximate the distance, ds, that an object falls from $t = 10.00$ s to $t = 10.03$ s.

25. The voltage V in volts varies according to $V = 10P^{2/3}$, where P is the power in watts. Find the change dV when the power changes from 125 W to 128 W.

26. The impedance Z in an ac circuit varies according to $Z = \sqrt{R^2 + X^2}$, where R is the resistance and X is the reactance. If $R = 300\ \Omega$ and $X = 225\ \Omega$, find dZ when R changes to $310\ \Omega$.

CHAPTER SUMMARY

1. *Algebraic curve sketching aids:*

(a) *Intercepts:*

(i) To find the x-intercepts, substitute $y = 0$ into the given equation and solve for x. Each resulting value of x is an x-intercept.

(ii) To find the y-intercepts, substitute $x = 0$ into the given equation and solve for y. Each resulting value of y is a y-intercept.

(b) *Location of the curve above and below the x-axis:*
   (i) Solve the given equation for y.
   (ii) Factor the result from Step (i) as much as possible.
   (iii) Set each factor from Step (ii) equal to zero.
   (iv) Find the sign of y in each region by choosing a test point. Substitute this test point value into each factor of the given equation and determine whether each factor is positive or negative in the region. If the sign of y is *positive*, the curve is *above* the x-axis in this region. If the sign of y is *negative*, the curve is *below* the x-axis in this region.

(c) *Symmetry:*
   (i) *Symmetry about the y-axis:* The graph of an equation is symmetric about the y-axis if replacement of x by $-x$ in the original equation results in an equivalent equation.
   (ii) *Symmetry about the x-axis:* The graph of an equation is symmetric about the x-axis if replacement of y by $-y$ in the original equation results in an equivalent equation.
   (iii) *Symmetry about the origin:* The graph of an equation is symmetric about the origin if replacement of both x by $-x$ and y by $-y$ in the original equation results in an equivalent equation.
   (iv) If any two of the three symmetry tests are satisfied, then the graph contains all three symmetries.

(d) *Asymptotes (for rational functions):*
   (i) *Vertical:* Set the denominator equal to zero and solve for x. Each solution, a, gives the equation of a vertical asymptote, $x = a$.
   (ii) *Horizontal:* $y = 0$ if the degree of the numerator is less than the degree of the denominator.
   $y = a_n/b_n$ if the degree of the numerator equals the degree of the denominator of the rational function in the form

   $$f(x) = \frac{a_n x^n + \cdots + a_0}{b_n x^n + \cdots + b_0}$$

   (iii) *Slant:* If the degree of the numerator is one greater than the degree of the denominator, the slant asymptote is found by dividing the numerator by the denominator. (Drop the remainder.) Then $y =$ *the quotient* is the equation of the slant asymptote.

(e) *Restricted domains:* Check for any restrictions in the domain of certain functions, such as factors that make the denominator zero in a rational function, values of the variable that make the radicand negative in an even root function, values that make the logarithm function undefined, and so forth.

2. *Increasing and decreasing functions:*
   (a) A function $y = f(x)$ is *increasing* in an interval if $f(x_2) > f(x_1)$ for any two points $x_2 > x_1$ in the interval.
   (b) A function $y = f(x)$ is *decreasing* in an interval if $f(x_2) < f(x_1)$ for any two points $x_2 > x_1$ in the interval.
   (c) $f(x)$ is increasing when $f'(x) > 0$.
   (d) $f(x)$ is decreasing when $f'(x) < 0$.

3. *Critical points:* Critical points are points of continuity $(x, f(x))$ on $y = f(x)$ where
   (a) $f'(x) = 0$, or
   (b) $f'(x)$ is undefined.
   That is, where $f'(x)$ is neither positive nor negative.

4. *First derivative test:* Given $(c, f(c))$ is a critical point of $y = f(x)$.
   (a) If $f'(x)$ changes from positive to negative at $x = c$, then $f(x)$ is changing from increasing to decreasing, and the point $(c, f(c))$ is a relative maximum.
   (b) If $f'(x)$ changes from negative to positive at $x = c$, then $f(x)$ is changing from decreasing to increasing, and the point $(c, f(c))$ is a relative minimum.
   (c) If $f'(x)$ does not change sign at $x = c$, then the point $(c, f(c))$ is neither a relative maximum nor a relative minimum.

5. *Calculus curve sketching aids:* To sketch $y = f(x)$:
   (a) Find all critical points for $y = f(x)$. That is, find all values of x where $f'(x) = 0$ or where $f'(x)$ is undefined.
   (b) Knowing the critical points, determine where the function is increasing and where it is decreasing.

$$y = f(x) \quad \text{is increasing where} \quad f'(x) > 0$$
$$y = f(x) \quad \text{is decreasing where} \quad f'(x) < 0$$

   (c) Find the relative maximum and minimum values. That is, find all values of x where $f'(x) = 0$ or is undefined. Check the critical points for possible relative maximums and relative minimums using the information obtained in Step (b).
   (d) Find the x-coordinate of each possible point of inflection. That is, find all values of x where $f''(x) = 0$ or is undefined.
   (e) From the possible points of inflection, determine the concavity of $y = f(x)$.

$$y = f(x) \quad \text{is concave upward where} \quad f''(x) > 0$$
$$y = f(x) \quad \text{is concave downward where} \quad f''(x) < 0$$

   (f) Find which points from Step (d) are points of inflection using the information obtained in Step (e). Note that it is possible for the second derivative to be zero at $x = a$ and $f''(a) = 0$, while $f(x)$ does not have a point of inflection at $x = a$. That is, the curve does not necessarily change concavity where $f''(x) = 0$. This must be checked in Step (e).
   (g) Sketch the curve.

6. *Second derivative test:*
   (a) If $f'(a) = 0$ and $f''(a) < 0$, then $M = f(a)$ is a relative maximum.
   (b) If $f'(a) = 0$ and $f''(a) > 0$, then $m = f(a)$ is a relative minimum.

7. *Maximum and minimum problems:*
   (a) Make a sketch of the conditions given in the problem whenever possible.
   (b) Label appropriate quantities and assign symbols for variable quantities.
   (c) Determine which variable quantity is to be maximized or minimized.
   (d) Write an equation that expresses the quantity identified in Step (c) as a function of one variable. You may need to combine different equations to obtain the desired function in one variable. These equations are obtained either from given conditions or from your understanding of the relationship between quantities involved in the problem.
   (e) Differentiate the function obtained from Step (d).
   (f) Set the derivative from Step (e) equal to zero. Solve for the values of the variable that make the derivative zero or undefined.
   (g) Determine which values of the variable from Step (f) provide the desired maximum or minimum.

8. *Related rate problems:*
   (a) Determine the equation for the variables whose rates are related. This equation may or may not be directly stated in the problem.
   (b) Differentiate each side of the resulting equation with respect to time (or other desired variable).

(c) *After* differentiating in Step (b), substitute all given rates and given values of variables into the equation.

(d) Solve for the unknown rate of change quantity.

9. *Differential:* $dy = f'(x)\,dx$

10. Errors:

(a) Relative error in $x = \dfrac{\text{error in } x}{x} = \dfrac{dx}{x}$

(b) Percentage error in $x = \dfrac{\text{error in } x}{x} \cdot 100\% = \dfrac{dx}{x} \cdot 100\%$

CHAPTER 3 REVIEW

Graph each equation.

1. $y = x^3 - 16x$

2. $y = \sqrt{-2 - 4x}$

3. $y = (x + 2)(25 - x^2)$

4. $y = (x^2 + 4x)(x - 1)^2$

5. $y = y^2 + 4$

6. $y = \dfrac{x - 2}{(x + 4)(x - 1)}$

7. $y = \dfrac{2x^2}{x^2 - 1}$

8. $y = \dfrac{x^2 + x - 12}{x + 3}$

9. $y = \dfrac{2x}{x^2 + 9}$

10. $y^2 = \dfrac{x}{(1 - x)(x + 4)}$

*For each function find (**a**) intervals for which it is increasing and decreasing, (**b**) relative maximums and relative minimums, (**c**) intervals for which it is concave upward and concave downward, and (**d**) points of inflection, and (**e**) sketch the curve.*

11. $y = 6x - x^3$

12. $y = x^2 - 3x - 4$

13. $y = x^3 - 7$

14. $y = 2x^3 - 9x^2 - 24x - 2$

15. $y = \dfrac{1}{(x + 1)^2}$

16. $y = \dfrac{x^2 - 1}{x^2 + 4}$

17. $y = \dfrac{10}{x^2 + 1}$

18. $y = \dfrac{x + 1}{x^2}$

19. The height of a missile fired vertically upward is given by $y = 240t - 16t^2$. Find the maximum height (in feet) that the missile will reach.

20. Find the dimensions of a right triangle with maximum area with hypotenuse 20.

21. The electrical power (in watts, W) produced by a source is given by $p = 3r - r^3$ where r is the resistance (in ohms, Ω) in the circuit. Find the value of r that provides the maximum power.

22. Find the area of the largest rectangle that can be inscribed in the right triangle in Fig. 3.35.

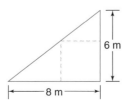

6 m

8 m

Figure 3.35

23. A long rectangular sheet of metal, 12 cm wide, is bent lengthwise to form a V-shaped trough. Find the depth of the trough that gives the maximum cross-sectional area and hence the greatest volume of flow.

24. Find the point on the curve $y = \sqrt{x}$ nearest the point $(1, 0)$.

25. A hot air balloon (spherical) is being inflated at the rate of 2 ft³/s. Find the rate at which the radius is increasing when the radius is 12 ft.

26. The voltage V produces a current i in a wire of radius r given by $V = 0.06i/r^2$. Find the rate at which the voltage is increasing when the current is increasing at the rate of 0.03 A/s in a wire with radius 0.01 in.

27. The current i through a circuit with resistance R and a battery whose voltage is E and whose internal resistance is r is given by $i = \dfrac{E}{R + r}$. The circuit has a variable resistor changing at the rate of 0.4 Ω/min with a 1.5-V battery whose resistance is 0.3 Ω. Find how fast the current is changing when the resistance of the variable resistor is 8 Ω.

28. A circular plate is being cooled. Find the rate of decrease in the area of the plate as it contracts if the radius is decreasing 0.05 cm/min when the radius is 8 cm.

29. Oil is leaking from an offshore well into a circular slick. If the area of the oil slick is increasing at a rate of 4 km²/day, at what rate is the radius increasing when the radius is 2.5 km?

30. A ground TV camera is 12 km from where a rocket is to be launched vertically. At what rate is the distance from the camera to the rocket increasing when the rocket is at an altitude of 10 km and rising at a rate of 3 km/s?

Find the differential for each expression.

31. $y = 4x^5 - 6x^3 + 2x$
32. $y = (3x - 5)^{-2/3}$
33. $s = \dfrac{3t^2 - 4}{5t + 1}$
34. Find dy: $(x^2 + y^2)^2 = y + 2x$

Using a differential expression, find the change in each expression for the given change in the given independent variable.

35. $s = 3t^2 - 5t + 6$ from $t = 9.50$ to 9.55

36. $y = (8x + 3)^{-3/4}$ from $x = 10.00$ to 10.06

37. Find the increase in volume if the radius of a sphere increases from 6.0 in. to 6.1 in. using differentials.

38. A particle moves along a straight line according to $s = \frac{1}{3}t^3 - 3t + 5$, where s is the distance, in metres, and t is the time, in seconds. Find the approximate distance covered by the particle between 3.00 s and 3.05 s.

39. Using differentials, approximate how much paint is needed to paint a cube 16 ft on a side if the thickness of the paint is to be 0.02 in. (One gallon contains 231 in³.)

40. A metal sphere is 8.00 cm in diameter. How much nickel is needed to plate the sphere with a thickness of 0.4 mm?

41. The attractive force between two unlike charged particles is given by $F = k/x^2$, where k is a constant and x is the distance between the particles. If x increases from 0.030 m to 0.031 m, find the approximate decrease in F (in newtons) using differentials.

4

Derivatives of Transcendental Functions

INTRODUCTION

Many problems that deal with maximizing or minimizing a quantity are expressed using trigonometric, logarithmic, or exponential functions. In addition, current in a circuit may be expressed using the derivative of the voltage across a capacitor or the derivative of the charge of a capacitor. For example, the equation $q = e^{-0.2t}(0.04 \cos 2t - 0.5 \sin 3t)$ represents the charge of a capacitor at time t. The current is represented by $i = \dfrac{dq}{dt}$. Thus we need to learn how to differentiate transcendental functions to expand our problem-solving abilities.

Objectives

- Use trigonometric identities to simplify trigonometric expressions or change a trigonometric expression to an equivalent expression.
- Differentiate trigonometric functions.
- Find an algebraic expression for a trigonometric function whose argument is an inverse trigonometric function.
- Find the derivative of an inverse trigonometric function.
- Find the derivative of a logarithmic function.
- Find the derivative of an exponential function.
- Use derivatives to solve application problems.

4.1   THE TRIGONOMETRIC FUNCTIONS

Before we begin to develop the calculus of the trigonometric functions, we briefly review some of the basics of trigonometry.

An angle is in **standard position** when its vertex is located at the origin and its initial side is lying on the positive x-axis. An angle resulting from a counterclockwise rotation, as indicated by the direction of the arrow, is a **positive angle.** But if the rotation is clockwise, the angle is **negative** (see Fig. 4.1).

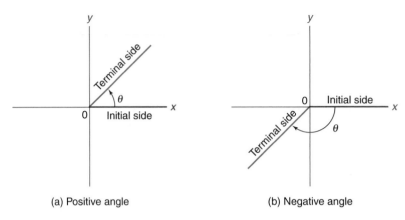

(a) Positive angle   (b) Negative angle

Figure 4.1

There are six trigonometric functions associated with angle θ in standard position. They can be expressed in terms of the coordinates of the point $P(x, y)$ as ratios where point P is on the terminal side of angle θ as in Fig. 4.2.

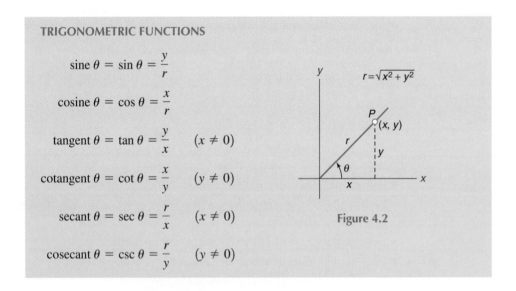

TRIGONOMETRIC FUNCTIONS

$$\text{sine } \theta = \sin \theta = \frac{y}{r}$$

$$\text{cosine } \theta = \cos \theta = \frac{x}{r}$$

$$\text{tangent } \theta = \tan \theta = \frac{y}{x} \quad (x \neq 0)$$

$$\text{cotangent } \theta = \cot \theta = \frac{x}{y} \quad (y \neq 0)$$

$$\text{secant } \theta = \sec \theta = \frac{r}{x} \quad (x \neq 0)$$

$$\text{cosecant } \theta = \csc \theta = \frac{r}{y} \quad (y \neq 0)$$

$r = \sqrt{x^2 + y^2}$

Figure 4.2

We know from algebra that $\dfrac{y}{r}$ and $\dfrac{r}{y}$ are reciprocals of each other; that is,

$$\sin \theta = \frac{y}{r} = \frac{1}{\frac{r}{y}} = \frac{1}{\csc \theta}$$

For this reason, $\sin \theta$ and $\csc \theta$ are called **reciprocal trigonometric functions.** In much the same way, we can complete the following table using the defining ratios.

RECIPROCAL TRIGONOMETRIC FUNCTIONS

$$\sin \theta = \frac{1}{\csc \theta} \qquad \csc \theta = \frac{1}{\sin \theta}$$

$$\cos \theta = \frac{1}{\sec \theta} \qquad \sec \theta = \frac{1}{\cos \theta}$$

$$\tan \theta = \frac{1}{\cot \theta} \qquad \cot \theta = \frac{1}{\tan \theta}$$

We must be careful to watch for angles where the point P on the terminal side has its abscissa, x, or its ordinate, y, equal to zero. Since we cannot divide by zero, $\tan \theta$ and $\sec \theta$ do not exist when $x = 0$. Likewise, $\cot \theta$ and $\csc \theta$ do not exist when $y = 0$.

Two or more angles in standard position are *coterminal* if they have the same terminal side. For example, $60°$, $420°$, and $-300°$ are coterminal.

A **quadrantal angle** is one which, when in standard position, has its terminal side coinciding with one of the axes.

EXAMPLE 1

Find the values of $\sin \theta$, $\cos \theta$, and $\tan \theta$ if θ is in standard position and its terminal side passes through the point $(-3, 4)$ (see Fig. 4.3).

$$r = \sqrt{x^2 + y^2} = \sqrt{9 + 16} = 5$$

$$\sin \theta = \frac{y}{r} = \frac{4}{5}$$

$$\cos \theta = \frac{x}{r} = \frac{-3}{5} = -\frac{3}{5}$$

$$\tan \theta = \frac{y}{x} = \frac{4}{-3} = -\frac{4}{3}$$

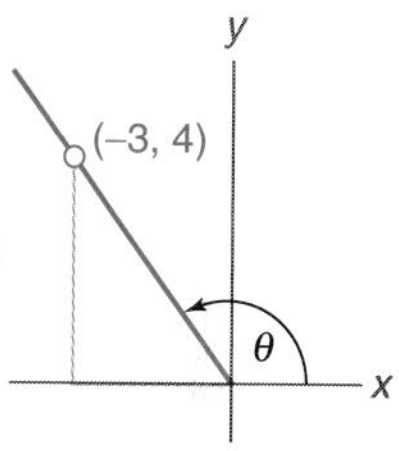

Figure 4.3

The **reference angle,** α, of any nonquadrantal angle, θ, in standard position is the *acute* angle between the terminal side of θ and the x-axis. Angle α is always considered to be a positive angle less than $90°$.

EXAMPLE 2

Find the reference angle α for each given angle θ (see Fig. 4.4).

Note that if angle θ is in standard position and

1. $0° < \theta < 90°$, then $\alpha = \theta$.
2. $90° < \theta < 180°$, then $\alpha = 180° - \theta$.
3. $180° < \theta < 270°$, then $\alpha = \theta - 180°$.
4. $270° < \theta < 360°$, then $\alpha = 360° - \theta$.

The reference angle α is often used to determine the angle θ when the value of the trigonometric function is known.

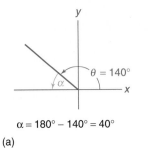

$\alpha = 180° - 140° = 40°$

(a)

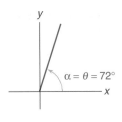

(b)

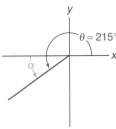

$\alpha = 215° - 180° = 35°$

(c)

$\alpha = 720° - 670° = 50°$

(d)

Figure 4.4

EXAMPLE 3

Given $\cos \theta = -\frac{1}{2}$, find θ for $0° \le \theta < 360°$.

Using a calculator, we find that $\alpha = 60°$. The cosine function is negative in quadrants II and III. The second-quadrant angle is $180° - 60° = 120°$. The third-quadrant angle is $180° + 60° = 240°$ (see Fig. 4.5).

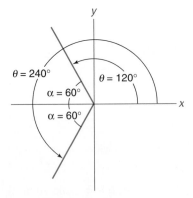

Figure 4.5

Angles are commonly given in degree measure and in radian measure. Since

$$1 \text{ revolution} = 360°$$

and

$$1 \text{ revolution} = 2\pi \text{ rad}$$

the basic relationship between radians and degrees is

$$\pi \text{ rad} = 180°$$

The degree and radian measures of several common angles are given in Table 4.1.

TABLE 4.1 Values of Common Angles

Radians	0	$\dfrac{\pi}{6}$	$\dfrac{\pi}{4}$	$\dfrac{\pi}{3}$	$\dfrac{\pi}{2}$	$\dfrac{2\pi}{3}$	$\dfrac{3\pi}{4}$	$\dfrac{5\pi}{6}$	π	$\dfrac{3\pi}{2}$	2π
Degrees	0°	30°	45°	60°	90°	120°	135°	150°	180°	270°	360°
$\sin\theta$	0	$\dfrac{1}{2}$	$\dfrac{\sqrt{2}}{2}$	$\dfrac{\sqrt{3}}{2}$	1	$\dfrac{\sqrt{3}}{2}$	$\dfrac{\sqrt{2}}{2}$	$\dfrac{1}{2}$	0	-1	0
$\cos\theta$	1	$\dfrac{\sqrt{3}}{2}$	$\dfrac{\sqrt{2}}{2}$	$\dfrac{1}{2}$	0	$-\dfrac{1}{2}$	$-\dfrac{\sqrt{2}}{2}$	$-\dfrac{\sqrt{3}}{2}$	-1	0	1
$\tan\theta$	0	$\dfrac{\sqrt{3}}{3}$	1	$\sqrt{3}$	undefined	$-\sqrt{3}$	-1	$-\dfrac{\sqrt{3}}{3}$	0	undefined	0
$\cot\theta$	undefined	$\sqrt{3}$	1	$\dfrac{\sqrt{3}}{3}$	0	$-\dfrac{\sqrt{3}}{3}$	-1	$-\sqrt{3}$	undefined	0	undefined
$\sec\theta$	1	$\dfrac{2\sqrt{3}}{3}$	$\sqrt{2}$	2	undefined	-2	$-\sqrt{2}$	$-\dfrac{2\sqrt{3}}{3}$	-1	undefined	1
$\csc\theta$	undefined	2	$\sqrt{2}$	$\dfrac{2\sqrt{3}}{3}$	1	$\dfrac{2\sqrt{3}}{3}$	$\sqrt{2}$	2	undefined	-1	undefined

Listed inside the back cover are the trigonometric identities commonly used in any trigonometry course, which we assume you have successfully completed. Trigonometric identities will be used to simplify trigonometric expressions, to change a given trigonometric expression into a different but equivalent expression that is easier to differentiate or integrate, and to compare results of integration by different methods to see that they are equivalent.

EXAMPLE 4

Prove $\sin^2\theta + \sin^2\theta \tan^2\theta = \tan^2\theta$.

$$\begin{aligned} \sin^2\theta + \sin^2\theta \tan^2\theta &= \sin^2\theta(1 + \tan^2\theta) \\ &= \sin^2\theta \sec^2\theta \\ &= \sin^2\theta\left(\frac{1}{\cos^2\theta}\right) \\ &= \tan^2\theta \end{aligned}$$

Therefore, $\sin^2\theta + \sin^2\theta \tan^2\theta = \tan^2\theta$.

EXAMPLE 5

Prove $\dfrac{\cos x}{1 + \sin x} = \dfrac{1 - \sin x}{\cos x}$.

Multiply the numerator and the denominator of the right-hand side by $1 + \sin x$.

$$\begin{aligned} \frac{1 - \sin x}{\cos x} \cdot \frac{1 + \sin x}{1 + \sin x} &= \frac{1 - \sin^2 x}{\cos x(1 + \sin x)} \\ &= \frac{\cos^2 x}{\cos x(1 + \sin x)} \\ &= \frac{\cos x}{1 + \sin x} \end{aligned}$$

Therefore, $\dfrac{\cos x}{1 + \sin x} = \dfrac{1 - \sin x}{\cos x}$.

EXAMPLE 6

Prove $\dfrac{1 - \cos 2x}{\sin 2x} = \tan x$.

$$\frac{1 - \cos 2x}{\sin 2x} = \frac{1 - (1 - 2\sin^2 x)}{2\sin x \cos x}$$

$$= \frac{2\sin^2 x}{2\sin x \cos x}$$

$$= \frac{\sin x}{\cos x}$$

$$= \tan x$$

Therefore, $\dfrac{1 - \cos 2x}{\sin 2x} = \tan x$.

EXAMPLE 7

Simplify $2\sin 3x \cos 3x$.

Using Formula 24 of the Common Trigonometric Identities listed inside the back cover,

$$2\sin \theta \cos \theta = \sin 2\theta$$

$$2\sin 3x \cos 3x = \sin 2(3x) = \sin 6x$$

EXAMPLE 8

Simplify $1 - 2\cos^2 5x$.

Using Formula 25b of the Common Trigonometric Identities listed inside the back cover,

$$1 - 2\cos^2 5x = -\cos 2(5x) = -\cos 10x$$

EXAMPLE 9

Simplify $\cos 2\theta \cos 3\theta - \sin 2\theta \sin 3\theta$.

By Formula 20 of the Common Trigonometric Identities listed inside the back cover,

$$\cos 2\theta \cos 3\theta - \sin 2\theta \sin 3\theta = \cos (2\theta + 3\theta) = \cos 5\theta$$

Fig. 4.6 shows the graphs of each of the six basic trigonometric functions. Recall that each is periodic. The *period* is the length of each cycle; that is, the period is the horizontal distance between any point on the curve and the next corresponding point in the next cycle where the graph begins to repeat itself.

The *amplitude* of the sine and cosine functions is one-half the distance difference between the maximum and minimum values of the function. Recall the following graphing facts and relationships for the six trigonometric functions.

$$\text{Period} = \frac{2\pi}{b} \begin{cases} \left.\begin{array}{l} y = a\sin(bx + c) \\ y = a\cos(bx + c) \end{array}\right\} \text{amplitude} = |a| \\ y = a\sec(bx + c) \\ y = a\csc(bx + c) \end{cases}$$

$$\text{Period} = \frac{\pi}{b} \begin{cases} y = a\tan(bx + c) \\ y = a\cot(bx + c) \end{cases}$$

Phase shift $= \dfrac{c}{b}$

to the *left* if $\dfrac{c}{b} > 0$ and

to the *right* if $\dfrac{c}{b} < 0$

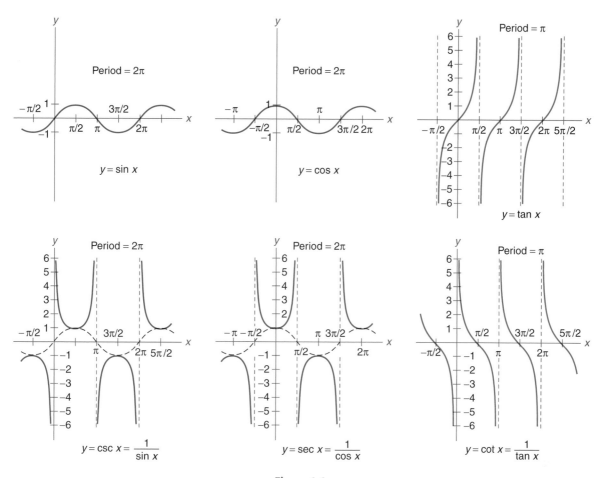

Figure 4.6

EXAMPLE 10

Graph $y = 2 \cos 3x$.

The amplitude is 2. The period is $P = \dfrac{2\pi}{b} = \dfrac{2\pi}{3}$ (see Fig. 4.7).

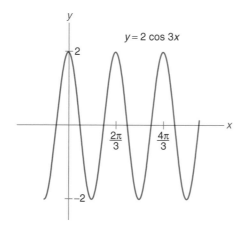

Figure 4.7

EXAMPLE 11

Graph $y = -3 \sin \frac{1}{2} x$.

The amplitude is 3. The period is $\dfrac{2\pi}{b} = \dfrac{2\pi}{\frac{1}{2}} = 4\pi$.

The effect of the negative sign is to flip, or invert, the curve $y = 3 \sin \frac{1}{2} x$ (see Fig. 4.8).

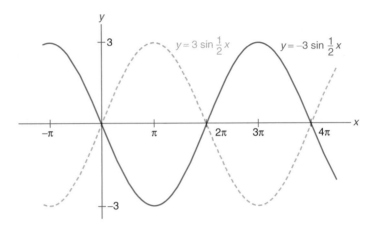

Figure 4.8

EXAMPLE 12

Graph $y = 3 \cos\left(6x + \dfrac{\pi}{4}\right)$.

The amplitude is 3. The period is $\dfrac{2\pi}{b} = \dfrac{2\pi}{6} = \dfrac{\pi}{3}$. The phase shift is $\dfrac{c}{b} = \dfrac{\pi/4}{6} = \dfrac{\pi}{24}$, or $\dfrac{\pi}{24}$ to the left (see Fig. 4.9).

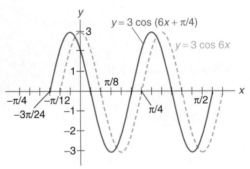

Each division on the x-axis is $\pi/24$.

Figure 4.9

Exercises 4.1

Prove each identity.

1. $\cos x \tan x = \sin x$

2. $(1 - \cos^2 x) \csc^2 x = 1$

3. $(\cot^2 x + 1) \tan^2 x = \sec^2 x$

4. $\cos \theta (\csc \theta - \sec \theta) = \cot \theta - 1$

5. $\tan^2 \theta - \tan^2 \theta \sin^2 \theta = \sin^2 \theta$

6. $\dfrac{\sin^2 x}{1 + \cos x} = 1 - \cos x$

7. $\dfrac{\cos x \tan x + \sin x}{\tan x} = 2 \cos x$

8. $\cos^4 \theta - \sin^4 \theta = 2 \cos^2 \theta - 1$

9. $\dfrac{1 + \sec x}{\csc x} = \sin x + \tan x$

10. $\dfrac{1 + \tan^2 x}{\tan^2 x} = \csc^2 x$

11. $(\sec x - \tan x)(\csc x + 1) = \cot x$

12. $\dfrac{1 - \cos \theta}{\cos \theta \tan \theta} = \dfrac{\sin \theta}{1 + \cos \theta}$

13. $\dfrac{1 - \tan x}{1 + \tan x} = \dfrac{\cot x - 1}{\cot x + 1}$

14. $\dfrac{1 - \sin^2 x}{1 - \cos^2 x} = \cot^2 x$

15. $\sin(x + \pi) = -\sin x$

16. $\cos(x + 180°) = -\cos x$

17. $\sin(x + 2\pi) = \sin x$

18. $\tan(x + \pi) = \tan x$

19. $\tan(\pi - x) = -\tan x$

20. $\sin(90° - \theta) = \cos \theta$

21. $\cos\left(\dfrac{\pi}{2} - \theta\right) = \sin \theta$

22. $\cos\left(\dfrac{\pi}{2} + \theta\right) = -\sin \theta$

23. $\cos(x + y)\cos(x - y) = \cos^2 x - \sin^2 y$

24. $\sin(x + y)\sin(x - y) = \sin^2 x - \sin^2 y$

25. $(\sin x + \cos x)^2 = 1 + \sin 2x$

26. $\sin 2x = \dfrac{2 \tan x}{1 + \tan^2 x}$

27. $\dfrac{1 - \tan^2 x}{1 + \tan^2 x} = \cos 2x$

28. $2 \tan x \csc 2x = \sec^2 x$

29. $\cot 2x = \dfrac{\cot^2 x - 1}{2 \cot x}$

30. $\sec 2x = \dfrac{\sec^2 x}{2 - \sec^2 x}$

31. $\tan x + \cot 2x = \csc 2x$

32. $\sin^2 \dfrac{x}{2} = \dfrac{\sec x - 1}{2 \sec x}$

33. $2 \cos^2 \dfrac{\theta}{2} = \dfrac{1 + \sec \theta}{\sec \theta}$

34. $\sec^2 \dfrac{x}{2} = \dfrac{2}{1 + \cos x}$

35. $\tan \dfrac{x}{2} = \dfrac{\sin x}{1 + \cos x}$

36. $2 \cos \dfrac{x}{2} = (1 + \cos x)\sec \dfrac{x}{2}$

Simplify each expression.

37. $\sin \theta \cos 3\theta + \cos \theta \sin 3\theta$

38. $\cos 2\theta \cos \theta - \sin 2\theta \sin \theta$

39. $\cos 4\theta \cos 3\theta + \sin 4\theta \sin 3\theta$

40. $\sin 2\theta \cos 3\theta - \cos 2\theta \sin 3\theta$

41. $\dfrac{\tan 3\theta + \tan 2\theta}{1 - \tan 3\theta \tan 2\theta}$

42. $\dfrac{\tan \theta - \tan 2\theta}{1 + \tan \theta \tan 2\theta}$

43. $\sin(\theta + \phi) + \sin(\theta - \phi)$

44. $\cos(\theta + \phi) + \cos(\theta - \phi)$

45. $2 \sin \dfrac{x}{4} \cos \dfrac{x}{4}$

46. $20 \sin^2 x \cos^2 x$

47. $1 - 2 \sin^2 3x$

48. $\sqrt{\dfrac{1 - \cos 6\theta}{2}}$

49. $\sqrt{\dfrac{1 + \cos \dfrac{\theta}{4}}{2}}$

50. $\cos 2x + 2 \sin^2 x$

51. $\cos^2 \dfrac{x}{6} - \sin^2 \dfrac{x}{6}$

52. $2 \sin 3x \cos 3x$

53. $20 \sin 4\theta \cos 4\theta$

54. $1 - 2 \sin^2 7t$

55. $4 - 8 \sin^2 \theta$

56. $15 \sin \dfrac{x}{6} \cos \dfrac{x}{6}$

Sketch each curve.

57. $y = 2 \cos x$

58. $y = \cos 3x$

59. $y = 3 \cos 6x$

60. $y = -\dfrac{1}{2} \sin \dfrac{2}{3} x$

61. $y = 2 \sin 3\pi x$

62. $y = 3 \sin\left(x - \dfrac{\pi}{3}\right)$

63. $y = -\sin\left(4x - \dfrac{2\pi}{3}\right)$

64. $y = -\cos(4x + \pi)$

65. $y = 3\sin\left(\dfrac{1}{2}x - \dfrac{\pi}{4}\right)$

66. $y = 5\cos\left(\dfrac{2}{3}x + \pi\right)$

67. $y = \tan 3x$

68. $y = 4\sin\left(\dfrac{\pi x}{6} - \dfrac{\pi}{3}\right)$

4.2 DERIVATIVES OF SINE AND COSINE FUNCTIONS

So far we have differentiated only algebraic functions; that is, functions in the form $y = f(x)$ where $f(x)$ is an algebraic expression. But many important applications involve the use of nonalgebraic functions called *transcendental functions*. The trigonometric and logarithmic functions are examples of transcendental functions. We will first find the derivative of $y = \sin u$.

By the definition of the derivative we have

$$\frac{d}{du}(\sin u) = \frac{dy}{du} = \lim_{\Delta u \to 0} \frac{\Delta y}{\Delta u}$$

$$= \lim_{\Delta u \to 0} \frac{\sin(u + \Delta u) - \sin u}{\Delta u}$$

Using the trigonometric identity (33) found inside the back cover,

$$\sin A - \sin B = 2\sin\left(\frac{A - B}{2}\right)\cos\left(\frac{A + B}{2}\right)$$

and letting $A = u + \Delta u$ and $B = u$, we have

$$\sin(u + \Delta u) - \sin u = 2\sin\left(\frac{u + \Delta u - u}{2}\right)\cos\left(\frac{u + \Delta u + u}{2}\right)$$

$$= 2\sin\left(\frac{\Delta u}{2}\right)\cos\left(u + \frac{\Delta u}{2}\right)$$

So now,

$$\frac{dy}{du} = \lim_{\Delta u \to 0} \frac{2\sin\left(\dfrac{\Delta u}{2}\right)\cos\left(u + \dfrac{\Delta u}{2}\right)}{\Delta u}$$

Next, divide numerator and denominator by 2.

$$\frac{dy}{du} = \lim_{\Delta u \to 0} \frac{\sin\left(\dfrac{\Delta u}{2}\right)\cos\left(u + \dfrac{\Delta u}{2}\right)}{\dfrac{\Delta u}{2}}$$

$$= \lim_{\Delta u \to 0} \frac{\sin\left(\dfrac{\Delta u}{2}\right)}{\dfrac{\Delta u}{2}} \cdot \lim_{\Delta u \to 0} \cos\left(u + \frac{\Delta u}{2}\right)$$

The right-hand factor, $\lim\limits_{\Delta u \to 0} \cos\left(u + \dfrac{\Delta u}{2}\right)$, is $\cos u$. We need, however, to determine

$$\lim_{\Delta u \to 0} \frac{\sin\left(\dfrac{\Delta u}{2}\right)}{\dfrac{\Delta u}{2}}$$

For this purpose we construct the geometric figure as shown in Fig. 4.10.

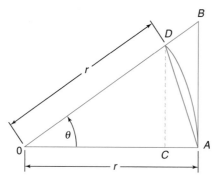

Figure 4.10

Angle θ is the central angle (measured in radians) of a circle of radius $r = OA = OD$. Note that

area of triangle OAD < area of sector OAD < area of right triangle OAB

$$\left(\tfrac{1}{2}r\right)(r \sin \theta) < \qquad \tfrac{1}{2}r^2\,\theta \qquad < \left(\tfrac{1}{2}r\right)(r \tan \theta)$$

$$\sin \theta < \qquad \theta \qquad < \tan \theta \qquad \text{(Divide each term by } \tfrac{1}{2}r^2.\text{)}$$

$$\sin \theta < \qquad \theta \qquad < \frac{\sin \theta}{\cos \theta} \qquad \left(\tan \theta = \frac{\sin \theta}{\cos \theta}\right)$$

Dividing each term now by $\sin \theta$, we have

$$1 < \qquad \frac{\theta}{\sin \theta} \qquad < \frac{1}{\cos \theta} \qquad (\sin \theta > 0)$$

By taking the reciprocal of each term, we have

$$1 > \qquad \frac{\sin \theta}{\theta} \qquad > \cos \theta$$

As θ approaches 0, $\cos \theta$ approaches 1, so we have

$$\lim_{\theta \to 0} 1 = 1 \quad \text{and} \quad \lim_{\theta \to 0} \cos \theta = 1$$

We conclude that $\dfrac{\sin \theta}{\theta}$, which is between, must *also* approach 1 as θ approaches 0, that is,

$$\lim_{\theta \to 0} \frac{\sin \theta}{\theta} = 1$$

So

$$\frac{d}{du}(\sin u) = \lim_{\Delta u \to 0} \frac{\sin\left(\dfrac{\Delta u}{2}\right)}{\dfrac{\Delta u}{2}} \cdot \lim_{\Delta u \to 0} \cos\left(u + \frac{\Delta u}{2}\right) = 1 \cdot \cos u = \cos u$$

Thus

$$\frac{d}{du}(\sin u) = \cos u$$

We are still unable, however, to find the derivative of a function such as

$$\frac{d}{dx}\sin(x^2 + 3x)$$

To find the derivative of $\sin u$ with respect to x when u is itself a function of x, we use the chain rule from Section 2.7. Recall that if $y = f(u)$ where $u = g(x)$, then $y = f[g(x)]$ is a function of x. Then,

CHAIN RULE

$$\frac{dy}{dx} = \frac{dy}{du} \cdot \frac{du}{dx}$$

EXAMPLE 1

Find the derivative of $y = \sin(x^2 + 3x)$.

Let $u = x^2 + 3x$ and $y = \sin u$ so that

$$\frac{dy}{du} = \cos u \quad \text{and} \quad \frac{du}{dx} = 2x + 3$$

and

$$\frac{dy}{dx} = \frac{dy}{du} \cdot \frac{du}{dx} = (\cos u)(2x + 3)$$

$$\frac{dy}{dx} = (2x + 3)\cos(x^2 + 3x)$$

EXAMPLE 2

Find $\dfrac{d}{dx}(\sin 2x)$.

Let $u = 2x$ and $y = \sin u$.

$$\frac{dy}{dx} = \frac{dy}{du} \cdot \frac{du}{dx}$$

$$= (\cos u)(2)$$

$$= 2\cos 2x$$

EXAMPLE 3

Find the derivative of $y = \sin(3x - 5)^2$.

Let $u = (3x - 5)^2$ and $y = \sin u$.

$$\frac{dy}{du} = \cos u \quad \text{and} \quad \frac{du}{dx} = 2(3x - 5)(3) = 6(3x - 5)$$

$$\frac{dy}{dx} = \frac{dy}{du} \cdot \frac{du}{dx}$$

$$= (\cos u)[6(3x - 5)]$$

$$= 6(3x - 5)\cos(3x - 5)^2$$

EXAMPLE 4

Find the derivative of $y = \sin^3 (2x - 3)$.

The chain rule is also used to find the derivative of a function, which is a chain of more than two functions of functions.

$$\frac{dy}{dx} = \frac{d}{dx}[\sin^3 (2x - 3)]$$

$$= 3 \sin^2 (2x - 3)\frac{d}{dx}[\sin (2x - 3)] \qquad \left[\frac{d}{dx}(u^n) = nu^{n-1}\frac{du}{dx}\right]$$

$$= 3 \sin^2 (2x - 3) \cos (2x - 3)\frac{d}{dx}(2x - 3) \qquad \left[\frac{d}{dx}(\sin u) = \cos u \frac{du}{dx}\right]$$

$$= 3 \sin^2 (2x - 3) \cos (2x - 3)(2)$$

$$= 6 \sin^2 (2x - 3) \cos (2x - 3)$$

To find the derivative of $y = \cos u$, first recall that $\cos u = \sin\left(\dfrac{\pi}{2} - u\right)$ and let $w = \dfrac{\pi}{2} - u$. Then $y = \cos u = \sin\left(\dfrac{\pi}{2} - u\right) = \sin w$. Using the chain rule,

$$\frac{dy}{du} = \frac{dy}{dw} \cdot \frac{dw}{du}$$

and noting that

$$\frac{dy}{dw} = \frac{d}{dw}(\sin w) = \cos w \quad \text{and} \quad \frac{dw}{du} = \frac{d}{du}\left(\frac{\pi}{2} - u\right) = -1$$

then we have

$$\frac{dy}{du} = (\cos w)(-1) = -\cos w$$

$$= -\cos\left(\frac{\pi}{2} - u\right)$$

$$= -\sin u \qquad \left[\text{since } \cos\left(\frac{\pi}{2} - u\right) = \sin u\right]$$

Thus

$$\frac{d}{du}(\cos u) = -\sin u$$

EXAMPLE 5

Find the derivative of $y = \cos (5x^2 + x)$.

Let $u = 5x^2 + x$ and $y = \cos u$.

$$\frac{dy}{du} = -\sin u \quad \text{and} \quad \frac{du}{dx} = 10x + 1$$

$$\frac{dy}{dx} = \frac{dy}{du} \cdot \frac{du}{dx}$$

$$= (-\sin u)(10x + 1)$$

$$= -(10x + 1) \sin (5x^2 + x)$$

EXAMPLE 6

Find the derivative of $y = \cos(x^2 - 1)^5$.

$$\frac{dy}{dx} = -\sin(x^2 - 1)^5 [5(x^2 - 1)^4](2x)$$

$$= -10x(x^2 - 1)^4 \sin(x^2 - 1)^5$$

EXAMPLE 7

Find the derivative of $y = \sin(7x^2 + 2) \cos 4x$.
Use the product rule for differentiation, followed by the chain rule.

$$\frac{dy}{dx} = \sin(7x^2 + 2)\frac{d}{dx}(\cos 4x) + \cos 4x \frac{d}{dx}\left[\sin(7x^2 + 2)\right]$$

$$= \sin(7x^2 + 2)(-\sin 4x)\frac{d}{dx}(4x) + \cos 4x \cos(7x^2 + 2)\frac{d}{dx}(7x^2 + 2)$$

$$= \sin(7x^2 + 2)(-\sin 4x)(4) + \cos 4x \cos(7x^2 + 2)(14x)$$

$$= -4\sin(7x^2 + 2)\sin 4x + 14x \cos 4x \cos(7x^2 + 2)$$

In summary,

$$\frac{d}{dx}(\sin u) = \cos u \frac{du}{dx}$$

$$\frac{d}{dx}(\cos u) = -\sin u \frac{du}{dx}$$

Exercises 4.2

Find the derivative of each function.

1. $y = \sin 7x$ **2.** $y = 5\sin 2x$ **3.** $y = 2\cos 5x$

4. $y = 4\cos 6x$ **5.** $y = 2\sin x^3$ **6.** $y = -3\cos x^4$

7. $y = 3\cos 4x^2$ **8.** $y = 5\sin 8x^3$ **9.** $y = 4\sin(1 - x)$

10. $y = 6\cos(1 - 3x)$ **11.** $y = 3\sin(x^2 + 4)$ **12.** $y = \sin(x^3 + 2x^2 - 4)$

13. $y = 4\cos(5x^2 + x)$ **14.** $y = 2\cos(x - 3)$ **15.** $y = \cos(x^4 - 2x^2 + 3)$

16. $y = 3\cos(x^2 - 4)$ **17.** $y = \cos^2(3x - 1)$ **18.** $y = \sin^2(1 - 2x)$

19. $y = \sin^3(2x + 3)$ **20.** $y = \cos^4(x^2 - 1)$ **21.** $y = \sin(2x - 5)^2$

22. $y = \cos(5x + 4)^3$ **23.** $y = \cos(x^3 - 4)^4$ **24.** $y = 4\sin(x^2 + x - 3)^6$

25. $y = \sin x \cos 3x$ **26.** $y = \sin x^2 \cos 5x$

27. $y = \sin 5x \cos 6x$ **28.** $y = \sin(3x - 1)\cos(4x + 3)$

29. $y = \cos 4x \cos 7x$ **30.** $y = \sin 3x \sin 5x$

31. $y = \sin(x^2 + 2x)\cos x^3$ **32.** $y = \sin(x + 2)^3 \cos(x^2 - 3)$

33. $y = (x^2 + 3x)\sin(5x - 2)$ **34.** $y = \sqrt{4x - 3}\cos(x^2 + 2)$

35. $y = \dfrac{\sin 5x}{x}$ **36.** $y = \dfrac{\cos 6x}{3x^2}$

37. $y = \dfrac{x^2 - 1}{\cos 3x}$ **38.** $y = \dfrac{4x - 5}{\sin(2x - 1)}$

39. $y = \sin 5x + \cos 6x$ **40.** $y = 2\sin 3x - 5\sin 6x$

41. $y = \sin(x^2 - 3x) + \cos 4x$

42. $y = x^2 - 3 \sin^3 2x$

43. $y = \dfrac{\sin x}{\cos x}$

44. $y = \dfrac{\cos x}{\sin x}$

45. Find $\dfrac{d^2y}{dx^2}$ for $y = \cos x$.

46. Find $\dfrac{d^2y}{dx^2}$ for $y = \sin x$.

47. Find $\dfrac{d^3y}{dx^3}$ for $y = \sin x$.

48. Find $\dfrac{d^4y}{dx^4}$ for $y = \cos x$.

49. Find $\dfrac{d^2y}{dx^2}$ for Exercise 39.

50. Find $\dfrac{d^2y}{dx^2}$ for Exercise 40.

51. Find the slope of the tangent line to $y = 4 \sin 3x$ at $x = \pi/18$.

52. Find the slope of the tangent line to $y = -4 \cos 2x$ at $x = 5\pi/12$.

53. Find the equation of the tangent line to $y = 2 \cos 5x$ at $x = \pi/10$.

54. Find the equation of the tangent line to $y = 6 \cos 4x$ at $x = 5\pi/24$.

4.3 DERIVATIVES OF OTHER TRIGONOMETRIC FUNCTIONS

Using $\tan u = \dfrac{\sin u}{\cos u}$ and the quotient rule for differentiation, we have

$$\frac{d}{du}(\tan u) = \frac{d}{du}\left(\frac{\sin u}{\cos u}\right)$$

$$= \frac{\cos u \dfrac{d}{du}(\sin u) - \sin u \dfrac{d}{du}(\cos u)}{\cos^2 u}$$

$$= \frac{(\cos u)(\cos u) - (\sin u)(-\sin u)}{\cos^2 u}$$

$$= \frac{\cos^2 u + \sin^2 u}{\cos^2 u}$$

$$= \frac{1}{\cos^2 u} \qquad (\cos^2 u + \sin^2 u = 1)$$

$$= \sec^2 u \qquad \left(\frac{1}{\cos u} = \sec u\right)$$

and

$$\frac{d}{dx}(\tan u) = \sec^2 u \frac{du}{dx} \qquad \text{(using the chain rule)}$$

In a similar manner it can be shown that

$$\frac{d}{du}(\cot u) = -\csc^2 u$$

and

$$\frac{d}{dx}(\cot u) = -\csc^2 u \frac{du}{dx}$$

Since $\sec u = \dfrac{1}{\cos u}$, we have

$$\frac{d}{du}(\sec u) = \frac{d}{du}(\cos u)^{-1}$$

$$= -(\cos u)^{-2}\frac{d}{du}(\cos u)$$

$$= \frac{-1}{\cos^2 u}(-\sin u)$$

$$= \frac{\sin u}{\cos^2 u}$$

$$= \frac{1}{\cos u} \cdot \frac{\sin u}{\cos u}$$

$$= \sec u \tan u$$

and using the chain rule,

$$\frac{d}{dx}(\sec u) = \sec u \tan u \frac{du}{dx}$$

In a similar manner you can show that

$$\frac{d}{du}(\csc u) = -\csc u \cot u$$

and

$$\frac{d}{dx}(\csc u) = -\csc u \cot u \frac{du}{dx}$$

EXAMPLE 1

Find the derivative of $y = \tan(x^2 + 3)$.

$$\frac{dy}{dx} = \sec^2(x^2 + 3)\frac{d}{dx}(x^2 + 3) \qquad (u = x^2 + 3)$$

$$= \sec^2(x^2 + 3)(2x)$$

$$= 2x\sec^2(x^2 + 3)$$

EXAMPLE 2

Find the derivative of $y = \sec^3 5x$.

First, use the power rule for differentiation.

$$\frac{dy}{dx} = 3\sec^2 5x\frac{d}{dx}(\sec 5x)$$

$$= (3\sec^2 5x)(\sec 5x \tan 5x)\frac{d}{dx}(5x) \qquad (u = 5x)$$

$$= 15\sec^3 5x \tan 5x$$

Using a calculator,

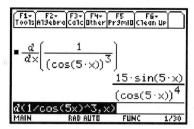

Note that the TI-89 uses $1/\cos(x)$ for sec x and that the derivative is obtained in terms of sine and cosine.

EXAMPLE 3

Find the derivative of $y = \cot \sqrt{3x + 1}$.

$$\frac{dy}{dx} = -\csc^2 \sqrt{3x + 1}\, \frac{d}{dx}(3x + 1)^{1/2} \qquad (u = \sqrt{3x + 1})$$

$$= (-\csc^2 \sqrt{3x + 1})\left[\frac{1}{2}(3x + 1)^{-1/2}(3)\right]$$

$$= \frac{-3\csc^2 \sqrt{3x + 1}}{2\sqrt{3x + 1}}$$

EXAMPLE 4

Find the derivative of $y = \sec 3x \tan 4x$.

$$\frac{dy}{dx} = (\sec 3x)(\sec^2 4x)(4) + (\tan 4x)(\sec 3x \tan 3x)(3)$$

$$= 4 \sec 3x \sec^2 4x + 3 \sec 3x \tan 3x \tan 4x$$

EXAMPLE 5

Find the derivative of $y = (x^2 + \cot^3 4x)^6$.

$$\frac{dy}{dx} = 6(x^2 + \cot^3 4x)^5[2x + 3(\cot^2 4x)(-\csc^2 4x)(4)]$$

$$= 6(x^2 + \cot^3 4x)^5[2x - 12 \cot^2 4x \csc^2 4x]$$

$$= 12(x^2 + \cot^3 4x)^5(x - 6 \cot^2 4x \csc^2 4x)$$

EXAMPLE 6

Find the derivative of $y = \csc (\sin 5x)$.

$$\frac{dy}{dx} = [-\csc (\sin 5x) \cot (\sin 5x)]\frac{d}{dx}(\sin 5x) \qquad (u = \sin 5x)$$

$$= -5 \csc (\sin 5x) \cot (\sin 5x) \cos 5x$$

When you prove trigonometric identities, you often have a choice of which identities to use. Similarly, here in finding the derivative of a trigonometric expression, you may use a trigonometric identity to change the form of the original expression before finding the derivative, to simplify the result, or to change the result to another form. When finding the derivative or later when finding the integral of a trigonometric expression, you will often need to use trigonometric identities to show that your answer is equivalent to the given answer.

EXAMPLE 7

Find the derivative of $y = \dfrac{\sin x}{\cot^2 x}$

Method 1: First, find the derivative and then change the trigonometric functions to sine and cosine.

$$\frac{dy}{dx} = \frac{(\cot^2 x)(\cos x) - (\sin x)[(2 \cot x)(-\csc^2 x)]}{\cot^4 x}$$

$$= \frac{\cot x \, (\cot x \cos x + 2 \sin x \csc^2 x)}{\cot^4 x}$$

$$= \frac{\dfrac{\cos x}{\sin x} \cos x + 2 \sin x \dfrac{1}{\sin^2 x}}{\cot^3 x}$$

$$= \frac{\dfrac{\cos^2 x + 2}{\sin x}}{\dfrac{\cos^3 x}{\sin^3 x}}$$

$$= \frac{\sin^2 x (\cos^2 x + 2)}{\cos^3 x}$$

Method 2: Change the expression to sine and cosine first and then find the derivative.

$$y = \frac{\sin x}{\cot^2 x} = \frac{\sin x}{\dfrac{\cos^2 x}{\sin^2 x}} = \frac{\sin^3 x}{\cos^2 x}$$

$$\frac{dy}{dx} = \frac{(\cos^2 x)[(3 \sin^2 x)(\cos x)] - (\sin^3 x)[(2 \cos x)(-\sin x)]}{\cos^4 x}$$

$$= \frac{\cos x \sin^2 x [3 \cos^2 x + 2 \sin^2 x]}{\cos^4 x}$$

$$= \frac{\sin^2 x [\cos^2 x + 2(\cos^2 x + \sin^2 x)]}{\cos^3 x}$$

$$= \frac{\sin^2 x (\cos^2 x + 2)}{\cos^3 x}$$

In summary, the derivatives of the trigonometric functions are as follows:

$$\frac{d}{dx}(\sin u) = \cos u \frac{du}{dx}$$

$$\frac{d}{dx}(\cos u) = -\sin u \frac{du}{dx}$$

$$\frac{d}{dx}(\tan u) = \sec^2 u \frac{du}{dx}$$

$$\frac{d}{dx}(\cot u) = -\csc^2 u \frac{du}{dx}$$

$$\frac{d}{dx}(\sec u) = \sec u \tan u \frac{du}{dx}$$

$$\frac{d}{dx}(\csc u) = -\csc u \cot u \frac{du}{dx}$$

Exercises 4.3

Find the derivative of each function.

1. $y = \tan 3x$

2. $y = \cot 2x$

3. $y = \sec 7x$

4. $y = \csc 4x$

5. $y = \cot(3x^2 - 7)$

6. $y = \tan(x^3 + 4)$

7. $y = 3 \csc(3x - 4)$

8. $y = 4 \sec\left(x - \dfrac{\pi}{3}\right)$

9. $y = \tan^2(5x - 2)$

10. $y = \sec^4 7x$

11. $y = 4 \cot^3 2x$

12. $y = \csc^3 x^4$

13. $y = \sec \sqrt{x^2 + x}$

14. $y = \sqrt{\tan 5x}$

15. $y = \dfrac{\csc x}{3x}$

16. $y = \dfrac{\cot 2x}{x}$

17. $y = \tan 3x - \sec(x^2 + 1)$

18. $y = \tan 2x + \cot 3x$

19. $y = \sec x \tan x$

20. $y = \sin x \sec x$

21. $y = \sin^2 x \cot x$

22. $y = \tan^2 x \sin x$

23. $y = x \sec x$

24. $y = x^2 \tan^2 x$

25. $y = x^2 + x^2 \tan^2 x$

26. $y = x^2 \csc^2 x - x^2$

27. $y = \csc 3x \cot 3x$

28. $y = \cot 2x \cos 4x$

29. $y = \csc^2 3x \sin 3x$

30. $y = \cos 2x \csc^2 x$

31. $y = (\sin x - \cos x)^2$

32. $y = \sec^2 x - \tan^2 x$

33. $y = (x + \sec^2 3x)^4$

34. $y = (\sin x - \tan^2 x)^3$

35. $y = (\sec x + \tan x)^3$

36. $y = (1 + \cot^3 x)^4$

37. $y = \sin(\tan x)$

38. $y = \sec(2 \cos x)$

39. $y = \tan(\cos x)$

40. $y = \cos(\cot x)$

41. $y = \sin^2(\cos x)$

42. $y = \tan^2(\sin x)$

43. $y = \dfrac{\tan x}{\cos x}$

44. $y = \dfrac{\sec x}{\cot x}$

45. $y = \dfrac{\sin^2 x}{\tan^2 x}$

46. $y = \dfrac{\csc^2 x}{\tan x}$

47. $y = \dfrac{\sin x}{1 + \tan x}$

48. $y = \dfrac{\sin x}{1 + \cos x}$

Find the second derivative of each function.

49. $y = \tan 3x$

50. $y = \sec 2x$

51. $y = x \cot x$

52. $y = \dfrac{\tan x}{x}$

53. Find the slope of the tangent line to the curve $y = \tan x$ at $x = \pi/4$.

54. Find the slope of the tangent line to the curve $y = \sec^2 x$ at $x = \pi/4$.

55. Show that $\dfrac{d}{du}(\cot u) = -\csc^2 u$.

56. Show that $\dfrac{d}{du}(\csc u) = -\csc u \cot u$.

4.4 THE INVERSE TRIGONOMETRIC FUNCTIONS

In mathematics, a *relation* is a set of ordered pairs of the form (x, y), usually written as an equation. The *inverse* of a given equation is the resulting equation when the x and y variables are interchanged. For example,

The inverse of	Is
$y = x^2$	$x = y^2$
$y = 3x^2 + 4x + 7$	$x = 3y^2 + 4y + 7$
$y = \dfrac{x + 4}{3x - 1}$	$x = \dfrac{y + 4}{3y - 1}$

Likewise, each basic trigonometric equation has an inverse.

The inverse of	Is
$y = \sin x$	$x = \sin y$
$y = \cos x$	$x = \cos y$
$y = \tan x$	$x = \tan y$
$y = \cot x$	$x = \cot y$
$y = \sec x$	$x = \sec y$
$y = \csc x$	$x = \csc y$

We find it necessary to write the inverse trigonometric equations solved for y. There are two common forms of the inverse trigonometric equations solved for y.

The inverse of	Is	Solved for y Is	Is
$y = \sin x$	$x = \sin y$	$y = \arcsin x$†	$y = \sin^{-1} x$‡
$y = \cos x$	$x = \cos y$	$y = \arccos x$	$y = \cos^{-1} x$
$y = \tan x$	$x = \tan y$	$y = \arctan x$	$y = \tan^{-1} x$
$y = \cot x$	$x = \cot y$	$y = \text{arccot } x$	$y = \cot^{-1} x$
$y = \sec x$	$x = \sec y$	$y = \text{arcsec } x$	$y = \sec^{-1} x$
$y = \csc x$	$x = \csc y$	$y = \text{arccsc } x$	$y = \csc^{-1} x$

†This is read, "y equals the arc sine of x" and means that y is the angle whose sine is x.

‡This notation will not be used here because of the confusion caused by the fact that -1 is not an exponent.

EXAMPLE 1

What is the meaning of each equation?
(a) $y = \arctan x$ (a) y is the angle whose tangent is x.
(b) $y = \arccos 3x$ (b) y is the angle whose cosine is $3x$.
(c) $y = 4 \text{ arccsc } 5x$ (c) y is four times the angle whose cosecant is $5x$.
Remember that $x = \sin y$ and $y = \arcsin x$ express the same relationship. The first form expresses the relationship in terms of the function (sine) of the angle; the second form expresses the relationship in terms of the angle itself.

EXAMPLE 2

Given the equation $y = \arcsin x$, find y when $x = \frac{1}{2}$. (Give the answer in radians.)
Substituting $x = \frac{1}{2}$, we have

$$y = \arcsin \tfrac{1}{2}$$

which means y is the angle whose sine is $\frac{1}{2}$. We know that $\sin \dfrac{\pi}{6} = \dfrac{1}{2}$, so $y = \dfrac{\pi}{6}$. But we also know that

$$\sin \frac{5\pi}{6} = \frac{1}{2}, \quad \text{so} \quad y = \frac{5\pi}{6}$$

$$\sin \frac{13\pi}{6} = \frac{1}{2}, \quad \text{so} \quad y = \frac{13\pi}{6} \qquad \left(\frac{13\pi}{6} = \frac{\pi}{6} + 2\pi \right)$$

$$\sin \frac{17\pi}{6} = \frac{1}{2}, \quad \text{so} \quad y = \frac{17\pi}{6} \qquad \left(\frac{17\pi}{6} = \frac{5\pi}{6} + 2\pi\right)$$

$$\vdots \qquad \qquad \vdots$$

$$\sin\left(\frac{-7\pi}{6}\right) = \frac{1}{2}, \quad \text{so} \quad y = \frac{-7\pi}{6}$$

$$\sin\left(\frac{-11\pi}{6}\right) = \frac{1}{2}, \quad \text{so} \quad y = \frac{-11\pi}{6}$$

$$\vdots \qquad \qquad \vdots$$

Thus there are infinitely many angles whose sine is $\frac{1}{2}$. We saw this when we graphed $y = \sin x$. That is,

$$y = \frac{\pi}{6} + n \cdot 2\pi$$

$$y = \frac{5\pi}{6} + n \cdot 2\pi$$

for every integer n.

EXAMPLE 3

Given the equation $y = \arctan x$, find y when $x = -1$ for $0 \le y < 2\pi$.
   Substituting $x = -1$, we have

$$y = \arctan(-1)$$

which means y is the angle whose tangent is -1. We know that

$$\tan \frac{3\pi}{4} = -1, \quad \text{so} \quad y = \frac{3\pi}{4}$$

and

$$\tan \frac{7\pi}{4} = -1, \quad \text{so} \quad y = \frac{7\pi}{4}$$

for $0 \le y < 2\pi$.

EXAMPLE 4

Find $\operatorname{arccot} \dfrac{1}{\sqrt{3}}$ for all angles in radians.

   Let $y = \operatorname{arccot} \dfrac{1}{\sqrt{3}}$, which means y is the angle whose cotangent is $\dfrac{1}{\sqrt{3}}$. We know that

$$\cot \frac{\pi}{3} = \frac{1}{\sqrt{3}} \quad \text{and} \quad \cot \frac{4\pi}{3} = \frac{1}{\sqrt{3}}$$

for $0 \le y < 2\pi$. Thus

$$\operatorname{arccot} \frac{1}{\sqrt{3}} = \begin{cases} \dfrac{\pi}{3} + 2n\pi \\[2mm] \dfrac{4\pi}{3} + 2n\pi \end{cases}$$

for every integer n.

EXAMPLE 5

Solve the equation $y = \cos 2x$ for x.

This equation is equivalent to

$$\arccos y = 2x$$

So

$$x = \frac{1}{2} \arccos y$$

EXAMPLE 6

Solve the equation $2y = \arctan 3x$ for x.

This equation is equivalent to

$$\tan 2y = 3x$$

So

$$x = \frac{1}{3} \tan 2y$$

EXAMPLE 7

Solve the equation $y = \frac{1}{3} \text{arcsec } 2x$ for x.

First, multiply both sides by 3.

$$3y = \text{arcsec } 2x$$

This equation is equivalent to

$$\sec 3y = 2x$$

Thus

$$x = \frac{1}{2} \sec 3y$$

To graph the inverse trigonometric relation

$$y = \arcsin x$$

first solve for x; that is,

$$x = \sin y$$

Then, plot ordered pairs of solutions of this equation as in Fig. 4.11.

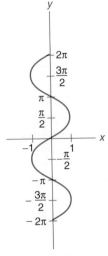

$y = \arcsin x$ **Figure 4.11**

In a similar manner, we can graph all six inverse trigonometric relations (see Fig. 4.12).

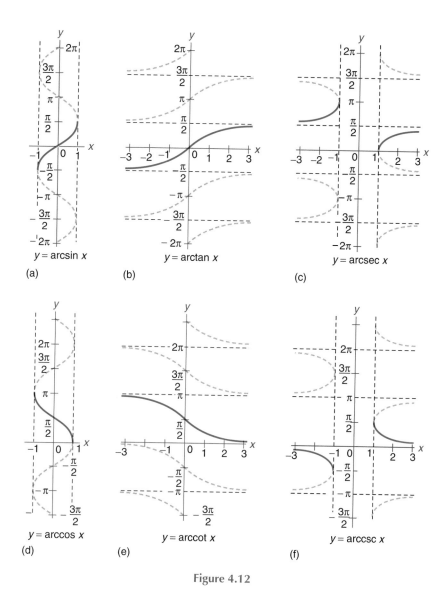

Figure 4.12

A *function* is a special relation: a set of ordered pairs in which no two distinct ordered pairs have the same first element. As you can see from the graphs, each *x*-value in each domain corresponds to (infinitely) many values of *y*. Thus none of the inverse trigonometric relations is a function. However, if we restrict the *y*-values of each inverse trigonometric relation, we can define an inverse that *is* a function. Although this could be done in any of several ways, it is customary to restrict the *y*-values as follows. *Note:* The inverse trigonometric *functions* are *capitalized* to distinguish them from the inverse trigonometric relations.

$$y = \text{Arcsin } x, \qquad -\frac{\pi}{2} \le y \le \frac{\pi}{2}$$

$$y = \text{Arccos } x, \qquad 0 \le y \le \pi$$

$$y = \text{Arctan } x, \qquad -\frac{\pi}{2} < y < \frac{\pi}{2}$$

$$y = \text{Arccot } x, \qquad 0 < y < \pi$$

$$y = \text{Arcsec } x, \qquad 0 \le y \le \pi, y \ne \frac{\pi}{2}$$

$$y = \text{Arccsc } x, \qquad -\frac{\pi}{2} \le y \le \frac{\pi}{2}, y \ne 0$$

Look once again at the graphs of the inverse trigonometric relations in Fig. 4.12. The solid lines indicate the portions of the graphs that correspond to the inverse trigonometric *functions* (see Fig. 4.13).

Note: The three inverse trigonometric functions on calculators are programmed to these same restricted ranges. When using a calculator to find the value of Arccot x, Arcsec x, or Arccsc x, use the following:

1. $\text{Arccot } x = \begin{cases} \text{Arctan } \dfrac{1}{x} & \text{if} \quad x > 0 \\[2mm] \pi + \text{Arctan } \dfrac{1}{x} & \text{if} \quad x < 0 \end{cases}$

2. $\text{Arcsec } x = \text{Arccos } \dfrac{1}{x} \quad$ where $\quad x \ge 1 \quad$ or $\quad x \le -1$

3. $\text{Arccsc } x = \text{Arcsin } \dfrac{1}{x} \quad$ where $\quad x \ge 1 \quad$ or $\quad x \le -1$

EXAMPLE 8

Find $\text{Arcsin}\left(\dfrac{1}{2}\right)$.

$$\text{Arcsin}\left(\frac{1}{2}\right) = \frac{\pi}{6}$$

This is the only value in the defined range of $-\dfrac{\pi}{2} \le y \le \dfrac{\pi}{2}$.

EXAMPLE 9

Find $\text{Arctan}\,(-1)$.

$$\text{Arctan}\,(-1) = -\frac{\pi}{4}$$

This is the only value in the defined range of $-\dfrac{\pi}{2} < y < \dfrac{\pi}{2}$.

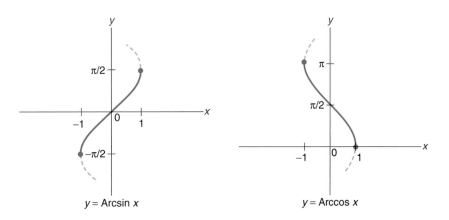

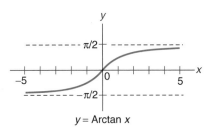

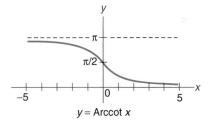

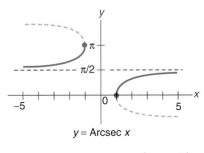

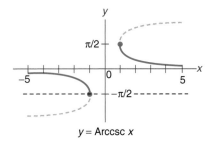

Inverse trigonometric functions

Figure 4.13

EXAMPLE 10

Find $\text{Arccos}\left(-\dfrac{1}{2}\right)$.

$$\text{Arccos}\left(-\frac{1}{2}\right) = \frac{2\pi}{3}$$

This is the only value in the defined range of $0 \leq y \leq \pi$.

EXAMPLE 11

Find $\tan\left[\text{Arccos}\left(-1\right)\right]$.

$$\tan\left[\text{Arccos}\left(-1\right)\right] = \tan\pi = 0$$

EXAMPLE 12

Find $\cos\left(\text{Arcsec } 2\right)$.

$$\cos\left(\text{Arcsec } 2\right) = \cos\frac{\pi}{3} = \frac{1}{2}$$

EXAMPLE 13

Find $\sin\left[\text{Arctan}\left(-\frac{1}{\sqrt{3}}\right)\right]$.

$$\sin\left[\text{Arctan}\left(-\frac{1}{\sqrt{3}}\right)\right] = \sin\left(-\frac{\pi}{6}\right) = -\frac{1}{2}$$

EXAMPLE 14

Find an algebraic expression for $\sin\left(\text{Arccos } x\right)$.

Let $\theta = \text{Arccos } x$. Then

$$\cos\theta = x = \frac{x}{1}$$

Draw a right triangle with θ as an acute angle, x as the adjacent side, and 1 as the hypotenuse as in Fig. 4.14(a).

Using the Pythagorean theorem,

$$c^2 = a^2 + b^2$$
$$1^2 = x^2 + (\text{side opposite } \theta)^2$$
$$\text{side opposite } \theta = \sqrt{1 - x^2} \quad \text{(Fig. 4.14b)}$$

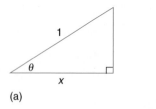
(a)

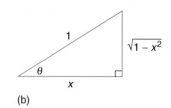
(b)

Figure 4.14

Now we see that

$$\sin\left(\text{Arccos } x\right) = \sin\theta$$
$$= \frac{\text{side opposite } \theta}{\text{hypotenuse}}$$

$$= \frac{\sqrt{1 - x^2}}{1}$$

$$= \sqrt{1 - x^2}$$

Using a calculator,

2nd SIN green diamond COS⁻¹ x)) ENTER

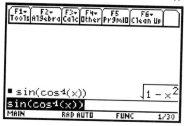

EXAMPLE 15

Find an algebraic expression for sec (Arctan x).

Let $\theta =$ Arctan x. Then

$$\tan \theta = x = \frac{x}{1}$$

Draw a right triangle with θ as an acute angle, x as the opposite side, and 1 as the adjacent side as in Fig. 4.15. Using the Pythagorean theorem, we find the hypotenuse is $\sqrt{x^2 + 1}$.

$$\begin{aligned}
\sec (\text{Arctan } x) &= \sec \theta \\
&= \frac{\text{hypotenuse}}{\text{side adjacent to } \theta} \\
&= \frac{\sqrt{x^2 + 1}}{1} \\
&= \sqrt{x^2 + 1}
\end{aligned}$$

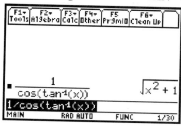

Figure 4.15

To use a calculator, you must write secant in terms of cosine.

1/ 2nd COS green diamond TAN⁻¹ x)) ENTER

EXAMPLE 16

Find an algebraic expression for cos (2 Arcsin x).

Let $\theta =$ Arcsin x. Then

$$\sin \theta = \frac{x}{1}$$

Draw a right triangle with θ as an acute angle, x as the opposite side, and 1 as the hypotenuse as in Fig. 4.16. Using the Pythagorean theorem, we find the side adjacent to θ is $\sqrt{1 - x^2}$.

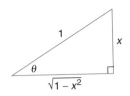

$$\cos{(2 \text{ Arcsin } x)} = \cos{2\theta}$$
$$= 1 - 2\sin^2{\theta}$$
$$= 1 - 2x^2$$

(Formula 25c of the Common Trigonometric Identities listed inside the back cover.)

Figure 4.16

Exercises 4.4

Solve each equation for $0 \leq y < 2\pi$.

1. $y = \arcsin{\left(\dfrac{\sqrt{3}}{2}\right)}$

2. $y = \arccos{\left(\dfrac{1}{2}\right)}$

3. $y = \arctan{1}$

4. $y = \text{arccot}{(-1)}$

5. $y = \text{arcsec}{\left(-\dfrac{2}{\sqrt{3}}\right)}$

6. $y = \text{arccsc}{\sqrt{2}}$

7. $y = \arccos{\left(-\dfrac{1}{2}\right)}$

8. $y = \arcsin{0}$

9. $y = \arccos{\left(-\dfrac{1}{\sqrt{2}}\right)}$

10. $y = \arctan{(-\sqrt{3})}$

11. $y = \text{arccot}{\left(\dfrac{1}{\sqrt{3}}\right)}$

12. $y = \arcsin{\left(-\dfrac{1}{\sqrt{2}}\right)}$

13. $y = \text{arcsec}{1}$

14. $y = \arcsin{(-1)}$

15. $y = \arctan{(1.963)}$

16. $y = \arccos{(-0.9063)}$

Find all angles in radians for each expression.

17. $\arctan{(-\sqrt{3})}$

18. $\arcsin{(-1)}$

19. $\arccos{\left(-\dfrac{\sqrt{3}}{2}\right)}$

20. $\text{arcsec}{\left(\dfrac{2}{\sqrt{3}}\right)}$

21. $\text{arccsc}{2}$

22. $\arctan{0}$

23. $\text{arcsec}{(-1)}$

24. $\arcsin{\left(\dfrac{1}{\sqrt{2}}\right)}$

25. $\text{arccot}{\sqrt{3}}$

26. $\arcsin{\left(-\dfrac{\sqrt{3}}{2}\right)}$

27. $\text{arcsec}{\left(\dfrac{2}{\sqrt{3}}\right)}$

28. $\arctan{(-\sqrt{3})}$

29. $\arccos{1}$

30. $\arcsin{1}$

31. $\arcsin{(-0.8572)}$

32. $\text{arccot}{(-0.8195)}$

Solve each equation for x.

33. $y = \sin{3x}$

34. $y = \tan{4x}$

35. $y = 4\cos{x}$

36. $y = 3\sec{x}$

37. $y = 5\tan{\dfrac{x}{2}}$

38. $y = \dfrac{1}{2}\cos{3x}$

39. $y = \dfrac{3}{2}\cot{\dfrac{x}{4}}$

40. $y = \dfrac{5}{2}\sin{\dfrac{2x}{3}}$

41. $y = 3\sin{(x - 1)}$

42. $y = 4\tan{(2x + 1)}$

43. $y = \dfrac{1}{2}\cos{(3x + 1)}$

44. $y = \dfrac{1}{3}\sec{(1 - 4x)}$

Find the value of each expression in radians.

45. $\text{Arcsin}\left(\dfrac{\sqrt{3}}{2}\right)$

46. $\text{Arccos}\left(\dfrac{1}{2}\right)$

47. $\text{Arctan}\left(-\dfrac{1}{\sqrt{3}}\right)$

48. $\text{Arcsin}\left(-\dfrac{1}{2}\right)$

49. $\text{Arccos}\left(-\dfrac{\sqrt{3}}{2}\right)$

50. $\text{Arctan}\left(\dfrac{1}{\sqrt{3}}\right)$

51. $\text{Arcsec}\,(-2)$

52. $\text{Arccot}\,(-1)$

53. $\text{Arccsc}\,\sqrt{2}$

54. $\text{Arcsin}\,(-1)$

55. $\text{Arctan}\,\sqrt{3}$

56. $\text{Arccos}\,0$

57. $\text{Arccos}\left(\dfrac{1}{\sqrt{2}}\right)$

58. $\text{Arcsin}\,1$

59. $\text{Arcsin}\left(-\dfrac{\sqrt{3}}{2}\right)$

60. $\text{Arctan}\,(-\sqrt{3})$

Find the value of each expression.

61. $\cos\,(\text{Arctan}\,\sqrt{3})$

62. $\tan\left[\text{Arcsin}\left(\dfrac{1}{\sqrt{2}}\right)\right]$

63. $\sin\left[\text{Arccos}\left(-\dfrac{1}{\sqrt{2}}\right)\right]$

64. $\sin\,[\text{Arctan}\,(-1)]$

65. $\tan\,[\text{Arccos}\,(-1)]$

66. $\sec\left[\text{Arccos}\left(-\dfrac{1}{2}\right)\right]$

67. $\sin\left[\text{Arcsin}\left(\dfrac{\sqrt{3}}{2}\right)\right]$

68. $\tan\,[\text{Arctan}\,(-\sqrt{3})]$

69. $\cos\left[\text{Arcsin}\left(\dfrac{3}{5}\right)\right]$

70. $\tan\left[\text{Arcsin}\left(\dfrac{12}{13}\right)\right]$

71. $\tan\,[\text{Arcsin}\,(-0.1560)]$

72. $\sin\,[\text{Arccot}\,(1.635)]$

Find an algebraic expression for each.

73. $\cos\,(\text{Arcsin}\,x)$

74. $\tan\,(\text{Arccos}\,x)$

75. $\sin\,(\text{Arcsec}\,x)$

76. $\cot\,(\text{Arcsec}\,x)$

77. $\sec\,(\text{Arccos}\,x)$

78. $\sin\,(\text{Arctan}\,x)$

79. $\tan\,(\text{Arctan}\,x)$

80. $\sin\,(\text{Arcsin}\,x)$

81. $\cos\,(\text{Arcsin}\,2x)$

82. $\tan\,(\text{Arccos}\,3x)$

83. $\sin\,(2\,\text{Arcsin}\,x)$

84. $\cos\,(2\,\text{Arctan}\,x)$

4.5 DERIVATIVES OF INVERSE TRIGONOMETRIC FUNCTIONS

To differentiate $y = \text{Arcsin}\,u$, first differentiate its inverse—the function $u = \sin y$ for $-\dfrac{\pi}{2} \le y \le \dfrac{\pi}{2}$. Then

$$\frac{du}{dx} = \frac{d}{dx}\,(\sin y) = \cos y\,\frac{dy}{dx}$$

Solving for $\dfrac{dy}{dx}$,

$$\frac{dy}{dx} = \frac{1}{\cos y} \cdot \frac{du}{dx}$$

Now express $\cos y$ in terms of $\sin y$ using the identity $\sin^2 y + \cos^2 y = 1$. That is,

$$\cos y = \sqrt{1 - \sin^2 y}$$

Note that $\cos y = +\sqrt{1 - \sin^2 y}$ because $-\dfrac{\pi}{2} \le \text{Arcsin } u \le \dfrac{\pi}{2}$ and $\cos y > 0$ in quadrants I and IV. So

$$\frac{dy}{dx} = \frac{1}{\sqrt{1 - \sin^2 y}} \frac{du}{dx}$$

$$= \frac{1}{\sqrt{1 - u^2}} \frac{du}{dx} \qquad (\text{since } u = \sin y)$$

EXAMPLE 1

Find the derivative of $y = \text{Arcsin } 2x$.
   Let $u = 2x$, then

$$\frac{dy}{dx} = \frac{1}{\sqrt{1 - (2x)^2}}(2) = \frac{2}{\sqrt{1 - 4x^2}}$$

To differentiate $y = \text{Arctan } u$, begin with its inverse:

$$u = \tan y$$

$$\frac{du}{dx} = \frac{d}{dx}(\tan y) = \sec^2 y \frac{dy}{dx}$$

Solving for $\dfrac{dy}{dx}$,

$$\frac{dy}{dx} = \frac{1}{\sec^2 y} \frac{du}{dx}$$

$$= \frac{1}{1 + \tan^2 y} \frac{du}{dx} \qquad (\sec^2 y = 1 + \tan^2 y)$$

$$= \frac{1}{1 + u^2} \frac{du}{dx} \qquad (u = \tan y)$$

EXAMPLE 2

Find the derivative of $y = \text{Arctan } 3x^2$.
   Let $u = 3x^2$, then

$$\frac{dy}{dx} = \frac{1}{1 + (3x^2)^2}(6x) = \frac{6x}{1 + 9x^4}$$

Formulas for the derivatives of the other trigonometric functions are found in a similar manner and are left as exercises. We now list the formulas for the derivatives of the six inverse trigonometric functions.

DERIVATIVES OF THE INVERSE TRIGONOMETRIC FUNCTIONS

$$\frac{d}{dx}(\text{Arcsin } u) = \frac{1}{\sqrt{1 - u^2}} \frac{du}{dx}$$

$$\frac{d}{dx}(\text{Arccos } u) = -\frac{1}{\sqrt{1 - u^2}} \frac{du}{dx}$$

$$\frac{d}{dx}(\text{Arctan } u) = \frac{1}{1 + u^2}\frac{du}{dx}$$

$$\frac{d}{dx}(\text{Arccot } u) = -\frac{1}{1 + u^2}\frac{du}{dx}$$

$$\frac{d}{dx}(\text{Arcsec } u) = \frac{1}{|u|\sqrt{u^2 - 1}}\frac{du}{dx}$$

$$\frac{d}{dx}(\text{Arccsc } u) = -\frac{1}{|u|\sqrt{u^2 - 1}}\frac{du}{dx}$$

EXAMPLE 3

Find the derivative of $y = \text{Arccos}^3\ 6x$.

Find the derivative of a power u^3, where $u = \text{Arccos } 6x$.

$$\frac{dy}{dx} = (3\ \text{Arccos}^2\ 6x)\left(-\frac{1}{\sqrt{1 - (6x)^2}}\right)(6)$$

$$= \frac{-18\ \text{Arccos}^2\ 6x}{\sqrt{1 - 36x^2}}$$

EXAMPLE 4

Find the derivative of $y = x^2\ \text{Arcsec}(1 - 3x)$.

$$\frac{dy}{dx} = x^2\left(\frac{1}{|1 - 3x|\ \sqrt{(1 - 3x)^2 - 1}}\right)(-3) + 2x\ \text{Arcsec}(1 - 3x)$$

$$= \frac{-3x^2}{|1 - 3x|\ \sqrt{9x^2 - 6x}} + 2x\ \text{Arcsec}(1 - 3x)$$

Exercises 4.5

Find the derivative of each.

1. $y = \text{Arcsin } 5x$ **2.** $y = \text{Arccos } 6x$ **3.** $y = \text{Arctan } 3x$

4. $y = \text{Arcsec } 4x$ **5.** $y = \text{Arccsc}(1 - x)$ **6.** $y = \text{Arccot } \sqrt{x}$

7. $y = 3\ \text{Arccos}(x - 1)$ **8.** $y = 4\ \text{Arcsin}(1/x)$ **9.** $y = 2\ \text{Arccot } 3x^2$

10. $y = 3\ \text{Arctan}(x^2 - 1)$ **11.** $y = 5\ \text{Arcsec } x^3$ **12.** $y = 6\ \text{Arccsc}(x/2)$

13. $y = \text{Arcsin}^3 x$ **14.** $y = \text{Arctan}^2 5x$ **15.** $y = 2\ \text{Arccos}^2 3x$

16. $y = 4\ \text{Arcsec}^3 2x$ **17.** $y = 3\ \text{Arctan}^4 \sqrt{x}$ **18.** $y = 6\ \text{Arcsin}^2(1 - x)$

19. $y = \text{Arcsin } x + \text{Arccos } x$ **20.** $y = x - \text{Arctan } x$ **21.** $y = \sqrt{1 - x^2} + \text{Arcsin } x$

22. $y = \sqrt{1 - x^2} + \text{Arccos } x$ **23.** $y = x\ \text{Arcsin } 3x$ **24.** $y = x^2\ \text{Arccos } x$

25. $y = x\ \text{Arctan } x$ **26.** $y = x\ \text{Arcsin } x^2$ **27.** $y = x\ \text{Arcsin } x + \sqrt{1 - x^2}$

28. $y = \dfrac{x}{\sqrt{1 - x^2}} - \text{Arcsin } x$ **29.** $y = \dfrac{x}{\text{Arcsin } x}$ **30.** $y = \dfrac{\text{Arctan } x}{x}$

Find the slope of the tangent line to each curve at the given value.

31. $y = x\ \text{Arcsin } x$ at $x = \frac{1}{2}$ **32.** $y = \text{Arctan } x$ at $x = 1$

33. $y = x\ \text{Arctan } x$ at $x = -1$ **34.** $y = \text{Arccos}^2 x$ at $x = \dfrac{\sqrt{3}}{2}$

35. Show that $\dfrac{d}{dx}\text{Arccos }u = -\dfrac{1}{\sqrt{1-u^2}}\dfrac{du}{dx}$.

36. Show that $\dfrac{d}{dx}\text{Arccot }u = -\dfrac{1}{1+u^2}\dfrac{du}{dx}$.

37. Show that $\dfrac{d}{dx}\text{Arcsec }u = \dfrac{1}{|u|\sqrt{(u^2-1)}}\dfrac{du}{dx}$.

38. Show that $\dfrac{d}{dx}\text{Arccsc }u = -\dfrac{1}{|u|\sqrt{(u^2-1)}}\dfrac{du}{dx}$.

4.6   EXPONENTIAL AND LOGARITHMIC FUNCTIONS

We have considered equations with a *constant* exponent in the form

$$y = x^n$$

These are called **power functions.** Two examples are $y = x^2$ and $y = x^{4/3}$.

> **EXPONENTIAL FUNCTION**
>
> Equations with a variable exponent in the form
>
> $$y = b^x$$
>
> where $b > 0$ and $b \neq 1$ are called **exponential functions.**

Two examples are $y = 2^x$ and $y = \left(\tfrac{3}{4}\right)^x$.

EXAMPLE 1

Graph $y = 2^x$ by plotting points (see Fig. 4.17).

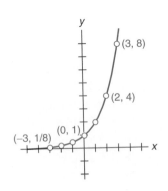

x	y	$y = 2^x$
0	1	$y = 2^0 = 1$
1	2	$y = 2^1 = 2$
2	4	$y = 2^2 = 4$
3	8	$y = 2^3 = 8$
-1	$\dfrac{1}{2}$	$y = 2^{-1} = \dfrac{1}{2}$
-2	$\dfrac{1}{4}$	$y = 2^{-2} = \dfrac{1}{4}$
-3	$\dfrac{1}{8}$	$y = 2^{-3} = \dfrac{1}{8}$

Figure 4.17

In general, for $b > 1$, $y = b^x$ is an **increasing** function. That is, as x increases, y increases.

EXAMPLE 2

Graph $y = \left(\tfrac{1}{2}\right)^x$ by plotting points (see Fig. 4.18).

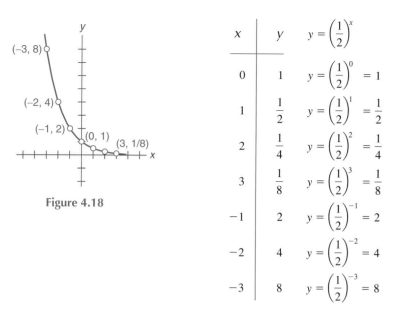

Figure 4.18

x	y	$y = \left(\dfrac{1}{2}\right)^x$
0	1	$y = \left(\dfrac{1}{2}\right)^0 = 1$
1	$\dfrac{1}{2}$	$y = \left(\dfrac{1}{2}\right)^1 = \dfrac{1}{2}$
2	$\dfrac{1}{4}$	$y = \left(\dfrac{1}{2}\right)^2 = \dfrac{1}{4}$
3	$\dfrac{1}{8}$	$y = \left(\dfrac{1}{2}\right)^3 = \dfrac{1}{8}$
-1	2	$y = \left(\dfrac{1}{2}\right)^{-1} = 2$
-2	4	$y = \left(\dfrac{1}{2}\right)^{-2} = 4$
-3	8	$y = \left(\dfrac{1}{2}\right)^{-3} = 8$

In general, for $0 < b < 1$, $y = b^x$ is a **decreasing** function. That is, as x increases, y decreases.

Some basic laws of exponents are given below for $a \neq 0$, $b \neq 0$.

LAWS OF EXPONENTS

1. $a^m \cdot a^n = a^{m+n}$

2. $\dfrac{a^m}{a^n} = a^{m-n}$

3. $(a^m)^n = a^{mn}$

4. $(ab)^n = a^n b^n$

5. $\left(\dfrac{a}{b}\right)^n = \dfrac{a^n}{b^n}$

6. $a^0 = 1$

When the values of x and y are interchanged in an equation, the resulting equation is called the **inverse** of the given equation. The inverse of the exponential equation $y = b^x$ is the exponential equation $x = b^y$. We define this inverse equation to be the **logarithmic** equation. The middle and right equations below show how to express this logarithmic equation in either exponential form or logarithmic form:

Exponential equation	Logarithmic equation in exponential form	Logarithmic equation in logarithmic form
$y = b^x$	$x = b^y$	$y = \log_b x$

That is, $x = b^y$ and $y = \log_b x$ are equivalent equations for $b > 0$ but $b \neq 1$.

The logarithm of a number is the *exponent* indicating the power to which the base must be raised to equal that number. The expression $\log_b x$ is read "the logarithm of x to the base b."

Remember: A logarithm is an exponent.

EXAMPLE 3

Write each equation in logarithmic form.

	Exponential form	Logarithmic form
(a)	$2^3 = 8$	$\log_2 8 = 3$
(b)	$5^2 = 25$	$\log_5 25 = 2$
(c)	$4^{-2} = \dfrac{1}{16}$	$\log_4\left(\frac{1}{16}\right) = -2$
(d)	$36^{1/2} = 6$	$\log_{36} 6 = \frac{1}{2}$
(e)	$p^q = r$	$\log_p r = q$

EXAMPLE 4

Write each equation in exponential form.

	Logarithmic form	Exponential form
(a)	$\log_7 49 = 2$	$7^2 = 49$
(b)	$\log_4 64 = 3$	$4^3 = 64$
(c)	$\log_{10} 0.01 = -2$	$10^{-2} = 0.01$
(d)	$\log_{27} 3 = \frac{1}{3}$	$27^{1/3} = 3$
(e)	$\log_m p = n$	$m^n = p$

EXAMPLE 5

Graph $y = \log_2 x$ by plotting points.

First, change the equation from logarithmic form to exponential form. That is, $y = \log_2 x$ is equivalent to $x = 2^y$. Then choose values for y and compute values for x (see Fig. 4.19).

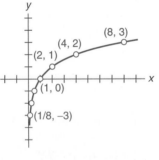

x	y	$x = 2^y$
1	0	$x = 2^0 = 1$
2	1	$x = 2^1 = 2$
4	2	$x = 2^2 = 4$
8	3	$x = 2^3 = 8$
$\dfrac{1}{2}$	-1	$x = 2^{-1} = \dfrac{1}{2}$
$\dfrac{1}{4}$	-2	$x = 2^{-2} = \dfrac{1}{4}$
$\dfrac{1}{8}$	-3	$x = 2^{-3} = \dfrac{1}{8}$

Figure 4.19

EXAMPLE 6

If $\log_3 81 = x$, find x.

The exponential form of $\log_3 81 = x$ is

$$3^x = 81$$

We know that

$$3^4 = 81$$

Therefore,

$$x = 4$$

EXAMPLE 7

If $\log_3 x = -2$, find x.

$$\log_3 x = -2 \quad \text{or} \quad 3^{-2} = x$$

Therefore,

$$x = \frac{1}{9}$$

EXAMPLE 8

If $\log_x 32 = \dfrac{5}{3}$, find x.

$$\log_x 32 = \frac{5}{3} \quad \text{or} \quad x^{5/3} = 32$$

$$x^{1/3} = 2 \qquad \text{(Take the fifth root of each side.)}$$
$$x = 8 \qquad \text{(Cube each side.)}$$

Or begin with

$$x^{5/3} = 32$$
$$(x^{5/3})^{3/5} = 32^{3/5} \qquad \text{(Raise each side to the } \tfrac{3}{5} \text{ power.)}$$
$$x = 8$$

The following are three basic logarithmic properties.

    1. *Multiplication:* If M and N are positive real numbers,

$$\log_a(M \cdot N) = \log_a M + \log_a N$$
where $a > 0$ and $a \neq 1$.

That is, the logarithm of a product equals the sum of the logarithms of its factors.

    2. *Division:* If M and N are positive real numbers,

$$\log_a\!\left(\frac{M}{N}\right) = \log_a M - \log_a N$$
where $a > 0$ and $a \neq 1$.

That is, the logarithm of a quotient equals the difference of the logarithms of its factors.

    3. *Powers:* If M is a positive real number and n is any real number,

$$\log_a M^n = n \log_a M$$
where $a > 0$ and $a \neq 1$.

That is, the logarithm of a power of a number equals the product of the exponent times the logarithm of the number.

    There are three special cases of the power property that are helpful.

(a) *Roots:* If M is any positive real number and n is any positive integer,

$$\log_a \sqrt[n]{M} = \frac{1}{n} \cdot \log_a M$$

Note that this is a special case of property 3, since $\sqrt[n]{M} = M^{1/n}$. That is, the logarithm of the root of a number equals the logarithm of the number divided by the index of the root.

(b) For $n = 0$,

$$\log_a M^0 = \log_a 1 \qquad (M^0 = 1)$$
$$\log_a M^0 = 0 \cdot \log_a M \qquad \text{(Property 3)}$$
$$= 0$$

Therefore,

$$\boxed{\log_a 1 = 0}$$

That is, the logarithm of one to any base is zero.

(c) For $n = -1$,

$$\log_a M^{-1} = \log_a \frac{1}{M} \qquad \left(M^{-1} = \frac{1}{M}\right)$$
$$\log_a M^{-1} = (-1)\log_a M \qquad \text{(Property 3)}$$

Therefore,

$$\boxed{\log_a \frac{1}{M} = -\log_a M}$$

That is, the logarithm of the reciprocal of a number is the negative of the logarithm of the number.

EXAMPLE 9

Write $\log_4 2x^5 y^2$ as a sum and multiple of logarithms of a single variable.

$$\log_4 2x^5 y^2 = \log_4 2 + \log_4 x^5 + \log_4 y^2 \qquad \text{(Property 1)}$$
$$= \log_4 2 + 5\log_4 x + 2\log_4 y \qquad \text{(Property 3)}$$

EXAMPLE 10

Write $\log_3 \dfrac{\sqrt{x(x-2)}}{(x+3)^2}$ as a sum, difference, or multiple of the logarithms of x, $x - 2$, and $x + 3$.

$$\log_3 \frac{\sqrt{x(x-2)}}{(x+3)^2} = \log_3 \frac{[x(x-2)]^{1/2}}{(x+3)^2}$$
$$= \log_3 [x(x-2)]^{1/2} - \log_3 (x+3)^2 \qquad \text{(Property 2)}$$
$$= \frac{1}{2}\log_3 [x(x-2)] - 2\log_3 (x+3) \qquad \text{(Property 3)}$$
$$= \frac{1}{2}\left[\log_3 x + \log_3 (x-2)\right] - 2\log_3 (x+3) \qquad \text{(Property 1)}$$
$$= \frac{1}{2}\log_3 x + \frac{1}{2}\log_3 (x-2) - 2\log_3 (x+3)$$

EXAMPLE 11

Write $3 \log_2 x + 4 \log_2 y - 2 \log_2 z$ as a single logarithmic expression.

$$\begin{aligned} 3 \log_2 x + 4 \log_2 y - 2 \log_2 z &= \log_2 x^3 + \log_2 y^4 - \log_2 z^2 \quad &\text{(Property 3)}\\ &= \log_2 (x^3 y^4) - \log_2 z^2 \quad &\text{(Property 1)}\\ &= \log_2 \frac{x^3 y^4}{z^2} \quad &\text{(Property 2)} \end{aligned}$$

EXAMPLE 12

Write $3 \log_{10} (x - 1) - \dfrac{1}{3} \log_{10} x - \log_{10} (2x + 3)$ as a single logarithmic expression.

$$3 \log_{10} (x - 1) - \frac{1}{3} \log_{10} x - \log_{10} (2x + 3)$$

$$\begin{aligned} &= \log_{10} (x - 1)^3 - \log_{10} x^{1/3} - \log_{10} (2x + 3) \quad &\text{(Property 3)}\\ &= \log_{10} (x - 1)^3 - [\log_{10} x^{1/3} + \log_{10} (2x + 3)] \\ &= \log_{10} (x - 1)^3 - [\log_{10} (x^{1/3})(2x + 3)] \quad &\text{(Property 1)}\\ &= \log_{10} \frac{(x - 1)^3}{x^{1/3}(2x + 3)} \quad &\text{(Property 2)}\\ &= \log_{10} \frac{(x - 1)^3}{\sqrt[3]{x}(2x + 3)} \end{aligned}$$

There are two other logarithmic properties that are useful in simplifying expressions.

$$\log_a a^x = x \quad \text{and} \quad a^{\log_a x} = x$$

To show the first one, we begin with the identity

$$(a^x) = a^x$$

Then, writing this exponential equation in logarithmic form, we have

$$\log_a (a^x) = x$$

To show the second one, we begin with the identity

$$\log_a x = (\log_a x)$$

Then, writing this logarithmic equation in exponential form, we have

$$a^{(\log_a x)} = x$$

EXAMPLE 13

Find the value of $\log_2 16$.

$$\begin{aligned} \log_2 16 &= \log_2 2^4 \quad &\text{(Note that this simplification is}\\ &= 4 \quad &\text{possible because 16 is a power of 2.)} \end{aligned}$$

EXAMPLE 14

Find the value of $\log_{10} 0.01$.

$$\begin{aligned} \log_{10} 0.01 &= \log_{10} 10^{-2} \quad &\text{(Note that } 0.01 = \tfrac{1}{100} = 10^{-2}.)\\ &= -2 \end{aligned}$$

EXAMPLE 15

Find the value of $5^{\log_5 8}$.

$$5^{\log_5 8} = 8$$

One other useful law is for change of base

$$\log_b x = \frac{\log_a x}{\log_a b}$$

Although all the laws of exponents and logarithms hold for any base b ($b > 0$ and $b \neq 1$), there are two standard bases for logarithms:

1. *Common logarithms:* Base 10

   Usual notation: $\log_{10} x = \log x$

2. *Natural logarithms:* Base e

   Usual notation: $\log_e x = \ln x$

As we shall see, the most convenient base for use in calculus is base e, which is defined as

$$e = \lim_{x \to 0} (1 + x)^{1/x} \approx 2.7182818$$

Exercises 4.6

Graph each equation.

1. $y = 4^x$ **2.** $y = 3^x$ **3.** $y = \left(\frac{1}{3}\right)^x$

4. $y = \left(\frac{1}{4}\right)^x$ **5.** $y = 4^{-x}$ **6.** $y = \left(\frac{2}{3}\right)^x$

7. $y = \left(\frac{4}{3}\right)^{-x}$ **8.** $y = 4^{2x}$

Write each equation in logarithmic form.

9. $3^2 = 9$ **10.** $7^2 = 49$ **11.** $5^3 = 125$

12. $10^3 = 1000$ **13.** $9^{1/2} = 3$ **14.** $4^0 = 1$

15. $10^{-5} = 0.00001$ **16.** $d^e = f$

Write each equation in exponential form.

17. $\log_5 25 = 2$ **18.** $\log_8 64 = 2$ **19.** $\log_{25} 5 = \frac{1}{2}$

20. $\log_{27} 3 = \frac{1}{3}$ **21.** $\log_2 \left(\frac{1}{4}\right) = -2$ **22.** $\log_2 \left(\frac{1}{8}\right) = -3$

23. $\log_{10} 0.01 = -2$ **24.** $\log_g h = k$

Graph each equation.

25. $y = \log_4 x$ **26.** $y = \log_3 x$ **27.** $y = \log_{10} x$

28. $y = \log_5 x$ **29.** $y = \log_{1/4} x$ **30.** $y = \log_{1/3} x$

Solve for x.

31. $\log_4 x = 3$ **32.** $\log_2 x = -1$ **33.** $\log_9 3 = x$

34. $\log_6 36 = x$ **35.** $\log_2 8 = x$ **36.** $\log_3 27 = x$

37. $\log_{25} 5 = x$ **38.** $\log_{27} 3 = x$ **39.** $\log_x 25 = 2$

40. $\log_x \left(\frac{1}{27}\right) = -3$ **41.** $\log_{1/2} \left(\frac{1}{8}\right) = x$ **42.** $\log_x 3 = \frac{1}{2}$

43. $\log_{12} x = 2$ **44.** $\log_8 \left(\frac{1}{64}\right) = x$ **45.** $\log_x 9 = \frac{2}{3}$

46. $\log_x 64 = \frac{3}{2}$ **47.** $\log_x \left(\frac{1}{8}\right) = -\frac{3}{2}$ **48.** $\log_x \left(\frac{1}{27}\right) = -\frac{3}{4}$

Write each expression as a sum, difference, or multiple of single logarithms.

49. $\log_2 5x^3 y$ **50.** $\log_3 \dfrac{8x^2 y^3}{z^4}$ **51.** $\log_b \dfrac{y^3 \sqrt{x}}{z^2}$

52. $\log_b \dfrac{7xy}{\sqrt[3]{z}}$ **53.** $\log_b \sqrt[3]{\dfrac{x^2}{y}}$ **54.** $\log_5 \sqrt[4]{xy^2 z}$

55. $\log_2 \dfrac{1}{x}\sqrt{\dfrac{y}{z}}$ **56.** $\log_b \dfrac{1}{z^2}\sqrt[3]{\dfrac{x^2}{y}}$ **57.** $\log_b \dfrac{z^3 \sqrt{x}}{\sqrt[3]{y}}$

58. $\log_b \dfrac{\sqrt{y\sqrt{x}}}{z^2}$ **59.** $\log_b \dfrac{x^2(x+1)}{\sqrt{x+2}}$ **60.** $\log_b \dfrac{\sqrt{x}(x+4)}{x^2}$

Write each as a single logarithmic expression.

61. $\log_b x + 2 \log_b y$ **62.** $2 \log_b z - 3 \log_b x$

63. $\log_b x + 2 \log_b y - 3 \log_b z$ **64.** $3 \log_7 x - 4 \log_7 y - 5 \log_7 z$

65. $\log_3 x + \frac{1}{3} \log_3 y - \frac{1}{2} \log_3 z$ **66.** $\frac{1}{2} \log_2 x - \frac{1}{3} \log_2 y - \log_2 z$

67. $2 \log_{10} x - \frac{1}{2} \log_{10} (x - 3) - \log_{10} (x + 1)$

68. $\log_3 (x + 1) + \frac{1}{2} \log_3 (x + 2) - 3 \log_3 (x - 1)$

69. $5 \log_b x + \frac{1}{3} \log_b (x - 1) - \log_b (x + 2)$

70. $\log_b (x + 1) + \frac{1}{3} \log_b (x - 7) - 2 \log_b x$

71. $\log_{10} x + 2 \log_{10} (x - 1) - \frac{1}{3}[\log_{10} (x + 2) + \log_{10} (x - 5)]$

72. $\frac{1}{2} \log_b (x + 1) - 3[\log_b x + \log_b (x - 1) + \log_b (2x - 1)]$

Find the value of each expression.

73. $\log_b b^3$ **74.** $\log_2 2^5$ **75.** $\log_3 9$

76. $\log_2 16$ **77.** $\log_5 125$ **78.** $\log_4 64$

79. $\log_2 \frac{1}{4}$ **80.** $\log_3 \frac{1}{27}$ **81.** $\log_{10} 0.001$

82 $\log_{10} 0.1$ **83.** $\log_3 1$ **84.** $\log_{10} 1$

85. $6^{\log_6 5}$ **86.** $3^{\log_3 9}$ **87.** $25^{\log_5 6}$

88. $27^{\log_3 2}$ **89.** $4^{\log_2 (1/5)}$ **90.** $8^{\log_2 (1/3)}$

4.7   DERIVATIVES OF LOGARITHMIC FUNCTIONS

We now find the derivative of a logarithmic function in the form $y = \log_b u$.

$$\frac{dy}{du} = \lim_{\Delta u \to 0} \frac{\Delta y}{\Delta u}$$

where

$$\frac{\Delta y}{\Delta u} = \frac{\log_b (u + \Delta u) - \log_b u}{\Delta u}$$

Recall that the difference of two logarithms is the logarithm of their quotient, so

$$\frac{\Delta y}{\Delta u} = \frac{\log_b\left(\dfrac{u + \Delta u}{u}\right)}{\Delta u}$$

$$= \frac{1}{u} \cdot \frac{u}{\Delta u} \log_b\left(\frac{u + \Delta u}{u}\right) \qquad \text{(Multiply numerator and denominator by } u.)$$

Also recall $n \log_b a = \log_b a^n$. We then have

$$\frac{\Delta y}{\Delta u} = \frac{1}{u} \log_b\left(\frac{u + \Delta u}{u}\right)^{u/\Delta u}$$

$$= \frac{1}{u} \log_b\left(1 + \frac{\Delta u}{u}\right)^{u/\Delta u}$$

and then

$$\frac{dy}{du} = \lim_{\Delta u \to 0} \frac{1}{u} \cdot \lim_{\Delta u \to 0} \log_b\left(1 + \frac{\Delta u}{u}\right)^{u/\Delta u}$$

The limit of the left-hand factor is

$$\lim_{\Delta u \to 0} \frac{1}{u} = \frac{1}{u} \qquad \text{(As } \Delta u \to 0, u \text{ itself is not affected.)}$$

Next, to determine the limit of the right-hand factor, one can show that the function $(1 + h)^{1/h}$ approaches a limiting number as h approaches 0. We denote this number by e, that is,

$$\lim_{h \to 0} (1 + h)^{1/h} = e$$

This is an irrational number whose decimal value can only be approximated. An approximation of e to seven decimal places is 2.7182818.

Note that if we let $h = \dfrac{\Delta u}{u}$, then h approaches 0 as Δu approaches 0. Then

$$\frac{dy}{du} = \frac{1}{u} \cdot \lim_{\Delta u \to 0} \log_b\left(1 + \frac{\Delta u}{u}\right)^{(1/\Delta u)/u} \qquad \left(\frac{u}{\Delta u} = \frac{1}{\dfrac{\Delta u}{u}}\right)$$

$$= \frac{1}{u} \log_b e$$

Using the chain rule, we also have

$$\frac{dy}{dx} = \frac{1}{u} \log_b e \, \frac{du}{dx}$$

Thus,

$$\frac{d}{du}(\log_b u) = \frac{1}{u} \log_b e$$

$$\frac{d}{dx}(\log_b u) = \frac{1}{u} \log_b e \, \frac{du}{dx}$$

Since $\log_e e = 1$ and since we denote $\log_e u$ by ln u (called the natural logarithm of u), we have

$$\frac{d}{du}(\ln u) = \frac{1}{u}$$

$$\frac{d}{dx}(\ln u) = \frac{1}{u}\frac{du}{dx}$$

Note: u must always be positive since $\log_b u$ or $\ln u$ is defined only for positive values of u.

EXAMPLE 1

Find the derivative of $y = \ln 5x$.

$$\frac{dy}{dx} = \frac{1}{5x}\frac{d}{dx}(5x) = \frac{5}{5x} = \frac{1}{x} \qquad (u = 5x)$$

Alternate method: Use the property of logarithms from Section 4.6: the logarithm of a product equals the sum of the logarithms of its factors.

$$y \quad = \ln 5x = \ln 5 + \ln x$$

$$\frac{dy}{dx} = \qquad 0 + \frac{1}{x} = \frac{1}{x}$$

EXAMPLE 2

Find the derivative of $y = \log (3x + 1)$.

$$\frac{dy}{dx} = \left(\frac{1}{3x + 1}\right)(\log e)\left[\frac{d}{dx}(3x + 1)\right] \qquad (u = 3x + 1)$$

$$= \frac{3 \log e}{3x + 1} \quad \text{or} \quad \frac{3(0.4343)}{3x + 1} = \frac{1.3029}{3x + 1}$$

EXAMPLE 3

Find the derivative of $y = \ln (\sin 3x)$.

$$\frac{dy}{dx} = \frac{1}{\sin 3x}\frac{d}{dx}(\sin 3x) \qquad (u = \sin 3x)$$

$$= \frac{3 \cos 3x}{\sin 3x}$$

$$= 3 \cot 3x$$

You will need to make simplifications using the properties of logarithms.

EXAMPLE 4

Find the derivative of $y = \ln\left(\dfrac{x^3}{2x - 5}\right)$

$$y = \ln\left(\frac{x^3}{2x - 5}\right) = \ln x^3 - \ln (2x - 5)$$

$$= 3 \ln x - \ln (2x - 5)$$

Then

$$\frac{dy}{dx} = 3\left(\frac{1}{x}\right) - \frac{1}{2x - 5}\frac{d}{dx}(2x - 5)$$

$$= \frac{3}{x} - \frac{2}{2x - 5}$$

$$= \frac{4x - 15}{x(2x - 5)}$$

EXAMPLE 5

Find the derivative of $y = \ln(x\sqrt{4 + x})$.

Rewriting we have

$$y = \ln x + \tfrac{1}{2}\ln(4 + x)$$

$$\frac{dy}{dx} = \frac{1}{x} + \frac{1}{2}\left(\frac{1}{4 + x}\right)(1)$$

$$= \frac{8 + 3x}{2x(4 + x)}$$

For more complicated algebraic expressions, consider the following method, called *logarithmic differentiation*.

EXAMPLE 6

Find the derivative of $y = \sqrt[3]{\dfrac{x^2(3x + 4)}{1 - 2x^2}}$.

First, take the natural logarithm of both sides.

$$\ln y = \ln\left(\frac{x^2(3x + 4)}{1 - 2x^2}\right)^{1/3}$$

$$\ln y = \frac{1}{3}\ln\left(\frac{x^2(3x + 4)}{1 - 2x^2}\right)$$

$$\ln y = \frac{1}{3}[2\ln x + \ln(3x + 4) - \ln(1 - 2x^2)]$$

Next, find the derivative of both sides.

$$\frac{1}{y}\frac{dy}{dx} = \frac{1}{3}\left[\frac{2}{x} + \frac{3}{3x + 4} - \frac{-4x}{1 - 2x^2}\right]$$

Solve for $\dfrac{dy}{dx}$:

$$\frac{dy}{dx} = \frac{y}{3}\left[\frac{2}{x} + \frac{3}{3x + 4} + \frac{4x}{1 - 2x^2}\right]$$

$$= \frac{1}{3}\sqrt[3]{\frac{x^2(3x + 4)}{1 - 2x^2}}\left[\frac{2}{x} + \frac{3}{3x + 4} + \frac{4x}{1 - 2x^2}\right]$$

Logarithmic differentiation is especially helpful for functions having both a variable base and a variable exponent.

EXAMPLE 7

Find the derivative of $y = x^{4x}$.

First, take the natural logarithm of both sides.

$$\ln y = \ln x^{4x}$$

$$\ln y = 4x \ln x$$

Then find the derivative of both sides.

$$\frac{1}{y}\frac{dy}{dx} = 4x\frac{1}{x} + 4\ln x$$

$$\frac{dy}{dx} = 4y(1 + \ln x)$$

$$= 4x^{4x}(1 + \ln x)$$

Exercises 4.7

Find the derivative of each function.

1. $y = \log(4x - 3)$ **2.** $y = \log(x^2 - 1)$ **3.** $y = \log_2 3x$

4. $y = \log_7(5x + 2)$ **5.** $y = \ln(2x^3 - 3)$ **6.** $y = \ln(x^2 - 4)$

7. $y = \ln(\tan 3x)$ **8.** $y = \ln(\sec x^2)$ **9.** $y = \ln(x \sin x)$

10. $y = \ln(x^3 \cos 2x)$ **11.** $y = \ln\sqrt{3x - 2}$ **12.** $y = \ln(x^2 - 1)^{1/3}$

13. $y = \ln\dfrac{x^3}{x^2 + 1}$ **14.** $y = \ln\dfrac{5x}{1 - x^2}$ **15.** $y = \tan(\ln x)$

16. $y = (\ln x)(\sin x)$ **17.** $y = \ln(\ln x)$ **18.** $y = x^2 + \ln(3x - 2)$

19. $y = \text{Arctan}(\ln x^2)$ **20.** $y = \text{Arcsin}(\ln x)$ **21.** $y = \ln(\text{Arccos}^2 x)$

22. $y = \ln(\text{Arctan}^3 x)$

Find $\dfrac{dy}{dx}$ using logarithmic differentiation.

23. $y = (3x + 2)(6x - 1)^2(x - 4)$ **24.** $y = \sqrt{x(2x + 1)(1 - 5x)}$

25. $y = \dfrac{(x + 1)(2x + 1)}{(3x - 4)(1 - 8x)}$ **26.** $y = \dfrac{x(2x - 1)^{3/2}}{\sqrt[3]{x + 1}}$

27. $y = x^x$ **28.** $y = x^{x+1}$ **29.** $y = x^{2/x}$ **30.** $y = (1 + x)^{1/x}$

31. $y = (\sin x)^x$ **32.** $y = (\tan x)^x$ **33.** $y = (1 + x)^{x^2}$ **34.** $y = x^{\sin x}$

Find the equation of the tangent line at the given value.

35. $y = \ln x$ at $x = 1$ **36.** $y = \ln x^2$ at $x = 1$

37. $y = \ln(\sin x)$ at $x = \pi/6$ **38.** $y = \tan(\ln x)$ at $x = 1$

4.8 DERIVATIVES OF EXPONENTIAL FUNCTIONS

An *exponential function* is a function in the form $y = b^u$ where b is a positive constant, $b \neq 1$, and u is a variable. If $y = b^u$, then $u = \log_b y$ in logarithmic form. Then from the preceding section,

$$\frac{du}{dx} = \frac{1}{y} \cdot \log_b e \, \frac{dy}{dx}$$

Solving for $\dfrac{dy}{dx}$, we have

$$\frac{dy}{dx} = \frac{y}{\log_b e} \frac{du}{dx}$$

$$= \frac{b^u}{\log_b e} \frac{du}{dx} \qquad (y = b^u)$$

That is,

$$\frac{d}{dx}(b^u) = \frac{b^u}{\log_b e} \frac{du}{dx}$$

When the base $b = e$, the previous expression simplifies to

$$\frac{d}{dx}(e^u) = e^u \frac{du}{dx} \qquad (\log_e e = 1)$$

EXAMPLE 1

Find the derivative of $y = e^{3x}$.

$$\frac{dy}{dx} = e^{3x} \frac{d}{dx}(3x) = 3e^{3x} \qquad (u = 3x)$$

EXAMPLE 2

Find the derivative of $y = 10^{2x}$.

$$\frac{dy}{dx} = \frac{10^{2x}}{\log e} \cdot \frac{d}{dx}(2x) \qquad (u = 2x)$$

$$= \frac{(2)10^{2x}}{\log e}$$

EXAMPLE 3

Find the derivative of $y = e^{x^2+3}$.

$$\frac{dy}{dx} = e^{x^2+3}(2x) \qquad (u = x^2 + 3)$$

$$= 2xe^{x^2+3}$$

EXAMPLE 4

Find the derivative of $y = e^{\tan x}$.

$$\frac{dy}{dx} = e^{\tan x}(\sec^2 x) \qquad (u = \tan x)$$

EXAMPLE 5

Find the derivative of $y = x^2 e^{-5x}$.
First apply the product rule for differentiation.

$$\frac{dy}{dx} = x^2 \frac{d}{dx}(e^{-5x}) + e^{-5x} \frac{d}{dx}(x^2)$$

$$= x^2 e^{-5x}(-5) + e^{-5x}(2x)$$

$$= xe^{-5x}(2 - 5x)$$

EXAMPLE 6

Find the derivative of $y = \text{Arcsin } e^{3x}$.
Let $u = e^{3x}$, then

$$\frac{dy}{dx} = \frac{1}{\sqrt{1 - (e^{3x})^2}}(3e^{3x}) = \frac{3e^{3x}}{\sqrt{1 - e^{6x}}}$$

Using a calculator,

F3 1 green diamond SIN^{-1} green diamond e^x 3x)),x) ENTER

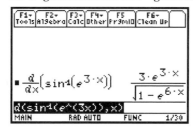

Exercises 4.8

Find the derivative of each function.

1. $y = e^{5x}$ **2.** $y = 2e^{3x+1}$ **3.** $y = 4e^{x^3}$

4. $y = e^{x^2+3x}$ **5.** $y = 10^{3x}$ **6.** $y = 4^{x^2}$

7. $y = \dfrac{1}{e^{6x}}$ **8.** $y = e^{-x^3}$ **9.** $y = e^{\sqrt{x}}$

10. $y = e^{\sqrt{3x-2}}$ **11.** $y = e^{\sin x}$ **12.** $y = 5e^{\cos x^2}$

13. $y = 6xe^{x^2-1}$ **14.** $y = xe^{x^3}$ **15.** $y = (\cos x)e^{3x^2}$

16. $y = 3e^{\tan x^2}$ **17.** $y = \ln(\cos e^{5x})$ **18.** $y = \tan e^{4x}$

19. $y = e^x - e^{-x}$ **20.** $y = \dfrac{e^x}{x}$ **21.** $y = \dfrac{2x^2}{3e^x - x}$

22. $y = \dfrac{e^{x^2}}{x-2}$ **23.** $y = 2\,\text{Arctan}\, e^{3x}$ **24.** $y = \text{Arcsin}\, e^{4x}$

25. $y = \text{Arccos}^3\, e^{-2x}$ **26.** $y = \text{Arctan}\, e^{x^2}$ **27.** $y = xe^x - e^x$

28. $y = (e^x + e^{-x})^3$ **29.** $y = \ln e^{x^2}$ **30.** $y = \ln(x - e^x)$

4.9 APPLICATIONS

The previous calculus techniques for curve sketching, finding maximums and minimums, solving problems of motion, and so on, apply also to the transcendental functions. Some of these applications are found in this section.

EXAMPLE 1

Sketch the curve $y = xe^x$.

The x- and y-intercepts are both $(0, 0)$. To find where the curve is above and below the x-axis:

Signs of factors in each interval:			
x	$-$		$+$
e^x	$+$		$+$

$$\xleftarrow{\hspace{3cm}} \underset{0}{\mid} \xrightarrow{\hspace{3cm}} x$$

Sign of y in each interval:		
	$-$	$+$
Function is:	(below)	(above)

Note: $e^x > 0$. This factor, being always positive, does not affect the sign in any interval. So we will not list it after this.

$$f'(x) = xe^x + e^x = e^x(x + 1)$$

To find where the curve is increasing and decreasing:

Sign of factor
in each interval: $(x + 1)$ $-$ $+$

$\longleftarrow \quad\quad\quad\quad\quad | \quad\quad\quad\quad\quad\quad \longrightarrow x$

Sign of $f'(x)$: $-$ -1 $+$
Function is: (decreasing)   (increasing)

Since $f(x)$ changes from decreasing to increasing at $x = -1$, $(-1, -1/e)$ is a relative minimum.

$$f''(x) = xe^x + 2e^x = e^x(x + 2)$$

To find where the curve is concave upward and concave downward:

Sign of $f''(x)$ in
each interval: $(x + 2)$ $-$ $+$

$\longleftarrow \quad\quad\quad\quad\quad | \quad\quad\quad\quad\quad\quad \longrightarrow x$

-2

Function is: concave   concave
                            down   up

So $(-2, -2/e^2)$ is a point of inflection.
With this information we now sketch the curve in Fig. 4.20.

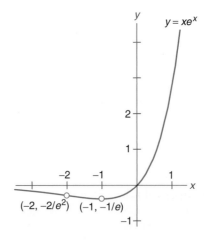

Figure 4.20

EXAMPLE 2

The insulation resistance of a shielded cable is given by

$$R = \frac{\rho}{2\pi} \ln\left(\frac{r_2}{r_1}\right) \Omega$$

where ρ is the resistivity of the insulation material, r_1 is the inner radius (in mm) of the insulation, and r_2 is the outer radius. Find the rate of change of the insulation resistance R with respect to the outer radius r_2.

$$\frac{dR}{dr_2} = \frac{d}{dr_2}\left[\left(\frac{\rho}{2\pi}\right)\ln\left(\frac{r_2}{r_1}\right)\right] \quad (r_1 \text{ is a constant.})$$

$$= \left(\frac{\rho}{2\pi}\right)\frac{d}{dr_2}\left[\ln\left(\frac{r_2}{r_1}\right)\right]$$

$$= \left(\frac{\rho}{2\pi}\right)\left(\frac{1}{r_2/r_1}\right)\frac{d}{dr_2}\left(\frac{r_2}{r_1}\right)$$

$$= \left(\frac{\rho}{2\pi}\right)\left(\frac{r_1}{r_2}\right)\left(\frac{1}{r_1}\right)$$

$$= \frac{\rho}{2\pi r_2}\ \Omega/\text{mm}$$

EXAMPLE 3

The current in an electrical circuit is given by $i = 10(1 - e^{-20t})$. Find the maximum value of the current.

$$\frac{di}{dt} = \frac{d}{dt}[10(1 - e^{-20t})]$$

$$= 10\frac{d}{dt}(1 - e^{-20t})$$

$$= 10(0 + 20e^{-20t})$$

$$= 200e^{-20t}$$

Since $\frac{di}{dt}$ is never zero and always positive, there is no maximum value. The current increases with time.

EXAMPLE 4

Find the equation of the line tangent to the curve $y = \ln x^2$ at the point $(1, 0)$.

$$\frac{dy}{dx} = \frac{d}{dx}(\ln x^2) = \frac{1}{x^2}(2x) = \frac{2}{x}$$

Alternate method: Use the property of logarithms from Section 4.6: the logarithm of a power of a number equals the product of the exponent and the logarithm of the number.

$$y = \ln x^2 = 2\ln x$$

$$\frac{dy}{dx} = 2 \cdot \frac{1}{x} = \frac{2}{x}$$

The slope of the tangent line is

$$m = \frac{dy}{dx}\bigg|_{x=1} = \frac{2}{1} = 2$$

Using the point-slope formula, we have

$$y - y_1 = m(x - x_1)$$

$$y - 0 = 2(x - 1)$$

$$y = 2x - 2$$

Exercises 4.9

Sketch each curve. Find any maximum, minimum, and inflection points.

1. $y = \sin x + \cos x$

2. $y = x \ln x, x > 0$

3. $y = \dfrac{\ln x}{x}, x > 0$

4. $y = \ln(\sin x), 0 < x < \pi$

5. $y = e^x + e^{-x}$ **6.** $y = \ln x - x, x > 0$ **7.** $y = x^2 e^{-x}$

8. $y = \ln\left(\dfrac{1}{x^2 - 1}\right)$ **9.** $y = \dfrac{1}{1 - e^x}$ **10.** $y = \dfrac{x}{\ln x}, x > 0$

Find any relative maximums or minimums of each function.

11. $y = xe^{-2x}$ **12.** $y = x^2 e^{2x}$ **13.** $y = \dfrac{x^2}{e^{2x}}$

14. $y = \dfrac{e^x}{x}$ **15.** $y = \dfrac{\ln x}{x^2}$ **16.** $y = \dfrac{x}{\ln^2 x}$

17. Find the equation of the tangent line to $y = \sin 2x$ at the point $(\pi/2, 0)$.

18. Find the equation of the tangent line to $y = xe^x$ at the point $(1, \ e)$.

19. Find the equation of the tangent line to $y = x^2 + x \ln x$ at $x = 1$.

20. Find the equation of the tangent line to $y = e^{\cos x}$ at $x = \pi/2$.

Find $\dfrac{d^4 y}{dx^4}$ for each function:

21. $y = e^x \sin x$ **22.** $y = e^{-x} \cos x$

23. The current in a circuit is given by $i = 50 \cos 2t$. Find the equation of the voltage V_L across a 3-henry (H) inductor $\left(V_L = L \dfrac{di}{dt}, \text{ where } L \text{ is the inductance}\right)$.

24. The current in a circuit is given by $i = 40(1 - 2e^{-20t})$. Find the equation of the voltage V_L across a 5-H inductor.

25. The voltage in a circuit is given by $V = 25e^{0.4t}$. Find $\dfrac{dV}{dt}$ at $t = 3$.

26. The current in a circuit is given by $i = 4t^2 e^{8/t}$. Find the value of t when $\dfrac{di}{dt} = 0$.

27. The apparent power p_a of a circuit is given by $p_a = p \sec \theta$ where p is the power and θ is the impedance phase angle. Find $\dfrac{dp_a}{dt}$ at $t = 1$ s if $\dfrac{d\theta}{dt} = 0.1$ rad/s and $\theta = \dfrac{\pi}{4}$. The power of the circuit, p, is 15 watts (W).

28. If work in a circuit is given by $W = 25 \sin^2 t$, find the equation for the power $p \left(p = \dfrac{dW}{dt}\right)$.

29. The charge q at any time t on a capacitor is given by $q = e^{-0.02t}(0.05 \cos 2t)$. Find the current in the circuit $\left(i = \dfrac{dq}{dt}\right)$.

30. The voltage across a capacitor at any time t is given by $V = e^{-0.001t}$. If the capacitance C is 2×10^{-5} farad (F), find the current in the circuit $\left(i = C \dfrac{dV}{dt}\right)$.

31. A particle moves along a straight line according to $s = 2e^{3t} + 5e^{-3t}$. Find expressions for its velocity and acceleration.

32. A particle moves along a straight line according to $s = \ln (t^2 + 1)$. Find expressions for its velocity and acceleration.

33. A particle moves along a straight line according to $s = e^{-t/2} \tan \dfrac{\pi t}{2}$. Find its velocity and acceleration at $t = 4$.

34. An object moves along a straight line according to $s = \sin 2t + e^t$. Find its velocity and acceleration at $t = 2$.

35. The charge of a capacitor at time t is given by $q = e^{-0.2t}(0.04 \cos 2t - 0.5 \sin 3t)$. Find the current $\left(i = \dfrac{dq}{dt} \right)$ at $t = \pi/2$.

CHAPTER SUMMARY

1. Trigonometric functions (see Fig. 4.21).

$$\sin \theta = \frac{y}{r} \qquad \cot \theta = \frac{x}{y}$$

$$\cos \theta = \frac{x}{r} \qquad \sec \theta = \frac{r}{x}$$

$$\tan \theta = \frac{y}{x} \qquad \csc \theta = \frac{r}{y}$$

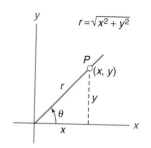

Figure 4.21

2. Graphing the six trigonometric functions

$$\text{Period} = \frac{2\pi}{b} \left\{ \begin{array}{l} y = a \sin(bx + c) \\ y = a \cos(bx + c) \\ y = a \sec(bx + c) \\ y = a \csc(bx + c) \end{array} \right\} \text{amplitude} = |a|$$

$$\text{Period} = \frac{\pi}{b} \left\{ \begin{array}{l} y = a \tan(bx + c) \\ y = a \cot(bx + c) \end{array} \right.$$

Phase shift $= \dfrac{c}{b}$

to the *left* if $\dfrac{c}{b} > 0$

and

to the *right* if $\dfrac{c}{b} < 0$

3. A list of trigonometric identities is given inside the back cover.
4. Derivatives of trigonometric functions

(a) $\dfrac{d}{dx}(\sin u) = \cos u \dfrac{du}{dx}$

(b) $\dfrac{d}{dx}(\cos u) = -\sin u \dfrac{du}{dx}$

(c) $\dfrac{d}{dx}(\tan u) = \sec^2 u \dfrac{du}{dx}$

(d) $\dfrac{d}{dx}(\cot u) = -\csc^2 u \dfrac{du}{dx}$

(e) $\dfrac{d}{dx}(\sec u) = \sec u \tan u \dfrac{du}{dx}$

(f) $\dfrac{d}{dx}(\csc u) = -\csc u \cot u \dfrac{du}{dx}$

5. Inverse trigonometric functions

(a) $y = \text{Arcsin } x, \ -\dfrac{\pi}{2} \le y \le \dfrac{\pi}{2}$

(b) $y = \text{Arccos } x, \ 0 \le y \le \pi$

(c) $y = \text{Arctan } x, \ -\dfrac{\pi}{2} < y < \dfrac{\pi}{2}$

(d) $y = \text{Arccot } x, \ 0 < y < \pi$

(e) $y = \text{Arcsec } x, \ 0 \le y \le \pi, y \ne \dfrac{\pi}{2}$

(f) $y = \text{Arccsc } x, \ -\dfrac{\pi}{2} \le y \le \dfrac{\pi}{2}, y \ne 0$

6. Derivatives of inverse trigonometric functions

(a) $\dfrac{d}{dx}(\text{Arcsin } u) = \dfrac{1}{\sqrt{1-u^2}}\dfrac{du}{dx}$

(b) $\dfrac{d}{dx}(\text{Arccos } u) = -\dfrac{1}{\sqrt{1-u^2}}\dfrac{du}{dx}$

(c) $\dfrac{d}{dx}(\text{Arctan } u) = \dfrac{1}{1+u^2}\dfrac{du}{dx}$

(d) $\dfrac{d}{dx}(\text{Arccot } u) = -\dfrac{1}{1+u^2}\dfrac{du}{dx}$

(e) $\dfrac{d}{dx}(\text{Arcsec } u) = \dfrac{1}{|u|\sqrt{u^2-1}}\dfrac{du}{dx}$

(f) $\dfrac{d}{dx}(\text{Arccsc } u) = -\dfrac{1}{|u|\sqrt{u^2-1}}\dfrac{du}{dx}$

7. Exponential function

$$y = b^x \quad \text{where} \quad b > 0 \quad \text{and} \quad b \neq 1$$

8. Laws of exponents

$$a^m \cdot a^n = a^{m+n}$$

$$\frac{a^m}{a^n} = a^{m-n}$$

$$(a^m)^n = a^{mn}$$

$$(ab)^n = a^n b^n$$

$$\left(\frac{a}{b}\right)^n = \frac{a^n}{b^n}$$

$$a^0 = 1$$

9. Logarithmic function

In exponential form	In logarithmic form	Is inverse of
$x = b^y$	$y = \log_b x$	$y = b^x$

Note: $b > 0$ and $b \neq 1$.

10. Logarithmic properties (for $a > 0$ and $a \neq 1$)

$$\log_a (M \cdot N) = \log_a M + \log_a N$$

$$\log_a \left(\frac{M}{N}\right) = \log_a M - \log_a N$$

$$\log_a M^n = n \log_a M$$

$$\log_a \sqrt[n]{M} = \frac{1}{n} \cdot \log_a M$$

$$\log_a 1 = 0$$

$$\log_a \frac{1}{M} = -\log_a M$$

$$\log_a (a^x) = x$$

$$a^{\log_a x} = x$$

$$\log_b x = \frac{\log_a x}{\log_a b}$$

11. Derivatives of logarithmic functions

$$\frac{d}{dx}(\ln u) = \frac{1}{u}\frac{du}{dx}$$

$$\frac{d}{dx}(\log_b u) = \frac{1}{u}\log_b e\,\frac{du}{dx}$$

12. Derivatives of exponential functions

$$\frac{d}{dx}(b^u) = \frac{b^u}{\log_b e}\frac{du}{dx}$$

$$\frac{d}{dx}(e^u) = e^u\frac{du}{dx}$$

CHAPTER 4 REVIEW

Prove each identity.

1. $\sec x \cot x = \csc x$

2. $\sec^2\theta + \tan^2\theta + 1 = \dfrac{2}{\cos^2\theta}$

3. $\dfrac{\cos\theta}{\cos\theta + \sin\theta} = \dfrac{\cot\theta}{1 + \cot\theta}$

4. $\cos\left(\theta - \dfrac{3\pi}{2}\right) = -\sin\theta$

5. $\left(\sin\dfrac{1}{2}x + \cos\dfrac{1}{2}x\right)^2 = 1 + \sin x$

6. $2\cos^2\dfrac{\theta}{2} = \dfrac{1 + \sec\theta}{\sec\theta}$

7. $\dfrac{2\cot\theta}{1 + \cot^2\theta} = \sin 2\theta$

8. $\csc x - \cot x = \tan\dfrac{1}{2}x$

9. $\tan 2x = \dfrac{2\cos x}{\csc x - 2\sin x}$

10. $\tan^2\dfrac{x}{2} + 1 = 2\tan\dfrac{x}{2}\csc x$

Simplify each expression.

11. $\sin\theta\cos\theta$

12. $\cos^2 3\theta - \sin^2 3\theta$

13. $\dfrac{1 + \cos 4\theta}{2}$

14. $1 - 2\sin^2\dfrac{\theta}{3}$

15. $\cos 2x\cos 3x - \sin 2x\sin 3x$

16. $\sin 2x\cos x - \cos 2x\sin x$

Sketch each curve.

17. $y = 4\cos 6x$

18. $y = -2\sin\dfrac{1}{3}x$

19. $y = 3\sin\left(x - \dfrac{\pi}{4}\right)$

20. $y = \cos\left(2x + \dfrac{2\pi}{3}\right)$

21. $y = 4\sin\left(\pi x + \dfrac{\pi}{2}\right)$

22. $y = \tan 5x$

23. $y = -\cot 3x$

24. $y = 2\sec 4x$

Find the derivative of each function.

25. $y = \sin(x^2 + 3)$

26. $y = \cos 8x$

27. $y = \cos^3(5x - 1)$

28. $y = \sin 3x\cos 2x$

29. $y = \tan(3x - 2)$

30. $y = \sec(4x + 3)$

31. $y = \cot 6x^2$

32. $y = \csc^2(8x^2 + x)$

33. $y = \sec^2 x\sin x$

34. $y = x^2 - \csc^2 x$

35. $y = \tan(\sec x)$

36. $y = \dfrac{\cos x}{1 + \sin x}$

37. $y = (1 - \sin x)^3$

38. $y = (1 + \sec 4x)^2$

Solve each equation for $0 \le y < 2\pi$.

39. $y = \arcsin\left(\dfrac{1}{2}\right)$

40. $y = \arctan\left(-\dfrac{1}{\sqrt{3}}\right)$

41. $y = \arccos\left(\dfrac{1}{\sqrt{2}}\right)$

42. $y = \operatorname{arcsec}(-2)$

Find all angles in radians for each expression.

43. $\arctan(-\sqrt{3})$

44. $\arcsin(-1)$

Solve each equation for x.

45. $y = \dfrac{1}{2}\sin\dfrac{3x}{4}$

46. $y = 5\tan(1 - 2x)$

Find the value of each expression in radians.

47. $\operatorname{Arcsin}\left(\dfrac{1}{\sqrt{2}}\right)$

48. $\operatorname{Arctan}\left(-\dfrac{1}{\sqrt{3}}\right)$

49. $\operatorname{Arcsec}(-1)$

50. $\operatorname{Arccos}\left(-\dfrac{1}{2}\right)$

Find the value of each expression.

51. $\sin\left[\operatorname{Arccos}\left(-\dfrac{1}{2}\right)\right]$

52. $\tan(\operatorname{Arctan}\sqrt{3})$

53. Find an algebraic expression for $\sin(\operatorname{Arccot} x)$.

Find the derivative of each function.

54. $y = \operatorname{Arcsin} x^3$

55. $y = \operatorname{Arctan} 3x$

56. $y = 3\operatorname{Arccos}\dfrac{1}{2x}$

57. $y = 2\operatorname{Arcsec} 4x$

58. $y = \operatorname{Arcsin}^2 3\sqrt{x}$

59. $y = x\operatorname{Arcsin} x$

Graph each equation.

60. $y = 3^x$

61. $y = \log_3 x$

Solve for x.

62. $\log_9 x = 2$

63. $\log_x 8 = 3$

64. $\log_2 32 = x$

Write each expression as a sum, difference, or multiple of single logarithms.

65. $\log_4 6x^2 y$

66. $\log_3 \dfrac{5x\sqrt{y}}{z^3}$

67. $\log_{10} \dfrac{x^2(x+1)^3}{\sqrt{x-4}}$

68. $\ln \dfrac{[x(x-1)]^3}{\sqrt{x+1}}$

Write each expression as a single logarithmic expression.

69. $\log_2 x + 3\log_2 y - 2\log_2 z$

70. $\dfrac{1}{2}\log_{10}(x+1) - 3\log_{10}(x-2)$

71. $4\ln x - 5\ln(x+1) - \ln(x+2)$

72. $\dfrac{1}{2}[\ln x + \ln(x+2)] - 2\ln(x-5)$

Simplify.

73. $\log_{10} 1000$
74. $\log_{10} 10^{x^2}$

75. $\ln e^2$
76. $\ln e^x$

Find the derivative of each function.

77. $y = \ln(2x^3 - 4)$
78. $y = \log_3(4x + 1)$
79. $y = \ln\left(\dfrac{x^2}{x^2 + 3}\right)$

80. $y = \cos(\ln x)$
81. $y = \dfrac{\sqrt{x + 1}\,(3x - 4)}{x^2(x + 2)}$
82. $y = x^{1-x}$

83. $y = e^{x^2 + 5}$
84. $y = 8^{3x}$
85. $y = \cot e^{2x}$

86. $y = e^{\sin x^2}$
87. $y = \text{Arcsin}\, e^{-4x}$
88. $y = x^3 e^{-4x}$

Sketch the graph of each equation.

89. $y = xe^{-x}$
90. $y = e^{-x^2}$

91. The current in a circuit is given by $i = 100 \cos 5t$. Find the equation of the voltage V_L across a 2-H inductor $\left(V_L = L\,\dfrac{di}{dt}, \text{ where } L \text{ is the inductance}\right)$.

92. The work done in a circuit is given by $W = 60 \cos^2 3t$. Find the equation for the power p $\left(p = \dfrac{dW}{dt}\right)$.

93. Find the equation of the tangent line to $y = \ln x^2$ at the point $(1, 0)$.

94. Find the velocity of an object moving along a straight line according to $s = e^{\sin t}$.

95. A particle moves along a straight line according to $s = 1 - 4e^{-t}$. Find its acceleration at $t = \frac{1}{2}$.

5

The Integral

INTRODUCTION

A large class of extremely important problems in science, engineering, and technology involves the notion of finding the "sum" of an "infinite number" of "infinitesimal" quantities. The integral calculus develops the mathematics needed to solve such problems.

Objectives

- Integrate polynomials.
- Find the constant of integration.
- Find the area under a curve.
- Evaluate definite integrals.

5.1 THE INDEFINITE INTEGRAL

As we have seen, many applications involve finding the derivative of a function. Following are some examples:

Function	Derivative
Position or displacement	Velocity
Velocity	Acceleration
Function	Slope of tangent line
Amount or quantity	Rate of increase or decrease
Work	Power

Often we have the derivative function and need to find the original function, which requires that we perform the inverse operation of differentiation. Now consider finding the derivative in reverse; that is, given the derivative of a function, find the function. This is called *antidifferentiation* or *integration*. That is, we look for an unknown function whose derivative is known.

> **ANTIDERIVATIVE**
>
> If $f(x)$ is a function, then $F(x)$ is an *antiderivative* of $f(x)$ if $F'(x) = f(x)$.

EXAMPLE 1

Find a function whose derivative is $10x^4$.

We need a function $y = f(x)$ where $\dfrac{dy}{dx} = 10x^4$. Recall that when differentiating a polynomial in x, the exponent of x is reduced by one. This means that the original function $y = f(x)$ must have a power of x with exponent 5, x^5. However, when we differentiate x^5, we obtain $5x^4$. If we rewrite $\dfrac{dy}{dx} = 10x^4 = 2(5x^4)$, observe that the given derivative is a multiple of two times the derivative of x^5.

Since $\dfrac{d}{dx}(2x^5) = 2\dfrac{d}{dx}(x^5) = 2(5x^4)$, we conclude that $y = 2x^5$ is a solution.

Example 1 immediately leads us into difficulty. While $y = 2x^5$ is a solution, so is $y = 2x^5 + 9$ since

$$\frac{d}{dx}(2x^5 + 9) = 10x^4 + 0 = 10x^4$$

In fact, any function of the form $y = 2x^5 + C$, where C is a constant, is a solution.

The process of antidifferentiation, unlike differentiation, does not lead to unique solutions. There are, in fact, an infinite number of solutions. Each solution depends on the choice of the value for C.

EXAMPLE 2

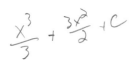

Find a function whose derivative is $\dfrac{dy}{dx} = x^2 + 3x$.

The term x^2 requires that the desired function have a term involving x^3, since differentiation decreases the power of x by one. Since $\dfrac{d}{dx}(x^3) = 3x^2$, the coefficient of x^3 must be changed to $\dfrac{1}{3}$. Then

$$\frac{d}{dx}\left(\frac{x^3}{3}\right) = 3\frac{(x^2)}{3} = x^2$$

Similarly, the term $3x$ indicates that the desired function must also have a term involving x^2. Observe that the coefficient of x^2 must be $\dfrac{3}{2}$ so that $\dfrac{d}{dx}\left(\dfrac{3}{2}x^2\right) = \left(\dfrac{3}{2}\right)(2x) = 3x$. The desired function is

$$y = \frac{x^3}{3} + \frac{3x^2}{2}$$

Note:

$$y = \frac{x^3}{3} + \frac{3x^2}{2} + 11$$

is also a solution for Example 2. The general solution is

$$y = \frac{x^3}{3} + \frac{3x^2}{2} + C$$

where C is a constant.

The process of antidifferentiation can be easily checked. Differentiate the solution obtained and compare the result with the original given function. They should be equal. In Example 2,

$$\frac{d}{dx}\left(\frac{x^3}{3} + \frac{3x^2}{2}\right) = x^2 + 3x$$

Any solution $y = F(x)$ resulting from performing the integration process is called an *antiderivative*. In Example 1, $y = 2x^5 + 9$ was found to be an antiderivative of $10x^4$. That is, $F(x)$ is an antiderivative for a given function $f(x)$ if $\dfrac{d}{dx}[F(x)] = f(x)$.

Using any antiderivative, $F(x)$, for a given function $f(x)$, $F(x) + C$ is a general solution; that is, $\dfrac{d}{dx}[F(x) + C] = f(x)$. This general solution is called the *indefinite integral* of $f(x)$ where $f(x)$ is called the *integrand*. The symbolism for the indefinite integral is

$$\int f(x)\, dx = F(x) + C$$

where $\int$ is called the *integral sign* and C is called the *constant of integration*. $F(x) + C$ can be considered to be the family of all curves whose derivatives for any given x are all equal to each other. This means that for any given x, the slopes of any two of these curves are equal.

EXAMPLE 3

Find the indefinite integral for $f(x) = 2x$.

Since $\dfrac{d}{dx}(x^2) = 2x$, we have

$$\int f(x)\, dx = \int 2x\, dx = x^2 + C$$

This is a family of parabolas $y = x^2 + C$ (see Fig. 5.1). At $x = 1$, each curve has slope

$$\frac{d}{dx}(x^2 + C)\bigg|_{x=1} = 2x\bigg|_{x=1} = 2(1) = 2$$

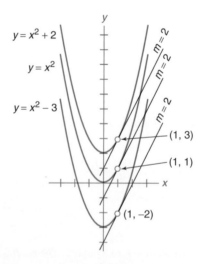

Figure 5.1 Family of parabolas $y = x^2 + C$

Just as there are basic differentiation formulas, there are basic integration formulas that can be developed. These make the integration process easier.

First,

$$\int x^n\, dx = \frac{x^{n+1}}{n + 1} + C \qquad (n \neq -1)$$

This is verified as follows:

$$\frac{d}{dx}\left(\frac{x^{n+1}}{n+1}+C\right)=\frac{(n+1)x^{(n+1)-1}}{n+1}+0$$

$$=x^n$$

The case for $n=-1$ is discussed in Section 7.2.

EXAMPLE 4

Integrate $y=x^{16}$.

$$\int x^{16}\,dx=\frac{x^{16+1}}{16+1}+C=\frac{x^{17}}{17}+C$$

EXAMPLE 5

Find $\int dx$.

$$\int dx=\int x^0\,dx \qquad\qquad (\text{since } x^0=1)$$

$$=\frac{x^{0+1}}{0+1}+C=x+C$$

EXAMPLE 6

Find $\int\dfrac{dx}{x^3}$.

$$\int\frac{dx}{x^3}=\int x^{-3}\,dx$$

$$=\frac{x^{-3+1}}{-3+1}+C$$

$$=\frac{x^{-2}}{-2}+C$$

$$=-\frac{1}{2x^2}+C$$

EXAMPLE 7

Find $\int\sqrt{x}\,dx$.

$$\int\sqrt{x}\,dx=\int x^{1/2}\,dx$$

$$=\frac{x^{(1/2)+1}}{\frac{1}{2}+1}+C$$

$$=\frac{x^{3/2}}{\frac{3}{2}}+C$$

$$=\frac{2}{3}x^{3/2}+C$$

Next,

$$\int[f(x)+g(x)]\,dx=\int f(x)\,dx+\int g(x)\,dx$$

This follows from the fact that the derivative of the sum of two functions equals the sum of their derivatives.

EXAMPLE 8

Find $\int (x^3 + x^2)\, dx$.

$$\int (x^3 + x^2)\, dx = \int x^3\, dx + \int x^2\, dx$$

$$= \frac{x^{3+1}}{3+1} + \frac{x^{2+1}}{2+1} + C$$

$$= \frac{x^4}{4} + \frac{x^3}{3} + C$$

Finally,

$$\int k f(x)\, dx = k \int f(x)\, dx \qquad \text{where } k \text{ is a constant}$$

This follows from the fact that the derivative of a constant times a function equals that constant times the derivative of the function.

EXAMPLE 9

Find $\int 12x^3\, dx$.

$$\int 12x^3\, dx = 12 \int x^3\, dx$$

$$= 12\frac{x^{3+1}}{3+1} + C$$

$$= 12\frac{x^4}{4} + C$$

$$= 3x^4 + C$$

EXAMPLE 10

Find $\int \left(3x^4 + 2x + \frac{1}{x^2} \right) dx$.

$$\int \left(3x^4 + 2x + \frac{1}{x^2} \right) dx = \int 3x^4\, dx + \int 2x\, dx + \int \frac{1}{x^2}\, dx$$

$$= 3\int x^4\, dx + 2\int x\, dx + \int x^{-2}\, dx$$

$$= 3\frac{x^{4+1}}{4+1} + 2\frac{x^{1+1}}{1+1} + \frac{x^{-2+1}}{-2+1} + C$$

$$= \frac{3x^5}{5} + x^2 - \frac{1}{x} + C$$

EXAMPLE 11

Find $\int 5(x^3 + 4)^4(3x^2)\, dx$.

This does not seem to fit any of the formulas developed so far. While we could perform the indicated multiplication $5(x^3 + 4)^4(3x^2)$ and integrate the resulting sum of terms, there is an easier method.

Observe that $3x^2$ is the derivative of $x^3 + 4$. Also recall that when differentiating the power formula u^{n+1}, we obtain $(n+1)u^n \dfrac{du}{dx}$. Then $5(x^3 + 4)^4(3x^2)$ is seen to be the derivative of $(x^3 + 4)^5$. That is,

$$\int 5(x^3 + 4)^4(3x^2)\,dx = (x^3 + 4)^5 + C$$

In Example 11, if we let $u = x^3 + 4$, then $du = 3x^2\,dx$. Then

$$\int 5(x^3 + 4)^4(3x^2)\,dx = \int 5u^4\,du$$

We must see the function to be integrated as a product of two factors. One factor is a power of u, that is, u^n. The other factor is the differential of u, that is, du. Then use the formula

$$\int u^n\,du = \frac{u^{n+1}}{n+1} + C$$

EXAMPLE 12

Find $\displaystyle\int 6x^2 \sqrt{2x^3 + 1}\,dx$.

Observe that $6x^2$ is the *derivative* of $2x^3 + 1$ (or $6x^2\,dx$ is the *differential* of $2x^3 + 1$). Setting $u = 2x^3 + 1$ then $du = 6x^2\,dx$, we have

$$\int 6x^2 \sqrt{2x^3 + 1}\,dx = \int \sqrt{u}\,du = \int u^{1/2}\,du$$

$$= \frac{2}{3}u^{3/2} + C$$

$$= \frac{2}{3}(2x^3 + 1)^{3/2} + C$$

Using a calculator,

F3   **2**   6x^2 **2nd** $\sqrt{}$ 2x^3+1),x)   **ENTER**

 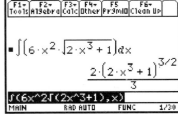

EXAMPLE 13

Find $\displaystyle\int x^3 \sqrt{3x^4 + 2}\,dx$.

If we let $u = 3x^4 + 2$, then $du = 12x^3\,dx$.

Then

$$\sqrt{u}\,du = 12x^3 \sqrt{3x^4 + 2}\,dx$$

Note that the only difference between the integrand $x^3 \sqrt{3x^4 + 2}$ and $\sqrt{u}\,du$ is the factor 12.

So write

$$x^3 \sqrt{3x^4 + 2}\, dx = \sqrt{3x^4 + 2}\left(\frac{12x^3\, dx}{12}\right) = \sqrt{u}\left(\frac{du}{12}\right)$$

and

$$\int x^3 \sqrt{3x^4 + 2}\, dx = \int \frac{\sqrt{u}\, du}{12}$$

$$= \frac{1}{12} \int (u)^{1/2}\, du$$

$$= \frac{1}{12} \frac{u^{3/2}}{\frac{3}{2}} + C$$

$$= \frac{1}{18}(3x^4 + 2)^{3/2} + C$$

$$\boxed{\begin{array}{l} u = 3x^4 + 2 \\ du = 12x^3\, dx \end{array}}$$

Note: We will continue to use a box to show the appropriate substitutions for u and du whenever the formula $\int u^n\, du$ is applied. The box will appear at the right of the integral in which the substitutions have been made.

In summary,

1. $\displaystyle\int du = u + C$

2. $\displaystyle\int u^n\, du = \frac{u^{n+1}}{n+1} + C \qquad (n \neq -1)$

3. $\displaystyle\int k f(x)\, dx = k \int f(x)\, dx \qquad$ where k is a constant

4. $\displaystyle\int [f(x) + g(x)]\, dx = \int f(x)\, dx + \int g(x)\, dx$

Exercises 5.1

Integrate.

1. $\displaystyle\int x^7\, dx$

2. $\displaystyle\int x^{24}\, dx$

3. $\displaystyle\int 3x^8\, dx$

4. $\displaystyle\int 200x^9\, dx$

5. $\displaystyle\int 4\, dx$

6. $\displaystyle\int 8\sqrt{x^3}\, dx$

7. $\displaystyle\int 9\sqrt[6]{x^5}\, dx$

8. $\displaystyle\int \frac{3\, dx}{x^2}$

9. $\displaystyle\int \frac{6\, dx}{x^3}$

10. $\displaystyle\int \frac{dx}{\sqrt[6]{x^5}}$

11. $\displaystyle\int (5x^2 - 12x + 8)\, dx$

12. $\displaystyle\int (7x^{10} + 8x^4 - 11x^3)\, dx$

13. $\displaystyle\int \left(3x^2 - x + \frac{5}{x^3}\right) dx$

14. $\displaystyle\int \left(2x^3 + 3x - \frac{5}{x^4}\right) dx$

15. $\displaystyle\int (2x^2 - 3)^2\, dx$

16. $\displaystyle\int (x^2 - 5)^2\, dx$

17. $\displaystyle\int \sqrt{6x + 2}\, dx$

18. $\displaystyle\int \sqrt[3]{8x - 1}\, dx$

19. $\displaystyle\int 8x(x^2 + 3)^3\, dx$

20. $\displaystyle\int 18x^2(x^3 + 2)^5\, dx$

21. $\displaystyle\int x\sqrt[3]{5x^2 - 1}\, dx$

22. $\displaystyle\int 3x\sqrt{6x^2 + 5}\, dx$

23. $\displaystyle\int x(x^2 - 1)^4\, dx$

24. $\displaystyle\int x^4(x^5 + 3)^7\, dx$

25. $\displaystyle\int \frac{2x\,dx}{\sqrt{x^2+1}}$ **26.** $\displaystyle\int \frac{24x\,dx}{\sqrt{8x^2-1}}$ **27.** $\displaystyle\int (3x^2+2)(x^3+2x)^3\,dx$

28. $\displaystyle\int (4x^3-6x)(x^4-3x^2)^4\,dx$ **29.** $\displaystyle\int \frac{x^2\,dx}{(x^3-4)^2}$ **30.** $\displaystyle\int \frac{(6x+1)\,dx}{(3x^2+x-2)^3}$

31. $\displaystyle\int (10x-1)\sqrt{5x^2-x}\,dx$ **32.** $\displaystyle\int (3x^2-2x)\sqrt{x^3-x^2+2}\,dx$

33. $\displaystyle\int \frac{(2x+1)\,dx}{\sqrt{x^2+x}}$ **34.** $\displaystyle\int \frac{(x-4)\,dx}{\sqrt{x^2-8x+3}}$ **35.** $\displaystyle\int (2x+3)^2\,dx$

36. $\displaystyle\int (4x-5)^2\,dx$ **37.** $\displaystyle\int (2x-1)^4\,dx$ **38.** $\displaystyle\int (5x+3)^6\,dx$

39. $\displaystyle\int (x^2+1)^3\,dx$ **40.** $\displaystyle\int (x^2-4)^2\,dx$ **41.** $\displaystyle\int 4x(x^2+1)^3\,dx$

42. $\displaystyle\int 6x(x^2-4)^2\,dx$ **43.** $\displaystyle\int 30x^2(5x^3+1)^4\,dx$ **44.** $\displaystyle\int 12x^3(6x^4-1)^5\,dx$

45. $\displaystyle\int \frac{6x^2\,dx}{\sqrt{x^3+1}}$ **46.** $\displaystyle\int \frac{24x^3\,dx}{\sqrt{2x^4-1}}$ **47.** $\displaystyle\int \frac{(6x^2+6)\,dx}{\sqrt[3]{x^3+3x}}$

48. $\displaystyle\int \frac{(60x^3+36x)\,dx}{\sqrt[4]{5x^4+6x^2}}$ **49.** $\displaystyle\int \frac{(x-1)\,dx}{x^3}$ **50.** $\displaystyle\int \frac{(4-3x^2)\,dx}{6x^4}$

5.2 THE CONSTANT OF INTEGRATION

When additional information about the antiderivative is known, we can determine the constant of integration C.

Note: This section is actually a brief introduction to the study of differential equations, which are more extensively developed in Chapters 11 and 12.

EXAMPLE 1

Find the antiderivative of $\dfrac{dy}{dx} = 3x^2$ where $y = 14$ when $x = 2$.

$$\int 3x^2\,dx = F(x) + C$$
$$= x^3 + C$$

We need that function $y = x^3 + C$ where $y = 14$ when $x = 2$. So

$$y = x^3 + C$$
$$(14) = (2)^3 + C$$
$$6 = C$$

The antiderivative is then $y = x^3 + 6$.

EXAMPLE 2

Find the equation of the curve that passes through $(4, -1)$ and whose slope is given by $m = \dfrac{dy}{dx} = 2x - 3$.

$$y = \int (2x-3)\,dx$$
$$y = x^2 - 3x + C$$

$$(-1) = (4)^2 - 3(4) + C$$
$$-5 = C$$

So the equation is $y = x^2 - 3x - 5$.

EXAMPLE 3

Find the equation describing the motion of an object moving along a straight line with constant acceleration 2 m/s² when the velocity at time $t = 3$ s is 10 m/s and when the object has traveled 70 m from the origin at $t = 5$ s.

Since $\dfrac{dv}{dt} = a$, we have

$$v = \int a\, dt = \int 2\, dt = 2t + C_1$$

since

$$v = 2t + C_1 \quad \text{and} \quad v = 10 \quad \text{when} \quad t = 3$$
$$(10) = 2(3) + C_1$$
$$4 = C_1$$

so

$$v = 2t + 4$$

Also, since $\dfrac{ds}{dt} = v$,

$$s = \int v\, dt = \int (2t + 4)\, dt$$

we have

$$s = t^2 + 4t + C_2 \quad \text{and} \quad s = 70 \quad \text{when} \quad t = 5$$
$$(70) = (5)^2 + 4(5) + C_2$$
$$25 = C_2$$

The equation is $s = t^2 + 4t + 25$.

EXAMPLE 4

A bullet is fired vertically from a gun with an initial velocity of 250 m/s. (a) Find the equation that describes its motion. (b) How long does it take to reach its maximum height? (c) How high does the bullet go?

In a problem involving a freely falling body, the acceleration is

$$a = g = -32 \text{ ft/s}^2 = -9.80 \text{ m/s}^2$$

An upward direction is positive, and a downward direction is negative.

(a) Since $\dfrac{dv}{dt} = a$, we have

$$v = \int a\, dt = \int -9.80\, dt = -9.80t + C_1$$

At $t = 0$, $v = 250$. Substituting, we have

$$250 = -9.80(0) + C_1$$
$$250 = C_1$$

So

$$v = -9.80t + 250$$

and since $\dfrac{ds}{dt} = v$,

$$s = \int v\, dt = \int (-9.80t + 250)\, dt = -4.90t^2 + 250t + C_2$$

At $t = 0$, $s = 0$. Substituting, we have

$$0 = -4.90(0)^2 + 250(0) + C_2$$
$$0 = C_2$$

So

$$s = -4.90t^2 + 250t$$

(b) At the bullet's maximum height, $v = 0$.
So

$$v = -9.80t + 250 = 0$$
$$t = \frac{250}{9.80} = 25.5 \text{ s}$$

(c) At $t = 25.5$ s,

$$s = -4.90t^2 + 250t$$
$$s = -4.90(25.5)^2 + (250)(25.5) = 3190 \text{ m}$$

In general, the position of an object moving freely (no air resistance) along a straight line with constant acceleration a, initial velocity v_0, and initial position s_0 may be written

$$s = \frac{1}{2}at^2 + v_0t + s_0$$

This equation is derived in Exercise 23.

The voltage V_C across a capacitor at any time t is given by

$$V_C = \frac{1}{C}\int i\, dt$$

where C is the capacitance in farads and i is the current in amperes.

EXAMPLE 5

A 1-microfarad capacitor $[1 \text{ microfarad} = 10^{-6} \text{ farad (F)}]$ has a voltage of 86 volts (V) across it. At a given instant $(t = 0)$, we connect this capacitor to a source that sends a current $i = 3t^2$ amperes (A) through the circuit. Find the voltage across the capacitor when $t = 0.1$ s.

$$V_C = \frac{1}{C}\int i\, dt = \frac{1}{10^{-6}}\int 3t^2\, dt$$
$$V_C = \frac{1}{10^{-6}} \cdot \frac{3t^3}{3} + C = 10^6 \cdot t^3 + C$$

When $t = 0$, $V_C = 86$ V, so

$$(86) = 10^6 \cdot (0)^3 + C$$
$$86 = C$$

We then have

$$V_C = 10^6 \cdot t^3 + 86$$
$$V_C = 10^6 \cdot (0.1)^3 + 86$$
$$= 10^6 \cdot 10^{-3} + 86$$
$$= 1086 \text{ V}$$

The current i in a circuit at any instant is given by

$$i = \frac{dq}{dt}$$

where q is the charge. That is,

$$q = \int i\, dt$$

EXAMPLE 6

The current in a circuit is given by $i = t^3 + 3t^2 - 4$ amperes (A). Find the charge in coulombs (C) that passes a given point in the circuit after 2 s.

$$q = \int i\, dt = \int (t^3 + 3t^2 - 4)\, dt$$

$$q = \frac{t^4}{4} + t^3 - 4t + C$$

If we assume that $q = 0$ when $t = 0$, then

$$0 = 0 + 0 - 0 + C$$
$$0 = C$$

and

$$q = \frac{t^4}{4} + t^3 - 4t$$

At $t = 2$ s,

$$q = \frac{(2)^4}{4} + (2)^3 - 4(2)$$

$$= 4 + 8 - 8 = 4\,\text{C}$$

Exercises 5.2

Find the equation of the curve $y = f(x)$ satisfying the given conditions.

1. $\dfrac{dy}{dx} = 3x$, passing through $(0, 1)$

2. $\dfrac{dy}{dx} = 5x^2$, passing through $\left(1, -\dfrac{1}{3}\right)$

3. $\dfrac{dy}{dx} = 3x^2 + 3$, passing through $(-1, 2)$

4. $\dfrac{dy}{dx} = 4x^3 - 2x + 2$, passing through $(1, -2)$

5. $\dfrac{dy}{dx} = x(x^2 - 3)^2$, passing through $\left(2, \dfrac{7}{6}\right)$

6. $\dfrac{dy}{dx} = \dfrac{x}{\sqrt{x^2 - 1}}$, passing through $(-3, 2\sqrt{2})$

7. Find the equation describing the motion of an object moving along a straight line when the acceleration is $a = 3t$, when the velocity at $t = 4$ s is 40 m/s, and when the object has traveled 86 m from the origin at $t = 2$ s.

8. Find the equation describing the motion of an object moving along a straight line when the acceleration is $a = 4t - 2$, when the velocity at $t = 5$ s is 25 m/s, and when the object has traveled 238 m from the origin at $t = 12$ s.

9. A stone is dropped from a height of 100 ft. For a free-falling object, the acceleration is $a = -32$ ft/s^2 (the effect of gravity). Find the distance the stone has traveled after 2 s. Note that the initial velocity is 0 because the stone was dropped, not thrown down. Find also the velocity of the stone when it hits the ground.

10. An object is dropped from a stationary balloon at 500 m. **(a)** Express the object's height above the ground as a function of time. **(b)** How long does it take to hit the ground? ($a = -9.80$ m/s^2.)

11. An airplane starts from rest and travels 3600 ft down a runway with constant acceleration before lifting off in 30 s. Find its velocity at the moment of lift-off. *Hint:* $s = \frac{1}{2}at^2$.

12. A ball rolls from a rest position down a 200-cm inclined plane in 4 s. Find its acceleration in cm/s^2.

13. A stone is hurled straight up from the ground at a velocity of 25 m/s. **(a)** Find the maximum height that the stone reaches. **(b)** How long does it take for the stone to hit the ground? **(c)** Find the speed at which the stone hits the ground.

14. A ball is thrown vertically upward with an initial velocity of 40 ft/s. **(a)** Find the maximum height of the ball. **(b)** How long does it take for the ball to hit the ground? **(c)** Find the speed at which the ball hits the ground.

15. A stone is thrown vertically upward from the roof of a 200-ft-tall building with an initial velocity of 30 ft/s. **(a)** Find the equation describing the altitude of the stone from the ground. **(b)** How long does it take for the stone to hit the ground?

16. A stone is thrown straight down from an 80-m-tall building with an initial velocity of 10 m/s. **(a)** Find the equation describing the height of the stone from the ground. **(b)** How long does it take for the stone to hit the ground?

17. A flywheel is turning at a rate given by $\omega = 80 - 12t + 3t^2$ where ω is the angular speed in revolutions per minute (rpm). Find the number of revolutions that the flywheel makes in the first 3 s. $\omega = \dfrac{d\theta}{dt}$ (Assume that $\theta = 0$ when $t = 0$.)

18. The power in a system equals the rate at which energy (work) is expended. That is, $p = \dfrac{dW}{dt}$. At $t = 0$, the energy in an electrical system is 4 joules (J). Find the energy after 4 s if $p = 2\sqrt{t}$ watts.

19. A capacitor with capacitance 10^{-4} F has a voltage of 100 V across it. At a given instant ($t = 0$) the capacitor is connected to a source that sends a current $i = \frac{1}{2}\sqrt{t} + 0.2$ A through the circuit. Find the voltage across the capacitor when $t = 0.16$ s.

20. A 0.1-F capacitor measures 150 V across it. At $t = 0$ the capacitor is connected to a source that sends current $i = \dfrac{16t}{\sqrt{4t^2 + 9}}$ amperes through the circuit. Find the voltage across the capacitor when $t = 2$ s.

21. The current in a circuit is given by $i = t\sqrt{t^2 + 1}$ A. Find the charge (in coulombs) that passes a given point in the circuit after 1 s. (Assume that $q = 0$ when $t = 0$.)

22. The current in a circuit is given by $i = t^{3/2} + 4t$ amperes. Find the charge in coulombs that passes a given point in the circuit after 4 s. (Assume that $q = 0$ when $t = 0$.)

23. In general, the position of an object moving freely (no air resistance) along a straight line with constant acceleration a, initial velocity v_0, and initial position s_0 may be written

$$s = \tfrac{1}{2}at^2 + v_0 t + s_0$$

For freely falling bodies ($a = g = -32$ ft/s$^2 = -9.80$ m/s^2), this equation becomes

$$s = -16t^2 + v_0 t + s_0 = -4.90t^2 + v_0 t + s_0$$

Derive this equation.

5.3 AREA UNDER A CURVE

Another application of integration is finding the area under a curve, which is actually the geometric interpretation of the integral. We can find the area of regions which would be impossible to determine if we had to rely only on regular geometric methods. The area under a curve $y = f(x)$ refers to the area of the region bounded by the curves $y = f(x)$, $x = a$, $x = b$, and the x-axis ($y = 0$) (see Fig. 5.2). We could approximate this area by forming the rectangles in Fig. 5.3. The sum of the areas of the rectangles could then be used as an approximation of the desired area.

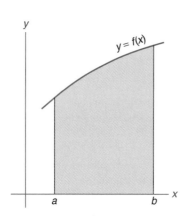

Figure 5.2   Area under a curve.

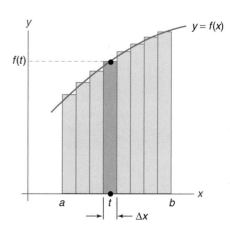

Figure 5.3   Approximating the area under a curve using rectangles.

The height of each rectangle is the value of the function $f(x)$ for some point t along the base of the rectangle. We have chosen Δx to be the base, or the width of each rectangle. The described area is then approximately equal to the sum

$$S_n = f(t_1)\,\Delta x + f(t_2)\,\Delta x + f(t_3)\,\Delta x + \cdots + f(t_k)\,\Delta x + \cdots + f(t_n)\,\Delta x$$

when using n rectangles with base Δx and t_k as a point along the base of the kth rectangle. The smaller we choose the width Δx, the better the approximation for the area under the curve. Observe that as $\Delta x \to 0$, the number of terms n in the approximating sum S_n will increase. In fact, as $\Delta x \to 0$, $n \to \infty$ and the sums S_n appear to approach the exact area A under the curve.

This summation process is symbolized as follows:

$$\lim_{n \to \infty} S_n = A$$

Although this method can be used to find area as illustrated in Example 1, the method is generally difficult to use. In practice, the approach shown after Example 1 for finding area is easier.

EXAMPLE 1

Find the area under the curve $y = 3x$ from $x = 0$ to $x = 2$.

Form the approximating sum S_n by letting the width of each rectangle be

$$\Delta x = \frac{b - a}{n} = \frac{2 - 0}{n} = \frac{2}{n}$$

In Fig. 5.4, $n = 4$, so $\Delta x = \frac{2}{4} = \frac{1}{2}$ and $t_1 = a = 0$, $t_2 = \frac{1}{2}$, $t_3 = 1$, and $t_4 = \frac{3}{2}$. Then

$$S_4 = f(0)\,\Delta x + f\!\left(\frac{1}{2}\right)\Delta x + f(1)\,\Delta x + f\!\left(\frac{3}{2}\right)\Delta x$$

$$= (0)\!\left(\frac{1}{2}\right) + \left(\frac{3}{2}\right)\!\left(\frac{1}{2}\right) + (3)\!\left(\frac{1}{2}\right) + \left(\frac{9}{2}\right)\!\left(\frac{1}{2}\right)$$

$$= \frac{18}{4} = \frac{9}{2}$$

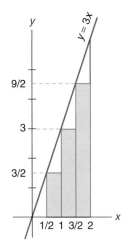

Figure 5.4

In general, since $f(x) = 3x$,

$$S_n = f(0)\,\Delta x \; + f\!\left(\frac{2}{n}\right)\Delta x + f\!\left(\frac{4}{n}\right)\Delta x$$

$$+ f\!\left(\frac{6}{n}\right)\Delta x + \cdots + f\!\left(\frac{2(n-1)}{n}\right)\Delta x$$

$$= 3(0)\!\left(\frac{2}{n}\right) + 3\!\left(\frac{2}{n}\right)\!\left(\frac{2}{n}\right) + 3\!\left(\frac{4}{n}\right)\!\left(\frac{2}{n}\right)$$

$$+ 3\!\left(\frac{6}{n}\right)\!\left(\frac{2}{n}\right) + \cdots + 3\!\left(\frac{2(n-1)}{n}\right)\!\left(\frac{2}{n}\right)$$

$$= \frac{12}{n^2}\big[1 + 2 + 3 + \cdots + (n-1)\big]$$

Since $1 + 2 + 3 + \cdots + (n-1)$ is the sum of an arithmetic progression whose last term is $n - 1$, we have

$$1 + 2 + 3 + \cdots + (n-1) = \frac{n-1}{2}\big[1 + (n-1)\big]$$

$$= \frac{(n-1)n}{2} \qquad \text{(See Section 9.1)}$$

Then

$$S_n = \frac{12}{n^2}\!\left(\frac{(n-1)n}{2}\right)$$

$$= \frac{6(n-1)}{n}$$

$$= 6\!\left(1 - \frac{1}{n}\right)$$

So

$$A = \lim_{n \to \infty} S_n = \lim_{n \to \infty} 6\left(1 - \frac{1}{n}\right)$$

$$= 6(1 - 0) = 6$$

Note that this is the same area as obtained by using the formula for the area of a triangle with height $h = 6$ and base $b = 2$:

$$A = \frac{1}{2}bh$$

$$= \frac{1}{2}(2)(6) = 6$$

To avoid the complicated summation process, let us now consider only the area under the curve $y = f(x)$ from a to x in Fig. 5.5. Note that the area increases as x increases. Consider the area determined by x as a function of x which we will denote by $A(x)$. The difference ΔA between $A(x + \Delta x)$ and $A(x)$ is the incremental change in the area for an incremental change Δx.

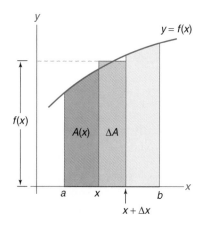

Figure 5.5

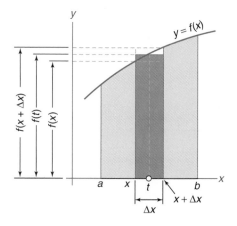

Figure 5.6

From Fig. 5.6 we see that

$$f(x) \, \Delta x \leq \Delta A \leq f(x + \Delta x) \, \Delta x$$

Dividing by Δx,

$$f(x) \leq \frac{\Delta A}{\Delta x} \leq f(x + \Delta x)$$

We can interpret $\dfrac{\Delta A}{\Delta x}$ as the average rate of change of the area as the area increases from x to $x + \Delta x$. As $\Delta x \to 0, f(x + \Delta x) \to f(x)$. Since $\dfrac{\Delta A}{\Delta x}$ is squeezed between $f(x)$ and $f(x + \Delta x)$, we have $\dfrac{\Delta A}{\Delta x} \to f(x)$ as $\Delta x \to 0$. That is

$$\lim_{\Delta x \to 0} \frac{\Delta A}{\Delta x} = f(x)$$

Formally, this limit $\lim\limits_{\Delta x \to 0} \dfrac{\Delta A}{\Delta x}$ represents the instantaneous rate of change of the area $A(x)$ at x. Since $\lim\limits_{\Delta x \to 0} \dfrac{\Delta A}{\Delta x} = f(x)$, we have

$$\frac{dA}{dx} = f(x)$$

To find $A(x)$ we integrate $\dfrac{dA}{dx} = f(x)$:

$$A(x) = \int f(x)\,dx = F(x) + C$$

where $F(x)$ is an antiderivative of $f(x)$. To determine the appropriate value for the constant of integration C, note that $A(a) = 0$. (When $x = a$, there is no area under the curve.) Then

$$0 = F(a) + C \quad \text{or} \quad C = -F(a)$$

The desired expression for $A(x)$ is then $A(x) = F(x) - F(a)$.

Finally, the area A under the curve, $y = f(x)$ from $x = a$ to $x = b$ is the value $A = F(b) - F(a)$. That is,

$$A = F(b) - F(a) \quad \text{where} \quad F(x) = \int f(x)\,dx$$

The constant of integration is not involved in the formula for A. Any antiderivative $F(x)$ for $f(x)$ can be used.

EXAMPLE 2

Find the area under the curve $y = x^2$ from $x = 1$ to $x = 2$.

$$\int x^2\,dx = \frac{x^3}{3} + C$$

Using the antiderivative $F(x) = x^3/3$, we have the desired area in Fig. 5.7.

$$A = F(2) - F(1) = \frac{(2)^3}{3} - \frac{(1)^3}{3} = \frac{8}{3} - \frac{1}{3} = \frac{7}{3}$$

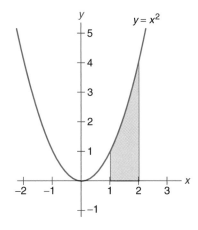

Figure 5.7

EXAMPLE 3

Revisiting the problem of Example 1, we can see geometrically how an antiderivative function is related to the area under the curve. Using Fig. 5.8, find a formula for the area under $y = 3x$ from $x = 0$ to an arbitrary value of x. Just think of the straight line $y = 3x$ as all points of the form $(x, 3x)$. A formula for the area of this triangle is

$$A = \frac{1}{2}bh = \frac{1}{2}(x)(3x) = \frac{3x^2}{2}$$

by comparison, $\int 3x\, dx = \dfrac{3x^2}{2} + C$

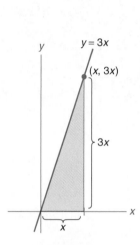 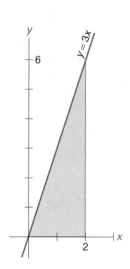

Figure 5.8   Figure 5.9

EXAMPLE 4

Find the area under the curve $y = 3x$ from $x = 0$ to $x = 2$ as in Fig. 5.9.

Note: We have already computed this area by the summation method in Example 1.

$$\int 3x\, dx = \frac{3x^2}{2} + C$$

Using the antiderivative $F(x) = \dfrac{3x^2}{2}$, we have

$$A = F(2) - F(0) = \frac{3(2)^2}{2} - \frac{3(0)^2}{2} = 6 - 0 = 6$$

EXAMPLE 5

Find the area bounded by $y = 2 + x - x^2$ and the x-axis.

First, graph the equation by finding the x-intercepts (see Fig. 5.10):

$$y = 2 + x - x^2 = 0$$
$$(2 - x)(1 + x) = 0$$
$$x = 2, -1$$

Then

$$\int (2 + x - x^2)\, dx = 2x + \frac{x^2}{2} - \frac{x^3}{3} + C$$

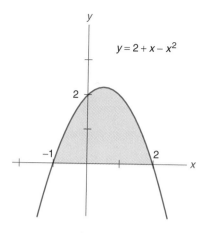

Figure 5.10

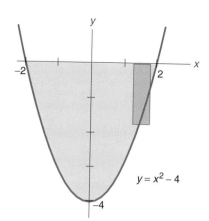

Figure 5.11

and

$$A = F(2) - F(-1) = \left[2(2) + \frac{2^2}{2} - \frac{2^3}{3} \right] - \left[2(-1) + \frac{(-1)^2}{2} - \frac{(-1)^3}{3} \right]$$

$$= \left(4 + 2 - \frac{8}{3} \right) - \left(-2 + \frac{1}{2} + \frac{1}{3} \right) = \frac{9}{2}$$

EXAMPLE 6

Find the area bounded by $y = x^2 - 4$ and the x-axis.

First, graph the equation (see Fig. 5.11).

Then

$$\int (x^2 - 4) \, dx = \frac{x^3}{3} - 4x + C$$

and

$$A = F(2) - F(-2) = \left[\frac{(2)^3}{3} - 4(2) \right] - \left[\frac{(-2)^3}{3} - 4(-2) \right]$$

$$= \left[\frac{8}{3} - 8 \right] - \left[-\frac{8}{3} + 8 \right]$$

$$= -\frac{32}{3}$$

Note: An area below the x-axis consists of approximating rectangles whose heights, $f(x) < 0$, and whose widths, $\Delta x > 0$, yield negative products. In such cases, the area is found by finding the absolute value of this result. The area in Example 6 is thus $\left| -\frac{32}{3} \right|$ or $\frac{32}{3}$. As we shall see in Section 6.1, complications occur when finding the area between the x-axis and a curve that crosses the x-axis.

Exercises 5.3

Find the area under each curve.

1. $y = x$ from $x = 0$ to $x = 2$

2. $y = 5x$ from $x = 1$ to $x = 4$

3. $y = 2x^2$ from $x = 1$ to $x = 3$

4. $y = x^2 + 3$ from $x = 0$ to $x = 2$

5. $y = 3x^2 - 2x$ from $x = 1$ to $x = 2$

6. $y = 4 - 3x^2$ from $x = 0$ to $x = 1$

7. $y = \dfrac{3}{x^2}$ from $x = 1$ to $x = 2$ **8.** $y = \dfrac{1}{x^3}$ from $x = 2$ to $x = 3$

9. $y = \sqrt{3x - 2}$ from $x = 1$ to $x = 2$ **10.** $y = \sqrt{2x + 3}$ from $x = 0$ to $x = 3$

11. $y = 4x - x^3$ from $x = 0$ to $x = 2$ **12.** $y = x^3 - 16x$ from $x = -4$ to $x = 0$

13. $y = 1 - x^4$ from $x = -1$ to $x = 1$ **14.** $y = 16 - x^4$ from $x = -2$ to $x = 2$

15. $y = \sqrt{2x + 1}$, $x = 4$ to $x = 12$

16. $y = \sqrt{4 - 3x}$ bounded by the coordinate axes

17. $y = \sqrt[3]{x - 1}$ bounded by the coordinate axes

18. $y = \sqrt[3]{8x + 27}$ bounded by the coordinate axes

19. $y = 1/x^2$ from $x = 1$ to $x = 5$ **20.** $y = x^4$ from $x = 1$ to $x = 3$

21. $y = \dfrac{1}{(2x - 1)^2}$ from $x = -3$ to y-axis   **22.** $y = \dfrac{1}{(2x + 3)^3}$ from $x = 3$ to y-axis

Find the area bounded by each curve and the x-axis.

23. $y = 9 - x^2$ **24.** $y = 12 - 3x^2$ **25.** $y = 2x - x^2$ **26.** $y = 5x - 2x^2$

27. $y = x^2 - x^3$ **28.** $y = 4x^2 - x^3$ **29.** $y = x^2 - x^4$ **30.** $y = 9x^2 - x^4$

5.4 THE DEFINITE INTEGRAL

In Section 5.3, we first found the area A under the curve $y = f(x)$ from $x = a$ to $x = b$ by the summation method, that is,

$$A = \lim_{n \to \infty} S_n$$

where $S_n = f(t_1)\,\Delta x + f(t_2)\,\Delta x + \cdots + f(t_k)\,\Delta x + \cdots + f(t_n)\,\Delta x$, where Δx is the width of each rectangle used to approximate the area under the curve, and $f(t_k)$ is the height of the kth rectangle where t_k is some point along its base.

We then found that we could determine this same area A using integration, that is,

$$A = F(b) - F(a)$$

where

$$\int f(x)\,dx = F(x) \qquad [F'(x) = f(x)]$$

so that

$$A = \lim_{n \to \infty} S_n = F(b) - F(a)$$

The symbol $\int$ indicates a type of summation. The summation process can be applied to situations not involving areas. With few restrictions on the function $y = f(x)$, we can ask whether $\lim_{n \to \infty} S_n$ exists when we consider values of x between a and b. If the limit does exist (unlike our examples in Section 5.3, it need not exist), then we make the following definition: the *definite integral* of a function $y = f(x)$ from $x = a$ to $x = b$ is the number $\lim_{n \to \infty} S_n$. We symbolize this number by

$$\int_a^b f(x)\,dx$$

That is,

$$\int_a^b f(x)\,dx = \lim_{n \to \infty} S_n$$

Note: The definite integral $\int_a^b f(x)\,dx$ is a number and is not to be confused with the indefinite integral $\int f(x)\,dx$ which is a family of functions. The number a is called the *lower limit* of the integral, and b is called the *upper limit.*

A fundamental result of the calculus (as found in Section 5.3) states that

DEFINITE INTEGRAL

$$\int_a^b f(x)\,dx = F(b) - F(a) \quad \text{where} \quad F'(x) = f(x)$$

Thus the technique for evaluating a definite integral is the same as for finding the area under a curve.

EXAMPLE 1

Evaluate the integral $\displaystyle\int_1^3 x^3\,dx$.

Since $\displaystyle\int x^3\,dx = \frac{x^4}{4} + C$, then $F(x) = \dfrac{x^4}{4}$, and

$$\int_1^3 x^3\,dx = F(3) - F(1)$$

$$= \frac{(3)^4}{4} - \frac{(1)^4}{4}$$

$$= \frac{81}{4} - \frac{1}{4}$$

$$= 20$$

We now introduce a shorthand notation for evaluating $F(b) - F(a)$:

$$F(x)\Big|_a^b = F(b) - F(a)$$

Remember that $F(x)$ can be any antiderivative of a given function $f(x)$, since $[F(b) + C] - [F(a) + C] = F(b) - F(a)$.

EXAMPLE 2

Evaluate $\displaystyle\int_1^2 (x^2 + 3x)\,dx$.

Since $\displaystyle\int (x^2 + 3x)\,dx = \frac{x^3}{3} + \frac{3x^2}{2} + C$, we have

$$\int_1^2 (x^2 + 3x)\,dx = \left(\frac{x^3}{3} + \frac{3x^2}{2}\right)\Bigg|_1^2$$

$$= \left[\frac{(2)^3}{3} + \frac{3(2)^2}{2}\right] - \left[\frac{(1)^3}{3} + \frac{3(1)^2}{2}\right]$$

$$= \left(\frac{8}{3} + \frac{12}{2}\right) - \left(\frac{1}{3} + \frac{3}{2}\right)$$

$$= \frac{41}{6}$$

EXAMPLE 3

Evaluate $\int_1^3 \frac{dx}{x^3}$.

$$\int_1^3 \frac{dx}{x^3} = \int_1^3 x^{-3}\, dx$$

$$= \frac{x^{-3+1}}{-3+1}\bigg|_1^3 = -\frac{1}{2x^2}\bigg|_1^3$$

$$= \left[-\frac{1}{2(3)^2}\right] - \left[-\frac{1}{2(1)^2}\right]$$

$$= -\frac{1}{18} + \frac{1}{2}$$

$$= \frac{8}{18} = \frac{4}{9}$$

EXAMPLE 4

Evaluate $\int_{-2}^1 x^2\, dx$.

$$\int_{-2}^1 x^2\, dx = \frac{x^3}{3}\bigg|_{-2}^1 = \frac{(1)^3}{3} - \frac{(-2)^3}{3} = \frac{1}{3} + \frac{8}{3} = 3$$

EXAMPLE 5

Evaluate $\int_1^0 x(x^2 + 1)^2\, dx$.

Let $u = x^2 + 1$, then

$$du = \frac{d}{dx}(x^2 + 1)\, dx = 2x\, dx$$

$$\boxed{\begin{array}{l} u = x^2 + 1 \\ du = 2x\, dx \end{array}}$$

So

$$\int x(x^2 + 1)^2\, dx = \int u^2 \frac{du}{2}$$

$$= \frac{1}{2} \cdot \frac{u^3}{3} + C$$

$$= \frac{u^3}{6} + C = \frac{(x^2 + 1)^3}{6} + C$$

Then

$$\int_1^0 x(x^2 + 1)^2\, dx = \frac{(x^2 + 1)^3}{6}\bigg|_1^0$$

$$= \left[\frac{(0 + 1)^3}{6}\right] - \left[\frac{(1 + 1)^3}{6}\right]$$

$$= \frac{1}{6} - \frac{8}{6} = -\frac{7}{6}$$

Note that in Example 5, the definite integral is negative. There are no restrictions on the value of a definite integral. The choice of upper and lower limits depends on the application of the definite integral. One can show, however, that

$$\int_a^b f(x)\, dx = -\int_b^a f(x)\, dx$$

EXAMPLE 6

Evaluate

$$\int_0^1 \frac{2x + 3}{(x^2 + 3x - 7)^2}\, dx$$

$$\int \frac{2x + 3}{(x^2 + 3x - 7)^2}\, dx = \int \frac{du}{u^2}$$

$$\boxed{\begin{aligned} u &= x^2 + 3x - 7 \\ du &= (2x + 3)\, dx \end{aligned}}$$

$$= \int u^{-2}\, du$$

$$= -u^{-1} + C$$

$$= -\frac{1}{x^2 + 3x - 7} + C$$

So

$$\int_0^1 \frac{2x + 3}{(x^2 + 3x - 7)^2}\, dx = -\frac{1}{x^2 + 3x - 7}\bigg|_0^1$$

$$= \left(-\frac{1}{1 + 3 - 7}\right) - \left(-\frac{1}{-7}\right)$$

$$= \frac{1}{3} - \frac{1}{7}$$

$$= \frac{4}{21}$$

2nd 7 $(2x+3)/(x^2+3x-7)^2,x,0,1)$ **ENTER**

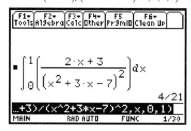

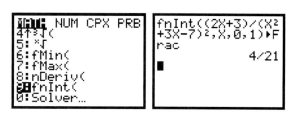

MATH 9 $(2x+3)/(x\ x^2+3x-7)\ x^2,x,0,1)$ **MATH 1 ENTER**

Exercises 5.4

Evaluate each definite integral.

1. $\displaystyle\int_0^1 5x\, dx$

2. $\displaystyle\int_0^1 x^4\, dx$

3. $\displaystyle\int_1^2 (x^2 + 3)\, dx$

4. $\displaystyle\int_2^4 (3x^2 + x - 1)\, dx$

5. $\displaystyle\int_2^0 (x^3 + 1)\, dx$

6. $\displaystyle\int_1^0 (3x^4 - 8)\, dx$

7. $\displaystyle\int_{-1}^{1} (x^2 + x + 2)\, dx$ **8.** $\displaystyle\int_{-2}^{0} (2x^3 - 4x)\, dx$ **9.** $\displaystyle\int_{4}^{9} (3x^{1/2} + x^{-1/2})\, dx$

10. $\displaystyle\int_{8}^{27} (5x^{2/3} + 4x^{-1/3})\, dx$ **11.** $\displaystyle\int_{1}^{9} \frac{x+3}{\sqrt{x}}\, dx$ **12.** $\displaystyle\int_{1}^{4} \frac{x^2 + 3x - 2}{\sqrt{x}}\, dx$

13. $\displaystyle\int_{1}^{2} (3x + 4)^4\, dx$ **14.** $\displaystyle\int_{0}^{1} (2x + 1)^5\, dx$ **15.** $\displaystyle\int_{0}^{16} \sqrt{2x + 4}\, dx$

16. $\displaystyle\int_{1}^{44} \sqrt[3]{5x - 4}\, dx$ **17.** $\displaystyle\int_{1}^{2} 4x(x^2 - 3)^3\, dx$ **18.** $\displaystyle\int_{0}^{1} 6x^2(x^3 + 1)^4\, dx$

19. $\displaystyle\int_{-1}^{0} x(1 - x^2)^{2/3}\, dx$ **20.** $\displaystyle\int_{-2}^{0} x\sqrt{4 - x^2}\, dx$ **21.** $\displaystyle\int_{0}^{1} x\sqrt{x^2 + 1}\, dx$

22. $\displaystyle\int_{0}^{2} x^2\sqrt{x^3 + 2}\, dx$ **23.** $\displaystyle\int_{0}^{4} \frac{6x}{\sqrt{x^2 + 9}}\, dx$ **24.** $\displaystyle\int_{-1}^{1} \frac{16x\, dx}{\sqrt[3]{3x^2 + 5}}$

25. $\displaystyle\int_{1}^{2} \frac{3x^2 + 1}{\sqrt{x^3 + x}}\, dx$ **26.** $\displaystyle\int_{2}^{3} \frac{2x + 1}{\sqrt{x^2 + x - 2}}\, dx$

CHAPTER SUMMARY

1. If $f(x)$ is a function, then $F(x)$ is the antiderivative of $f(x)$ if $F'(x) = f(x)$.

2. *Integration formulas:*

(a) $\displaystyle\int x^n\, dx = \frac{x^{n+1}}{n+1} + C \qquad (n \neq -1)$

(b) $\displaystyle\int [f(x) + g(x)]\, dx = \int f(x)\, dx + \int g(x)\, dx$

(c) $\displaystyle\int k f(x)\, dx = k\int f(x)\, dx$, where k is a constant

3. The area under the curve $y = f(x)$ in Fig. 5.12 from $x = a$ to $x = b$ is $A = F(b) - F(a)$.

4. $\displaystyle\int_{a}^{b} f(x)\, dx = F(b) - F(a)$

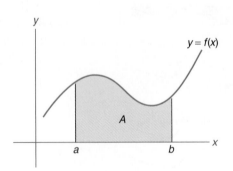

Figure 5.12

CHAPTER 5 REVIEW

Integrate.

1. $\displaystyle\int (5x^2 - x)\, dx$

2. $\displaystyle\int (3x^7 + 2x + 4)\, dx$

3. $\displaystyle\int 6\sqrt{x^7}\, dx$

4. $\displaystyle\int 4x^{2/3}\, dx$

5. $\displaystyle\int \frac{3\, dx}{x^5}$

6. $\displaystyle\int \frac{dx}{\sqrt{x^3}}$

7. $\displaystyle\int (6x^3 + 1)(3x^4 + 2x - 1)^3\, dx$

8. $\displaystyle\int (7x + 4)(7x^2 + 8x + 2)^{3/5}\, dx$

9. $\displaystyle\int \frac{2x + 5}{\sqrt{x^2 + 5x}}\, dx$

10. $\displaystyle\int (15x^2 + 4)(5x^3 + 4x)^{-2/3}\, dx$

11. Find the equation of the curve $y = f(x)$ passing through $(1, -3)$ whose slope is given by $\dfrac{dy}{dx} = 3x^2$.

12. A stone is thrown vertically upward from a cliff 100 ft high with an initial velocity of 25 ft/s. Find the equation describing the altitude of the stone from the ground below $(a = -32 \text{ ft/s}^2)$.

13. The rate of change of resistance with respect to temperature of an electrical resistor is given by $\dfrac{dR}{dT} = 0.009T^2 + 0.02T - 0.7$. Find the resistance when the temperature is $30°C$ if $R = 0.2\ \Omega$ when $T = 0°C$.

14. The current in an electrical circuit is given by $i = \dfrac{3t^2 + 1}{\sqrt{t^3 + t + 2}}$ amperes. Find the charge (in coulombs) that passes a given point in the circuit after 0.2 s. (Assume $q = 2\sqrt{2}$ at $t = 0$.)

Find the area under each curve.

15. $y = x^2 + 1$ from $x = 0$ to $x = 2$

16. $y = 8 - 6x^2$ from $x = 0$ to $x = 1$

17. $y = \dfrac{1}{x^5}$ from $x = 1$ to $x = 2$

18. $y = \dfrac{4x}{(x^2 + 1)^2}$ from $x = 0$ to $x = 1$

19. $y = \sqrt{5x + 6}$ from $x = 0$ to $x = 6$

20. $y = x\sqrt{x^2 + 4}$ from $x = 0$ to $x = 2$

Evaluate each definite integral.

21. $\displaystyle\int_0^1 (x^3 + 2x^2 + x)\, dx$

22. $\displaystyle\int_1^2 (3x^4 - 2x^3 + 7x)\, dx$

23. $\displaystyle\int_0^2 x(x^2 + 1)^2\, dx$

24. $\displaystyle\int_1^2 \frac{5x^2}{(x^3 + 2)^2}\, dx$

25. $\displaystyle\int_1^2 (2x + 1)\sqrt{x^2 + x}\, dx$

26. $\displaystyle\int_0^3 \frac{x^2}{\sqrt{x^3 + 1}}\, dx$

27. $\displaystyle\int_2^1 \left(3x^2 - x + \frac{1}{x^2}\right) dx$

28. $\displaystyle\int_0^{1/2} \frac{3x}{\sqrt{2x^2 + \frac{1}{2}}}\, dx$

6

Applications of Integration

INTRODUCTION

Kurt's basement was flooded with 3.5 feet of water after a heavy rainfall. The amount of *work* required to pump the water out of the basement can be calculated using integrals. In this chapter we study some of the technical applications of integrals.

Objectives

- Find an area bounded by two or more curves.
- Find a volume of revolution.
- Find the center of mass of a linear system.
- Find the centroid of a region bounded by given curves.
- Find moments of inertia.
- Use integration to solve work, fluid pressure, and average value problems.

6.1 AREA BETWEEN CURVES

In Chapter 5 we studied the process of computing the area between a given curve and the *x*-axis. Now consider the problem of finding the area *between* two given curves. First, observe that the area between one given curve $y = f(x)$ and the *x*-axis between $x = a$ and $x = b$ is a definite integral:

$$A = \int_a^b f(x)\, dx = F(b) - F(a)$$

where $F'(x) = f(x)$.

Further recall that definite integration as applied to area is a summation process. That is, the definite integral is the limit of sums of approximating rectangles where the area of a typical rectangle is $f(t)\, \Delta x$ as shown in Fig. 6.1. The typical rectangle, which is shaded, is called an *element* of the area. Note the correspondence between the form of the expression for the area of the element, $f(t)\, \Delta x$, and for the differential, $f(x)\, dx$, which is to be integrated:

$$f(t)\, \Delta x \leftrightarrow f(x)\, dx$$

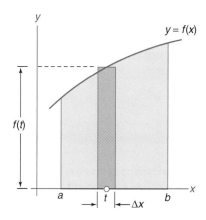

Figure 6.1 Area under the curve $y = f(x)$.

Noting this correspondence will be a visual aid in setting up the appropriate integrals for computing areas between curves. The correct form for $f(x)\,dx$ can be found by finding the appropriate expression $f(t)\,\Delta x$ from viewing a sketch of the area.

Now let's determine the area bounded by the curves $y = f(x)$, $y = g(x)$, $x = a$, and $x = b$ as shown in Fig. 6.2.

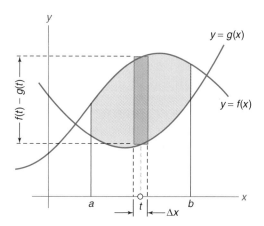

Figure 6.2 Area between two curves.

Between a and b, $g(x) \le f(x)$; that is, the curve $y = g(x)$ lies below the curve $y = f(x)$. The area of the element shown, used in approximating the area between the two curves, is

$$[f(t) - g(t)]\,\Delta x$$

This corresponds to the differential

$$[f(x) - g(x)]\,dx$$

We therefore use the definite integral

$$\int_a^b [f(x) - g(x)]\,dx$$

to find the desired area.

EXAMPLE 1

Find the area between the curves $y = 8 - x^2$ and $y = x + 2$.

The area is bounded by a parabola and a straight line. First, find the points where the two curves intersect by solving the two equations simultaneously.

$$8 - x^2 = x + 2$$
$$0 = x^2 + x - 6$$
$$0 = (x - 2)(x + 3)$$
$$x = 2, x = -3$$

The curves intersect at the points $(-3, -1)$ and $(2, 4)$. Note that between $x = -3$ and $x = 2$, the line $y = x + 2$ is below the parabola $y = 8 - x^2$. The length of the element shown in Fig. 6.3 is the difference between the upper curve $y = 8 - x^2$ and the lower curve $y = x + 2$ at a given value of x.

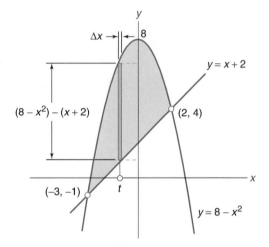

Figure 6.3

The area A between the curves is the value of the definite integral.

$$A = \int_{-3}^{2} [(8 - x^2) - (x + 2)] \, dx$$
$$= \int_{-3}^{2} (6 - x - x^2) \, dx$$
$$= \left(6x - \frac{x^2}{2} - \frac{x^3}{3} \right)\Big|_{-3}^{2}$$
$$= \left(12 - 2 - \frac{8}{3} \right) - \left(-18 - \frac{9}{2} + 9 \right) = 20\frac{5}{6}$$

The limits of integration are determined by the points where the curves intersect. The lower limit is the smallest value of x where the curves intersect, and the upper limit is the largest value of x where the curves intersect.

EXAMPLE 2

Find the area between the line $y = -x + 2$ and the parabola $x = 4 - y^2$.

The points of intersection of these two curves are $(0, 2)$ and $(3, -1)$. In this example, we run into a problem. If we use vertical elements to approximate the area as in Fig. 6.4, then

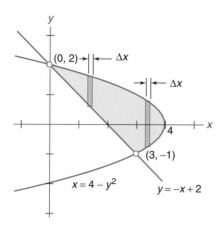

Figure 6.4

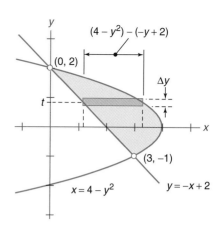

Figure 6.5

between $x = 0$ and $x = 3$ the height of the element is the difference between the curve $x = 4 - y^2$ $(y = \sqrt{4-x})$ and $y = -x + 2$. And between $x = 3$ and $x = 4$ the height of the element is the difference between $y = \sqrt{4-x}$ and $y = -\sqrt{4-x}$. Because of the change in boundaries at $x = 3$, we would have to find the desired area A by separately computing the areas

$$A_1 = \int_0^3 [(\sqrt{4-x}) - (-x + 2)]\, dx$$

and

$$A_2 = \int_3^4 [(\sqrt{4-x}) - (-\sqrt{4-x})]\, dx$$

Then $A = A_1 + A_2$. By integrating, we find $A_1 = \frac{19}{6}$ and $A_2 = \frac{4}{3}$. So

$$A = \frac{19}{6} + \frac{4}{3} = \frac{9}{2}$$

Sometimes, as in this example, it is easier to set up the problem using horizontal elements as in Fig. 6.5. Then express the given curves as functions of y instead of x: $x = 4 - y^2$ and $x = -y + 2$, and integrate with respect to the y variable and use limits of integration based on the y-coordinates of the points of intersection: $y = -1$ and $y = 2$. With respect to *the y-axis*, the curve $x = 4 - y^2$ lies above $x = -y + 2$ from $y = -1$ to $y = 2$. The length of a typical horizontal element is then $(4 - y^2) - (-y + 2)$.

$$A = \int_{-1}^2 [(4 - y^2) - (-y + 2)]\, dy$$

$$= \int_{-1}^2 (2 + y - y^2)\, dy$$

$$= \left(2y + \frac{y^2}{2} - \frac{y^3}{3}\right)\Big|_{-1}^2$$

$$= \left(4 + 2 - \frac{8}{3}\right) - \left(-2 + \frac{1}{2} + \frac{1}{3}\right) = \frac{9}{2}$$

Note: Using horizontal elements and working in terms of dy is usually simpler than using vertical elements and working in terms of dx when one of the curves does not represent a function (contains a y^2-term, for instance).

In summary, to find the area between two given curves between $x = a$ and $x = b$:

1. Find the points of intersection of the two curves, if necessary.
2. Sketch the two curves:
   (a) Determine whether to use vertical elements with the curves expressed as functions of x or horizontal elements with curves expressed as functions of y.
   (b) Determine which curve lies above the other.
3. Find the height of a typical element based in Step 2(b).
4. Write the definite integral:

$$\int_a^b [f(x) - g(x)]\, dx$$

where $f(x) - g(x)$ is the length of vertical elements between $x = a$ and $x = b$ with $a < b$, or

$$\int_c^d [f(y) - g(y)]\, dy$$

where $f(y) - g(y)$ is the length of horizontal elements between $y = c$ and $y = d$ with $c < d$.

Note that the choice of using vertical or horizontal elements depends on the difficulty of the resulting definite integral. Also, the curves do not need to lie above the x- or y-axis in order to find the area between them.

EXAMPLE 3

Find the area between the curves $x = y^3$ and $x = -y^2$.

The points of intersection are $(0, 0)$ and $(-1, -1)$. We could use either vertical or horizontal elements. We will use horizontal elements as in Fig. 6.6. Since the curve $x = y^3$ lies *above* $x = -y^2$ (*in the positive direction along the x-axis*) from $y = -1$ to $y = 0$, we have

$$A = \int_{-1}^0 [y^3 - (-y^2)]\, dy$$

$$= \left(\frac{y^4}{4} + \frac{y^3}{3}\right)\Big|_{-1}^0$$

$$= 0 - \left(\frac{1}{4} - \frac{1}{3}\right) = \frac{1}{12}$$

Using a calculator to evaluate, we have

2nd 7 y^3-(-y^2),y,-1,0) **ENTER**

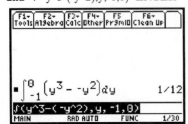 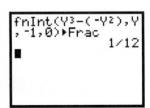

MATH 9 ALPHA Y MATH 3 -(-ALPHA Y x^2), ALPHA Y ,-1,0) MATH 1 ENTER

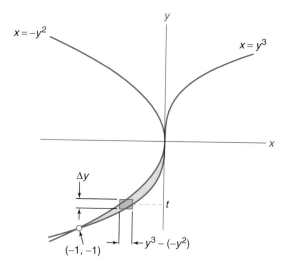

Figure 6.6

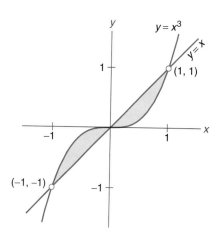

Figure 6.7

EXAMPLE 4

Find the area between the curves $y = x$ and $y = x^3$.

There are three points of intersection: $(-1, -1), (0, 0),$ and $(1, 1)$. Between $x = -1$ and $x = 0, y = x^3$ lies above $y = x$. Between $x = 0$ and $x = 1, y = x$ lies above $y = x^3$. We need to compute the two areas as in Fig. 6.7.

$$A_1 = \int_{-1}^{0} (x^3 - x)\, dx \qquad \text{and} \qquad A_2 = \int_{0}^{1} (x - x^3)\, dx$$

$$A_1 = \left(\frac{x^4}{4} - \frac{x^2}{2}\right)\Big|_{-1}^{0} = \frac{1}{4} \quad \text{and} \quad A_2 = \left(\frac{x^2}{2} - \frac{x^4}{4}\right)\Big|_{0}^{1} = \frac{1}{4}$$

The desired area $A = A_1 + A_2 = \frac{1}{4} + \frac{1}{4} = \frac{1}{2}$.

Using the symmetry of the two areas, note that $A_1 = A_2$. We could also compute

$$A = 2\int_{0}^{1} (x - x^3)\, dx$$

EXAMPLE 5

Find the area between the x-axis and $y = x^2 - 4$.

The desired region lies between $x = -2$ and $x = 2$. Note that in this region the curve $y = x^2 - 4$ lies below the x-axis in Fig. 6.8. Since the x-axis is the curve $y = 0$, we have

$$A = \int_{-2}^{2} [0 - (x^2 - 4)]\, dx$$

$$= \int_{-2}^{2} (4 - x^2)\, dx$$

$$= \left(4x - \frac{x^3}{3}\right)\Big|_{-2}^{2}$$

$$= \left(8 - \frac{8}{3}\right) - \left(-8 + \frac{8}{3}\right) = \frac{32}{3}$$

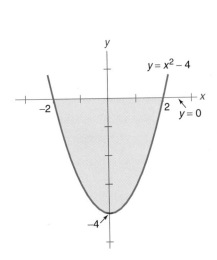

Figure 6.8

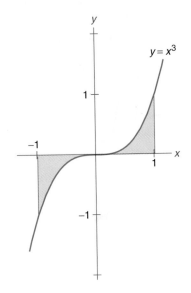

Figure 6.9

EXAMPLE 6

Find the area between the x-axis and $y = x^3$ from $x = -1$ to $x = 1$.

You may be tempted to simply form the integral $\displaystyle\int_{-1}^{1} x^3\, dx$. But

$$\int_{-1}^{1} x^3\, dx = \frac{x^4}{4}\bigg|_{-1}^{1} = \frac{1}{4} - \frac{1}{4} = 0$$

We have incorrectly obtained the value zero since we failed to observe that at $x = 0$ the curve $y = x^3$ crosses the x-axis as in Fig. 6.9. To the left of $x = 0$ the curve is below the x-axis, but to the right of $x = 0$ the curve is above the x-axis. We must therefore separate the computation into two integrals as in Example 4:

$$A = \int_{-1}^{0} [0 - (x^3)]\, dx + \int_{0}^{1} [(x^3) - 0]\, dx$$

$$= -\int_{-1}^{0} x^3\, dx + \int_{0}^{1} x^3\, dx$$

$$= -\left(\frac{x^4}{4}\right)\bigg|_{-1}^{0} + \frac{x^4}{4}\bigg|_{0}^{1}$$

$$= -\left(-\frac{1}{4}\right) + \frac{1}{4} = \frac{1}{2}$$

Exercises 6.1

Find each area bounded by the curves.

1. $y = x^2$, $y = 0$, and $x = 1$

2. $y = x^2$, $y = 0$, $x = 1$, and $x = 2$

3. $y = 1 - x$, $x = 0$, and $y = 0$

4. $y = 2x$, $y = 0$, $x = 1$, and $x = 2$

5. $y = 2 - x^2$ and $y + x = 0$

6. $y = x^2 - 2x$ and $y = 3$

7. $y^2 = x$ and $x = 4$

8. $y^2 = 4x$, $x = 0$, $y = -1$, and $y = 4$

9. $y = x^2$ and $y = x$

10. $y = 2x^2$ and $y^2 = 4x$

11. $x = y^2 - 2y$ and $y = x$

12. $y = x^3, y = 2 - x^2, x = 0,$ and $x = 1$

13. $y = x^3 - x, y = 0, x = -1,$ and $x = 1$

14. $y^3 = x, y = 1,$ and $x = -1$

15. $x = y + 1$ and $x = 3 - y^2$

16. $x = y^2$ and $x = y + 2$

17. $y = 4 - 4x^2$ and $y = 1 - x^2$

18. $y = x^2$ and $y = 8 - x^2$

19. $x = y^4$ and $x = 2 - y^2$

20. $x = y^2$ and $x = 4 + 2y - y^2$

21. $y = x^2 - 3x - 4$ and $y = 6$

22. $y = x^2$ and $y = \sqrt{x}$

23. $x^2 y = 8, y = x, x = 5,$ and $y = 0$

24. $y = x, x + y = 6,$ and $2y = x$

25. $y = x(x - 1)(x - 3)$ and the x-axis

26. $y = x(x + 3)(x - 2)$ and the x-axis

6.2 VOLUMES OF REVOLUTION: DISK METHOD

Another application of integration is finding the volume of a solid resulting from rotating an area about an axis. For example, consider the region bounded by the curves $y = f(x), x = a, x = b,$ and the x-axis. Revolving this region about the x-axis determines a solid figure as in Fig. 6.10(a).

The area ΔA of a typical rectangle used in Chapter 5 to approximate the area under the curve $y = f(x)$ is $\Delta A = f(t)\,\Delta x$, where t is a point at the base of the rectangle, Δx is the width, and $f(t)$ is the height. If we now rotate this area about the x-axis as in Fig. 6.10(b), we obtain a cylindrical disk with volume

$$\Delta V = \pi r^2 h$$

where r, the radius, is $f(t)$ and h, the width, is Δx. So

$$\Delta V = \pi [f(t)]^2\,\Delta x$$

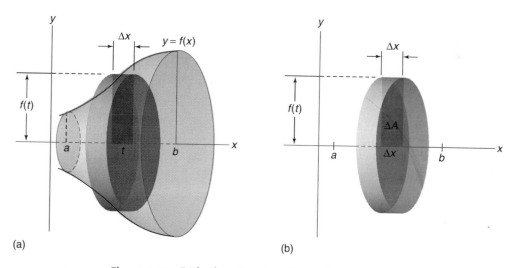

(a)

(b)

Figure 6.10 Disk of a solid of revolution about the x-axis.

By a method similar to approximating the area under a curve by rectangles, we approximate the volume of revolution by using the sum of the volumes of differential disks (see Fig. 6.11).

For areas, the integral

$$A = \int_a^b f(x)\,dx$$

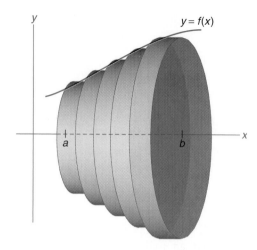

Figure 6.11 Approximating the volume of revolution about the *x*-axis by using the sum of the volumes of differential disks.

gives the exact area for a region that is approximated by summing the areas of rectangles:

$$\Delta A = f(t)\, \Delta x$$

In a similar manner, the integral

$$V = \pi \int_a^b [f(x)]^2\, dx = \pi \int_a^b y^2\, dx$$

gives the exact volume for the solid of revolution about the *x*-axis which is approximated by summing the volumes of disks: $\Delta V = \pi[f(t)]^2\, \Delta x$. This method of computing the volume of a solid is called the *disk method*.

CIRCULAR DISK METHOD

$V =$ sum of circular disks

$$\overset{\text{radius}^2}{\downarrow} \quad \overset{\text{thickness}}{\downarrow}$$

$$V = \pi \int_a^b [f(x)]^2 \quad dx \qquad \text{(revolved about } x\text{-axis)}$$

$$V = \pi \int_c^d [f(y)]^2 \quad dy \qquad \text{(revolved about } y\text{-axis)}$$

EXAMPLE 1

Find the volume of the solid formed by revolving the curve $y = x$ from $x = 0$ to $x = 2$ about the *x*-axis.

The area rotated about the *x*-axis is a triangle as in Fig. 6.12(a). The resulting solid is a cone. The volume of a typical differential disk in Fig. 6.12(b) is

$$\Delta V = \pi[f(t)]^2\, \Delta x = \pi(t)^2\, \Delta x$$

So the integral giving the exact volume is

$$V = \pi \int_a^b [f(x)]^2\, dx$$

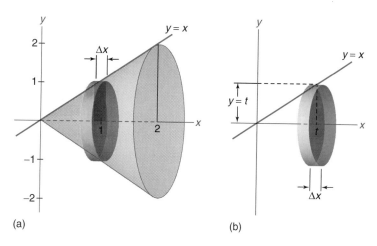

Figure 6.12

$$= \pi \int_0^2 x^2 \, dx \qquad (y = x)$$

$$= \pi \left. \frac{x^3}{3} \right|_0^2$$

$$= \pi \left[\frac{8}{3} - \frac{0}{3} \right]$$

$$= \frac{8\pi}{3}$$

This same volume could also be found using the formula for the volume of a cone from geometry, $V = \frac{1}{3}\pi r^2 h$. In this example, $r = 2$ (radius of the base of the cone) and $h = 2$ (the altitude). So

$$V = \frac{1}{3}\pi(2)^2(2) = \frac{8\pi}{3}$$

Although this problem could also be solved using a geometrical formula, this is not always the case. In the following example, integration provides the only solution.

EXAMPLE 2

Find the volume of the solid obtained by revolving the region bounded by the curves $y = x^2$, $y = 0$, and $x = 2$ about the x-axis.

The volume of a differential disk in Fig. 6.13 is

$$\Delta V = \pi [f(t)]^2 \, \Delta x = \pi [t^2]^2 \, \Delta x$$

The exact volume is then

$$V = \pi \int_a^b [f(x)]^2 \, dx = \pi \int_0^2 (x^2)^2 \, dx \qquad (y = x^2)$$

$$= \pi \int_0^2 x^4 \, dx$$

$$= \pi \left. \frac{x^5}{5} \right|_0^2$$

$$= \frac{32\pi}{5}$$

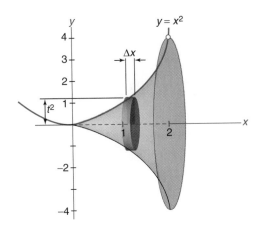

Figure 6.13

When a solid is formed by revolving an area about the y-axis, the integral giving the volume is

$$V = \pi \int_c^d [f(y)]^2\, dy = \pi \int_c^d x^2\, dy$$

Here express the radius of a differential disk as a distance x from the y-axis and the width of the disk as an increment of y, Δy (see Fig. 6.14).

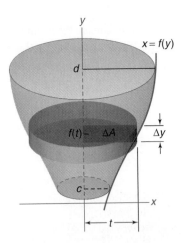

Figure 6.14   Disk of a solid of revolution about the y-axis.

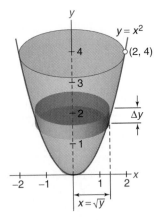

Figure 6.15

EXAMPLE 3

Find the volume of the solid obtained by revolving the region bounded by the curves $y = x^2$, $x = 0$, and $y = 4$ about the y-axis.

When revolving the area about the y-axis as in Fig. 6.15, the boundary curve $(y = x^2)$ must be determined in a manner that expresses x as a function of y. That is,

$$x = \sqrt{y} \quad (\text{for } 0 \le y \le 4)$$

Note that the boundary curve determines the radius of the differential disks.

The volume of the solid is then given by

$$V = \pi \int_c^d [f(y)]^2\, dy = \pi \int_0^4 (\sqrt{y})^2\, dy \qquad (x = \sqrt{y})$$

$$= \pi \int_0^4 y\, dy$$

$$= \pi \frac{y^2}{2}\Big|_0^4 = \pi \left(\frac{16}{2} - 0\right)$$

$$= 8\pi$$

EXAMPLE 4

Find the volume of the solid formed by revolving the region bounded by the curves $y = 4 - x^2$, $x = 0$, and $y = 0$ about the y-axis.

Expressing the radius of a differential disk in terms of x, we have

$$y = 4 - x^2$$
$$x^2 = 4 - y$$
$$x = \sqrt{4 - y}$$

The desired volume in Fig. 6.16 is then

$$V = \pi \int_c^d [f(y)]^2\, dy$$

$$= \pi \int_0^4 (\sqrt{4 - y})^2\, dy \qquad (x = \sqrt{4 - y})$$

$$= \pi \int_0^4 (4 - y)\, dy$$

$$= \pi \left(4y - \frac{y^2}{2}\right)\Big|_0^4$$

$$= \pi \left[\left(16 - \frac{16}{2}\right) - (0)\right]$$

$$= 8\pi$$

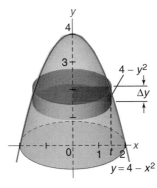

Figure 6.16

When the region between two curves is rotated as shown in Fig. 6.17 about the x-axis, the rotation results in a *washer*-type solid. If $y = f(x)$ is the outer radius and $y = g(x)$ is the inner radius of the region being revolved, the volume of the resulting solid is

$$V = \pi \int_a^b \{[f(x)]^2 - [g(x)]^2\}\, dx$$

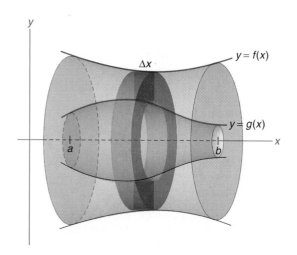

Figure 6.17 Washer-type solid of revolution about the *x*-axis.

EXAMPLE 5

Find the volume of the solid formed by revolving the region bounded by $y = x^2$ and $y = x$ about the *x*-axis.

From Fig. 6.18, we have

$$V = \pi \int_0^1 [(x)^2 - (x^2)^2]\, dx$$

$$= \pi \int_0^1 (x^2 - x^4)\, dx$$

$$= \pi \left(\frac{x^3}{3} - \frac{x^5}{5} \right) \Big|_0^1$$

$$= \pi \left[\left(\frac{1}{3} - \frac{1}{5} \right) - (0) \right]$$

$$= \frac{2\pi}{15}$$

EXAMPLE 6

Drill a hole of radius 3 in. through the center of a metal sphere of radius 5 in. Find the volume of the resulting ring.

First, rotate the shaded portion of the circle $x^2 + y^2 = 25$ about the *x*-axis as shown in Fig. 6.19. The outer radius is $y = \sqrt{25 - x^2}$ while its inner radius (radius of hole) is $y = 3$. Thus

$$V = \pi \int_{-4}^4 [(\sqrt{25 - x^2})^2 - (3)^2]\, dx$$

$$= \pi \int_{-4}^4 (16 - x^2)\, dx$$

$$= \pi \left(16x - \frac{x^3}{3} \right) \Big|_{-4}^4$$

$$= \pi \left[\left(64 - \frac{64}{3} \right) - \left(-64 + \frac{64}{3} \right) \right]$$

$$= \frac{256\pi}{3} \text{ in}^3$$

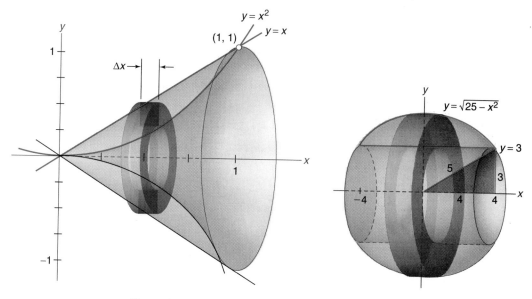

Figure 6.18

Figure 6.19

EXAMPLE 7

Find the volume of the solid obtained by revolving the region bounded by the curves $y = x^2$, $x = 2$, and the x-axis about the line $x = 2$.

Note that the radius of the differential disk in Fig. 6.20(b) is $2 - x$ and its thickness is Δy with $0 \le y \le 4$.

$$V = \pi \int_0^4 (2 - x)^2 \, dy$$

$$= \pi \int_0^4 (4 - 4x + x^2) \, dy$$

$$= \pi \int_0^4 (4 - 4\sqrt{y} + y) \, dy \qquad (x^2 = y \text{ and } x = \sqrt{y})$$

$$= \pi \left(4y - \frac{8}{3} y^{3/2} + \frac{y^2}{2} \right) \Big|_0^4$$

$$= \pi \left[\left(16 - \frac{8}{3} \cdot 8 + 8 \right) - (0) \right]$$

$$= \frac{8\pi}{3}$$

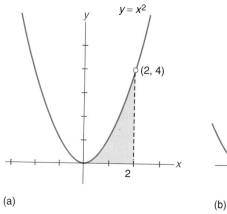

(a)

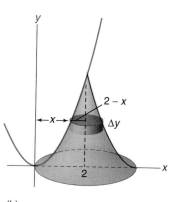

(b)

Figure 6.20

Exercises 6.2

Find the volume of each solid formed by revolving the region bounded by the given curves about the given line.

1. $y = x + 1$, $y = 0$, $x = 0$, and $x = 2$ about the x-axis
2. $y = x$, $y = 0$, $x = 2$, and $x = 4$ about the x-axis
3. $y = x^2 + 1$, $y = 0$, $x = 1$, and $x = 2$ about the x-axis
4. $y = \sqrt{x}$, $y = 0$, $x = 1$, and $x = 4$ about the x-axis
5. $y = x - 1$, $x = 0$, and $y = 1$ about the y-axis
6. $y^2 = 2x$, $x = 0$, and $y = 2$ about the y-axis
7. $y = 4x^2$, $x = 0$, and $y = 4$ about the y-axis
8. $y = 4 - x^2$, $x = 0$, $y = 1$, and $y = 2$ about the y-axis
9. $y = x$, $x = 1$, and $y = 0$ about $x = 1$
10. $y = x^2$, $x = 2$, and $y = 0$ about $x = 2$
11. $y = x$, $x = 1$, $x = 2$, and $y = 1$ about $y = 1$
12. $y = x^2$, $x = 1$, $x = 2$, and $y = 1$ about $y = 1$
13. $4y = x^2$, $y = 0$, and $x = 2$ about the y-axis
14. $y^2 = x$, $y = 0$, and $x = 4$ about the y-axis
15. $y^2 = x$, $y = 0$, and $x = 4$ about the x-axis
16. $y = x^3$ and $y = x$ from $x = 0$ to $x = 1$ about the x-axis
17. $y = x^3$ and $y = x$ from $x = 0$ to $x = 1$ about the y-axis
18. $y = 2 - x^2$, $y = x$, and $x = 0$ about the x-axis
19. $y = 2 - x^2$, $y = x$, and $x = 0$ about the y-axis
20. $2y = x$ and $y^2 = x$ about the y-axis
21. $2y = x$ and $y^2 = x$ about the x-axis
22. $x = y^2$ and $x = 2 - y^2$ about the y-axis
23. $y = 3 - x^2$ and $y = x^2 + 1$ about the x-axis
24. $x = 2y^2 + 1$ and $x = 4 - y^2$ about the y-axis
25. The area bounded by the first and second quadrants of the ellipse $9x^2 + 25y^2 = 225$ is revolved about the x-axis. Find the volume.
26. The area bounded by the first and fourth quadrants of the ellipse $9x^2 + 25y^2 = 225$ is revolved about the y-axis. Find the volume.
27. Drill a hole of radius 2 in. through the center of the solid (along the x-axis) described in Exercise 25. Find the volume of the resulting solid.
28. Drill a hole of radius 2 in. through the center of the solid (along the y-axis) described in Exercise 26. Find the volume of the resulting solid.
29. Use the disk method to verify that the volume of a sphere of radius r is $V = \frac{4}{3}\pi r^3$.
30. Use the disk method to verify that the volume of a right circular cone is $V = \frac{1}{3}\pi r^2 h$, where r is the radius of the base and h is the height.

6.3 VOLUMES OF REVOLUTION: SHELL METHOD

A second method of obtaining the volume of a solid uses cylindrical shells instead of disks. Let's use this method to find the volume of the solid described in Example 4 of Section 6.2.

The volume ΔV of a typical shell in Fig. 6.21(a) is

$$\Delta V = \pi r_2^2 h - \pi r_1^2 h$$

where

$$r_1 = t, \qquad \text{(radius of inside wall of the shell)}$$
$$r_2 = t + \Delta x \qquad \text{(radius of outside wall of the shell)}$$

Now

$$\begin{aligned}
\Delta V &= \pi r_2^2 h - \pi r_1^2 h \\
&= \pi h (r_2^2 - r_1^2) \\
&= \pi h (r_2 + r_1)(r_2 - r_1) \\
&= 2\pi h \left(\frac{r_2 + r_1}{2}\right)(r_2 - r_1) \qquad \text{(Multiply numerator and denominator by 2.)}
\end{aligned}$$

$$\underbrace{\phantom{2\pi h \left(\frac{r_2 + r_1}{2}\right)}}_{\text{average radius}}$$

or

$$\begin{aligned}
\Delta V &= 2\pi y \left(\frac{2t + \Delta x}{2}\right)(\Delta x) \qquad [r_2 + r_1 = (t + \Delta x) + t = 2t + \Delta x] \\
&= 2\pi y \left(t + \frac{\Delta x}{2}\right)\Delta x \\
&= 2\pi f(x)\, x\, \Delta x \qquad \left(\text{where we let } x = t + \frac{\Delta x}{2}\right)
\end{aligned}$$

This expression for ΔV is the product of the circumference of the shell of radius x, height y, and thickness Δx. By taking the sum of the volumes of all such approximating shells as in Fig. 6.21(b), we obtain another approximation for the desired volume.

Again using the methods of Chapter 5, this approximation leads to the integral

$$V = 2\pi \int_a^b x f(x)\, dx$$

which gives the exact volume of the solid.

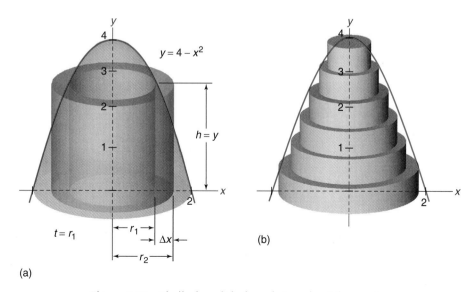

Figure 6.21   Shell of a solid of revolution about the y-axis.

Expressing y as a function of x we have $y = f(x) = 4 - x^2$ and

$$V = 2\pi \int_0^2 x(4 - x^2)\, dx$$

$$= 2\pi \int_0^2 (4x - x^3)\, dx$$

$$= 2\pi \left(2x^2 - \frac{x^4}{4}\right)\Big|_0^2 = 2\pi[(8 - 4) - (0)] = 8\pi$$

This method of computing the volume of a solid is called the *shell method.*

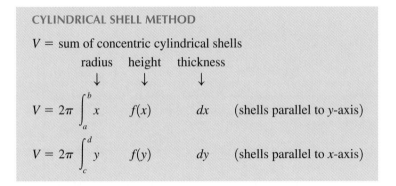

CYLINDRICAL SHELL METHOD

$V =$ sum of concentric cylindrical shells

radius   height   thickness

$\downarrow$ $\downarrow$ $\downarrow$

$$V = 2\pi \int_a^b x \quad f(x) \quad dx \quad \text{(shells parallel to y-axis)}$$

$$V = 2\pi \int_c^d y \quad f(y) \quad dy \quad \text{(shells parallel to x-axis)}$$

EXAMPLE 1

Find the volume of the solid formed by the region under $y = x^2 + 1$ from $x = 0$ to $x = 2$ revolved about the y-axis. Note that the disk method would not readily solve this problem (some pieces would be disks and others would be washers, so there is no typical unit of volume for that approach). Instead, revolve vertical rectangles about the y-axis to form cylindrical shells (see Fig. 6.22).

$$\Delta V = 2\pi(x)(y)\Delta x$$

where x is the radius of the shell, y is the height, and Δx is the thickness. The integral for the volume is

$$V = 2\pi \int_0^2 x y\, dx = 2\pi \int_0^2 x(x^2 + 1)\, dx = 2\pi \int_0^2 (x^3 + x)\, dx$$

$$= 2\pi \left(\frac{x^4}{4} + \frac{x^2}{2}\right)\Big|_0^2 = 2\pi \left(\frac{2^4}{4} + \frac{2^2}{2} - 0\right) = 12\pi$$

The volume is $12\,\pi$ cubic units.

EXAMPLE 2

Find the volume of the solid formed by revolving the region bounded by $y = x^2$, $y = 0$, and $x = 1$ about the line $x = 1$.

In this example the shell method is more convenient than the disk method. The volume of a typical shell in Fig. 6.23 is

$$\Delta V = 2\pi(1 - x)y\, \Delta x$$

where $1 - x$ is the radius of the shell, y is the height, and Δx is its thickness. The integral for V then becomes

$$V = 2\pi \int_0^1 (1 - x)y\, dx$$

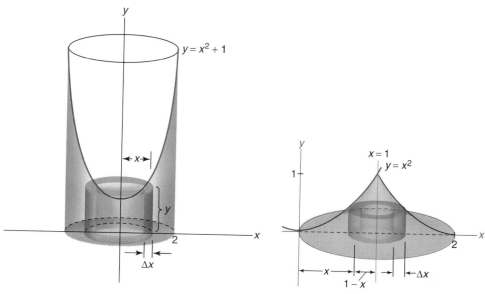

Figure 6.22

Figure 6.23

$$V = 2\pi \int_0^1 (1 - x)(x^2)\, dx \qquad (y = x^2)$$

$$= 2\pi \int_0^1 (x^2 - x^3)\, dx$$

$$= 2\pi\left(\frac{x^3}{3} - \frac{x^4}{4}\right)\Big|_0^1$$

$$= 2\pi\left[\left(\frac{1}{3} - \frac{1}{4}\right) - (0)\right]$$

$$= \frac{\pi}{6}$$

EXAMPLE 3

Find the volume of the solid formed by revolving the region bounded by $y = x^2$ and $y = x$ about the x-axis.

The volume of a typical shell in Fig. 6.24 is

$$\Delta V = 2\pi y(x_2 - x_1)\, \Delta y$$

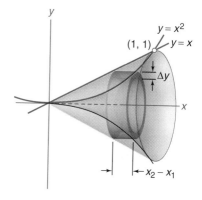

Figure 6.24

where y is the radius of the shell, $x_2 - x_1$ is the height of the shell, and Δy is its thickness. The height of the shell, $x_2 - x_1$, is found by subtracting the curve $x_1 = f(y) = y$ from curve $x_2 = g(y) = \sqrt{y}$.

The integral for V then becomes

$$V = 2\pi \int_c^d y[g(y) - f(y)]\, dy$$

$$= 2\pi \int_0^1 y(x_2 - x_1)\, dy$$

$$= 2\pi \int_0^1 y(\sqrt{y} - y)\, dy$$

$$= 2\pi \int_0^1 (y^{3/2} - y^2)\, dy$$

$$= 2\pi \left(\frac{2}{5}y^{5/2} - \frac{y^3}{3}\right)\Big|_0^1$$

$$= 2\pi \left[\left(\frac{2}{5} - \frac{1}{3}\right) - (0)\right]$$

$$= \frac{2\pi}{15}$$

Exercises 6.3

Find the volume of each solid formed by revolving the region bounded by the given curves about the given line using the shell method.

1. $y = 4x^2$, $x = 0$, and $y = 4$ about the y-axis
2. $y = 4x^2$, $x = 0$, and $y = 4$ about the x-axis
3. $4y = x^2$, $y = 0$, and $x = 2$ about the y-axis
4. $y^2 = x$, $y = 0$, and $x = 4$ about the y-axis
5. $y^2 = 2x$, $x = 0$, and $y = 2$ about the x-axis
6. $y = x^3$, $y = 0$, and $x = 2$ about the y-axis
7. $y = x^3$, $y = 0$, and $x = 2$ about the x-axis
8. $y = \sqrt{x}$, $x = 0$, and $y = 2$ about the x-axis
9. $y = 2x - x^2$ and the x-axis about the y-axis
10. $x = 3y - y^2$ and the y-axis about the x-axis
11. $y = x$, $x = 1$, and $y = 0$ about $x = 1$
12. $y = x^2$, $x = 2$, and $y = 0$ about $x = 2$
13. $x = y^2$, $y = 1$, and $x = 0$ about $y = 2$
14. $y = x^2$ and $y = 4$ about $y = -2$
15. $y = x^3$ and $y = x$ from $x = 0$ to $x = 1$ about the x-axis
16. $y = x^3$ and $y = x$ from $x = 0$ to $x = 1$ about the y-axis
17. $y = x^2 - 3x + 2$ and $y = 0$ about the y-axis
18. $x = y^2 - 6y + 8$ and $x = 0$ about the x-axis
19. $y = x(x - 2)^2$ and $y = 0$ about $x = 2$
20. $y = x(x - 2)^2$ and $y = 0$ about the y-axis

6.4 CENTER OF MASS OF A SYSTEM OF PARTICLES

The next application of integration involves finding the center of mass, which is discussed in the next two sections. We find the center of mass of a system of particles in this section and the center of mass of a thin plate and of a solid of revolution in the next section. Finding the center of mass is of fundamental importance in the study of mechanics. The *center of mass or center of gravity* of an object or system of objects is the point at which the object or system balances or at which the entire mass can be considered to be concentrated.

Before finding the center of mass of a system of particles, we must first introduce the concept of a moment. The *moment* about a point P produced by some mass m is given by

$$\text{moment} = md$$

where d is the length of the *moment arm,* which is the distance between the mass and point P (see Fig. 6.25).

Suppose we have a 15-kg sign hanging from a support 0.8 m from a building as shown in Fig. 6.26. The length of the moment arm is 0.8 m, the distance from the mass to P. The mass is 15 kg. Therefore,

$$\text{moment} = md = (15 \text{ kg})(0.8 \text{ m}) = 12 \text{ kg m}$$

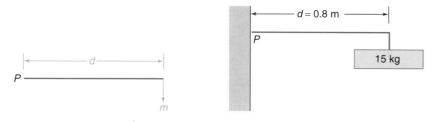

Figure 6.25   Length of moment arm.   **Figure 6.26**

Consider the mobile that is balanced by the five weights as shown in Fig. 6.27. Let the masses m_1, m_2, m_3, m_4, and m_5 be at distances d_1, d_2, d_3, d_4, and d_5, respectively, from a point P. The moment of the system about P is

$$\text{moment} = m_1 d_1 + m_2 d_2 + m_3 d_3 + m_4 d_4 + m_5 d_5$$

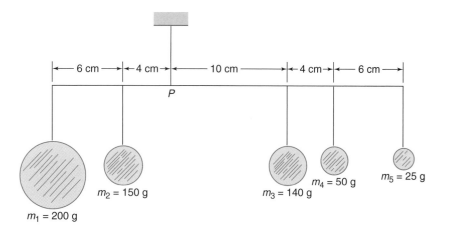

Figure 6.27

If we let P be the origin of the x-axis, $d_1 = -10$, $d_2 = -4$, $d_3 = 10$, $d_4 = 14$, and $d_5 = 20$. Then the moment about P is

$$\begin{aligned} \text{moment} &= (200)(-10) + (150)(-4) + (140)(10) + (50)(14) + (25)(20) \\ &= -2000 - 600 + 1400 + 700 + 500 \\ &= 0 \end{aligned}$$

If the moment is zero, the system is in equilibrium; that is, the mobile balances.

Note: Point P may be any point from which the lengths of the moment arms are measured. We usually choose point P to be the center of mass or the pivot point.

MOMENT IN A LINEAR SYSTEM ALONG THE X-AXIS

Let $\bar{x}$ be the center of mass or balancing point of a linear system along the x-axis with n masses. That is, $\bar{x}$ is the point where all the mass seems to be concentrated. So

$$(m_1 + m_2 + m_3 + \cdots + m_n)\bar{x} = m_1 x_1 + m_2 x_2 + m_3 x_3 + \cdots + m_n x_n$$

Then

$$\bar{x} = \frac{m_1 x_1 + m_2 x_2 + m_3 x_3 + \cdots + m_n x_n}{m_1 + m_2 + m_3 + \cdots + m_n}$$

If we let M_0 be the moment about the origin, $\bar{x}$ be the center of mass, and m be the total mass of the system, then

$$\bar{x} = \frac{M_0}{m}$$

EXAMPLE 1

Find the center of mass of the linear system $m_1 = 10$, $x_1 = -4$; $m_2 = 25$, $x_2 = 2$; $m_3 = 40$, $x_3 = 5$; $m_4 = 15$, $x_4 = 10$.

The moment about the origin is

$$\begin{aligned} M_0 &= m_1 x_1 + m_2 x_2 + m_3 x_3 + m_4 x_4 \\ &= (10)(-4) + (25)(2) + (40)(5) + (15)(10) \\ &= -40 + 50 + 200 + 150 = 360 \end{aligned}$$

The total mass of the system is

$$m = 10 + 25 + 40 + 15 = 90$$

The center of mass is

$$\bar{x} = \frac{M_0}{m} = \frac{360}{90} = 4$$

To extend these concepts to two dimensions as in Fig. 6.28:

MOMENTS OF A TWO-DIMENSIONAL SYSTEM

Consider n masses m_1, m_2, $\ldots$, m_n located in the xy-plane at points $(x_1, y_1)(x_2, y_2)$, $\ldots$, (x_n, y_n), respectively. Their moments with respect to the x-axis and the y-axis are defined as follows: The moment about the y-axis, M_y, is

$$M_y = m_1 x_1 + m_2 x_2 + \cdots + m_n x_n$$

and the moment about the x-axis, M_x, is

$$M_x = m_1 y_1 + m_2 y_2 + \cdots + m_n y_n$$

If we let m be the total mass of the system, the center of mass $(\bar{x}, \bar{y})$ is given by

$$\bar{x} = \frac{M_y}{m} \quad \text{and} \quad \bar{y} = \frac{M_x}{m}$$

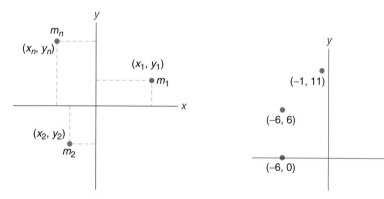

Figure 6.28 Moments of a two-dimensional system.

Figure 6.29

The quantities $m\bar{x}$ and $m\bar{y}$ are regarded as the moments about the y-axis and the x-axis, respectively, of a mass m located at $(\bar{x}, \bar{y})$. That is, $(\bar{x}, \bar{y})$ is the point where the total mass that would give the same moments M_y and M_x seems to be concentrated.

EXAMPLE 2

Find the center of mass of the system $m_1 = 10$ at $(9, 3)$; $m_2 = 6$ at $(-1, 11)$; $m_3 = 8$ at $(-6, 6)$; $m_4 = 12$ at $(-6, 0)$.

From Fig. 6.29,

$$m = 10 + 6 + 8 + 12 = 36$$
$$M_y = (10)(9) + (6)(-1) + (8)(-6) + (12)(-6) = -36$$
$$M_x = (10)(3) + (6)(11) + (8)(6) + (12)(0) = 144$$

Then

$$\bar{x} = \frac{M_y}{m} = \frac{-36}{36} = -1$$

$$\bar{y} = \frac{M_x}{m} = \frac{144}{36} = 4$$

So the center of mass of this system is $(-1, 4)$.

Exercises 6.4

Find the center of mass of each linear system.

1. $m_1 = 3$, $x_1 = -5$; $m_2 = 7$, $x_2 = 3$; $m_3 = 4$, $x_3 = 6$

2. $m_1 = 6$, $x_1 = -12$; $m_2 = 3$, $x_2 = -3$; $m_3 = 10$, $x_3 = 0$; $m_4 = 5$, $x_4 = 9$

3. $m_1 = 24, x_1 = -15; m_2 = 15, x_2 = -9; m_3 = 12, x_3 = 3; m_4 = 9, x_4 = 6$

4. $m_1 = 8, x_1 = -15; m_2 = 15, x_2 = -9; m_3 = 7, x_3 = -1; m_4 = 20, x_4 = 8; m_5 = 24, x_5 = 12$

5. There is a mass of 6 at $(9, 0)$ and a mass of 18 at $(-2, 0)$. Find where a mass of 3 should be placed on the x-axis so that the origin is the center of mass.

6. There is a mass of 30 at $(-4, 0)$, a mass of 9 at $(2, 0)$, and a mass of 15 at $(8, 0)$. Find where a mass of 3 should be placed on the x-axis so that the origin is the center of mass.

7. There is a mass of 24 at $(-8, 0)$ and a mass of 36 at $(12, 0)$. Find where a mass of 9 should be placed on the x-axis so that $(3, 0)$ is the center of mass.

8. There is a mass of 15 at $(-8, 0)$, a mass of 5 at $(-4, 0)$, and a mass of 12 at $(3, 0)$. Find where a mass of 8 should be placed on the x-axis so that $(-3, 0)$ is the center of mass.

9. There is a mass of 6 at $(-3, 0)$ and a mass of 9 at $(12, 0)$. Find what mass should be placed at $(-6, 0)$ so that the origin is the center of mass.

10. There is a mass of 4 at $(-5, 0)$, a mass of 16 at $(3, 0)$, and a mass of 24 at $(8, 0)$. Find what mass should be placed at $(-4, 0)$ so that the origin is the center of mass.

11. There is a mass of 25 at $(-6, 0)$, a mass of 45 at $(8, 0)$, and a mass of 40 at $(10, 0)$. Find what mass should be placed at $(-4, 0)$ so that $(3, 0)$ is the center of mass.

12. There is a mass of 18 at $(3, 0)$, a mass of 54 at $(9, 0)$, a mass of 24 at $(12, 0)$, and a mass of 36 at $(15, 0)$. Find what mass should be placed at $(4, 0)$ so that $(6, 0)$ is the center of mass.

13. A straight road connects Flatville (population 75,000), Pleasant Hill (population 50,000), and Harristown (population 25,000). Pleasant Hill is 18 mi north of Flatville while Harristown is 30 mi north of Flatville. Where is the best place to locate an airport to serve these three communities?

14. A straight road connects Leadville (population 1750), Branburg (population 2800), Princeton (population 970), and Four Oaks (population 480). The distance from Leadville to Branburg is 4 mi, Princeton is 10 mi, and Four Oaks is 13 mi. Where is the best place to locate a hospital to serve these four communities?

Find the center of mass of each two-dimensional system.

15. $m_1 = 6$ at $(1, 4)$; $m_2 = 3$ at $(6, 2)$; $m_3 = 12$ at $(3, 3)$

16. $m_1 = 20$ at $(-5, 10)$; $m_2 = 15$ at $(-10, -15)$; $m_3 = 40$ at $(0, -5)$

17. $m_1 = 8$ at $(8, 12)$; $m_2 = 16$ at $(-12, 8)$; $m_3 = 20$ at $(-16, -4)$; $m_4 = 36$ at $(4, -20)$

18. $m_1 = 9$ at $(3, 6)$; $m_2 = 12$ at $(-6, 12)$; $m_3 = 18$ at $(0, -9)$; $m_4 = 30$ at $(15, 0)$; $m_5 = 15$ at $(-12, -9)$

19. There is a mass of 6 at $(4, 2)$ and a mass of 9 at $(-5, 8)$. Find where a mass of 10 should be placed so that the origin is the center of mass.

20. There is a mass of 18 at $(-4, 6)$, a mass of 12 at $(1, -2)$, and a mass of 9 at $(8, 0)$. Find where a mass of 6 should be placed so that $(2, -3)$ is the center of mass.

21. There is a mass of 15 at $(10, 3)$, a mass of 25 at $(-6, -1)$, and a mass of 40 at $(8, -2)$. Find what mass should be placed at $(-5, -3)$ so that $(-1, -2)$ is the center of mass.

22. There is a mass of 4 at $(-5, -3)$, a mass of 16 at $(-4, 3)$, and a mass of 12 at $(6, -4)$. Find what mass should be placed at $(2, 2)$ so that the origin is the center of mass.

23. Three towns plan to build a new health clinic to serve all three communities. Town B (population 8200) is 6 mi east and 3 mi south of Town A (population 12,500). Town C (population 5200) is 2 mi west and 8 mi south of Town A. Find the best location for the new health clinic. Do not consider new roads in determining the best location.

24. Four cities plan to build a new airport to serve all four communities. City B (population 180,000) is 4 mi north and 3 mi west of City A (population 75,000). City C (population 240,000) is 6 mi east and 12 mi south of City A. City D (population 105,000) is 15 mi due south of City A. Find the best location for the airport. Do not consider new roads in determining the best location.

6.5 CENTER OF MASS OF CONTINUOUS MASS DISTRIBUTIONS

As we saw in Section 6.4, the center of mass of any system of finite particles may be found arithmetically by summing the moments and the masses and dividing. Recall that the center of mass of a linear system along the x-axis with n masses was given by

$$\bar{x} = \frac{M_0}{m} = \frac{m_1 x_1 + m_2 x_2 + m_3 x_3 + \cdots + m_n x_n}{m_1 + m_2 + m_3 + \cdots + m_n}$$

For a continuous mass distribution, the center of mass is found by integration. For example, to find the center of mass of a straight thin wire of constant density ρ, place the wire on the x-axis as shown in Fig. 6.30. Then, subdivide the wire into n equal lengths, each of length Δx and mass Δm. The mass of the ith length is

$$\text{mass} = (\text{density})(\text{length})$$
$$m = \rho \quad \Delta x$$

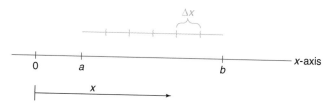

Figure 6.30   Center of mass of a straight thin wire of constant density.

The total mass of the wire is the integral

$$m = \rho \int_a^b dx$$

Next, find the moment of the ith length:

$$\text{moment} = (\text{mass})(\text{length of moment arm})$$
$$= (\rho \, \Delta x)(x)$$

So, the moment of the entire wire about the origin is the integral

$$M_0 = \rho \int_a^b x \, dx$$

Then, the center of mass is

> **CENTER OF MASS OF A CONTINUOUS THIN UNIFORM MASS**
>
> $$\dot{x} = \frac{M_0}{m} = \frac{\displaystyle\int_a^b x \, dx}{\displaystyle\int_a^b dx}$$

Note: The density ρ cancels in all such cases when the density is constant or uniform. For any homogeneous mass distribution having constant density (constant mass per unit length, per unit area, or per unit volume), the *centroid* is the same as the center of mass. The centroid often refers to the geometric center.

EXAMPLE 1

Find the center of mass of a straight wire 12 cm long and of uniform density.

$$
\bar{x} = \frac{\displaystyle\int_a^b x \, dx}{\displaystyle\int_a^b dx} = \frac{\displaystyle\int_0^{12} x \, dx}{\displaystyle\int_0^{12} dx} = \frac{\left.\dfrac{x^2}{2}\right|_0^{12}}{\left. x \right|_0^{12}} = \frac{72}{12} = 6
$$

This result should be no surprise. For a uniform linear object, the center of mass is at its center, the point at which the object can be supported. For example, a metre stick is supported on one's finger at the 50-cm mark as in Fig. 6.31.

Figure 6.31 The center of mass, or centroid, of a metre stick is the 50-cm mark, the point at which it can be balanced on one's finger.

For a nonuniform object, the center of mass is usually not at its geometric center, but at the point at which it can be supported in equilibrium by a single force or the point about which it spins if allowed to spin freely in space as in Fig. 6.32.

When the density of a continuous thin mass is not uniform, the center of mass is found as follows:

CENTER OF MASS OF A CONTINUOUS THIN MASS OF VARIABLE DENSITY

$$
\bar{x} = \frac{M_0}{m} = \frac{\displaystyle\int_a^b \rho(x)\, x \, dx}{\displaystyle\int_a^b \rho(x)\, dx}
$$

where $\rho(x)$ is the density expressed as a function of x. Density is mass per unit length.

EXAMPLE 2

Find the center of mass of a straight wire 16 cm long and whose density is given by $\rho(x) = 4\sqrt{x}$, where x is the distance from one end of the wire.

$$
\bar{x} = \frac{M_0}{m} = \frac{\displaystyle\int_a^b \rho(x)\, x \, dx}{\displaystyle\int_a^b \rho(x)\, dx} = \frac{\displaystyle\int_0^{16} 4\sqrt{x}\, x \, dx}{\displaystyle\int_0^{16} 4\sqrt{x}\, dx} = \frac{\displaystyle\int_0^{16} x^{3/2} \, dx}{\displaystyle\int_0^{16} x^{1/2} \, dx} = \frac{\left.\dfrac{2}{5}x^{5/2}\right|_0^{16}}{\left.\dfrac{2}{3}x^{3/2}\right|_0^{16}}
$$

$$
= \frac{\frac{2}{5}[16^{5/2} - 0]}{\frac{2}{3}[16^{3/2} - 0]} = \frac{\frac{2}{5}(1024)}{\frac{2}{3}(64)} = \frac{48}{5} = 9.6 \text{ cm from the lighter end}
$$

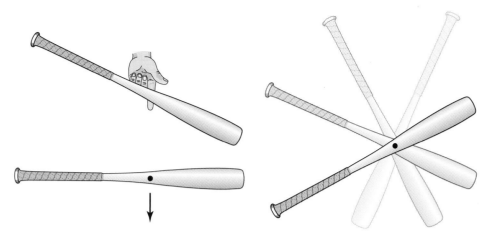

Figure 6.32 The center of mass of any object is the point about which it spins freely in space.

The center of mass of a two-dimensional thin plate is the point at which the plate can be supported as in Fig. 6.33.

If the thin plate is in a regular geometric shape, its center of mass is its geometric center because of its symmetry. Examples of four common geometric figures with each corresponding center of mass are shown in Fig. 6.34.

The center of mass of a more complex but uniformly thin object can be found by subdividing the object into combinations of simpler figures. Find the center of each simpler figure. Consider the mass of each simpler figure to be concentrated at its center and proceed using the method for moments of a two-dimensional system used in Example 2 of Section 6.4.

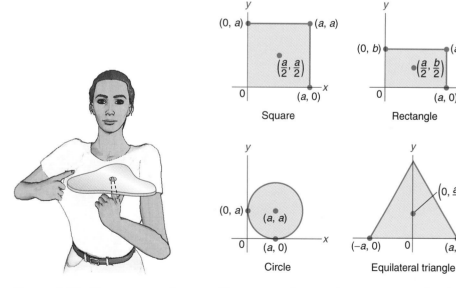

Figure 6.33 The center of mass, or centroid, of a thin plate is the point at which the plate can be supported.

Figure 6.34 Centers of mass, or centroids, of some common geometric shapes placed in the xy-plane.

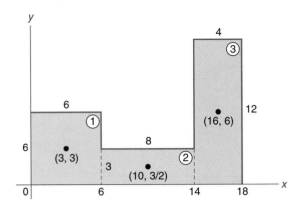

Figure 6.35

EXAMPLE 3

Find the center of mass of the uniform thin plate in Fig. 6.35.

The center of region 1 (square) is $(3, 3)$, of region 2 (rectangle) is $(10, \frac{3}{2})$, and of region 3 (rectangle) is $(16, 6)$. Since the plate is uniform, the mass in each region is proportional to its area. The area of region 1 is 36 units, of region 2 is 24 units, and of region 3 is 48 units. So

$$m = 36 + 24 + 48 = 108$$

$$
\begin{aligned}
M_y &= m_1 x_1 + m_2 x_2 + m_3 x_3 \\
&= (36)(3) + (24)(10) + (48)(16) \\
&= 1116
\end{aligned}
$$

$$
\begin{aligned}
M_x &= m_1 y_1 + m_2 y_2 + m_3 y_3 \\
&= (36)(3) + (24)(\tfrac{3}{2}) + (48)(6) \\
&= 432
\end{aligned}
$$

Then

$$\bar{x} = \frac{M_y}{m} = \frac{1116}{108} = 10\tfrac{1}{3}$$

$$\bar{y} = \frac{M_x}{m} = \frac{432}{108} = 4$$

The center of mass of the plate is $(10\tfrac{1}{3}, 4)$.

Note that in Example 3 the center of mass is not on the surface of the plate.

Next, let's find the centroid of an irregular-shaped area (or thin plate) of constant density ρ between the curves shown in Fig. 6.36. First divide the area into n rectangles, each of width Δx. Let (x_i, y_i) be the center of mass of the ith rectangle. The y-value of the geometric center of the ith rectangle is

$$y_i = \frac{f(x_i) + g(x_i)}{2}$$

The area of the ith rectangle is $[f(x_i) - g(x_i)]\,\Delta x$. The mass of the ith rectangle is

$$
\begin{aligned}
\text{mass} &= (\text{density})(\text{area}) \\
&= \rho[f(x_i) - g(x_i)]\,\Delta x
\end{aligned}
$$

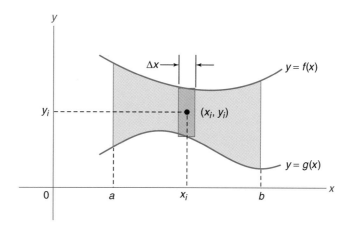

Figure 6.36 Finding the centroid of an irregular-shaped area or thin plate of constant density.

The total mass is the integral

$$m = \rho \int_a^b [f(x) - g(x)]\, dx = \rho A$$

where A is the area of the region.

Next, find the moment of the ith rectangle about the x-axis:

$$
\begin{aligned}
\text{moment} &= (\text{mass})(\text{moment arm}) \\
&= \rho[f(x_i) - g(x_i)]\, \Delta x \cdot y_i \\
&= \rho[f(x_i) - g(x_i)]\, \Delta x \cdot \frac{f(x_i) + g(x_i)}{2} \\
&= \frac{\rho}{2}\{[f(x_i)]^2 - [g(x_i)]^2\}\, \Delta x
\end{aligned}
$$

The moment about the x-axis is the integral

$$M_x = \frac{\rho}{2} \int_a^b \{[f(x)]^2 - [g(x)]^2\}\, dx$$

Similarly, the moment about the y-axis is found to be the integral

$$M_y = \rho \int_a^b x[f(x) - g(x)]\, dx$$

MOMENTS AND CENTER OF MASS OF A PLANE AREA OR THIN PLATE

Let $g(x) \le f(x)$ be continuous functions on $a \le x \le b$ for the area of uniform density ρ bounded by $y = f(x)$, $y = g(x)$, $x = a$, and $x = b$. The moments about the x-axis and the y-axis are

$$M_x = \frac{\rho}{2} \int_a^b \{[f(x)]^2 - [g(x)]^2\}\, dx \quad \text{and} \quad M_y = \rho \int_a^b x[f(x) - g(x)]\, dx$$

Its mass is given by

$$m = \rho \int_a^b [f(x) - g(x)] \, dx$$

and its center of mass $(\overline{x}, \overline{y})$ is

$$\overline{x} = \frac{M_y}{m} \quad \text{and} \quad \overline{y} = \frac{M_x}{m}$$

EXAMPLE 4

Find the center of mass of the uniformly thin plate of density ρ bounded by $y = x^2$ and $y = x + 2$.

First, graph the equations and find the points of intersection as in Fig. 6.37.

$$x^2 = x + 2$$
$$x^2 - x - 2 = 0$$
$$(x - 2)(x + 1) = 0$$
$$x = 2, -1$$

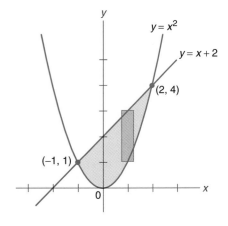

Figure 6.37

$$M_x = \frac{\rho}{2} \int_a^b \{[f(x)]^2 - [g(x)]^2\} \, dx$$

$$= \frac{\rho}{2} \int_{-1}^{2} [(x + 2)^2 - (x^2)^2] \, dx$$

$$= \frac{\rho}{2} \int_{-1}^{2} [x^2 + 4x + 4 - x^4] \, dx$$

$$= \frac{\rho}{2} \left(\frac{x^3}{3} + 2x^2 + 4x - \frac{x^5}{5} \right)\Big|_{-1}^{2}$$

$$= \frac{\rho}{2} \left[\left(\frac{8}{3} + 8 + 8 - \frac{32}{5} \right) - \left(-\frac{1}{3} + 2 - 4 + \frac{1}{5} \right) \right]$$

$$= \frac{\rho}{2} \left(\frac{72}{5} \right)$$

$$= \frac{36\rho}{5}$$

$$M_y = \rho \int_a^b x[f(x) - g(x)] \, dx$$

$$= \rho \int_{-1}^{2} x(x + 2 - x^2) \, dx$$

$$= \rho \int_{-1}^{2} (x^2 + 2x - x^3) \, dx$$

$$= \rho\left(\frac{x^3}{3} + x^2 - \frac{x^4}{4}\right)\Big|_{-1}^{2}$$

$$= \rho\left[\left(\frac{8}{3} + 4 - 4\right) - \left(-\frac{1}{3} + 1 - \frac{1}{4}\right)\right]$$

$$= \frac{9\rho}{4}$$

$$m = \rho \int_a^b [f(x) - g(x)]\, dx$$

$$= \rho \int_{-1}^{2} (x + 2 - x^2)\, dx$$

$$= \rho\left(\frac{x^2}{2} + 2x - \frac{x^3}{3}\right)\Big|_{-1}^{2}$$

$$= \rho\left[\left(2 + 4 - \frac{8}{3}\right) - \left(\frac{1}{2} - 2 + \frac{1}{3}\right)\right]$$

$$= \frac{9\rho}{2}$$

Then

$$\bar{x} = \frac{M_y}{m} = \frac{9\rho/4}{9\rho/2} = \frac{1}{2}$$

$$\bar{y} = \frac{M_x}{m} = \frac{36\rho/5}{9\rho/2} = \frac{8}{5}$$

The center of mass is $\left(\frac{1}{2}, \frac{8}{5}\right)$.

Note that the density ρ cancels in both $\bar{x}$ and $\bar{y}$. That is, the center of mass of a thin plate or area of uniform density depends only on its shape and not its density. Thus we may find the centroid as follows:

CENTROID OF A PLANE REGION OR THIN PLATE

Let $g(x) \leq f(x)$ be continuous functions on $a \leq x \leq b$. The centroid $(\bar{x}, \bar{y})$ of the region bounded by $y = f(x)$, $y = g(x)$, $x = a$, and $y = b$ is

$$\bar{x} = \frac{\displaystyle\int_a^b x[f(x) - g(x)]\, dx}{A}$$

and

$$\bar{y} = \frac{\dfrac{1}{2}\displaystyle\int_a^b \{[f(x)]^2 - [g(x)]^2\}\, dx}{A}$$

where A is the area of the region.

EXAMPLE 5

Find the centroid of the region bounded by $y = x^4$ and $y = x$.

First, graph the equations and find the points of intersection (see Fig. 6.38).

$$A = \int_0^1 (x - x^4)\, dx$$

$$= \left(\frac{x^2}{2} - \frac{x^5}{5} \right)\Bigg|_0^1$$

$$= \left(\frac{1}{2} - \frac{1}{5} \right) - (0) = \frac{3}{10}$$

$$\bar{x} = \frac{\displaystyle\int_a^b x[f(x) - g(x)]\, dx}{A}$$

$$= \frac{\displaystyle\int_0^1 x(x - x^4)\, dx}{3/10}$$

$$= \frac{10}{3} \int_0^1 (x^2 - x^5)\, dx$$

$$= \frac{10}{3} \left(\frac{x^3}{3} - \frac{x^6}{6} \right)\Bigg|_0^1$$

$$= \frac{10}{3} \left[\left(\frac{1}{3} - \frac{1}{6} \right) - (0) \right]$$

$$= \frac{10}{3} \left(\frac{1}{6} \right) = \frac{5}{9}$$

$$\bar{y} = \frac{\dfrac{1}{2} \displaystyle\int_a^b \{[f(x)]^2 - [g(x)]^2\}\, dx}{A}$$

$$= \frac{\dfrac{1}{2} \displaystyle\int_0^1 [(x)^2 - (x^4)^2]\, dx}{3/10}$$

$$= \frac{5}{3} \int_0^1 (x^2 - x^8)\, dx$$

$$= \frac{5}{3} \left(\frac{x^3}{3} - \frac{x^9}{9} \right)\Bigg|_0^1$$

$$= \frac{5}{3} \left[\left(\frac{1}{3} - \frac{1}{9} \right) - (0) \right]$$

$$= \frac{5}{3} \left(\frac{2}{9} \right)$$

$$= \frac{10}{27}$$

The centroid is $\left(\frac{5}{9}, \frac{10}{27} \right)$.

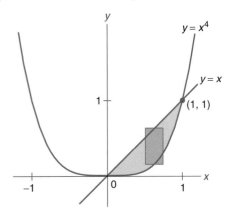

Figure 6.38

The most general case of finding the center of mass of a two-dimensional mass distribution requires double integration and is not treated in this text.

A solid of revolution of constant density has its centroid on its axis of revolution. Let the area bounded by $y = f(x)$, $x = a$, and $x = b$ be revolved about the x-axis. We have drawn a typical disk in Fig. 6.39. Its center of mass is x units from the y-axis; therefore, the length of its moment arm is x. Its volume is $\pi y^2\,dx$. Since the solid is of constant density ρ, its mass is proportional to its volume; that is, $m = \rho\pi y^2\,dx$. Thus the moment about the y-axis is

$$md = (\rho\pi y^2\,dx)x = \rho\pi xy^2\,dx$$

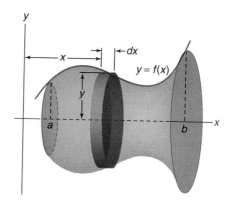

Figure 6.39   Finding the center of mass, or centroid, of a solid of revolution about the x-axis.

The sum of the moments of all such disks may be expressed by the integral

$$M_y = \rho\pi \int_a^b xy^2\,dx$$

The mass of the solid may be expressed by the integral

$$m = \rho\pi \int_a^b y^2\,dx$$

CENTROID OF A SOLID OF REVOLUTION ABOUT THE X-AXIS

$$\bar{x} = \frac{M_y}{m} = \frac{\displaystyle\int_a^b xy^2\,dx}{\displaystyle\int_a^b y^2\,dx} \quad \text{and} \quad \bar{y} = 0$$

Note: The ρ and π factors cancel.
In a similar manner, we can show:

CENTROID OF A SOLID OF REVOLUTION ABOUT THE Y-AXIS

$$\bar{y} = \frac{M_x}{m} = \frac{\displaystyle\int_c^d yx^2\,dy}{\displaystyle\int_c^d x^2\,dy} \quad \text{and} \quad \bar{x} = 0$$

EXAMPLE 6

Find the centroid of the solid formed by revolving the region bounded by $y = x^2$, $x = 1$, and $y = 0$ about the x-axis (see Fig. 6.40).

$$\overline{x} = \frac{\displaystyle\int_a^b xy^2\,dx}{\displaystyle\int_a^b y^2\,dx}$$

$$= \frac{\displaystyle\int_0^1 x(x^2)^2\,dx}{\displaystyle\int_0^1 (x^2)^2\,dx}$$

$$= \frac{\displaystyle\int_0^1 x^5\,dx}{\displaystyle\int_0^1 x^4\,dx}$$

$$= \frac{\left.\dfrac{x^6}{6}\right|_0^1}{\left.\dfrac{x^5}{5}\right|_0^1} = \frac{\dfrac{1}{6}}{\dfrac{1}{5}} = \frac{5}{6}$$

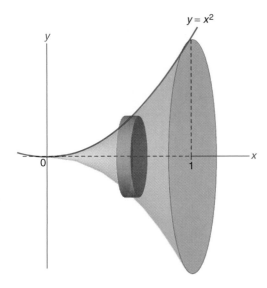

Figure 6.40

The centroid is $\left(\frac{5}{6}, 0\right)$.

EXAMPLE 7

Find the centroid of the solid formed by revolving the region bounded by $y = 4 - x^2$, $x = 0$, and $y = 0$ about the y-axis (see Fig. 6.41).

$$\overline{y} = \frac{\displaystyle\int_c^d yx^2\,dy}{\displaystyle\int_c^d x^2\,dy}$$

$$= \frac{\displaystyle\int_0^4 y(4 - y)\,dy}{\displaystyle\int_0^4 (4 - y)\,dy} \qquad (\textit{Note: } x^2 = 4 - y.)$$

$$= \frac{\displaystyle\int_0^4 (4y - y^2)\,dy}{\displaystyle\int_0^4 (4 - y)\,dy}$$

$$= \frac{\left.\left(2y^2 - \dfrac{y^3}{3}\right)\right|_0^4}{\left.\left(4y - \dfrac{y^2}{2}\right)\right|_0^4} = \frac{\left(32 - \dfrac{64}{3}\right) - (0)}{(16 - 8) - (0)} = \frac{32/3}{8} = \frac{4}{3}$$

The centroid is $\left(0, \frac{4}{3}\right)$.

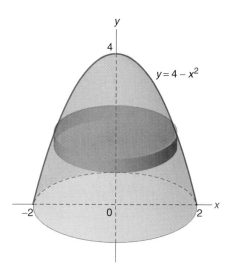

 placed above — figure graph.

Figure 6.41

The most general case of finding the center of mass of a three-dimensional mass distribution requires triple integration and is not treated in this text.

Exercises 6.5

1. Find the center of mass of a straight wire 20 cm long and of uniform density.

2. Show that the center of mass of a straight wire of length L and of uniform density is at its midpoint.

3. Find the center of mass measured from the lighter end of a straight wire 10 cm long and whose density is given by $\rho(x) = 0.1x$, where x is the distance from one end.

4. Find the center of mass measured from the lighter end of a straight wire 8 cm long and whose density is given by $\rho(x) = 0.1x^2$, where x is the distance from one end.

5. Find the center of mass measured from the lighter end of a straight wire 12 cm long and whose density is given by $\rho(x) = 4 + x^2$, where x is the distance from one end.

6. Find the center of mass measured from the lighter end of a straight wire 9 cm long and whose density is given by $\rho(x) = 3 - \sqrt{x}$, where x is the distance from one end.

7. The density of a straight wire 6 cm long is directly proportional to the distance from one end. Find its center of mass.

8. The density of a straight wire 9 cm long varies inversely as the square root of the distance from one end. Find its center of mass.

Find the center of mass of each uniform thin plate.

9.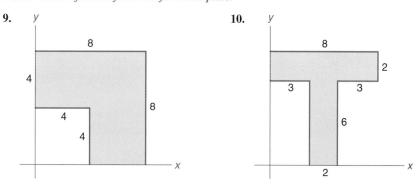
10.

11.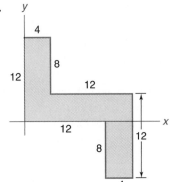

12.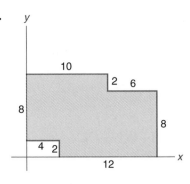

13.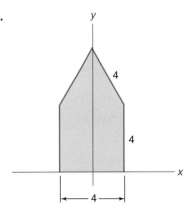

14.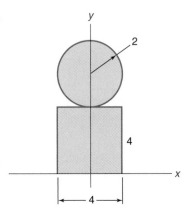

Find the centroid of each region bounded by the given curves.

15. $y = \sqrt{x}$, $y = 0$, and $x = 9$

16. $y = x^2$ and $y = 4$

17. $y = x^2 - 2x$ and $y = 0$

18. $y = 4 - x^2$ and $y = x^2 - 4$

19. $y = 4 - x^2$ and $y = 0$

20. $y = 4 - x^2$ and $y = x + 2$

21. $y = x^3$ and $y = x$ (first quadrant)

22. $y = x^3$, $x = 0$, and $y = 1$

23. Find the centroid of the semicircle of radius 1 with center at the origin lying in the first and second quadrants.

24. Find the centroid of the quarter circle of radius 1 with center at the origin lying in the first quadrant.

Find the centroid of the solid formed by revolving each region bounded by the given curves about the given axis.

25. $y = x^3$, $y = 0$, and $x = 1$ about the x-axis

26. $y = x^3$, $y = 0$, and $x = 1$ about the y-axis

27. $y = 3 - x$, $x = 0$, and $y = 0$ about the x-axis

28. $y = 3 - 2x$, $x = 0$, and $y = 0$ about the y-axis

29. $y = x^2$, $x = 1$, and $y = 0$ about the y-axis

30. $x^2 + y^2 = 1$, $x = 0$, and $y = 0$ about the x-axis

6.6 MOMENTS OF INERTIA

In Sections 6.4 and 6.5 we used moments to find centers of mass and centroids. Each moment was the product of the mass and its distance from some line. This case is called the first moment.

The second moment, called the *moment of inertia* about a line, is defined as the product of a mass m and the square of its distance d from a given line; that is,

$$I = md^2$$

Inertia is a property of an object that resists a change in its motion. That is, inertia is a property of an object that causes it to remain at rest if it is at rest or to continue moving with constant velocity.

MOMENT OF INERTIA OF A SYSTEM

Let masses $m_1, m_2, m_3, \ldots, m_n$ be at distance $d_1, d_2, d_3, \ldots, d_n$, respectively, rotating about some axis. The moment of inertia, I, of the system is

$$I = m_1 d_1^2 + m_2 d_2^2 + m_3 d_3^2 + \cdots + m_n d_n^2$$

Let m be the sum of all the masses in the systems and let R be the distance from the axis of rotation that gives the same total moment of inertia

$$I = mR^2 = m_1 d_1^2 + m_2 d_2^2 + m_3 d_3^2 + \cdots + m_n d_n^2$$

R is called the *radius* of *gyration.* It tells how far from the axis of rotation the entire mass would be concentrated to have the same moment of inertia. This is a convenient way to express the moment of inertia of the mass of a body in terms of its mass and a length.

EXAMPLE 1

Find the moment of inertia and the radius of gyration about the y-axis of the system $m_1 = 8$ at $(2, -4)$; $m_2 = 3$ at $(-9, 8)$; and $m_3 = 6$ at $(-5, 2)$.

$$I_y = m_1 x_1^2 + m_2 x_2^2 + m_3 x_3^2$$
$$I_y = 8(2)^2 + 3(-9)^2 + 6(-5)^2 = 425$$
$$m = m_1 + m_2 + m_3 = 8 + 3 + 6 = 17$$
$$I_y = mR^2$$
$$R = \sqrt{\frac{I_y}{m}} = \sqrt{\frac{425}{17}} = 5$$

Now, let's find the moment of inertia of an area of constant density ρ about the y-axis as shown in Fig. 6.42. First, divide the area into n rectangles, each of width Δx. Let (x_i, y_i) be the center of mass of the ith rectangle. As we saw in the preceding sections, the y-value of the geometric center of the ith rectangle is

$$y_i = \frac{f(x_i) + g(x_i)}{2}$$

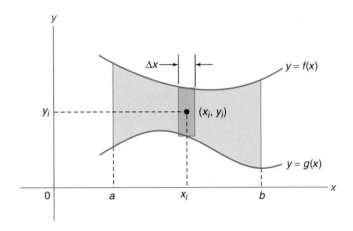

Figure 6.42   Finding the moment of inertia of an area of constant density about the *y*-axis.

Its area is $[f(x_i) - g(x_i)]\,\Delta x$, and its mass is $\rho[f(x_i) - g(x_i)]\,\Delta x$. The total mass is the integral

$$m = \rho \int_a^b [f(x) - g(x)]dx = \rho A$$

where A is the area of the region.

The distance of the center of the *i*th rectangle from the *y*-axis is x_i. Its moment of inertia, its second moment, is then

$$(\text{mass})(\text{moment arm})^2 = \rho[f(x_i) - g(x_i)]\,\Delta x \cdot (x_i)^2$$

Summing the moments of all such rectangles, the moment of inertia of the region about the *y*-axis is the integral

$$I_y = \rho \int_a^b x^2[f(x) - g(x)]\,dx$$

Similarly, the moment of inertia of an area about the *x*-axis is found to be

$$I_x = \rho \int_c^d y^2[f(y) - g(y)]\,dy$$

The radius of gyration for each moment is

About *y*-axis	About *x*-axis
$I_y = mR^2$	$I_x = mR^2$
$R = \sqrt{\dfrac{I_y}{m}}$	$R = \sqrt{\dfrac{I_x}{m}}$

MOMENTS OF INERTIA OF AN AREA OF CONSTANT DENSITY ABOUT THE X- AND Y-AXES

Let $g(x) \le f(x)$ be continuous functions on $a \le x \le b$ for the area of constant density ρ bounded by $y = f(x)$, $y = g(x)$, $x = a$, and $x = b$. The moment of inertia about the *y*-axis is

$$I_y = \rho \int_a^b x^2[f(x) - g(x)]\, dx \qquad \left(R = \sqrt{\frac{I_y}{m}} \right)$$

Similarly, the moment of inertia about the *x*-axis is

$$I_x = \rho \int_c^d y^2[f(y) - g(y)]\, dy \qquad \left(R = \sqrt{\frac{I_x}{m}} \right)$$

EXAMPLE 2

Find the moment of inertia and the radius of gyration about the *y*-axis of the region bounded by $y = x^2$, $y = 0$, and $x = 3$, where the region has a constant density of 5 (see Fig. 6.43).

$$I_y = \rho \int_a^b x^2[f(x) - g(x)]\, dx$$

$$= 5 \int_0^3 x^2[x^2 - 0]\, dx$$

$$= 5 \int_0^3 x^4\, dx$$

$$= 5 \frac{x^5}{5} \Big|_0^3$$

$$= 243$$

$$m = \rho \int_a^b [f(x) - g(x)]\, dx$$

$$= 5 \int_0^3 (x^2 - 0)\, dx$$

$$= 5 \cdot \frac{x^3}{3} \Big|_0^3$$

$$= 45$$

$$R = \sqrt{\frac{I_y}{m}} = \sqrt{\frac{243}{45}} = 2.32$$

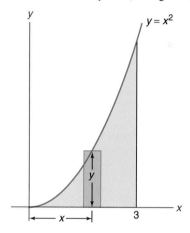

Figure 6.43

EXAMPLE 3

Find the moment of inertia and the radius of gyration about the *x*-axis of the region described in Example 2 (see Fig. 6.44).

$$I_x = \rho \int_c^d y^2[f(y) - g(y)]\, dy$$

$$= 5 \int_0^9 y^2[3 - \sqrt{y}]\, dy \qquad (3 - x = 3 - \sqrt{y})$$

$$= 5 \int_0^9 [3y^2 - y^{5/2}]\, dy$$

$$= 5 \left[y^3 - \frac{y^{7/2}}{7/2} \right] \Big|_0^9$$

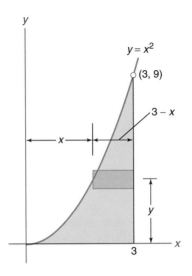

Figure 6.44

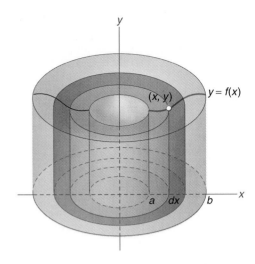

Figure 6.45   Finding the moment of inertia of a solid of revolution with respect to its y-axis of revolution.

$$= 5\left[9^3 - \frac{9^{7/2}}{7/2}\right] - 5[0] = 521 \qquad \text{(three significant digits)}$$

$$R = \sqrt{\frac{I_x}{m}} = \sqrt{\frac{521}{45}} = 3.40$$

To find the moment of inertia of a solid of revolution with respect to its axis of revolution, it is most convenient to use the shell method. Let the area bounded by $y = f(x), y = 0, x = a$, and $x = b$ be revolved about the y-axis. We have drawn a typical shell in Fig. 6.45. The mass of the solid is proportional to its volume $m = \rho V$:

$$m = 2\pi\rho \int_a^b x f(x) \, dx$$

and x^2 is the square of its distance from the y-axis. Summing all such shells gives the following results:

MOMENTS OF INERTIA OF A SOLID OF REVOLUTION

The moment of inertia of a solid of revolution about the y-axis is given by the integral

$$I_y = 2\pi\rho \int_a^b x^3 f(x) \, dx$$

Similarly, let the area bounded by $x = f(y), x = 0, y = c$, and $y = d$ be revolved about the x-axis as in Fig. 6.46. Its moment of inertia about the x-axis is given by the integral

$$I_x = 2\pi\rho \int_c^d y^3 f(y) \, dy$$

Note: $f(x)$ and $f(y)$ correspond to the height of the shell.

Each radius of gyration is found in a similar way as that for a plane region; that is,

$$R = \sqrt{\frac{I_y}{m}} \quad \text{or} \quad R = \sqrt{\frac{I_x}{m}}$$

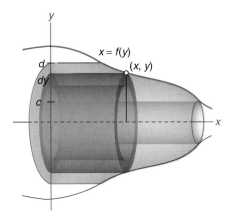

Figure 6.46 Finding the moment of inertia of a solid of revolution with respect to its x-axis of revolution.

EXAMPLE 4

Find the moment of inertia and the radius of gyration of the solid formed by revolving the region bounded by $y = 2x$, $y = 0$, and $x = 3$ about the y-axis. Assume that $\rho = 5$ (see Fig. 6.47).

$$I_y = 2\pi\rho \int_a^b x^3 f(x)\, dx$$

$$= 2\pi(5) \int_0^3 x^3 (2x)\, dx$$

$$= 20\pi \int_0^3 x^4\, dx$$

$$= 20\pi \cdot \frac{x^5}{5}\Big|_0^3 = 972\pi$$

$$m = 2\pi\rho \int_a^b x f(x)\, dx$$

$$= 2\pi(5) \int_0^3 x(2x)\, dx$$

$$= 20\pi \int_0^3 x^2\, dx$$

$$= 20\pi \cdot \frac{x^3}{3}\Big|_0^3 = 180\pi$$

$$R = \sqrt{\frac{I_y}{m}} = \sqrt{\frac{972\pi}{180\pi}} = 2.32$$

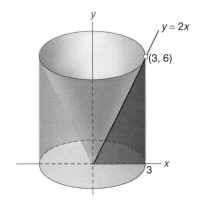

Figure 6.47

EXAMPLE 5

Find the moment of inertia and the radius of gyration of the solid formed by revolving the region bounded by $y = x^2$, $y = 0$, and $x = 2$ about the x-axis. Assume that $\rho = 1$ (see Fig. 6.48).

$$I_x = 2\pi\rho \int_c^d y^3 f(y)\, dy$$

$$= 2\pi(1) \int_0^4 y^3(2 - \sqrt{y})\, dy \qquad (2 - x = 2 - \sqrt{y})$$

$$= 2\pi \int_0^4 (2y^3 - y^{7/2})\, dy$$

$$= 2\pi\left(\frac{y^4}{2} - \frac{y^{9/2}}{9/2}\right)\Big|_0^4$$

$$= 2\pi\left(128 - \frac{1024}{9}\right) = \frac{256\pi}{9}$$

$$m = 2\pi\rho \int_c^d y f(y)\, dy$$

$$= 2\pi(1) \int_0^4 y(2 - \sqrt{y})\, dy$$

$$= 2\pi \int_0^4 (2y - y^{3/2})\, dy$$

$$= 2\pi\left(y^2 - \frac{y^{5/2}}{5/2}\right)\Big|_0^4$$

$$= 2\pi\left(16 - \frac{64}{5}\right) = \frac{32\pi}{5}$$

$$R = \sqrt{\frac{I_x}{m}} = \sqrt{\frac{256\pi/9}{32\pi/5}} = 2.11$$

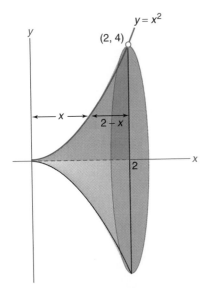

Figure 6.48

Exercises 6.6

Find the moment of inertia and the radius of gyration about the y-axis of each system.

1. $m_1 = 9$ at $(3, -2)$; $m_2 = 12$ at $(5, 4)$; $m_3 = 15$ at $(3, 7)$
2. $m_1 = 3$ at $(-6, 2)$; $m_2 = 9$ at $(-5, 8)$; $m_3 = 8$ at $(7, -2)$
3. $m_1 = 15$ at $(3, -9)$; $m_2 = 10$ at $(6, -4)$; $m_3 = 18$ at $(9, 2)$; $m_4 = 12$ at $(1, 3)$
4. $m_1 = 24$ at $(-6, 3)$; $m_2 = 36$ at $(7, -9)$; $m_3 = 15$ at $(4, 4)$; $m_4 = 12$ at $(-2, -5)$; $m_5 = 18$ at $(8, 6)$

Find the moment of inertia and the radius of gyration about the x-axis of each system.

5. The system in Exercise 1.
6. $m_1 = 8$ at $(-3, 9)$; $m_2 = 16$ at $(8, -6)$; $m_3 = 14$ at $(4, 2)$
7. $m_1 = 9$ at $(-5, -9)$; $m_2 = 5$ at $(8, -2)$; $m_3 = 8$ at $(5, 0)$; $m_4 = 10$ at $(-3, 1)$
8. The system in Exercise 4.

Find the moment of inertia and the radius of gyration of each region bounded by the given curves about the given axis.

9. $y = x^2$, $x = 0$, and $y = 4$ about the y-axis ($\rho = 5$).
10. Region of Exercise 9 about the x-axis.

11. $y = x^2$, $y = x$, and $x > 0$ about the x-axis ($\rho = 4$).

12. Region of Exercise 11 about the y-axis.

13. $y = 5 - x^2$, $y = 1$, and $x = 0$ about the y-axis ($\rho = 3$).

14. $x = 1 + y^2$, $x = 10$, and $y = 0$ about the x-axis ($\rho = 5$).

15. $y = 1/x^2$, $y = 0$, $x = 1$, and $x = 2$ about the y-axis ($\rho = 2$).

16. $y = 4x - x^2$ and $y = 0$ about the y-axis ($\rho = 15$).

Find the moment of inertia and the radius of gyration of the solid formed by revolving the region bounded by the given curves about the given axis.

17. $y = 3x$, $y = 0$, and $x = 2$ about the y-axis ($\rho = 15$).

18. Region of Exercise 17 about the x-axis.

19. $y = 4x^2$, $y = 0$, and $x = 2$ about the x-axis ($\rho = 1$).

20. Region of Exercise 19 about the y-axis.

21. $y = 9 - x^2$, $y = 0$, and $x = 0$ about the y-axis ($\rho = 12$).

22. $x = 4 - y^2$, $y = 0$, and $x = 0$ about the x-axis ($\rho = 6$).

23. $y = 4x - x^2$ and $y = 0$ about the y-axis ($\rho = 15$).

24. $y = 1/x^3$, $x = 0$, $y = 1$, and $y = 8$ about the x-axis ($\rho = 2$).

6.7 WORK, FLUID PRESSURE, AND AVERAGE VALUE

Although there are still many more technical applications of the integral, we will consider only three more in this section: work, fluid pressure, and average value.

Work

When a constant force F is applied to an object, moving it through a distance s, the technical term *work* is defined to be the product $F \cdot s$. That is, work W is the product of the force and the distance through which the force acts.

EXAMPLE 1

A 70-lb container is lifted 8 ft above the floor. Find the work done.

$$W = F \cdot s$$
$$= (70 \text{ lb})(8 \text{ ft})$$
$$= 560 \text{ ft-lb}$$

This formula for work is appropriate when the force remains constant. However, if an object is moved from a to b by a variable force F, then we can approximate the work done as follows: Divide the interval from a to b into intervals, each of width Δx. Let F_k represent the value of the force acting on the object somewhere in the kth interval (see Fig. 6.49).

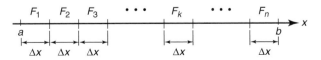

Figure 6.49 The work done by a variable force in moving an object from a to b is the sum of the work done in each interval of width Δx.

Then $\Delta W_k = F_k \cdot \Delta x$ is an approximation for the work done in the kth interval. If there are n intervals, each of width Δx, from a to b, then

$$W_{\text{approx}} = F_1 \cdot \Delta x + F_2 \cdot \Delta x + F_3 \cdot \Delta x + \cdots + F_n \cdot \Delta x$$

is an approximation for the work done moving the object from a to b. If the variable force can be expressed as a function of the distance traveled by the object, then it is possible to use integration techniques to find the actual work done. Suppose that $F = f(x)$ expresses the force as a function of the distance traveled by the object. Then

$$W_{\text{approx}} = f(x_1) \cdot \Delta x + f(x_2) \cdot \Delta x + f(x_3) \cdot \Delta x + \cdots$$
$$+ f(x_k) \cdot \Delta x + \cdots + f(x_n) \cdot \Delta x$$

is an approximation for W, the actual work done, where x_k is a number in the kth interval.

The smaller we choose Δx, the better approximation we obtain for W_{approx}. In fact, if we let Δx approach 0, then W_{approx} approaches the actual work done W. Then

> **WORK**
>
> $$W = \int_a^b f(x)\, dx$$

since this is how the definite integral of $f(x)$ from a to b was described in Chapter 5.

EXAMPLE 2

Find the work done by a force F moving an object from $x = 1$ to $x = 2$ according to $F = f(x) = x^2$.

$$W = \int_1^2 x^2\, dx$$

$$= \left.\frac{x^3}{3}\right|_1^2 = \frac{8}{3} - \frac{1}{3} = \frac{7}{3}$$

EXAMPLE 3

Hooke's law states that the force required to stretch a spring is directly proportional to the amount that it is stretched. Find the work done in stretching a spring 3 in. if it requires 12 lb of force to stretch it 10 in.

Let x represent the distance stretched by the force $F = f(x)$. Then by Hooke's law

$$f(x) = kx$$

At $x = 10$ we know that $f(10) = 12$, so

$$12 = k(10)$$

$$k = \frac{6}{5}$$

Then

$$f(x) = \frac{6}{5}x$$

and

$$W = \int_a^b f(x)\, dx$$

$$= \int_0^3 \frac{6}{5} x \, dx$$

$$= \frac{6}{5} \left(\frac{x^2}{2} \right) \Big|_0^3 = \frac{6}{5} \left(\frac{9}{2} - 0 \right)$$

$$= \frac{27}{5} \text{ in.-lb}$$

EXAMPLE 4

Two charged particles separated by a distance x (in metres) attract each other with a force $F = 4.65 \times 10^{-20} x^{-2}$ newton (N). Find the work done (in joules) in separating them over an interval from $x = 0.01$ m to $x = 0.1$ m.

$$W = \int_{0.01}^{0.1} 4.65 \times 10^{-20} x^{-2} \, dx$$

$$= 4.65 \times 10^{-20} \int_{0.01}^{0.1} x^{-2} \, dx$$

$$= 4.65 \times 10^{-20} \left(\frac{x^{-1}}{-1} \Big|_{0.01}^{0.1} \right)$$

$$= 4.65 \times 10^{-20} \left(\frac{-1}{0.1} - \frac{-1}{0.01} \right)$$

$$= 4.65 \times 10^{-20} (-10 + 100)$$

$$= 4.19 \times 10^{-18} \text{ N-m}$$

$$= 4.19 \times 10^{-18} \text{ J}$$

Note: The metric system unit of work is the joule (J). 1 J = 1 N-m.

EXAMPLE 5

A chain 50 ft long and weighing 4 lb/ft is hanging from a pulley. (a) How much work is needed to pull 30 ft of the chain to the top? (b) How much work is needed to pull all of the chain to the top?

The force needed to lift the chain at any one time equals the weight of the chain hanging down at that time. If x feet of chain are hanging down and the chain weighs 4 lb/ft, the force is

$$f(x) = F = 4x$$

(a) $W = \int_{20}^{50} 4x \, dx$ (b) $W = \int_0^{50} 4x \, dx$

$\quad = 2x^2 \Big|_{20}^{50} = 4200$ ft-lb $\quad = 2x^2 \Big|_0^{50} = 5000$ ft-lb

EXAMPLE 6

A cylindrical tank 10 ft in diameter and 12 ft high is full of water. How much work is needed to pump all the water out over the top? The density of water $\rho = 62.4$ lb/ft^3.

First, divide the tank into n layers, each of thickness Δx as in Fig. 6.50. Let x be the distance that each layer travels as it is pumped to the top. The force is the weight of each layer of water. $F = \rho V$.

Each layer is in the shape of a cylinder, whose volume is given by

$$V = \pi r^2 h$$

So

$$F = \rho V = 62.4 \pi (5^2) \, \Delta x$$

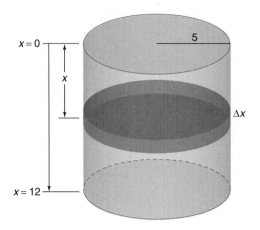

Figure 6.50

is the force of each layer. Since each layer travels a distance of x feet, the work needed to move one layer to the top is

$$W = F \cdot s = 62.4\pi (5^2)\Delta x \cdot x$$
$$= 1560\pi x \, \Delta x$$

Summing the work of such layers gives

$$W = \int_0^{12} 1560\pi x \, dx$$
$$= 780\pi \, x^2 \Big|_0^{12}$$
$$= 112{,}320\pi \text{ ft-lb}$$

Fluid Pressure

As a body goes deeper under water, the pressure on it increases because the water's weight increases with depth. Fluids are different in this respect from solids in that, where solids exert only a downward force due to gravity, the force exerted by fluids is in all directions. *Hydrostatic pressure* is the pressure at any given depth in a fluid due to its weight and may be expressed by

$$p = \rho g h$$

where p is the pressure, ρ is the mass density of the fluid, g is the force of gravity, and h is the height or depth of the fluid. For example, if you are in a swimming pool 12 ft below the surface, the pressure you feel is

$$p = \rho g h$$
$$p = (62.4 \text{ lb/ft}^3)(12 \text{ ft}) \qquad (\rho g = 62.4 \text{ lb/ft}^3)$$
$$= 748.8 \text{ lb/ft}^2$$

The total force is given by

$$F = pA$$

One main interest in fluid pressure is determining the total force exerted by a fluid on the walls of its container. If the container has vertical sides, it is simple to calculate the total force on the *bottom* of the container; that is,

$$F = \rho g h A$$

For example, the total force on the bottom of a rectangular swimming pool 12 ft × 30 ft when the water is 8 ft deep is

$$F = (62.4 \text{ lb/ft}^3)(8 \text{ ft})(12 \text{ ft} \times 30 \text{ ft})$$
$$F = 180{,}000 \text{ lb (approx.)}$$

The more difficult problem is finding the total force against the *vertical sides* of a container because the pressure is not constant. The pressure increases as the depth increases.

Let a vertical plane region be submerged into a fluid of constant density ρ as shown in Fig. 6.51. We need to find the total force against this region from depth $h - a$ to $h - b$. First, divide the interval $a \le y \le b$ into n rectangles each of width Δy. The ith rectangle has length L_i, area $L_i \, \Delta y$, and depth $h - y_i$. The force on the ith rectangle is

$$\Delta F_i = \rho g(h - y_i)L_i \, \Delta y$$

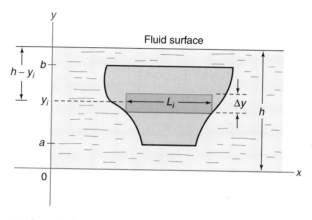

Figure 6.51 Finding the force exerted by a fluid against a submerged vertical plane.

Summing the forces on all such rectangles gives the following integral.

FORCE EXERTED BY A FLUID

The force F exerted by a fluid of constant mass density ρ against a submerged vertical plane region from $y = a$ to $y = b$ is given by

$$F = \rho g \int_a^b (h - y)L \, dy$$

where h is the total depth of the fluid and L is the horizontal length of the region at y.
    Note: In the metric system, the mass density, ρ, must be known and $g = 9.80 \text{ m/s}^2$. In the English system, the weight density, ρg, must be known.

EXAMPLE 7

A vertical gate in a dam is in the shape of an isosceles trapezoid 12 ft across the top and 8 ft across the bottom, with a height of 10 ft. Find the total force against the gate if the water surface is at the top of the gate.

The solution can be simplified if we position the trapezoid in the plane as shown in Fig. 6.52. The equation of the line through $(4, 0)$ and $(6, 10)$ is

$$y - 0 = 5(x - 4)$$
$$x = \frac{y + 20}{5}$$

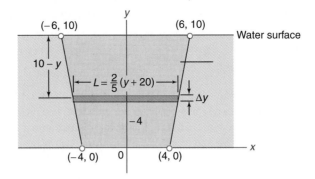

Figure 6.52

The width of the ith rectangle is Δy; its length L may be written as

$$L = 2x = 2\left(\frac{y + 20}{5}\right) = \frac{2}{5}(y + 20)$$

and its depth is $10 - y$. So

$$F = \rho g \int_a^b (h - y)L\, dy$$

$$F = 62.4 \int_0^{10} (10 - y)\frac{2}{5}(y + 20)\, dy$$

$$= 24.96 \int_0^{10} (10 - y)(y + 20)\, dy$$

$$= 24.96 \int_0^{10} (200 - 10y - y^2)\, dy$$

$$= 24.96\left(200y - 5y^2 - \frac{y^3}{3}\right)\Big|_0^{10}$$

$$= 24.96\left(2000 - 500 - \frac{1000}{3}\right)$$

$$= 29{,}120 \text{ lb}$$

EXAMPLE 8

A vertical gate in a dam is semicircular and has a diameter of 8 m. Find the total force against the gate if the water level is 1 m from the top of the gate.

Here place the semicircle with center at the origin as in Fig. 6.53. Its equation is $x^2 + y^2 = 16$. Other positions make the integration more difficult.

$$L = 2x = 2\sqrt{16 - y^2}$$

The depth of the rectangle is $0 - y$, or $-y$, $\rho = 1000 \text{ kg/m}^3$, and $g = 9.80 \text{ m/s}^2$.

$$F = \rho g \int_a^b (h - y)\, L\, dy$$

$$F = 9800 \int_{-4}^{-1} (-y)\, 2\sqrt{16 - y^2}\, dy$$

$$= -19{,}600 \int_{-4}^{-1} y(16 - y^2)^{1/2}\, dy$$

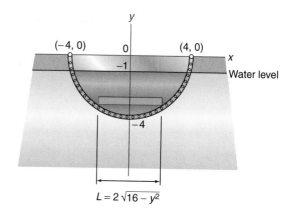

Figure 6.53

$$= -19,600 \int yu^{1/2} \frac{du}{-2y}$$

$$\boxed{\begin{array}{l} u = 16 - y^2 \\ du = -2y \, dy \end{array}}$$

$$= 9800 \int u^{1/2} \, du$$

$$= 9800 \cdot \frac{u^{3/2}}{3/2}$$

$$= \frac{19,600}{3}(16 - y^2)^{3/2} \Big|_{-4}^{-1}$$

$$= \frac{19,600}{3}(15^{3/2} - 0)$$

$$= 379,600 \text{ N} \qquad (4 \text{ significant digits})$$

EXAMPLE 9

A swimming pool is 15 ft wide and 20 ft long. The bottom is flat but sloped so that the water is 4 ft deep at one end and 12 ft deep at the other end. Find the force of the water on one 20-ft side.

First, let's position the vertical side in the plane as shown in Fig. 6.54. Notice that the right ends of the horizontal strips sometimes are on the vertical line $x = 20$ and sometimes on the line through $(0, 0)$ and $(20, 8)$, whose equation is $y = 2x/5$. Thus we need two integrals. For each integral $L = x$.

$$F = 62.4 \int_0^8 (12 - y)x \, dy + 62.4 \int_8^{12} (12 - y)x \, dy$$

$$= 62.4 \int_0^8 (12 - y)\left(\frac{5}{2}y\right) dy + 62.4 \int_8^{12} (12 - y)(20) \, dy$$

$$= 156 \int_0^8 (12y - y^2) \, dy + 1248 \int_8^{12} (12 - y) \, dy$$

$$= 156\left(6y^2 - \frac{y^3}{3}\right)\Big|_0^8 + 1248\left(12y - \frac{y^2}{2}\right)\Big|_8^{12}$$

$$= 33,280 + 9984$$

$$= 43,264 \text{ lb}$$

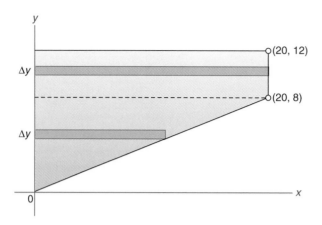

Figure 6.54

Note: The position of the vertical side in the *xy*-plane has no effect on the final result. You should repeat this problem using other positions.

Average Value

The average value of a function, y_{av}, is another application of integration. The sum

$$y_{av} = \frac{f(x_1) + f(x_2) + f(x_3) + \cdots + f(x_n)}{n}$$

is the average value of a function for the given values of x: $x_1, x_2, x_3, \ldots, x_n$.

For example, if $f(x_1) = 3, f(x_2) = 5$, and $f(x_3) = 4$, then

$$y_{av} = \frac{3 + 5 + 4}{3} = 4$$

is the average of the three given values.

Now

$$
\begin{aligned}
y_{av} &= \frac{f(x_1) + f(x_2) + f(x_3) + \cdots + f(x_n)}{n} \\
&= \frac{[f(x_1) + f(x_2) + f(x_3) + \cdots + f(x_n)]\, \Delta x}{n\, \Delta x} \quad \text{(Multiply numerator and denominator by } \Delta x.\text{)} \\
&= \frac{1}{n\, \Delta x}[f(x_1)\, \Delta x + f(x_2)\, \Delta x + f(x_3)\, \Delta x + \cdots + f(x_n)\, \Delta x]
\end{aligned}
$$

In Fig. 6.55, $n\, \Delta x = b - a$, so that

$$y_{av} = \frac{1}{b - a}[f(x_1)\, \Delta x + f(x_2)\, \Delta x + f(x_3)\, \Delta x + \cdots + f(x_n)\, \Delta x]$$

As we let Δx approach 0, the right-hand factor approaches the integral

$$\int_a^b f(x)\, dx$$

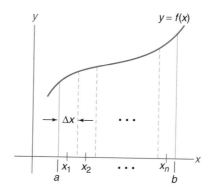

Figure 6.55 The average value of $y = f(x)$.

We then consider

AVERAGE VALUE

The *average value* of the function $y = f(x)$ over the interval $x = a$ to $x = b$ is

$$y_{av} = \frac{1}{b - a} \int_a^b f(x)\, dx$$

A geometrical interpretation can be given to y_{av} as the height of a rectangle with base, $b - a$, having the same area as the area bounded by the curve $y = f(x)$, $x = a$, $x = b$, and the x-axis as shown in Fig. 6.56.

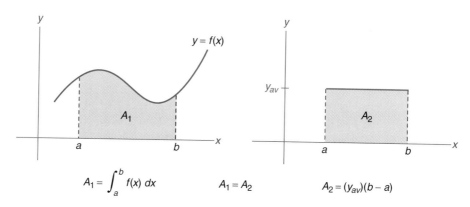

$$A_1 = \int_a^b f(x)\, dx \qquad A_1 = A_2 \qquad A_2 = (y_{av})(b - a)$$

Figure 6.56 Geometrically, y_{av} is the height of a rectangle with base $b - a$, having the same area as the area under the curve $y = f(x)$ from $x = a$ to $x = b$.

EXAMPLE 10

Find the average value of the function $y = 4 - x^2$ from $x = 0$ to $x = 2$.

$$y_{av} = \frac{1}{b - a} \int_a^b f(x)\, dx = \frac{1}{2 - 0} \int_0^2 (4 - x^2)\, dx$$

$$= \frac{1}{2}\left(4x - \frac{x^3}{3}\right)\Big|_0^2 = \frac{1}{2}\left[\left(8 - \frac{8}{3}\right) - (0)\right] = \frac{8}{3}$$

A rectangle with height $\frac{8}{3}$ and width 2 has area also equal to A: $\left(\frac{8}{3}\right)(2) = \frac{16}{3}$ (see Fig. 6.57).

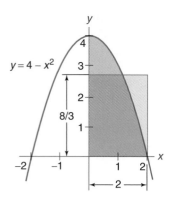

Figure 6.57

EXAMPLE 11

The power developed in a resistor is given by $p = 0.5i^3$. Find the average power (in watts) as the current changes from 1 A to 3 A.

$$p_{av} = \frac{1}{3-1} \int_1^3 0.5i^3 \, di$$

$$= \frac{0.5}{2} \cdot \frac{i^4}{4} \Big|_1^3$$

$$= \frac{0.5}{2} \left(\frac{81}{4} - \frac{1}{4} \right)$$

$$= 5.0 \text{ W}$$

Exercises 6.7

1. A force moves an object from $x = 0$ to $x = 3$ according to $F = x^3 - x$. Find the work done.

2. A force moves an object from $x = 1$ to $x = 100$ according to $F = \sqrt{x}$. Find the work done.

3. Find the work done in stretching a spring 5 in. if it requires a 20-lb force to stretch it 10 in.

4. Find the work done in stretching a spring 12 in. if it requires a 10-lb force to stretch it 8 in.

5. A spring with a natural length of 8 cm measures 12 cm after a 150-N weight is attached. Find the work required to stretch the spring 6 cm from its natural length.

6. A spring with a natural length of 10 cm measures 14 cm after a 60-N weight is attached. Find the work done in stretching the spring 10 cm from its natural length.

7. Two charged particles separated by a distance x attract each other with a force $F = 3.62 \times 10^{-16} x^{-2}$ N. Find the work done (in joules) in separating them over an interval from $x = 0.01$ m to $x = 0.05$ m.

8. Two charged particles separated by a distance x attract each other with a force $F = 4.52 \times 10^{-18} x^{-2}$ N. Find the work done (in joules) in separating them over an interval from $x = 0.02$ m to $x = 0.08$ m.

9. A chain 50 ft long and weighing 2 lb/ft is hanging over a pulley. How much work is needed to pull (a) 10 ft of the chain to the top? (b) half of the chain to the top? (c) all of the chain to the top?

10. Find the amount of work done in winding all of a 300-ft hanging cable that weighs 120 lb.

11. A cylindrical tank 8 ft in diameter and 12 ft high is full of water. How much work is needed to pump all the water out over the top?

12. How much work is needed in Exercise 11 to pump half the water out over the top of the tank?

13. Suppose that the tank in Exercise 11 is placed on a 10-ft platform. How much work is needed to fill the tank from the ground when the water is pumped in through a hole in the bottom of the tank?

14. How much work is needed to fill the tank in Exercise 13 if the water is pumped into the top of the tank?

15. A conical tank (inverted right circular cone) filled with water is 10 ft across the top and 12 ft high. How much work is needed to pump all the water out over the top?

16. How much work is needed in Exercise 15 to pump 4 ft of water out over the top (a) when the tank is full? (b) when the tank has 4 ft of water in it?

17. A dam contains a vertical rectangular gate 10 ft high and 8 ft wide. The top of the gate is at the water's surface. Find the force on the gate.

18. A dam contains a vertical rectangular gate 6 ft high and 8 ft wide. The top of the gate is 4 ft below the water's surface. Find the force on the gate.

19. A rectangular tank is 8 m wide and 4 m deep. If the tank is $\frac{3}{4}$ full of water, find the force against the side.

20. A cylindrical tank is lying on its side and is half-filled with water. If its diameter is 6 m, find the force against an end.

21. A cylindrical tank of oil is half full of oil ($\rho = 870$ kg/m^3) and lying on its side. If the diameter of the tank is 10 m, find the force on an end of the tank.

22. A rectangular porthole on a vertical side of a ship is 1 ft square. Find the total force on the porthole if its top is 20 ft below the water's surface.

23. A trough is 12 ft long and 2 ft high. Vertical cross sections are isosceles right triangles with the hypotenuse horizontal. Find the force on one end if the trough is filled with water.

24. A trough is 12 m long and 1 m high. Vertical cross sections are equilateral triangles with the top side horizontal. Find the force on one end if the trough is filled with alcohol ($\rho = 790$ kg/m^3).

25. A dam has a vertical gate in the shape of an isosceles trapezoid with upper base 10 ft, lower base 16 ft, and height 6 ft. Find the force on the gate if the upper base is 8 ft below the water's surface.

26. Find the force on the gate in Exercise 25 if the gate is inverted.

27. A swimming pool is 12 ft wide and 18 ft long. The bottom is flat but sloped so that the water is 3 ft deep at one end and 9 ft deep at the other end. Find the force on one 18-ft side.

28. A dam is in the shape of a parabola 12 ft high and 8 ft across its top. Find the force on it when the water's surface is at the top.

Find the average value of each function.

29. $y = x^2$ from $x = 1$ to $x = 3$

30. $y = \sqrt{x}$ from $x = 1$ to $x = 4$

31. $y = \dfrac{1}{\sqrt{x-1}}$ from $x = 5$ to $x = 10$

32. $y = x^2 - 1/x^2$ from $x = 1$ to $x = 3$

33. The electric current for a certain circuit is given by $i = 6t - t^2$. Find the average value of the current (in amperes) over the interval from $t = 0.1$ s to $t = 0.5$ s.

34. The power developed in a resistor is given by $P = 0.28i^3$. Find the average power (in watts) as the current changes from 1 A to 4 A.

CHAPTER SUMMARY

1. The area between two curves $y = f(x)$ and $y = g(x)$ between $x = a$ and $x = b$ is

$$\int_a^b [f(x) - g(x)]\, dx$$

for $f(x) \geq g(x)$ and $a \leq x \leq b$. The area between two curves $x = f(y)$ and $x = g(y)$ between $y = c$ and $y = d$ is

$$\int_c^d [f(y) - g(y)]\, dy$$

for $f(y) \geq g(y)$ and $c \leq y \leq d$.

2. *Volume of revolution, disk method:*

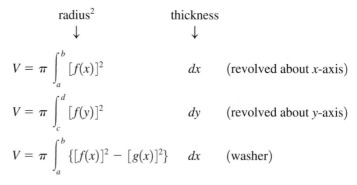

radius² ↓   thickness ↓

$$V = \pi \int_a^b [f(x)]^2 \qquad dx \qquad \text{(revolved about x-axis)}$$

$$V = \pi \int_c^d [f(y)]^2 \qquad dy \qquad \text{(revolved about y-axis)}$$

$$V = \pi \int_a^b \{[f(x)]^2 - [g(x)]^2\} \qquad dx \qquad \text{(washer)}$$

3. *Volume of revolution, shell method:*

radius ↓   height ↓   thickness ↓

$$V = 2\pi \int_a^b \quad x \quad f(x) \qquad dx \qquad \text{(shells parallel to y-axis)}$$

$$V = 2\pi \int_c^d \quad y \quad f(y) \qquad dy \qquad \text{(shells parallel to x-axis)}$$

4. *Moment in a linear system along the x-axis:* Let $\bar{x}$ be the center of mass of a linear system along the x-axis with n masses; then

$$\bar{x} = \frac{m_1 x_1 + m_2 x_2 + m_3 x_3 + \cdots + m_n x_n}{m_1 + m_2 + m_3 + \cdots + m_n}$$

If we let M_0 be the moment about the origin and m be the total mass of the system, then

$$\bar{x} = \frac{M_0}{m}$$

5. *Moments of a two-dimensional system:* Consider n masses $m_1, m_2, \ldots, m_n$ located at points $(x_1, y_1), (x_2, y_2), \ldots, (x_n, y_n)$, respectively. The moment about the y-axis, M_y, is

$$M_y = m_1 x_1 + m_2 x_2 + \cdots + m_n x_n$$

and the moment about the x-axis, M_x, is

$$M_x = m_1 y_1 + m_2 y_2 + \cdots + m_n y_n.$$

If we let m be the total mass of the system, the center of mass $(\bar{x}, \bar{y})$ is

$$\bar{x} = \frac{M_y}{m} \quad \text{and} \quad \bar{y} = \frac{M_x}{m}$$

6. The center of mass of a continuous thin mass of variable density from $x = a$ to $x = b$ is given by

$$\bar{x} = \frac{M_0}{m} = \frac{\displaystyle\int_a^b \rho(x)\, x\, dx}{\displaystyle\int_a^b \rho(x)\, dx}$$

where $\rho(x)$ is the density expressed as a function of x. If the density is constant, ρ will cancel.

7. *Moments and center of mass of a plane area or thin plate:* Let $g(x) \leq f(x)$ be continuous functions on $a \leq x \leq b$ for the area of uniform density ρ bounded by $y = f(x)$, $y = g(x)$, $x = a$, and $x = b$, the moments about the x-axis and the y-axis are

$$M_x = \frac{\rho}{2} \int_a^b \{[f(x)]^2 - [g(x)]^2\}\, dx \quad \text{and} \quad M_y = \rho \int_a^b x[f(x) - g(x)]\, dx$$

Its mass is given by

$$m = \rho \int_a^b [f(x) - g(x)]\, dx$$

and its center of mass is $(\bar{x}, \bar{y})$ where

$$\bar{x} = \frac{M_y}{m} \quad \text{and} \quad \bar{y} = \frac{M_x}{m}$$

8. *Centroid of a plane region or thin plate:* Let $g(x) \leq f(x)$ be continuous functions on $a \leq x \leq b$. The centroid $(\bar{x}, \bar{y})$ of the region bounded by $y = f(x)$, $y = g(x)$, $x = a$, and $x = b$ is

$$\bar{x} = \frac{\displaystyle\int_a^b x[f(x) - g(x)]\, dx}{A} \quad \text{and} \quad \bar{y} = \frac{\dfrac{1}{2} \displaystyle\int_a^b \{[f(x)]^2 - [g(x)]^2\}\, dx}{A}$$

where A is the area of the region.

9. *Centroid of a solid of revolution:*

$$\bar{x} = \frac{M_y}{m} = \frac{\displaystyle\int_a^b xy^2\, dx}{\displaystyle\int_a^b y^2\, dx} \quad \text{and} \quad \bar{y} = 0 \qquad \text{(revolved about } x\text{-axis)}$$

$$\bar{y} = \frac{M_x}{m} = \frac{\displaystyle\int_c^d yx^2\, dy}{\displaystyle\int_c^d x^2\, dy} \quad \text{and} \quad \bar{x} = 0 \qquad \text{(revolved about } y\text{-axis)}$$

10. *Moment of inertia of a system:* Let masses $m_1, m_2, m_3, \ldots, m_n$ be at distances $d_1, d_2, d_3, \ldots, d_n$, respectively, rotating about some axis. The moment of inertia, I, of the system is

$$I = m_1 d_1^2 + m_2 d_2^2 + m_3 d_3^2 + \cdots + m_n d_n^2$$

Let m be the sum of all the masses in the system and let R be the distance from the axis of rotation that gives the same total moment of inertia

$$I = mR^2$$

R is called the radius of gyration.

11. *Moments of intertia of an area of constant density about the x- and y-axes:* Let $g(x) \leq f(x)$ be continuous functions on $a \leq x \leq b$ for the area of constant density ρ bounded by $y = f(x)$, $y = g(x)$, $x = a$, and $x = b$. The moment of inertia about the y-axis is

$$I_y = \rho \int_a^b x^2[f(x) - g(x)]\, dx \quad \text{and} \quad R = \sqrt{\frac{I_y}{m}}$$

Similarly, the moment of inertia about the x-axis is

$$I_x = \rho \int_c^d y^2[f(y) - g(y)]\, dy \quad \text{and} \quad R = \sqrt{\frac{I_x}{m}}$$

12. *Moment of inertia of a solid of revolution:* Let the area bounded by $y = f(x)$, $y = 0$, $x = a$, and $x = b$ be revolved about the y-axis. The moment of inertia about the y-axis is

$$I_y = 2\pi\rho \int_a^b x^3 f(x)\, dx$$

Its radius of gyration is $R = \sqrt{\dfrac{I_y}{m}}$, where $m = 2\pi\rho \displaystyle\int_a^b x f(x)\, dx$.

Similarly, let the area bounded by $x = f(y)$, $x = 0$, $y = c$, and $y = d$ be revolved about the x-axis. Its moment of inertia about the x-axis is

$$I_x = 2\pi\rho \int_c^d y^3 f(y)\, dy$$

Its radius of gyration is $R = \sqrt{\dfrac{I_x}{m}}$, where $m = 2\pi\rho \displaystyle\int_c^d y f(y)\, dy$.

Note: $f(x)$ and $f(y)$ correspond to the height of the shell.

13. *Work of a variable force $f(x)$ acting through the distance from $x = a$ to $x = b$.*

$$W = \int_a^b f(x)\, dx$$

14. *Force exerted by a fluid:* The force F exerted by a fluid of constant density ρ against a submerged vertical plane region from $y = a$ to $y = b$ is

$$F = \rho g \int_a^b (h - y)\, L\, dy$$

where h is the total depth of the fluid and L is the horizontal length of the region at y.

15. *Average value:* The average value of the function $y = f(x)$ over the interval $x = a$ to $x = b$ is

$$y_{av} = \frac{1}{b - a} \int_a^b f(x)\, dx$$

CHAPTER 6 REVIEW

Find each area bounded by the curves.

1. $y = x^2 + 3$, $y = 0$, $x = 1$, and $x = 2$.

2. $y = 1 - x^2$, $y = 0$, and $x = 0$.

3. $x = y^2 - y^3$ and the *y*-axis.

4. $y = 3x^2 - 12x + 9$, $y = 0$, $x = 0$, and $x = 4$.

5. $x = y^4 - 2y^2$ and $x = 2y^2$.

6. $x = y^2$ and $x = 9$.

Find the volume of each solid formed by revolving the region bounded by the given curves about the given line.

7. $y = \sqrt{x}$, $y = 0$, and $x = 4$ about the *x*-axis (shell method).

8. $y = \sqrt{x}$, $y = 0$, and $x = 4$ about the *x*-axis (disk method).

9. $y = x - x^2$ and $y = 0$ about the *x*-axis.

10. $y = x$ and $y = 3x - x^2$ about the *y*-axis.

11. $y = 3x^2 - x^3$ and $y = 0$ about the *y*-axis.

12. $y = x^2 + 1$, $y = 0$, $x = 0$, and $x = 3$ about the *y*-axis.

13. $x = y^2$ and $x = 4$ about *y*-axis.

14. $x = 4y - y^2$, $x = 0$, and $y = 3$ about *x*-axis.

15. Find the center of mass of the linear system $m_1 = 12$, $x_1 = -4$; $m_2 = 20$, $x_2 = 9$; $m_3 = 24$, $x_3 = 12$.

16. Find the center of mass of the system $m_1 = 24$ at $(11, -3)$; $m_2 = 36$ at $(-4, -15)$; $m_3 = 30$ at $(-7, 0)$.

17. Find the center of mass of the uniform thin plate in Fig. 6.58.

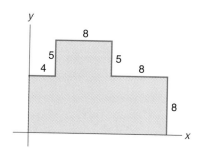

Figure 6.58

Find the centroid of each region bounded by the given curves.

18. $y = 5x$, $x = 4$, and $y = 0$.

19. $y = 6x - x^2$ and $y = 3x$.

20. $y = x^2 - 6x$ and $y = 0$.

Find the centroid of the solid formed by revolving each region bounded by the given curves about the given axis.

21. $y = 2x$, $x = 0$, and $y = 2$ about the *y*-axis. **22.** $y = x^2$, $x = 0$, and $y = 1$ about the *x*-axis.

23. $x = y^2 - 4y$ and $x = 0$ about the *y*-axis.

24. Find the moment of inertia and the radius of gyration about the *x*-axis of the system: $m_1 = 10$ at $(3, 2)$; $m_2 = 6$ at $(5, 7)$; $m_3 = 8$ at $(8, -4)$.

Find the moment of inertia and the radius of gyration of each region bounded by the given curves about the given axis.

25. $y = 3x$, $x = 4$, and $y = 0$ about the *y*-axis ($\rho = 1$).

26. Region in Exercise 25 about the x-axis.

27. $x = 1 - y^2$, $y = 0$, and $x = 0$ about the x-axis ($\rho = 4$).

Find the moment of inertia and the radius of gyration of the solid formed by revolving the region bounded by the given curves about the given axis.

28. $y = x^3$, $y = 0$, and $x = 1$ about the x-axis ($\rho = 4$).

29. Region of Exercise 28 about the y-axis.

30. $y = 1/x$, $y = 0$, $x = 1$, and $x = 4$ about the y-axis ($\rho = 3$).

31. Find the work done in stretching a spring 10 in. if it requires a 16-lb force to stretch it 4 in.

32. Two charged particles separated by a distance x attract each other with a force $F = 5.24 \times 10^{-18} x^{-2}$ N. Find the work done (in joules) in separating them over an interval from $x = 0.01$ m to $x = 0.02$ m.

33. A cable weighs 4 lb/ft and has a 250-lb weight attached in a hole 200 ft below the ground. Find the amount of work needed to pull the cable and weight to ground level.

34. A dam contains a vertical rectangular gate 8 ft high and 10 ft wide. The top of the gate is 6 ft below the water's surface. Find the force on the gate.

35. A cylindrical tank, 10 m in diameter, is lying on its side and is half full of water. Find the force against an end.

36. Find the average value of the voltage, V_{av}, in an electrical circuit from $t = 0$ s to $t = 3$ s if $V = t^2 + 3t + 2$.

37. Find the average value of the current, i_{av}, in an electrical circuit from $t = 4$ s to $t = 9$ s if $i = 4t^{3/2}$.

38. The power in a circuit varies according to $p = 2t^3$. Find the average power p_{av} (in watts) from $t = 1$ s to $t = 3$ s.

7

Methods
of Integration

INTRODUCTION

In the previous chapter we studied several applications of the integral of a function. We are now ready to learn more sophisticated methods of integration so that we may solve a wider variety of problems.

Objectives

- Integrate products and quotients of algebraic and exponential functions.
- Integrate expressions that are derivatives of logarithmic functions.
- Integrate trigonometric functions.
- Integrate expressions that are derivatives of inverse trigonometric functions.
- Use the partial fraction technique to express a fraction as the sum of two or more simpler fractions.
- Use the method of integration by parts.
- Use the method of integration by trigonometric substitution.
- Use tables to find integrals.
- Use the trapezoidal rule and Simpson's rule to approximate definite integrals.
- Find the area of a region defined with polar coordinates.

7.1 THE GENERAL POWER FORMULA

In this chapter we develop techniques that can be used to integrate more complicated functions. The first technique is based on the use of the general power formula developed in Chapter 5:

$$\int u^n \, du = \frac{u^{n+1}}{n+1} + C \qquad (n \neq -1)$$

This formula can be effectively used to integrate numerous functions involving either transcendental or algebraic functions. As seen in Chapter 5, the effective use of this formula depends on the proper choice of u and du.

305

EXAMPLE 1

Integrate $\int \sin^5 2x \cos 2x \, dx$.

If we choose $u = \sin 2x$, then $du = 2 \cos 2x \, dx$. We can then apply the general power formula with $n = 5$.

$$\int \sin^5 2x \cos 2x \, dx = \int u^5 \frac{du}{2}$$

$$= \frac{1}{2} \int u^5 \, du$$

$$= \frac{1}{2} \cdot \frac{u^6}{6} + C$$

$$= \frac{1}{12} \sin^6 2x + C$$

$$\boxed{\begin{array}{l} u = \sin 2x \\ du = 2 \cos 2x \, dx \\ n = 5 \end{array}}$$

EXAMPLE 2

Integrate $\int e^{3x}(2 - e^{3x})^4 \, dx$.

$$\int e^{3x}(2 - e^{3x})^4 \, dx = \int u^4 \left(-\frac{du}{3}\right)$$

$$= -\frac{1}{3} \int u^4 \, du$$

$$= -\frac{1}{3} \cdot \frac{u^5}{5} + C$$

$$= -\frac{1}{15}(2 - e^{3x})^5 + C$$

$$\boxed{\begin{array}{l} u = 2 - e^{3x} \\ du = -3e^{3x} \, dx \\ n = 4 \end{array}}$$

EXAMPLE 3

Integrate $\int \dfrac{\sec^2 x \, dx}{\sqrt{4 + \tan x}}$.

$$\int \frac{\sec^2 x \, dx}{\sqrt{4 + \tan x}} = \int (4 + \tan x)^{-1/2} \sec^2 x \, dx$$

$$= \int u^{-1/2} \, du$$

$$= \frac{u^{1/2}}{\frac{1}{2}} + C$$

$$= 2\sqrt{4 + \tan x} + C$$

$$\boxed{\begin{array}{l} u = 4 + \tan x \\ du = \sec^2 x \, dx \\ n = -\dfrac{1}{2} \end{array}}$$

EXAMPLE 4

Evaluate $\int_1^e \dfrac{\ln x}{x} dx$.

$$\int \frac{\ln x}{x} dx = \int u \, du$$

$$= \frac{u^2}{2} + C$$

$$= \frac{1}{2} \ln^2 x + C$$

$$\boxed{\begin{array}{l} u = \ln x \\ du = \dfrac{1}{x} dx \\ n = 1 \end{array}}$$

So

$$\int_1^e \frac{\ln x}{x}\, dx = \frac{1}{2}\ln^2 x \Big|_1^e = \frac{1}{2}\ln^2 e - \frac{1}{2}\ln^2 1 = \frac{1}{2}(1)^2 - \frac{1}{2}(0)^2 = \frac{1}{2}$$

An appropriate change in the form of the integral is needed to integrate some functions. In Example 5 we need to multiply the function $\dfrac{\tan x}{\sec^3 x}$ by 1 in the form $\dfrac{\sec x}{\sec x}$.

EXAMPLE 5

Integrate $\displaystyle\int \frac{\tan x}{\sec^3 x}\, dx.$

$$\int \frac{\tan x}{\sec^3 x}\, dx = \int \left(\frac{\sec x}{\sec x}\right)\left(\frac{\tan x}{\sec^3 x}\right) dx$$

$$\boxed{\begin{aligned} u &= \sec x \\ du &= \sec x \tan x\, dx \\ n &= -4 \end{aligned}}$$

$$= \int \frac{\sec x \tan x}{\sec^4 x}\, dx$$

$$= \int \frac{du}{u^4} = \int u^{-4}\, du$$

$$= \frac{u^{-3}}{-3} + C$$

$$= -\frac{1}{3}\frac{1}{\sec^3 x} + C$$

$$= -\frac{1}{3}\cos^3 x + C$$

EXAMPLE 6

Evaluate $\displaystyle\int_{1/6}^{1/3} \frac{\text{Arccos}^2\, 3x}{\sqrt{1 - 9x^2}}\, dx.$

First, find the indefinite integral.

$$\int \frac{\text{Arccos}^2\, 3x}{\sqrt{1 - 9x^2}}\, dx = \int \frac{u^2}{-3}\, du$$

$$\boxed{\begin{aligned} u &= \text{Arccos}\, 3x \\ du &= \frac{-3}{\sqrt{1 - 9x^2}}\, dx \\ n &= 2 \end{aligned}}$$

$$= -\frac{1}{3}\cdot\frac{u^3}{3}$$

$$= -\frac{1}{9}\text{Arccos}^3\, 3x + C$$

Thus

$$\int_{1/6}^{1/3} \frac{\text{Arccos}^2\, 3x}{\sqrt{1 - 9x^2}}\, dx = -\frac{1}{9}\text{Arccos}^3\, 3x \Big|_{1/6}^{1/3}$$

$$= -\frac{1}{9}\left[0 - \left(\frac{\pi}{3}\right)^3\right]$$

$$= \frac{\pi^3}{243}$$

Using a calculator, we have

2nd 7 green diamond COS⁻¹ 3x)^2/2nd √ 1-9x^2),x,1/6,1/3) ENTER

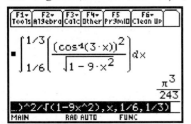

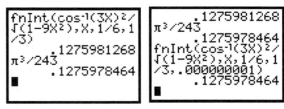

MATH 9 2nd COS⁻¹ 3x) x²/ 2nd √ 1-9x x²),x,1/6,1/3) ENTER
Note that the numerical integrator (**fnInt**) of the TI-83 Plus originally calculated 6 significant digits. The last frame shows that greater accuracy can be obtained by specifying the optional error tolerance.

Exercises 7.1

Integrate.

1. $\displaystyle\int \sqrt{3x+2}\,dx$

2. $\displaystyle\int \sqrt[3]{2x-1}\,dx$

3. $\displaystyle\int \frac{dx}{\sqrt{4+x}}$

4. $\displaystyle\int \frac{dx}{(2x+1)^3}$

5. $\displaystyle\int (x+2)(x^2+4x)^{3/4}\,dx$

6. $\displaystyle\int x\sqrt{x^2+4}\,dx$

7. $\displaystyle\int \cos^3 x \sin x\,dx$

8. $\displaystyle\int \frac{\cos 2x\,dx}{\sqrt{\sin 2x}}$

9. $\displaystyle\int \tan^3 4x \sec^2 4x\,dx$

10. $\displaystyle\int \cos x \sin x\,dx$

11. $\displaystyle\int \sin 4x\,(\cos 4x+1)\,dx$

12. $\displaystyle\int \cos 2x\,(1-\sin 2x)^2\,dx$

13. $\displaystyle\int \sqrt{9+\sec x}\,\sec x \tan x\,dx$

14. $\displaystyle\int \frac{\csc^2 2x\,dx}{\sqrt{8-\cot 2x}}$

15. $\displaystyle\int \sqrt{1+e^{2x}}\,e^{2x}\,dx$

16. $\displaystyle\int (10-e^{-2x})^3\,e^{-2x}\,dx$

17. $\displaystyle\int \frac{xe^{x^2}\,dx}{\sqrt{1+e^{x^2}}}$

18. $\displaystyle\int \frac{e^{\tan x}\sec^2 x\,dx}{\sqrt{2+e^{\tan x}}}$

19. $\displaystyle\int \frac{\ln(3x-5)}{3x-5}\,dx$

20. $\displaystyle\int \frac{\ln 4x}{x}\,dx$

21. $\displaystyle\int \frac{dx}{x\ln^2 x}$

22. $\displaystyle\int \frac{x\ln(x^2+1)\,dx}{x^2+1}$

23. $\displaystyle\int \frac{\operatorname{Arcsin}3x}{\sqrt{1-9x^2}}\,dx$

24. $\displaystyle\int \frac{\operatorname{Arctan}2x}{1+4x^2}\,dx$

25. $\displaystyle\int \frac{\cot x}{\csc^4 x}\,dx$

26. $\displaystyle\int \frac{x\sec^2 x^2\,dx}{\sqrt{9+\tan x^2}}$

27. $\displaystyle\int \frac{\operatorname{Arctan}^2 x}{1+x^2}\,dx$

28. $\displaystyle\int \frac{\ln^2 4x}{2x}\,dx$

29. $\displaystyle\int_3^5 x\sqrt{x^2-9}\,dx$

30. $\displaystyle\int_0^6 3x\sqrt{x^2+1}\,dx$

31. $\displaystyle\int_0^1 \frac{e^{3x}\,dx}{\sqrt{1+e^{3x}}}$

32. $\displaystyle\int_0^{\pi/12} \sin^3 6x \cos 6x\,dx$

33. $\displaystyle\int_1^2 \frac{\ln(2x-1)}{2x-1}\,dx$

34. $\displaystyle\int_0^{\sqrt{2}/4} \frac{\operatorname{Arcsin}2x}{\sqrt{1-4x^2}}\,dx$

35. Find the area bounded by $y=\sin^2 x \cos x$ from $x=0$ to $x=\pi/2$ and $y=0$.

36. Find the area bounded by $y=\dfrac{\ln^2 x}{x}$ from $x=1$ to $x=e$ and $y=0$.

Since integration is the inverse of differentiation and $\dfrac{d}{du}(\ln u) = \dfrac{1}{u}$, we find

$$\int \frac{du}{u} = \ln |u| + C$$

The absolute value of u is necessary because logarithms are defined only for positive numbers. If $u > 0$, then $\int \dfrac{du}{u} = \ln u + C$ and if $u < 0$, then $\int \dfrac{du}{u} = \int \dfrac{-du}{-u} = \ln(-u) + C$ so that in both cases we obtain $\ln |u| + C$. *Note:* This form is the general power formula for $n = -1$.

EXAMPLE 1

Integrate $\int \dfrac{dx}{x - 1}$.

$$\int \frac{dx}{x - 1} = \int \frac{du}{u}$$
$$= \ln |u| + C$$
$$= \ln |x - 1| + C$$

$$\boxed{\begin{aligned} u &= x - 1 \\ du &= dx \end{aligned}}$$

EXAMPLE 2

Integrate $\int \dfrac{x \, dx}{x^2 + 6}$.

$$\int \frac{x \, dx}{x^2 + 6} = \int \frac{1}{u} \left(\frac{du}{2} \right)$$
$$= \frac{1}{2} \int \frac{du}{u}$$
$$= \frac{1}{2} \ln |u| + C$$
$$= \frac{1}{2} \ln |x^2 + 6| + C$$

$$\boxed{\begin{aligned} u &= x^2 + 6 \\ du &= 2x \, dx \end{aligned}}$$

Use the integral formula $\int \dfrac{du}{u} = \ln |u| + C$ whenever the integrand is a quotient and the derivative of the denominator is a constant multiple of the numerator.

EXAMPLE 3

Integrate $\int \dfrac{\sec 3x \tan 3x \, dx}{2 + \sec 3x}$.

$$\int \frac{\sec 3x \tan 3x \, dx}{2 + \sec 3x} = \int \frac{1}{u} \left(\frac{du}{3} \right)$$
$$= \frac{1}{3} \int \frac{du}{u}$$
$$= \frac{1}{3} \ln |u| + C$$
$$= \frac{1}{3} \ln |2 + \sec 3x| + C$$

$$\boxed{\begin{aligned} u &= 2 + \sec 3x \\ du &= 3 \sec 3x \tan 3x \, dx \end{aligned}}$$

EXAMPLE 4

Evaluate $\displaystyle\int_2^3 \frac{x^2\,dx}{1-x^3}$.

$$\int \frac{x^2\,dx}{1-x^3} = \int \frac{1}{u}\left(\frac{du}{-3}\right)$$

$$= -\frac{1}{3}\int \frac{du}{u}$$

$$= -\frac{1}{3}\ln |u| + C$$

$$= -\frac{1}{3}\ln |1-x^3| + C$$

$$\boxed{\begin{array}{l} u = 1-x^3 \\ du = -3x^2\,dx \end{array}}$$

So

$$\int_2^3 \frac{x^2\,dx}{1-x^3} = -\frac{1}{3}\ln |1-x^3|\,\Big|_2^3$$

$$= -\frac{1}{3}\ln |-26| + \frac{1}{3}\ln |-7|$$

$$= \frac{1}{3}(\ln 7 - \ln 26)$$

$$= \frac{1}{3}\ln\left(\frac{7}{26}\right) = -0.437$$

From the derivative formula $\dfrac{d}{du}\left(e^u\right) = e^u$, we find

$$\int e^u\,du = e^u + C$$

EXAMPLE 5

Integrate $\displaystyle\int e^{5x}\,dx$.

$$\int e^{5x}\,dx = \int e^u\left(\frac{du}{5}\right)$$

$$= \frac{1}{5}\int e^u\,du$$

$$= \frac{1}{5}e^u + C$$

$$= \frac{1}{5}e^{5x} + C$$

$$\boxed{\begin{array}{l} u = 5x \\ du = 5\,dx \end{array}}$$

EXAMPLE 6

Integrate $\displaystyle\int \frac{dx}{e^{2x}}$.

$$\int \frac{dx}{e^{2x}} = \int e^{-2x}\,dx = \int e^u\left(\frac{du}{-2}\right)$$

$$= -\frac{1}{2}\int e^u\,du$$

$$= -\frac{1}{2}e^u + C$$

$$= -\frac{1}{2}e^{-2x} + C$$

$$\boxed{\begin{array}{l} u = -2x \\ du = -2\,dx \end{array}}$$

EXAMPLE 7

Evaluate $\displaystyle\int_0^1 x^2 e^{\,x^3+1} dx$.

First,

$$\int x^2 e^{x^3+1}\,dx = \int e^u\left(\frac{du}{3}\right)$$

$$= \frac{1}{3}\int e^u\,du$$

$$= \frac{1}{3}e^u + C$$

$$= \frac{1}{3}e^{\,x^3+1} + C$$

$$\boxed{\begin{array}{l} u = x^3 + 1 \\ du = 3x^2\,dx \end{array}}$$

So

$$\int_0^1 x^2 e^{\,x^3+1}\,dx = \left.\frac{1}{3}e^{\,x^3+1}\right|_0^1$$

$$= \frac{1}{3}e^2 - \frac{1}{3}e$$

$$= \frac{1}{3}e(e - 1)$$

For bases other than e, we have

$$\int a^u\,du = \frac{a^u}{\ln a} + C \qquad (a > 0,\, a \neq 1)$$

EXAMPLE 8

Find the area bounded by $xy = 1$, $x = 1$, $x = 2$, and $y = 0$.

This area is given by the definite integral (see Fig. 7.1).

$$\int_1^2 \frac{dx}{x} = \ln|x|\,\Big|_1^2 = \ln 2 - \ln 1 = \ln 2 = 0.6931$$

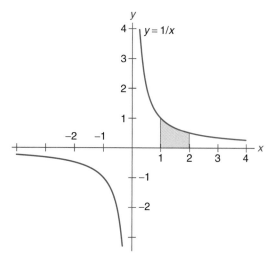

Figure 7.1

Exercises 7.2

Integrate.

1. $\displaystyle\int \frac{dx}{3x + 2}$

2. $\displaystyle\int \frac{dx}{x + 5}$

3. $\displaystyle\int \frac{dx}{1 - 4x}$

4. $\displaystyle\int \frac{dx}{4x + 2}$

5. $\displaystyle\int \frac{4x\,dx}{1 - x^2}$

6. $\displaystyle\int \frac{3x\,dx}{x^2 + 1}$

7. $\displaystyle\int \frac{x^3\,dx}{x^4 - 1}$

8. $\displaystyle\int \frac{(x + 1)\,dx}{x^2 + 2x - 3}$

9. $\displaystyle\int \frac{\csc^2 x\,dx}{\cot x}$

10. $\displaystyle\int \frac{\sin x\,dx}{\cos x}$

11. $\displaystyle\int \frac{\sec^2 3x\,dx}{1 + \tan 3x}$

12. $\displaystyle\int \frac{\cos 2x}{1 + \sin 2x}\,dx$

13. $\displaystyle\int \frac{\csc x \cot x}{1 + \csc x}\,dx$

14. $\displaystyle\int \frac{\csc^2 4x}{1 - \cot 4x}\,dx$

15. $\displaystyle\int \frac{\cos x}{1 + \sin x}\,dx$

16. $\displaystyle\int \frac{(x - \sin 2x)\,dx}{x^2 + \cos 2x}$

17. $\displaystyle\int \frac{dx}{x \ln x}$

18. $\displaystyle\int \frac{dx}{x(4 + \ln x)}$

19. $\displaystyle\int e^{2x}\,dx$

20. $\displaystyle\int e^{3x - 1}\,dx$

21. $\displaystyle\int \frac{dx}{e^{4x}}$

22. $\displaystyle\int \frac{dx}{e^{2x+3}}$

23. $\displaystyle\int xe^{x^2}\,dx$

24. $\displaystyle\int x^3 e^{x^4 - 1}\,dx$

25. $\displaystyle\int \frac{x\,dx}{e^{x^2+9}}$

26. $\displaystyle\int xe^{-x^2}\,dx$

27. $\displaystyle\int (\sin x)\,e^{\cos x}\,dx$

28. $\displaystyle\int (\sec^2 x)\,e^{2\tan x}\,dx$

29. $\displaystyle\int_0^2 xe^{x^2+2}\,dx$

30. $\displaystyle\int_0^{\pi/2} (\cos x)\,e^{\sin x}\,dx$

31. $\displaystyle\int 4^x\,dx$

32. $\displaystyle\int 10^{2x}\,dx$

33. $\displaystyle\int \frac{2e^x}{e^x + 4}\,dx$

34. $\displaystyle\int \frac{e^{-x}}{1 - e^{-x}}\,dx$

35. $\displaystyle\int_0^1 \frac{x\,dx}{x^2 + 1}$

36. $\displaystyle\int_{\pi/4}^{\pi/2} \frac{\cos x\,dx}{\sin x}$

37. $\displaystyle\int_1^5 \frac{4\,dx}{2x - 1}$

38. $\displaystyle\int_0^6 \frac{dx}{5x + 4}$

39. $\displaystyle\int_0^2 e^{x/2}\,dx$

40. $\displaystyle\int_{-2}^0 e^{-4x}\,dx$

41. $\displaystyle\int_0^1 x^2 e^{x^3}\,dx$

42. $\displaystyle\int_1^2 xe^{x^2}\,dx$

43. $\displaystyle\int_0^{\pi/6} \frac{\cos x\,dx}{1 - \sin x}$

44. $\displaystyle\int_{\pi/3}^{\pi/2} \frac{\sin x}{1 + \cos x}\,dx$

45. Find the area bounded by $y = 1/(1 + 2x)$, $x = 0$, $x = 1$, and $y = 0$.

46. Find the area bounded by $xy = 1$, $x = 1$, $x = 3$, and $y = 0$.

47. Find the area bounded by $y = e^{2x}$, $x = 0$, $x = 4$, and $y = 0$.

48. Find the area bounded by $y = e^{-x}$, $x = 0$, $x = 5$, and $y = 0$.

7.3 BASIC TRIGONOMETRIC FORMS

We have found that

$$\frac{d}{du}(\sin u) = \cos u \quad \text{and} \quad \frac{d}{du}(\cos u) = -\sin u$$

Now, since integration is the inverse of differentiation, we have

$$\int \cos u \, du = \sin u + C$$

$$\int \sin u \, du = -\cos u + C$$

EXAMPLE 1

Integrate $\int \cos 3x \, dx$.

$$\int \cos 3x \, dx = \int (\cos u)\left(\frac{du}{3}\right) \qquad \boxed{\begin{array}{l} u = 3x \\ du = 3 \, dx \end{array}}$$

$$= \frac{1}{3} \int \cos u \, du$$

$$= \frac{1}{3} \sin u + C$$

$$= \frac{1}{3} \sin 3x + C$$

EXAMPLE 2

Integrate $\int \sin(4x + 3) \, dx$.

$$\int \sin(4x + 3) \, dx = \int (\sin u)\left(\frac{du}{4}\right) \qquad \boxed{\begin{array}{l} u = 4x + 3 \\ du = 4 \, dx \end{array}}$$

$$= \frac{1}{4} \int \sin u \, du$$

$$= \frac{1}{4}(-\cos u) + C$$

$$= -\frac{1}{4} \cos(4x + 3) + C$$

EXAMPLE 3

Integrate $\int x^2 \sin(x^3 + 2) \, dx$.

$$\int x^2 \sin(x^3 + 2) \, dx = \int (\sin u)\left(\frac{du}{3}\right) \qquad \boxed{\begin{array}{l} u = x^3 + 2 \\ du = 3x^2 \, dx \end{array}}$$

$$= \frac{1}{3} \int \sin u \, du$$

$$= \frac{1}{3}(-\cos u) + C$$

$$= -\frac{1}{3} \cos(x^3 + 2) + C$$

EXAMPLE 4

Find $\int_0^{\pi/4} \cos 2x \, dx$.

$$\int \cos 2x \, dx = \int \cos u \left(\frac{du}{2}\right) \qquad \boxed{\begin{aligned} u &= 2x \\ du &= 2 \, dx \end{aligned}}$$

$$= \frac{1}{2} \int \cos u \, du$$

$$= \frac{1}{2} \sin u + C$$

$$= \frac{1}{2} \sin 2x + C$$

So

$$\int_0^{\pi/4} \cos 2x \, dx = \frac{1}{2} \sin 2x \Big|_0^{\pi/4} = \frac{1}{2} \sin \frac{\pi}{2} - \frac{1}{2} \sin 0$$

$$= \frac{1}{2}(1) - \frac{1}{2}(0) = \frac{1}{2}$$

From the derivatives of $\tan u$, $\cot u$, $\sec u$, and $\csc u$ and noting that integration is the inverse of differentiation, we have:

$$\int \sec^2 u \, du = \tan u + C$$

$$\int \csc^2 u \, du = -\cot u + C$$

$$\int \sec u \tan u \, du = \sec u + C$$

$$\int \csc u \cot u \, du = -\csc u + C$$

EXAMPLE 5

Integrate $\displaystyle\int \sec^2 3x \, dx$

$$\int \sec^2 3x \, dx = \int (\sec^2 u)\left(\frac{du}{3}\right) \qquad \boxed{\begin{aligned} u &= 3x \\ du &= 3 \, dx \end{aligned}}$$

$$= \frac{1}{3} \int \sec^2 u \, du$$

$$= \frac{1}{3} \tan u + C$$

$$= \frac{1}{3} \tan 3x + C$$

Using a calculator,

2nd 7 1/ 2nd COS 3x)^2,x) **ENTER**

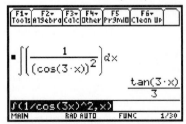

EXAMPLE 6

Integrate $\int \sec(2x+5)\tan(2x+5)\,dx$.

$$\int \sec(2x+5)\tan(2x+5)\,dx = \int (\sec u \tan u)\left(\frac{du}{2}\right) \qquad \boxed{\begin{array}{l} u = 2x+5 \\ du = 2\,dx \end{array}}$$

$$= \frac{1}{2}\int \sec u \tan u\,du$$

$$= \frac{1}{2}\sec u + C$$

$$= \frac{1}{2}\sec(2x+5) + C$$

We determine $\int \tan u\,du$ as follows:

$$\int \tan u\,du = \int \frac{\sin u}{\cos u}\,du \qquad \boxed{\begin{array}{l} w = \cos u \\ dw = -\sin u\,du \end{array}}$$

$$= \int \frac{1}{w}(-dw)$$

$$= -\int \frac{dw}{w}$$

$$= -\ln|w| + C$$

$$= -\ln|\cos u| + C$$

In a similar manner,

$$\int \cot u\,du = \int \frac{\cos u}{\sin u}\,du \qquad \boxed{\begin{array}{l} w = \sin u \\ dw = \cos u\,du \end{array}}$$

$$= \int \frac{1}{w}\,dw$$

$$= \ln|w| + C$$

$$= \ln|\sin u| + C$$

To find $\int \sec u\,du$, first multiply $\sec u$ by 1 in the form

$$\frac{\sec u + \tan u}{\sec u + \tan u}$$

That is,

$$\int \sec u\,du = \int \sec u\left(\frac{\sec u + \tan u}{\sec u + \tan u}\right)du$$

$$= \int \frac{\sec^2 u + \sec u \tan u}{\sec u + \tan u}\,du$$

$$= \int \frac{dw}{w} \qquad \boxed{\begin{array}{l} w = \sec u + \tan u \\ dw = (\sec u \tan u + \sec^2 u)\,du \end{array}}$$

$$= \ln|w| + C$$

$$= \ln|\sec u + \tan u| + C$$

In a similar manner, we find that

$$\int \csc u\,du = \ln|\csc u - \cot u| + C$$

We now can integrate all six basic trigonometric functions:

$$\int \sin u \, du = -\cos u + C$$

$$\int \cos u \, du = \sin u + C$$

$$\int \tan u \, du = -\ln|\cos u| + C$$

$$\int \cot u \, du = \ln|\sin u| + C$$

$$\int \sec u \, du = \ln|\sec u + \tan u| + C$$

$$\int \csc u \, du = \ln|\csc u - \cot u| + C$$

EXAMPLE 7

Integrate $\displaystyle\int \tan 3x \, dx$.

$$\int \tan 3x \, dx = \int \tan u \left(\frac{du}{3}\right) \qquad \boxed{\begin{array}{l} u = 3x \\ du = 3 \, dx \end{array}}$$

$$= \frac{1}{3} \int \tan u \, du$$

$$= \frac{1}{3}(-\ln|\cos u|) + C$$

$$= -\frac{1}{3}\ln|\cos 3x| + C$$

EXAMPLE 8

Integrate $\displaystyle\int x \sec 3x^2 \, dx$.

$$\int x \sec 3x^2 \, dx = \int \sec u \left(\frac{du}{6}\right) \qquad \boxed{\begin{array}{l} u = 3x^2 \\ du = 6x \, dx \end{array}}$$

$$= \frac{1}{6} \int \sec u \, du$$

$$= \frac{1}{6}\ln|\sec u + \tan u| + C$$

$$= \frac{1}{6}\ln|\sec 3x^2 + \tan 3x^2| + C$$

EXAMPLE 9

Evaluate $\displaystyle\int_{\pi/8}^{\pi/4} \cot 2x \, dx$.

$$\int \cot 2x \, dx = \int \cot u \left(\frac{du}{2}\right) \qquad \boxed{\begin{array}{l} u = 2x \\ du = 2 \, dx \end{array}}$$

$$= \frac{1}{2} \int \cot u \, du$$

$$= \frac{1}{2} \ln |\sin u| + C$$

$$= \frac{1}{2} \ln |\sin 2x| + C$$

So

$$\int_{\pi/8}^{\pi/4} \cot 2x \, dx = \frac{1}{2} \ln |\sin 2x| \Big|_{\pi/8}^{\pi/4} = \frac{1}{2} \ln \left| \sin \frac{\pi}{2} \right| - \frac{1}{2} \ln \left| \sin \frac{\pi}{4} \right|$$

$$= \frac{1}{2} \ln 1 - \frac{1}{2} \ln \frac{1}{\sqrt{2}}$$

$$= 0 - \frac{1}{2} \ln 2^{-1/2} = \frac{1}{4} \ln 2 = 0.173$$

Algebraic simplifications often lead to less complicated integrals as shown in Example 10.

EXAMPLE 10

Integrate $\int \dfrac{2 - \cos x}{\sin x} \, dx$.

$$\int \frac{2 - \cos x}{\sin x} \, dx = \int \frac{2 \, dx}{\sin x} - \int \frac{\cos x}{\sin x} \, dx$$

$$= 2 \int \csc x \, dx - \int \cot x \, dx$$

$$= 2 \ln |\csc x - \cot x| - \ln |\sin x| + C$$

Exercises 7.3

Integrate.

1. $\displaystyle\int \sin 5x \, dx$ **2.** $\displaystyle\int \cos 6x \, dx$ **3.** $\displaystyle\int \cos (3x - 1) \, dx$

4. $\displaystyle\int \sin (2x + 7) \, dx$ **5.** $\displaystyle\int x \sin (x^2 + 5) \, dx$ **6.** $\displaystyle\int x^3 \cos x^4 \, dx$

7. $\displaystyle\int (3x^2 - 2x) \cos (x^3 - x^2) \, dx$ **8.** $\displaystyle\int (x - 1) \sin (x^2 - 2x + 3) \, dx$

9. $\displaystyle\int \csc^2 5x \, dx$ **10.** $\displaystyle\int \sec^2 2x \, dx$

11. $\displaystyle\int \sec 3x \tan 3x \, dx$ **12.** $\displaystyle\int \csc 7x \cot 7x \, dx$

13. $\displaystyle\int \sec^2 (4x + 3) \, dx$ **14.** $\displaystyle\int \csc^2 (3x - 2) \, dx$

15. $\displaystyle\int \csc (2x - 3) \cot (2x - 3) \, dx$ **16.** $\displaystyle\int \sec \frac{x}{2} \tan \frac{x}{2} \, dx$

17. $\displaystyle\int x \sec^2 (x^2 + 3) \, dx$ **18.** $\displaystyle\int x^2 \csc^2 x^3 \, dx$

19. $\displaystyle\int x^2 \csc (x^3 - 1) \cot (x^3 - 1) \, dx$ **20.** $\displaystyle\int x \sec 3x^2 \tan 3x^2 \, dx$

21. $\displaystyle\int \tan 4x \, dx$ **22.** $\displaystyle\int x \cot x^2 \, dx$ **23.** $\displaystyle\int \sec 5x \, dx$

24. $\displaystyle\int x^2 \csc x^3 \, dx$ **25.** $\displaystyle\int e^x \cot e^x \, dx$ **26.** $\displaystyle\int \frac{\sec (\ln x)}{x} \, dx$

27. $\displaystyle\int (1 + \sec x)^2 \, dx$ **28.** $\displaystyle\int (1 + \tan 3x)^2 \, dx$ **29.** $\displaystyle\int \frac{5 + \sin x}{\cos x} \, dx$

30. $\displaystyle\int \frac{\tan x - \cos x}{\sin x} \, dx$ **31.** $\displaystyle\int_0^{\pi/4} \sin 2x \, dx$ **32.** $\displaystyle\int_0^{\pi/6} 2 \cos 3x \, dx$

33. $\displaystyle\int_0^{\pi/2} 3 \cos \left(x - \frac{\pi}{2} \right) dx$ **34.** $\displaystyle\int_0^{\pi/2} \sin \left(x - \frac{\pi}{4} \right) dx$ **35.** $\displaystyle\int_0^{\pi/4} \sec^2 x \, dx$

36. $\displaystyle\int_{\pi/4}^{\pi/2} \csc^2 x \, dx$ **37.** $\displaystyle\int_0^{\pi/8} \sec 2x \tan 2x \, dx$ **38.** $\displaystyle\int_{1/4}^{1/2} \csc \pi x \cot \pi x \, dx$

Find the area of each region bounded by the given curves.

39. $y = \sin x$ from $x = 0$ to $x = \pi$ and $y = 0$.

40. $y = 2 \cos x$ from $x = 0$ to $x = \pi/2$ and $y = 0$.

41. $y = \sec^2 x$, $x = 0$, $x = \pi/4$, and $y = 0$.

42. $y = \sec x \tan x$, $x = 0$, $x = \pi/4$, and $y = 0$.

43. $y = \tan x$, $x = 0$, $x = \pi/4$, and $y = 0$.

44. $y = \sin x + \cos x$, $x = 0$, and $y = 0$.

45. Find the volume of the solid formed by revolving the region bounded by $y = \sec x$ from $x = 0$ to $x = \pi/4$ and $y = 0$ about the x-axis.

7.4 OTHER TRIGONOMETRIC FORMS

To integrate powers of trigonometric functions, use trigonometric identities as shown in the following methods.

Odd Powers of Sines or Cosines

Various forms of the trigonometric identity $\sin^2 x + \cos^2 x = 1$ are needed for this case.

EXAMPLE 1

Integrate $\displaystyle\int \sin^5 x \, dx$.

$$\int \sin^5 x \, dx = \int \sin^4 x \sin x \, dx = \int (\sin^2 x)^2 \sin x \, dx$$

$$= \int (1 - \cos^2 x)^2 \sin x \, dx \qquad (\sin^2 x = 1 - \cos^2 x)$$

$$= \int (1 - 2\cos^2 x + \cos^4 x) \sin x \, dx$$

$$= \int (1 - 2u^2 + u^4)(-du) \qquad \boxed{\begin{array}{l} u = \cos x \\ du = -\sin x \, dx \end{array}}$$

$$= -u + 2\frac{u^3}{3} - \frac{u^5}{5} + C$$

$$= -\cos x + \frac{2}{3} \cos^3 x - \frac{1}{5} \cos^5 x + C$$

EXAMPLE 2

Integrate $\int \sin^2 x \cos^3 x \, dx$.

$$\int \sin^2 x \cos^3 x \, dx = \int \sin^2 x \cos^2 x \cos x \, dx$$

$$= \int \sin^2 x (1 - \sin^2 x) \cos x \, dx \qquad (\cos^2 x = 1 - \sin^2 x)$$

$$= \int u^2(1 - u^2) \, du$$

$$= \int (u^2 - u^4) \, du$$

$$\boxed{\begin{array}{l} u = \sin x \\ du = \cos x \, dx \end{array}}$$

$$= \frac{u^3}{3} - \frac{u^5}{5} + C$$

$$= \frac{1}{3} \sin^3 x - \frac{1}{5} \sin^5 x + C$$

In each of the two previous examples the method used involved making a substitution so that the function to be integrated becomes a product of a power of sine (or cosine) and the first power only of cosine (or sine).

Even Powers of Sines and Cosines

For this case we use the identities:

$$\sin^2 x = \frac{1}{2}(1 - \cos 2x)$$

$$\cos^2 x = \frac{1}{2}(1 + \cos 2x)$$

EXAMPLE 3

Integrate $\int \cos^2 x \, dx$.

$$\int \cos^2 x \, dx = \int \frac{1}{2}(1 + \cos 2x) \, dx \qquad \left[\cos^2 x = \frac{1}{2}(1 + \cos 2x)\right]$$

$$= \frac{1}{2} \int dx + \frac{1}{2} \int \cos 2x \, dx$$

$$= \frac{1}{2}x + \frac{1}{4} \sin 2x + C \qquad \left[\int \cos nx \, dx = \frac{1}{n} \sin nx + C\right]$$

EXAMPLE 4

Integrate $\int \sin^4 x \, dx$.

$$\int \sin^4 x \, dx = \int (\sin^2 x)^2 \, dx$$

$$= \int \left[\frac{1}{2}(1 - \cos 2x)\right]^2 dx \qquad \left[\sin^2 x = \frac{1}{2}(1 - \cos 2x)\right]$$

$$= \int \left(\frac{1}{4} - \frac{1}{2} \cos 2x + \frac{1}{4} \cos^2 2x\right) dx$$

$$= \frac{1}{4} \int dx - \frac{1}{2} \int \cos 2x \, dx + \frac{1}{4} \int \cos^2 2x \, dx$$

$$= \frac{1}{4}x - \frac{1}{4} \sin 2x + \frac{1}{4} \int \cos^2 2x \, dx.$$

Now find

$$\int \cos^2 2x \, dx = \int \frac{1}{2}(1 + \cos 4x) \, dx \qquad \left[\cos^2 2x = \frac{1}{2}(1 + \cos 4x)\right]$$

$$= \frac{1}{2} \int dx + \frac{1}{2} \int \cos 4x \, dx$$

$$= \frac{1}{2}x + \frac{1}{8} \sin 4x + C_1$$

So

$$\int \sin^4 x \, dx = \frac{1}{4}x - \frac{1}{4} \sin 2x + \frac{1}{4}\left(\frac{1}{2}x + \frac{1}{8} \sin 4x + C_1\right)$$

$$= \frac{1}{4}x - \frac{1}{4} \sin 2x + \frac{1}{8}x + \frac{1}{32} \sin 4x + \frac{1}{4}C_1$$

$$= \frac{3}{8}x - \frac{1}{4} \sin 2x + \frac{1}{32} \sin 4x + C$$

Powers of Other Trigonometric Functions

The identities $1 + \tan^2 x = \sec^2 x$ and $1 + \cot^2 x = \csc^2 x$ are used to integrate even powers of $\sec x$ and $\csc x$, and powers of $\tan x$ and $\cot x$.

EXAMPLE 5

Integrate $\displaystyle\int \sec^4 x \, dx$.

$$\int \sec^4 x \, dx = \int (1 + \tan^2 x) \sec^2 x \, dx$$

$$= \int \sec^2 x \, dx + \int \tan^2 x \sec^2 x \, dx$$

$$= \tan x + \frac{1}{3}\tan^3 x + C$$

EXAMPLE 6

Integrate $\displaystyle\int \tan^5 x \, dx$.

$$\int \tan^5 x \, dx = \int \tan^3 x \tan^2 x \, dx$$

$$= \int \tan^3 x \, (\sec^2 x - 1) \, dx \qquad\qquad (\tan^2 x = \sec^2 x - 1)$$

$$= \int \tan^3 x \sec^2 x \, dx - \int \tan^3 x \, dx$$

$$= \frac{1}{4} \tan^4 x - \int \tan x \tan^2 x \, dx$$

$$= \frac{1}{4} \tan^4 x - \int \tan x \, (\sec^2 x - 1) \, dx$$

$$= \frac{1}{4} \tan^4 x - \int \tan x \sec^2 x \, dx + \int \tan x \, dx$$

$$= \frac{1}{4} \tan^4 x - \frac{1}{2} \tan^2 x - \ln |\cos x| + C$$

EXAMPLE 7

Integrate $\int \csc^6 5x \, dx$.

$$\int \csc^6 5x \, dx = \int \csc^4 5x \csc^2 5x \, dx$$

$$= \int (1 + \cot^2 5x)^2 \csc^2 5x \, dx \qquad (\csc^2 5x = 1 + \cot^2 5x)$$

$$= \int (1 + u^2)^2 \left(\frac{du}{-5} \right) \qquad \boxed{\begin{array}{l} \text{let } u = \cot 5x \\ du = (-\csc^2 5x)\, 5 \, dx \end{array}}$$

$$= -\frac{1}{5} \int (1 + 2u^2 + u^4) \, du$$

$$= -\frac{1}{5} \left(u + \frac{2}{3} u^3 + \frac{1}{5} u^5 \right) + C$$

$$= -\frac{1}{5} \cot 5x - \frac{2}{15} \cot^3 5x - \frac{1}{25} \cot^5 5x + C$$

Exercises 7.4

Integrate.

1. $\int \sin^3 x \, dx$ **2.** $\int \cos^3 x \, dx$ **3.** $\int \cos^5 x \, dx$

4. $\int \sin^7 x \, dx$ **5.** $\int \sin^2 x \cos x \, dx$ **6.** $\int \cos^2 x \sin x \, dx$

7. $\int \frac{\sin x \, dx}{\cos^3 x}$ **8.** $\int \sin^3 x \cos x \, dx$ **9.** $\int \sin^3 x \cos^2 x \, dx$

10. $\int \sin^2 3x \cos^3 3x \, dx$ **11.** $\int \sin^2 x \, dx$ **12.** $\int \cos^2 4x \, dx$

13. $\int \cos^4 3x \, dx$ **14.** $\int \sin^4 7x \, dx$ **15.** $\int \sin^2 x \cos^2 x \, dx$

16. $\int \sin^4 x \cos^2 x \, dx$ **17.** $\int \sin^2 x \cos^4 x \, dx$ **18.** $\int \sin^4 2x \cos^4 2x \, dx$

19. $\int \tan^3 x \, dx$ **20.** $\int \cot^5 4x \, dx$ **21.** $\int \cot^4 2x \, dx$

22. $\int \sec^4 7x \, dx$ **23.** $\int \sec^6 x \, dx$ **24.** $\int \csc^4 3x \, dx$

25. $\int \tan^4 2x \, dx$ **26.** $\int \cot^6 x \, dx$

27. Find the area of the region bounded by $y = \sin^2 x$, $x = 0$, $x = \pi$, and $y = 0$.

28. Find the volume of the solid formed by revolving the region bounded by $y = \cos^3 x$ from $x = 0$ to $x = \pi/2$ and $y = 0$ about the x-axis.

7.5 INVERSE TRIGONOMETRIC FORMS

A major use of the inverse trigonometric functions is to integrate certain algebraic functions. Two important integral formulas are based on the derivatives of the inverse sine and inverse tangent functions:

$$\int \frac{du}{\sqrt{a^2 - u^2}} = \text{Arcsin } \frac{u}{a} + C$$

since

$$\frac{d}{du}\left(\text{Arcsin } \frac{u}{a}\right) = \frac{1}{\sqrt{1 - \left(\frac{u}{a}\right)^2}} \cdot \frac{1}{a} = \frac{a}{\sqrt{a^2 - u^2}} \cdot \frac{1}{a} = \frac{1}{\sqrt{a^2 - u^2}}$$

$$\int \frac{du}{a^2 + u^2} = \frac{1}{a}\text{Arctan } \frac{u}{a} + C$$

since

$$\frac{d}{du}\left(\text{Arctan } \frac{u}{a}\right) = \frac{1}{1 + \left(\frac{u}{a}\right)^2} \cdot \frac{1}{a} = \frac{a^2}{a^2 + u^2} \cdot \frac{1}{a} = \frac{a}{a^2 + u^2}$$

so that

$$\frac{d}{du}\left(\frac{1}{a}\text{Arctan } \frac{u}{a}\right) = \frac{1}{a} \cdot \frac{d}{du}\left(\text{Arctan } \frac{u}{a}\right) = \frac{1}{a^2 + u^2}$$

EXAMPLE 1

Integrate $\displaystyle\int \frac{dx}{\sqrt{16 - x^2}}$.

$$\int \frac{dx}{\sqrt{16 - x^2}} = \int \frac{du}{\sqrt{a^2 - u^2}}$$

$$= \text{Arcsin } \frac{u}{a} + C$$

$$= \text{Arcsin } \frac{x}{4} + C$$

$$\boxed{\begin{array}{l} u = x \\ du = dx \\ a = 4 \end{array}}$$

EXAMPLE 2

Evaluate $\displaystyle\int_0^2 \frac{dx}{x^2 + 4}$

$$\int \frac{dx}{x^2 + 4} = \int \frac{du}{a^2 + u^2} = \frac{1}{a}\text{Arctan } \frac{u}{a} + C = \frac{1}{2}\text{Arctan } \frac{x}{2} + C$$

$$\boxed{\begin{array}{l} u = x \\ du = dx \\ a = 2 \end{array}}$$

So,

$$\int_0^2 \frac{dx}{x^2 + 4} = \frac{1}{2}\text{Arctan } \frac{x}{2}\Big|_0^2 = \frac{1}{2}\text{Arctan } 1 - \frac{1}{2}\text{Arctan } 0 = \frac{1}{2}\left(\frac{\pi}{4}\right) - 0 = \frac{\pi}{8}$$

EXAMPLE 3

Integrate $\int \dfrac{dx}{4x^2 + 9}$.

$$\int \frac{dx}{4x^2 + 9} = \int \frac{1}{a^2 + u^2} \cdot \frac{du}{2} = \frac{1}{2} \int \frac{du}{a^2 + u^2}$$

$$= \frac{1}{2}\left(\frac{1}{a} \text{ Arctan } \frac{u}{a}\right) + C$$

$$= \frac{1}{2}\left(\frac{1}{3} \text{ Arctan } \frac{2x}{3}\right) + C$$

$$= \frac{1}{6} \text{ Arctan } \frac{2x}{3} + C$$

$$\boxed{\begin{array}{l} u = 2x \\ du = 2\,dx \\ a = 3 \end{array}}$$

EXAMPLE 4

Integrate $\int \dfrac{dx}{x^2 + 2x + 5}$.

Here we must complete the square in the denominator:

$$x^2 + 2x + 5 = (x + 1)^2 + 4 = (x + 1)^2 + (2)^2$$

Then let $u = x + 1$ and $a = 2$.

$$\int \frac{dx}{x^2 + 2x + 5} = \int \frac{dx}{(x + 1)^2 + (2)^2} = \int \frac{du}{u^2 + a^2}$$

$$= \frac{1}{a} \text{ Arctan } \frac{u}{a} + C$$

$$= \frac{1}{2} \text{ Arctan } \left(\frac{x + 1}{2}\right) + C$$

$$\boxed{\begin{array}{l} u = x + 1 \\ du = dx \\ a = 2 \end{array}}$$

EXAMPLE 5

Evaluate $\int_1^2 \dfrac{dx}{\sqrt{25 - 4x^2}}$.

$$\int \frac{dx}{\sqrt{25 - 4x^2}} = \int \frac{1}{\sqrt{a^2 - u^2}} \cdot \frac{du}{2}$$

$$= \frac{1}{2} \int \frac{du}{\sqrt{a^2 - u^2}}$$

$$= \frac{1}{2} \text{ Arcsin } \frac{u}{a} + C$$

$$= \frac{1}{2} \text{ Arcsin } \frac{2x}{5} + C$$

$$\boxed{\begin{array}{l} u = 2x \\ du = 2\,dx \\ a = 5 \end{array}}$$

So

$$\int_1^2 \frac{dx}{\sqrt{25 - 4x^2}} = \frac{1}{2} \text{ Arcsin } \frac{2x}{5}\bigg|_1^2 = \frac{1}{2}\left(\text{Arcsin } \frac{4}{5} - \text{Arcsin } \frac{2}{5}\right)$$

$$= \frac{1}{2}(0.93 - 0.41) = 0.26 \qquad \text{(radian measure)}$$

Exercises 7.5

Integrate.

1. $\displaystyle\int \frac{dx}{\sqrt{1 - 9x^2}}$

2. $\displaystyle\int \frac{dx}{\sqrt{1 - x^2}}$

3. $\displaystyle\int \frac{dx}{\sqrt{9 - x^2}}$

4. $\displaystyle\int \frac{dx}{\sqrt{144 - 25x^2}}$

5. $\displaystyle\int \frac{dx}{x^2 + 25}$

6. $\displaystyle\int \frac{dx}{x^2 + 4}$

7. $\displaystyle\int \frac{dx}{9x^2 + 4}$

8. $\displaystyle\int \frac{dx}{16 + 25x^2}$

9. $\displaystyle\int \frac{dx}{\sqrt{36 - 25x^2}}$

10. $\displaystyle\int \frac{dx}{\sqrt{5 - 6x^2}}$

11. $\displaystyle\int \frac{dx}{\sqrt{3 - 12x^2}}$

12. $\displaystyle\int \frac{dx}{\sqrt{3 - 4x^2}}$

13. $\displaystyle\int \frac{dx}{4 + (x - 1)^2}$

14. $\displaystyle\int \frac{dx}{16 + (x - 3)^2}$

15. $\displaystyle\int \frac{dx}{x^2 + 6x + 25}$

16. $\displaystyle\int \frac{dx}{x^2 + 4x + 13}$

17. $\displaystyle\int \frac{e^x \, dx}{\sqrt{1 - e^{2x}}}$

18. $\displaystyle\int \frac{\sec x \tan x \, dx}{1 + \sec^2 x}$

19. $\displaystyle\int \frac{\sin x \, dx}{1 + \cos^2 x}$

20. $\displaystyle\int \frac{\cos x \, dx}{\sqrt{4 - \sin^2 x}}$

21. $\displaystyle\int_0^1 \frac{dx}{1 + x^2}$

22. $\displaystyle\int_0^2 \frac{dx}{\sqrt{4 - x^2}}$

23. $\displaystyle\int_0^1 \frac{dx}{\sqrt{25 - 9x^2}}$

24. $\displaystyle\int_0^1 \frac{dx}{25x^2 + 9}$

25. A force is acting on an object according to the equation

$$F = \frac{100}{1 + 4x^2}$$

where x is measured in metres and F is measured in newtons. Find the work done in moving the object from $x = 1$ m to $x = 2$ m.

26. Find an equation for the distance traveled by an object moving along a straight line if the velocity at time t is given by

$$v = \frac{1}{\sqrt{9 - t^2}}$$

The object was 10 m from the point of reference at $t = 0$ s.

7.6) PARTIAL FRACTIONS

We add two fractions such as

$$\frac{5}{x + 1} + \frac{6}{x - 2} = \frac{5(x - 2)}{(x + 1)(x - 2)} + \frac{6(x + 1)}{(x + 1)(x - 2)}$$
$$= \frac{5x - 10 + 6x + 6}{(x + 1)(x - 2)}$$
$$= \frac{11x - 4}{(x + 1)(x - 2)}$$

At times, we need to express a fraction as the sum of two or more fractions that are each simpler than the original; that is, we reverse the operation. Such simpler fractions whose numerators are of lower degree than their denominators are called **partial fractions**.

We separate our study of partial fractions into four cases. In each case we assume that the given fraction is expressed in lowest terms and the degree of each numerator is less than the degree of its denominator.

EXAMPLE 1

Find the partial fractions of $\dfrac{11x - 4}{(x + 1)(x - 2)}$.

The possible partial fractions are $\dfrac{A}{x + 1}$ and $\dfrac{B}{x - 2}$, so we have

$$\frac{11x - 4}{(x + 1)(x - 2)} = \frac{A}{x + 1} + \frac{B}{x - 2}$$

Multiply each side of this equation by the L.C.D.: $(x + 1)(x - 2)$.

$$11x - 4 = A(x - 2) + B(x + 1)$$

Removing parentheses and rearranging terms, we have

$$11x - 4 = Ax - 2A + Bx + B$$
$$11x - 4 = Ax + Bx - 2A + B$$
$$11x - 4 = (A + B)x - 2A + B$$

Next, the coefficients of x must be equal and the constant terms must be equal. This gives the following system of linear equations:

$$A + B = 11$$
$$-2A + B = -4$$

Subtracting the two equations gives

$$3A = 15$$
$$A = 5$$

Substituting $A = 5$ into either of the preceding equations gives

$$B = 6$$

Then

$$\frac{11x - 4}{(x + 1)(x - 2)} = \frac{5}{x + 1} + \frac{6}{x - 2}$$

EXAMPLE 2

Find the partial fractions of $\dfrac{3x^2 - 27x - 12}{x(2x + 1)(x - 4)}$.

The possible partial fractions are

$$\frac{A}{x} \qquad \frac{B}{2x + 1} \quad \text{and} \quad \frac{C}{x - 4}$$

So we have

$$\frac{3x^2 - 27x - 12}{x(2x + 1)(x - 4)} = \frac{A}{x} + \frac{B}{2x + 1} + \frac{C}{x - 4}$$

Now multiply each side of this equation by the L.C.D.: $x(2x + 1)(x - 4)$.

$$3x^2 - 27x - 12 = A(2x + 1)(x - 4) + Bx(x - 4) + Cx(2x + 1)$$

Removing parentheses and rearranging terms, we have

$$3x^2 - 27x - 12 = 2Ax^2 - 7Ax - 4A + Bx^2 - 4Bx + 2Cx^2 + Cx$$
$$3x^2 - 27x - 12 = (2A + B + 2C)x^2 + (-7A - 4B + C)x - 4A$$

Then the coefficients of x^2 must be equal, the coefficients of x must be equal, and the constant terms must be equal. This gives the following system of linear equations:

$$2A + B + 2C = 3$$
$$-7A - 4B + C = -27$$
$$-4A = -12$$

Note that $A = 3$ from the third equation. Substituting $A = 3$ into the first two equations gives

$$6 + B + 2C = 3$$
$$-21 - 4B + C = -27$$

or

$$B + 2C = -3$$
$$-4B + C = -6$$

Multiplying the second equation by 2 gives

$$B + 2C = -3$$
$$-8B + 2C = -12$$

Subtracting these two equations gives

$$9B = 9$$
$$B = 1$$

Then $C = -2$ and

$$\frac{3x^2 - 27x - 12}{x(2x + 1)(x - 4)} = \frac{3}{x} + \frac{1}{2x + 1} - \frac{2}{x - 4}$$

CASE 2: REPEATED LINEAR DENOMINATOR FACTORS

For every factor $(ax + b)^k$ of the denominator of a given fraction, there correspond the possible partial fractions

$$\frac{A_1}{ax + b}, \frac{A_2}{(ax + b)^2}, \frac{A_3}{(ax + b)^3}, \ldots, \frac{A_k}{(ax + b)^k}$$

where $A_1, A_2, A_3, \ldots, A_k$ are constants.

EXAMPLE 3

Find the partial fractions of $\dfrac{-x^2 - 8x + 27}{x(x - 3)^2}$.

The possible partial fractions are

$$\frac{A}{x} \qquad \frac{B}{x - 3} \quad \text{and} \quad \frac{C}{(x - 3)^2}$$

So we have

$$\frac{-x^2 - 8x + 27}{x(x - 3)^2} = \frac{A}{x} + \frac{B}{x - 3} + \frac{C}{(x - 3)^2}$$

Then multiply each side of this equation by the L.C.D.: $x(x - 3)^2$.

$$-x^2 - 8x + 27 = A(x - 3)^2 + Bx(x - 3) + Cx$$

Removing parentheses and rearranging terms, we have

$$-x^2 - 8x + 27 = Ax^2 - 6Ax + 9A + Bx^2 - 3Bx + Cx$$
$$-x^2 - 8x + 27 = (A + B)x^2 + (-6A - 3B + C)x + 9A$$

Equating coefficients, we have

$$A + B = -1$$
$$-6A - 3B + C = -8$$
$$9A = 27$$

From the third equation, we have $A = 3$. Substituting $A = 3$ into the first equation gives $B = -4$. Then substituting $A = 3$ and $B = -4$ into the second equation gives $C = -2$. Thus,

$$\frac{-x^2 - 8x + 27}{x(x - 3)^2} = \frac{3}{x} - \frac{4}{x - 3} - \frac{2}{(x - 3)^2}$$

The TI-89 uses the **expand** command to produce partial fraction expansions.

F2   3   (-x^2-8x+27)/(x*(x-3)^2)) **ENTER**

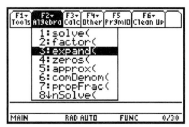 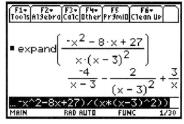

EXAMPLE 4

Find the partial fractions of $\dfrac{3x^2 - 12x + 17}{(x - 2)^3}$.

Since $x - 2$ is repeated as a linear factor three times, the possible partial fractions are

$$\frac{A}{x - 2} \qquad \frac{B}{(x - 2)^2} \quad \text{and} \quad \frac{C}{(x - 2)^3}$$

So we have

$$\frac{3x^2 - 12x + 17}{(x - 2)^3} = \frac{A}{x - 2} + \frac{B}{(x - 2)^2} + \frac{C}{(x - 2)^3}$$

Then multiply each side of this equation by the L.C.D.: $(x - 2)^3$.

$$3x^2 - 12x + 17 = A(x - 2)^2 + B(x - 2) + C$$

Removing parentheses and rearranging terms, we have

$$3x^2 - 12x + 17 = Ax^2 - 4Ax + 4A + Bx - 2B + C$$
$$3x^2 - 12x + 17 = Ax^2 + (-4A + B)x + 4A - 2B + C$$

Equating coefficients, we have the system

$$A = 3$$
$$-4A + B = -12$$
$$4A - 2B + C = 17$$

Substituting $A = 3$ into the second equation, we have $B = 0$. Then substituting $A = 3$ and $B = 0$ into the third equation, we have $C = 5$. Thus,

$$\frac{3x^2 - 12x + 17}{(x - 2)^3} = \frac{3}{x - 2} + \frac{5}{(x - 2)^3}$$

CASE 3: NONREPEATED QUADRATIC DENOMINATOR FACTORS

For every nonrepeated factor $ax^2 + bx + c$ of the denominator of a given fraction, there corresponds the partial fraction

$$\frac{Ax + B}{ax^2 + bx + c}$$

where A and B are constants.

EXAMPLE 5

Find the partial fractions of $\dfrac{11x^2 + 8x - 12}{(2x^2 + x + 2)(x + 1)}$.

The possible partial fractions are

$$\frac{Ax + B}{2x^2 + x + 2} \quad \text{and} \quad \frac{C}{x + 1}$$

So we have

$$\frac{11x^2 + 8x - 12}{(2x^2 + x + 2)(x + 1)} = \frac{Ax + B}{2x^2 + x + 2} + \frac{C}{x + 1}$$

Then multiply each side of this equation by the L.C.D.: $(2x^2 + x + 2)(x + 1)$.

$$11x^2 + 8x - 12 = (Ax + B)(x + 1) + C(2x^2 + x + 2)$$

Removing parentheses and rearranging terms, we have

$$11x^2 + 8x - 12 = Ax^2 + Ax + Bx + B + 2Cx^2 + Cx + 2C$$
$$11x^2 + 8x - 12 = (A + 2C)x^2 + (A + B + C)x + B + 2C$$

Equating coefficients, we have

$$A + 2C = 11$$
$$A + B + C = 8$$
$$B + 2C = -12$$

The solution of this system of linear equations is $A = 17, B = -6, C = -3$. Then

$$\frac{11x^2 + 8x - 12}{(2x^2 + x + 2)(x + 1)} = \frac{17x - 6}{2x^2 + x + 2} - \frac{3}{x + 1}$$

EXAMPLE 6

Find the partial fractions of $\dfrac{2x^3 - 2x^2 + 8x + 7}{(x^2 + 1)(x^2 + 4)}$.

The possible partial fractions are

$$\frac{Ax + B}{x^2 + 1} \quad \text{and} \quad \frac{Cx + D}{x^2 + 4}$$

So we have

$$\frac{2x^3 - 2x^2 + 8x + 7}{(x^2 + 1)(x^2 + 4)} = \frac{Ax + B}{x^2 + 1} + \frac{Cx + D}{x^2 + 4}$$

Then multiply each side of this equation by the L.C.D.: $(x^2 + 1)(x^2 + 4)$.

$$2x^3 - 2x^2 + 8x + 7 = (Ax + B)(x^2 + 4) + (Cx + D)(x^2 + 1)$$

Removing parentheses and rearranging terms, we have

$$2x^3 - 2x^2 + 8x + 7 = Ax^3 + Bx^2 + 4Ax + 4B + Cx^3 + Dx^2 + Cx + D$$
$$2x^3 - 2x^2 + 8x + 7 = (A + C)x^3 + (B + D)x^2 + (4A + C)x + 4B + D$$

Equating coefficients, we have

$$A + C = 2$$
$$B + D = -2$$
$$4A + C = 8$$
$$4B + D = 7$$

The solution of this system of linear equations is $A = 2, B = 3, C = 0, D = -5$. Then

$$\frac{2x^3 - 2x^2 + 8x + 7}{(x^2 + 1)(x^2 + 4)} = \frac{2x + 3}{x^2 + 1} - \frac{5}{x^2 + 4}$$

CASE 4: REPEATED QUADRATIC DENOMINATOR FACTORS

For every factor $(ax^2 + bx + c)^k$ of the denominator of a given fraction, there correspond the possible partial fractions

$$\frac{A_1x + B_1}{ax^2 + bx + c}, \frac{A_2x + B_2}{(ax^2 + bx + c)^2}, \frac{A_3x + B_3}{(ax^2 + bx + c)^3}, \cdots, \frac{A_kx + B_k}{(ax^2 + bx + c)^k}$$

where $A_1, A_2, A_3, \cdots, A_k, B_1, B_2, B_3, \cdots, B_k$ are constants.

EXAMPLE 7

Find the partial fractions of $\dfrac{5x^4 - x^3 + 44x^2 - 5x + 75}{x(x^2 + 5)^2}$.

The possible partial fractions are

$$\frac{A}{x} \quad \frac{Bx + C}{x^2 + 5} \quad \text{and} \quad \frac{Dx + E}{(x^2 + 5)^2}$$

So we have

$$\frac{5x^4 - x^3 + 44x^2 - 5x + 75}{x(x^2 + 5)^2} = \frac{A}{x} + \frac{Bx + C}{x^2 + 5} + \frac{Dx + E}{(x^2 + 5)^2}$$

Then multiply each side of this equation by the L.C.D.: $x(x^2 + 5)^2$.

$$5x^4 - x^3 + 44x^2 - 5x + 75 = A(x^2 + 5)^2 + (Bx + C)(x^2 + 5)(x) + (Dx + E)x$$

Removing parentheses and rearranging terms, we have

$$5x^4 - x^3 + 44x^2 - 5x + 75 = Ax^4 + 10Ax^2 + 25A + Bx^4 + Cx^3$$
$$+ 5Bx^2 + 5Cx + Dx^2 + Ex$$
$$5x^4 - x^3 + 44x^2 - 5x + 75 = (A + B)x^4 + Cx^3 + (10A + 5B + D)x^2$$
$$+ (5C + E)x + 25A$$

Equating coefficients, we have

$$A + B = 5$$
$$C = -1$$
$$10A + 5B + D = 44$$

$$5C + E = -5$$
$$25A = 75$$

The solution of this system of linear equations is $A = 3, B = 2, C = -1, D = 4, E = 0$. Then

$$\frac{5x^4 - x^3 + 44x^2 - 5x + 75}{x(x^2 + 5)^2} = \frac{3}{x} + \frac{2x - 1}{x^2 + 5} + \frac{4x}{(x^2 + 5)^2}$$

If the degree of the numerator is greater than or equal to the degree of the denominator of the original fraction, you must first divide the numerator by the denominator using long division. Then find the partial fractions of the resulting remainder.

EXAMPLE 8

Find the partial fractions of $\dfrac{x^3 + 3x^2 + 7x + 4}{x^2 + 2x}$.

Since the degree of the numerator is greater than the degree of the denominator, divide as follows:

$$
\begin{array}{r}
x + 1 \\
x^2 + 2x \overline{)x^3 + 3x^2 + 7x + 4} \\
x^3 + 2x^2 \\
\hline
x^2 + 7x \\
x^2 + 2x \\
\hline
5x + 4
\end{array}
$$

or

$$\frac{x^3 + 3x^2 + 7x + 4}{x^2 + 2x} = x + 1 + \frac{5x + 4}{x^2 + 2x}$$

Now factor the denominator and find the partial fractions of the remainder.

$$\frac{5x + 4}{x(x + 2)} = \frac{A}{x} + \frac{B}{x + 2}$$
$$5x + 4 = A(x + 2) + Bx$$
$$5x + 4 = Ax + 2A + Bx$$
$$5x + 4 = (A + B)x + 2A$$

Then

$$A + B = 5$$
$$2A = 4$$

So $A = 2$ and $B = 3$ and

$$\frac{x^3 + 3x^2 + 7x + 4}{x^2 + 2x} = x + 1 + \frac{2}{x} + \frac{3}{x + 2}$$

By calculator,

F2   3   (x^3+3x^2+7x+4)/(x^2+2x)) **ENTER**

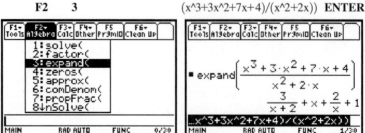

Exercises 7.6

Find the partial fractions of each expression.

1. $\dfrac{8x - 29}{(x + 2)(x - 7)}$

2. $\dfrac{10x - 34}{(x - 4)(x - 2)}$

3. $\dfrac{-x - 18}{2x^2 - 5x - 12}$

4. $\dfrac{17x - 18}{3x^2 + x - 2}$

5. $\dfrac{61x^2 - 53x - 28}{x(3x - 4)(2x + 1)}$

6. $\dfrac{11x^2 - 7x - 42}{(2x + 3)(x^2 - 2x - 3)}$

7. $\dfrac{x^2 + 7x + 10}{(x + 1)(x + 3)^2}$

8. $\dfrac{3x^2 - 18x + 9}{(2x - 1)(x - 1)^2}$

9. $\dfrac{48x^2 - 20x - 5}{(4x - 1)^3}$

10. $\dfrac{x^2 + 8x}{(x + 4)^3}$

11. $\dfrac{11x^2 - 18x + 3}{x(x - 1)^2}$

12. $\dfrac{6x^2 + 4x + 4}{x^3 + 2x^2}$

13. $\dfrac{-x^2 - 4x + 3}{(x^2 + 1)(x^2 - 3)}$

14. $\dfrac{-6x^3 + 2x^2 - 3x + 10}{(2x^2 + 1)(x^2 + 5)}$

15. $\dfrac{4x^3 - 21x - 6}{(x^2 + x + 1)(x^2 - 5)}$

16. $\dfrac{x^3 + 6x^2 + 2x - 2}{(3x^2 - x - 1)(x^2 + 4)}$

17. $\dfrac{4x^3 - 16x^2 - 93x - 9}{(x^2 + 5x + 3)(x^2 - 9)}$

18. $\dfrac{12x^2 + 8x - 72}{(x^2 + x - 1)(x^2 - 16)}$

19. $\dfrac{8x^4 - x^3 + 13x^2 - 6x + 5}{x(x^2 + 1)^2}$

20. $\dfrac{-4x^4 + 6x^3 + 8x^2 - 19x + 17}{(x - 1)(x^2 - 3)^2}$

21. $\dfrac{x^5 - 2x^4 - 8x^2 + 4x - 8}{x^2(x^2 + 2)^2}$

22. $\dfrac{3x^5 + x^4 + 24x^3 + 10x^2 + 48x + 16}{x^2(x^2 + 4)^2}$

23. $\dfrac{6x^2 + 108x + 54}{x^4 - 81}$

24. $\dfrac{x^6 + 2x^4 + 3x^2 + 1}{x^2(x^2 + 1)^3}$

25. $\dfrac{x^3}{x^2 - 1}$

26. $\dfrac{x^4 + x^2}{(x + 1)(x - 2)}$

27. $\dfrac{x^3 - x^2 + 8}{x^2 - 4}$

28. $\dfrac{2x^3 - 2x^2 + 8x - 3}{x(x - 1)}$

29. $\dfrac{3x^4 - 2x^3 - 2x + 5}{x(x^2 + 1)}$

30. $\dfrac{x^5 - x^4 - 3x^3 + 7x^2 + 3x + 20}{(x + 2)(x^2 + 2)}$

7.7 INTEGRATION USING PARTIAL FRACTIONS

Integrals of the form $\displaystyle\int \dfrac{P(x)}{Q(x)}\,dx$, where $P(x)$ and $Q(x)$ are polynomials, may be integrated by first writing the rational expression as a sum of partial fractions. (See Section 7.6.) Then integrate each term.

EXAMPLE 1

Integrate $\displaystyle\int \dfrac{3x + 1}{x^2 - x - 6}\,dx.$

First, write $\dfrac{3x + 1}{x^2 - x - 6}$ as a sum of partial fractions.

$$\frac{3x + 1}{(x - 3)(x + 2)} = \frac{A}{x - 3} + \frac{B}{x + 2}$$

Multiply each side of this equation by the L.C.D.: $(x - 3)(x + 2)$.

$$3x + 1 = A(x + 2) + B(x - 3)$$

Removing parentheses and rearranging terms, we have

$$3x + 1 = Ax + 2A + Bx - 3B$$
$$3x + 1 = Ax + Bx + 2A - 3B$$
$$3x + 1 = (A + B)x + 2A - 3B$$

Next, equate the coefficients of x and the constant terms, which gives the following system of linear equations.

$$3 = A + B$$
$$1 = 2A - 3B$$

Solving this system, we have

$$A = 2 \qquad B = 1$$

So we write

$$\frac{3x + 1}{x^2 - x - 6} = \frac{2}{x - 3} + \frac{1}{x + 2}$$

So

$$\int \frac{3x + 1}{x^2 - x - 6}\,dx = \int \left(\frac{2}{x - 3} + \frac{1}{x + 2} \right) dx$$
$$= 2 \int \frac{dx}{x - 3} + \int \frac{dx}{x + 2}$$
$$= 2 \ln|x - 3| + \ln|x + 2| + C$$

Using the properties of logarithms, we can write the result above as

$$\int \frac{3x + 1}{x^2 - x - 6}\,dx = \ln|(x - 3)^2(x + 2)| + C$$

We thus see how a complicated integral can be written as a sum of less complicated integrals by using partial fractions.

EXAMPLE 2

Integrate $\displaystyle\int \frac{x^3 + 3x^2 + 7x + 4}{x^2 + 2x}\,dx.$

This rational expression was written as a sum of partial fractions in Example 8 in Section 7.6.

$$\int \frac{x^3 + 3x^2 + 7x + 4}{x^2 + 2x}\,dx = \int \left(x + 1 + \frac{2}{x} + \frac{3}{x + 2} \right) dx$$
$$= \int x\,dx + \int dx + 2 \int \frac{dx}{x} + 3 \int \frac{dx}{x + 2}$$
$$= \frac{x^2}{2} + x + 2 \ln|x| + 3 \ln|x + 2| + C$$
$$= \frac{x^2}{2} + x + \ln|x^2(x + 2)^3| + C$$

Since the method of writing a rational expression as a sum of partial fractions was presented in detail in Section 7.6, the next examples will concentrate on the integration. The process of finding the partial fractions is left to the student.

EXAMPLE 3

Integrate $\int \dfrac{x^2 - x + 2}{x^3 - 2x^2 + x} \, dx$.

$$\int \frac{x^2 - x + 2}{x^3 - 2x^2 + x} \, dx = \int \left[\frac{2}{x} - \frac{1}{x-1} + \frac{2}{(x-1)^2} \right] dx$$

$$= 2 \int \frac{dx}{x} - \int \frac{dx}{x-1} + 2 \int (x-1)^{-2} \, dx$$

$$= 2 \ln |x| - \ln |x-1| - \frac{2}{x-1} + C$$

$$= \ln \left| \frac{x^2}{x-1} \right| - \frac{2}{x-1} + C$$

EXAMPLE 4

Integrate $\int \dfrac{x^2 + x - 1}{x^3 + x} \, dx$.

$$\int \frac{x^2 + x - 1}{x^3 + x} \, dx = \int \left(\frac{2x+1}{x^2+1} - \frac{1}{x} \right) dx$$

$$= \int \frac{2x+1}{x^2+1} \, dx - \int \frac{dx}{x}$$

$$= \int \frac{2x}{x^2+1} \, dx + \int \frac{dx}{x^2+1} - \int \frac{dx}{x}$$

$$= \ln |x^2 + 1| + \text{Arctan } x - \ln |x| + C$$

$$= \ln \left| \frac{x^2 + 1}{x} \right| + \text{Arctan } x + C$$

EXAMPLE 5

Integrate $\int \dfrac{4x^2 - 3x + 2}{x^3 - x^2 - 2x} \, dx$.

$$\int \frac{4x^2 - 3x + 2}{x^3 - x^2 - 2x} \, dx = \int \left(-\frac{1}{x} + \frac{2}{x-2} + \frac{3}{x+1} \right) dx$$

$$= -\int \frac{dx}{x} + 2 \int \frac{dx}{x-2} + 3 \int \frac{dx}{x+1}$$

$$= -\ln |x| + 2 \ln |x-2| + 3 \ln |x+1| + C$$

$$= \ln \left| \frac{(x-2)^2 (x+1)^3}{x} \right| + C$$

Exercises 7.7

Integrate.

1. $\displaystyle\int \frac{dx}{1 - x^2}$

2. $\displaystyle\int \frac{dx}{x^2 - 5x + 6}$

3. $\displaystyle\int \frac{dx}{x^2 + 2x - 8}$

4. $\displaystyle\int \frac{dx}{x^2 + x}$

5. $\displaystyle\int \frac{x \, dx}{x^2 - 3x + 2}$

6. $\displaystyle\int \frac{x \, dx}{x^3 - 3x^2 + 2x}$

7. $\displaystyle\int \frac{x + 1}{x^2 + 4x - 5} \, dx$

8. $\displaystyle\int \frac{3x - 4}{2 - x - x^2} \, dx$

9. $\displaystyle\int \frac{dx}{x(x + 1)^2}$

10. $\displaystyle\int \frac{7x - 4}{(x - 1)^2(x + 2)}\,dx$ **11.** $\displaystyle\int \frac{2x^2 + x + 3}{x^2(x + 3)}\,dx$ **12.** $\displaystyle\int \frac{x^2 - x + 2}{x^2(x + 2)}\,dx$

13. $\displaystyle\int \frac{x^3\,dx}{x^2 + 3x + 2}$ **14.** $\displaystyle\int \frac{x^2\,dx}{x^2 + 2x + 1}$ **15.** $\displaystyle\int \frac{x^2 - 2}{(x^2 + 1)x}\,dx$

16. $\displaystyle\int \frac{5x^2 - x + 11}{(x^2 + 4)(x - 1)}\,dx$ **17.** $\displaystyle\int \frac{x^3 + 2x^2 - 9}{x^2(x^2 + 9)}\,dx$ **18.** $\displaystyle\int \frac{2x^3 + 3x}{(x^2 + 1)(x^2 + 2)}\,dx$

19. $\displaystyle\int \frac{x^3\,dx}{(x^2 + 1)^2}$ **20.** $\displaystyle\int \frac{x^2 - 2x + 1}{(x^2 + 1)^2}\,dx$ **21.** $\displaystyle\int_2^3 \frac{3\,dx}{1 - x^2}$

22. $\displaystyle\int_2^3 \frac{5x + 1}{x^2 + x - 2}\,dx$ **23.** $\displaystyle\int_2^4 \frac{x\,dx}{x^2 + 4x - 5}$ **24.** $\displaystyle\int_1^3 \frac{x - 2}{x^3 + x^2}\,dx$

25. Find the area of the region bounded by $y = \dfrac{4x}{x^2 + 2x - 3}$, $x = 2$, $x = 4$, and $y = 0$.

26. Find the area of the region bounded by $y = \dfrac{x + 1}{(x + 2)^2}$, $x = 0$, $x = 1$, and $y = 0$.

7.8 INTEGRATION BY PARTS

Integration by parts is another method of transforming integrals into a form that can be integrated by using familiar integration formulas. This method is based on the formula for differentiating the product of two functions u and v:

$$\frac{d}{dx}(u \cdot v) = u\frac{dv}{dx} + v\frac{du}{dx}$$

The differential form is then

$$d(u \cdot v) = u\,dv + v\,du$$

Integrating each side, we have

$$\int d(u \cdot v) = \int (u\,dv + v\,du)$$

$$u \cdot v = \int u\,dv + \int v\,du$$

Solving for $\displaystyle\int u\,dv$, we can write this equation as

> **INTEGRATION BY PARTS**
>
> $$\int u\,dv = u \cdot v - \int v\,du$$

We now demonstrate the method of integration by parts.

EXAMPLE 1

Integrate $\displaystyle\int xe^{2x}\,dx$.

Let $u = x$ and $dv = e^{2x}\,dx$, then

$$\int (x)(e^{2x}\,dx) = \int u\,dv$$

which is the left-hand side of the formula for integration by parts. Note that what we choose to call dv must contain the factor dx.

If $u = x$, then $du = dx$ and if $dv = e^{2x} dx$, then $v = \int dv = \int e^{2x} dx = \frac{1}{2}e^{2x} + C_1$.

Since

$$\int u\, dv = \qquad uv \qquad - \int v\, du$$

we have

$$\int (x)(e^{2x}\, dx) = (x)\left(\frac{1}{2}e^{2x} + C_1\right) - \int\left(\frac{1}{2}e^{2x} + C_1\right)dx$$

$$= \frac{1}{2}xe^{2x} + C_1x \quad - \int \frac{1}{2}e^{2x}\, dx - C_1 \int dx$$

$$= \frac{1}{2}xe^{2x} + C_1x \quad - \frac{1}{4}e^{2x} - C_1x + C$$

$$= \frac{1}{2}xe^{2x} - \frac{1}{4}e^{2x} + C$$

Note: The constant of integration C_1 that arose from integrating dv disappears in the final result. This will *always* occur when using the method of integration by parts. Because of this, we will omit C_1 when integrating dv.

The decision on what to choose for u and dv is not always clear. A trial-and-error approach may be necessary. For example, we could have chosen $u = e^{2x}$ and $dv = x\, dx$ in Example 1. We would then have had $du = 2e^{2x}\, dx$ and $v = \frac{1}{2}x^2$. So,

$$\int e^{2x}(x\, dx) = \frac{1}{2}x^2e^{2x} - \int x^2e^{2x}\, dx$$

and the right-hand integral is more complicated than the original integral. When this occurs, you should try another choice for u and dv.

EXAMPLE 2

Integrate $\int x \sin x\, dx$.

$$\int u\, dv \qquad\qquad\qquad \int v\, du$$

$$\boxed{\begin{array}{l} u = x \\ dv = \sin x\, dx \end{array}} \longrightarrow \boxed{\begin{array}{l} du = dx \\ v = -\cos x \end{array}}$$

$$\int u\, dv = \quad uv \quad - \int v\, du$$

$$\int x \sin x\, dx = -x \cos x - \int (-\cos x)\, dx$$

$$= -x \cos x + \int \cos x\, dx$$

$$= -x \cos x + \sin x + C$$

EXAMPLE 3

Integrate $\displaystyle\int x^3 \ln x \, dx$.

$$\int u \, dv \qquad\qquad\qquad \int v \, du$$

$$\boxed{\begin{array}{l} u = \ln x \\ dv = x^3 \, dx \end{array}} \qquad\longrightarrow\qquad \boxed{\begin{array}{l} du = \dfrac{1}{x} dx \\[2mm] v = \dfrac{x^4}{4} \end{array}}$$

$$\int u \, dv = \quad uv \quad - \int v \, du$$

$$\int (\ln x)(x^3 \, dx) = (\ln x)\left(\frac{x^4}{4}\right) - \int \left(\frac{x^4}{4}\right)\left(\frac{1}{x} dx\right)$$

$$= \frac{1}{4}x^4 \ln x \quad - \frac{1}{4}\int x^3 \, dx$$

$$= \frac{1}{4}x^4 \ln x \quad - \frac{1}{16}x^4 + C$$

EXAMPLE 4

Integrate $\displaystyle\int \text{Arcsin } x \, dx$.

$$\int u \, dv \qquad\qquad\qquad \int v \, du$$

$$\boxed{\begin{array}{l} u = \text{Arcsin } x \\ dv = dx \end{array}} \qquad\longrightarrow\qquad \boxed{\begin{array}{l} du = \dfrac{dx}{\sqrt{1 - x^2}} \\[2mm] v = x \end{array}}$$

$$\int u \, dv = \quad uv \quad - \int v \, du$$

$$\int (\text{Arcsin } x)(dx) = (\text{Arcsin } x)(x) - \int (x)\left(\frac{dx}{\sqrt{1 - x^2}}\right)$$

$$= x \, \text{Arcsin } x \quad - \int (1 - x^2)^{-1/2} x \, dx$$

$$= x \, \text{Arcsin } x \quad + \sqrt{1 - x^2} + C$$

EXAMPLE 5

Integrate $\displaystyle\int \sec^3 x \, dx$.

$$\int u \, dv \qquad\qquad\qquad \int v \, du$$

$$\boxed{\begin{array}{l} u = \sec x \\ dv = \sec^2 x \, dx \end{array}} \qquad\longrightarrow\qquad \boxed{\begin{array}{l} du = \sec x \tan x \, dx \\ v = \tan x \end{array}}$$

$$\int u\,dv = \qquad uv \qquad - \int v\,du$$

$$\int (\sec x)(\sec^2 x\,dx) = (\sec x)(\tan x) - \int (\tan x)(\sec x \tan x\,dx)$$

$$= \sec x \tan x \quad - \int \sec x \tan^2 x\,dx$$

$$= \sec x \tan x \quad - \int (\sec x)(\sec^2 x - 1)\,dx$$

$$= \sec x \tan x \quad - \int \sec^3 x\,dx + \int \sec x\,dx$$

or

$$\int \sec^3 x\,dx = \sec x \tan x - \int \sec^3 x\,dx + \ln|\sec x + \tan x| + C_1$$

$$2\int \sec^3 x\,dx = \sec x \tan x + \ln|\sec x + \tan x| + C_1 \qquad \text{(Add } \int \sec^3 x\,dx \text{ to each side.)}$$

$$\int \sec^3 x\,dx = \frac{1}{2}[\sec x \tan x + \ln|\sec x + \tan x|] + C$$

Note: Whenever a multiple (not equal to 1) of the original integral appears on the right-hand side, it may be combined with the left-hand integral to complete the integration process.

EXAMPLE 6

Evaluate $\int_0^{\pi/2} e^{2x} \sin x\,dx$.

$$\int u\,dv \qquad\qquad\qquad \int v\,du$$

$$\boxed{\begin{array}{l} u = e^{2x} \\ dv = \sin x\,dx \end{array}} \qquad \longrightarrow \qquad \boxed{\begin{array}{l} du = 2e^{2x}\,dx \\ v = -\cos x \end{array}}$$

$$\int u\,dv = \qquad uv \qquad - \int v\,du$$

$$\int (e^{2x})(\sin x\,dx) = (e^{2x})(-\cos x) - \int (-\cos x)(2e^{2x}\,dx)$$

$$= -e^{2x}\cos x \quad + 2\int e^{2x}\cos x\,dx \qquad\qquad (1)$$

In this example, we need to repeat the integration by parts process for the integral $\int e^{2x}\cos x\,dx$.

$$\int u'\,dv' \qquad\qquad\qquad \int v'\,du'$$

$$\boxed{\begin{array}{l} u' = e^{2x} \\ dv' = \cos x\,dx \end{array}} \qquad \longrightarrow \qquad \boxed{\begin{array}{l} du' = 2e^{2x}\,dx \\ v' = \sin x \end{array}}$$

$$\int u'\,dv' = \qquad u'v' \qquad - \int v'\,du'$$

$$\int (e^{2x})(\cos x \, dx) = (e^{2x})(\sin x) - \int (\sin x)(2e^{2x} \, dx)$$

$$= e^{2x} \sin x - 2 \int e^{2x} \sin x \, dx$$

So, substituting this result in Equation (1), we have

$$\int e^{2x} \sin x \, dx = -e^{2x} \cos x + 2\left(e^{2x} \sin x - 2 \int e^{2x} \sin x \, dx \right)$$

$$\int e^{2x} \sin x \, dx = -e^{2x} \cos x + 2e^{2x} \sin x - 4 \int e^{2x} \sin x \, dx$$

Adding $4 \int e^{2x} \sin x \, dx$ to each side and including a constant of integration, we have

$$5 \int e^{2x} \sin x \, dx = -e^{2x} \cos x + 2e^{2x} \sin x + C_1$$

$$\int e^{2x} \sin x \, dx = \frac{e^{2x}}{5} (2 \sin x - \cos x) + C$$

$$\int_0^{\pi/2} e^{2x} \sin x \, dx = \frac{e^{2x}}{5} (2 \sin x - \cos x) \Big|_0^{\pi/2}$$

$$= \frac{e^{\pi}}{5} (2 - 0) - \frac{1}{5}(0 - 1)$$

$$= \frac{1}{5}(2e^{\pi} + 1)$$

Exercises 7.8

Integrate.

1. $\displaystyle\int \ln x \, dx$

2. $\displaystyle\int x \cos x \, dx$

3. $\displaystyle\int xe^x \, dx$

4. $\displaystyle\int xe^{-x} \, dx$

5. $\displaystyle\int \sqrt{x} \ln x \, dx$

6. $\displaystyle\int x^2 \ln x \, dx$

7. $\displaystyle\int \ln x^2 \, dx$

8. $\displaystyle\int \frac{\ln x}{\sqrt{x}} dx$

9. $\displaystyle\int \text{Arccos } x \, dx$

10. $\displaystyle\int \text{Arctan } x \, dx$

11. $\displaystyle\int e^x \cos x \, dx$

12. $\displaystyle\int x^2 \sin x \, dx$

13. $\displaystyle\int x^2 \cos x \, dx$

14. $\displaystyle\int x^3 \ln x \, dx$

15. $\displaystyle\int x \sec^2 x \, dx$

16. $\displaystyle\int x^2 e^{2x} \, dx$

17. $\displaystyle\int (\ln x)^2 \, dx$

18. $\displaystyle\int x \, \text{Arctan } x \, dx$

19. $\displaystyle\int x \sec x \tan x \, dx$

20. $\displaystyle\int x^3 e^{3x} \, dx$

21. $\displaystyle\int_0^1 xe^{3x} \, dx$

22. $\displaystyle\int_0^1 \text{Arcsin } x \, dx$

23. $\displaystyle\int_1^2 x\sqrt{x-1} \, dx$

24. $\displaystyle\int_0^3 x\sqrt{x+1} \, dx$

25. $\displaystyle\int_1^2 \ln(x+1) \, dx$

26. $\displaystyle\int_0^{\pi/2} x^2 \cos x \, dx$

27. Find the area of the region bounded by $y = \ln 2x$, $x = \frac{1}{2}$, $x = 1$, and $y = 0$.

28. Find the area of the region bounded by $y = x^2 e^x$, $x = 0$, $x = 2$, and $y = 0$.

29. Find the volume of the solid generated by revolving the region bounded by $y = e^x$, $y = 0$, $x = 0$, and $x = 1$ about the y-axis.

30. Find the volume of the solid generated by revolving the region bounded by $y = \sin x$, $y = 0$, $x = 0$, and $x = \pi$ about the y-axis.

7.9 INTEGRATION BY TRIGONOMETRIC SUBSTITUTION

Certain algebraic functions cannot yet be integrated by the methods presented so far. Appropriate trigonometric substitutions often lead to an integration solution. Algebraic functions involving the expressions $\sqrt{a^2 - u^2}$, $\sqrt{u^2 - a^2}$, or $\sqrt{u^2 + a^2}$ can often be integrated by substitutions based on the diagrams in Fig. 7.2.

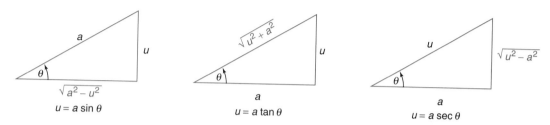

Figure 7.2 Reference triangles used for integrating expressions in the form $\sqrt{a^2 - u^2}$, $\sqrt{u^2 + a^2}$, and $\sqrt{u^2 - a^2}$, respectively.

The following examples illustrate the use of these substitutions.

EXAMPLE 1

Integrate $\displaystyle\int \frac{x^2\, dx}{\sqrt{9 - x^2}}$ (see Fig. 7.3).

Since $\sqrt{9 - x^2}$ appears in the integral, we will use the substitution

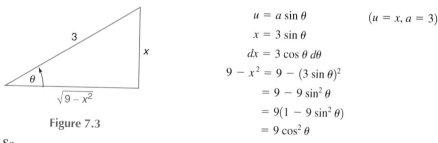

Figure 7.3

$$u = a \sin \theta \qquad (u = x, a = 3)$$
$$x = 3 \sin \theta$$
$$dx = 3 \cos \theta\, d\theta$$
$$9 - x^2 = 9 - (3 \sin \theta)^2$$
$$= 9 - 9 \sin^2 \theta$$
$$= 9(1 - 9 \sin^2 \theta)$$
$$= 9 \cos^2 \theta$$

So
$$\sqrt{9 - x^2} = 3 \cos \theta$$

Then
$$\int \frac{x^2\, dx}{\sqrt{9 - x^2}} = \int \frac{(3 \sin \theta)^2 (3 \cos \theta\, d\theta)}{3 \cos \theta}$$
$$= 9 \int \sin^2 \theta\, d\theta$$
$$= 9 \int \frac{1}{2}(1 - \cos 2\theta)\, d\theta \qquad \left[\sin^2 \theta = \frac{1}{2}(1 - \cos 2\theta)\right]$$
$$= \frac{9}{2} \int d\theta - \frac{9}{2} \int \cos 2\theta\, d\theta$$

$$= \frac{9}{2}\theta - \frac{9}{4}\sin 2\theta + C$$

$$= \frac{9}{2}\theta - \frac{9}{2}\sin\theta\cos\theta + C \qquad (\sin 2\theta = 2\sin\theta\cos\theta)$$

From Fig. 7.3

$$\sin\theta = \frac{x}{3} \quad \text{so} \quad \theta = \text{Arcsin}\frac{x}{3}$$

$$\cos\theta = \frac{\sqrt{9-x^2}}{3}$$

Making these substitutions, we have

$$\int \frac{x^2\,dx}{\sqrt{9-x^2}} = \frac{9}{2}\text{Arcsin}\frac{x}{3} - \frac{9}{2}\cdot\frac{x}{3}\cdot\frac{\sqrt{9-x^2}}{3} + C = \frac{9}{2}\text{Arcsin}\frac{x}{3} - \frac{x\sqrt{9-x^2}}{2} + C$$

Using a calculator,

2nd 7 x^2/ **2nd** $\sqrt{}$ 9-x^2),x) **ENTER**

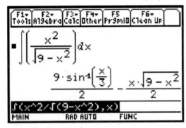

EXAMPLE 2

Integrate $\displaystyle\int \frac{dx}{\sqrt{x^2+4}}$ (see Fig. 7.4).

Since $\sqrt{x^2+4}$ appears in the integral, we use the substitutions:

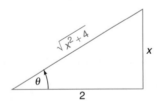

Figure 7.4

$$u = a\tan\theta \qquad (u = x, a = 2)$$
$$x = 2\tan\theta$$
$$dx = 2\sec^2\theta\,d\theta$$
$$x^2 + 4 = (2\tan\theta)^2 + 4$$
$$= 4\tan^2\theta + 4$$
$$= 4(\tan^2\theta + 1)$$
$$= 4\sec^2\theta$$

So

$$\sqrt{x^2+4} = 2\sec\theta$$

Then

$$\int \frac{dx}{\sqrt{x^2+4}} = \int \frac{2\sec^2\theta\,d\theta}{2\sec\theta}$$

$$= \int \sec\theta\,d\theta$$

$$= \ln|\sec\theta + \tan\theta| + C$$

From Fig. 7.4 we have

$$\sec \theta = \frac{\sqrt{x^2 + 4}}{2}$$

$$\tan \theta = \frac{x}{2}$$

Making these substitutions, we have

$$\int \frac{dx}{\sqrt{x^2 + 4}} = \ln \left| \frac{\sqrt{x^2 + 4}}{2} + \frac{x}{2} \right| + C$$

$$= \ln \left| \frac{\sqrt{x^2 + 4} + x}{2} \right| + C$$

$$= \ln \left| \sqrt{x^2 + 4} + x \right| - \ln 2 + C$$

$$= \ln \left| \sqrt{x^2 + 4} + x \right| + C' \qquad \text{(since } -\ln 2 \text{ is also a } constant\text{)}$$

EXAMPLE 3

Integrate $\displaystyle\int \frac{dx}{\sqrt{(4x^2 - 25)^3}}$ (see Fig. 7.5).

The denominator can be written as $\sqrt{(4x^2 - 25)^3} = (4x^2 - 25)\sqrt{4x^2 - 25}$. This suggests the substitutions:

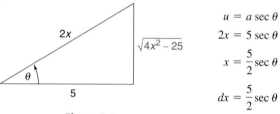

Figure 7.5

$$u = a \sec \theta \qquad (u = 2x, a = 5)$$

$$2x = 5 \sec \theta$$

$$x = \frac{5}{2} \sec \theta$$

$$dx = \frac{5}{2} \sec \theta \tan \theta \, d\theta$$

$$4x^2 - 25 = 4\left(\frac{25}{4} \sec^2\theta\right) - 25$$

$$= 25 \sec^2 \theta - 25$$

$$= 25(\sec^2 \theta - 1)$$

$$= 25 \tan^2 \theta$$

so

$$\sqrt{4x^2 - 25} = 5 \tan \theta$$

Then

$$\int \frac{dx}{\sqrt{(4x^2 - 25)^3}} = \int \frac{\frac{5}{2} \sec \theta \tan \theta \, d\theta}{(5 \tan \theta)^3}$$

$$= \frac{1}{50} \int \frac{\sec \theta \tan \theta \, d\theta}{\tan^3 \theta}$$

$$= \frac{1}{50} \int \frac{\sec \theta}{\tan^2 \theta} \, d\theta$$

$$= \frac{1}{50} \int \frac{\dfrac{1}{\cos \theta}}{\dfrac{\sin^2 \theta}{\cos^2 \theta}} \, d\theta$$

$$= \frac{1}{50} \int \frac{\cos \theta}{\sin^2 \theta} \, d\theta$$

$$= \frac{1}{50}\left(-\frac{1}{\sin\theta}\right) + C$$

$$= -\frac{1}{50}\csc\theta + C \qquad \left(\csc\theta = \frac{2x}{\sqrt{4x^2 - 25}}\text{ from Fig. 7.5}\right)$$

$$= \frac{-x}{25\sqrt{4x^2 - 25}} + C$$

EXAMPLE 4

Evaluate $\displaystyle\int_0^4 \frac{dx}{(\sqrt{x^2+9})^3}$ (see Fig. 7.6).

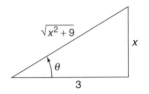

Figure 7.6

$$\boxed{\begin{array}{l} x = 3\tan\theta \\ dx = 3\sec^2\theta\, d\theta \end{array}}$$

$$\int \frac{dx}{(\sqrt{x^2+9})^3} = \int \frac{3\sec^2\theta\, d\theta}{(\sqrt{9\tan^2\theta + 9})^3} = \int \frac{3\sec^2\theta\, d\theta}{(\sqrt{9\sec^2\theta})^3} = \int \frac{3\sec^2\theta\, d\theta}{27\sec^3\theta}$$

$$= \frac{1}{9}\int \frac{1}{\sec\theta}\, d\theta = \frac{1}{9}\int \cos\theta\, d\theta$$

$$= \frac{1}{9}\sin\theta + C = \frac{1}{9}\cdot\frac{x}{\sqrt{x^2+9}} + C = \frac{x}{9\sqrt{x^2+9}} + C$$

So,

$$\int_0^4 \frac{dx}{(\sqrt{x^2+9})^3} = \frac{x}{9\sqrt{x^2+9}}\Big|_0^4 = \frac{4}{9\sqrt{25}} - 0 = \frac{4}{45}$$

Using a calculator,

2nd 7 1/ 2nd $\sqrt{}$ **x^2+9)^3,x,0,4) ENTER**

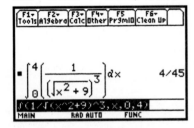

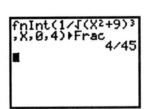

Math 9 1/ 2nd $\sqrt{}$ **x x² +9) MATH 3 ,x,0,4) MATH 1 ENTER**

Exercises 7.9

Integrate.

1. $\displaystyle\int \frac{dx}{\sqrt{9 + 4x^2}}$

2. $\displaystyle\int \frac{dx}{\sqrt{9 - 16x^2}}$

3. $\displaystyle\int \frac{x^2}{\sqrt{4 - 9x^2}}\, dx$

4. $\displaystyle\int \frac{x^2}{\sqrt{1 - 16x^2}}\, dx$

5. $\displaystyle\int_0^1 \frac{dx}{\sqrt{9 - x^2}}$

6. $\displaystyle\int_0^2 \frac{dx}{\sqrt{16 + x^2}}$

7. $\displaystyle\int_0^1 \frac{dx}{\sqrt{(4 - x^2)^3}}$

8. $\displaystyle\int_2^3 \frac{dx}{\sqrt{4x^2 - 9}}$

9. $\displaystyle\int \frac{dx}{x\sqrt{x^2 + 4}}$

10. $\displaystyle\int \frac{dx}{x\sqrt{9 - x^2}}$

11. $\displaystyle\int \frac{\sqrt{x^2 - 9}}{x^2}\,dx$

12. $\displaystyle\int \frac{dx}{\sqrt{(x^2 + 4)^3}}$

13. $\displaystyle\int \frac{dx}{x^2\sqrt{16 - x^2}}$

14. $\displaystyle\int \frac{dx}{x\sqrt{x^2 + 9}}$

15. $\displaystyle\int \frac{\sqrt{9 + x^2}}{x}\,dx$

16. $\displaystyle\int \frac{\sqrt{4 + 9x^2}}{x}\,dx$

17. $\displaystyle\int \frac{dx}{(25 - x^2)^{3/2}}$

18. $\displaystyle\int \frac{dx}{(x^2 - 5)^{3/2}}$

19. $\displaystyle\int \frac{dx}{\sqrt{x^2 - 9}}$

20. $\displaystyle\int \frac{dx}{x\sqrt{1 - x^2}}$

21. $\displaystyle\int \frac{x^3\,dx}{\sqrt{9x^2 + 4}}$

22. $\displaystyle\int \frac{x^2\,dx}{(4 + x^2)^2}$

23. $\displaystyle\int \frac{dx}{\sqrt{x^2 - 6x + 8}}$

(*Hint:* complete the square under the radical.)

24. $\displaystyle\int \frac{dx}{\sqrt{-9x^2 + 18x - 5}}$

25. $\displaystyle\int \frac{dx}{(x^2 + 8x + 15)^{3/2}}$

26. $\displaystyle\int \frac{dx}{\sqrt{9x^2 + 36x + 52}}$

27. Find the area of the region bounded by $y = \dfrac{1}{\sqrt{x^2 + 4}}$, $x = 0$, $x = 2$, and $y = 0$.

28. Find the area of the region bounded by $y = \sqrt{4 - x^2}$, $x = 0$, $x = 2$, and $y = 0$.

7.10 INTEGRATION USING TABLES

The Table of Integrals in Appendix B lists some standard integration formulas for integrating selected functions. These formulas have been developed using the techniques of this chapter and other methods of integrating various complicated functions. A more extensive list of integration formulas (usually more than 400) can be found in most standard handbooks of mathematical tables.

We now illustrate how to use these tables. In the Table of Integrals (Appendix B), u represents a function of x.

EXAMPLE 1

Integrate $\displaystyle\int x\sqrt{3 + 4x}\,dx$.

If we let $u = x$, $a = 3$, and $b = 4$, this integral is in the form of Formula 11 in Appendix B. Substituting these values of u, a, and b in this formula, we have

$$\int x\sqrt{3 + 4x}\,dx = \frac{-2[2(3) - 3(4)(x)][(3) + (4)(x)]^{3/2}}{15(4)^2} + C$$

$$= \frac{(2x - 1)(3 + 4x)^{3/2}}{20} + C$$

Using a calculator,

2nd 7 x **2nd** $\sqrt{}$ 3+4x),x) **ENTER**

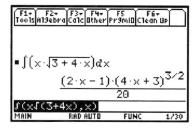

EXAMPLE 2

Integrate $\int \dfrac{dx}{x\sqrt{9 + 5x}}$.

If we let $u = x$, $a = 9$, and $b = 5$, then the integral is in the form of Formula 15. Substituting these values of u, a, and b in this formula, we have

$$\int \frac{dx}{x\sqrt{9 + 5x}} = \frac{1}{3}\ln\left|\frac{\sqrt{9 + 5x} - 3}{\sqrt{9 + 5x} + 3}\right| + C$$

EXAMPLE 3

Integrate $\int x^{100} \ln x \, dx$.

If we let $u = x$ and $n = 100$, then this integral is in the form of Formula 57.

$$\int x^{100} \ln x \, dx = \frac{x^{101} \ln x}{101} - \frac{x^{101}}{(101)^2} + C$$

EXAMPLE 4

Integrate $\int 2^{\sin x} \cos x \, dx$.

If we let $u = \sin x$ and $a = 2$, then since $du = \cos x \, dx$, this integral is in the form of Formula 52.

$$\int 2^{\sin x} \cos x \, dx = \int 2^u \, du$$
$$= \frac{2^u}{\ln 2} + C$$
$$= \frac{2^{\sin x}}{\ln 2} + C$$

$$\boxed{\begin{array}{l} u = \sin x \\ du = \cos x \, dx \end{array}}$$

EXAMPLE 5

Integrate $\int \sin^5 3x \, dx$.

If we let $u = 3x$ and $n = 5$, then the integral is in the form of Formula 83.

$$\int \sin^5 3x \, dx = \int \sin^5 u \, \frac{du}{3}$$
$$= \frac{1}{3}\int \sin^5 u \, du$$
$$= \frac{1}{3}\left(-\frac{1}{5}\sin^{5-1} u \cos u + \frac{4}{5}\int \sin^3 u \, du\right)$$
$$= -\frac{1}{15}\sin^4 u \cos u + \frac{4}{15}\int \sin^3 u \, du$$

$$\boxed{\begin{array}{l} u = 3x \\ du = 3 \, dx \end{array}}$$

The solution is not complete. We still need to integrate $\int \sin^3 u \, du$, which can be integrated by using Formula 83 again.

$$\int \sin^3 u \, du = -\frac{1}{3}\sin^2 u \cos u + \frac{2}{3}\int \sin u \, du$$
$$= -\frac{1}{3}\sin^2 u \cos u - \frac{2}{3}\cos u$$

So

$$\int \sin^5 3x\, dx = -\frac{1}{15}\sin^4 u \cos u + \frac{4}{15}\left(-\frac{1}{3}\sin^2 u \cos u - \frac{2}{3}\cos u\right) + C$$

$$= -\frac{1}{15}\sin^4 3x \cos 3x - \frac{4}{45}\sin^2 3x \cos 3x - \frac{8}{45}\cos 3x + C$$

Exercises 7.10

Integrate using the Table of Integrals in Appendix B. Give the number of the formula used.

1. $\displaystyle\int \frac{dx}{x\sqrt{x+5}}$ **2.** $\displaystyle\int \frac{x\, dx}{2+7x}$ **3.** $\displaystyle\int \frac{dx}{\sqrt{x^2-4}}$

4. $\displaystyle\int \frac{dx}{\sqrt{6+x^2}}$ **5.** $\displaystyle\int \frac{x\, dx}{\sqrt{2x+3}}$ **6.** $\displaystyle\int x\sqrt{4x+7}\, dx$

7. $\displaystyle\int \sin 7x \sin 3x\, dx$ **8.** $\displaystyle\int \cos 5x \cos 2x\, dx$ **9.** $\displaystyle\int \frac{x^2\, dx}{\sqrt{9-x^2}}$

10. $\displaystyle\int \frac{dx}{x\sqrt{16-x^2}}$ **11.** $\displaystyle\int \frac{dx}{x(1+9x)^2}$ **12.** $\displaystyle\int x^2 e^{4x}\, dx$

13. $\displaystyle\int \frac{dx}{x^2-25}$ **14.** $\displaystyle\int \frac{dx}{x\sqrt{9-16x^2}}$ **15.** $\displaystyle\int \sqrt{x^2+4}\, dx$

16. $\displaystyle\int \frac{dx}{x(2+5x)^2}$ **17.** $\displaystyle\int \frac{x\, dx}{(3+4x)^2}$ **18.** $\displaystyle\int \frac{5x\, dx}{\sqrt{2+4x}}$

19. $\displaystyle\int \frac{dx}{x\sqrt{9x^2-16}}$ **20.** $\displaystyle\int xe^{5x}\, dx$ **21.** $\displaystyle\int e^{3x} \sin 4x\, dx$

22. $\displaystyle\int e^{2x} \cos 5x\, dx$ **23.** $\displaystyle\int (2x-3)\sin(2x-3)\, dx$

24. $\displaystyle\int x^2 \cos x^2\, dx$ **25.** $\displaystyle\int \sin^4 x\, dx$ **26.** $\displaystyle\int \cos^5 x\, dx$

27. $\displaystyle\int \frac{\sqrt{9x^2-16}}{x}\, dx$ **28.** $\displaystyle\int \frac{dx}{(25-4x^2)^{3/2}}$

7.11 NUMERICAL METHODS OF INTEGRATION

Despite the numerous integration techniques and formulas available, there are still many functions that are difficult to integrate. In fact, the value of some integrals cannot be exactly determined by any known method of integration. However, several numerical techniques have been developed for approximating the value of an integral. These numerical methods can easily be used with the help of a calculator or a computer.

The *trapezoidal rule* can be demonstrated by considering $\int_a^b f(x)\, dx$ as representing the area bounded by the curves $y = f(x)$, $x = a$, $x = b$, and the x-axis. The trapezoidal rule is based on approximating this area by the sum of the areas of selected trapezoids.

The line segment from a to b is divided into n intervals, each of width $\Delta x = \dfrac{b-a}{n}$ as shown in Fig. 7.7. $\left(\text{Here } n = 4, \text{ so } \Delta x = \dfrac{b-a}{4}.\right)$ This process determines $(n+1)$

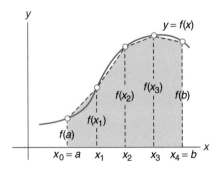

Figure 7.7 The area under a curve may be estimated by adding the areas of the trapezoids.

numbers on the x-axis: $x_0 = a, x_1 = a + \Delta x, x_2 = a + 2\,\Delta x, x_3 = a + 3\,\Delta x, \ldots,$ $x_n = a + n\,\Delta x = b$. Then forming the $n = 4$ trapezoids as shown in Fig. 7.7, we have

$$\frac{1}{2}[f(a) + f(x_1)]\Delta x \qquad \text{as the area of the first trapezoid}$$

$$\frac{1}{2}[f(x_1) + f(x_2)]\Delta x \qquad \text{as the area of the second trapezoid}$$

$$\frac{1}{2}[f(x_2) + f(x_3)]\Delta x \qquad \text{as the area of the third trapezoid}$$

$$\frac{1}{2}[f(x_3) + f(b)]\Delta x \qquad \text{as the area of the fourth trapezoid}$$

The sum of these four areas is then used as an approximation for the area $\int_a^b f(x)\,dx$. In summing these four areas, we have

$$\frac{\Delta x}{2}[f(a) + 2f(x_1) + 2f(x_2) + 2f(x_3) + f(b)]$$

In general,

TRAPEZOIDAL RULE

$$\int_a^b f(x)\,dx \cong \frac{\Delta x}{2}[f(a) + 2f(x_1) + 2f(x_2) + 2f(x_3) + \cdots + 2f(x_{n-1}) + f(b)]$$

The symbol $\cong$ means "approximately equals."
Note: The pattern of coefficients is 1, 2, 2, 2, ..., 2, 2, 1.

EXAMPLE 1

Use the trapezoidal rule with $n = 4$ to find the approximate value of $\int_1^2 \frac{dx}{x}$. (See Fig. 7.8.)

Since $n = 4$, $\Delta x = \dfrac{b - a}{n} = \dfrac{2 - 1}{4} = \dfrac{1}{4}$ and

$$x_0 = a = 1 \qquad\qquad f(a) = \frac{1}{1} = 1$$

$$x_1 = 1 + \frac{1}{4} = \frac{5}{4} \qquad\qquad f(x_1) = \frac{1}{\frac{5}{4}} = \frac{4}{5}$$

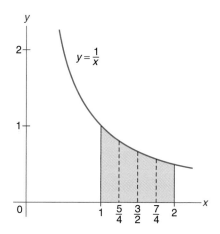

Figure 7.8

$$x_2 = 1 + 2\left(\frac{1}{4}\right) = \frac{3}{2} \qquad f(x_2) = \frac{1}{\frac{3}{2}} = \frac{2}{3}$$

$$x_3 = 1 + 3\left(\frac{1}{4}\right) = \frac{7}{4} \qquad f(x_3) = \frac{1}{\frac{7}{4}} = \frac{4}{7}$$

$$x_4 = b = 1 + 4\left(\frac{1}{4}\right) = 2 \qquad f(b) = \frac{1}{2}$$

$$\int_1^2 \frac{dx}{x} \cong \frac{\Delta x}{2}[f(a) + 2f(x_1) + 2f(x_2) + 2f(x_3) + f(b)]$$

$$= \frac{\frac{1}{4}}{2}\left[1 + 2\left(\frac{4}{5}\right) + 2\left(\frac{2}{3}\right) + 2\left(\frac{4}{7}\right) + \frac{1}{2}\right]$$

$$= \frac{1}{8}\left(1 + \frac{8}{5} + \frac{4}{3} + \frac{8}{7} + \frac{1}{2}\right) = 0.6970$$

For comparison, in Section 7.2, Example 8, we found the exact value to be

$$\int_1^2 \frac{dx}{x} = \ln x \Big|_1^2 = \ln 2 - \ln 1 = \ln 2 - 0 = \ln 2 = 0.6931.$$

EXAMPLE 2

Use the trapezoidal rule with $n = 8$ to find the approximate value of $\displaystyle\int_1^5 x\sqrt{x-1}\,dx$.

(See Fig. 7.9.) First, $\Delta x = \dfrac{b-a}{n} = \dfrac{5-1}{8} = \dfrac{4}{8} = \dfrac{1}{2} = 0.5$

$x_0 = a = 1$	$f(a) = f(1) = 0$
$x_1 = 1.5$	$f(x_1) = f(1.5) = 1.061$
$x_2 = 2$	$f(x_2) = f(2) = 2$
$x_3 = 2.5$	$f(x_3) = f(2.5) = 3.062$
$x_4 = 3$	$f(x_4) = f(3) = 4.243$
$x_5 = 3.5$	$f(x_5) = f(3.5) = 5.534$
$x_6 = 4$	$f(x_6) = f(4) = 6.928$
$x_7 = 4.5$	$f(x_7) = f(4.5) = 8.419$
$x_8 = b = 5$	$f(b) = f(5) = 10$

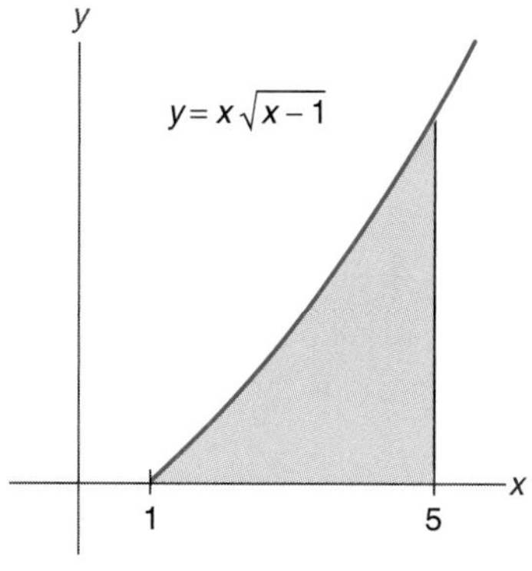

$y = x\sqrt{x-1}$

Figure 7.9

So

$$\int_1^5 x\sqrt{x-1}\,dx \cong \frac{\Delta x}{2}[f(a) + 2f(x_1) + 2f(x_2) + 2f(x_3) + 2f(x_4) + 2f(x_5)$$
$$+ 2f(x_6) + 2f(x_7) + f(b)]$$
$$= \frac{0.5}{2}[0 + 2(1.061) + 2(2) + 2(3.062) + 2(4.243)$$
$$+ 2(5.534) + 2(6.928) + 2(8.419) + 10]$$
$$= 18.12$$

Note: Slight variations in the results may occur due to rounding.

Numerical methods of integration are especially helpful when the function to be integrated is not completely known.

EXAMPLE 3

Find the work done in moving an object along a straight line for 10 ft if the following measurements were made.

Distance moved (ft)	0	2	4	6	8	10
Force (lb)	12	9	7	5	4	2

The formula for work W done in moving an object from a to b is $W = \int_a^b f(x)\,dx$, where $f(x)$ is the force exerted on the object at x.

In this example we do not know what the function $f(x)$ looks like, but we do know its values at 2-ft intervals ($\Delta x = 2$). Here $x_0 = a = 0$, $x_1 = 2$, $x_2 = 4$, $x_3 = 6$, $x_4 = 8$, and $x_5 = b = 10$. So,

$$W = \int_0^{10} f(x)\,dx \cong \frac{\Delta x}{2}[f(a) + 2f(x_1) + 2f(x_2) + 2f(x_3) + 2f(x_4) + f(b)]$$
$$= \frac{2}{2}[12 + 2(9) + 2(7) + 2(5) + 2(4) + 2]$$
$$= 64 \text{ foot-pounds (ft-lb)}$$

A second method of approximating the value of an integral is called *Simpson's rule*, which uses parabolic areas instead of trapezoidal areas and is usually more accurate. As before, divide the line segment from a to b into n intervals, each of width $\Delta x = \dfrac{b-a}{n}$ (n must be even here). Then fit parabolas to adjacent triples of points as shown in Fig. 7.10.

The areas under the parabolic segments are then summed. The development of this formula is rather difficult and can be found in more advanced texts.

SIMPSON'S RULE

$$\int_a^b f(x)\,dx \cong \frac{\Delta x}{3}[f(a) + 4f(x_1) + 2f(x_2) + 4f(x_3) + 2f(x_4)$$
$$+ \cdots + 4f(x_{n-1}) + f(b)]$$

(n must be an even integer)

Note: The pattern of coefficients is 1, 4, 2, 4, 2, 4, ..., 2, 4, 1.

We now use Simpson's rule to obtain another approximation for $\int_1^2 \dfrac{dx}{x}$ (see Example 1).

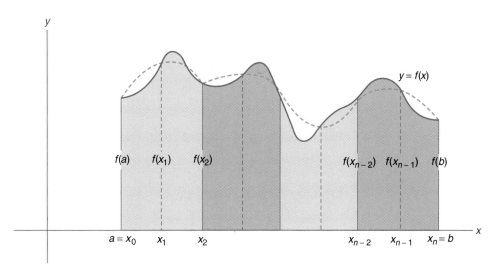

y

y = f(x)

f(a)   f(x₁)   f(x₂)

f(x_{n-2}) f(x_{n-1})   f(b)

a = x₀   x₁   x₂

x_{n-2}   x_{n-1}   x_n = b

x

Figure 7.10   The area under a curve may be estimated by adding the areas under parabolic segments, instead of trapezoidal strips.

$$\int_1^2 \frac{dx}{x} \cong \frac{\frac{1}{4}}{3}\left[1 + 4\left(\frac{4}{5}\right) + 2\left(\frac{2}{3}\right) + 4\left(\frac{4}{7}\right) + \frac{1}{2}\right]$$

$$= \frac{1}{12}\left(1 + \frac{16}{5} + \frac{4}{3} + \frac{16}{7} + \frac{1}{2}\right) = 0.6933$$

Note that the exact answer is ln 2, which is approximately 0.693147.

EXAMPLE 4

Use Simpson's rule with $n = 6$ to find the approximate value of $\int_1^4 \frac{dx}{x+2}$.

First, $\Delta x = \dfrac{b-a}{n} = \dfrac{4-1}{6} = \dfrac{3}{6} = \dfrac{1}{2} = 0.5$.

$$x_0 = a = 1 \qquad f(a) = f(1) = \frac{1}{3}$$

$$x_1 = 1.5 \qquad f(x_1) = f(1.5) = \frac{1}{3.5}$$

$$x_2 = 2 \qquad f(x_2) = f(2) = \frac{1}{4}$$

$$x_3 = 2.5 \qquad f(x_3) = f(2.5) = \frac{1}{4.5}$$

$$x_4 = 3 \qquad f(x_4) = f(3) = \frac{1}{5}$$

$$x_5 = 3.5 \qquad f(x_5) = f(3.5) = \frac{1}{5.5}$$

$$x_6 = b = 4 \qquad f(b) = f(4) = \frac{1}{6}$$

Then

$$\int_1^4 \frac{dx}{x+2} \cong \frac{\Delta x}{3}\left[f(a) + 4f(x_1) + 2f(x_2) + 4f(x_3) + 2f(x_4) + 4f(x_5) + f(b)\right]$$

$$= \frac{0.5}{3}\left[\frac{1}{3} + 4\left(\frac{1}{3.5}\right) + 2\left(\frac{1}{4}\right) + 4\left(\frac{1}{4.5}\right) + 2\left(\frac{1}{5}\right) + 4\left(\frac{1}{5.5}\right) + \frac{1}{6}\right]$$

$$= 0.6932$$

EXAMPLE 5

Estimate $\int_2^5 f(x)\, dx$ using Simpson's rule and the following table of values of $f(x)$. Note that since Simpson's rule requires the number of intervals to be *even,* the number of function values must be *odd* (for example, these 7 equally spaced x-values have 6 spaces between them).

x	2	2.5	3	3.5	4	4.5	5
f(x)	3	6	5	4	7	8	9

$$\int_2^5 f(x)\, dx \cong \frac{0.5}{3}[(3) + 4(6) + 2(5) + 4(4) + 2(7) + 4(8) + (9)] = 18$$

Calculator Programs: Trapezoidal Rule, Simpson's Rule

The following programs are written for a Texas Instruments TI-83 or TI-83 Plus Graphing Calculator. Sample output is shown in Figure 7.11.

Exercises 7.11

Use the trapezoidal rule to approximate the value of each integral.

1. $\int_1^3 \frac{dx}{x}, n = 4$ **2.** $\int_1^2 \frac{dx}{x}, n = 10$ **3.** $\int_0^1 \frac{dx}{1 + x^2}, n = 4$

4. $\int_0^2 \frac{dx}{4 + x}, n = 6$ **5.** $\int_0^1 \sqrt{4 - x^2}\, dx, n = 10$ **6.** $\int_0^2 2^x\, dx, n = 4$

7. $\int_0^3 \sqrt{1 + x^3}\, dx, n = 6$ **8.** $\int_0^2 \frac{dx}{\sqrt{1 + x^3}}, n = 4$ **9.** $\int_0^{\pi/3} \cos x^2\, dx, n = 8$

10. $\int_0^{\pi/4} \tan x^2\, dx, n = 4$

11. Use the trapezoidal rule to find the work done in moving an object along a straight line for 14 ft if the following measurements were made:

Distance moved (ft)	0	2	4	6	8	10	12	14
Force (lb)	24	21	18	17	15	12	10	9

12. Use the trapezoidal rule to find the distance traveled in 6 s by an object moving along a straight line if the following data were recorded:

Time (s)	0	1	2	3	4	5	6
Velocity (m/s)	20	30	50	60	40	30	10

```
Program:   TRAPZOID
------------------------------
:ClrHome
:Output(1,4,"NUMERICAL")
:Output(2,3,"INTEGRATION")
:Output(4,1,"TRAPEZOIDAL RULE")
:Output(8,1,"* PRESS  ENTER *")
:Pause
:Radian
:ClrHome
:Menu("INTEGRATING:","THE CURRENT Y_0",1,"WRITE A NEW Y_0",A)
:Lbl A
:ClrHome
:Disp "ENTER YOUR NEW","FUNCTION.",""
:Input "Y_0=",Str0
:String►Equ(Str0,Y_0)
:FnOff 0
:Lbl 1
:ClrHome
:Disp "LIMITS OF","INTEGRATION:",""
:Input "LOWER=",A
:Input "UPPER=",B
:Disp "","N SUBINTERVALS:"
:Input "N=",N
:Output(1,1,"PLEASE WAIT ...")
:0→S
:0→K
:B-A→D
:For(J,1,N)
:Y_0(A+KD/N)+S→S
:Y_0(A+(K+1)D/N)+S→S
:K+1→K
:End
:SD/(2N)→I
:Disp "TRAPEZ. INTEGRAL",I
:Output(3,1,"Y_0=")
:Equ►String(Y_0,Str0)
:Output(3,4,Str0)
```

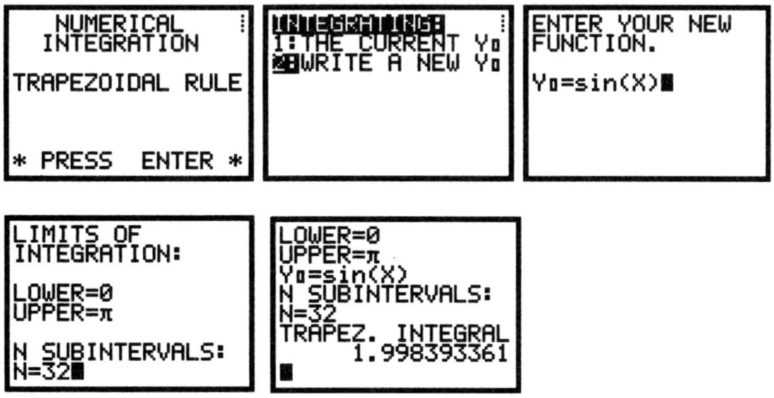

Figure 7.11a

Use the trapezoidal rule to find the approximate area under the curve through each set of points.

13.

x	3	6	9	12	15	18	21
y	12	11	18	25	19	6	10

14.

x	-4	0	4	8	12	16	20
y	11	9	3	10	21	30	40

```
Program:   SIMPSON
------------------------------
:ClrHome
:Output(1,4,"NUMERICAL")
:Output(2,3,"INTEGRATION")
:Output(4,2,"SIMPSON'S RULE")
:Output(8,1,"* PRESS  ENTER *")
:Pause
:Radian
:ClrHome
:Menu("INTEGRATING:","THE CURRENT Y₀",1,"WRITE A NEW Y₀",A)
:Lbl A
:ClrHome
:Disp "ENTER YOUR NEW","FUNCTION.",""
:Input "Y₀=",Str0
:String▶Equ(Str0,Y₀)
:FnOff 0
:Lbl 1
:ClrHome
:Disp "LIMITS OF","INTEGRATION:",""
:Input "LOWER=",A
:Input "UPPER=",B
:Lbl 2
:Disp "","N SUBINTERVALS:"
:Input "N=",N
:If N-2int(N/2)≠0:Then
:Disp "N MUST BE EVEN,","TRY AGAIN.","",""
:Goto 2
:End
:Output(1,1,"PLEASE WAIT ...")
:0▶S
:0▶K
:B-A▶D
:For(J,1,N/2)
:Y₀(A+KD/N)+S▶S
:4Y₀(A+(K+1)D/N)+S▶S
:Y₀(A+(K+2)D/N)+S▶S
:K+2▶K
:End
:SD/(3N)▶I
:Disp "SIMPSON INTEGRAL",I
:Output(3,1,"Y₀=")
:Equ▶String(Y₀,Str0)
:Output(3,4,Str0)
```

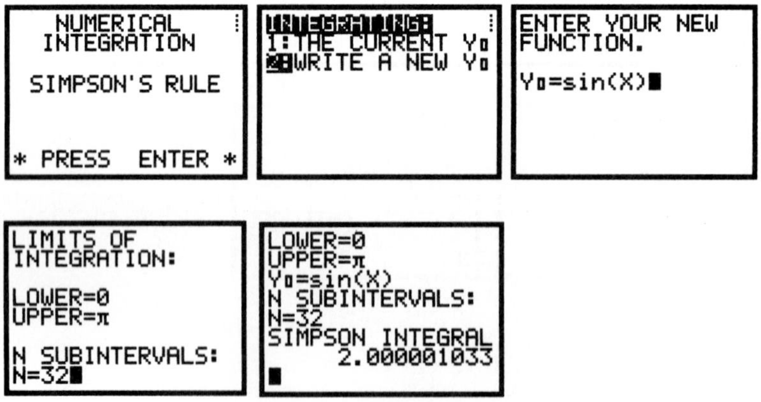

```
┌─────────────────────┐
│     NUMERICAL       │
│    INTEGRATION      │
│                     │
│   SIMPSON'S RULE    │
│                     │
│                     │
│ * PRESS   ENTER *   │
└─────────────────────┘
```
```
┌─────────────────────┐
│ INTEGRATING:        │
│ 1:THE CURRENT Y₀    │
│ 2:WRITE A NEW Y₀    │
│                     │
└─────────────────────┘
```
```
┌─────────────────────┐
│ ENTER YOUR NEW      │
│ FUNCTION.           │
│                     │
│ Y₀=sin(X)■          │
└─────────────────────┘
```
```
┌─────────────────────┐
│ LIMITS OF           │
│ INTEGRATION:        │
│                     │
│ LOWER=0             │
│ UPPER=π             │
│                     │
│ N SUBINTERVALS:     │
│ N=32■               │
└─────────────────────┘
```
```
┌─────────────────────┐
│ LOWER=0             │
│ UPPER=π             │
│ Y₀=sin(X)           │
│ N SUBINTERVALS:     │
│ N=32                │
│ SIMPSON INTEGRAL    │
│        2.000001033  │
│ ■                   │
└─────────────────────┘
```

Figure 7.11b

Use Simpson's rule to approximate the value of each integral.

15. $\displaystyle\int_0^2 \frac{dx}{\sqrt{1+x^2}}, n = 4$

16. $\displaystyle\int_0^2 \sqrt[3]{8-x^2}\, dx, n = 4$

17. $\displaystyle\int_2^6 \frac{dx}{1+x^3}, n = 8$

18. $\displaystyle\int_1^3 x^x\, dx, n = 8$

19. $\displaystyle\int_0^1 e^{x^2}\,dx,\ n=4$

20. $\displaystyle\int_0^1 e^{-x^2}\,dx,\ n=4$

21. $\displaystyle\int_0^{\pi/4} x\,\tan x\,dx,\ n=4$

22. $\displaystyle\int_1^2 \sqrt{x}\,\sin x\,dx,\ n=4$

23. $\displaystyle\int_0^{\pi/2} \sqrt{1+\cos^2 x}\,dx,\ n=6$

24. $\displaystyle\int_{\pi/6}^{\pi/3} \csc x\,dx,\ n=4$

Approximate the value of each integral by (a) using the trapezoidal rule and (b) using Simpson's rule.

25. $\displaystyle\int_0^3 \sqrt{9-x^2}\,dx,\ n=6$

26. $\displaystyle\int_0^\pi \frac{\sin x}{1+x}\,dx,\ n=6$

27. $\displaystyle\int_0^1 xe^x\,dx,\ n=4$

28. $\displaystyle\int_0^1 xe^{x^2}\,dx,\ n=4$

29. $\displaystyle\int_0^{\pi/2} \cos\sqrt{x}\,dx,\ n=6$

30. $\displaystyle\int_0^{\pi/2} \cos^2 x\,dx,\ n=6$

31. Find the approximate area of the field in Fig. 7.12 **(a)** using the trapezoidal rule and **(b)** using Simpson's rule.

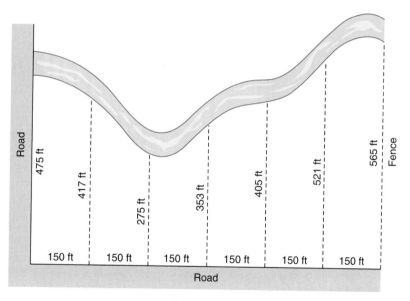

Figure 7.12

7.12   AREAS IN POLAR COORDINATES

Finding areas in polar coordinates is similar to finding the area of a region in rectangular coordinates. Suppose that we have a region bounded by $r=f(\theta)$ and the terminal sides of angles α and β in standard position, where $\alpha < \beta$ (see Fig. 7.13a). Instead of rectangles as in rectangular coordinates, let's subdivide the angular region into n subintervals of circular sectors as follows:

$$\alpha = \theta_0,\ \theta_1,\ \theta_2,\ \ldots,\ \theta_n = \beta$$

Let r_i be any ray within the ith sector. Then $\Delta\theta_i$ is the central angle of the ith sector and $r_i = f(\theta_i)$ is the radius of the ith sector.

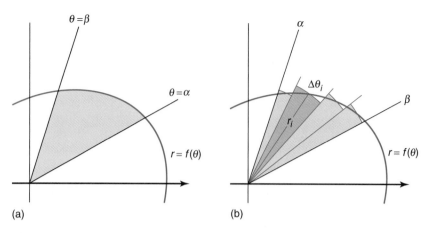

Figure 7.13 (a) The region shown is bounded by the polar curve $r = f(\theta)$ and the rays $\theta = \alpha$ and $\theta = \beta$. (b) The area of the region shown can be estimated by adding the areas of the small circular sectors.

Recall that the area of a sector of a circle whose central angle is θ (in radians) and radius r is

$$A = \frac{1}{2} r^2 \theta$$

The area of the ith sector in Fig. 7.13(b) is

$$\frac{1}{2} r_i^2 \, \Delta\theta_i = \frac{1}{2} [f(\theta_i)]^2 \, \Delta\theta_i$$

The total area is

$$\frac{1}{2} r_1^2 \, \Delta\theta_1 + \frac{1}{2} r_2^2 \, \Delta\theta_2 + \frac{1}{2} r_3^2 \, \Delta\theta_3 + \cdots + \frac{1}{2} r_n^2 \, \Delta\theta_n$$

which can be represented by the integral

> **AREA OF A REGION IN POLAR COORDINATES**
>
> $$A = \frac{1}{2} \int_\alpha^\beta r^2 \, d\theta = \frac{1}{2} \int_\alpha^\beta [f(\theta)]^2 \, d\theta$$

Due to the squaring of r, you should review Section 7.4 and the integration of even powers of sine and cosine functions. Recall that the identities used are

$$\sin^2 x = \frac{1}{2}(1 - \cos 2x)$$

$$\cos^2 x = \frac{1}{2}(1 + \cos 2x)$$

EXAMPLE 1

Find the area of the region inside $r^2 = 4 \sin 2\theta$.

First, graph $r^2 = 4 \sin 2\theta$ (see Fig. 7.14).

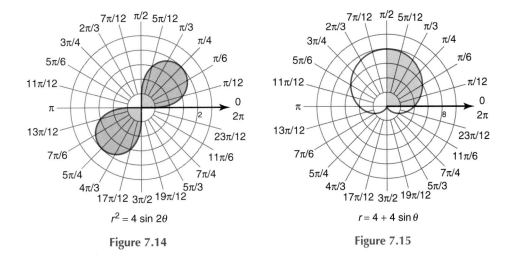

$r^2 = 4 \sin 2\theta$

Figure 7.14

$r = 4 + 4 \sin \theta$

Figure 7.15

Due to symmetry, the area may be expressed by

$$A = 2\left[\frac{1}{2}\int_0^{\pi/2} r^2\, d\theta\right]$$

$$= \int_0^{\pi/2} 4 \sin 2\theta\, d\theta$$

$$= 4\left(-\frac{1}{2}\cos 2\theta\right)\Big|_0^{\pi/2}$$

$$= -2[-1-1] = 4$$

EXAMPLE 2

Find the area of the region in the first quadrant within $r = 4 + 4 \sin \theta$.

First, graph $r = 4 + 4 \sin \theta$ (see Fig. 7.15).

$$A = \frac{1}{2}\int_\alpha^\beta r^2\, d\theta$$

$$= \frac{1}{2}\int_0^{\pi/2} (4 + 4 \sin \theta)^2\, d\theta$$

$$= 8\int_0^{\pi/2} (1 + \sin \theta)^2\, d\theta$$

$$= 8\int_0^{\pi/2} (1 + 2 \sin \theta + \sin^2 \theta)\, d\theta$$

$$= 8\int_0^{\pi/2} \left[1 + 2 \sin \theta + \frac{1}{2}(1 - \cos 2\theta)\right] d\theta$$

$$= 8\int_0^{\pi/2} \left[\frac{3}{2} + 2 \sin \theta - \frac{1}{2}\cos 2\theta\right] d\theta$$

$$= 8\left(\frac{3\theta}{2} - 2 \cos \theta - \frac{1}{4}\sin 2\theta\right)\Big|_0^{\pi/2}$$

$$= 8\left[\left(\frac{3}{2}\cdot\frac{\pi}{2} - 2\cdot 0 - \frac{1}{4}\cdot 0\right) - \left(\frac{3}{2}\cdot 0 - 2\cdot 1 - \frac{1}{4}\cdot 0\right)\right]$$

$$= 8\left(\frac{3\pi}{4} + 2\right) = 6\pi + 16$$

Using a calculator,

1/2* 2nd 7 (4+4 2nd SIN green diamond ^))^2, green diamond ^ ,0, 2nd π /2) ENTER

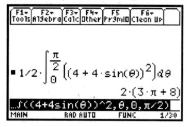   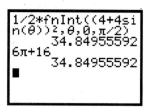

1/2* MATH 9 (4+4 SIN ALPHA 3)) x², ALPHA 3 ,0, 2nd π /2) ENTER

EXAMPLE 3

Find the area within the inner loop of $r = 2 + 4\cos\theta$.

First, graph $r = 2 + 4\cos\theta$ (see Fig. 7.16). The end points of the inner loop are found by noting that $r = 0$ at the endpoints.

$$r = 2 + 4\cos\theta$$

$$0 = 2 + 4\cos\theta$$

$$\cos\theta = -\frac{1}{2}$$

$$\theta = \frac{2\pi}{3}, \frac{4\pi}{3}$$

So the inner loop is determined by $\frac{2\pi}{3} \le \theta \le \frac{4\pi}{3}$.

$$A = \frac{1}{2}\int_{2\pi/3}^{4\pi/3} r^2\,d\theta$$

$$= \frac{1}{2}\int_{2\pi/3}^{4\pi/3} (2 + 4\cos\theta)^2\,d\theta$$

$$= 2\int_{2\pi/3}^{4\pi/3} (1 + 4\cos\theta + 4\cos^2\theta)\,d\theta$$

$$= 2\int_{2\pi/3}^{4\pi/3} \left[1 + 4\cos\theta + 4\cdot\frac{1}{2}(1 + \cos 2\theta)\right]d\theta$$

$$= 2\int_{2\pi/3}^{4\pi/3} (3 + 4\cos\theta + 2\cos 2\theta)\,d\theta$$

$$= 2[3\theta + 4\sin\theta + \sin 2\theta]\Big|_{2\pi/3}^{4\pi/3}$$

$$= 2\left\{\left[3\cdot\frac{4\pi}{3} + 4\left(\frac{-\sqrt{3}}{2}\right) + \frac{\sqrt{3}}{2}\right] - \left[3\cdot\frac{2\pi}{3} + 4\cdot\frac{\sqrt{3}}{2} + \left(-\frac{\sqrt{3}}{2}\right)\right]\right\}$$

$$= 2\left(4\pi - 2\sqrt{3} + \frac{\sqrt{3}}{2} - 2\pi - 2\sqrt{3} + \frac{\sqrt{3}}{2}\right)$$

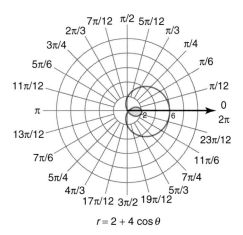

$r = 2 + 4\cos\theta$

Figure 7.16

$$= 2(2\pi - 3\sqrt{3})$$
$$= 4\pi - 6\sqrt{3}$$

The area between two polar curves may be expressed by

$$A = \frac{1}{2}\int_{\alpha}^{\beta}\left\{[f(\theta)]^2 - [g(\theta)]^2\right\}d\theta$$

EXAMPLE 4

Find the area of the region inside $r = 1$ and outside $r = 1 + \sin\theta$.

First, graph the two equations (see Fig. 7.17).

Use the substitution method of solving equations simultaneously to find the points of intersection.

$$1 + \sin\theta = 1$$
$$\sin\theta = 0$$
$$\theta = 0, \pi, 2\pi$$

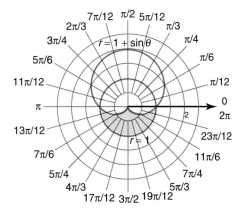

Figure 7.17

Due to symmetry with respect to the y-axis, the area may be expressed by

$$A = 2\left[\frac{1}{2}\int_{\pi}^{3\pi/2}\left[1^2 - (1 + \sin\theta)^2\right]d\theta\right]$$

$$= \int_{\pi}^{3\pi/2}(1 - 1 - 2\sin\theta - \sin^2\theta)\,d\theta$$

$$= \int_{\pi}^{3\pi/2}\left(-2\sin\theta - \frac{1}{2}(1 - \cos 2\theta)\right)d\theta$$

$$= \int_{\pi}^{3\pi/2}\left(-2\sin\theta - \frac{1}{2} + \frac{1}{2}\cos 2\theta\right)d\theta$$

$$= \left(2\cos\theta - \frac{\theta}{2} + \frac{1}{4}\sin 2\theta\right)\Big|_{\pi}^{3\pi/2}$$

$$= \left(2\cdot 0 - \frac{3\pi/2}{2} + \frac{1}{4}\cdot 0\right) - \left(2(-1) - \frac{\pi}{2} + \frac{1}{4}\cdot 0\right)$$

$$= -\frac{3\pi}{4} + 2 + \frac{\pi}{2}$$

$$= 2 - \frac{\pi}{4}$$

EXAMPLE 5

A microphone with a cardioid pickup pattern is often the choice of sound engineers recording live performances. The cardioid pattern offers good frontal sensitivity, while suppressing audience noise in back of the microphone. In Fig. 7.18, find the area on the stage that lies within the optimal pickup range of the microphone, if the boundaries of this region are given by the cardioid $r = 20 + 20\cos\theta$ and the vertical line $x = 15$ (the microphone has been placed 15 feet from the edge of the stage).

First rewrite $x = 15$ in polar coordinates.

$$x = 15$$

$$r\cos\theta = 15 \qquad \text{(recall that in polar form, } x = r\cos\theta\text{)}$$

$$r = \frac{15}{\cos\theta}$$

$$r = 15\sec\theta$$

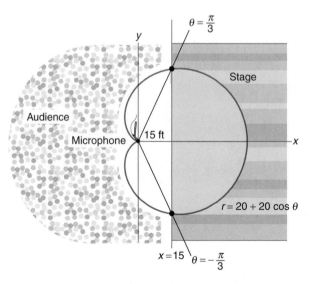

Figure 7.18

Next, find the θ-values of the points of intersection by equating the two expressions for r.

$$20 + 20 \cos \theta = \frac{15}{\cos \theta}$$

$$20 \cos \theta + 20 \cos^2 \theta = 15 \qquad \text{(multiplying both sides by } \cos \theta)$$

$$20 \cos^2 \theta + 20 \cos \theta - 15 = 0$$

$$4 \cos^2 \theta + 4 \cos \theta - 3 = 0 \qquad \text{(dividing both sides by 5)}$$

$$(2 \cos \theta + 3)(2 \cos \theta - 1) = 0$$

$$\cos \theta = -\frac{3}{2} \quad \text{or} \quad \cos \theta = \frac{1}{2}$$

$$\theta = \pm \frac{\pi}{3}$$

Note: $\cos \theta = -\dfrac{3}{2}$ is impossible, since $-1 \leq \cos \theta \leq 1$.

The integral for the area of the region is

$$\frac{1}{2} \int_{-\pi/3}^{\pi/3} \left[(20 + 20 \cos \theta)^2 - (15 \sec \theta)^2 \right] d\theta$$

$$= 2 \cdot \frac{1}{2} \int_{0}^{\pi/3} \left[(20 + 20 \cos \theta)^2 - (15 \sec \theta)^2 \right] d\theta \qquad \text{(using } x\text{-axis symmetry)}$$

$$= \int_{0}^{\pi/3} \left[400 + 800 \cos \theta + 400 \cos^2 \theta - 225 \sec^2 \theta \right] d\theta$$

$$= \int_{0}^{\pi/3} \left[400 + 800 \cos \theta + 400 \left(\frac{1 + \cos 2\theta}{2} \right) - 225 \sec^2 \theta \right] d\theta$$

$$= \int_{0}^{\pi/3} \left[600 + 800 \cos \theta + 200 \cos 2\theta - 225 \sec^2 \theta \right] d\theta$$

$$= \left. \left(600 \theta + 800 \sin \theta + 100 \sin 2\theta - 225 \tan \theta \right) \right|_{0}^{\pi/3}$$

$$= 600 \left(\frac{\pi}{3} \right) + 800 \left(\frac{\sqrt{3}}{2} \right) + 100 \left(\frac{\sqrt{3}}{2} \right) - 225 \sqrt{3} - 0$$

$$= 200 \pi + 225 \sqrt{3}$$

$$= 1018 \text{ ft}^2$$

Using a calculator (please be sure it is set in *radians*),

1/2 **2nd 7** (20+20 **2nd COS green diamond** $\wedge$))^2-(15/ **2nd COS green diamond** $\wedge$))^2, **green diamond** $\wedge$,-π/3,π/3) **ENTER**   up arrow *twice* **2nd** right arrow   down arrow **ENTER green diamond ENTER**

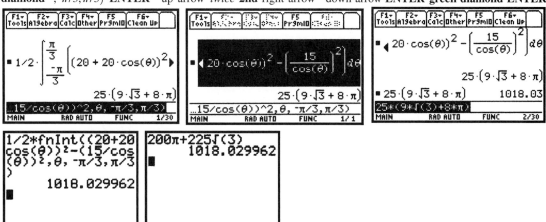

1/2* **MATH 9** (20+20   **COS ALPHA 3**)) x^2 -(15/ **COS ALPHA 3**)) x^2, **ALPHA 3** ,-π/3,π/3) **ENTER**

Exercises 7.12

Find the area of each region.

1. $r = 2, 0 \le \theta \le \dfrac{2\pi}{3}$ **2.** $r = 4, \dfrac{\pi}{3} \le \theta \le \dfrac{7\pi}{6}$ **3.** Inside $r = 3 \cos \theta$

4. Inside $r = 6 \sin \theta$ **5.** Inside $r^2 = 9 \sin 2\theta$ **6.** Inside $r^2 = \cos 2\theta$

7. Inside $r = 1 + \cos \theta$ **8.** Inside $r = 2 - \sin \theta$ **9.** Inside $r = 4 \sin 2\theta$

10. Inside $r = 2 \sin 3\theta$ **11.** Inside $r^2 = 16 \cos^2 \theta$ **12.** Inside $r^2 = 4 \sin^2 \theta$

13. Inside $r = 4 + 3 \cos \theta$ **14.** Inside $r = 3 + 3 \sin \theta$

15. $r = e^\theta, 0 \le \theta \le \pi$ **16.** $r = 2^\theta, 0 \le \theta \le \pi$

17. Inside the inner loop of $r = 1 - 2 \cos \theta$

18. Inside the outer loop and outside the inner loop of $r = 2 - 4 \sin \theta$

19. Inside $r = 2 \sin \theta + 2 \cos \theta$ **20.** Inside $r = 2 \sin \theta - 2 \cos \theta$

21. Inside $r = \sin \theta$ and $r = \sin 2\theta$ **22.** Inside $r = \sin \theta$ and $r = \cos \theta$

23. Inside $r = 2 \cos 2\theta$ and outside $r = 1$

24. Inside $r^2 = 8 \cos 2\theta$ and outside $r = 2$

25. Inside $r = 2 + \sin \theta$ and outside $r = 5 \sin \theta$

26. Inside $r = -6 \cos \theta$ and outside $r = 2 - 2 \cos \theta$

27. Inside $r = 3 + 3 \cos \theta$ and outside $r = 3 + 3 \sin \theta$

28. Inside the resulting inner loops of $r = 3 + 3 \sin \theta$ and $r = 3 - 3 \sin \theta$

CHAPTER SUMMARY

1. *Basic integration formulas:*

(a) $\displaystyle \int u^n \, du = \dfrac{u^{n+1}}{n+1} + C \qquad (n \ne -1)$

(b) $\displaystyle \int \dfrac{du}{u} = \ln |u| + C$

(c) $\displaystyle \int e^u \, du = e^u + C$

(d) $\displaystyle \int a^u \, du = \dfrac{a^u}{\ln a} + C$

(e) $\displaystyle \int \ln u \, du = u \ln u - u + C$

(f) $\displaystyle \int \sin u \, du = -\cos u + C$

(g) $\displaystyle \int \cos u \, du = \sin u + C$

(h) $\displaystyle \int \tan u \, du = -\ln |\cos u| + C$

(i) $\displaystyle \int \cot u \, du = \ln |\sin u| + C$

(j) $\displaystyle \int \sec u \, du = \ln |\sec u + \tan u| + C$

(k) $\displaystyle\int \csc u \, du = \ln|\csc u - \cot u| + C$

(l) $\displaystyle\int \sec^2 u \, du = \tan u + C$

(m) $\displaystyle\int \csc^2 u \, du = -\cot u + C$

(n) $\displaystyle\int \sec u \tan u \, du = \sec u + C$

(o) $\displaystyle\int \csc u \cot u \, du = -\csc u + C$

(p) $\displaystyle\int \frac{du}{\sqrt{a^2 - u^2}} = \text{Arcsin} \, \frac{u}{a} + C$

(q) $\displaystyle\int \frac{du}{a^2 + u^2} = \frac{1}{a} \text{Arctan} \, \frac{u}{a} + C$

(r) $\displaystyle\int u \, dv = uv - \int v \, du$ (integration by parts)

Other integrals can often be written in terms of these basic integrals by using the methods of partial fractions and trigonometric substitution.

2. *Partial fractions:*

(a) *Case 1: Nonrepeated linear denominator factors.* For every nonrepeated factor $ax + b$ of the denominator of a given fraction, there corresponds the partial fraction $\dfrac{A}{ax + b}$, where A is a constant.

(b) *Case 2: Repeated linear denominator factors.* For every factor $(ax + b)^k$ of the denominator of the given fraction, there correspond the possible partial fractions

$$\frac{A_1}{ax + b}, \frac{A_2}{(ax + b)^2}, \frac{A_3}{(ax + b)^3}, \ldots, \frac{A_k}{(ax + b)^k}$$

where $A_1, A_2, A_3, \ldots, A_k$ are constants.

(c) *Case 3: Nonrepeated quadratic denominator factors.* For every nonrepeated factor $ax^2 + bx + c$ of the denominator of the given fraction, there corresponds the partial fraction

$$\frac{Ax + B}{ax^2 + bx + c}$$

where A and B are constants.

(d) *Case 4: Repeated quadratic denominator factors.* For every factor $(ax^2 + bx + c)^k$ of the denominator of the given fraction, there correspond the possible partial fractions

$$\frac{A_1 x + B_1}{ax^2 + bx + c}, \frac{A_2 x + B_2}{(ax^2 + bx + c)^2}, \frac{A_3 x + B_3}{(ax^2 + bx + c)^3}, \ldots, \frac{A_k x + B_k}{(ax^2 + bx + c)^k}$$

where $A_1, A_2, A_3, \ldots, A_k, B_1, B_2, B_3, \ldots, B_k$ are constants.

3. *Trigonometric substitutions* (See Fig. 7.19.):

4. *Trapezoidal rule for approximating the value of an integral:*

$$\int_a^b f(x) \, dx \cong \frac{\Delta x}{2}[f(a) + 2 f(x_1) + 2 f(x_2) + 2 f(x_3) + \cdots + 2 f(x_{n-1}) + f(b)]$$

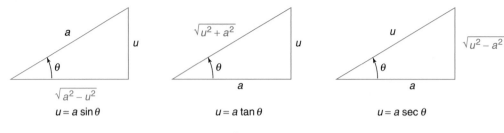

$$u = a\sin\theta \qquad\qquad u = a\tan\theta \qquad\qquad u = a\sec\theta$$

Figure 7.19

5. *Simpson's rule for approximating the value of an integral:*

$$\int_a^b f(x)\,dx \cong \frac{\Delta x}{3}[f(a) + 4f(x_1) + 2f(x_2) + 4f(x_3) + 2f(x_4) + \cdots$$
$$+ 4f(x_{n-1}) + f(b)] \qquad n \text{ must be an even integer}$$

6. *Area of a region in polar coordinates:*

$$A = \frac{1}{2}\int_\alpha^\beta r^2\,d\theta = \frac{1}{2}\int_\alpha^\beta [f(\theta)]^2\,d\theta$$

7. *Area between polar curves:*

$$A = \frac{1}{2}\int_\alpha^\beta \{[f(\theta)]^2 - [g(\theta)]^2\}\,d\theta$$

CHAPTER 7 REVIEW

Integrate without using a table of integrals.

1. $\displaystyle\int \frac{\cos 3x\,dx}{\sqrt{2 + \sin 3x}}$

2. $\displaystyle\int (5 + \tan 2x)^3 \sec^2 2x\,dx$

3. $\displaystyle\int \cos 3x\,dx$

4. $\displaystyle\int \frac{x\,dx}{x^2 - 5}$

5. $\displaystyle\int xe^{3x^2}\,dx$

6. $\displaystyle\int \frac{dx}{9 + 4x^2}$

7. $\displaystyle\int \frac{dx}{\sqrt{16 - x^2}}$

8. $\displaystyle\int \sec^2(7x + 2)\,dx$

9. $\displaystyle\int \frac{\sec^2 x\,dx}{3 + 5\tan x}$

10. $\displaystyle\int x^2 \sin(x^3 + 4)\,dx$

11. $\displaystyle\int_0^1 \frac{dx}{16 + 9x^2}$

12. $\displaystyle\int_0^{1/2} \sin \pi x\,dx$

13. $\displaystyle\int_0^{\pi/4} \frac{\sin x\,dx}{\cos x}$

14. $\displaystyle\int_0^1 \frac{dx}{\sqrt{9 - 4x^2}}$

15. $\displaystyle\int \frac{dx}{x\sqrt{16x^2 - 9}}$

16. $\displaystyle\int \frac{\tan 3x}{\sec^4 3x}\,dx$

17. $\displaystyle\int \frac{\text{Arctan } 3x}{1 + 9x^2}\,dx$

18. $\displaystyle\int \sec 5x\,dx$

19. $\displaystyle\int x \tan x^2\,dx$

20. $\displaystyle\int \sin^5 2x \cos^2 2x\,dx$

21. $\displaystyle\int \cos^4 3x \sin^2 3x\,dx$

22. $\displaystyle\int \frac{dx}{x^2 + 2x - 3}$

23. $\displaystyle\int e^{3x} \cos 4x\,dx$

24. $\displaystyle\int \sin x \ln(\cos x)\,dx$

25. $\displaystyle\int \tan^4 x\,dx$

26. $\displaystyle\int \cos^2 5x\,dx$

27. $\displaystyle\int \frac{x - 3}{6x^2 - x - 1}\,dx$

28. $\displaystyle\int \sqrt{x}\ln x\,dx$

29. $\displaystyle\int e^{-x}(\cos^2 e^{-x})(\sin e^{-x})\,dx$ **30.** $\displaystyle\int \frac{4\,dx}{x^2 - 4}$

31. $\displaystyle\int x^2 \sin x\,dx$ **32.** $\displaystyle\int \frac{\text{Arcsin } 5x}{\sqrt{1 - 25x^2}}\,dx$ **33.** $\displaystyle\int_2^3 \frac{dx}{x\sqrt{x^2 - 1}}$

34. $\displaystyle\int_0^1 xe^{4x}\,dx$ **35.** $\displaystyle\int_0^{\pi/8} 4\tan 2x\,dx$ **36.** $\displaystyle\int_0^1 \frac{8\,dx}{4 - x^2}$

37. $\displaystyle\int \frac{3x^2 - 11x + 12}{x^3 - 4x^2 + 4x}\,dx$ **38.** $\displaystyle\int \frac{dx}{x\sqrt{16 - 9x^2}}$

Find the area bounded by the given curves.

39. $y(x - 1) = 1$, $x = 2$, $x = 4$, and $y = 0$.   **40.** $y = e^{x+2}$, $x = 1$, $x = 3$, and $y = 0$.

41. $y = e^{2x}$, $x = 0$, $x = 1$, and $y = 0$.   **42.** $y(1 + x^2) = 1$, $x = 0$, $x = 1$, and $y = 0$.

43. $y^2(4 - x^2) = 1$, $x = 0$, $x = 1$, and $y = 0$.   **44.** $xy = 1$, $y = x$, $y = 0$, and $x = 2$.

45. $y = \sec^2 x$, $y = 0$, $x = 0$, and $x = \pi/4$.   **46.** $y(4 + x^2) = 1$, $x = 0$, $x = 2$, and $y = 0$.

47. $y = \tan x$, $x = 0$, $x = \pi/4$, and $y = 0$.

48. Find the volume of the solid formed by revolving the region bounded by $y = \ln x$, $x = 1$, $x = 2$, and $y = 0$ about the x-axis.

49. Find the volume of the solid formed by revolving the region bounded by $y = e^x$, $x = 0$, $x = 1$, and $y = 0$ about the x-axis.

50. The current i in an electric circuit varies according to the equation $i = 4t \sin 3t$ amperes. Find an equation for the charge q (in coulombs, C) transferred as a function of time ($q = \int i\,dt$).

51. The current i in an electric circuit varies according to the equation $i = \dfrac{5t + 1}{t^2 + t - 2}$. Find an equation for the charge q (in coulombs) transferred as a function of time.

52. A force is acting on an object according to the equation $F = xe^{x^2}$ where x is measured in metres and F is measured in newtons. Find the work done in moving the object from $x = 1$ m to $x = 2$ m.

Integrate using the Table of Integrals in Appendix B. Give the number of the formula used.

53. $\displaystyle\int \frac{dx}{x(5 + 3x)}$ **54.** $\displaystyle\int \frac{\sqrt{16 - x^2}}{x^2}\,dx$ **55.** $\displaystyle\int \frac{dx}{\sqrt{3 + 6x + x^2}}$

56. $\displaystyle\int e^x \sin 2x\,dx$ **57.** $\displaystyle\int \frac{\sqrt{4x^2 - 9}}{x}\,dx$ **58.** $\displaystyle\int \cos^4 3x \sin 3x\,dx$

59. $\displaystyle\int \frac{\sqrt{9 + 4x}}{x}\,dx$ **60.** $\displaystyle\int \tan^6 x\,dx$

Use the trapezoidal rule to approximate the value of each integral.

61. $\displaystyle\int_1^4 \frac{dx}{2x - 1}$, $n = 6$ **62.** $\displaystyle\int_1^3 \frac{dx}{9 + x^2}$, $n = 4$

63. $\displaystyle\int_0^4 \sqrt{16 - x^2}\,dx$, $n = 8$ **64.** $\displaystyle\int_1^4 \sqrt[3]{18 - x^2}\,dx$, $n = 6$

Use the trapezoidal rule to find the approximate area under the curve through each set of points.

65.

x	1	3	5	7	9
y	2.3	2.8	3.4	2.7	2.1

66.

x	1.6	1.8	2.0	2.2	2.4	2.6
y	0.9	0.7	0.8	1.1	1.3	0.9

Use Simpson's rule to approximate the value of each integral.

67. $\displaystyle\int_0^4 \sqrt{16 + x^2}\, dx, n = 8$

68. $\displaystyle\int_0^{12} \frac{4}{1 + x^2}\, dx, n = 6$

69. $\displaystyle\int_0^{\pi/2} \frac{dx}{2 + \sin x}, n = 4$

70. $\displaystyle\int_0^8 \frac{x^3}{\sqrt{1 + x^3}}\, dx, n = 8$

Find the area of each region.

71. Inside $r^2 = 4 \cos 2\theta$

72. Inside $r = 1 - \sin \theta$

73. Inside $r = 3 - 2 \sin \theta$

74. Inside $r = \frac{1}{2} \cos 3\theta$

75. Inside $r = \frac{1}{2}(\theta + \pi)$ and outside $r = \theta$ for $0 \le \theta \le \pi$

76. Inside $r = 2 \cos \theta$ and outside $r = 1$

77. Inside the inner loop of $r = 2 - 4 \cos \theta$

78. Inside $r = 1$ and outside $r = 1 + \sin \theta$

79. Inside $r^2 = 8 \sin 2\theta$ and outside $r = 2$

80. Inside both $r^2 = \cos 2\theta$ and $r^2 = \sin 2\theta$

8

Three-Space
Partial Derivatives and Double Integrals

INTRODUCTION

We have been discussing functions of only one variable whose graphs lie in a plane, which is often called the Euclidian plane or two-space. Now, we need to extend our discussion to functions of two variables whose graphs require three dimensions, sometimes called three-space.

Objectives

- Sketch graphs in three-space.
- Find the equation of a surface given appropriate information.
- Find first and second partial derivatives.
- Given a surface, find the slope of the tangent lines parallel to the xz- and yz-planes through a given point.
- Find the total differential of a function of two variables.
- Find relative maximum and minimum points of a function of two variables.
- Evaluate double integrals.

8.1 FUNCTIONS IN THREE-SPACE

Consider three mutually perpendicular number lines (x-, y-, and z-axes) with their zero points intersecting at point O, called the origin. Although these axes may be oriented in any way, we will use the orientation as shown in Fig. 8.1(a), which is often called the *right-handed system*.

Think of the y- and z-axes lying in the plane of the paper with the positive y-direction to the right and the positive z-direction upward. The x-axis is perpendicular to the paper, and its positive direction is toward us. This is called a right-handed system because if the fingers on the right hand are cupped so they curve from the positive x-axis to the positive y-axis, the thumb points along the positive z-axis as shown in Fig. 8.1(b).

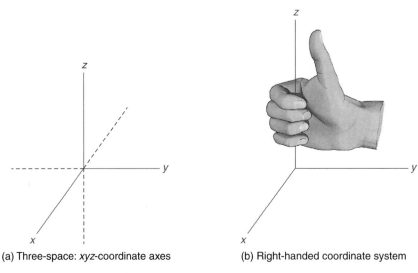

(a) Three-space: *xyz*-coordinate axes   (b) Right-handed coordinate system

Figure 8.1

The three axes determine three planes (*xy*-, *xz*-, and *yz*-planes), which divide three-space into octants as shown in Fig. 8.2.

Each point in three-space is represented by an ordered triple of numbers in the form (*x*, *y*, *z*), which indicate the directed distances from the three planes as shown in Fig. 8.3.

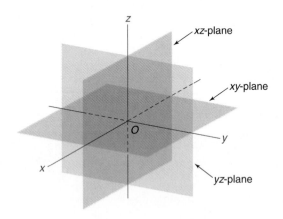

Figure 8.2   Coordinate planes.

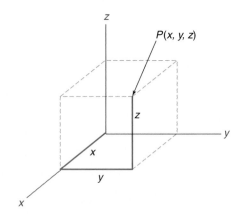

Figure 8.3   An ordered triple of numbers in the form *P*(*x*, *y*, *z*) represents a point in three-space.

EXAMPLE 1

Plot the points $(2, -4, 5)$ and $(-3, 5, -4)$ (see Fig. 8.4).

The graph of an equation expressed in three variables is normally a surface. We shall now discuss the graphs of some surfaces on a case-by-case basis.

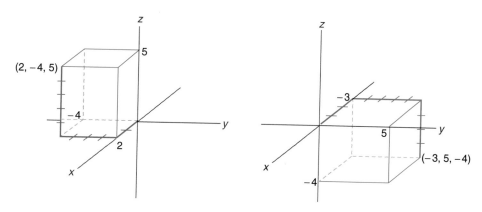

Figure 8.4 Plotting points in three-space.

> **PLANE**
>
> The graph of an equation in the form
>
> $$Ax + By + Cz + D = 0$$
>
> is a plane.

EXAMPLE 2

Sketch $3x + 6y + 2z = 12$.

If the plane does not pass through the origin, its sketch may be found by graphing the intercepts. To find the x-intercept, set y and z equal to zero and solve for x: $x = 4$, which corresponds to the point $(4, 0, 0)$. Similarly, the y- and z-intercepts are $(0, 2, 0)$ and $(0, 0, 6)$, respectively. These three points determine the plane. The lines through pairs of these points are in the coordinate planes and are called *traces*, as shown in Fig. 8.5.

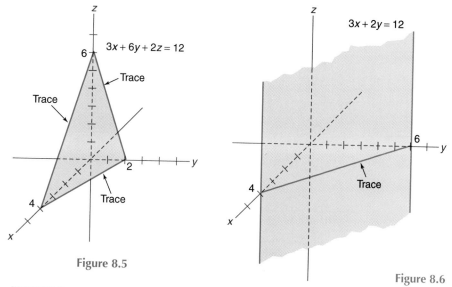

Figure 8.5

Figure 8.6

EXAMPLE 3

Sketch $3x + 2y = 12$ in three-space.

The x- and y-intercepts are $(4, 0, 0)$ and $(0, 6, 0)$, respectively, which determine the trace in the xy-plane. Note that the plane never crosses the z-axis because x and y cannot both be zero. Thus the plane is parallel to the z-axis, as shown in Fig. 8.6.

EXAMPLE 4

Sketch $x = 3$ in three-space.

Its x-intercept is $(3, 0, 0)$. The plane is parallel to both the y- and z-axes, as shown in Fig. 8.7.

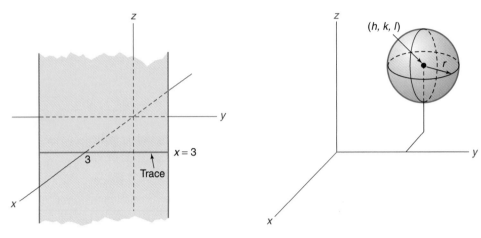

Figure 8.7

Figure 8.8   Graph of sphere with center at (h, k, l) and radius r.

The formula for the distance between two points in three-space is given in Exercise 41. Using this formula, the equation of a sphere is as follows:

SPHERE

A sphere with center at (h, k, l) and radius r (see Fig. 8.8) is given by the equation

$$(x - h)^2 + (y - k)^2 + (z - l)^2 = r^2$$

EXAMPLE 5

Find the center and radius of the sphere whose equation is

$$x^2 + y^2 + z^2 - 6x + 10y - 16z + 82 = 0$$

and sketch its graph.

First, complete the square on each variable.

$$(x^2 - 6x \quad\;\,) + (y^2 + 10y \quad\;\,) + (z^2 - 16z \quad\;\,) = -82$$
$$(x^2 - 6x + 9) + (y^2 + 10y + 25) + (z^2 - 16z + 64) = -82 + 9 + 25 + 64$$
$$(x - 3)^2 + (y + 5)^2 + (z - 8)^2 = 16$$

The center is at $(3, -5, 8)$ and the radius is 4 (see Fig. 8.9).

In general, graphing more complex surfaces can be quite complicated. One helpful technique involves graphing the intersections of the surface with the coordinate planes. Such intersections of the surfaces and the coordinate planes are called *traces*.

CYLINDRICAL SURFACE

The graph in three-space of an equation expressed in only two variables is a cylindrical surface. The surface is parallel to the coordinate axis of the missing variable.

EXAMPLE 6

Sketch the graph of $y^2 = 4x$ in three-space.

First, graph the trace $y^2 = 4x$ in the xy-plane. Then extend the cylindrical surface parallel to the z-axis as in Fig. 8.10.

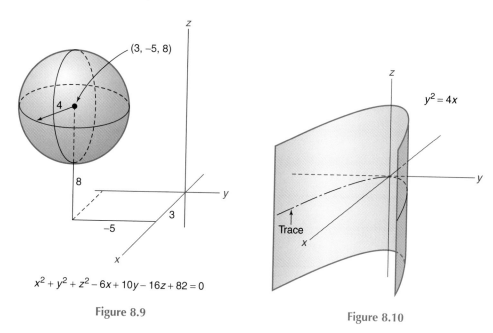

$$x^2 + y^2 + z^2 - 6x + 10y - 16z + 82 = 0$$

Figure 8.9

Figure 8.10

EXAMPLE 7

Sketch the graph of $z = \sin y$ in three-space.

First, graph the trace $z = \sin y$ in the yz-plane. Then extend the cylindrical surface parallel to the x-axis as in Fig. 8.11.

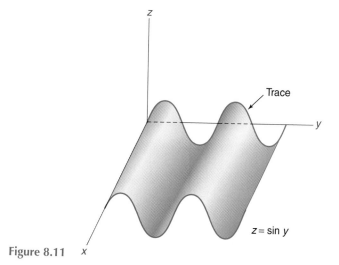

Figure 8.11

QUADRIC SURFACE

The graph in three-space of a second-degree equation is a quadric surface. Plane sections, intersections, or slices of the given surface with planes are conics.

Surface	Equation	Traces
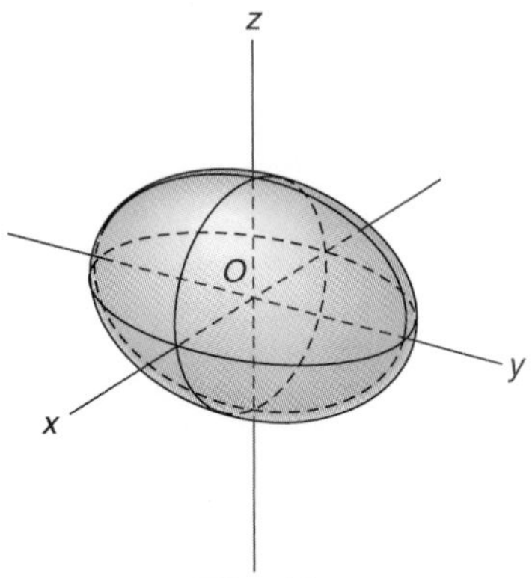 \ \ Ellipsoid	$$\frac{x^2}{a^2} + \frac{y^2}{b^2} + \frac{z^2}{c^2} = 1$$ *Note:* If $a = b = c$, the surface is a sphere.	xy-plane: ellipse\ xz-plane: ellipse\ yz-plane: ellipse\ All traces in planes parallel to the coordinate planes are ellipses.
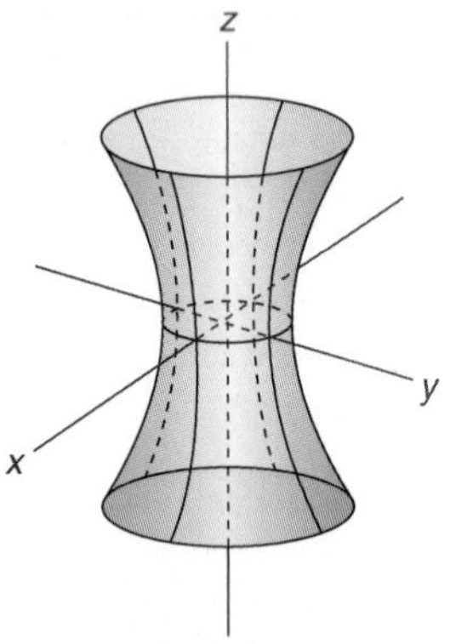 \ \ Hyperboloid of one sheet	$$\frac{x^2}{a^2} + \frac{y^2}{b^2} - \frac{z^2}{c^2} = 1$$ *Note:* If $a = b$, the surface may be generated by rotating a hyperbola about its conjugate axis.	xy-plane: ellipse\ xz-plane: hyperbola\ yz-plane: hyperbola\ All traces parallel to the xy-plane are ellipses.
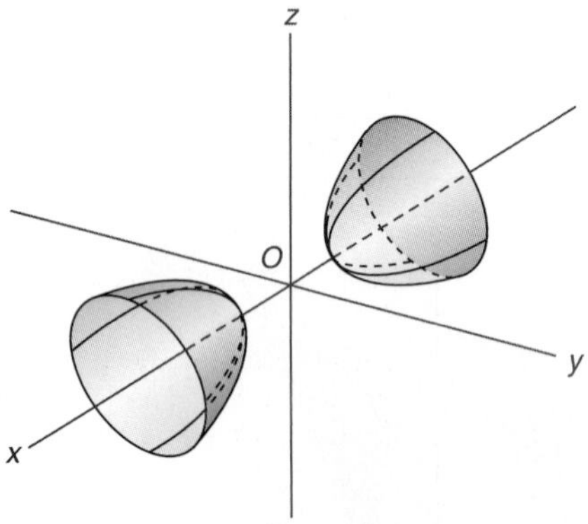 \ \ Hyperboloid of two sheets	$$\frac{x^2}{a^2} - \frac{y^2}{b^2} - \frac{z^2}{c^2} = 1$$ *Note:* If $b = c$, the surface may be generated by rotating a hyperbola about its transverse axis.	xy-plane: hyperbola\ xz-plane: hyperbola\ yz-plane: no trace\ All traces parallel to the yz-plane that intersect the surface are ellipses.

Figure 8.12 Six basic types of quadric surfaces.

Surface	Equation	Traces
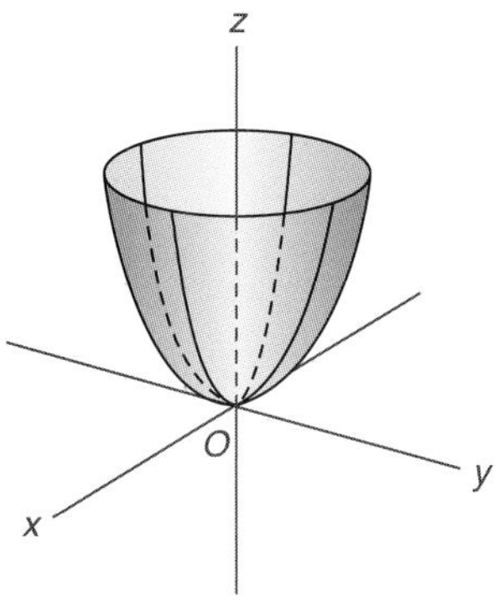 Elliptic paraboloid	$\dfrac{x^2}{a^2} + \dfrac{y^2}{b^2} = \dfrac{z}{c}$ *Note:* If $a = b$, the surface may be generated by rotating a parabola about its axis.	xy-plane: point xz-plane: parabola yz-plane: parabola If $c > 0$, all traces parallel to and above the xy-plane are ellipses; below the xy-plane, there is no intersection. If $c < 0$, the situation is reversed.
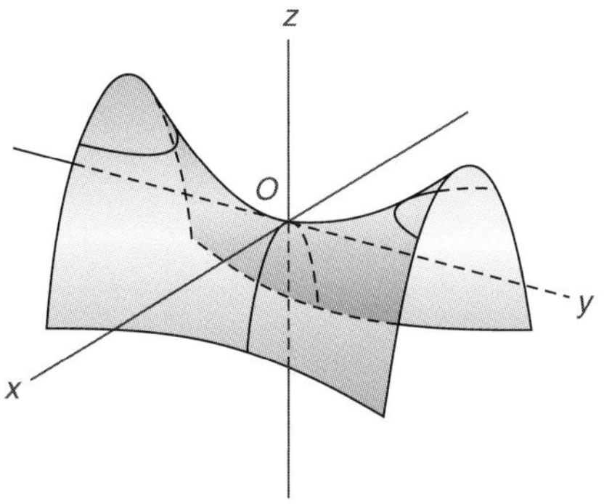 Hyperbolic paraboloid	$\dfrac{x^2}{a^2} - \dfrac{y^2}{b^2} = \dfrac{z}{c}$ *Note:* The origin is called a saddle point because this surface looks like a saddle.	xy-plane: pair of intersecting lines xz-plane: parabola $\left.\right\}$ one opens up, yz-plane: parabola $\left.\right\}$ one opens down Traces parallel to the xy-plane are hyperbolas.
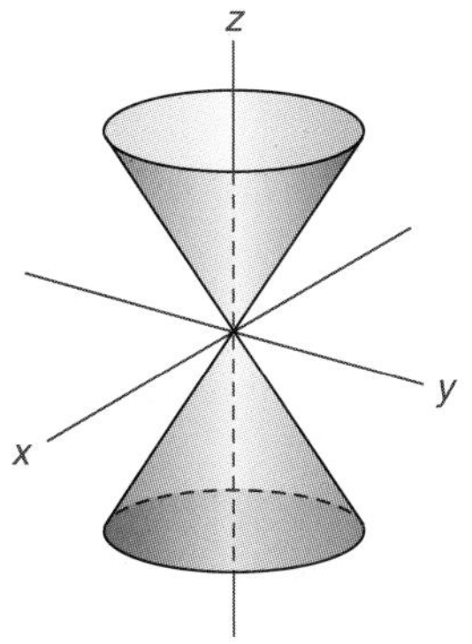 Elliptic cone	$\dfrac{x^2}{a^2} + \dfrac{y^2}{b^2} - \dfrac{z^2}{c^2} = 0$	xy-plane: point xz-plane: pair of intersecting lines yz-plane: pair of intersecting lines Traces parallel to the xy-plane are ellipses. Traces parallel to the xz- and yz-planes are hyperbolas.

Figure 8.12 *(continued)*

The second-degree equation in three-space has the form

$$Ax^2 + By^2 + Cz^2 + Dxy + Exz + Fyz + Gx + Hy + Iz + J = 0$$

The equations of the basic or central quadrics after translations and/or rotations can be expressed in one of the two general forms:

$$Ax^2 + By^2 + Cz^2 + J = 0$$

or

$$Ax^2 + By^2 + Iz = 0$$

Figure 8.12 shows six basic types of quadric surfaces.

EXAMPLE 8

Name and sketch the graph of $\dfrac{x^2}{16} + \dfrac{y^2}{9} - \dfrac{z^2}{36} = 1$.

First, find the traces in the three coordinate planes:

xy-plane: Set $z = 0$, $\dfrac{x^2}{16} + \dfrac{y^2}{9} = 1$, an ellipse.

xz-plane: Set $y = 0$, $\dfrac{x^2}{16} - \dfrac{z^2}{36} = 1$, a hyperbola.

yz-plane: Set $x = 0$, $\dfrac{y^2}{9} - \dfrac{z^2}{36} = 1$, a hyperbola

This is a hyperboloid of one sheet (see Fig. 8.13).

EXAMPLE 9

Name and sketch the graph of $2x^2 + 2y^2 = 8z$.

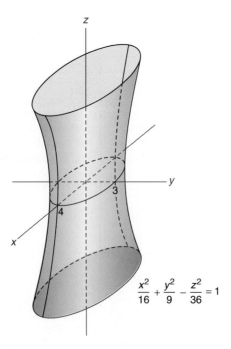

$$\frac{x^2}{16} + \frac{y^2}{9} - \frac{z^2}{36} = 1$$

Figure 8.13

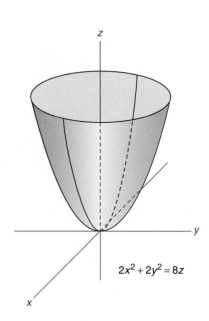

$$2x^2 + 2y^2 = 8z$$

Figure 8.14

First, divide both sides by 2.

$$x^2 + y^2 = 4z$$

Then, find the traces in the three coordinate planes:

xy-plane: Set $z = 0$, $x^2 + y^2 = 0$, the origin (degenerate circle).

xz-plane: Set $y = 0$, $x^2 = 4z$, a parabola.

yz-plane: Set $x = 0$, $y^2 = 4z$, a parabola.

(see Fig. 8.14). This is an elliptic paraboloid.

Exercises 8.1

Plot each point.

1. $(2, 4, 5)$ **2.** $(3, 5, 2)$ **3.** $(3, -2, 4)$

4. $(-2, 0, 3)$ **5.** $(-1, 4, -3)$ **6.** $(3, -2, -4)$

Name and sketch the graph of each equation in three-space. Sketch the traces in the coordinate planes when appropriate.

7. $2x + 3y + 8z = 24$ **8.** $6x - 3y + 9z = 18$ **9.** $x + 3y - z = 9$

10. $-4x + 8y + 3z = 12$ **11.** $2x + 5y = 20$ **12.** $4x - 3y = 12$

13. $y = 4$ **14.** $z = -5$

15. $x^2 + y^2 + z^2 = 25$ **16.** $(x - 3)^2 + y^2 + (z - 2)^2 = 36$

17. $x^2 + y^2 + z^2 - 8x + 6y - 4z + 4 = 0$

18. $x^2 + y^2 + z^2 + 6x - 6y + 9 = 0$

19. $z^2 = 8y$ **20.** $x^2 + z^2 = 16$ **21.** $9x^2 + 4y^2 = 36$

22. $y^2 - z^2 = 4$ **23.** $4x^2 + 9y^2 + z^2 = 36$ **24.** $x^2 - y^2 = 16z$

25. $x^2 + y^2 - 4z = 0$ **26.** $y = e^x$

27. $9x^2 - 36y^2 - 4z^2 = 36$ **28.** $16x^2 + 9y^2 - 36z^2 = 144$

29. $y = \cos x$ **30.** $4x^2 + 9y^2 + 9z^2 = 36$

31. $y^2 - z^2 = 8x$ **32.** $y^2 + z^2 = 4x$

33. $4x^2 + 9y^2 - 9z^2 = 0$ **34.** $9x^2 - 16y^2 - 16z^2 = 144$

35. $81y^2 + 36z^2 - 4x^2 = 324$ **36.** $x^2 + y^2 - z^2 = 0$

37. Find the equation of the sphere with center at $(3, -2, 4)$ and radius 6.

38. Find the equation of the sphere with center at $(3, -2, 4)$ and tangent to the xz-plane.

39. Find the equation of the sphere with center at $(3, -2, 4)$ and tangent to the xy-plane.

40. Find the equation of the sphere of radius 4 that is tangent to the three coordinate planes and whose center is in the first octant.

41. Given two points $P_1(x_1, y_1, z_1)$ and $P_2(x_2, y_2, z_2)$, show that the distance between them is given by

$$d = P_1P_2 = \sqrt{(x_2 - x_1)^2 + (y_2 - y_1)^2 + (z_2 - z_1)^2}$$

Find the distance between each pair of points.

42. $(3, 6, 8)$ and $(7, -2, 7)$ **43.** $(5, -3, 2)$ and $(3, 1, -2)$ **44.** $(-4, 0, 6)$ and $(2, 5, -2)$

45. Show that $(5, 6, 3)$, $(2, 8, 4)$, and $(3, 5, 6)$ are vertices of an equilateral triangle.

46. Show that $(1, 1, 2)$, $(3, 1, 0)$, and $(5, 3, 2)$ are vertices of a right triangle. (*Hint:* Use the Pythagorean theorem.)

8.2 PARTIAL DERIVATIVES

Let $z = f(x, y)$ represent a function z with independent variables x and y. That is, there exists a unique value of z for every pair of x- and y-values.

PARTIAL DERIVATIVES

If $z = f(x, y)$ is a function of two variables, the partial derivative of z with respect to x (y is treated as a constant) is defined by

$$\frac{\partial z}{\partial x} = \lim_{\Delta x \to 0} \frac{f(x + \Delta x, y) - f(x, y)}{\Delta x}$$

provided that this limit exists.

If $z = f(x, y)$ is a function of two variables, the partial derivative of z with respect to y (x is treated as a constant) is defined by

$$\frac{\partial z}{\partial y} = \lim_{\Delta y \to 0} \frac{f(x, y + \Delta y) - f(x, y)}{\Delta y}$$

provided that this limit exists.

Other common notations for partial derivatives include

$$\frac{\partial z}{\partial x} \quad \text{or} \quad \frac{\partial f}{\partial x} \quad \text{or} \quad f_x \quad \text{or} \quad f_x(x, y)$$

$$\frac{\partial z}{\partial y} \quad \text{or} \quad \frac{\partial f}{\partial y} \quad \text{or} \quad f_y \quad \text{or} \quad f_y(x, y)$$

Note that in the definition of $\dfrac{\partial z}{\partial x}$, y is treated as a constant (held fixed) and only x is allowed to vary. Think of z as a function of only one variable x and find the usual derivative with respect to x, treating y as a constant. For $\dfrac{\partial z}{\partial y}$, find the usual derivative with respect to y, treating x as a constant.

EXAMPLE 1

If $z = 5x^2 - 4x^2y + y^3$, find $\dfrac{\partial z}{\partial x}$ and $\dfrac{\partial z}{\partial y}$.

$$\frac{\partial z}{\partial x} = 10x - 8xy$$

$$\frac{\partial z}{\partial y} = -4x^2 + 3y^2$$

EXAMPLE 2

If $z = x^2 \ln y - e^{xy}$, find $\dfrac{\partial z}{\partial x}$ and $\dfrac{\partial z}{\partial y}$.

$$\frac{\partial z}{\partial x} = 2x \ln y - ye^{xy}$$

$$\frac{\partial z}{\partial y} = \frac{x^2}{y} - xe^{xy}$$

EXAMPLE 3

If $z = e^{-x} \sin y + \ln(x^2 + y^2)$, find $\dfrac{\partial z}{\partial x}$ and $\dfrac{\partial z}{\partial y}$.

$$\frac{\partial z}{\partial x} = -e^{-x} \sin y + \frac{2x}{x^2 + y^2}$$

$$\frac{\partial z}{\partial y} = e^{-x} \cos y + \frac{2y}{x^2 + y^2}$$

Using a calculator,

2nd 8 green diamond e^x -x) 2nd SIN y)+ 2nd LN x^2+y^2),x) ENTER right arrow left arrow $\leftarrow$ y **ENTER**

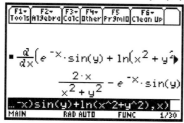
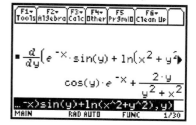

Now, let's consider a graphical interpretation of partial derivatives. Let $z = f(x, y)$ be a function of x and y. Suppose that we are considering the partial derivative of z with respect to x at the point (x_0, y_0, z_0). We treat y as a constant and take the usual derivative with respect to x; this is graphically equivalent to finding the slope at $x = x_0$ of the curve, which is determined by the intersection of the plane $y = y_0$ and the surface $z = f(x, y)$ (see Fig. 8.15a).

Likewise for the partial derivative of z with respect to y; this is graphically equivalent to finding the slope at $y = y_0$ of the curve, which is determined by the intersection of the plane $x = x_0$ and the surface $z = f(x, y)$ (see Fig. 8.15b).

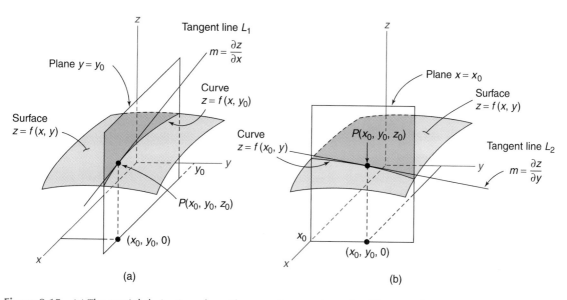

(a) (b)

Figure 8.15 (a) The partial derivative of z with respect to x at the point $P(x_0, y_0, z_0)$ is graphically equivalent to the slope of L_1, the tangent line to the surface through P and parallel to the xz-plane. (b) The partial derivative of z with respect to y at the point $P(x_0, y_0, z_0)$ is graphically equivalent to the slope of L_2, the tangent line to the surface through P and parallel to the yz-plane.

EXAMPLE 4

Given the surface $z = 4x^2 + y^2$, find the slope of the tangent lines parallel to the xz- and yz-planes and through the point $(2, -1, 9)$.

First,

$$\frac{\partial z}{\partial x} = 8x \quad \text{and} \quad \frac{\partial z}{\partial y} = 2y$$

The slope of each tangent line is the value of each partial derivative evaluated at the point $(2, -1, 9)$. Thus

$$\left.\frac{\partial z}{\partial x}\right|_{(2,-1,9)} = 16 \quad \text{and} \quad \left.\frac{\partial z}{\partial y}\right|_{(2,-1,9)} = -2$$

So the slope of the tangent line parallel to the xz-plane is 16, and the slope of the tangent line parallel to the yz-plane is -2.

EXAMPLE 5

In an RC-circuit, the current I may be given by

$$I = \frac{E}{R} e^{-t/(RC)}$$

Assume that all quantities are constant except I and R. Find $\dfrac{\partial I}{\partial R}$.

$$\frac{\partial I}{\partial R} = \frac{E}{R} e^{-t/(RC)}\left(\frac{-t}{C} \cdot \frac{-1}{R^2}\right) + e^{-t/(RC)}\left(\frac{-E}{R^2}\right)$$

$$= \frac{tE}{CR^3} e^{-t/(RC)} - \frac{E}{R^2} e^{-t/(RC)}$$

$$= \frac{E}{R^2} e^{-t/(RC)}\left(\frac{t}{RC} - 1\right)$$

Exercises 8.2

Find (a) $\dfrac{\partial z}{\partial x}$ *and (b)* $\dfrac{\partial z}{\partial y}$ *for each function.*

1. $z = 4x^3y^2$

2. $z = 8\pi x^2\sqrt{y}$

3. $z = 6x^2y^4 + 2xy^2$

4. $z = 3xy^5 - 4x^2y^2$

5. $z = \sqrt{x^2 + y^2}$

6. $z = \ln\sqrt{x^2 + y^2}$

7. $z = \dfrac{x^2 - y^2}{2xy}$

8. $z = \dfrac{1}{3}\pi x^2 y$

9. $z = axy$

10. $z = e^{xy} - e^{-x}$

11. $z = \ln\dfrac{x}{y}$

12. $z = (x^2 - y^2)^{-3}$

13. $z = \tan(x - y)$

14. $z = \sin(x - y)$

15. $z = e^{3x}\sin xy$

16. $z = \tan\dfrac{x}{y}$

17. $z = \dfrac{e^{xy}}{x\sin y}$

18. $z = ye^{\sin x}$

19. $z = 2\sin y\cos x$

20. $z = x^2\cos 2y$

21. $z = xy\tan xy$

22. $z = x^2y\cos y$

The following formulas are used in electronics and physics. Find the indicated partial derivative.

23. $P = I^2R; \dfrac{\partial P}{\partial I}$

24. $P = \dfrac{V^2}{R}; \dfrac{\partial P}{\partial R}$

25. $I = \dfrac{E}{R + r}; \dfrac{\partial I}{\partial R}$ **26.** $R = \dfrac{R_1 R_2}{R_1 + R_2}; \dfrac{\partial R}{\partial R_1}$

27. $Z = \sqrt{R^2 + X_L^2}; \dfrac{\partial Z}{\partial R}$ **28.** $Z = \sqrt{R^2 + (X_L - X_C)^2}; \dfrac{\partial Z}{\partial X_L}$

29. $e = E \sin 2\pi ft; \dfrac{\partial e}{\partial t}$ **30.** $E = I_2 R_2 + I_3 R_3; \dfrac{\partial E}{\partial I_2}$

31. $E = I_2 R_2 + I_2 R_3; \dfrac{\partial E}{\partial I_2}$ **32.** $I = \dfrac{E_2 R_1 + E_2 R_3 - E_1 R_3}{R_1 R_2 + R_1 R_3 + R_2 R_3}; \dfrac{\partial I}{\partial R_1}$

33. $I = \dfrac{E}{R} e^{-t/(RC)}; \dfrac{\partial I}{\partial t}$ **34.** $I = \dfrac{E}{R} e^{-t/(RC)}; \dfrac{\partial I}{\partial C}$

35. $q = CEe^{-t/(RC)}; \dfrac{\partial q}{\partial C}$ **36.** $q = CE(1 - e^{-t/(RC)}); \dfrac{\partial q}{\partial C}$

37. $\tan \phi = \dfrac{X_L}{R}; \dfrac{\partial \phi}{\partial R}$ **38.** $\tan \phi = \dfrac{X_L - X_C}{R}; \dfrac{\partial \phi}{\partial X_L}$

*For each surface, find the slope of the tangent lines parallel to **(a)** the xz-plane and **(b)** the yz-plane and through the given point.*

39. $z = 9x^2 + 4y^2; (1, -2, 25)$ **40.** $z = 8y^2 - x^2; (-2, 1, 4)$

41. $z = \sqrt{25x^2 + 36y^2 + 164}; (1, -1, 15)$

42. $z = 6x^2 + 2xy + 4y^2 - 6x + 8y - 2; (1, -2, -6)$

43. The ideal gas law is $PV = nRT$, where P is the pressure, V is the volume, T is the absolute temperature, n is the number of moles of gas, and R is a constant. Show that

$$\frac{\partial P}{\partial V} \cdot \frac{\partial V}{\partial T} \cdot \frac{\partial T}{\partial P} = -1$$

44. Find the only point on the surface $z = x^2 + 3xy + 4y^2 - 10x - 8y + 4 = 0$ at which its tangent plane is horizontal.

45. The volume of a right circular cylinder is given by $V = \pi r^2 h$. If the height h is fixed at 8 cm, find the rate of change of the volume V with respect to the radius r when $r = 6$ cm.

46. Given that $z = \dfrac{e^{x+y}}{e^x + e^y}$, show that $\dfrac{\partial z}{\partial x} + \dfrac{\partial z}{\partial y} = z$.

47. Given $w = t^2 + \tan t e^{1/s}$, show that $s^2 \dfrac{\partial w}{\partial s} + t \dfrac{\partial w}{\partial t} = 2t^2$.

8.3 APPLICATIONS OF PARTIAL DERIVATIVES

Earlier applications of the derivative of a function of one variable may be extended to functions of two variables by using partial derivatives. In Section 3.6 we defined the differential of a function of one variable as

$$dy = f'(x)\, dx$$

Similarly, we define the total differential of a function of two variables as

---

TOTAL DIFFERENTIAL OF $z = f(x, y)$

$$dz = \frac{\partial f}{\partial x}\, dx + \frac{\partial f}{\partial y}\, dy$$

The same type of definition may be extended to functions of three or more variables.

EXAMPLE 1

Find dz for $z = x^3 + 5x^2y - 4xy^2$.

$$dz = \frac{\partial z}{\partial x}\,dx + \frac{\partial z}{\partial y}\,dy$$
$$= (3x^2 + 10xy - 4y^2)\,dx + (5x^2 - 8xy)\,dy$$

In Section 3.6 we used differential approximations to find changes in the function as small changes were made in the independent variable. Now we can use the total differential to find changes in a function of two variables as small changes are made in one or both variables.

EXAMPLE 2

The height of a right circular cylinder measures 20.00 cm with a possible error of 0.10 cm, while its radius measures 8.00 cm with a possible error of 0.05 cm. Find the maximum error in its volume.

$$V = \pi r^2 h$$
$$dV = \frac{\partial V}{\partial r}\,dr + \frac{\partial V}{\partial h}\,dh$$
$$= 2\pi rh\,dr + \pi r^2\,dh$$
$$= 2\pi(8.00\text{ cm})(20.00\text{ cm})(0.05\text{ cm}) + \pi(8.00\text{ cm})^2(0.10\text{ cm})$$
$$= 70.4\text{ cm}^3$$

EXAMPLE 3

The angle of elevation to the top of a monument when measured 150 ft ($\pm$ 0.5 ft) from its base is 31.00° with a maximum error of 0.05°. Find the maximum error in measuring its height (see Fig. 8.16).

$$\tan\theta = \frac{y}{x}$$
$$y = x\tan\theta$$
$$dy = \frac{\partial y}{\partial x}\,dx + \frac{\partial y}{\partial \theta}\,d\theta$$
$$= \tan\theta\,dx + x\sec^2\theta\,d\theta$$
$$= (\tan 31.00°)(0.5\text{ ft}) + (150\text{ ft})(\sec 31.00°)^2\left(0.05° \times \frac{\pi\text{ rad}}{180°}\right)$$
$$= 0.479\text{ ft}$$

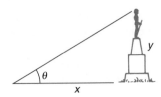

Figure 8.16

The second partial derivatives of a function of two variables are calculated by taking partial derivatives of the first partials. There are four possibilities. The second partial derivative with respect to x is notated $f_{xx}(x, y)$ or $\frac{\partial^2 f}{\partial x^2}$, the second partial with respect to y is notated $f_{yy}(x, y)$ or $\dfrac{\partial^2 f}{\partial y^2}$, and the mixed partials $f_{xy}(x, y)$ and $f_{yx}(x, y)$ can also be notated as $\dfrac{\partial^2 f}{\partial y \partial x}$ and $\dfrac{\partial^2 f}{\partial x \partial y}$ respectively. While the order of differentiation in the mixed partials may look significant, a theorem of advanced calculus guarantees that $f_{xy}(x, y) = f_{yx}(x, y)$ if the original function, its first partials, and these mixed partials are all continuous functions of two variables (note the equality of the mixed partials in the following example).

EXAMPLE 4

Find the second partial derivatives of $f(x, y) = x^3y^5 + x \cos y$.
The first partials are

$$f_x(x, y) = \frac{\partial f}{\partial x} = 3x^2y^5 + \cos y$$

$$f_y(x, y) = \frac{\partial f}{\partial y} = 5x^3y^4 - x \sin y$$

so the second partials are

$$f_{xx}(x, y) = 6xy^5 \qquad \text{[the partial with respect to } x \text{ of } f_x(x, y)\text{]}$$
$$f_{yy}(x, y) = 20x^3y^3 - x \cos y \qquad \text{[the partial with respect to } y \text{ of } f_y(x, y)\text{]}$$
$$f_{xy}(x, y) = 15x^2y^4 - \sin y \qquad \text{[the partial with respect to } y \text{ of } f_x(x, y)\text{]}$$
$$f_{yx}(x, y) = 15x^2y^4 - \sin y \qquad \text{[the partial with respect to } x \text{ of } f_y(x, y)\text{]}$$

One principal application of the derivative of a function in one variable is finding relative maximums and minimums. This application may also be extended to functions of two variables by using partial derivatives. Geometrically, let's start this discussion with a surface $z = f(x, y)$ as shown in Fig. 8.17. If point P is a relative maximum (or minimum), then every tangent line through P must be parallel to the xy-plane. That is, $\dfrac{\partial z}{\partial x} = 0$ and $\dfrac{\partial z}{\partial y} = 0$.

Note: Both partial derivatives must equal zero. However, this result gives only critical points, that is, only possible relative maximum or minimum values. *A saddle point is a*

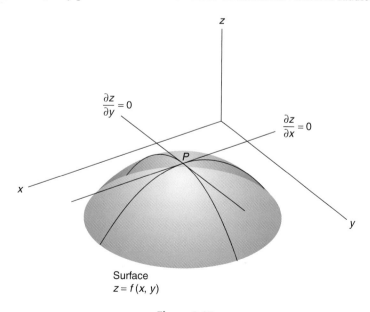

Figure 8.17

critical point that is neither a maximum nor a minimum. The name "saddle point" refers to the center point on a horse's saddle, which has this property. For a drawing of a saddle point, see the hyperbolic paraboloid in Fig. 8.12.

> **DISCRIMINANT**
>
> Let $f(x, y)$ be a function that has continuous second partial derivatives; then its discriminant $D(x, y)$ is given by $D(x, y) = f_{xx}(x, y)f_{yy}(x, y) - [f_{xy}(x, y)]^2$.

The Second Partials Test

Let $f(x, y)$ be a function that has continuous second partial derivatives and a critical point $(a, b, f(a, b))$:

1. If $D(a, b) < 0$, then the point $(a, b, f(a, b))$ is a saddle point (it's neither a maximum nor a minimum).
2. If $D(a, b) > 0$ and $f_{xx}(a, b) < 0$, then $(a, b, f(a, b))$ is a relative maximum point.
3. If $D(a, b) > 0$ and $f_{xx}(a, b) > 0$, then $(a, b, f(a, b))$ is a relative minimum point.
4. Nothing can be concluded if $D(a, b) = 0$.

Though this is a rather advanced theorem, just think of the discriminant's job as telling us whether or not our critical point corresponds to an extreme point on the surface. A *negative* discriminant means *no,* it's not a relative extreme (our critical point is a saddle point). A *positive* discriminant means *yes,* our critical point is either a relative maximum or minimum. To distinguish a maximum from a minimum, $f_{xx}(a, b)$ is evaluated to measure the concavity of the surface (using a trace parallel to the xz-plane). If $f_{xx}(a, b)$ is negative, the surface is opening downward, so our critical point would be a relative maximum. Similarly, if $f_{xx}(a, b)$ is positive, the surface is opening upward, so our critical point would be a relative minimum. As noted in Part (4), a zero discriminant does not allow any conclusion to be made. An interested student with access to a computer graphics utility may wish to investigate $f(x, y) = 1 + x^4 - y^4$, $g(x, y) = 1 - x^4 - y^4$, and $h(x, y) = 1 + x^4 + y^4$, which all have zero discriminants at the critical point $(0, 0, 1)$. This critical point turns out to be a saddle point of f, a relative maximum of g, and a relative minimum of h, so any of the three cases are possible with a zero discriminant.

EXAMPLE 5

Find any relative maximum or minimum points for $z = f(x, y) = 3x - x^2 - xy - y^2 - 2$.
 The first partial derivatives are

$$\frac{\partial z}{\partial x} = f_x(x, y) = 3 - 2x - y$$

$$\frac{\partial z}{\partial y} = f_y(x, y) = -x - 2y$$

Set each expression equal to zero and solve the resulting system of equations.

$$3 - 2x - y = 0$$
$$-x - 2y = 0$$

Multiply the second equation by 2 and add.

$$2x + y = 3$$
$$\underline{-2x - 4y = 0}$$
$$-3y = 3$$
$$y = -1 \quad \text{and} \quad x = 2$$

Then

$$z = f(2, -1) = 3(2) - (2)^2 - (2)(-1) - (-1)^2 - 2 = 1$$

Next we need to determine if the point $(2, -1, 1)$ is a maximum or a minimum.

$$f_{xx}(x, y) = -2$$
$$f_{yy}(x, y) = -2$$
$$f_{xy}(x, y) = -1 \quad \text{(note that quadric surfaces yield } constant \text{ second partials)}$$
$$D(x, y) = f_{xx}(x, y) f_{yy}(x, y) - [f_{xy}(x, y)]^2$$
$$D(x, y) = \quad (-2)(-2) \quad - [-1]^2 = 3$$

$D(2, -1) = 3 > 0$ and $f_{xx}(2, -1) = -2 < 0$, so $(2, -1, 1)$ is a relative maximum point. A graph of this surface can be obtained as follows:

MODE right arrow **5 ENTER**   **green diamond Y=** 3x-x^2-x*y-y^2-2   **ENTER   F1 9 (Format)** etc.

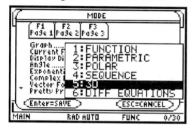
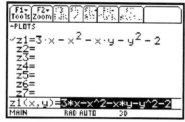
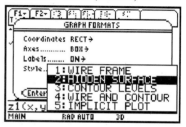

Set Axes to **BOX** and Style to **HIDDEN SURFACE**.

F2   6

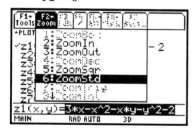
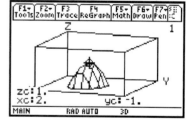

F3 (Trace) 2 ENTER -1 ENTER

the relative maximum point $(2, -1, 1)$

A professionally drawn sketch is shown in Fig. 8.18.

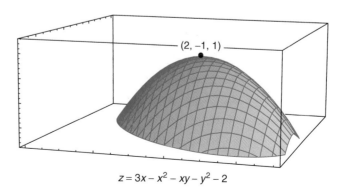

Figure 8.18 $z = 3x - x^2 - xy - y^2 - 2$

EXAMPLE 6

Find any relative maximum or minimum points for $z = f(x, y) = x^3 + y^3 - 3xy + 4$.

The first partial derivatives are

$$\frac{\partial z}{\partial x} = f_x(x, y) = 3x^2 - 3y$$

$$\frac{\partial z}{\partial y} = f_y(x, y) = 3y^2 - 3x$$

Set each expression equal to zero and solve the resulting system of equations.

$$3x^2 - 3y = 0$$
$$3y^2 - 3x = 0$$

The first equation implies that $y = x^2$. Substitute this into the second equation.

$$3(x^2)^2 - 3x = 0$$
$$3x^4 - 3x = 0$$
$$3x(x^3 - 1) = 0$$
$$3x = 0 \quad \text{or} \quad x^3 = 1$$
$$x = 0 \quad \text{or} \quad x = 1$$

If $x = 0$, $y = 0^2 = 0$ and $z = f(0, 0) = (0)^3 + (0)^3 - 3(0)(0) + 4 = 4$
If $x = 1$, $y = 1^2 = 1$ and $z = f(1, 1) = (1)^3 + (1)^3 - 3(1)(1) + 4 = 3$
So $(0, 0, 4)$ and $(1, 1, 3)$ are critical points.

The second partials are

$$f_{xx}(x, y) = 6x$$
$$f_{yy}(x, y) = 6y$$
$$f_{xy}(x, y) = -3$$
$$D(x, y) = f_{xx}(x, y)f_{yy}(x, y) - [f_{xy}(x, y)]^2$$
$$D(x, y) = (6x)(6y) - [-3]^2 = 36xy - 9$$

$D(0, 0) = 36(0)(0) - 9 = -9 < 0$, so $(0, 0, 4)$ is a saddle point.
$D(1, 1) = 36(1)(1) - 9 = 27 > 0$ and $f_{xx}(1, 1) = 6(1) = 6 > 0$, so $(1, 1, 3)$ is a relative minimum point.

The graph of this function has the topography of a mountain lake with a waterfall. The top of the waterfall is the saddle point and the bottom of the lake is the relative minimum point. For instructions on setting the TI-89 in **3D** mode, with **HIDDEN SURFACE** format style and **BOX** axes, see the previous example or Appendix Section D.16.

green diamond Y= x^3+y^3-3x*y+4 **ENTER green diamond WINDOW** -70 **ENTER** 66 **ENTER**, etc.

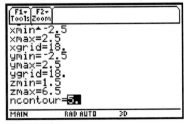

green diamond GRAPH **multiplication sign F3 (Trace)** 0 **ENTER** 0 **ENTER**   1 **ENTER** 1 **ENTER**

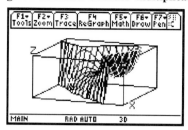 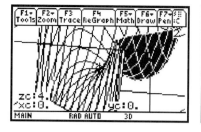 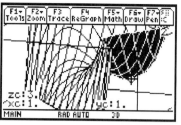

the saddle point $(0, 0, 4)$ the relative minimum point $(1, 1, 3)$

Two different views of professionally drawn sketches are shown in Fig. 8.19.

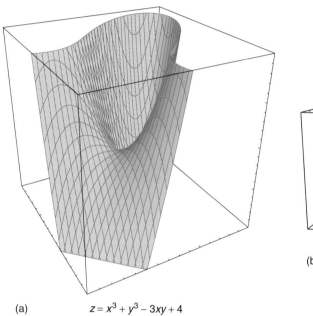

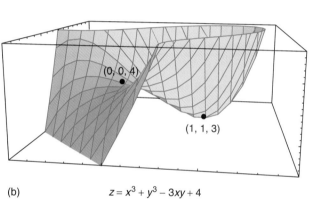

(b) $z = x^3 + y^3 - 3xy + 4$

(a) $z = x^3 + y^3 - 3xy + 4$

Figure 8.19

EXAMPLE 7

Find the dimensions of a rectangular box, with no top, having a volume of 32 m^3 and using the least amount of material.

The area or amount of the material is $A = lw + 2wh + 2lh$. The volume of the box is

$V = lwh = 32$ (see Fig. 8.20). Next, solve this volume equation for h, $h = \dfrac{32}{lw}$, and substitute into the area equation.

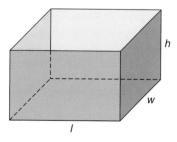

Figure 8.20

$$A = lw + 2w\left(\frac{32}{lw}\right) + 2l\left(\frac{32}{lw}\right)$$

$$A = lw + \frac{64}{l} + \frac{64}{w}$$

which expresses A as a function of two variables. Now find the partial derivatives.

$$\frac{\partial A}{\partial l} = w - \frac{64}{l^2} \quad \text{and} \quad \frac{\partial A}{\partial w} = l - \frac{64}{w^2}$$

Setting each partial derivative equal to zero, we have

$$w - \frac{64}{l^2} = 0 \quad \text{and} \quad l - \frac{64}{w^2} = 0$$

$$w = \frac{64}{l^2}$$

Substitute $w = 64/l^2$ into the second equation.

$$l - \frac{64}{w^2} = 0$$

$$l - \frac{64}{(64/l^2)^2} = 0$$

$$l - \frac{l^4}{64} = 0$$

$$l\left(1 - \frac{l^3}{64}\right) = 0$$

Then $l = 0$ or $l = 4$. (*Note:* $l = 0$ is impossible.) When $l = 4$,

$$w = \frac{64}{l^2} = \frac{64}{16} = 4$$

and

$$h = \frac{32}{lw} = \frac{32}{4 \cdot 4} = 2$$

So the dimensions of the box are 4 m × 4 m × 2 m.

Exercises 8.3

Find the total differential for each function.

1. $z = 3x^2 + 4xy + y^3$ **2.** $z = x^2y + 4x^2y^2 - 5x^3y^3$

3. $z = x^2 \cos y$ **4.** $z = \frac{y}{x} + \ln x$ **5.** $z = \frac{x - y}{xy}$

6. $z = \text{Arctan } xy$ **7.** $z = \ln \sqrt{1 + xy}$ **8.** $z = e^{xy} + \sin xy$

9. The sides of a box measure 26.00 cm × 26.00 cm × 12.00 cm with a maximum error of 0.15 cm on each side. Use a differential to approximate the change in volume.

10. A 6.00-in.-radius cylindrical rod is 2 ft long. Use a differential to approximate how much nickel (in in^3) is needed to coat the entire rod with a thickness of 0.12 in.

11. The total resistance R of two resistors R_1 and R_2, connected in parallel, is $R = \frac{R_1 R_2}{R_1 + R_2}$.
R_1 measures 400 Ω, with a maximum error of 25 Ω. R_2 measures 600 Ω with a maximum error of 50 Ω. Use a differential to approximate the change in R.

12. The angle of elevation to the top of a hill is 18.00°. The distance to the top of the hill is 450 ± 15 ft. If the maximum error in measuring the angle is 0.5°, find the maximum error in calculating the height of the hill.

13. The height of a right circular cone increases from 21.00 cm to 21.10 cm while the radius decreases from 12.00 cm to 11.85 cm. Use a total differential to approximate the change in volume.

14. The electric current in a circuit containing a variable resistor R is given by $i = 25(1 - e^{-Rt/25})$. Use a total differential to approximate the change in current as R changes from 8.00 Ω to 8.25 Ω and t changes from 4.0 s to 4.1 s.

Find any relative maximum or minimum points or saddle points for each function.

15. $z = 9 + 6x - 8y - 3x^2 - 2y^2$

16. $z = 4x^2 + 3y^2 - 16x - 24y + 5$

17. $z = \dfrac{1}{x} + \dfrac{1}{y} + xy$

18. $z = 60x + 60y - xy - x^2 - y^2$

19. $z = x^2 - y^2 - 2x - 4y - 4$

20. $z = x^2 + y^2 + 2x + 3y + 3$

21. $z = x^2 - y^2 - 6x + 4y$

22. $z = x^3 + y^3 + 3xy + 4$

23. $z = 4x^3 + y^2 - 12x^2 - 36x - 2y$

24. $z = x^3 + y^3 - 3x^2 - 9y^2 - 24x$

25. Find the dimensions of a rectangular box, with no top, having a volume of 500 cm^3 and using the least amount of material.

26. A rectangular box with no top and a volume of 6 ft^3 needs to be built from material that costs \$6/ft^2 for the bottom, \$2/ft^2 for the front and back, and \$1/ft^2 for the sides. Find the dimensions of the box in order to minimize the cost.

27. Find three positive numbers whose sum is 30 and whose product is a maximum.

28. Find three positive numbers whose sum is 30 and whose sum of squares is a minimum.

8.4 DOUBLE INTEGRALS

In Section 8.2 on partial differentiation, we showed how to differentiate functions of two variables by differentiating with respect to one variable at a time while holding the other variable constant. Next, we need to consider the inverse of partial differentiation of a function of two variables. That is, if $z = f(x, y)$, integrate by first holding either variable x or y constant and integrate with respect to the other. Then we integrate with respect to the variable first held constant. This is called *double integration* or a *double integral.*

> **DOUBLE INTEGRAL**
>
> $$\int_a^b \left[\int_{g(x)}^{G(x)} f(x, y)\, dy \right] dx = \int_a^b \int_{g(x)}^{G(x)} f(x, y)\, dy\, dx$$
>
> $$\int_c^d \left[\int_{h(y)}^{H(y)} f(x, y)\, dx \right] dy = \int_c^d \int_{h(y)}^{H(y)} f(x, y)\, dx\, dy$$

Note: The brackets are normally not included when writing a double integral.

To integrate the double integral, $\displaystyle\int_a^b \int_{g(x)}^{G(x)} f(x, y)\, dy\, dx$:

1. Integrate $f(x, y)$ with respect to y, holding x constant.

2. Evaluate this integral by substituting the limits on the inner or right integral. These limits are either numbers or functions of x.

3. Integrate the result from step 2 with respect to x.

4. Evaluate this integral by substituting the remaining limits (those on the outer or left integral), which are numerical values.

EXAMPLE 1

Evaluate $\displaystyle\int_0^1 \int_x^{3x^2} (6xy - x)\, dy\, dx.$

First, integrate with respect to y, holding x constant.

$$\int_0^1 \int_x^{3x^2} (6xy - x)\, dy\, dx = \int_0^1 (3xy^2 - xy)\Big|_x^{3x^2} dx$$

$$= \int_0^1 \{[3x(3x^2)^2 - x(3x^2)] - [3x(x)^2 - x(x)]\}\, dx$$

$$= \int_0^1 (27x^5 - 3x^3 - 3x^3 + x^2)\, dx$$

$$= \int_0^1 (27x^5 - 6x^3 + x^2)\, dx$$

$$= \left(\frac{9x^6}{2} - \frac{3x^4}{2} + \frac{x^3}{3}\right)\Big|_0^1$$

$$= \left(\frac{9}{2} - \frac{3}{2} + \frac{1}{3}\right) - (0)$$

$$= \frac{10}{3}$$

EXAMPLE 2

Evaluate $\displaystyle\int_0^4 \int_{-y}^{3y} \sqrt{x + y}\, dx\, dy.$

First, integrate with respect to x, holding y constant.

$$\int_0^4 \int_{-y}^{3y} \sqrt{x + y}\, dx\, dy = \int_0^4 \frac{2}{3}(x + y)^{3/2}\Big|_{-y}^{3y} dy$$

$$= \int_0^4 \left[\frac{2}{3}(3y + y)^{3/2} - \frac{2}{3}(-y + y)^{3/2}\right] dy$$

$$= \int_0^4 \frac{2}{3}(4y)^{3/2}\, dy$$

$$= \int_0^4 \frac{16}{3} y^{3/2}\, dy \qquad (4^{3/2} = 8)$$

$$= \frac{16}{3} \cdot \frac{y^{5/2}}{5/2}\Big|_0^4$$

$$= \frac{16}{3} \cdot \frac{2}{5}(4^{5/2} - 0)$$

$$= \frac{16}{3} \cdot \frac{2}{5} \cdot 32$$

$$= \frac{1024}{15}$$

Geometrically, a double integral may be interpreted as the volume under a surface. Let $z = f(x, y)$ be a continuous function whose surface is shown in Fig. 8.21. Let R be a region in the xy-plane bounded by $x = a$, $x = b$, $y = G(x)$, and $y = g(x)$.

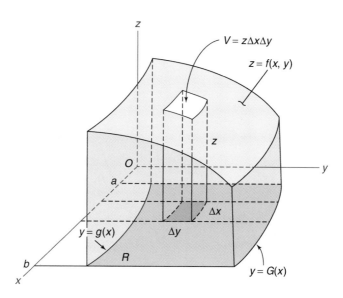

Figure 8.21 The volume of the solid is approximated by summing rectangular solids.

First, divide region R into n rectangles, each of which has dimensions Δx and Δy or dx and dy. Note that we have chosen to draw a typical rectangle. Next, construct a box with the rectangle as the base and z or $f(x, y)$ as its height. The volume of this typical volume element can be approximated by

$$z \, \Delta x \, \Delta y$$

Now let $n \to \infty$, where n is the number of such rectangles and the number of volume elements. Then we have

$$V = \int_a^b \int_{g(x)}^{G(x)} f(x, y) \, dy \, dx$$

VOLUME OF A SOLID

Let R be a region in the xy-plane bounded by $x = a$, $x = b$, $y = G(x)$, and $y = g(x)$. The volume of the solid between R and the surface $z = f(x, y)$ is

$$\int_a^b \int_{g(x)}^{G(x)} f(x, y) \, dy \, dx$$

Similarly, let R be a region in the xy-plane bounded by $y = c$, $y = d$, $x = H(y)$, and $x = h(y)$. The volume of the solid between R and the surface $z = f(x, y)$ is

$$\int_c^d \int_{h(y)}^{H(y)} f(x, y) \, dx \, dy$$

EXAMPLE 3

Find the volume of the solid bounded by the plane $2x + 3y + z = 6$ and the coordinate planes.
    First, make a sketch as in Fig. 8.22.

The region R is bounded in the xy-plane by $x = 0$, $y = 0$, and $2x + 3y = 6$ or $y = \dfrac{6 - 2x}{3}$.

Next, determine the variable and constant limits for the double integral.

constant limits for x: $\quad 0 \le x \le 3$

variable limits for y: $\quad 0 \le y \le \dfrac{6 - 2x}{3}$

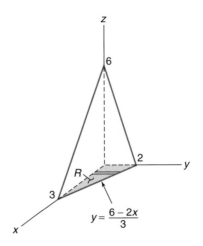

Figure 8.22

The volume is then

$$V = \int_0^3 \int_0^{(6-2x)/3} (6 - 2x - 3y) \, dy \, dx$$

$$= \int_0^3 \left(6y - 2xy - \frac{3y^2}{2} \right) \Big|_0^{(6-2x)/3} dx$$

$$= \int_0^3 \left\{ \left[6\left(\frac{6-2x}{3} \right) - 2x\left(\frac{6-2x}{3} \right) - \frac{3}{2}\left(\frac{6-2x}{3} \right)^2 \right] - (0) \right\} dx$$

$$= \int_0^3 \left(12 - 4x - 4x + \frac{4}{3}x^2 - 6 + 4x - \frac{2}{3}x^2 \right) dx$$

$$= \int_0^3 \left(6 - 4x + \frac{2}{3}x^2 \right) dx$$

$$= \left(6x - 2x^2 + \frac{2}{9}x^3 \right) \Big|_0^3$$

$$= (18 - 18 + 6) - (0)$$

$$= 6$$

Note: Since we have a pyramid, we can check this result using the formula

$$V = \frac{1}{3}Bh \qquad \text{(where } B \text{ is the area of the base and } h \text{ is the height of the pyramid)}$$

$$B = \frac{1}{2}(3)(2) = 3 \qquad \text{(area of triangular base)}$$

$$V = \frac{1}{3}(3)(6) = 6$$

What happens if you reverse the order of the integration in Example 3?

constant limits for y: $0 \le y \le 2$

variable limits for x: $0 \le x \le \dfrac{6 - 3y}{2}$

Then the double integral is

$$V = \int_0^2 \int_0^{(6-3y)/2} (6 - 2x - 3y)\, dx\, dy$$

Show that this double integral also gives 6 as the result.

Using a calculator,

2nd 7 2nd 7 6-2x-3y,y,0,(6-2x)/3),x,0,3) **ENTER 2nd 7 2nd 7** 6-2x-3y,x,0,(6-3y)/2),y,0,2) **ENTER**

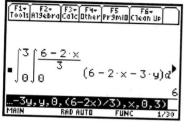 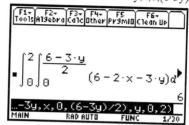

EXAMPLE 4

Find the volume of the solid inside the cylinder $x^2 + y^2 = 4$, below the plane $y = z$, and above the xy-plane.

Region R as in Fig. 8.23 is bounded in the xy-plane by $x^2 + y^2 = 4$ and $y = 0$.

constant limits for x: $-2 \le x \le 2$

variable limits for y: $0 \le y \le \sqrt{4 - x^2}$

Note that the height of a typical volume element is $z = y$.

The volume is then

$$V = \int_{-2}^2 \int_0^{\sqrt{4-x^2}} y\, dy\, dx$$

$$= \int_{-2}^2 \frac{y^2}{2} \Big|_0^{\sqrt{4-x^2}} dx$$

$$= \int_{-2}^2 \left(\frac{4 - x^2}{2} \right) dx$$

$$= \frac{1}{2} \left(4x - \frac{x^3}{3} \right) \Big|_{-2}^2$$

$$= \frac{1}{2}\left(8 - \frac{8}{3} \right) - \frac{1}{2}\left(-8 + \frac{8}{3} \right)$$

$$= \frac{8}{3} + \frac{8}{3} = \frac{16}{3}$$

Using a calculator,

2nd 7 2nd 7 y,y,0, **2nd** $\sqrt{}$ 4-x^2)),x,-2,2) **ENTER**

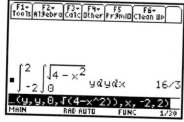

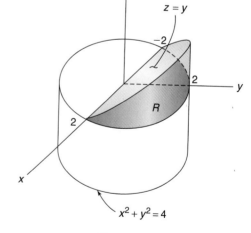

Figure 8.23

Exercises 8.4

Evaluate each double integral.

1. $\displaystyle\int_0^2 \int_0^{3x} (x + y)\, dy\, dx$

2. $\displaystyle\int_0^3 \int_0^{1-y} (x + y)\, dx\, dy$

3. $\displaystyle\int_0^1 \int_y^{2y} (3x^2 + xy)\, dx\, dy$

4. $\displaystyle\int_0^2 \int_0^x (4x^2 - 3xy^2 + 8y^3)\, dy\, dx$

5. $\displaystyle\int_{-1}^1 \int_x^{x^2} (4xy + 9x^2 + 6y)\, dy\, dx$

6. $\displaystyle\int_0^1 \int_{x^2}^{3x+1} x\, dy\, dx$

7. $\displaystyle\int_0^1 \int_0^{\sqrt{1-y^2}} (x + y)\, dx\, dy$

8. $\displaystyle\int_0^2 \int_0^{\sqrt{4-y^2}} \frac{2}{\sqrt{4 - y^2}}\, dx\, dy$

9. $\displaystyle\int_0^1 \int_0^x e^{x+y}\, dy\, dx$

10. $\displaystyle\int_0^1 \int_0^{x^3} e^{y/x}\, dy\, dx$

11. $\displaystyle\int_0^3 \int_0^{4y} \sqrt{y^2 + 16}\, dx\, dy$

12. $\displaystyle\int_0^1 \int_0^{4x} \frac{1}{x^2 + 1}\, dy\, dx$

13. $\displaystyle\int_0^{\pi/2} \int_0^x \cos x \sin y\, dy\, dx$

14. $\displaystyle\int_0^{\pi} \int_0^x x \sin y\, dy\, dx$

Find the volume of each solid bounded by the given surfaces.

15. $3x + y + 6z = 12$ and the coordinate planes

16. $3x + 4y + 2z = 12$ and the coordinate planes

17. $z = xy, y = x, x = 2$ (first octant)

18. $x^2 + y^2 = 4, z = x + y$ (first octant)

19. $x^2 + y^2 = 4, z = 2x + 3y$, and above the xy-plane

20. $x^2 + y^2 = 1$ and $x^2 + z^2 = 1$

CHAPTER SUMMARY

1. The graph of an equation expressed in three variables or in the form $z = f(x, y)$ is usually a surface.

2. The graph of an equation in the form

$$Ax + By + Cz + D = 0$$

is a plane.

3. A sphere with center at (h, k, l) and radius r is given by the equation

$$(x - h)^2 + (y - k)^2 + (z - l)^2 = r^2$$

4. The graph in three-space of an equation expressed in only two variables is a cylindrical surface. The surface is parallel to the coordinate axis of the missing variable.

5. The graph in three-space of a second-degree equation is a quadric surface. Plane sections, intersections, or slices of the given surface with planes are conics. The graphs, equations, and trace information of six basic quadric surfaces are given on pages 370 and 371.

6. The distance between two points $P_1(x_1, y_1, z_1)$ and $P_2(x_2, y_2, z_2)$ is given by

$$d = P_1 P_2 = \sqrt{(x_2 - x_1)^2 + (y_2 - y_1)^2 + (z_2 - z_1)^2}$$

7. *Partial derivatives:* If $z = f(x, y)$ is a function of two variables, the partial derivative of z with respect to x (y is treated as a constant) is defined by

$$\frac{\partial z}{\partial x} = \lim_{\Delta x \to 0} \frac{f(x + \Delta x, y) - f(x, y)}{\Delta x}$$

provided that this limit exists.

If $z = f(x, y)$ is a function of two variables, the partial derivative of z with respect to y (x is treated as a constant) is defined by

$$\frac{\partial z}{\partial y} = \lim_{\Delta y \to 0} \frac{f(x, y + \Delta y) - f(x, y)}{\Delta y}$$

provided that this limit exists.

8. *Total differential of $z = f(x, y)$:*

$$dz = \frac{\partial f}{\partial x} \, dx + \frac{\partial f}{\partial y} \, dy$$

9. To find a relative maximum or minimum of a function of two variables in the form $z = f(x, y)$, set both

$$\frac{\partial f}{\partial x} = 0 \quad \text{and} \quad \frac{\partial f}{\partial y} = 0$$

which will determine the critical points. Using the discriminant

$$D(x, y) = f_{xx}(x, y)f_{yy}(x, y) - [f_{xy}(x, y)]^2$$

and the Second Partials Test, check each critical point $(a, b, f(a, b))$ as follows:
(a) If $D(a, b) < 0$, then the point $(a, b, f(a, b))$ is a saddle point (it's neither a maximum nor a minimum).
(b) If $D(a, b) > 0$ and $f_{xx}(a, b) < 0$, then $(a, b, f(a, b))$ is a relative maximum point.
(c) If $D(a, b) > 0$ and $f_{xx}(a, b) > 0$, then $(a, b, f(a, b))$ is a relative minimum point.
(d) Nothing can be concluded if $D(a, b) = 0$.

10. To integrate the double integral $\displaystyle\int_a^b \int_{g(x)}^{G(x)} f(x, y) \, dy \, dx$:

(a) Integrate $f(x, y)$, with respect to y, holding x constant.
(b) Evaluate this integral by substituting the limits on the inner or right integral. These limits are either numbers or functions of x.
(c) Integrate the result from Step (b) with respect to x.
(d) Evaluate this integral by substituting the remaining limits (those on the outer or left integral), which are numerical values.

11. *Volume of a solid:* Let R be a region in the xy-plane bounded by $x = a$, $x = b$, $y = G(x)$, and $y = g(x)$. The volume of the solid between R and the surface $z = f(x, y)$ is

$$\int_a^b \int_{g(x)}^{G(x)} f(x, y) \, dy \, dx$$

Similarly, let R be a region in the xy-plane bounded by $y = c$, $y = d$, $x = H(y)$, and $x = h(y)$. The volume of the solid between R and the surface $z = f(x, y)$ is

$$\int_c^d \int_{h(y)}^{H(y)} f(x, y) \, dx \, dy$$

CHAPTER 8 REVIEW

Name and sketch the graph of each equation in three-space. Sketch the traces in the coordinate planes when appropriate.

1. $3x + 6y + 4z = 36$

2. $9x^2 + 9y^2 = 3z$

3. $36y^2 + 9z^2 - 16x^2 = 144$

4. $16x^2 + 9y^2 + 36z^2 = 144$

5. $x^2 = -12y$

6. $9x^2 - 9y^2 = 3z$

7. $9y^2 - 36x^2 - 16z^2 = 144$

8. $9x^2 + 36y^2 - 16z^2 = 0$

9. Find the equation of the sphere with the center at $(-3, 2, 1)$ and tangent to the yz-plane.

10. Find the distance between the points $(1, 2, 9)$ and $(4, -2, 4)$.

Find (a) $\dfrac{\partial z}{\partial x}$ and (b) $\dfrac{\partial z}{\partial y}$ for each function.

11. $z = x^3 + 3x^2y + 2y^2$

12. $z = 3x^2 e^{2y}$

13. $z = \ln(3x^2 y)$

14. $z = \sin 3x \sin 3y$

15. $z = \dfrac{y\, e^{x^2}}{x \ln y}$

16. $z = \dfrac{e^x \sin y}{y \sin x}$

Find the indicated partial derivative.

17. $v = \sqrt{\dfrac{P}{w}}; \dfrac{\partial v}{\partial w}$

18. $U = \dfrac{1}{2} LI^2; \dfrac{\partial U}{\partial I}$

19. $Z = \sqrt{R^2 + X_C^2}; \dfrac{\partial Z}{\partial X_C}$

20. $V = \dfrac{RE}{R + r}; \dfrac{\partial V}{\partial R}$

21. Find the slope of the tangent lines to $z = y^2 + xy + 3x^2 - 4x$ parallel to (a) the xz-plane and (b) the yz-plane and through the point $(2, 0, 4)$.

Find the total differential for each function.

22. $z = \ln \sqrt{x^2 + xy}$

23. $z = \dfrac{x - y}{x + y}$

24. A cylindrical tank has a radius of 2.000 m and a length of 10.000 m. Use a total differential to approximate the amount of paint (in litres) needed to paint one coat a thickness of 3 mm over its total exterior.

25. An electric current varies according to $i = 50(1 - e^{-t/50R})$. Use a total differential to approximate the change in current as R changes from 150 Ω to 160 Ω and t changes from 6.0 s to 6.5 s.

Find any relative maximum or minimum points or saddle points for each function.

26. $z = x^2 + 2xy - y^2 - 14x - 6y + 8$

27. $z = y^2 - xy - x^2 + 4x - 3y - 6$

28. $z = -x^2 + xy - y^2 + 4x - 8y + 9$

29. A rectangular box with no top needs to be built to hold 8 m³. The bottom costs twice as much as the material for the sides. Find the dimensions of the box in order to minimize the cost.

Evaluate each double integral.

30. $\displaystyle \int_0^1 \int_0^{x^2} (x - 2y)\, dy\, dx$

31. $\displaystyle \int_0^2 \int_y^{4y} (4xy + 6x^2 - 9y^2)\, dx\, dy$

32. $\displaystyle \int_1^3 \int_0^{\ln y} ye^x\, dx\, dy$

33. $\displaystyle \int_0^{\pi/4} \int_0^x \sec^2 y\, dy\, dx$

Find the volume of each solid bounded by the given surfaces.

34. $ax + by + cz = d, a > 0, b > 0, c > 0$, and the coordinate planes.

35. $x^2 + y^2 = 1, z = 2x$ (first octant)

36. $z = 1 - y - x^2$ (first octant)

9

Progressions and the Binomial Theorem

INTRODUCTION

The fact that the sum of an infinite number of numbers could be finite is somewhat startling, yet we use this concept when we consider a rational number written as a repeating decimal. For example

$$\frac{2}{3} = 0.666 \ldots$$

$$= \frac{6}{10} + \frac{6}{100} + \frac{6}{1000} + \cdots$$

The numbers $\frac{6}{10}, \frac{6}{100}, \frac{6}{1000}, \ldots$ form a progression or sequence, and their sum is called a series. In this chapter we study progressions and series as well as the binomial theorem.

Objectives

- Find the nth term of an arithmetic progression.
- Find the sum of the first n terms of an arithmetic series.
- Find the nth term of a geometric progression.
- Find the sum of the first n terms of a geometric series.
- Use the binomial theorem to expand a binomial raised to the nth power.
- Find a specified term of a binomial expansion.

9.1 ARITHMETIC PROGRESSIONS

Many technical applications make use of infinite series. One in particular, the Fourier series, is very useful in the field of electronics. An *infinite series* is the summation of an infinite sequence of numbers. We will first look, however, at sums of finite sequences.

A *finite sequence* consists of a succession of quantities $a_1, a_2, a_3, \ldots, a_n$, where the three dots represent the quantities or terms between a_3 and a_n. The term a_n is called the *n*th term or last term. An *infinite sequence* is a succession of quantities which continues indefinitely and may be written

$$a_1, a_2, a_3, \ldots, a_n, \ldots \quad \text{or} \quad a_1, a_2, a_3, \ldots$$

A basic example of an infinite sequence is an arithmetic progression. An *arithmetic progression* is a sequence of terms where each term differs from the immediately preceding term by a fixed number, d, which is called the *common difference* of the progression. For example, the sequence 5, 10, 15, 20, 25, 30, . . . is an arithmetic progression with a common difference of 5. The sequence $1, \frac{1}{3}, -\frac{1}{3}, -1, -\frac{5}{3}, \ldots$ is an arithmetic progression with a common difference of $-\frac{2}{3}$.

In any arithmetic progression if a is the first term and d is the common difference, then the second term would be $a + d$. Likewise, the third term would be $a + 2d$, and the fourth term would be $a + 3d$. Continuing in this manner we could express the arithmetic progression as follows:

$$a, a + d, a + 2d, a + 3d, \ldots$$

Note that the first n terms of an arithmetic progression, an example of a finite sequence, can be written as

$$a, a + d, a + 2d, a + 3d, \ldots, a + (n - 1)d$$

The *n*th or last term l of such a finite arithmetic progression is then given by

$$l = a + (n - 1)\,d$$

EXAMPLE 1

Find the 6th term of the arithmetic progression 5, 9, 13, ...

The common difference is the difference between *any* term and the preceding term, that is,

$$d = 9 - 5 = 13 - 9 = 4, \qquad a = 5 \quad \text{and} \quad n = 6$$

We then have

$$l = a + (n - 1)\,d$$
$$= 5 + (6 - 1)(4)$$
$$= 25$$

EXAMPLE 2

Find the 22nd term of the arithmetic progression $1, \frac{1}{3}, -\frac{1}{3}, \ldots$.

Since $d = \frac{1}{3} - 1 = -\frac{2}{3}, a = 1$, and $n = 22$, we have

$$l = a + (n - 1)d$$
$$= 1 + (22 - 1)(-\tfrac{2}{3})$$
$$= -13$$

To find the sum S_n of the first n terms of an arithmetic progression, note that the first n terms can be written as an expression involving l instead of a:

$$a, \ldots, l - 2d, l - d, l$$

In fact, we can indicate the sum S_n of these terms by

$$S_n = a + (a + d) + (a + 2d) + \cdots + (l - 2d) + (l - d) + l$$

or

$$S_n = l + (l - d) + (l - 2d) + \cdots + (a + 2d) + (a + d) + a$$

where the terms of the last equation are written in reverse order.

If we add these two equations for S_n, we have

$$2S_n = (a + l) + (a + l) + (a + l) + \cdots + (a + l) + (a + l) + (a + l)$$

since, term by term, the multiples of d add to zero. Note that we obtain a sum of n terms of the form $(a + l)$. That is,

$$2S_n = n(a + l)$$

or

$$S_n = \frac{n}{2}(a + l)$$

EXAMPLE 3

Find the sum of the first 12 terms of the arithmetic progression $6, 11, 16, \ldots$.
Since $d = 11 - 6 = 5$, $a = 6$, and $n = 12$, we have

$$l = a + (n - 1)d$$
$$l = 6 + (12 - 1)(5)$$
$$= 61$$
$$S_n = \frac{n}{2}(a + l)$$
$$= \frac{12}{2}(6 + 61)$$
$$= 402$$

EXAMPLE 4

Find the sum of the first 500 positive integers.
Since $a = 1$, $d = 1$, $n = 500$, and $l = 500$, we have

$$S_n = \frac{n}{2}(a + l)$$
$$= \frac{500}{2}(1 + 500)$$
$$= 125{,}250$$

Exercises 9.1

Find the nth term of each arithmetic progression.

1. $2, 5, 8, \ldots, n = 6$ **2.** $-3, -7, -11, \ldots, n = 7$ **3.** $3, 4\frac{1}{2}, 6, \ldots, n = 15$

4. $-2, \frac{1}{5}, 2\frac{2}{5}, \ldots, n = 8$ **5.** $4, -5, -14, \ldots, n = 12$ **6.** $10, 50, 90, \ldots, n = 9$

7–12. Find the sum of the first n terms of the progressions in Exercises 1–6.

Write the first five terms of each arithmetic progression whose first term is a and whose common difference is d.

13. $a = 2, d = -3$ **14.** $a = -4, d = 2$ **15.** $a = 5, d = \frac{2}{3}$ **16.** $a = 3, d = -\frac{1}{2}$

17. Find the first term of an arithmetic progression whose 10th term is 12 and whose sum of the first ten terms is 80.

18. Find the common difference of an arithmetic progression whose first term is 7 and whose 8th term is 16.

19. Find the sum of the first 1000 odd positive integers.

20. Find the sum of the first 500 even positive integers.

21. A man is employed at an initial salary of $24,000. If he receives an annual raise of $800, what is his salary for the tenth year?

22. Equipment purchased at an original value of $1360 is depreciated $120 per year for 10 years. Find the depreciated value after four years. Find the scrap value (depreciated value after 10 years).

9.2 GEOMETRIC PROGRESSIONS

A geometric progression is another example of a sequence. A *geometric progression* is a sequence of terms each of which can be obtained by multiplying the preceding term by a fixed number r, which is called the *common ratio*. For example, $1, \frac{1}{2}, \frac{1}{4}, \frac{1}{8}, \ldots$ is a geometric progression with a common ratio of $\frac{1}{2}$, and $-6, -18, -54, -162, \ldots$ is a geometric progression with a common ratio of 3.

For any geometric progression if a is the first term and r is the common ratio, then ar would be the second term. Likewise, $(ar)r = ar^2$ would be the third term, and $(ar^2)r = ar^3$ would be the fourth term. Continuing in this manner we could express a geometric progression as follows:

$$a, ar, ar^2, ar^3, \ldots, ar^{n-1}$$

where ar^{n-1} is the nth or last term, l, of the progression. That is,

$$l = ar^{n-1}$$

EXAMPLE 1

Find the 8th term of the geometric progression $1, \frac{1}{2}, \frac{1}{4}, \frac{1}{8}, \ldots$.

The common ratio is found by dividing *any* term by the preceding term. So

$$r = \frac{\frac{1}{2}}{1} = \frac{\frac{1}{4}}{\frac{1}{2}} = \frac{\frac{1}{8}}{\frac{1}{4}} = \frac{1}{2}$$

Since $a = 1$ and $n = 8$, we have

$$l = ar^{n-1}$$
$$= (1)(\tfrac{1}{2})^{8-1}$$
$$= \frac{1}{128}$$

EXAMPLE 2

Find the 10th term of the geometric progression $3, -6, 12, -24, \ldots$.

In this example $a = 3$, $r = -\dfrac{6}{3} = -2$, and $n = 10$.

$$l = ar^{n-1}$$
$$= (3)(-2)^{10-1}$$
$$= -1536$$

To find the sum S_n of the first n terms of the geometric progression

$$S_n = a + ar + ar^2 + ar^3 + \cdots + ar^{n-2} + ar^{n-1}$$

multiply each side of this equation by r and subtract as follows:

$$S_n = a + ar + ar^2 + ar^3 + \cdots + ar^{n-2} + ar^{n-1}$$

$$rS_n = \qquad ar + ar^2 + ar^3 + ar^4 + \cdots + ar^{n-1} + ar^n$$

$$\overline{S_n - rS_n = a \qquad\qquad\qquad\qquad\qquad\qquad - ar^n}$$

Solving for S_n, we have

$$S_n(1 - r) = a(1 - r^n)$$

or

$$S_n = \frac{a(1 - r^n)}{1 - r}$$

EXAMPLE 3

Find the sum of the first eight terms of the geometric progression $1, \frac{1}{2}, \frac{1}{4}, \frac{1}{8}, \ldots$.
Since $r = \frac{1}{2}$, $a = 1$, and $n = 8$, we have

$$S_n = \frac{a(1 - r^n)}{1 - r}$$

$$= \frac{(1)[1 - (\frac{1}{2})^8]}{1 - \frac{1}{2}}$$

$$= \frac{1 - \frac{1}{256}}{\frac{1}{2}} = \frac{255}{128}$$

EXAMPLE 4

Find the sum of the first five terms of the geometric progression $2, -\frac{2}{3}, \frac{2}{9}, -\frac{2}{27}, \ldots$.
In this example $a = 2$, $r = -\frac{1}{3}$, and $n = 5$.

$$S_n = \frac{a(1 - r^n)}{1 - r}$$

$$= \frac{(2)[1 - (-\frac{1}{3})^5]}{1 - (-\frac{1}{3})}$$

$$= \frac{(2)(1 + \frac{1}{243})}{\frac{4}{3}} = \frac{122}{81}$$

EXAMPLE 5

If \$3000 is deposited annually in a savings account at 8% interest compounded annually, find the total amount in this account after 4 years.

The total amount in the account is the sum of a geometric progression

$$(3000)(1.08) + (3000)(1.08)^2 + (3000)(1.08)^3 + (3000)(1.08)^4$$

since the value of each dollar in the account increases by 8% each year. Note that the first term $(3000)(1.08) = \$3240$ represents the amount in the account after one year. Thus $a = 3240$, $r = 1.08$, and $n = 4$.

$$S_n = \frac{a(1 - r^n)}{1 - r}$$

$$= \frac{(3240)(1 - 1.08^4)}{1 - 1.08}$$

$$= \$14,600$$

The term *series* is used to denote the sum of a sequence of terms. Each S_n is thus a finite series. The methods of computing the sums S_n of finite arithmetic and geometric series have already been shown.

An infinite series is the indicated sum of an infinite sequence of terms. For example, $1 + \frac{1}{2} + \frac{1}{4} + \frac{1}{8} + \cdots$ is an infinite series. Since it is the infinite summation of the terms of a geometric sequence, it is called an *infinite geometric series.*

In Example 3 we found that the sum of the first eight terms of the geometric progression $1, \frac{1}{2}, \frac{1}{4}, \frac{1}{8}, \ldots$ is $\frac{255}{128}$. The sum of the first nine terms can be shown to be $\frac{511}{256}$ and the sum of the first ten terms is $\frac{1023}{512}$. This last sum is close to the value 2. In fact, the sum of the first 50 terms is given by

$$S_n = \frac{(1)[1 - (\frac{1}{2})^{50}]}{1 - \frac{1}{2}} = \frac{1 - (\frac{1}{2})^{50}}{\frac{1}{2}} = 2[1 - (\frac{1}{2})^{50}]$$

But since $\left(\frac{1}{2}\right)^{50} = \dfrac{1}{1,125,899,906,842,624}$, which is practically zero, we conclude that the sum S_n is very close to the value 2. The sum S_n gets closer and closer to 2 as n is given a larger and larger value.

If we denote the sum of the first n terms of a geometric progression by S_n and S as the sum of the terms of an infinite geometric progression, then

$$S = 1 + \frac{1}{2} + \frac{1}{4} + \frac{1}{8} + \cdots = 2$$

Not every infinite series has a finite sum. For example, $3 + 6 + 12 + 24 + \cdots$ has no finite sum. When the sum exists, the series is said to *converge.* When the limit or sum does not exist, the series is said to *diverge.*

In general, if $|r| < 1$, the infinite geometric series

$$a + ar + ar^2 + \cdots + ar^{n-1} + \cdots$$

has sum

$$S = \frac{a}{1 - r}$$

If $|r| \geq 1$, the infinite geometric series has no sum. (It diverges.)

EXAMPLE 6

Find the sum of the infinite geometric series $3 + \frac{3}{5} + \frac{3}{25} + \cdots + 3(\frac{1}{5})^{n-1} + \cdots$
Since $r = \frac{1}{5}$ and $a = 3$, we have

$$S = \frac{a}{1 - r}$$

$$= \frac{3}{1 - \frac{1}{5}} = \frac{15}{4}$$

EXAMPLE 7

Find, if possible, the sum of the infinite geometric series

$$1 + 2 + (2)^2 + (2)^3 + \cdots + (2)^{n-1} + \cdots$$

Since $r = 2 \geq 1$, this infinite geometric series diverges.

EXAMPLE 8

Find a fraction that is equivalent to the decimal $0.232323\ldots.$

We can write this decimal as the infinite series

$$0.23 + 0.0023 + 0.000023 + \cdots$$

Then $a = 0.23$ and $r = 0.01$. Thus

$$S = \frac{a}{1 - r} = \frac{0.23}{1 - 0.01} = \frac{0.23}{0.99} = \frac{23}{99}$$

Note that any repeating decimal can be expressed as a fraction. The fractional form can be found by using the method shown in Example 8.

Exercises 9.2

Find the nth term of each geometric progression.

1. $20, \frac{20}{3}, \frac{20}{9}, \ldots, n = 8$ **2.** $\frac{1}{8}, -\frac{1}{4}, \frac{1}{2}, \ldots, n = 7$

3. $\sqrt{2}, 2, 2\sqrt{2}, \ldots, n = 6$ **4.** $6, 3, \frac{3}{2}, \ldots, n = 8$

5. $8, -4, 2, \ldots, n = 10$ **6.** $3, 12, 48, \ldots, n = 5$

7–12. Find the sum of the first n terms of the progressions in Exercises 1–6.

Write the first five terms of each geometric progression whose first term is a and whose common ratio is r.

13. $a = 3, r = \frac{1}{2}$ **14.** $a = -6, r = \frac{1}{3}$ **15.** $a = 5, r = -\frac{1}{4}$

16. $a = 2, r = -\frac{3}{2}$ **17.** $a = -4, r = 3$ **18.** $a = -5, r = -2$

19. Find the common ratio of a geometric progression whose first term is 6 and whose fourth term is $\frac{3}{4}$.

20. Find the first term of a geometric progression with common ratio $\frac{1}{3}$ if the sum of the first three terms is 13.

21. If \$1000 is deposited annually in an account at 10% interest compounded annually, find the total amount in the account after 10 years.

22. If \$200 is deposited quarterly in an account at 9% annual interest compounded quarterly, find the total amount in the account after 10 years.

23. A ball is dropped from a height of 12 ft. After each bounce, it rebounds to $\frac{1}{2}$ the height of the previous height from which it fell. Find the distance the ball rises after the fifth bounce.

24. The half-life of a chlorine isotope, ^{38}Cl, used in radioisotope therapy is 37 min. This means that half of a given amount will disintegrate in 37 min. This means also that three-fourths will have disintegrated after 74 min. Find how much will have disintegrated in 148 min.

25. A salt solution is being cooled so that the temperature decreases 20% each minute. Find the temperature of the solution after 8 min if the original temperature was 90°C.

26. A tank contains 400 gal of acid. Then 100 gal is drained out and refilled with water. Then 100 gal of the mixture is drained out and refilled with water. Assuming that this process continues, how much acid remains in the tank after five 100-gal units are drained out?

Find the sum, when possible, for each infinite geometric series.

27. $4 + \frac{4}{7} + \frac{4}{49} + \cdots + 4(\frac{1}{7})^{n-1} + \cdots$ **28.** $6 + \frac{6}{11} + \frac{6}{121} + \cdots + 6(\frac{1}{11})^{n-1} + \cdots$

29. $3 - \frac{3}{8} + \frac{3}{64} - \cdots + 3(-\frac{1}{8})^{n-1} + \cdots$ **30.** $1 - \frac{1}{9} + \frac{1}{81} - \cdots + (-\frac{1}{9})^{n-1} + \cdots$

31. $4 + 12 + 36 + \cdots + 4(3)^{n-1} + \cdots$ **32.** $-5 - \frac{5}{2} - \frac{5}{4} - \cdots - 5(\frac{1}{2})^{n-1} - \cdots$

33. $3 + 3.1 + 3.01 + \cdots + [3 + (0.1)^{n-1}] + \cdots$ **34.** $2 + 2 + 2 + \cdots + 2(1)^{n-1} + \cdots$

Find the fraction that is equivalent to each decimal.

35. $0.3333\ldots$ **36.** $0.135135135\ldots$ **37.** $0.0121212\ldots$

38. $0.6252525\ldots$ **39.** $0.86666\ldots$ **40.** $0.365365365\ldots$

9.3   THE BINOMIAL THEOREM

The *binomial theorem* provides us with a convenient means of expressing any power of a binomial as a sum of terms.

For small nonnegative integers n, we can find $(a + b)^n$ by actual multiplication.

$$n = 0: \quad (a + b)^0 = 1$$
$$n = 1: \quad (a + b)^1 = a + b$$
$$n = 2: \quad (a + b)^2 = a^2 + 2ab + b^2$$
$$n = 3: \quad (a + b)^3 = a^3 + 3a^2b + 3ab^2 + b^3$$
$$n = 4: \quad (a + b)^4 = a^4 + 4a^3b + 6a^2b^2 + 4ab^3 + b^4$$
$$n = 5: \quad (a + b)^5 = a^5 + 5a^4b + 10a^3b^2 + 10a^2b^3 + 5ab^4 + b^5$$

We could continue this process, but the multiplications become more complicated for larger values of n.

No matter what positive integral value of n is chosen, the following results:

1. $(a + b)^n$ has $n + 1$ terms.

2. The first term is a^n.

3. The second term is $na^{n-1}b$

4. The exponent of a decreases by 1 and the exponent of b increases by 1 for each successive term.

5. In each term, the sum of the exponents of a and b is n.

6. The last term is b^n.

The kth term is given by the formula

$$\frac{n(n - 1)(n - 2)\cdots(n - k + 2)}{(k - 1)!}a^{n-k+1}b^{k-1}$$

where $k!$ (k factorial) indicates the product of the first k positive integers. (For example, $3! = 3 \cdot 2 \cdot 1 = 6$; $5! = 5 \cdot 4 \cdot 3 \cdot 2 \cdot 1 = 120$; and $6! = 6 \cdot 5 \cdot 4 \cdot 3 \cdot 2 \cdot 1 = 720$.) The three dots in the numerator indicate that the multiplication of decreasing numbers is to continue until the number $n - k + 2$ is reached. For example, if $n = 8$ and $k = 4$, then the formula gives us

$$\frac{8 \cdot 7 \cdot 6}{3 \cdot 2 \cdot 1} = 56$$

BINOMIAL THEOREM

$$(a + b)^n = a^n + na^{n-1}b + \frac{n(n - 1)}{2!}a^{n-2}b^2 + \frac{n(n - 1)(n - 2)}{3!}a^{n-3}b^3 + \cdots$$
$$+ \frac{n(n - 1)(n - 2)\cdots(n - k + 2)}{(k - 1)!}a^{n-k+1}b^{k-1} + \cdots + b^n$$

where the three dots indicate that you are to complete the process of calculating the terms. The expression for the kth term is also given.

EXAMPLE 1

Expand $(x + 4y)^5$ by using the binomial theorem.
Let $a = x$, $b = 4y$, and $n = 5$.

$$(x + 4y)^5 = x^5 + 5x^{5-1}(4y) + \frac{5(5-1)}{2!}x^{5-2}(4y)^2 + \frac{5(5-1)(5-2)}{3!}x^{5-3}(4y)^3$$
$$+ \frac{5(5-1)(5-2)(5-3)}{4!}x^{5-4}(4y)^4 + (4y)^5$$
$$= x^5 + 20x^4y + 160x^3y^2 + 640x^2y^3 + 1280xy^4 + 1024y^5$$

For small values of n, it is possible to determine the coefficients of each term of the expansion by the use of *Pascal's triangle* as shown below:

$$
\begin{array}{ccccccccccc}
n = 0: & & & & & 1 & & & & & \\
n = 1: & & & & 1 & & 1 & & & & \\
n = 2: & & & 1 & & 2 & & 1 & & & \\
n = 3: & & 1 & & 3 & & 3 & & 1 & & \\
n = 4: & 1 & & 4 & & 6 & & 4 & & 1 & \\
n = 5: 1 & & 5 & & 10 & & 10 & & 5 & & 1 \\
\end{array}
$$

Observe the similarity of this triangle with the triangular format shown earlier when expanding $(a + b)^n$ for $n = 0, 1, 2, 3, 4,$ and 5. Each row gives the coefficients for all terms of the binomial expansion for a given integer n. Each successive row provides the coefficients for the next integer n. Each row is read from left to right. The first and last coefficients are always 1 as observed in the triangle. Beginning with the third row ($n = 2$), coefficients of terms other than the first and last are found by adding together the two nearest coefficients found in the row above. For example, the coefficient of the fourth term for the expansion with $n = 5$ is 10, which is the sum of 6 and 4. The numbers 6 and 4 appear just above 10 in Pascal's triangle.

We can enlarge Pascal's triangle to obtain a row for any desired integer n. However, this is not very practical for very large values of n.

EXAMPLE 2

Using Pascal's triangle, find the coefficients of the terms of the binomial expansion for $n = 7$.
We need two more rows of the triangle:

$$
\begin{array}{cccccccccccccc}
n = 6: & & 1 & & 6 & & 15 & & 20 & & 15 & & 6 & & 1 \\
n = 7: & 1 & & 7 & & 21 & & 35 & & 35 & & 21 & & 7 & & 1 \\
\end{array}
$$

The last row provides the desired coefficients.

EXAMPLE 3

Expand $(2m + k)^7$ using the binomial theorem.
Let $a = 2m$, $b = k$, and $n = 7$.

$$(2m + k)^7 = 1(2m)^7 + 7(2m)^{7-1}k^1 + 21(2m)^{7-2}k^2 + 35(2m)^{7-3}k^3$$
$$+ 35(2m)^{7-4}k^4 + 21(2m)^{7-5}k^5 + 7(2m)^{7-6}k^6 + (1)k^7$$
$$= 128m^7 + 448m^6k + 672m^5k^2 + 560m^4k^3$$
$$+ 280m^3k^4 + 84m^2k^5 + 14mk^6 + k^7$$

EXAMPLE 4

Find the 7th term of $(x^3 - 2y)^{10}$.

First, note that $k = 7$, $n = 10$, $a = x^2$, and $b = -2y$.

$$k\text{th term} = \frac{n(n-1)(n-2)\cdots(n-k+2)}{(k-1)!}a^{n-k+1}b^{k-1}$$

$$7\text{th term} = \frac{10 \cdot 9 \cdot 8 \cdot 7 \cdot 6 \cdot 5}{6!}(x^3)^4(-2y)^6$$

$$= 210(x^{12})(64y^6)$$

$$= 13{,}440x^{12}y^6$$

Exercises 9.3

Expand each binomial using the binomial theorem.

1. $(3x + y)^3$
2. $(2x - 3y)^4$
3. $(a - 2)^5$
4. $(5x + 1)^4$
5. $(2x - 1)^4$
6. $(1 + x)^7$
7. $(2a + 3b)^6$
8. $(a - 2b)^8$
9. $(\frac{2}{3}x - 2)^5$
10. $(\frac{3}{4}m + \frac{2}{3}k)^5$
11. $(a^{1/2} + 3b^2)^4$
12. $(x^{1/2} - y^{1/2})^4$
13. $\left(\dfrac{x}{y} - \dfrac{2}{z}\right)^4$
14. $\left(\dfrac{2x}{y} + \dfrac{3}{z}\right)^6$

Find the indicated term of each binomial expansion.

15. $(x - y)^9$; 6th term
16. $(4x + 2y)^5$; 3rd term
17. $(2x - y)^{13}$; 9th term
18. $(x^{1/2} + 2)^{10}$; 8th term
19. $(2x + y^2)^7$; 5th term
20. $(x^2 - y^3)^8$; middle term
21. $(3x + 2y)^6$; middle term
22. $(x - 2y)^{10}$; middle term
23. $(2x - 1)^{10}$; term containing x^5
24. $(x^2 + 1)^8$; term containing x^6

CHAPTER SUMMARY

1. An arithmetic progression is a sequence of terms where each term differs from the immediately preceding term by a fixed number, d, which is called the common difference. The general form of an arithmetic progression is written

$$a, a + d, a + 2d, a + 3d, \ldots, a + (n-1)d$$

(a) The nth or last term l of such a finite arithmetic progression is given by

$$l = a + (n-1)d$$

(b) The sum of the first n terms of a finite arithmetic progression is given by

$$S_n = \frac{n}{2}(a + l)$$

2. A geometric progression is a sequence of terms each of which can be obtained by multiplying the preceding term by a fixed number, r, which is called the common ratio. The general form of a geometric progression is given by

$$a, ar, ar^2, ar^3, \ldots, ar^{n-1}$$

(a) The nth or last term l of such a finite geometric progression is given by

$$l = ar^{n-1}$$

(b) The sum of the first n terms of a finite geometric progression is given by

$$S_n = \frac{a(1 - r^n)}{1 - r}$$

(c) The sum of the infinite geometric series, where $|r| < 1$, is

$$S = \frac{a}{1 - r}$$

3. $k!$ (k factorial) $= k(k - 1)(k - 2) \cdots 4 \cdot 3 \cdot 2 \cdot 1$

$$5! = 5 \cdot 4 \cdot 3 \cdot 2 \cdot 1$$

$$10! = 10 \cdot 9 \cdot 8 \cdot 7 \cdot 6 \cdot 5 \cdot 4 \cdot 3 \cdot 2 \cdot 1$$

4. *Binomial theorem*

$$(a + b)^n = a^n + na^{n-1}b + \frac{n(n - 1)}{2!}a^{n-2}b^2 + \frac{n(n - 1)(n - 2)}{3!}a^{n-3}b^3 + \cdots$$

$$+ \frac{n(n - 1)(n - 2) \cdots (n - k + 2)}{(k - 1)!}a^{n-k+1}b^{k-1} + \cdots + b^n$$

5. The kth term of a binomial expansion is given by the formula

$$\frac{n(n - 1)(n - 2) \cdots (n - k + 2)}{(k - 1)!}a^{n-k+1}b^{k-1}$$

CHAPTER 9 REVIEW

Find the nth term of each progression.

1. $3, 7, 11, 15, \ldots, n = 12$

2. $4, 2, 1, \frac{1}{2}, \ldots, n = 7$

3. $\sqrt{3}, -3, 3\sqrt{3}, -9, \ldots, n = 8$

4. $4, -2, -8, -14, \ldots, n = 12$

5. $6, 2, \frac{2}{3}, \frac{2}{9}, \ldots, n = 6$

6. $5, 15, 25, 35, \ldots, n = 10$

7–12. Find the sum of the first n terms of the progressions in Exercises 1–6.

13. Find the sum of the first 1000 even positive integers.

14. If $500 is deposited annually in a savings account at 6% interest compounded annually, find the total amount in the account after five years.

Find the sum, when possible, for each infinite geometric series.

15. $3 + 6 + 12 + \cdots$

16. $5 + \frac{5}{7} + \frac{5}{49} + \cdots$

17. $2 - \frac{2}{3} + \frac{2}{9} - \cdots$

18. $3 + \frac{9}{2} + \frac{27}{4} + \cdots$

19. Find the fraction equivalent to $0.454545 \ldots$.

20. Find the fraction equivalent to $0.9212121 \ldots$.

Expand each binomial using the binomial theorem.

21. $(a - b)^6$ **22.** $(2x^2 - 1)^5$ **23.** $(2x + 3y)^4$ **24.** $(1 + x)^8$

Find the indicated term of each binomial expansion.

25. $(1 - 3x)^5$; 3rd term **26.** $(a + 4b)^6$; 4th term

27. $(x + 2b^2)^{10}$; middle term **28.** $(3x^2 - 1)^{12}$; term containing x^{16}

10
Series

INTRODUCTION

The functions e^x, sin x, cos x, ln x, and many others may be written as polynomials with an infinite number of terms called *power series*. When you use your calculator to evaluate one of these functions, the calculator is evaluating a polynomial with a finite number of terms to give an approximation of the functional value you have indicated. In this chapter we study the conditions needed for writing a power series.

Objectives

- Use the Σ notation.
- Know the convergence conditions for a *p*-series.
- Use the comparison test for convergence.
- Use the limit comparison test for convergence.
- Use the ratio test for convergence.
- Use the integral test for convergence.
- Use the alternating series test for convergence.
- Know the definitions for absolute and conditional convergence.
- Find the interval of convergence of a series.
- Find a Maclaurin series expansion for a given function.
- Find a Taylor series expansion for a given function for a given value of *a*.
- Find a Fourier series that represents a given wave function.

10.1 SERIES AND CONVERGENCE

We now begin a more general study of series. *Sigma notation* is used for writing series. The Greek letter Σ (sigma) is used to indicate that the given expression is a sum. The term following Σ represents the general form of each term. For example,

$$\sum_{n=1}^{6} n^2 = 1^2 + 2^2 + 3^2 + 4^2 + 5^2 + 6^2$$

where the general form of each term is n^2. The numbers below and above Σ must be integers and indicate the values of *n* to be used for the first and last terms, respectively, of the series. The other terms are found by replacing n by the consecutive integers between 1 and 6. The finite sum

$$1 + 2 + 3 + \cdots + n$$

can be represented by $\sum_{k=1}^{n} k$, while the infinite sum

$$\frac{2}{3} + \frac{2}{9} + \frac{2}{27} + \cdots$$

can be represented by $\sum_{n=1}^{\infty} 2\left(\frac{1}{3}\right)^n$

The following example illustrates the use of sigma notation.

EXAMPLE 1

Σ notation	Expanded form of sum
(a) $\sum_{k=1}^{6} (2k)$	$2 + 4 + 6 + 8 + 10 + 12$
(b) $\sum_{k=5}^{9} (3k - 4)$	$11 + 14 + 17 + 20 + 23$
(c) $\sum_{k=1}^{n} (k - 2)$	$-1 + 0 + 1 + 2 + 3 + \cdots + (n - 2)$
(d) $\sum_{k=0}^{n} 2^k$	$1 + 2 + 2^2 + 2^3 + \cdots + 2^n$
(e) $\sum_{k=1}^{n} A_k$	$A_1 + A_2 + A_3 + \cdots + A_n$

EXAMPLE 2

Write the expanded form of $\sum_{n=1}^{5} \frac{4}{2n + 1}$.

$$\sum_{n=1}^{5} \frac{4}{2n + 1} = \frac{4}{2 \cdot 1 + 1} + \frac{4}{2 \cdot 2 + 1} + \frac{4}{2 \cdot 3 + 1} + \frac{4}{2 \cdot 4 + 1} + \frac{4}{2 \cdot 5 + 1}$$

$$= \frac{4}{3} + \frac{4}{5} + \frac{4}{7} + \frac{4}{9} + \frac{4}{11}$$

Using a calculator,

F3   4

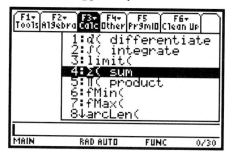

4/(2 **alpha** N+1), **alpha** N ,1,5) **ENTER**

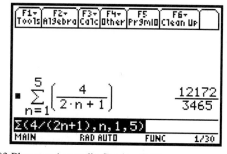

For calculator examples using the TI-83 Plus, see Appendix Section C.12.

EXAMPLE 3

Write the sum $\dfrac{6}{9} + \dfrac{6}{16} + \dfrac{6}{25} + \cdots + \dfrac{6}{121}$ using sigma notation.

Build the general form of each term. Note that 6 appears in the numerator of each term while the denominator changes in a pattern of perfect squares beginning with 3^2. Do you see that this series has nine terms?

$$\sum_{n=3}^{11} \frac{6}{n^2} \quad \text{or} \quad \sum_{n=1}^{9} \frac{6}{(n+2)^2}$$

are acceptable forms.

Next, we need to study convergence and divergence of infinite series. To begin this process, consider the following series:

DEFINITION OF CONVERGENCE AND DIVERGENCE

$$\sum_{n=1}^{\infty} a_n = a_1 + a_2 + a_3 + \cdots$$

Then

$$S_1 = a_1$$
$$S_2 = a_1 + a_2$$
$$S_3 = a_1 + a_2 + a_3$$
$$\vdots$$
$$S_n = a_1 + a_2 + a_3 + \cdots + a_n$$
$$\vdots$$

where S_1 is the sum of the first term, S_2 is the sum of the first two terms, S_3 is the sum of the first three terms, $\ldots$, S_n is the sum of the first n terms (sometimes called the *n*th partial sum) . . . .

(a) If $\lim\limits_{n \to \infty} S_n = S$ (where S is finite), the series $\sum\limits_{n=1}^{\infty} a_n$ *converges,* and S is the *sum of the infinite series.*

(b) If $\lim\limits_{n \to \infty} S_n$ does not exist, the series $\sum\limits_{n=1}^{\infty} a_n$ *diverges.*

The geometric series $1 + \dfrac{1}{2} + \dfrac{1}{4} + \dfrac{1}{8} + \cdots$ from Section 9.2 converges because $\lim\limits_{n \to \infty} S_n = 2$. The arithmetic series $1 + 2 + 3 + 4 + \cdots$ diverges because $\lim\limits_{n \to \infty} S_n = \infty$.

In the remainder of this section and Sections 10.2 to 10.4, we seek answers to the following two questions regarding infinite series:

1. Does a given series converge or diverge?

2. If the series converges, to what value does it converge?

The answers to these questions are not always easy, especially the second one. There is no general method for finding the sum of a convergent infinite series. We will begin the study of these two questions with a simple test for divergence.

> ### nth TERM TEST FOR DIVERGENCE OF A SERIES
>
> If $\lim_{n \to \infty} a_n \neq 0$, then the series $\sum_{n=1}^{\infty} a_n$ diverges.

In other words, it is impossible for an infinite series to converge if the terms you are adding fail to get extremely small in absolute values as $n \to \infty$. Please note that this test will help you to identify some, but not all, divergent series (many infinite series diverge for a completely different reason).

EXAMPLE 4

Show that each series diverges by using the nth term test for divergence.

(a) $\sum_{n=1}^{\infty} 2^n$ (b) $\sum_{n=1}^{\infty} \dfrac{n}{n+1}$ (c) $1 - 1 + 1 - 1 + 1 - 1 + \cdots$

The series in Parts (a) and (b) diverge because

(a) $\lim_{n \to \infty} 2^n = \infty$ and (b) $\lim_{n \to \infty} \dfrac{n}{n+1} = \lim_{n \to \infty} \dfrac{1}{1 + \dfrac{1}{n}} = 1.$

The series in Part (c) also diverges because $\lim_{n \to \infty} a_n$ clearly is not zero.

Note of warning: If $\lim_{n \to \infty} a_n = 0$, there is no guarantee that the series converges. For example, the series

$$\sum_{n=1}^{\infty} \frac{1}{n} = 1 + \frac{1}{2} + \frac{1}{3} + \frac{1}{4} + \cdots \quad \text{(called the harmonic series)}$$

diverges even though $\lim_{n \to \infty} \dfrac{1}{n} = 0.$

We now present some tests to determine whether a given series of positive terms converges or diverges. The next special series that we need to consider is the p-series. We can show that the conditions for convergence and divergence of the p-series are the following:

> ### CONVERGENCE AND DIVERGENCE OF A p-SERIES
>
> Any series in the form
>
> $$\sum_{n=1}^{\infty} \frac{1}{n^p} = \frac{1}{1^p} + \frac{1}{2^p} + \frac{1}{3^p} + \cdots$$
>
> where p is a real number, is called a p-series. The p-series
>
> **(a)** coverges for $p > 1$, and
>
> **(b)** diverges for $p \leq 1$.

EXAMPLE 5

Determine whether each p-series converges or diverges.

(a) $1 + \dfrac{1}{2^3} + \dfrac{1}{3^3} + \dfrac{1}{4^3} + \cdots$ (b) $1 + \dfrac{1}{2} + \dfrac{1}{3} + \dfrac{1}{4} + \cdots$

(c) $1 + \dfrac{1}{\sqrt{2}} + \dfrac{1}{\sqrt{3}} + \cdots$

(a) This p-series $(p = 3)$ converges.
(b) This p-series $(p = 1)$ diverges. Recall that this series is called the *harmonic series*.
(c) This p-series $(p = 1/2)$ diverges.

The next test for convergence is called the *comparison test*.

COMPARISON TEST FOR CONVERGENCE AND DIVERGENCE

Let N be a positive integer, $\displaystyle\sum_{n=1}^{\infty} a_n$ and $\displaystyle\sum_{n=1}^{\infty} b_n$ be series of positive terms, and $0 \le a_n \le b_n$ for all $n > N$; then

1. If $\displaystyle\sum_{n=1}^{\infty} b_n$ converges, then $\displaystyle\sum_{n=1}^{\infty} a_n$ also converges.

2. If $\displaystyle\sum_{n=1}^{\infty} a_n$ diverges, then $\displaystyle\sum_{n=1}^{\infty} b_n$ also diverges.

In other words, the comparison test says

1. A series of positive terms that is term by term smaller than a known convergent series must also converge.

2. A series of positive terms that is term by term larger than a known divergent series must also diverge.

EXAMPLE 6

Use the comparison test to determine whether each series converges or diverges.

(a) $\displaystyle\sum_{n=1}^{\infty} \dfrac{1}{2^n + 1}$ (b) $\displaystyle\sum_{n=2}^{\infty} \dfrac{1}{\sqrt{n} - 1}$

(a) We know that the geometric series $\displaystyle\sum_{n=1}^{\infty} \dfrac{1}{2^n}$ converges, and

$$\dfrac{1}{2^n + 1} \le \dfrac{1}{2^n} \quad \text{for} \quad n \ge 1$$

Then by the comparison test, $\displaystyle\sum_{n=1}^{\infty} \dfrac{1}{2^n + 1}$ also converges.

(b) We know that the p-series $\displaystyle\sum_{n=2}^{\infty} \dfrac{1}{\sqrt{n}} \; (p = \tfrac{1}{2})$ diverges, and

$$\dfrac{1}{\sqrt{n} - 1} \ge \dfrac{1}{\sqrt{n}} \quad \text{for } n \ge 2$$

Then by the comparison test, $\displaystyle\sum_{n=2}^{\infty} \dfrac{1}{\sqrt{n} - 1}$ also diverges.

A test that is often easier to apply than the comparison test is the *limit comparison test*.

First, a definition:

Let $\sum\limits_{n=1}^{\infty} a_n$ and $\sum\limits_{n=1}^{\infty} b_n$ be two series of positive terms.

(a) Then $\sum\limits_{n=1}^{\infty} a_n$ and $\sum\limits_{n=1}^{\infty} b_n$ have the *same order of magnitude* if $\lim\limits_{n \to \infty} \dfrac{a_n}{b_n} = L$, where L is a real number and $L \neq 0$.

(b) The series $\sum\limits_{n=1}^{\infty} a_n$ has a *lesser order of magnitude* than $\sum\limits_{n=1}^{\infty} b_n$ if $\lim\limits_{n \to \infty} \dfrac{a_n}{b_n} = 0$.

(c) The series $\sum\limits_{n=1}^{\infty} a_n$ has a *greater order of magnitude* than $\sum\limits_{n=1}^{\infty} b_n$ if $\lim\limits_{n \to \infty} \dfrac{a_n}{b_n} = \infty$.

EXAMPLE 7

Compare the orders of magnitude of each pair of series.

(a) $\sum\limits_{n=1}^{\infty} (2n)$ and $\sum\limits_{n=1}^{\infty} (n - 4)$

(b) $\sum\limits_{n=1}^{\infty} (3n)$ and $\sum\limits_{n=1}^{\infty} (n^2 + 1)$

(c) $\sum\limits_{n=1}^{\infty} \left(\dfrac{1}{n}\right)$ and $\sum\limits_{n=1}^{\infty} \dfrac{1}{n^2 + 1}$

(a) $\lim\limits_{n \to \infty} \dfrac{2n}{n - 4} = \lim\limits_{n \to \infty} \dfrac{2}{1 - \dfrac{4}{n}} = 2$. These two series have the same order of magnitude.

(b) $\lim\limits_{n \to \infty} \dfrac{3n}{n^2 + 1} = \lim\limits_{n \to \infty} \dfrac{3}{n + \dfrac{1}{n}} = 0$. So $\sum\limits_{n=1}^{\infty} (3n)$ has a lesser order of magnitude than

$\sum\limits_{n=1}^{\infty} (n^2 + 1)$.

(c) $\lim\limits_{n \to \infty} \dfrac{\dfrac{1}{n}}{\dfrac{1}{n^2 + 1}} = \lim\limits_{n \to \infty} \dfrac{n^2 + 1}{n} = \lim\limits_{n \to \infty} \left(n + \dfrac{1}{n}\right) = \infty$. So $\sum\limits_{n=1}^{\infty} \left(\dfrac{1}{n}\right)$ has a greater order of

magnitude than $\sum\limits_{n=1}^{\infty} \dfrac{1}{n^2 + 1}$.

We can now give the limit comparison test:

LIMIT COMPARISON TEST

Let $\sum\limits_{n=1}^{\infty} a_n$ and $\sum\limits_{n=1}^{\infty} b_n$ be series of positive terms.

(a) If both series have the same order of magnitude, then either both series converge or both series diverge.

(b) If the series $\sum\limits_{n=1}^{\infty} a_n$ has a lesser order of magnitude than $\sum\limits_{n=1}^{\infty} b_n$ and $\sum\limits_{n=1}^{\infty} b_n$ is known to converge, then $\sum\limits_{n=1}^{\infty} a_n$ also converges.

(c) If $\sum\limits_{n=1}^{\infty} a_n$ has a greater order of magnitude than $\sum\limits_{n=1}^{\infty} b_n$ and $\sum\limits_{n=1}^{\infty} b_n$ is known to diverge, then $\sum\limits_{n=1}^{\infty} a_n$ also diverges.

EXAMPLE 8

Use the limit comparison test to determine whether each series converges or diverges.

(a) $\displaystyle\sum_{n=1}^{\infty} \frac{1}{n(3n+1)}$ (b) $\displaystyle\sum_{n=1}^{\infty} \frac{\ln n}{n}$

(a) Let's compare with $\displaystyle\sum_{n=1}^{\infty} \frac{1}{n^2}$ (p-series, $p = 2$), which converges.

$$\lim_{n\to\infty} \frac{\dfrac{1}{n(3n+1)}}{\dfrac{1}{n^2}} = \lim_{n\to\infty} \frac{n^2}{3n^2+n} = \lim_{n\to\infty} \frac{1}{3+\dfrac{1}{n}} = \frac{1}{3}$$

Since $\displaystyle\sum_{n=1}^{\infty} \frac{1}{n^2}$ converges and both series have the same order of magnitude, the series $\displaystyle\sum_{n=1}^{\infty} \frac{1}{n(3n+1)}$ also converges.

(b) Let's compare with $\displaystyle\sum_{n=1}^{\infty} \frac{1}{n}$ (harmonic series), which diverges.

$$\lim_{n\to\infty} \frac{\dfrac{\ln n}{n}}{\dfrac{1}{n}} = \lim_{n\to\infty} \ln n = \infty$$

Since $\displaystyle\sum_{n=1}^{\infty} \frac{1}{n}$ diverges and the series $\displaystyle\sum_{n=1}^{\infty} \frac{\ln n}{n}$ has a greater order of magnitude, the series $\displaystyle\sum_{n=1}^{\infty} \frac{\ln n}{n}$ also diverges.

Exercises 10.1

Write the expanded form of each series.

1. $\displaystyle\sum_{n=1}^{6} (4n+1)$ **2.** $\displaystyle\sum_{n=1}^{5} (1-n^2)$ **3.** $\displaystyle\sum_{n=3}^{8} (n^2+1)$

4. $\displaystyle\sum_{n=1}^{6} (n^2-4n)$ **5.** $\displaystyle\sum_{k=1}^{n} \frac{k^2}{k+1}$ **6.** $\displaystyle\sum_{k=1}^{n} \frac{4}{k(k+1)}$

7. $\displaystyle\sum_{n=1}^{\infty} (-1)^n \frac{1}{n^2}$ **8.** $\displaystyle\sum_{n=1}^{\infty} (-1)^{n+1} \frac{n}{n+1}$

Write each sum using sigma notation.

9. $1 + 2 + 3 + \cdots + 12$

10. $1 + 3 + 5 + \cdots + 43$

11. $2 + 4 + 6 + \cdots + 100$

12. $1^3 + 2^3 + 3^3 + \cdots + (n-1)^3$

13. $1 + 3 + 5 + \cdots + (2n-1)$

14. $\dfrac{1}{2} + \dfrac{1}{4} + \dfrac{1}{8} + \cdots + \dfrac{1}{2^n}$

15. $10 + 17 + 26 + 37 + \cdots + (n^2+1)$

16. $\dfrac{1}{2} + \dfrac{2}{3} + \dfrac{3}{4} + \cdots + \dfrac{n+1}{n+2}$

Determine whether each series converges or diverges.

17. $3 + 9 + 27 + \cdots + 3^n + \cdots$

18. $1 + 2 + 3 + \cdots + n + \cdots$

19. $\displaystyle\sum_{n=2}^{\infty} \frac{2n}{n-1}$

20. $\displaystyle\sum_{n=1}^{\infty} \frac{3n+1}{2n-1}$

21. $1 + \dfrac{1}{\sqrt[4]{2}} + \dfrac{1}{\sqrt[4]{3}} + \dfrac{1}{\sqrt[4]{4}} + \cdots + \dfrac{1}{\sqrt[4]{n}} + \cdots$

22. $1 + \dfrac{1}{2^4} + \dfrac{1}{3^4} + \dfrac{1}{4^4} + \cdots + \dfrac{1}{n^4} + \cdots$

23. $\displaystyle\sum_{n=1}^{\infty} \dfrac{1}{n^2}$ **24.** $\displaystyle\sum_{n=1}^{\infty} \dfrac{1}{\sqrt[3]{n}}$ **25.** $\displaystyle\sum_{n=1}^{\infty} \dfrac{1}{(n+1)^2}$ **26.** $\displaystyle\sum_{n=4}^{\infty} \dfrac{1}{n-3}$

27. $\dfrac{1}{1 \cdot 2} + \dfrac{1}{2 \cdot 3} + \dfrac{1}{3 \cdot 4} + \cdots + \dfrac{1}{n(n+1)} + \cdots$

28. $1 + \dfrac{1}{3} + \dfrac{1}{5} + \cdots + \dfrac{1}{2n-1} + \cdots$ **29.** $\dfrac{1}{2} + \dfrac{1}{4} + \dfrac{1}{6} + \dfrac{1}{8} + \cdots + \dfrac{1}{2n} + \cdots$

30. $\displaystyle\sum_{n=1}^{\infty} \dfrac{1}{n^2 + 3n}$ **31.** $\displaystyle\sum_{n=1}^{\infty} \dfrac{1}{(2n-1)^2}$ **32.** $\displaystyle\sum_{n=1}^{\infty} \dfrac{1}{n^2 + 1}$

33. $\displaystyle\sum_{n=1}^{\infty} \dfrac{1}{\sqrt{n^2 + 1}}$ **34.** $\displaystyle\sum_{n=2}^{\infty} \dfrac{1}{n\sqrt{n^2 - 1}}$ **35.** $\displaystyle\sum_{n=1}^{\infty} \dfrac{1}{\sqrt{n}(n+1)}$

36. $\displaystyle\sum_{n=1}^{\infty} \dfrac{1}{\sqrt[3]{n^2 + 1}}$ **37.** $\displaystyle\sum_{n=1}^{\infty} \dfrac{1}{2^n + 2n}$ **38.** $\displaystyle\sum_{n=3}^{\infty} \dfrac{1}{n^2 - 4}$

39. $\displaystyle\sum_{n=2}^{\infty} \dfrac{1}{\ln n}$ **40.** $\displaystyle\sum_{n=3}^{\infty} \dfrac{n^2}{n^2 - 4}$ **41.** $\displaystyle\sum_{n=1}^{\infty} \dfrac{1 + \sin n\pi}{n^2}$

42. $\displaystyle\sum_{n=1}^{\infty} \dfrac{3n}{(n+1)(n+2)}$ **43.** $\displaystyle\sum_{n=1}^{\infty} \dfrac{1}{\sqrt{n(n+1)}}$ **44.** $\displaystyle\sum_{n=1}^{\infty} \dfrac{1}{n^n}$

10.2   RATIO AND INTEGRAL TESTS

The tests for convergence and divergence in Section 10.1 work for some series but not others. As a result, we show two additional tests in this section.

> ### RATIO TEST FOR CONVERGENCE AND DIVERGENCE
>
> Let $\displaystyle\sum_{n=1}^{\infty} a_n$ be a series of positive terms and
>
> $$r = \lim_{n \to \infty} \dfrac{a_{n+1}}{a_n}$$
>
> **(a)** If $r < 1$, the series converges.
> **(b)** If $r > 1$ (including $r = \infty$), the series diverges.
> **(c)** If $r = 1$, the test fails. Some other test must be used.

EXAMPLE 1

Determine whether the series $\displaystyle\sum_{n=1}^{\infty} \dfrac{2n}{3^n}$ converges or diverges.

$$r = \lim_{n \to \infty} \dfrac{a_{n+1}}{a_n} = \lim_{n \to \infty} \dfrac{\dfrac{2(n+1)}{3^{n+1}}}{\dfrac{2n}{3^n}} = \lim_{n \to \infty} \dfrac{3^n \, 2(n+1)}{3^{n+1}(2n)} = \lim_{n \to \infty} \dfrac{n+1}{3n}$$

$$= \lim_{n \to \infty} \dfrac{1}{3}\left(1 + \dfrac{1}{n}\right) = \dfrac{1}{3} < 1$$

Since $r < 1$, the given series converges.

A calculator may actually show the sum.

F3 4 2 alpha N /3^ alpha N , alpha N ,1, green diamond ∞) ENTER

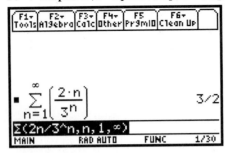

The ratio test is especially helpful in testing expressions involving factorials. Recall that

$$5! = 5 \cdot 4 \cdot 3 \cdot 2 \cdot 1$$
$$10! = 10 \cdot 9 \cdot 8 \cdot 7 \cdot 6 \cdot 5 \cdot 4 \cdot 3 \cdot 2 \cdot 1$$
$$n! = n(n - 1)(n - 2) \cdots 4 \cdot 3 \cdot 2 \cdot 1 \qquad (n \text{ must be a positive integer})$$

EXAMPLE 2

Determine whether the series $\displaystyle\sum_{n=1}^{\infty} \frac{2^n}{n!}$ converges or diverges.

$$r = \lim_{n \to \infty} \frac{a_{n+1}}{a_n} = \lim_{n \to \infty} \frac{\dfrac{2^{n+1}}{(n+1)!}}{\dfrac{2^n}{n!}} = \lim_{n \to \infty} \frac{n! \, 2^{n+1}}{(n+1)! \, 2^n}$$

$$= \lim_{n \to \infty} \frac{2}{n+1} = 0 < 1$$

Since $r < 1$, the given series converges.

EXAMPLE 3

Determine whether the series $\displaystyle\sum_{n=1}^{\infty} \frac{2^n}{n^2}$ converges or diverges.

$$r = \lim_{n \to \infty} \frac{a_{n+1}}{a_n} = \lim_{n \to \infty} \frac{\dfrac{2^{n+1}}{(n+1)^2}}{\dfrac{2^n}{n^2}} = \lim_{n \to \infty} \frac{n^2 \, 2^{n+1}}{(n+1)^2 \, 2^n}$$

$$= \lim_{n \to \infty} \frac{1(2)}{\left(1 + \dfrac{1}{n}\right)^2} = 2 > 1$$

Since $r > 1$, the given series diverges.

INTEGRAL TEST FOR CONVERGENCE AND DIVERGENCE

Let $\displaystyle\sum_{n=1}^{\infty} a_n$ be a series of positive terms and $f(x)$ be a continuous, decreasing function for $x \geq 1$ such that $f(n) = a_n$ for all positive integers n. Then $\displaystyle\sum_{n=1}^{\infty} a_n$ and $\displaystyle\int_{1}^{\infty} f(x)\, dx$ both converge or they both diverge.

Note: Before using the integral test, you must be certain that the function $f(x)$ is decreasing and continuous.

EXAMPLE 4

Determine whether the series $\displaystyle\sum_{n=2}^{\infty} \frac{1}{n \ln n}$ converges or diverges.

Note: This series begins with $n = 2$ because $\ln 1 = 0$.

Let $f(x) = \dfrac{1}{x \ln x}$ for $x \geq 2$; $f(x)$ is continuous and decreasing for $x \geq 2$.

$$\int_2^{\infty} f(x)\, dx = \int_2^{\infty} \frac{dx}{x \ln x} = \lim_{b \to \infty} \int_2^b \frac{dx}{x \ln x}$$

First, integrate $\displaystyle\int \frac{dx}{x \ln x}$.

$$\int \frac{dx}{x \ln x} = \int \frac{du}{u}$$
$$= \ln u$$
$$= \ln \ln x \quad \text{or} \quad \ln(\ln x)$$

$$\boxed{\begin{aligned} u &= \ln x \\ du &= \frac{1}{x} dx \end{aligned}}$$

Then

$$\lim_{b \to \infty} \int_2^b \frac{dx}{x \ln x} = \lim_{b \to \infty} \ln \ln x \Big|_2^b = \lim_{b \to \infty} (\ln \ln b - \ln \ln 2) = \infty$$

Since $\displaystyle\int_2^{\infty} f(x)\, dx$ diverges, the given series also diverges.

EXAMPLE 5

Determine whether the series $\displaystyle\sum_{n=1}^{\infty} \frac{n}{e^n}$ converges or diverges.

Let $f(x) = xe^{-x}$ for $x \geq 1$; $f(x)$ is continuous and decreasing for $x \geq 1$.

$$\int_1^{\infty} f(x)\, dx = \int_1^{\infty} xe^{-x}\, dx = \lim_{b \to \infty} \int_1^b xe^{-x}\, dx$$

Integrate $\displaystyle\int xe^{-x}\, dx$ using integration by parts.

$$\int u\, dv \qquad\qquad \int v\, du$$

$$\boxed{\begin{aligned} u &= x \\ dv &= e^{-x}\, dx \end{aligned}} \quad \longrightarrow \quad \boxed{\begin{aligned} du &= dx \\ v &= -e^{-x} \end{aligned}}$$

$$\int u\, dv = uv - \int v\, du$$
$$\int xe^{-x}\, dx = -xe^{-x} - \int (-e^{-x})\, dx$$
$$= -xe^{-x} - e^{-x}$$
$$= e^{-x}(-x - 1)$$

Then

$$\lim_{b \to \infty} \int_1^b xe^{-x}\, dx = \lim_{b \to \infty} \left[e^{-x}(-x - 1) \right]\Big|_1^b$$

$$= \lim_{b \to \infty} \left[e^{-b}(-b - 1) - e^{-1}(-1 - 1) \right]$$

$$= \lim_{b \to \infty} \left[\frac{-b - 1}{e^b} + \frac{2}{e} \right] = 0 + \frac{2}{e} = \frac{2}{e}$$

Since $\int_1^\infty f(x) \, dx$ converges (has a finite value), the given series also converges.

The integral test can be used to find the values of p for which the p-series converges. Let's begin by considering the p-series

$$\sum_{n=1}^\infty \frac{1}{n^p} \qquad \text{for } p > 0$$

Note: For $p \le 0$, $\lim_{n \to \infty} \frac{1}{n^p} \ne 0$, so the p-series diverges for $p \le 0$.

Let $f(x) = \frac{1}{x^p}$ for $p > 0$; $f(x)$ is continuous and decreasing for $x > 0$. We need two cases: For $p \ne 1$,

$$\int_1^\infty \frac{dx}{x^p} = \lim_{b \to \infty} \int_1^b x^{-p} \, dx$$

$$= \lim_{b \to \infty} \frac{x^{1-p}}{1 - p} \bigg|_1^b$$

$$= \lim_{b \to \infty} \left(\frac{b^{1-p}}{1 - p} - \frac{1}{1 - p} \right) = \begin{cases} \dfrac{1}{p - 1} & \text{if } p > 1 \\ \infty & \text{if } p < 1 \end{cases}$$

For $p = 1$,

$$\int_1^\infty \frac{dx}{x} = \lim_{b \to \infty} \int_1^b \frac{dx}{x} = \lim_{b \to \infty} \ln x \bigg|_1^b = \lim_{b \to \infty} (\ln b - \ln 1) = \infty$$

So the p-series converges for $p > 1$ and diverges for $p \le 1$.

Exercises 10.2

Use either the ratio test or the integral test to determine whether each series converges or diverges.

1. $\sum_{n=1}^\infty \dfrac{n + 1}{n \cdot 3^n}$ 2. $\sum_{n=1}^\infty \dfrac{2^{n+1}}{3^{n-1}}$ 3. $\sum_{n=1}^\infty \dfrac{1}{n!}$ 4. $\sum_{n=1}^\infty \dfrac{n + 2}{n!}$

5. $\sum_{n=1}^\infty \dfrac{n^2}{n!}$ 6. $\sum_{n=1}^\infty \dfrac{n^2}{2^n}$ 7. $\sum_{n=1}^\infty \dfrac{3^n}{n \cdot 2^n}$ 8. $\sum_{n=1}^\infty \dfrac{n!}{10^n}$

9. $\sum_{n=1}^\infty \dfrac{2n + 3}{2^n}$ 10. $\sum_{n=1}^\infty \dfrac{2^n}{n^2 + 1}$ 11. $\sum_{n=1}^\infty \dfrac{1}{2n + 1}$ 12. $\sum_{n=1}^\infty \dfrac{1}{n\sqrt{n}}$

13. $\sum_{n=2}^\infty \dfrac{1}{n\sqrt{\ln n}}$ 14. $\sum_{n=1}^\infty \dfrac{1}{\sqrt[3]{n}}$

15. $1 + \dfrac{1}{3} + \dfrac{1}{5} + \dfrac{1}{7} + \cdots$ 16. $\dfrac{1}{2} + \dfrac{1}{4} + \dfrac{1}{6} + \dfrac{1}{8} + \cdots$

17. $\sum_{n=1}^\infty \dfrac{n}{n^2 + 1}$ 18. $\sum_{n=2}^\infty \dfrac{\ln n}{n}$ 19. $\sum_{n=1}^\infty \dfrac{n^2}{e^n}$ 20. $\sum_{n=1}^\infty \dfrac{n}{\sqrt{n^2 + 1}}$

10.3 ALTERNATING SERIES AND CONDITIONAL CONVERGENCE

Up to now we have considered only series with all positive terms. An *alternating series* is a series whose terms are alternately positive and negative. Examples of alternating series are

$$\sum_{n=1}^{\infty} (-1)^{n+1} 2^n = 2 - 4 + 8 - 16 + \cdots$$

$$\sum_{n=1}^{\infty} (-1)^n \frac{3}{n} = -3 + \frac{3}{2} - \frac{3}{3} + \frac{3}{4} - \frac{3}{5} + \cdots$$

$$\sum_{n=1}^{\infty} (-1)^{n+1} a_n = a_1 - a_2 + a_3 - a_4 + \cdots \qquad (a_n > 0 \text{ for each } n)$$

We have a relatively simple test for convergence of alternating series.

ALTERNATING SERIES TEST

The alternating series

$$\sum_{n=1}^{\infty} (-1)^{n+1} a_n = a_1 - a_2 + a_3 - a_4 + \cdots \qquad (a_n > 0 \text{ for each } n)$$

converges provided that both of the following conditions are fulfilled:

(a) $0 < a_{n+1} \le a_n$ for $n \ge 1$ and

(b) $\lim_{n \to \infty} a_n = 0$

In addition, if S is the sum of the infinite series and S_n is the nth partial sum, then

$$|S - S_n| \le a_{n+1}$$

Note: Condition (a) guarantees that the terms decrease in absolute value as n increases.

EXAMPLE 1

Determine whether the alternating series $\sum_{n=1}^{\infty} \frac{(-1)^{n+1}}{n} = 1 - \frac{1}{2} + \frac{1}{3} - \frac{1}{4} + \cdots$ converges or diverges.

Since $a_{n+1} = \frac{1}{n+1} < \frac{1}{n} = a_n$ and $\lim_{n \to \infty} \frac{1}{n} = 0$, this alternating series converges.

We have already shown that the harmonic series

$$\sum_{n=1}^{\infty} \frac{1}{n} = 1 + \frac{1}{2} + \frac{1}{3} + \frac{1}{4} + \cdots$$

diverges, even though $a_{n+1} < a_n$ and $\lim_{n \to \infty} a_n = 0$. The alternating series in Example 1 is sometimes called the *alternating harmonic series*.

EXAMPLE 2

Determine whether the series $\sum_{n=2}^{\infty} \frac{(-1)^{n+1}}{n \ln n}$ converges or diverges.

Since

$$a_{n+1} = \frac{1}{(n+1)\ln(n+1)} < \frac{1}{n\ln n} = a_n \quad \text{and} \quad \lim_{n\to\infty}\frac{1}{n\ln n} = 0$$

this alternating series converges.

We have shown that the series of positive terms $\sum_{n=2}^{\infty} \frac{1}{n\ln n}$ diverges in Example 4 of Section 10.2.

ABSOLUTE AND CONDITIONAL CONVERGENCE

Suppose that $\sum_{n=1}^{\infty} a_n$ converges.

1. If $\sum_{n=1}^{\infty} |a_n|$ converges, then $\sum_{n=1}^{\infty} a_n$ *converges absolutely.*

2. If $\sum_{n=1}^{\infty} |a_n|$ diverges, then $\sum_{n=1}^{\infty} a_n$ *converges conditionally.*

For example, the series $\sum_{n=1}^{\infty} (-1)^{n+1}\frac{1}{n^2}$ converges absolutely because $\sum_{n=1}^{\infty} \frac{1}{n^2}$ converges and the series $\sum_{n=1}^{\infty} (-1)^{n+1}\frac{1}{n}$ converges conditionally because $\sum_{n=1}^{\infty} \frac{1}{n}$ diverges.

Exercises 10.3

Determine whether each alternating series converges or diverges. If it converges, find whether it converges absolutely or converges conditionally.

1. $\sum_{n=1}^{\infty} (-1)^{n+1}\frac{1}{2n+1}$

2. $\sum_{n=1}^{\infty} (-1)^{n+1}\frac{1}{2^n}$

3. $\sum_{n=1}^{\infty} (-1)^n\frac{1}{(2n)^2}$

4. $\sum_{n=1}^{\infty} (-1)^n\frac{n}{2n-1}$

5. $\sum_{n=1}^{\infty} (-1)^{n+1}\frac{2n}{2n-1}$

6. $\sum_{n=1}^{\infty} (-1)^{n-1}\frac{1}{n^2}$

7. $\sum_{n=2}^{\infty} \frac{(-1)^{n-1}}{\ln n}$

8. $\sum_{n=1}^{\infty} (-1)^{n+1}\frac{1}{\sqrt{n}}$

9. $\sum_{n=1}^{\infty} (-1)^n\frac{n^2}{2^n}$

10. $\sum_{n=1}^{\infty} (-1)^n\frac{1}{n!}$

11. $\sum_{n=1}^{\infty} (-1)^{n+1}\frac{n^2}{n^2+1}$

12. $\sum_{n=1}^{\infty} (-1)^{n+1}\frac{n}{e^n}$

13. $\sum_{n=1}^{\infty} (-1)^{n+1}\frac{n!}{3^n}$

14. $\sum_{n=1}^{\infty} (-1)^{n+1}\frac{n}{2^n}$

15. $\sum_{n=1}^{\infty} (-1)^n\frac{2n+1}{n^2}$

16. $\sum_{n=1}^{\infty} (-1)^n\frac{1}{\sqrt{n^2+1}}$

17. $\sum_{n=2}^{\infty} (-1)^{n+1}\frac{n}{\ln n}$

18. $\sum_{n=1}^{\infty} (-1)^{n+1}\frac{\sin \pi n}{n}$

19. $\sum_{n=1}^{\infty} (-1)^n\frac{\cos n}{n^2}$

20. $\sum_{n=1}^{\infty} (-1)^n\frac{n+1}{n\sqrt{n}}$

10.4 POWER SERIES

A more general example of an infinite series is the power series. A *power series* is an infinite series in the form

$$\sum_{n=0}^{\infty} a_n x^n = a_0 + a_1 x + a_2 x^2 + \cdots + a_n x^n + \cdots$$

Certain functions can be written as a power series. For example, the function $f(x) = \dfrac{1}{1 + x}$ can be expressed as a power series as follows:

$$
\begin{array}{r}
1 - x + x^2 - x^3 + \cdots \\
1 + x \overline{)\, 1 } \\
\underline{1 + x} \\
-x \\
\underline{-x - x^2} \\
x^2 \\
\underline{x^2 + x^3} \\
-x^3
\end{array}
$$

Thus

$$
f(x) = \frac{1}{1 + x} = 1 - x + x^2 - x^3 + \cdots
$$

Note however that the infinite division process as indicated is valid only for $|x| < 1$, because the right-hand side is a geometric series (Section 9.2) with $a = 1$ and $r = -x$, which has the sum $\dfrac{1}{1 - (-x)}$ only for $|x| < 1$. Thus, the equality is not valid for $|x| \geq 1$.

A more general series in the form

$$
\sum_{n=0}^{\infty} a_n(x - a)^n = a_0 + a_1(x - a) + a_2(x - a)^2 + a_3(x - a)^3 + \cdots + a_n(x - a)^n + \cdots
$$

is called a *power series centered at a.*

For each particular value of x (x is the variable in the power series), we have an infinite series of constants, which either converges or diverges. In many power series, the series converges for some values of x and diverges for other values of x. The ratio test (when it works) is usually the simplest test to use on a power series.

Let

$$
\lim_{n \to \infty} \left| \frac{u_{n+1}}{u_n} \right| = r(x)
$$

where u_{n+1} is the $(n + 1)$ term and u_n is the nth term of a power series. This ratio is most often a function of x, or $r(x)$. Then the series will

1. converge absolutely for $r(x) < 1$ and

2. diverge for $r(x) > 1$.

Recall that the ratio test is not valid for $r = 1$ or $r(x) = 1$. These values of x must be checked individually using other tests.

A power series always converges on an interval, which may vary from a single value of x to all real numbers x—the entire number line. This interval is called the *interval of convergence.* The interval of convergence may include both end points, only one end point, or neither end point. The interval of convergence of the power series

$$
\sum_{n=0}^{\infty} a_n(x - a)^n
$$

is always centered at $x = a$. The *radius of convergence* is the distance from the point $x = a$ to either end point of the interval. Thus the radius of convergence is one-half the length of the interval of convergence.

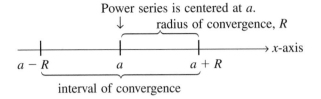

EXAMPLE 1

For what values of x does the series $\displaystyle\sum_{n=0}^{\infty} 5(x - 3)^n$ converge? That is, find the interval of convergence.

$$r(x) = \lim_{n\to\infty} \left|\frac{u_{n+1}}{u_n}\right| = \lim_{n\to\infty} \left|\frac{5(x - 3)^{n+1}}{5(x - 3)^n}\right| = \lim_{n\to\infty} |x - 3| = |x - 3|$$

This series converges for

$$|x - 3| < 1$$
$$-1 < x - 3 < 1$$
$$2 < x < 4$$

Next, check the end points. For $x = 2$, the series is

$$\sum_{n=0}^{\infty} (-1)^n(5) \qquad \text{which diverges.}$$

For $x = 4$, the series is

$$\sum_{n=0}^{\infty} 1^n(5) \qquad \text{which also diverges.}$$

Thus the interval of convergence is $2 < x < 4$, whose graph is

Note: the interval of convergence is centered at $x = 3$ and the radius of convergence is 1.

EXAMPLE 2

Find the interval of convergence of the series $\displaystyle\sum_{n=0}^{\infty} \frac{x^n}{n!}$.

$$r(x) = \lim_{n\to\infty} \left|\frac{u_{n+1}}{u_n}\right| = \lim_{n\to\infty} \left|\frac{\dfrac{x^{n+1}}{(n + 1)!}}{\dfrac{x^n}{n!}}\right| = \lim_{n\to\infty} \left|\frac{x}{n + 1}\right| = 0$$

Since $r(x) < 1$ for all values of x, this series converges for all real numbers or for all values of x, which may also be written

$$-\infty < x < \infty$$

EXAMPLE 3

Find the interval of convergence of the series $\displaystyle\sum_{n=0}^{\infty} n!\, x^n$.

$$r(x) = \lim_{n\to\infty} \left| \frac{u_{n+1}}{u_n} \right| = \lim_{n\to\infty} \left| \frac{(n+1)!\, x^{n+1}}{n!\, x^n} \right|$$

$$= \lim_{n\to\infty} |(n+1)x| = \begin{cases} 0 & \text{if } x = 0 \\ \infty & \text{if } x \neq 0 \end{cases}$$

This series converges only for $x = 0$. This interval of convergence consists of one point.

EXAMPLE 4

Find the interval of convergence of the series $\displaystyle\sum_{n=0}^{\infty} \frac{n^2}{2^n}(x-1)^n$.

$$r(x) = \lim_{n\to\infty} \left| \frac{u_{n+1}}{u_n} \right| = \lim_{n\to\infty} \left| \frac{\dfrac{(n+1)^2}{2^{n+1}}(x-1)^{n+1}}{\dfrac{n^2}{2^n}(x-1)^n} \right| = \lim_{n\to\infty} \left| \frac{(n+1)^2(x-1)}{2n^2} \right|$$

$$= \lim_{n\to\infty} \left| \frac{1}{2}\left(1 + \frac{1}{n}\right)^2 (x-1) \right| = \left| \frac{x-1}{2} \right|$$

This series converges for

$$\left| \frac{x-1}{2} \right| < 1$$

$$-1 < \frac{x-1}{2} < 1$$

$$-2 < x - 1 < 2$$

$$-1 < x < 3$$

Check the end points. For $x = -1$, the series is

$$\sum_{n=0}^{\infty} \frac{n^2(-2)^n}{2^n} = \sum_{n=0}^{\infty} (-1)^n n^2 \qquad \text{which diverges.}$$

For $x = 3$, the series is

$$\sum_{n=0}^{\infty} \frac{n^2\, 2^n}{2^n} = \sum_{n=0}^{\infty} n^2 \qquad \text{which also diverges.}$$

Thus the interval of convergence is $-1 < x < 3$.

$$-1 \qquad\qquad\qquad\qquad 3$$

EXAMPLE 5

Find the interval of convergence of the series $\displaystyle\sum_{n=0}^{\infty} \frac{nx^n}{(n+1)^2}$.

$$r(x) = \lim_{n\to\infty} \left| \frac{u_{n+1}}{u_n} \right| = \lim_{n\to\infty} \left| \frac{\dfrac{(n+1)x^{n+1}}{(n+2)^2}}{\dfrac{nx^n}{(n+1)^2}} \right| = \lim_{n\to\infty} \left| \frac{(n+1)^3\, x}{n(n+2)^2} \right| = |x|$$

Using a calculator,

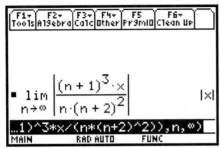

This series converges for $|x| < 1$ or $-1 < x < 1$.

Check the end points. For $x = -1$, the series is

$$\sum_{n=0}^{\infty} \frac{n(-1)^n}{(n+1)^2} \quad \text{which converges conditionally.}$$

For $x = 1$, the series is

$$\sum_{n=0}^{\infty} \frac{n \, 1^n}{(n+1)^2} \quad \text{which diverges.}$$

Thus the interval of convergence is $-1 \le x < 1$.

Exercises 10.4

Find the interval of convergence of each series.

1. $\displaystyle\sum_{n=0}^{\infty} \left(\frac{x}{2}\right)^n$

2. $\displaystyle\sum_{n=1}^{\infty} \frac{(-1)^n x^n}{n}$

3. $\displaystyle\sum_{n=0}^{\infty} (4n)! \left(\frac{x}{2}\right)^n$

4. $\displaystyle\sum_{n=0}^{\infty} nx^n$

5. $\displaystyle\sum_{n=1}^{\infty} \frac{(4x)^n}{(2n)!}$

6. $\displaystyle\sum_{n=1}^{\infty} \frac{x^n}{n}$

7. $\displaystyle\sum_{n=1}^{\infty} \frac{(-1)^{n+1} x^n}{(n+1)(n+2)}$

8. $\displaystyle\sum_{n=0}^{\infty} \frac{(-1)^{n+1} x^n}{4^n}$

9. $\displaystyle\sum_{n=0}^{\infty} \frac{nx^n}{(n+1)^2}$

10. $\displaystyle\sum_{n=0}^{\infty} 3^n x^n$

11. $\displaystyle\sum_{n=0}^{\infty} \frac{2^n x^n}{3^n}$

12. $\displaystyle\sum_{n=1}^{\infty} \frac{(-1)^{n+1}(x-1)^{n+1}}{n+1}$

13. $\displaystyle\sum_{n=1}^{\infty} \frac{(-1)^{n+1} x^n}{n \cdot 2^n}$

14. $\displaystyle\sum_{n=1}^{\infty} (-2)^n (x+1)^n$

15. $\displaystyle\sum_{n=1}^{\infty} \frac{(-1)^n (x-2)^n}{\sqrt{n}}$

16. $\displaystyle\sum_{n=1}^{\infty} \frac{(x+3)^n}{n^2 \cdot 2^n}$

17. $\displaystyle\sum_{n=1}^{\infty} \frac{x^n}{n^2}$

18. $\displaystyle\sum_{n=1}^{\infty} \frac{x^n}{\sqrt{n} \, 3^n}$

19. $\displaystyle\sum_{n=1}^{\infty} \frac{(-1)^n x^{2n}}{n!}$

20. $\displaystyle\sum_{n=1}^{\infty} \frac{n! \, x^n}{(2n)!}$

21. $\displaystyle\sum_{n=1}^{\infty} \frac{2^n x^{n+1}}{n(3^{n+1})}$

22. $\displaystyle\sum_{n=1}^{\infty} \frac{(-1)^n x^n}{n^n}$

23. $\displaystyle\sum_{n=1}^{\infty} \frac{(2x-5)^n}{n^2}$

24. $\displaystyle\sum_{n=1}^{\infty} \frac{\sin^n x}{n!}$

10.5   MACLAURIN SERIES

The Maclaurin series expansion of a function is a power series developed by differentiation. If a function can be differentiated repeatedly at $x = 0$, then it will have a Maclaurin series expansion. Thus we can write

$$f(x) = a_0 + a_1x + a_2x^2 + a_3x^3 + a_4x^4 + \cdots + a_nx^n + \cdots$$
$$f'(x) = a_1 + 2a_2x + 3a_3x^2 + 4a_4x^3 + \cdots + na_nx^{n-1} + \cdots$$
$$f''(x) = 2a_2 + 2 \cdot 3a_3x + 3 \cdot 4a_4x^2 + \cdots + (n-1)(n)a_nx^{n-2} + \cdots$$
$$f'''(x) = 2 \cdot 3a_3 + 2 \cdot 3 \cdot 4a_4x + \cdots + (n-2)(n-1)(n)a_nx^{n-3} + \cdots$$

$$\vdots \qquad \vdots$$

If we let $x = 0$, then

$$f(0) = a_0, \qquad f'(0) = a_1, \qquad f''(0) = 2a_2, \quad \text{and} \quad f'''(0) = 2 \cdot 3a_3$$

If we continue differentiating, the nth derivative at $x = 0$ is

$$f^{(n)}(0) = 1 \cdot 2 \cdot 3 \cdots (n-2)(n-1)(n)a_n = n!a_n$$

Rewriting, we have

$$a_0 = f(0)$$
$$a_1 = f'(0)$$
$$a_2 = \frac{f''(0)}{2!}$$
$$a_3 = \frac{f'''(0)}{3!}$$

$$\vdots$$

$$a_n = \frac{f^{(n)}(0)}{n!}$$

Replace the coefficients of the powers of x in the power series

$$f(x) = a_0 + a_1x + a_2x^2 + a_3x^3 + \cdots + a_nx^n + \cdots$$

by the preceding equivalent expressions to obtain

MACLAURIN SERIES EXPANSION

$$f(x) = f(0) + f'(0)x + \frac{f''(0)}{2!}x^2 + \frac{f'''(0)}{3!}x^3 + \cdots + \frac{f^{(n)}(0)}{n!}x^n + \cdots$$

The expansion is valid for all values of x for which the power series converges and for which the function $f(x)$ is repeatedly differentiable.

EXAMPLE 1

Find the first four terms of the Maclaurin expansion for $f(x) = \dfrac{1}{1+x}$.

$$f(x) = \frac{1}{1 + x} \qquad f(0) = 1$$

$$f'(x) = \frac{-1}{(1 + x)^2} \qquad f'(0) = -1$$

$$f''(x) = \frac{2}{(1 + x)^3} \qquad f''(0) = 2$$

$$f'''(x) = \frac{-6}{(1 + x)^4} \qquad f'''(0) = -6$$

So

$$f(x) = 1 - x + \frac{2}{2!}x^2 + \frac{-6}{3!}x^3 + \cdots$$

$$= 1 - x + x^2 - x^3 + \cdots$$

which is the same power series we obtained earlier by division.

EXAMPLE 2

Find the first five terms of the Maclaurin series expansion for $f(x) = \cos 3x$.

$$f(x) = \cos 3x \qquad f(0) = 1$$
$$f'(x) = -3 \sin 3x \qquad f'(0) = 0$$
$$f''(x) = -9 \cos 3x \qquad f''(0) = -9$$
$$f'''(x) = 27 \sin 3x \qquad f'''(0) = 0$$
$$f^{(4)}(x) = 81 \cos 3x \qquad f^{(4)}(0) = 81$$

Thus

$$f(x) = \cos 3x = 1 + 0x + \frac{-9}{2!}x^2 + 0x^3 + \frac{81}{4!}x^4 + \cdots$$

$$= 1 - \frac{9}{2}x^2 + \frac{27}{8}x^4 - \cdots$$

EXAMPLE 3

Find the Maclaurin series expansion for $f(x) = e^x$.

Since $\dfrac{d}{dx}(e^x) = e^x$, we have $f^{(n)}(x) = e^x$ for all n. So

$$f(0) = f'(0) = f''(0) = f'''(0) = \cdots = f^{(n)}(0) = e^0 = 1$$

Thus

$$f(x) = e^x = 1 + x + \frac{1}{2!}x^2 + \frac{1}{3!}x^3 + \frac{1}{4!}x^4 + \cdots + \frac{1}{n!}x^n + \cdots$$

$$= 1 + x + \frac{x^2}{2!} + \frac{x^3}{3!} + \frac{x^4}{4!} + \cdots + \frac{x^n}{n!} + \cdots$$

Note: This series converges (has a sum) and is a valid representation for $f(x) = e^x$ for all values of x. (See Example 2, Section 10.4.)

EXAMPLE 4

Find the first four nonzero terms of the Maclaurin expansion for $f(x) = e^x \cos x$.

$$f(x) = e^x \cos x \qquad\qquad f(0) = 1$$

$$f'(x) = e^x(-\sin x) + e^x \cos x$$
$$= e^x(\cos x - \sin x) \qquad f'(0) = 1$$

$$f''(x) = e^x(-\sin x - \cos x) + e^x(\cos x - \sin x)$$
$$= -2e^x \sin x \qquad\qquad f''(0) = 0$$

$$f'''(x) = -2e^x \cos x - 2e^x \sin x$$
$$= -2e^x(\cos x + \sin x) \qquad f'''(0) = -2$$

$$f^{(4)}(x) = -2e^x(-\sin x + \cos x) - 2e^x(\cos x + \sin x)$$
$$= -4e^x \cos x \qquad\qquad f^{(4)}(0) = -4$$

Thus

$$f(x) = 1 + x + \frac{0x^2}{2!} - \frac{2}{3!}x^3 - \frac{4}{4!}x^4$$

$$= 1 + x - \frac{1}{3}x^3 - \frac{1}{6}x^4$$

Using a calculator,

F3   9 $\qquad\qquad$ **green diamond e^x x) 2nd COS x),x,4) ENTER**

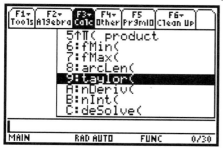 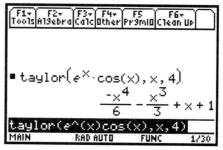

4th-degree Maclaurin expansion

Exercises 10.5

Find a Maclaurin series expansion for each function.

1. $f(x) = \sin x$ $\qquad$ **2.** $f(x) = \cos x$ $\qquad$ **3.** $f(x) = e^{-x}$

4. $f(x) = \dfrac{1}{1 - x}$ $\qquad$ **5.** $f(x) = \ln(1 + x)$ $\qquad$ **6.** $f(x) = e^{3x}$

7. $f(x) = \cos 2x$ $\qquad$ **8.** $f(x) = \sin 4x$ $\qquad$ **9.** $f(x) = xe^x$

10. $f(x) = x \sin x$ $\qquad$ **11.** $f(x) = \sqrt{4 - x}$ $\qquad$ **12.** $f(x) = \dfrac{1}{\sqrt{9 + x}}$

13. $f(x) = \sin\left(x - \dfrac{\pi}{2}\right)$ $\qquad$ **14.** $f(x) = \dfrac{1}{(1 + x)^2}$ $\qquad$ **15.** $f(x) = \dfrac{1}{(1 - x)^2}$

16. $f(x) = (1 + x)^{3/2}$ $\qquad$ **17.** $f(x) = (1 + x)^5$ $\qquad$ **18.** $f(x) = (2x - 1)^4$

19. $f(x) = e^{-x} \sin x$

20. Show that $(1 + x)^n = 1 + nx + \dfrac{n(n - 1)}{2!}x^2 + \dfrac{n(n - 1)(n - 2)}{3!}x^3 + \cdots$ by using the

Maclaurin expansion. This series is called the *binomial series,* which is valid for all real numbers n for $|x| < 1$. (See also Section 9.3.)

10.6 OPERATIONS WITH SERIES

We now summarize four important series that were developed previously as exercises.

$$e^x = 1 + x + \frac{x^2}{2!} + \frac{x^3}{3!} + \frac{x^4}{4!} + \cdots + \frac{x^n}{n!} + \cdots \quad (1)$$

$$\sin x = x - \frac{x^3}{3!} + \frac{x^5}{5!} - \cdots \quad (2)$$

$$\cos x = 1 - \frac{x^2}{2!} + \frac{x^4}{4!} - \cdots \quad (3)$$

$$\ln(1 + x) = x - \frac{x^2}{2} + \frac{x^3}{3} - \frac{x^4}{4} + \cdots \quad (4)$$

From these and similar basic power series expansions we can often obtain power series of other functions.

EXAMPLE 1

Find the Maclaurin series expansion for $f(x) = \cos 3x$.

Substituting $3x$ for x in Equation (3) we obtain

$$\cos 3x = 1 - \frac{(3x)^2}{2!} + \frac{(3x)^4}{4!} - \cdots$$

$$= 1 - \frac{9x^2}{2!} + \frac{81x^4}{4!} - \cdots$$

$$= 1 - \frac{9x^2}{2} + \frac{27x^4}{8} - \cdots$$

Compare this result with Example 2, Section 10.5.

EXAMPLE 2

Find a power series expansion for $f(x) = \dfrac{\ln(1 + x)}{x}$.

Dividing each side of Equation (4) by x, we have

$$\frac{\ln(1 + x)}{x} = \frac{x}{x} - \frac{x^2}{2x} + \frac{x^3}{3x} - \frac{x^4}{4x} + \cdots$$

$$= 1 - \frac{x}{2} + \frac{x^2}{3} - \frac{x^3}{4} + \cdots$$

EXAMPLE 3

Find the Maclaurin series expansion for $f(x) = e^{-3x}$.

Substituting $-3x$ for x in Equation (1), we have

$$e^{-3x} = 1 + (-3x) + \frac{(-3x)^2}{2!} + \frac{(-3x)^3}{3!} + \frac{(-3x)^4}{4!} + \cdots$$

$$= 1 - 3x + \frac{9x^2}{2} - \frac{9x^3}{2} + \frac{27x^4}{8} - \cdots$$

EXAMPLE 4

Evaluate $\displaystyle\int_0^1 \frac{\sin x}{x}\,dx$.

In Chapter 4 we used the function $\dfrac{\sin x}{x}$ to find the derivative of $y = \sin x$. We have no trouble differentiating this function, but none of the techniques of integration introduced in the preceding chapters leads to finding the integral $\displaystyle\int_0^1 \dfrac{\sin x}{x}\,dx$. However, if we divide each side of Equation (2) by x, we have

$$\frac{\sin x}{x} = \frac{x}{x} - \frac{x^3}{3!x} + \frac{x^5}{5!x} - \frac{x^7}{7!x} + \cdots$$

$$= 1 - \frac{x^2}{3!} + \frac{x^4}{5!} - \frac{x^6}{7!} + \cdots$$

Then

$$\int_0^1 \frac{\sin x}{x}\,dx = \int_0^1 \left(1 - \frac{x^2}{3!} + \frac{x^4}{5!} - \frac{x^6}{7!} + \cdots\right)dx$$

$$= \left(x - \frac{x^3}{3!3} + \frac{x^5}{5!5} - \frac{x^7}{7!7} + \cdots\right)\bigg|_0^1$$

$$= \left(1 - \frac{1}{18} + \frac{1}{600} - \frac{1}{35{,}280} + \cdots\right) - (0)$$

$$= 0.946083 \qquad \text{(sum of the first four nonzero terms)}$$

EXAMPLE 5

Evaluate $\displaystyle\int_{0.1}^1 \frac{e^{-x} - 1}{x}\,dx$.

Using Equation (1), we have

$$\frac{e^{-x}}{x} = \frac{1}{x} + \frac{-x}{x} + \frac{(-x)^2}{2!x} + \frac{(-x)^3}{3!x} + \frac{(-x)^4}{4!x} + \cdots$$

$$= \frac{1}{x} - 1 + \frac{x}{2!} - \frac{x^2}{3!} + \frac{x^3}{4!} - \cdots$$

So

$$\frac{e^{-x} - 1}{x} = \frac{e^{-x}}{x} - \frac{1}{x} = \left(\frac{1}{x} - 1 + \frac{x}{2!} - \frac{x^2}{3!} + \frac{x^3}{4!} - \cdots\right) - \frac{1}{x}$$

$$= -1 + \frac{x}{2!} - \frac{x^2}{3!} + \frac{x^3}{4!} - \cdots$$

$$\int_{0.1}^1 \frac{(e^{-x} - 1)}{x}\,dx = \int_{0.1}^1 \left(-1 + \frac{x}{2!} - \frac{x^2}{3!} + \frac{x^3}{4!} - \cdots\right)dx$$

$$= \left(-x + \frac{x^2}{4} - \frac{x^3}{18} + \frac{x^4}{96} - \cdots\right)\bigg|_{0.1}^1$$

$$= \left(-1 + \frac{1}{4} - \frac{1}{18} + \frac{1}{96} - \cdots\right)$$

$$\qquad - \left(-0.1 + \frac{0.01}{4} - \frac{0.001}{18} + \frac{0.0001}{96} - \cdots\right)$$

$$= -0.6976 \qquad \text{(using the first four terms)}$$

The exponential form of a complex number is based on the expression

$$e^{j\theta} = \cos\theta + j\sin\theta \qquad \text{where } j = \sqrt{-1}$$

We will now show that this is a valid identity. Recall from p. 424 that

$$e^x = 1 + x + \frac{x^2}{2!} + \frac{x^3}{3!} + \frac{x^4}{4!} + \frac{x^5}{5!} + \cdots \tag{1}$$

$$\cos x = 1 - \frac{x^2}{2!} + \frac{x^4}{4!} - \cdots \tag{3}$$

$$\sin x = x - \frac{x^3}{3!} + \frac{x^5}{5!} - \cdots \tag{2}$$

If we let $x = j\theta$ in Equation (1) and $x = \theta$ in Equations (3) and (2), then we have Equations (5), (6), and (7), respectively:

$$e^{j\theta} = 1 + j\theta + \frac{(j\theta)^2}{2!} + \frac{(j\theta)^3}{3!} + \frac{(j\theta)^4}{4!} + \frac{(j\theta)^5}{5!} + \cdots$$

$$= 1 + j\theta - \frac{\theta^2}{2!} - j\frac{\theta^3}{3!} + \frac{\theta^4}{4!} + j\frac{\theta^5}{5!} - \cdots \tag{5}$$

$$\cos\theta = 1 - \frac{\theta^2}{2!} + \frac{\theta^4}{4!} - \cdots \tag{6}$$

$$j\sin\theta = j\left(\theta - \frac{\theta^3}{3!} + \frac{\theta^5}{5!} - \cdots\right)$$

$$= j\theta - j\frac{\theta^3}{3!} + j\frac{\theta^5}{5!} - \cdots \tag{7}$$

Adding Equations (6) and (7), we have

$$\cos\theta + j\sin\theta = \left(1 - \frac{\theta^2}{2!} + \frac{\theta^4}{4!} - \cdots\right) + \left(j\theta - j\frac{\theta^3}{3!} + j\frac{\theta^5}{5!} - \cdots\right)$$

$$= 1 + j\theta - \frac{\theta^2}{2!} - j\frac{\theta^3}{3!} + \frac{\theta^4}{4!} + j\frac{\theta^5}{5!} - \cdots \qquad \text{(same as Equation 5)}$$

$$= e^{j\theta}$$

Exercises 10.6

Find a Maclaurin series expansion for each function.

1. $f(x) = e^{-x}$
2. $f(x) = \cos\sqrt{x}$
3. $f(x) = e^{x^2}$
4. $f(x) = \sin x^2$
5. $f(x) = \ln(1 - x)$
6. $f(x) = e^{-4x}$
7. $f(x) = \cos 5x^2$
8. $f(x) = \ln(1 + 3x)$
9. $f(x) = \sin x^3$
10. $f(x) = e^{-2x^2}$
11. $f(x) = xe^x$
12. $f(x) = x^2\sin x$
13. $f(x) = \dfrac{\cos x - 1}{x}$
14. $f(x) = \dfrac{e^x}{x - 1}$

15. Evaluate $\displaystyle\int_0^1 e^{-x^2}\,dx$. (Use first four nonzero terms.)

16. Evaluate $\displaystyle\int_0^1 \frac{\cos x - 1}{x}\,dx$. (Use first three nonzero terms.)

17. Evaluate $\displaystyle\int_2^3 \frac{e^{x-1}}{x - 1}\,dx$. (Use first three nonzero terms.)

18. Evaluate $\displaystyle\int_0^1 \cos x^2\, dx.$ (Use first four nonzero terms.)

19. Evaluate $\displaystyle\int_0^1 \sin \sqrt{x}\, dx.$ (Use first three nonzero terms.)

20. Evaluate $\displaystyle\int_0^{\pi/2} \sqrt{x} \cos x\, dx.$ (Use first six nonzero terms.)

21. The hyperbolic sine function is defined by $\sinh x = \frac{1}{2}(e^x - e^{-x})$. Find its Maclaurin series.

22. The hyperbolic cosine function is defined by $\cosh x = \frac{1}{2}(e^x + e^{-x})$. Find its Maclaurin series.

23. If $i = \sin t^2$ amperes, find the amount of charge q (in coulombs) transmitted by this current from $t = 0$ to $t = 0.5$ s. *Note:* $q = \displaystyle\int i\, dt.$

24. The current supplied to a capacitor is given by $i = \dfrac{1 - \cos t}{t}$ amperes. Find the voltage V across the capacitor after 0.1 s, where the capacitance $C = 1 \times 10^{-6}$ farad. *Note:*
$$V = \frac{1}{C}\int i\, dt.$$

Using power series for sin x, cos x, and e^x, show that

25. $\sin x = \dfrac{e^{jx} - e^{-jx}}{2j}$ 26. $\cos x = \dfrac{e^{jx} + e^{-jx}}{2}$

10.7 TAYLOR SERIES

When a function $f(x)$ is repeatedly differentiable at a number a and at x as well as all numbers between a and x, then the function usually has a Taylor series expansion that is a valid representation of the given function at x.

TAYLOR SERIES

A Taylor series expansion of a function $f(x)$ is a power series in the form

$$f(x) = f(a) + f'(a)(x - a) + \frac{f''(a)}{2!}(x - a)^2 + \frac{f'''(a)}{3!}(x - a)^3 + \cdots$$
$$+ \frac{f^{(n)}(a)}{n!}(x - a)^n + \cdots$$

Note that a Maclaurin series is a special case of a Taylor series with $a = 0$.

EXAMPLE 1

Find the Taylor series expansion for $f(x) = \ln x$ with $a = 2$.

$$
\begin{array}{ll}
f(x) = \ln x & f(2) = \ln 2 \\[2mm]
f'(x) = \dfrac{1}{x} & f'(2) = \dfrac{1}{2} \\[2mm]
f''(x) = -\dfrac{1}{x^2} & f''(2) = -\dfrac{1}{4} \\[2mm]
f'''(x) = \dfrac{2}{x^3} & f'''(2) = \dfrac{2}{8} = \dfrac{1}{4} \\[2mm]
f^{(4)}(x) = -\dfrac{6}{x^4} & f^{(4)}(2) = -\dfrac{6}{16} = -\dfrac{3}{8}
\end{array}
$$

So

$$f(x) = \ln x = \ln 2 + \frac{1}{2}(x - 2) + \frac{\left(-\frac{1}{4}\right)}{2!}(x - 2)^2 + \frac{\frac{1}{4}}{3!}(x - 2)^3 + \frac{\left(-\frac{3}{8}\right)}{4!}(x - 2)^4 + \cdots$$

$$= \ln 2 + \frac{1}{2}(x - 2) - \frac{1}{8}(x - 2)^2 + \frac{1}{24}(x - 2)^3 - \frac{1}{64}(x - 2)^4 + \cdots$$

EXAMPLE 2

Find the Taylor series expansion for $f(x) = e^x$ with $a = 1$.

Since $\dfrac{d}{dx}(e^x) = e^x$, we have

$$f(1) = f'(1) = f''(1) = f'''(1) = \cdots = f^{(n)}(1) = e^1 = e$$

So,

$$f(x) = e^x = e + e(x - 1) + \frac{e}{2!}(x - 1)^2 + \frac{e}{3!}(x - 1)^3 + \cdots + \frac{e}{n!}(x - 1)^n + \cdots$$

$$= e\left[1 + (x - 1) + \frac{1}{2!}(x - 1)^2 + \frac{1}{3!}(x - 1)^3 + \cdots + \frac{1}{n!}(x - 1)^n + \cdots\right]$$

EXAMPLE 3

Find the Taylor series expansion for $f(x) = \sin x$ at $a = \pi/2$.

$$f(x) = \sin x \qquad f\left(\frac{\pi}{2}\right) = 1$$

$$f'(x) = \cos x \qquad f'\left(\frac{\pi}{2}\right) = 0$$

$$f''(x) = -\sin x \qquad f''\left(\frac{\pi}{2}\right) = -1$$

$$f'''(x) = -\cos x \qquad f'''\left(\frac{\pi}{2}\right) = 0$$

$$f^{(4)}(x) = \sin x \qquad f^{(4)}\left(\frac{\pi}{2}\right) = 1$$

So,

$$f(x) = \sin x = 1 - \frac{1}{2!}\left(x - \frac{\pi}{2}\right)^2 + \frac{1}{4!}\left(x - \frac{\pi}{2}\right)^4 - \cdots$$

Using a calculator,

F3   9   **2nd SIN** x),x,4, **2nd** π/2) **ENTER**

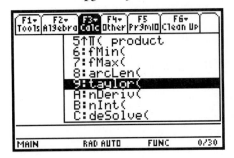
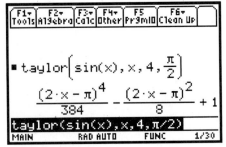

4th-degree Taylor expansion at $a = \pi/2$.

Exercises 10.7

Find the Taylor series expansion for each function for the given value of a.

1. $f(x) = \cos x, a = \dfrac{\pi}{2}$ **2.** $f(x) = \sin x, a = \dfrac{\pi}{4}$ **3.** $f(x) = e^x, a = 2$

4. $f(x) = \sqrt{x}, a = 4$ **5.** $f(x) = \sqrt{x}, a = 9$ **6.** $f(x) = \tan x, a = \dfrac{\pi}{4}$

7. $f(x) = \dfrac{1}{x}, a = 2$ **8.** $f(x) = e^{-x}, a = 1$ **9.** $f(x) = \ln x, a = 1$

10. $f(x) = x \ln x, a = 1$ **11.** $f(x) = \dfrac{1}{\sqrt{x}}, a = 1$ **12.** $f(x) = \dfrac{1}{1 + 2x}, a = 1$

13. $f(x) = \dfrac{1}{x^2}, a = 1$ **14.** $f(x) = \cos x, a = \dfrac{\pi}{3}$ **15.** $f(x) = \cos x, a = \pi$

16. $f(x) = e^{-x}, a = -3$

10.8 COMPUTATIONAL APPROXIMATIONS

One important use of power series expansions is to compute the numerical values of transcendental functions.

EXAMPLE 1

Calculate ln 1.1.

From Exercise 9 in Section 10.7 we found that

$$\ln x = (x - 1) - \frac{(x - 1)^2}{2} + \frac{(x - 1)^3}{3} - \frac{(x - 1)^4}{4} + \cdots$$

then

$$\ln 1.1 = (1.1 - 1) - \frac{(1.1 - 1)^2}{2} + \frac{(1.1 - 1)^3}{3} - \frac{(1.1 - 1)^4}{4} + \cdots$$

$$= 0.1 - \frac{(0.1)^2}{2} + \frac{(0.1)^3}{3} - \frac{(0.1)^4}{4} + \cdots$$

$$= 0.095308 \quad \text{(sum of the first four terms)}$$

Unlike the geometric series, it is difficult to compute the sum of a power series. Usually we must settle for an approximate value by simply evaluating only the first few terms of the series.

EXAMPLE 2

Calculate $e^{-0.2}$.

From

$$e^x = 1 + x + \frac{x^2}{2!} + \frac{x^3}{3!} + \cdots$$

we find

$$e^{-0.2} = 1 - 0.2 + \frac{(-0.2)^2}{2!} + \frac{(-0.2)^3}{3!} + \cdots$$

$$= 0.81867 \quad \text{(sum of the first four terms)}$$

Or,

F3 9 green diamond e^x x),x,3)|x=-.2 ENTER

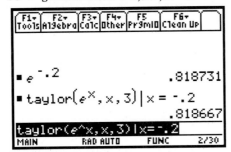

EXAMPLE 3

Calculate sin 3°.

From

$$\sin x = x - \frac{x^3}{3!} + \frac{x^5}{5!} - \cdots$$

and the first two terms, we have

$$\sin 3° = \sin \frac{\pi}{60} = \frac{\pi}{60} - \frac{\left(\frac{\pi}{60}\right)^3}{3!} + \cdots \qquad \left(3° = \frac{\pi}{60}\text{rad}\right)$$

$$= 0.05236 - 0.00002$$

$$= 0.05234$$

EXAMPLE 4

Calculate cos 32°.

The Taylor series expansion for $f(x) = \cos x$ at $a = \pi/6$ is found as follows:

$$f(x) = \cos x \qquad f\left(\frac{\pi}{6}\right) = \frac{\sqrt{3}}{2}$$

$$f'(x) = -\sin x \qquad f'\left(\frac{\pi}{6}\right) = -\frac{1}{2}$$

$$f''(x) = -\cos x \qquad f''\left(\frac{\pi}{6}\right) = -\frac{\sqrt{3}}{2}$$

$$f'''(x) = \sin x \qquad f'''\left(\frac{\pi}{6}\right) = \frac{1}{2}$$

So

$$\cos x = \frac{\sqrt{3}}{2} - \frac{1}{2}\left(x - \frac{\pi}{6}\right) - \frac{\sqrt{3}}{2}\frac{(x - \pi/6)^2}{2!} + \frac{1}{2}\frac{(x - \pi/6)^3}{3!} - \cdots$$

Note: We need to write $x = 32° = 30° + 2° = \frac{\pi}{6} + \frac{\pi}{90}$.

Then

$$x - a = x - \frac{\pi}{6} = \left(\frac{\pi}{6} + \frac{\pi}{90}\right) - \frac{\pi}{6} = \frac{\pi}{90}$$

and

$$\cos 32° = \frac{\sqrt{3}}{2} - \frac{1}{2}\left(\frac{\pi}{90}\right) - \frac{\sqrt{3}}{2}\frac{(\pi/90)^2}{2!} + \frac{1}{2}\frac{(\pi/90)^3}{3!} - \cdots$$

$$= 0.848048 \qquad \text{(using the first four nonzero terms)}$$

More advanced texts explain how many terms need to be used in approximating the value of a function to a desired accuracy. In the exercises that follow, the number of terms to be used will be specified.

If one desires to evaluate $e^{1.1}$, then it is better to use the Taylor expansion with $a = 1$:

$$e^x = e\left[1 + (x - 1) + \frac{1}{2!}(x - 1)^2 + \frac{1}{3!}(x - 1)^3 + \cdots\right]$$

rather than the Maclaurin series

$$e^x = 1 + x + \frac{x^2}{2!} + \frac{x^3}{3!} + \cdots$$

because the powers of $x - 1$ become smaller faster than do the powers of x when $x = 1.1$. Thus, an accurate approximation can be obtained with the Taylor expansion using fewer terms. This observation illustrates the importance of the Taylor series expansion.

Using a calculator,

green diamond e^x 1.1) **ENTER F3 9 green diamond** e^x x),x,3,1)|x=1.1) **ENTER** *erase* ",1" **ENTER**

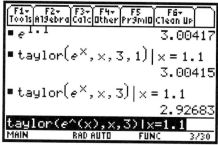

The 3rd degree Taylor expansion using $a = 1$ gives a much better estimate of $e^{1.1}$ than does the 3rd-degree Maclaurin expansion (using the default, $a = 0$).

Exercises 10.8

Calculate the value of each expression.

1. $e^{0.1}$ (Use first four nonzero terms.)
2. $e^{-0.3}$ (Use first four nonzero terms.)
3. $\cos 1°$ (Use first two nonzero terms.)
4. $\sin 2°$ (Use first two nonzero terms.)
5. $\ln 0.5$ (Use first four nonzero terms.)
6. $\ln 1.5$ (Use first four nonzero terms.)
7. $\sqrt{1.1}$ (Use first four nonzero terms.)
8. $\sqrt{0.9}$ (Use first four nonzero terms.)
9. $e^{1.3}$ (Use first four nonzero terms.)
10. $\sqrt{3.9}$ (Use first four nonzero terms.)
11. $\sin 29°$ (Use first three nonzero terms.)
12. $e^{0.9}$ (Use first four nonzero terms.)
13. Find the value of a current $i = \sin \omega t$ when $\omega t = 0.03$ rad.
14. Find the value of a current $i = 3e^{t^2}$ when $t = 0.1$ s.

10.9 FOURIER SERIES

One of the difficulties with Taylor series expansions is that, in general, they can be used to represent a given function only for values of x close to a [when expanded in powers of $(x - a)$]. A Fourier series expansion is often used when it is necessary to approximate a function over a larger interval of values of x.

The following expression is called a *Fourier series expansion* representing $f(x)$.

FOURIER SERIES

$$f(x) = a_0 + a_1 \cos x + a_2 \cos 2x + \cdots + a_n \cos nx + \cdots$$
$$+ b_1 \sin x + b_2 \sin 2x + \cdots + b_n \sin nx + \cdots$$

The coefficients are determined as follows:

$$a_0 = \frac{1}{2\pi} \int_0^{2\pi} f(x)\, dx$$

$$a_n = \frac{1}{\pi} \int_0^{2\pi} f(x) \cos nx\, dx \qquad (n = 1, 2, 3, \cdots)$$

$$b_n = \frac{1}{\pi} \int_0^{2\pi} f(x) \sin nx\, dx \qquad (n = 1, 2, 3, \cdots)$$

Note that n is a positive integer and that the Fourier series expansion is periodic with period 2π.

Four of the basic periodic waves that commonly occur in the analysis of electrical and mechanical systems are shown in Fig. 10.1.

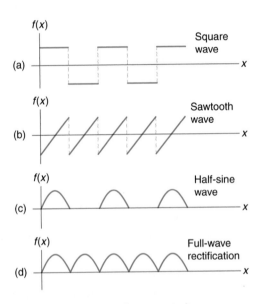

(a) Square wave

(b) Sawtooth wave

(c) Half-sine wave

(d) Full-wave rectification

Figure 10.1 Four basic periodic waves.

EXAMPLE 1

Find the Fourier series which represents the wave function $f(x) = x\ (0 \le x < 2\pi)$ with period 2π shown in Fig. 10.2.

Finding a_0: $\quad a_0 = \dfrac{1}{2\pi} \int_0^{2\pi} x\, dx = \dfrac{1}{2\pi} \dfrac{x^2}{2} \Big|_0^{2\pi} = \pi$

Finding a_n: $\quad a_n = \dfrac{1}{\pi} \int_0^{2\pi} x \cos nx\, dx$

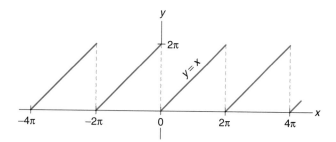

Figure 10.2

From Appendix B, Formula 81, we have

$$\int u \cos u \, du = \cos u + u \sin u + C$$

$$\int x \cos nx \, dx = \int \frac{u}{n} \cos u \, \frac{du}{n} \qquad \boxed{\begin{array}{l} u = nx \\ du = n \, dx \end{array}}$$

$$= \frac{1}{n^2} \int u \cos u \, du$$

$$= \frac{1}{n^2} (\cos u + u \sin u) + C$$

$$= \frac{1}{n^2} (\cos nx + nx \sin nx) + C$$

Thus

$$a_n = \frac{1}{\pi n^2} (\cos nx + nx \sin nx) \Big|_0^{2\pi}$$

$$= \frac{1}{\pi n^2} \{ (\cos 2n\pi + 2n\pi \sin 2n\pi) - [\cos 0 + n(0) \sin 0] \}$$

$$= \frac{1}{\pi n^2} (1 + 0 - 1 - 0) = 0 \qquad \text{(Recall that } n \text{ is a positive integer.)}$$

Finding b_n: $\quad b_n = \frac{1}{\pi} \int_0^{2\pi} x \sin nx \, dx$

From Appendix B, Formula 80, we have

$$\int u \sin u \, du = \sin u - u \cos u + C$$

So

$$\int x \sin nx \, dx = \int \frac{u}{n} \sin u \, \frac{du}{n} \qquad \boxed{\begin{array}{l} u = nx \\ du = n \, dx \end{array}}$$

$$= \frac{1}{n^2} \int u \sin u \, du$$

$$= \frac{1}{n^2} (\sin u - u \cos u) + C$$

$$= \frac{1}{n^2} (\sin nx - nx \cos nx) + C$$

Thus

$$b_n = \frac{1}{\pi n^2} (\sin nx - nx \cos nx) \Big|_0^{2\pi}$$

$$= \frac{1}{\pi n^2} \{(\sin 2n\pi - 2n\pi \cos 2n\pi) - [\sin 0 - n(0) \cos 0]\}$$

$$= \frac{1}{\pi n^2} (0 - 2n\pi - 0 + 0)$$

$$= -\frac{2}{n}$$

That is, $b_1 = -\frac{2}{1} = -2$, $b_2 = -\frac{2}{2} = -1$, $b_3 = -\frac{2}{3}, \ldots$, and the Fourier series is

$$f(x) = \pi - 2 \sin x - \sin 2x - \frac{2}{3} \sin 3x - \cdots$$

Note: There are no terms involving cosine because $a_1 = a_2 = a_3 = \cdots = a_n = 0$. The graphs of the sums of the first few terms are sketched in Fig. 10.3.

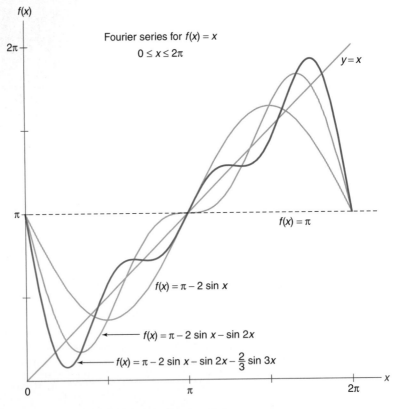

Figure 10.3

EXAMPLE 2

Find the Fourier series for the wave function given by

$$f(x) = \begin{cases} \pi & 0 \leq x < \pi \\ 2\pi - x & \pi \leq x < 2\pi \end{cases}$$

First, sketch several periods of $f(x)$ as in Fig. 10.4.

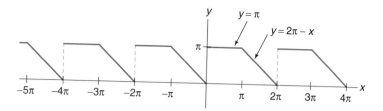

Figure 10.4

Finding a_0: $a_0 = \dfrac{1}{2\pi} \displaystyle\int_0^\pi \pi \, dx + \dfrac{1}{2\pi} \int_\pi^{2\pi} (2\pi - x) \, dx$

$$= \frac{1}{2\pi}(\pi x)\Big|_0^\pi + \frac{1}{2\pi}\left(2\pi x - \frac{x^2}{2}\right)\Big|_\pi^{2\pi}$$

$$= \frac{\pi}{2} + \frac{\pi}{4} = \frac{3\pi}{4}$$

Note: Two separate integrals must be used to determine the coefficients. This is because the function is defined differently on the two intervals $0 \le x < \pi$ and $\pi \le x < 2\pi$.

Finding a_n: $a_n = \dfrac{1}{\pi} \displaystyle\int_0^\pi \pi \cos nx \, dx + \dfrac{1}{\pi} \int_\pi^{2\pi} (2\pi - x) \cos nx \, dx$

$$= \int_0^\pi \cos nx \, dx + 2\int_\pi^{2\pi} \cos nx \, dx - \frac{1}{\pi}\int_\pi^{2\pi} x \cos nx \, dx$$

$$= \frac{1}{n}(\sin nx)\Big|_0^\pi + \frac{2}{n}(\sin nx)\Big|_\pi^{2\pi}$$

$$\quad - \frac{1}{\pi n^2}(\cos nx + nx \sin nx)\Big|_\pi^{2\pi} \qquad \text{(from Example 1)}$$

$$= \begin{cases} 0 + 0 - \dfrac{2}{\pi n^2} = -\dfrac{2}{\pi n^2} & \text{if } n \text{ is odd} \\ 0 & \text{if } n \text{ is even} \end{cases}$$

Finding b_n: $b_n = \dfrac{1}{\pi} \displaystyle\int_0^\pi \pi \sin nx \, dx + \dfrac{1}{\pi} \int_\pi^{2\pi} (2\pi - x) \sin nx \, dx$

$$= \int_0^\pi \sin nx \, dx + 2\int_\pi^{2\pi} \sin nx \, dx - \frac{1}{\pi}\int_\pi^{2\pi} x \sin nx \, dx$$

$$= \left(-\frac{1}{n}\right)(\cos nx)\Big|_0^\pi + \left(-\frac{2}{n}\right)(\cos nx)\Big|_\pi^{2\pi}$$

$$\quad - \frac{1}{\pi n^2}(\sin nx - nx \cos nx)\Big|_\pi^{2\pi} \qquad \text{(from Example 1)}$$

$$= \begin{cases} \dfrac{2}{n} - \dfrac{4}{n} + \dfrac{3}{n} = \dfrac{1}{n} & \text{if } n \text{ is odd} \\ 0 & \text{if } n \text{ is even} \end{cases}$$

We thus obtain the Fourier series

$$f(x) = \frac{3\pi}{4} - \frac{2}{\pi}\left(\cos x + \frac{1}{9}\cos 3x + \frac{1}{25}\cos 5x + \cdots\right)$$

$$+ \left(\sin x + \frac{1}{3}\sin 3x + \frac{1}{5}\sin 5x + \cdots\right)$$

We will now show how the formulas for the coefficients a_0, a_n, and b_n are obtained. Note that if we integrate each side of the equation

$$f(x) = a_0 + a_1 \cos x + a_2 \cos 2x + a_3 \cos 3x + \cdots$$
$$+ \, b_1 \sin x + b_2 \sin 2x + b_3 \sin 3x + \cdots$$

from 0 to 2π, then the integrals should be equal. That is,

$$\int_0^{2\pi} f(x)\, dx = \int_0^{2\pi} a_0\, dx + \int_0^{2\pi} a_1 \cos x\, dx + \int_0^{2\pi} a_2 \cos 2x\, dx + \cdots$$
$$+ \int_0^{2\pi} b_1 \sin x\, dx + \int_0^{2\pi} b_2 \sin 2x\, dx + \cdots$$

All terms on the right-hand side are zero except for $\int_0^{2\pi} a_0\, dx = 2\pi a_0$, so

$$\int_0^{2\pi} f(x)\, dx = 2\pi a_0$$

$$a_0 = \frac{1}{2\pi} \int_0^{2\pi} f(x)\, dx$$

Multiply each side of the preceding integral series equation by $\cos nx$. Then

$$\int_0^{2\pi} f(x) \cos nx\, dx = \int_0^{2\pi} a_0 \cos nx\, dx + \int_0^{2\pi} a_1 (\cos nx) \cos x\, dx$$
$$+ \int_0^{2\pi} a_2 (\cos nx) \cos 2x\, dx + \cdots$$
$$+ \int_0^{2\pi} b_1 (\cos nx) \sin x\, dx$$
$$+ \int_0^{2\pi} b_2 (\cos nx) \sin 2x\, dx + \cdots$$

All terms on the right-hand side are found to be zero except the term

$$\int_0^{2\pi} a_n (\cos nx)(\cos nx)\, dx = \pi a_n$$

So

$$\int_0^{2\pi} f(x) \cos nx\, dx = \pi a_n$$

$$a_n = \frac{1}{\pi} \int_0^{2\pi} f(x) \cos nx\, dx$$

In a similar manner (multiplying each side of the Fourier series equation by $\sin nx$), we can show that

$$b_n = \frac{1}{\pi} \int_0^{2\pi} f(x) \sin nx\, dx$$

Note: If the function to be analyzed ranges periodically from $-\pi$ to π, then the coefficients become

$$a_0 = \frac{1}{2\pi} \int_{-\pi}^{\pi} f(x)\, dx$$

$$a_n = \frac{1}{\pi} \int_{-\pi}^{\pi} f(x) \cos nx \, dx$$

$$b_n = \frac{1}{\pi} \int_{-\pi}^{\pi} f(x) \sin nx \, dx$$

If you need to find a Fourier series of a function defined over a still different interval, the coefficients become

$$a_0 = \frac{1}{2L} \int_{-L}^{L} f(x) \, dx$$

$$a_n = \frac{1}{L} \int_{-L}^{L} f(x) \cos \frac{n\pi x}{L} \, dx$$

$$b_n = \frac{1}{L} \int_{-L}^{L} f(x) \sin \frac{n\pi x}{L} \, dx$$

where the period of the function is $2L$. The Fourier series is

$$f(x) = a_0 + a_1 \cos \frac{\pi x}{L} + a_2 \cos \frac{2\pi x}{L} + a_3 \cos \frac{3\pi x}{L} + \cdots + a_n \cos \frac{n\pi x}{L} + \cdots +$$
$$b_1 \sin \frac{\pi x}{L} + b_2 \sin \frac{2\pi x}{L} + b_3 \sin \frac{3\pi x}{L} + \cdots + b_n \sin \frac{n\pi x}{L} + \cdots$$

Exercises 10.9

Sketch several periods of each given function and find its Fourier series expansion.

1. $f(x) = -x, 0 \le x < 2\pi$

2. $f(x) = 2x, 0 \le x < 2\pi$

3. $f(x) = \frac{1}{3}x, 0 \le x < 2\pi$

4. $f(x) = 2x, -\pi \le x < \pi$

5. $f(x) = \begin{cases} 0 & 0 \le x < \pi \\ 1 & \pi \le x < 2\pi \end{cases}$

6. $f(x) = \begin{cases} \pi & 0 \le x < \pi \\ 0 & \pi \le x < 2\pi \end{cases}$

7. $f(x) = \begin{cases} 1 & 0 \le x < \pi \\ -1 & \pi \le x < 2\pi \end{cases}$

8. $f(x) = \begin{cases} x & 0 \le x < \pi \\ \pi & \pi \le x < 2\pi \end{cases}$

9. $f(x) = \begin{cases} 0 & -5 \le x < 0 \\ 6 & 0 \le x < 5 \end{cases}$

10. $f(x) = \begin{cases} 1 & -4 \le x < 0 \\ -1 & 0 \le x < 4 \end{cases}$

11. $f(x) = \begin{cases} x & 0 \le x < \pi \\ 2\pi - x & \pi \le x < 2\pi \end{cases}$

12. $f(x) = \begin{cases} 0 & 0 \le x < \pi \\ x & \pi \le x < 2\pi \end{cases}$

13. $f(x) = e^x, 0 \le x < 2\pi$

14. $f(x) = e^{-2x}, 0 \le x < 2\pi$

15. $f(x) = \begin{cases} \sin x & 0 \le x < \pi \\ 0 & \pi \le x < 2\pi \end{cases}$

16. $f(x) = \begin{cases} \sin x & 0 \le x < \pi \\ -\sin x & \pi \le x < 2\pi \end{cases}$

CHAPTER SUMMARY

1. *Definition of convergence and divergence*

$$\sum_{n=1}^{\infty} a_n = a_1 + a_2 + a_3 + \cdots$$

Then

$$S_1 = a_1$$
$$S_2 = a_1 + a_2$$
$$S_3 = a_1 + a_2 + a_3$$
.
.
.
$$S_n = a_1 + a_2 + a_3 + \cdots + a_n$$
.
.
.

where S_1 is the sum of the first term, S_2 is the sum of the first two terms, S_3 is the sum of the first three terms, $\ldots$, S_n is the sum of the first n terms (or sometimes called the nth partial sum).

(a) If $\lim_{n \to \infty} S_n = S$ (where S is finite), the series $\sum_{n=1}^{\infty} a_n$ converges and S is the *sum of the infinite series.*

(b) If $\lim_{n \to \infty} S_n$ does not exist, the series $\sum_{n=1}^{\infty} a_n$ *diverges.*

2. *nth Term Test for divergence of a series:* If $\lim_{n \to \infty} a_n \neq 0$, then the series *diverges.* Stated another way, if the series $\sum_{n=1}^{\infty} a_n$ converges, then $\lim_{n \to \infty} a_n = 0$.

3. *Convergence and divergence of a p-series:* Any series in the form

$$\sum_{n=1}^{\infty} \frac{1}{n^p} = \frac{1}{1^p} + \frac{1}{2^p} + \frac{1}{3^p} + \cdots,$$ where p is a real number, is called a p-series. The p-series
(a) converges for $p > 1$ and
(b) diverges for $p \leq 1$.

4. *Comparison test for convergence and divergence:* Let N be a positive integer, $\sum_{n=1}^{\infty} a_n$ and $\sum_{n=1}^{\infty} b_n$ be series of positive terms, and $0 \leq a_n \leq b_n$ for all $n > N$.

(a) If $\sum_{n=1}^{\infty} b_n$ converges, then $\sum_{n=1}^{\infty} a_n$ also converges.

(b) If $\sum_{n=1}^{\infty} a_n$ diverges, then $\sum_{n=1}^{\infty} b_n$ also diverges.

In other words, the comparison test says
(a) A series of positive terms that is term by term smaller than a known convergent series must also converge.
(b) A series of positive terms that is term by term larger than a known divergent series must also diverge.

5. Let $\sum_{n=1}^{\infty} a_n$ and $\sum_{n=1}^{\infty} b_n$ be two series of positive terms.

(a) Then $\sum_{n=1}^{\infty} a_n$ and $\sum_{n=1}^{\infty} b_n$ have the *same order of magnitude* if $\lim_{n \to \infty} \frac{a_n}{b_n} = L$, where L is a real number and $L \neq 0$.

(b) The series $\displaystyle\sum_{n=1}^{\infty} a_n$ has a *lesser order of magnitude* than $\displaystyle\sum_{n=1}^{\infty} b_n$ if $\displaystyle\lim_{n\to\infty} \frac{a_n}{b_n} = 0$.

(c) The series $\displaystyle\sum_{n=1}^{\infty} a_n$ has a *greater order of magnitude* than $\displaystyle\sum_{n=1}^{\infty} b_n$ if $\displaystyle\lim_{n\to\infty} \frac{a_n}{b_n} = \infty$.

6. *Limit comparison test:* Let $\displaystyle\sum_{n=1}^{\infty} a_n$ and $\displaystyle\sum_{n=1}^{\infty} b_n$ be series of positive terms.

(a) If both series have the same order of magnitude, then either both series converge or both series diverge.

(b) If the series $\displaystyle\sum_{n=1}^{\infty} a_n$ has a lesser order of magnitude than $\displaystyle\sum_{n=1}^{\infty} b_n$ and $\displaystyle\sum_{n=1}^{\infty} b_n$ is known to converge, then $\displaystyle\sum_{n=1}^{\infty} a_n$ also converges.

(c) If $\displaystyle\sum_{n=1}^{\infty} a_n$ has a greater order of magnitude than $\displaystyle\sum_{n=1}^{\infty} b_n$ and $\displaystyle\sum_{n=1}^{\infty} b_n$ is known to diverge, then $\displaystyle\sum_{n=1}^{\infty} a_n$ also diverges.

7. *Ratio test for convergence and divergence:* Let $\displaystyle\sum_{n=1}^{\infty} a_n$ be a series of positive terms and

$$r = \lim_{n\to\infty} \frac{a_{n+1}}{a_n}$$

(a) If $r < 1$, the series converges.
(b) If $r > 1$ (including $r = \infty$), the series diverges.
(c) If $r = 1$, the test fails. Some other test must be used.

8. *Integral test for convergence and divergence:* Let $\displaystyle\sum_{n=1}^{\infty} a_n$ be a series of positive terms and $f(x)$ be a continuous, decreasing function for $x \geq 1$ such that $f(n) = a_n$ for all positive integers n. Then $\displaystyle\sum_{n=1}^{\infty} a_n$ and $\displaystyle\int_1^{\infty} f(x)\,dx$ both converge or both diverge.

9. *Alternating series test:* The alternating series

$$\sum_{n=1}^{\infty} (-1)^{n+1} a_n = a_1 - a_2 + a_3 - a_4 + \cdots \qquad (a_n > 0 \text{ for each } n)$$

converges provided that both of the following conditions are fulfilled:
(a) $0 < a_{n+1} \leq a_n$ for $n \geq 1$
(b) $\displaystyle\lim_{n\to\infty} a_n = 0$

In addition, if S is the sum of the infinite series and S_n is the nth partial sum, then

$$|S - S_n| \leq a_{n+1}$$

10. *Absolute and conditional convergence:* Suppose that $\displaystyle\sum_{n=1}^{\infty} a_n$ converges.

(a) If $\displaystyle\sum_{n=1}^{\infty} |a_n|$ converges, then $\displaystyle\sum_{n=1}^{\infty} a_n$ *converges absolutely.*

(b) If $\displaystyle\sum_{n=1}^{\infty} |a_n|$ diverges, then $\displaystyle\sum_{n=1}^{\infty} a_n$ *converges conditionally.*

11. A *power series* is an infinite series in the form

$$\sum_{n=0}^{\infty} a_n x^n = a_0 + a_1 x + a_2 x^2 + \cdots + a_n x^n + \cdots$$

A series in the form

$$\sum_{n=0}^{\infty} a_n(x - a)^n = a_0 + a_1(x - a) + a_2(x - a)^2 + a_3(x - a)^3$$
$$+ \cdots + a_n(x - a)^n + \cdots$$

is called a *power series centered at a.* Let

$$\lim_{n=\infty} \left| \frac{u_{n+1}}{u_n} \right| = r(x)$$

where u_{n+1} is the $(n + 1)$ term and u_n is the nth term of a power series. Note that the ratio will most often be a function of x, or $r(x)$. Then the series
(a) converges absolutely for $r(x) < 1$ and
(b) diverges for $r(x) > 1$.
Recall that the ratio test is not valid for $r = 1$ or $r(x) = 1$. These values of x must be checked individually by other methods.

12. A power series converges on an interval, called the *interval of convergence.* The interval of convergence may include both end points, only one end point, or neither end point. The interval of convergence of the power series

$$\sum_{n=0}^{\infty} a_n(x - a)^n$$

is centered at $x = a$. The *radius of convergence* is the distance from the point $x = a$ to either end point of the interval. Thus the radius of convergence is one-half the length of the interval of convergence.

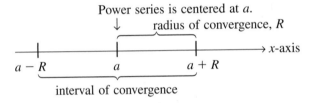

13. The Maclaurin series expansion of the function $f(x)$ is

$$f(x) = f(0) + f'(0)x + \frac{f''(0)}{2!}x^2 + \frac{f'''(0)}{3!}x^3 + \cdots + \frac{f^{(n)}(0)}{n!}x^n + \cdots$$

The expansion is valid for all values of x for which the power series converges and for which the function $f(x)$ is repeatedly differentiable.

14. A Taylor series expansion of $f(x)$ is a power series in the form

$$f(x) = f(a) + f'(a)(x - a) + \frac{f''(a)}{2!}(x - a)^2 + \frac{f'''(a)}{3!}(x - a)^3 + \cdots$$
$$+ \frac{f^{(n)}(a)}{n!}(x - a)^n + \cdots$$

Note that a Maclaurin series is a special case of a Taylor series with $a = 0$.

15. The Fourier series expansion of the function $f(x)$ is

$$f(x) = a_0 + a_1 \cos x + a_2 \cos 2x + \cdots + a_n \cos nx + \cdots$$
$$+ b_1 \sin x + b_2 \sin 2x + \cdots + b_n \sin nx + \cdots$$

The coefficients are determined as follows:

$$a_0 = \frac{1}{2\pi} \int_0^{2\pi} f(x) \, dx$$

$$a_n = \frac{1}{\pi} \int_0^{2\pi} f(x) \cos nx \, dx$$

$$b_n = \frac{1}{\pi} \int_0^{2\pi} f(x) \sin nx \, dx$$

Note that n is a positive integer and that the Fourier series expansion is periodic with period 2π.

CHAPTER 10 REVIEW

Write the expanded form of each series.

1. $\displaystyle\sum_{n=1}^{6} (1 - 3n)$

2. $\displaystyle\sum_{k=1}^{n} \frac{k+1}{k}$

Write each sum using sigma notation.

3. $\dfrac{1}{3} + \dfrac{1}{9} + \dfrac{1}{27} + \cdots + \dfrac{1}{2187}$

4. $\dfrac{1}{4} + \dfrac{2}{5} + \dfrac{3}{6} + \dfrac{4}{7} + \cdots + \dfrac{10}{13}$

Determine whether each series converges or diverges.

5. $\displaystyle\sum_{n=1}^{\infty} \frac{1}{n^3}$

6. $\displaystyle\sum_{n=1}^{\infty} \frac{1}{\sqrt[4]{n}}$

7. $\displaystyle\sum_{n=1}^{\infty} \frac{1}{6n^2 + 2}$

8. $\displaystyle\sum_{n=2}^{\infty} \frac{\sqrt{n}}{n^2 - 1}$

9. $\displaystyle\sum_{n=2}^{\infty} \frac{n}{\ln n}$

10. $\displaystyle\sum_{n=1}^{\infty} \frac{5n + 2}{(3n + 1)\,4^n}$

11. $\displaystyle\sum_{n=1}^{\infty} \frac{n^3}{2^n}$

12. $\displaystyle\sum_{n=1}^{\infty} \frac{3n + 1}{4n - 5}$

13. $\displaystyle\sum_{n=1}^{\infty} \frac{n + 1}{n^2 + 4n}$

14. $\displaystyle\sum_{n=1}^{\infty} \frac{\sin n}{n^2 + 1}$

Determine whether each alternating series converges or diverges. If it converges, find whether it converges absolutely or converges conditionally.

15. $\displaystyle\sum_{n=1}^{\infty} \frac{(-1)^{n+1}\, 3^n}{n!}$

16. $\displaystyle\sum_{n=1}^{\infty} \frac{(-1)^{n+1}\, 2^n}{5^n\,(n + 1)}$

17. $\displaystyle\sum_{n=2}^{\infty} (-1)^n \frac{n + 1}{n - 1}$

18. $\displaystyle\sum_{n=1}^{\infty} (-1)^{n+1} \frac{n + 2}{n(n + 3)}$

Find the interval of convergence of each series.

19. $\displaystyle\sum_{n=0}^{\infty} \frac{n}{2}(x - 2)^n$

20. $\displaystyle\sum_{n=0}^{\infty} n^2(x - 3)^n$

21. $\displaystyle\sum_{n=1}^{\infty} \frac{(x - 1)^n}{n!}$

22. $\displaystyle\sum_{n=1}^{\infty} \frac{3^n(x - 4)^n}{n^2}$

Find a Maclaurin series expansion for each function.

23. $f(x) = \dfrac{1}{1-x}$ **24.** $f(x) = \sqrt{x+1}$ **25.** $f(x) = \sin x + \cos x$

26. $f(x) = e^x \sin x$ **27.** $f(x) = \dfrac{1 - e^x}{x}$ **28.** $f(x) = \cos x^2$

29. $f(x) = \sin 3x$ **30.** $f(x) = e^{\sin x}$

31. Evaluate $\displaystyle\int_0^{0.1} \dfrac{\ln(x+1)}{x}\, dx.$ (Use first four nonzero terms.)

32. If $i = \dfrac{\sin t}{t}$ amperes, find the amount of charge q (in coulombs) transmitted by this current

from $t = 0$ to $t = 0.1$ s. *Note:* $q = \displaystyle\int i\, dt.$

Find the Taylor series expansions for each function for the given value of a.

33. $f(x) = \cos 2x,\ a = \dfrac{\pi}{6}$ **34.** $f(x) = \ln x,\ a = 4$

35. $f(x) = e^{x^2},\ a = 1$ **36.** $f(x) = \sin x,\ a = \dfrac{3\pi}{2}$

Calculate the value of each expression.

37. $\sin 31°$ (Use first three nonzero terms.)   **38.** $e^{1.2}$ (Use first four nonzero terms).

39. $\ln 1.2$ (Use first four nonzero terms.)   **40.** $\sqrt{4.1}$ (Use first four nonzero terms.)

Find the Fourier series for each function.

41. $f(x) = \begin{cases} 0 & 0 \le x < \pi \\ -1 & \pi \le x < 2\pi \end{cases}$ **42.** $f(x) = \begin{cases} x^2 & 0 \le x < \pi \\ 0 & \pi \le x < 2\pi \end{cases}$

11

First-Order Differential Equations

INTRODUCTION

Often in physics, engineering, and other technical areas, we need to search for an unknown function. In many cases, this search leads to an equation involving derivatives (or differentials) of the unknown function. Such equations involving derivatives (or differentials) are called differential equations.

Objectives

- Solve differential equations by the method of separation of variables.
- Solve differential equations by using an integrating factor.
- Use differential equations to solve applications problems.

11.1 SOLVING DIFFERENTIAL EQUATIONS

In Chapters 11 and 12 we present methods of solving differential equations. That is, we will find ways in which we use differential equations to determine an unknown function.

EXAMPLE 1

The following are examples of differential equations.

(a) $\dfrac{dy}{dx} = x^2 y$

(b) $\dfrac{dy}{dx} = \sin x$

(c) $\dfrac{d^2y}{dx^2} + x\,\dfrac{dy}{dx} + y = 0$

(d) $x^2\,\dfrac{d^3y}{dx^3} + 2y\,\dfrac{d^2y}{dx^2} - \left(\dfrac{dy}{dx}\right)^4 + 3 = 0$

(e) $e^x\,dy - x^2 y\,dx = 2$

The *order of a differential equation* is n, if n is the highest order derivative that appears in the equation.

443

EXAMPLE 2

Determine the order of the differential equation

$$\frac{d^2y}{dx^2} + 2\left(\frac{dy}{dx}\right)^3 + 5 = 0$$

The order is 2, since the second derivative $\frac{d^2y}{dx^2}$ (order 2 derivative) is the highest-order derivative appearing in the equation.

The *degree of a differential equation* is the highest power of the derivative of highest order.

EXAMPLE 3

Determine the degree and order of each differential equation:

$$\frac{d^2y}{dx^2} - 7\frac{dy}{dx} + \left(\frac{dy}{dx}\right)^3 = 0 \qquad \textbf{(1)}$$

$$\left(\frac{dy}{dx}\right)^2 - 3\frac{dy}{dx} + y = 0 \qquad \textbf{(2)}$$

Equation (1) is a first-degree differential equation of order 2, since $\frac{d^2y}{dx^2}$ is the highest-order derivative in the equation and is raised to the first power. Note that the third power of $\frac{dy}{dx}$ has no effect on the degree of Equation (1) because $\frac{dy}{dx}$ is of lesser order than $\frac{d^2y}{dx^2}$.

Equation (2) is a second-degree, first-order differential equation. $\frac{dy}{dx}$ is the highest-order derivative (order 1), and 2 is the highest power of $\frac{dy}{dx}$ appearing in the equation.

A *solution* of a differential equation is a function $y = f(x)$ that together with its derivatives satisfies the given differential equation.

EXAMPLE 4

Verify that $y = x^2 + 5x$ is a solution of the second-order, first-degree differential equation
$$x\frac{d^2y}{dx^2} - \frac{dy}{dx} + 5 = 0$$

First, $\frac{dy}{dx} = 2x + 5$ and $\frac{d^2y}{dx^2} = 2$. Then substitute in the differential equation:

$$x\frac{d^2y}{dx^2} - \frac{dy}{dx} + 5 = 0$$
$$x(2) - (2x + 5) + 5 = 0$$
$$2x - 2x - 5 + 5 = 0$$
$$0 = 0$$

EXAMPLE 5

Verify that $y = \dfrac{1}{x^2 + C}$ is a solution of the first-order, first-degree differential equation

$$\frac{dy}{dx} = -2xy^2$$

First, $\dfrac{dy}{dx} = -\dfrac{2x}{(x^2 + C)^2}$. Substitute this result in the given differential equation:

$$\frac{dy}{dx} = -2xy^2$$

$$\frac{-2x}{(x^2 + C)^2} = -2x\left(\frac{1}{x^2 + C}\right)^2$$

$$\frac{-2x}{(x^2 + C)^2} = \frac{-2x}{(x^2 + C)^2}$$

The solution $y = x^2 + 5x$ in Example 4 is an example of a *particular solution* of a differential equation. One can verify that $y = x^2 + 5x - 7$ is also a particular solution of the differential equation in Example 4. A differential equation can have infinitely many particular solutions.

A solution $y = f(x)$ of a differential equation of order n containing n arbitrary constants is called a *general solution*. Thus the solution $y = \dfrac{1}{x^2 + C}$ in Example 5 or $y = x^2 + 5x + C$ in Example 4 is an example of a general solution.

We will solve only first-degree equations. Differential equations that do not contain partial derivatives are called *ordinary differential equations*. We restrict our considerations to first-degree ordinary differential equations.

Recall the use of other notations for derivatives:

$$y' = \frac{dy}{dx} \qquad y'' = \frac{d^2y}{dx^2} \qquad y''' = \frac{d^3y}{dx^3} \cdots$$

EXAMPLE 6

Verify that $y = C_1 + C_2x + C_3e^x$ is a general solution of the differential equation $y''' = y''$.

First, find the first three derivatives of the given function:

$$y' = C_2 + C_3e^x$$
$$y'' = C_3e^x$$
$$y''' = C_3e^x$$

Substituting in the differential equation, we have

$$y''' = y''$$
$$C_3e^x = C_3e^x$$

Therefore, $y = C_1 + C_2x + C_3e^x$ is a general solution of the given differential equation because it has three distinct arbitrary constants.

Exercises 11.1

State the order and degree of each differential equation.

1. $\dfrac{dy}{dx} = x^2 - y^2$

2. $\left(\dfrac{dy}{dx}\right)^2 - 3x\dfrac{dy}{dx} + 2 = 0$

3. $\dfrac{d^2y}{dx^2} + 5xy\dfrac{dy}{dx} = x^2y$

4. $x^2\dfrac{dy}{dx} + y\left(\dfrac{dy}{dx}\right)^2 = 0$

5. $y''' - 4y'' + xy = 0$

6. $y' + x\cos x = 0$

7. $(y'')^3 - xy' + y'' = 0$

8. $y'' + e^xy = 2$

Verify that each function $y = f(x)$ is a solution of the differential equation.

9. $\dfrac{dy}{dx} = 3; y = 3x - 7$

10. $\dfrac{dy}{dx} + y + 2x + 4 = x^2; y = x^2 - 4x$

11. $x\dfrac{dy}{dx} - 2y = 4x; y = x^2 - 4x$

12. $\dfrac{d^2y}{dx^2} + y = 0; y = 2\sin x + 3\cos x$

13. $\dfrac{dy}{dx} + y = e^{-x}; y = (x + 2)e^{-x}$

14. $x\dfrac{dy}{dx} = x^2 + y; y = x^2 + Cx$

15. $\dfrac{d^2y}{dx^2} + 16y = 0; y = C_1\sin 4x + C_2\cos 4x$

16. $\dfrac{d^2y}{dx^2} = 20x^3; y = x^5 + 3x - 2$

17. $\dfrac{dy}{dx} + y - 2\cos x = 0; y = \sin x + \cos x - e^{-x}$

18. $\dfrac{d^2y}{dx^2} - y + x^2 = 2; y = e^{-x} + x^2$

19. $\left(\dfrac{d^2y}{dx^2}\right)^2 + 4\left(\dfrac{dy}{dx}\right)^2 = 4; y = \sin x \cos x$

20. $\dfrac{d^2y}{dx^2} = 9y; y = e^{3x}$

21. $\dfrac{d^2y}{dx^2} - 5\left(\dfrac{dy}{dx}\right) + 4y = 0; y = e^{4x}$

22. $\dfrac{d^2y}{dx^2} + 2\left(\dfrac{dy}{dx}\right) + y = 0; y = e^{-x}$

23. $\dfrac{d^2y}{dx^2} + 2\left(\dfrac{dy}{dx}\right) + y = 0; y = xe^{-x}$

24. $\dfrac{d^2y}{dx^2} - 2\left(\dfrac{dy}{dx}\right) + y = -\dfrac{e^{-x}}{x^2}; y = e^x \ln x$

11.2 SEPARATION OF VARIABLES

There are numerous methods for solving ordinary differential equations. We present a few of these methods. Certain first-order differential equations can be solved most easily by using the method of separation of variables.

A first-order differential equation is a relation involving the first derivative. That is, it can be written in the form

$$N(x, y)\frac{dy}{dx} + M(x, y) = 0 \tag{1}$$

or (by multiplying each side by the differential dx, where $dx \neq 0$)

$$M(x, y)\, dx + N(x, y)\, dy = 0 \tag{2}$$

where $M(x, y)$ and $N(x, y)$ are functions involving the variables x and y.

EXAMPLE 1

Rewrite the first-degree differential equation $x^2y' - e^{xy} = 0$ in the form of Equation (2).

$$x^2y' - e^{xy} = 0$$
$$x^2\frac{dy}{dx} - e^{xy} = 0$$
$$x^2\, dy - e^{xy}\, dx = 0 \quad \text{(Multiply each side by } dx.)$$
$$-e^{xy}\, dx + x^2\, dy = 0$$

In this example, $M(x, y) = -e^{xy}$ and $N(x, y) = x^2$.

Some first-degree equations in the form $M(x, y)\, dx + N(x, y)\, dy = 0$ can be rewritten in the form

$$f(x)\, dx + g(y)\, dy = 0 \tag{3}$$

where $f(x)$ is a function of x alone and $g(y)$ is a function of y alone.

EXAMPLE 2

Rewrite the first-degree differential equation $x^2yy' - 2xy^3 = 0$ in the form of Equation (3).

$$x^2yy' - 2xy^3 = 0$$

$$x^2y \frac{dy}{dx} - 2xy^3 = 0$$

$$x^2y\, dy - 2xy^3\, dx = 0 \qquad \text{(Multiply each side by } dx.)$$

$$\left(\frac{1}{x^2y^3}\right)(x^2y\, dy - 2xy^3\, dx) = (0)\left(\frac{1}{x^2y^3}\right) \qquad \text{(Divide each side by } x^2y^3.)$$

$$\frac{1}{y^2}\, dy - \frac{2}{x}\, dx = 0$$

or

$$-\frac{2}{x}\, dx + \frac{1}{y^2}\, dy = 0$$

The process demonstrated in Example 2 is called *separating the variables*. By appropriate multiplications and divisions, we separate the equation into terms such that each term involves only one variable and its corresponding differential. Because of the expression e^{xy} in Example 1, it is impossible to separate the variables.

When a first-order differential equation can be separated so that we can collect all y terms with dy and all x terms with dx, then the general solution can be obtained by integrating each term. If we separate the variables to each side of the equation, the general solution of a differential equation in the form

$$f(x)\, dx = g(y)\, dy$$

is

$$\int f(x)\, dx = \int g(y)\, dy$$

$$F(x) = G(y) + C$$

where $F(x)$ is the antiderivative of $f(x)$, $G(y)$ is the antiderivative of $g(y)$, and C is the constant of integration.

EXAMPLE 3

Find the general solution of the differential equation $x^2yy' - 2xy^3 = 0$.

In Example 2 we wrote $x^2yy' - 2xy^3 = 0$ as

$$-\frac{2}{x}\, dx + \frac{1}{y^2}\, dy = 0$$

$$\frac{1}{y^2}\, dy = \frac{2}{x}\, dx$$

$$\int \frac{1}{y^2}\, dy = \int \frac{2}{x}\, dx \qquad \text{(Integrate each side of the equation.)}$$

$$-\frac{1}{y} = 2 \ln x + C$$

$$y = -\frac{1}{2 \ln x + C}$$

Note: When the solution of a differential equation involves integrating a term in the form $\dfrac{du}{u}$, we will now write $\displaystyle\int \dfrac{du}{u} = \ln u + C$ rather than $\displaystyle\int \dfrac{du}{u} = \ln |u| + C$. We now assume that the solution is valid only when u is positive. Remember also to include the constant of integration C.

EXAMPLE 4

Solve the differential equation $y' = \dfrac{y}{x^2 + 1}$.

Rewriting, we have

$$\frac{dy}{dx} = \frac{y}{x^2 + 1}$$

$$dy = \frac{y}{x^2 + 1}\, dx \qquad \text{(Multiply each side by } dx.\text{)}$$

$$\frac{dy}{y} = \frac{dx}{x^2 + 1} \qquad \left(\text{Multiply each side by } \frac{1}{y}.\right)$$

$$\int \frac{dy}{y} = \int \frac{dx}{x^2 + 1} \qquad \text{(Integrate each side.)}$$

$$\ln y = \text{Arctan } x + C$$

Using a calculator,

F3   alpha C

y 2nd = =y/(x^2+1),x,y) ENTER

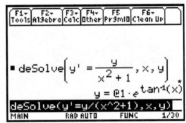

*see appendix D.17 for further information.

EXAMPLE 5

Solve the differential equation $x(1 + y^2) - y(1 + x^2)y' = 0$.

Rewriting, we have

$$x(1 + y^2) - y(1 + x^2)\frac{dy}{dx} = 0$$

$$x(1 + y^2)\, dx - y(1 + x^2)\, dy = 0$$

$$\frac{x}{1 + x^2}\, dx - \frac{y}{1 + y^2}\, dy = 0 \qquad [\text{Divide each side by } (1 + y^2)(1 + x^2).]$$

$$\frac{x}{1 + x^2}\, dx = \frac{y}{1 + y^2}\, dy$$

$$\int \frac{x}{1 + x^2}\, dx = \int \frac{y}{1 + y^2}\, dy$$

$$\frac{1}{2}\ln (1 + x^2) = \frac{1}{2}\ln (1 + y^2) + C$$

$$\frac{1}{2}\ln (1 + x^2) - \frac{1}{2}\ln (1 + y^2) = C$$

$$\frac{1}{2}\ln \left(\frac{1 + x^2}{1 + y^2}\right) = C$$

Since C is an arbitrary constant, we could rewrite this constant as $C = \frac{1}{2} \ln k$ where $k > 0$. Then we have

$$\frac{1}{2} \ln \left(\frac{1 + x^2}{1 + y^2} \right) = \frac{1}{2} \ln k$$

$$\frac{1 + x^2}{1 + y^2} = k$$

$$1 + x^2 = k + ky^2$$

$$x^2 - ky^2 + 1 - k = 0$$

This last equation is easier to work with since it no longer involves natural logarithms. The equations $\frac{1}{2} \ln \left(\frac{1 + x^2}{1 + y^2} \right) = C$ and $x^2 - ky^2 + 1 - k = 0$ are equivalent. They differ only in the form of the constant of integration. By working the exercises, you will gain experience in choosing the most appropriate form for this arbitrary constant.

EXAMPLE 6

Solve the differential equation $y' - 2x = 0$.

Rewriting, we have

$$\frac{dy}{dx} = 2x$$

$$dy = 2x \, dx$$

$$\int dy = \int 2x \, dx$$

$$y = x^2 + C$$

This general solution represents a family of functions where each function $y = f(x)$ is a particular solution of the differential equation $y' - 2x = 0$. In this case, the solution is a family of parabolas some of which are sketched in Fig. 11.1.

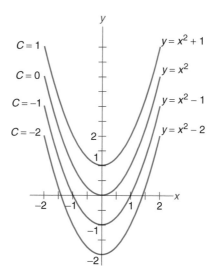

Figure 11.1 Family of parabolas $y = x^2 + C$.

A unique particular solution can be obtained if certain initial conditions are given. An *initial condition* of a differential equation is a condition that specifies a particular value of y, y_0, corresponding to a particular value of x, x_0. That is, if $y = f(x)$ is a solution of the differential equation, then the function must satisfy the condition $y_0 = f(x_0)$. A differential equation with initial conditions is called an *initial value problem.*

EXAMPLE 7

Solve the differential equation $y' - 2x = 0$ subject to the initial condition that $y = 1$ when $x = 2$.

From Example 6 the general solution is

$$y = x^2 + C$$

Substituting $y = 1$ and $x = 2$, we have

$$1 = (2)^2 + C$$
$$-3 = C$$

So the particular solution is

$$y = x^2 - 3$$

EXAMPLE 8

Solve the differential equation $y + xy' = 0$ subject to the initial condition that $y = 2$ when $x = 3$, which may also be written $y(3) = 2$.

Rewriting, we have

$$y + x\frac{dy}{dx} = 0$$

$$y\,dx + x\,dy = 0$$

$$\frac{dx}{x} + \frac{dy}{y} = 0$$

$$\frac{dx}{x} = -\frac{dy}{y}$$

$$\int \frac{dx}{x} = -\int \frac{dy}{y}$$

$$\ln x = -\ln y + C$$

$$\ln x + \ln y = C$$

$$\ln xy = C$$

or

$$\ln xy = \ln k \qquad \text{where } C = \ln k$$

$$xy = k$$

Substituting $y = 2$ when $x = 3$, we have

$$(3)(2) = k$$
$$6 = k$$

So the required particular solution is

$$xy = 6$$

or

$$y = \frac{6}{x}$$

Using a calculator,

F3 alpha C y+x*y **2nd** = =0,x,y) **ENTER** **F3 alpha C** y+x*y **2nd** = =0 **2nd MATH 8 8** y(3)=2,x,y) **ENTER**

Exercises 11.2

Solve each differential equation.

1. $x\,dy - y^2\,dx = 0$
2. $3x^3y^2\,dx - xy\,dy = 0$
3. $x\,dy + y\,dx = 0$

4. $\sec x\,dy + \csc y\,dx = 0$
5. $\dfrac{dy}{dx} = y^{3/2}$
6. $(y^2 - 4)\dfrac{dy}{dx} = 1$

7. $\dfrac{dy}{dx} = x^2 + x^2y^2$
8. $y\dfrac{dy}{dx} = \sin x$
9. $x\dfrac{dy}{dx} + y = 3$

10. $\dfrac{dy}{dx} = 1 - y$
11. $\dfrac{dy}{dx} = \dfrac{x^2}{y}$
12. $\dfrac{dy}{dx} = \dfrac{xy}{x^2 + 3}$

13. $\dfrac{dy}{dx} + y^3\cos x = 0$
14. $\dfrac{dy}{dx} - \dfrac{e^x}{y^2} = 0$
15. $e^{3x}\dfrac{dy}{dx} - e^x = 0$

16. $x\sqrt{1 - y^2}\,dx - 3\,dy = 0$
17. $(1 + x^2)\,dy - dx = 0$
18. $(1 + x^2)\,dy + x\,dx = 0$

19. $\dfrac{dy}{dx} = 1 + x^2 + y^2 + x^2y^2$
20. $\dfrac{dy}{dx} = e^{x+y}$
21. $y' = e^{x-y}$

22. $y' = xe^{x^2+2y}$
23. $(x + 1)y' = y^2 + 4$
24. $\sqrt{1 - 16y^2} = (4x^2 + 9)y'$

25. $(4xy + 12x)\,dx = (5x^2 + 5)\,dy$
26. $8x^3(1 - y^2) = 3y(1 + x^4)y'$

Find the particular solution of each differential equation subject to the given conditions.

27. $\dfrac{dy}{dx} = x^2y^4;\ y = 1$ when $x = 1$
28. $ye^{-x}\dfrac{dy}{dx} + 2 = 0;\ y = 2$ when $x = 0$

29. $\dfrac{dy}{dx} = \dfrac{2x}{y + x^2y};\ y = 4$ when $x = 0$
30. $x^2\,dy = y\,dx;\ y = 1$ when $x = 1$

31. $y\dfrac{dy}{dx} = e^x;\ y(0) = 6$
32. $y^2x\dfrac{dy}{dx} - 2x + 4 = 0;\ y(1) = 3$

33. $\sqrt{x} + \sqrt{y}\dfrac{dy}{dx} = 0;\ y(1) = 4$
34. $e^{-2y}\dfrac{dy}{dx} = x - 2;\ y(1) = 0$

35. $xy\dfrac{dy}{dx} = \ln x;\ y(1) = 0$
36. $\dfrac{dy}{dx} = e^{x-y};\ y(0) = 1$

11.3 USE OF INTEGRATING FACTORS

Not all differential equations can be solved by separating the variables. There are some differential equations, however, that can be solved by recognizing a combination of differentials that can be integrated. For example, the left-hand side of the differential equation

$$\frac{x\,dy - y\,dx}{x^2} = 3\,dx$$

we recognize as

$$d\left(\frac{y}{x}\right) = \frac{x\,dy - y\,dx}{x^2}$$

so we can integrate as follows:

$$\int \frac{x\,dy - y\,dx}{x^2} = \int 3\,dx$$

$$\int d\left(\frac{y}{x}\right) = \int 3\,dx$$

$$\frac{y}{x} = 3x + C$$

$$y = 3x^2 + Cx$$

is the general solution.

Some easily recognizable differentials are expressed below:

1. $d\left(\dfrac{x}{y}\right) = \dfrac{y\,dx - x\,dy}{y^2}$

2. $d\left(\dfrac{y}{x}\right) = \dfrac{x\,dy - y\,dx}{x^2} = -\left(\dfrac{y\,dx - x\,dy}{x^2}\right)$

3. $d\left(\text{Arctan}\,\dfrac{y}{x}\right) = \dfrac{x\,dy - y\,dx}{x^2 + y^2}$

4. $d\left(\text{Arctan}\,\dfrac{x}{y}\right) = \dfrac{y\,dx - x\,dy}{x^2 + y^2}$

5. $d(xy) = x\,dy + y\,dx$

6. $d(\ln\sqrt{x^2 + y^2}) = \dfrac{x\,dx + y\,dy}{x^2 + y^2}$

7. $d(x^2 + y^2) = 2(x\,dx + y\,dy)$

Thus, be on the lookout for the terms $x\,dy - y\,dx$, $x\,dy + y\,dx$, and $x\,dx + y\,dy$ appearing in a differential equation. It may be possible to integrate the differential equation after multiplying or dividing the equation by an appropriate expression, called the *integrating factor*.

EXAMPLE 1

Solve the differential equation

$$x\,dy - y\,dx = (x^2 + y^2)\,dy$$

Divide each side of the equation by $(x^2 + y^2)$.

$$\frac{x\,dy - y\,dx}{x^2 + y^2} = dy$$

$$d\left(\text{Arctan}\,\frac{y}{x}\right) = dy$$

$$\int d\left(\text{Arctan}\,\frac{y}{x}\right) = \int dy$$

Then

$$\text{Arctan} \frac{y}{x} = y + C$$

is the general solution.

EXAMPLE 2

Solve the differential equation

$$x \, dy + y \, dx = 3xy \, dx$$

Divide each side of the equation by xy.

$$\frac{x \, dy + y \, dx}{xy} = 3 \, dx$$

$$\frac{d(xy)}{xy} = 3 \, dx \qquad [d(xy) = x \, dy + y \, dx]$$

$$\int \frac{d(xy)}{xy} = \int 3 \, dx$$

Then

$$\ln xy = 3x + C$$

is the general solution.

EXAMPLE 3

Find the particular solution of the differential equation

$$y \, dx - x \, dy + dx = 4x^4 \, dx$$

subject to the initial condition that $y = \frac{1}{3}$ when $x = 1$.

Rewriting, we have

$$y \, dx - x \, dy = 4x^4 \, dx - dx$$

$$\frac{y \, dx - x \, dy}{x^2} = 4x^2 \, dx - \frac{dx}{x^2} \qquad \text{(Divide each side by } x^2.\text{)}$$

$$-d\left(\frac{y}{x}\right) = 4x^2 \, dx - \frac{dx}{x^2}$$

$$-\int d\left(\frac{y}{x}\right) = \int 4x^2 \, dx - \int \frac{dx}{x^2}$$

$$-\frac{y}{x} = \frac{4x^3}{3} + \frac{1}{x} + C$$

and $y = -\frac{4}{3}x^4 - 1 - Cx$ is the general solution. Substituting $y = \frac{1}{3}$ when $x = 1$, we have

$$\left(\frac{1}{3}\right) = -\frac{4}{3}(1)^4 - 1 - C(1)$$

$$\frac{1}{3} = -\frac{4}{3} - 1 - C$$

$$-\frac{8}{3} = C$$

So

$$y = -\frac{4}{3}x^4 + \frac{8}{3}x - 1$$

is the desired particular solution.

Exercises 11.3

Solve each differential equation.

1. $x \, dy + y \, dx = y^2 \, dy$

2. $x \, dx + y \, dy = (x^2 + y^2) \, dx$

3. $x \, dy - y \, dx = 5x^2 \, dy$

4. $x \, dx + y \, dy = x^3 \, dx$

5. $y \, dx - x \, dy + y^2 \, dx = 3 \, dy$

6. $x \, dy - y \, dx = x^4 \, dx + x^2 y^2 \, dx$

7. $x\sqrt{x^2 + y^2} \, dx - 2x \, dx = 2y \, dy$

8. $x \, dy - y \, dx + x^4 y^2 \, dx = 0$

9. $x \, dx + y \, dy = (x^3 + xy^2) \, dy + (x^2 y + y^3) \, dx$

10. $(y + x) \, dx + (y - x) \, dy = (x^2 + y^2) \, dx$

Find the particular solution of each differential equation for the given initial conditions.

11. $x \, dy + y \, dx = 2x \, dx + 2y \, dy$ for $y = 1$ when $x = 0$

12. $y \, dx - x \, dy = y^2 \, dx$ for $y = 3$ when $x = 1$

13. $x \, dy - 4 \, dx = (x^3 + xy^2) \, dy + (x^2 y + y^3) \, dx$ for $y = 2$ when $x = 2$

14. $x \, dy - y \, dx = x^5 \, dx$ for $y = 4$ when $x = 1$

11.4   LINEAR EQUATIONS OF FIRST ORDER

A differential equation where both the unknown function $y = f(x)$ and its derivatives are of first degree is called a *linear differential equation*. A linear differential equation of first order can be written in the form

$$\frac{dy}{dx} + y \, P(x) = Q(x) \tag{1}$$

or

$$dy + y \, P(x) \, dx = Q(x) \, dx \tag{2}$$

where $P(x)$ and $Q(x)$ are functions of x.

A method has been devised for solving linear equations of first order. *An integrating factor* is an expression that, when multiplied on both sides of a differential equation, gives a differential equation that can be integrated in order to find its solution. We will show that $e^{\int P(x) \, dx}$ is an integrating factor for these equations. Multiplying each side of Equation (2) by $e^{\int P(x) \, dx}$, we have

$$e^{\int P(x) \, dx} \, dy + e^{\int P(x) \, dx} \, y \, P(x) \, dx = e^{\int P(x) \, dx} \, Q(x) \, dx$$

First, observe that

$$d\left(y e^{\int P(x) \, dx}\right) = y \, d\left(e^{\int P(x) \, dx}\right) + e^{\int P(x) \, dx} \, dy$$

$$= y e^{\int P(x) \, dx} \, d\left[\int P(x) \, dx\right] + e^{\int P(x) \, dx} \, dy$$

$$= y e^{\int P(x) \, dx} \, P(x) \, dx + e^{\int P(x) \, dx} \, dy$$

$$= e^{\int P(x) \, dx} \left[y P(x) \, dx + dy\right]$$

$$= e^{\int P(x) \, dx} \, Q(x) \, dx \qquad \text{[From Equation (2)]}$$

Thus

$$d\left(y e^{\int P(x) \, dx}\right) = \left[Q(x) e^{\int P(x) \, dx}\right] dx$$

$$\int d\left(y e^{\int P(x) \, dx}\right) = \int \left[Q(x) e^{\int P(x) \, dx}\right] dx \qquad \text{(Integrate each side.)}$$

$$ye^{\int P(x)\,dx} = \int Q(x)e^{\int P(x)\,dx}\,dx$$

So $e^{\int P(x)\,dx}$ is an integrating factor. The general solution of the first-order linear differential equation

$$\frac{dy}{dx} + y\,P(x) = Q(x)$$

is

$$ye^{\int P(x)\,dx} = \int Q(x)e^{\int P(x)\,dx}\,dx$$

EXAMPLE 1

Solve $\dfrac{dy}{dx} + 2xy = 5x$.

Here $P(x) = 2x$, $Q(x) = 5x$, and $\displaystyle\int P(x)\,dx = \int 2x\,dx = x^2$. We do not write a constant of integration for $\displaystyle\int P(x)\,dx$ because we are merely obtaining an integrating factor. The solution is

$$ye^{\int P(x)\,dx} = \int Q(x)e^{\int P(x)\,dx}\,dx$$

$$ye^{x^2} = \int 5xe^{x^2}\,dx$$

$$ye^{x^2} = \int \frac{5}{2}e^{u}\,du$$

$$\boxed{\begin{array}{l} \text{let } u = x^2 \\ du = 2x\,dx \end{array}}$$

$$ye^{x^2} = \frac{5}{2}e^{u} + C$$

$$ye^{x^2} = \frac{5}{2}e^{x^2} + C$$

$$y = \frac{5}{2} + Ce^{-x^2}$$

EXAMPLE 2

Solve $y' + y = e^{-x}\cos x$.

Here $P(x) = 1$, $Q(x) = e^{-x}\cos x$, and $\displaystyle\int P(x)\,dx = \int dx = x$. The solution is

$$ye^{\int P(x)\,dx} = \int Q(x)e^{\int P(x)\,dx}\,dx$$

$$ye^{x} = \int (e^{-x}\cos x)(e^{x})\,dx$$

$$ye^{x} = \int \cos x\,dx$$

$$ye^{x} = \sin x + C$$

$$y = e^{-x}\sin x + Ce^{-x}$$

EXAMPLE 3

Solve $(2y - 6x^2)\,dx + x\,dy = 0$

Rewriting, this equation becomes

$$dy + \frac{2y}{x}\,dx = 6x\,dx$$

where $P(x) = \dfrac{2}{x}$, $Q(x) = 6x$, and

$$\int P(x)\,dx = \int \frac{2}{x}\,dx = 2 \ln x = \ln x^2$$

The solution is then

$$ye^{\int P(x)\,dx} = \int Q(x)e^{\int P(x)\,dx}\,dx$$

$$ye^{\ln x^2} = \int 6xe^{\ln x^2}\,dx$$

$$yx^2 = \int 6x^3\,dx \qquad\qquad (\text{since } e^{\ln x^2} = x^2)$$

$$yx^2 = \frac{3}{2}x^4 + C$$

$$y = \frac{3}{2}x^2 + \frac{C}{x^2}$$

Exercises 11.4

Solve each differential equation.

1. $\dfrac{dy}{dx} - 5y = e^{3x}$ 　　　　　　　**2.** $\dfrac{dy}{dx} + 3y = e^{-2x}$

3. $\dfrac{dy}{dx} + \dfrac{3y}{x} = x^3 - 2$ 　　　　**4.** $\dfrac{dy}{dx} - \dfrac{2y}{x} = x^2 + 5$

5. $y' + 2xy = e^{3x}(3 + 2x)$ 　　　　**6.** $y' - 3x^2y = e^x(3x^2 - 1)$

7. $dy - 4y\,dx = x^2e^{4x}\,dx$ 　　　　　**8.** $dy - 3x^2y\,dx = x^2\,dx$

9. $x\,dy - 5y\,dx = (x^6 + 4x)\,dx$ 　　**10.** $x\dfrac{dy}{dx} + 2y = (x^2 + 4)^3$

11. $(1 + x^2)\,dy + 2xy\,dx = 3x^2\,dx$ 　　**12.** $\dfrac{dy}{dx} - y \tan x = \sin x$

13. $x^2y' + 2xy = x^4 - 7$ 　　　　　**14.** $x^2y' - 2xy = x^3 + 5$

15. $\dfrac{dy}{dx} + 2y = e^{-x}$ 　　**16.** $(x + 1)\dfrac{dy}{dx} + 5y = 10$ 　　**17.** $x\dfrac{dy}{dx} - y = 3x^2$

18. $\dfrac{dy}{dx} + \dfrac{y}{3x - 1} = 8$ 　　**19.** $y' + y\cos x = \cos x$ 　　**20.** $\dfrac{dy}{dx} + y\sec^2 x = \sec^2 x$

Find a particular solution of each differential equation subject to the given initial conditions.

21. $\dfrac{dy}{dx} - 3y = e^{2x}$; $y = 2$ when $x = 0$ 　　**22.** $\dfrac{dy}{dx} - \dfrac{y}{x} = x^2 + 3$; $y = 3$ when $x = 1$

23. $\dfrac{dy}{dx} = \csc x - y \cot x$; $y\left(\dfrac{\pi}{2}\right) = \dfrac{3\pi}{2}$ 　　**24.** $dy = (x - 3y)\,dx$; $y(0) = 1$

25. $y' = e^x - y$; $y(0) = \dfrac{3}{2}$ 　　　　**26.** $y' + 8x^2y = 4x^2$; $y(0) = 2$

27. $x\dfrac{dy}{dx} + y = 3$; $y(1) = -2$ 　　**28.** $\dfrac{dy}{dx} - \dfrac{3y}{x - 4} = (x - 4)^2$; $y(5) = 3$

29. $xy' + y = 4x^3$; $y(2) = 3$ 　　　　**30.** $y' + y\sin x = \sin x$; $y(\pi/2) = 3$

11.5 APPLICATIONS OF FIRST-ORDER DIFFERENTIAL EQUATIONS

Many technical problems involve first-order, first-degree differential equations. In the applications that follow, observe that certain phenomena involve a rate of change (a derivative). The presence of a derivative often leads to a differential equation that describes the physical situation in mathematical terms. We solve the differential equation, and the mathematical solution is given a physical interpretation that provides a solution to the original technical problem.

As with all mathematical applications, we create a mathematical model of certain physical phenomena. The mathematical model is usually an approximation. The model is only as accurate as the interpretation given the physical phenomena.

EXAMPLE 1   RADIOACTIVE DECAY

Radioactive material decays at a rate proportional to the amount present. For a certain radioactive substance, approximately 10% of the original quantity decomposes in 25 years. Find the half-life of this radioactive material. That is, find the time that elapses for the quantity of material to decay to one-half of its original quantity.

Step 1: Set up the mathematical model.

Let Q_0 represent the original quantity present and Q the amount present at any time t (in years). The mathematical model describing the observed rate of decay process is

$$\frac{dQ}{dt} = kQ$$

where k is the constant of proportionality of decay.

Step 2: Solve the differential equation.

Separate the variables:

$$dQ = kQ \, dt$$
$$\frac{dQ}{Q} = k \, dt$$

then

$$\int \frac{dQ}{Q} = \int k \, dt$$
$$\ln Q = kt + \ln C$$

We have the initial condition that $Q = Q_0$ when $t = 0$. So

$$\ln Q_0 = k(0) + \ln C$$
$$\ln Q_0 = \ln C$$
$$C = Q_0$$

We then have

$$\ln Q = kt + \ln Q_0$$
$$\ln Q - \ln Q_0 = kt$$
$$\ln \frac{Q}{Q_0} = kt$$
$$\frac{Q}{Q_0} = e^{kt} \qquad \text{(Rewrite each side as a power of } e.\text{)}$$
$$Q = Q_0 e^{kt}$$

which is an expression for the amount of radioactive material present after t years where k is the constant of proportionality of decay.

Step 3: Solve the particular problem in question.

To find t, the material's half-life, first find k. We are given that after 25 years, 10% of the material has decayed. That is, at $t = 25$ years, $Q = (1 - 0.1)Q_0 = 0.9Q_0$. Thus

$$0.9Q_0 = Q_0 e^{25k}$$

$$0.9 = e^{25k}$$

$$\ln 0.9 = 25k$$

$$\frac{\ln 0.9}{25} = k$$

$$-0.00421 = k$$

Now determine the half-life. Since the **half-life** represents the time for the material to decay to half its original quantity (that is, $Q = Q_0/2$), the half-life can be determined by substituting $Q = Q_0/2$ in the equation $Q = Q_0 e^{kt}$.

$$\frac{Q_0}{2} = Q_0 e^{kt}$$

$$\frac{1}{2} = e^{kt}$$

$$\ln \frac{1}{2} = kt \qquad \text{(Take the ln of each side.)}$$

$$-\ln 2 = kt \qquad (\ln \tfrac{1}{2} = \ln 1 - \ln 2 = 0 - \ln 2)$$

$$\frac{-\ln 2}{k} = t \qquad \textbf{(half-life formula)}$$

$$\frac{-\ln 2}{-0.00421} = t$$

$$t = 165 \text{ years}$$

EXAMPLE 2   ELECTRICAL CIRCUIT

The current i in a series circuit with constant inductance L, constant resistance R, and a constant applied voltage V is described by the differential equation

$$L \frac{di}{dt} + Ri = V$$

Find an equation for the current i as a function of time t.

Rewrite the equation as

$$\frac{di}{dt} + \frac{R}{L} i = \frac{V}{L}$$

and apply the method described in Section 11.4:

$$P(t) = \frac{R}{L} \qquad Q(t) = \frac{V}{L}$$

and

$$\int P(t)\, dt = \int \frac{R}{L}\, dt = \frac{R}{L} t$$

and

$$i e^{(R/L)t} = \int \frac{V}{L} e^{(R/L)t}\, dt$$

$$= \frac{V}{L} \int e^{(R/L)t}\, dt$$

$$= \frac{V}{L} \cdot \frac{L}{R} e^{(R/L)t} + C$$

$$= \frac{V}{R} e^{(R/L)t} + C$$

If there is no initial current, then $i = 0$ when $t = 0$. So we have

$$0 \cdot e^{(R/L)(0)} = \frac{V}{R} e^{(R/L)(0)} + C$$

$$0 = \frac{V}{R} + C$$

$$C = -\frac{V}{R}$$

Finally, we solve for i:

$$i e^{(R/L)t} = \frac{V}{R} e^{(R/L)t} - \frac{V}{R}$$

$$i = \frac{\frac{V}{R}(e^{(R/L)t} - 1)}{e^{(R/L)t}}$$

$$i = \frac{V}{R}(1 - e^{-(R/L)t})$$

This shows that V/R is a limiting value for the current i. As $t \to \infty$,

$$e^{-(R/L)t} = \frac{1}{e^{(R/L)t}} \to 0$$

and

$$i = \frac{V}{R}\left(1 - \frac{1}{e^{(R/L)t}}\right) \to \frac{V}{R}(1 - 0) = \frac{V}{R} \qquad \text{(see Fig. 11.2)}$$

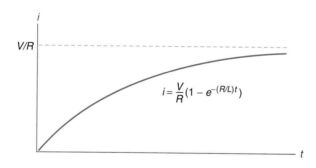

Figure 11.2 As $t \to \infty$, $i \to \dfrac{V}{R}$.

The equation for i involves two terms, V/R and $(V/L)e^{-(R/L)t}$. The limit V/R is called the *steady state solution* and represents the solution when no inductance is present ($Ri = V$); the other term, $(V/R)e^{-(R/L)t}$, represents the effect of inductance. The addition of this term gives what is called the *transient solution*. Thus the study of this type of differential equation enables us to understand the effect of inductance in an electrical circuit.

EXAMPLE 3 MIXTURES

Salt is being dissolved in a tank filled with 100 gallons (gal) of water. Originally 25 lb of salt was dissolved in the tank. Saltwater containing 1 lb of salt per gallon is poured in at the rate of

3 gal/min; the solution is kept well stirred, and the mixture is poured out at the same rate of 3 gal/min. Find an expression for the amount of salt Q in the tank at time t. How much salt remains after 1 h?

The rate of change of salt $\dfrac{dQ}{dt}$ is equal to the rate at which salt enters the tank minus the rate at which it leaves. That is,

$$\frac{dQ}{dt} = \text{rate of gain} - \text{rate of loss}$$

$$\text{rate of gain} = (1 \text{ lb/gal})(3 \text{ gal/min}) = 3 \text{ lb/min}$$

$$\text{rate of loss} = \left(\frac{Q \text{ lb}}{100 \text{ gal}}\right)(3 \text{ gal/min}) = \frac{3Q}{100} \text{ lb/min}$$

so

$$\frac{dQ}{dt} = 3 - \frac{3Q}{100}$$

$$\frac{dQ}{dt} = 3\left(1 - \frac{Q}{100}\right)$$

$$\frac{dQ}{dt} = 3\left(\frac{100 - Q}{100}\right)$$

Separating the variables, we have

$$\frac{dQ}{100 - Q} = \frac{3}{100} dt$$

or rewriting,

$$\int \frac{dQ}{Q - 100} = \int -\frac{3}{100} dt$$

$$\ln(Q - 100) = -\frac{3}{100}t + \ln C$$

$$\ln(Q - 100) - \ln C = -\frac{3}{100}t$$

$$\ln\left(\frac{Q - 100}{C}\right) = -\frac{3}{100}t$$

$$\frac{Q - 100}{C} = e^{-(3/100)t}$$

$$Q = Ce^{-(3/100)t} + 100$$

At $t = 0$, $Q = 25$, so

$$25 = Ce^{-(3/100)(0)} + 100$$

$$25 = C + 100$$

$$C = -75$$

So

$$Q = -75e^{-(3/100)t} + 100$$

After 1 h = 60 min, we have

$$Q = -75e^{-(3/100)(60)} + 100$$

$$= -75e^{-1.8} + 100$$

$$= -12.4 + 100$$

$$= 87.6 \text{ lb}$$

Using a calculator,

F3 alpha C alpha Q 2nd = =3-3 alpha Q /100 2nd MATH 8 8 alpha Q (0)=25,t, alpha Q) ENTER
up arrow **ENTER** |t=60 **green diamond ENTER (≈)**

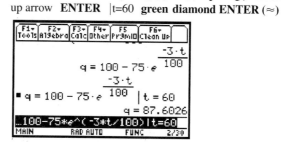

EXAMPLE 4   TEMPERATURE CHANGE

Under certain conditions the temperature of an object changes at a rate proportional to the difference between the temperature outside the object and the temperature of the object (Newton's law of cooling). A thermometer registering 75°F is taken outside where the temperature is 40°F. After 5 min, the thermometer registers 60°F. Find the temperature reading on the thermometer 7 min after having been outside.

Let T represent the temperature reading of the thermometer at any time t

$$\frac{dT}{dt} = k(T - T_{\text{outside}})$$

$$\frac{dT}{dt} = k(T - 40)$$

The equation is subject to the conditions $T = 75$ when $t = 0$ and $T = 60$ when $t = 5$.

Separating the variables, we obtain

$$\frac{dT}{T - 40} = k \, dt$$

$$\int \frac{dT}{T - 40} = \int k \, dt$$

$$\ln (T - 40) = kt + \ln C$$

or

$$T = Ce^{kt} + 40$$

Since $T = 75$ when $t = 0$, we can find C:

$$75 = Ce^{k(0)} + 40$$

$$75 - 40 = C$$

$$35 = C$$

So

$$T = 35e^{kt} + 40$$

Now determine k. Since $T = 60$ when $t = 5$, we have

$$60 = 35e^{k(5)} + 40$$

$$\frac{20}{35} = e^{5k}$$

$$\ln \frac{20}{35} = 5k$$

$$k = -0.11$$

So

$$T = 35e^{-0.11t} + 40$$

After 7 min, we have

$$T = 35e^{(-0.11)(7)} + 40$$
$$= 35e^{-0.77} + 40$$
$$= 56°F$$

Exercises 11.5

1. Find the rate of change of velocity with respect to time of a body moving along a straight line with acceleration $a = \dfrac{dv}{dt}$. A body is moving along a straight line with constant acceleration $a = 5$ m/s². If the body had an initial velocity of $v = 10$ m/s (when $t = 0$), find the equation for velocity. Find the velocity at $t = 3$ s.

2. Find the equation for the velocity of an object moving along a straight line with acceleration $a = t^2$ m/s², if the object started moving from rest. Find the velocity at $t = 2$ s. (See Exercise 1.)

3. Find the equation of the curve passing through the point $(1, 1)$ with slope $\dfrac{x^2 - y}{x}$.

4. Find the equation of the curve passing through the point $(-1, 0)$ with slope $\dfrac{x^2}{y^3}$.

5. Find the equation for the current i in a series circuit with inductance $L = 0.1$ H, resistance $R = 80$ Ω, and voltage $V = 120$ volts if the initial current $i_0 = 2$ amperes when $t = 0$. (See Example 2.)

6. Find the equation relating charge q in terms of time t in a series circuit containing only a resistance and a capacitance if the initial charge $q_0 = 3$ C when $t = 0$. This circuit is described by the differential equation $Ri + \dfrac{q}{C} = 0$ where $i = \dfrac{dq}{dt}$.

7. Find the equation for the current i in a series circuit with inductance $L = 0.1$ H, resistance $R = t$ ohms, and voltage $V = t$ volts. The initial current $i_0 = 0$ when $t = 0$.

8. Find the equation for the current i in a series circuit with inductance $L = 0.1$ H, resistance $R = t^2$ ohms, and voltage $V = t^2$ volts. The initial current $i_0 = 2$ amperes when $t = 0$.

9. The isotope ^{238}U of uranium has a half-life of 4.5×10^9 years. Determine the amount of ^{238}U left after t years if the initial quantity is 1 g. ^{238}U decays at a rate proportional to the amount present. (See Example 1.)

10. The isotope ^{234}U of uranium has a half-life of 2.7×10^5 years. Determine the amount of ^{234}U left after t years if the initial quantity is 2 g. ^{234}U decays at a rate proportional to the amount present.

11. A radioactive material with mass 5 g decays 10% after 36 h. The material decays at a rate proportional to the amount present. Find an equation expressing the amount of material present at any given time t. Find its half-life.

12. A radioactive material with an original mass of 10 g has a mass of 8 g after 50 years. The material decays at a rate proportional to the amount present. Find an expression for the amount of material present at any time t. Find its half-life.

13. Fifty gallons of brine originally containing 10 lb of salt is in a tank. Saltwater containing 2 lb of salt per gallon is poured in at the rate of 2 gal/min, the solution is kept well stirred, and the mixture is poured out at the same rate of 2 gal/min. Find the amount of salt remaining after 30 min. (See Example 3.)

14. One hundred gallons of brine originally containing 20 lb of salt is in a tank. Pure water is poured in at the rate of 1 gal/min, the solution is kept well stirred, and the mixture is poured out at the same rate of 1 gal/min. Find the amount of salt remaining after 1 h.

15. An object at 90°C is cooled in air, which is at 10°C. If the object is 70°C after 5 min, find the temperature of the object after 30 min. (See Example 4.)

16. A thermometer registering 80°F is taken outside where the temperature is 35°F. After 3 min, the thermometer registers 50°F. Find the temperature reading on the thermometer 5 min after having been outside.

17. Populations tend to have a growth rate that is proportional to the present population. That is, $y' = ky$, where y is the present population. Suppose that the population of a country has doubled in the last 20 years. Find the expected population in 80 years if the current population is 2,000,000. (*Hint:* First solve the differential equation $y' = ky$ before determining k.)

18. An amount of money earning compound interest continuously also satisfies the growth equation $y' = ky$ (see Exercise 17), where k is the rate of compound interest. Find the value of a deposit of $500 after 10 years if it is earning 9% interest compounded continuously.

19. A gas undergoing an adiabatic change has a rate of change of pressure with respect to volume that is directly proportional to the pressure and inversely proportional to the volume. Solve the resulting differential equation and express the pressure in terms of the volume.

20. A lake contains 5×10^7 litres (L) of water. Industrial waste in the form of a chemical compound begins to be dumped into the lake at the rate of 5 L/h. A freshwater stream feeds the lake at a rate of 345 L/h. Determine how long it will take for the lake in Fig. 11.3 to become polluted if the lake is considered polluted when the water contains 0.2% of the chemical compound. (*Hint:* $Q' = 5 - 7 \times 10^{-6}Q$, where Q is the amount of the chemical compound present in the water at any time t.)

Figure 11.3

CHAPTER SUMMARY

1. The order of a differential equation is n, if n is the highest order derivative that appears in the equation.

2. The degree of a differential equation is the highest power of the derivative of highest order.

3. A solution of a differential equation is a function $y = f(x)$ that together with its derivatives satisfies the given differential equation.

4. *Separation of variables method:* By appropriate multiplications and divisions, separate the differential equation into terms where each term involves only one variable and its differential.

5. Some easily recognizable differentials:

$$\text{(a) } d\left(\frac{x}{y}\right) = \frac{y\,dx - x\,dy}{y^2}$$

$$\text{(b) } d\left(\frac{y}{x}\right) = \frac{x\,dy - y\,dx}{x^2} = -\left(\frac{y\,dx - x\,dy}{x^2}\right)$$

$$\text{(c) } d\left(\text{Arctan}\,\frac{y}{x}\right) = \frac{x\,dy - y\,dx}{x^2 + y^2}$$

$$\text{(d) } d\left(\text{Arctan}\,\frac{x}{y}\right) = \frac{y\,dx - x\,dy}{x^2 + y^2}$$

$$\text{(e) } d(xy) = x\,dy + y\,dx$$

$$\text{(f) } d(\ln \sqrt{x^2 + y^2}) = \frac{x\,dx + y\,dy}{x^2 + y^2}$$

$$\text{(g) } d(x^2 + y^2) = 2(x\,dx + y\,dy)$$

6. *Linear differential equation of the first order:* A differential equation where both the unknown function $y = f(x)$ and its derivative are of first power. A method for solving such equations involves using the integrating factor, $e^{\int P(x)\,dx}$. The general solution of

$$\frac{dy}{dx} + y\,P(x) = Q(x) \text{ is}$$

$$ye^{\int P(x)\,dx} = \int Q(x)e^{\int P(x)\,dx}\,dx$$

CHAPTER 11 REVIEW

State the order and degree of each differential equation.

1. $\dfrac{d^2y}{dx^2} - 3x\dfrac{dy}{dx} + x^2y^3 = 0$

2. $\left(\dfrac{dy}{dx}\right)^2 - xy\dfrac{dy}{dx} = 3$

3. $(y'')^3 - 3y' + x^2y^5 = 2$

4. $\left(\dfrac{dy}{dx}\right)^2 + 3y^3 = 7$

Verify that each function $y = f(x)$ is a solution of the given differential equation.

5. $\dfrac{d^2y}{dx^2} + 3y = 3x^3;\ y = x^3 - 2x$

6. $y'' - 2y' + y = 4e^{-x};\ y = 2e^x - 3xe^x + e^{-x}$

7. $y'' + y = x^2 + 2;\ y = \sin x + x^2$

8. $\dfrac{dy}{dx} - 2y = 5x^4e^{2x};\ y = e^{2x}(x^5 - 1)$

Solve each differential equation.

9. $x^3\dfrac{dy}{dx} - y = 0$

10. $e^{2x}\dfrac{dy}{dx} - 3y = 0$

11. $(9 + x^2)y^2\,dy + x\,dx = 0$

12. $\dfrac{dy}{dx} = e^y \sec^2 x$

13. $x\,dy - y\,dx = 3x^4\,dx$

14. $\dfrac{dy}{dx} + 6x^2y = 12x^2$

15. $\dfrac{dy}{dx} - \dfrac{y}{x^2} = \dfrac{5}{x^2}$

16. $x\,dx + y\,dy = (x^2y + y^3)\,dy$

17. $x\,dy - y\,dx = (x^4 + x^2y^2)\,dx$

18. $x\,dy + y\,dx = 14x^5\,dx$

19. $dy + 3y\,dx = e^{-2x}\,dx$

20. $\dfrac{dy}{dx} - \dfrac{5y}{x} = x^3 + 7$

21. $x\,dy - 3y\,dx = (4x^3 - x^2)\,dx$

22. $\dfrac{dy}{dx} + y\cot x = \cos x$

Find a particular solution of each differential equation subject to the given initial conditions.

23. $3x^2y^3\,dx - xy\,dy = 0$; $y = -1$ when $x = -2$

24. $ye^{2x}\,dy + 3\,dx = 0$; $y = -2$ when $x = 0$

25. $x\,dy - y\,dx = x^5\,dx$; $y(1) = 1$

26. $x\,dy + y\,dx = x\ln x\,dx$; $y(1) = \frac{3}{4}$

27. $\dfrac{dy}{dx} - y = e^{5x}$; $y(0) = -3$

28. $\dfrac{dy}{dx} - \dfrac{2y}{x} = x^3 - 5x$; $y(1) = 3$

29. Find the equation for the current i in a series circuit with inductance $L = 0.2$ H, resistance $R = 60\ \Omega$, and voltage $V = 120$ volts if the initial current $i = 1$ ampere when $t = 0$.

30. The isotope ^{235}U of uranium has a half-life of 8.8×10^8 years. Determine the amount of ^{235}U left after t years if the initial quantity is 1 g. ^{235}U decays at a rate proportional to the amount present.

31. Two hundred litres of brine originally containing 5 kg of salt is in a tank. Pure water is poured in at the rate of 1 L/min, the solution is kept well stirred, and the mixture is poured out at the same rate of 1 L/min. Find the amount of salt remaining after 20 min.

32. A block of material registers 30°C. The block is cooled in air, which is at 12°C. If the block is at 25°C after 20 min, find the temperature of the material after 45 min.

33. Find the equation of the curve passing through the point $(0, 2)$ and having slope x^2y.

34. Find the equation for the velocity of an object moving along a straight line with acceleration $a = 4t^3$ m/s², if the object started moving from rest. Find the velocity after $t = 3$ s.

12

Second-Order Differential Equations

INTRODUCTION

In Chapter 11, we studied first-order differential equations. However, problems in mechanical vibrations, buoyancy, and electric circuits require the use of second-order differential equations, discussed in this chapter.

Objectives

- Solve homogeneous linear differential equations.
- Solve nonhomogeneous second-order linear differential equations.
- Use linear differential equations to solve application problems.
- Find Laplace transforms of functions using a table.
- Use Laplace transforms to solve linear differential equations subject to initial conditions.

12.1 HIGHER-ORDER HOMOGENEOUS DIFFERENTIAL EQUATIONS

We now consider higher-order linear differential equations—that is, consider equations that have derivatives of an unknown function with order higher than first order. These equations are called *linear* because they contain no powers of the unknown function and its derivatives higher than the first power. Thus a linear differential equation of order n is represented by the form

$$a_0 \frac{d^n y}{dx^n} + a_1 \frac{d^{n-1} y}{dx^{n-1}} + \cdots + a_{n-1} \frac{dy}{dx} + a_n y = b$$

where $a_0, a_1, a_2, \ldots, a_{n-1}, a_n$, and b can be either functions of x or constants. If $b = 0$, the linear differential equation is called *homogeneous*. If $b \neq 0$, the equation is called *nonhomogeneous*.

We have already seen linear differential equations in which the coefficients of y and $\frac{dy}{dx}$ were not constants. However, in this chapter we consider only equations in which the coefficients a_i are constants.

EXAMPLE 1

The following are examples of linear differential equations with constant coefficients and with b constant.

$$3\frac{d^4y}{dx^4} - 2\frac{d^2y}{dx^2} + 5\frac{dy}{dx} + 3y = 7 \tag{1}$$

$$y''' + 2y'' - 5y = 2 \tag{2}$$

$$4\frac{d^5y}{dx^5} - \frac{dy}{dx} + y = 0 \tag{3}$$

$$y'' - 3y' + 2y = 0 \tag{4}$$

Equations (1) and (2) are nonhomogeneous whereas Equations (3) and (4) are homogeneous.

We now present a method of solving homogeneous linear differential equations with constant coefficients. These equations arise in many practical situations. Although this method is applicable to equations with order higher than the second, we cover only second-order equations in the remaining sections.

We introduce a *differential operator D*, which operates on a function $y = f(x)$ as follows:

$$Dy = \frac{dy}{dx} \qquad D^2y = \frac{d^2y}{dx^2} \qquad D^3y = \frac{d^3y}{dx^3} \cdots$$

For example, if $y = 2x^3 + 5 \ln x$, then

$$Dy = D(2x^3 + 5 \ln x) = 6x^2 + \frac{5}{x}$$

and

$$D^2y = D(Dy) = D\left(6x^2 + \frac{5}{x}\right) = 12x - \frac{5}{x^2}$$

Operator D is an example of a linear operator. An algebraic expression in the form $ax + by$ where a and b are constants is called a *linear combination*. A *linear operator* is any process that after operating on a linear combination results in another linear combination. We thus observe that for the operator D, we have

$$D[a\, f(x) + b\, g(x)] = a\, D[f(x)] + b\, D[g(x)]$$

We can apply algebraic operations to linear operators. In particular, we can factor expressions involving the differential operator D. For example, $(D^2 - 2D - 3)y = (D - 3)(D + 1)y$, which can be used to solve differential equations. The method is explained in Example 2.

EXAMPLE 2

Solve the differential equation

$$y'' - 2y' - 3y = 0$$

Step 1: First rewrite this equation using the differential operator D.

$$y'' - 2y' - 3y = D^2y - 2Dy - 3y = 0$$
$$(D^2 - 2D - 3)y = 0$$

Factor:

$$(D - 3)(D + 1)y = 0$$

Step 2: Next, let $z = (D + 1)y = Dy + y$. Note that z is a function of x and is a linear combination of the functions $\dfrac{dy}{dx}$ and y: that is,

$$z = \frac{dy}{dx} + y$$

Step 3: Then $(D - 3)(D + 1)y = (D - 3)z = 0$. $(D - 3)z = 0$ is a linear differential equation of first order which can be solved by the method of separation of variables:

$$(D - 3)z = 0$$
$$Dz - 3z = 0$$
$$\frac{dz}{dx} - 3z = 0$$
$$dz = 3z\,dx$$
$$\frac{dz}{z} = 3\,dx$$
$$\int \frac{dz}{z} = \int 3\,dx$$
$$\ln z = 3x + \ln C_1 \qquad$$
$$\ln z - \ln C_1 = 3x$$
$$\ln \frac{z}{C_1} = 3x$$
$$\frac{z}{C_1} = e^{3x}$$
$$z = C_1 e^{3x}$$

(We use $\ln C_1$ instead of C_1 as our constant of integration in order to obtain a simpler expression for z.)

Step 4: Now replace z by $(D + 1)y$ and obtain

$$(D + 1)y = C_1 e^{3x}$$
$$\frac{dy}{dx} + y = C_1 e^{3x}$$

which is another differential equation of first order.

Step 5: Next, solve this differential equation by the method shown in Section 11.4.

$$P(x) = 1 \qquad Q(x) = C_1 e^{3x} \qquad \int P(x)\,dx = \int dx = x$$

Then

$$ye^x = \int (C_1 e^{3x})e^x\,dx$$
$$ye^x = \int C_1 e^{4x}\,dx$$
$$ye^x = \frac{C_1}{4}e^{4x} + C_2$$
$$y = \frac{C_1}{4}e^{3x} + C_2 e^{-x}$$

or

$$y = k_1 e^{3x} + k_2 e^{-x}$$

where

$$k_1 = C_1/4 \quad \text{and} \quad k_2 = C_2.$$

Let's consider a second, simpler method. The general solution $y = k_1e^{3x} + k_2e^{-x}$ of the differential equation $D^2y - 2Dy - 3y = 0$ in Example 2 suggests that $y = ke^{mx}$ is a particular solution of a second-order linear differential equation

$$a_0\,D^2y + a_1\,Dy + a_2y = 0.$$

Since $Dy = D(ke^{mx}) = mke^{mx}$ and $D^2y = D[D(ke^{mx})] = D(mke^{mx}) = m^2ke^{mx}$, we can write

$$a_0\,D^2y + a_1\,Dy + a_2y = 0$$
$$a_0(m^2ke^{mx}) + a_1(mke^{mx}) + a_2ke^{mx} = 0$$
$$(a_0m^2 + a_1m + a_2)ke^{mx} = 0$$

Note: $ke^{mx} \neq 0$ (e^{mx} is never zero and k cannot be zero if $y \neq 0$).
Thus $a_0m^2 + a_1m + a_2 = 0$ if $y = ke^{mx}$ is a solution.
The equation $a_0m^2 + a_1m + a_2 = 0$ is called the *auxiliary equation* (*characteristic equation*) for the differential equation

$$a_0\,D^2y + a_1\,Dy + a_2y = 0$$

Observe that $y = ke^{mx}$ is a solution if m satisfies the auxiliary equation $a_0m^2 + a_1m + a_2 = 0$. To find a solution of the differential equation, find the two roots, m_1 and m_2, of the auxiliary equation.

EXAMPLE 3

Solve the differential equation $y'' - 2y' - 3y = 0$ by solving its auxiliary equation.
    The auxiliary equation is $m^2 - 2m - 3 = 0$. Since this factors as $(m - 3)(m + 1) = 0$, the roots are $m_1 = 3$ and $m_2 = -1$. Thus

$$y = k_1e^{m_1x} = k_1e^{3x}$$

and

$$y = k_2e^{m_2x} = k_2e^{-x}$$

are both solutions of the differential equation. Since the sum of the solutions of a homogeneous linear differential equation can also be shown to be a solution, $y = k_1e^{3x} + k_2e^{-x}$ is the general solution. Note that this is the same solution as in Example 2, as should be the case.

GENERAL SOLUTION OF HOMOGENEOUS LINEAR DIFFERENTIAL EQUATION

1. Write the auxiliary equation for a given differential equation

$$a_0\,D^2y + a_1\,Dy + a_2y = 0$$
$$a_0m^2 + a_1m + a_2 = 0$$

Note that the power of m corresponds to the order of the derivative.

2. Find the two roots, m_1 and m_2, of the auxiliary equation.
3. Write the general solution as

$$y = k_1e^{m_1x} + k_2e^{m_2x}$$

EXAMPLE 4

Solve the differential equation $y'' + 3y' = 0$.

Step 1: Since $D^2y + 3\,Dy = 0$, the auxiliary equation is $m^2 + 3m = 0$.

Step 2: Solve for m:

$$m^2 + 3m = 0$$
$$m(m + 3) = 0$$
$$m_1 = 0 \quad \text{and} \quad m_2 = -3$$

Step 3: The general solution is

$$y = k_1 e^{m_1 x} + k_2 e^{m_2 x}$$
$$y = k_1 e^0 + k_2 e^{-3x}$$
$$y = k_1 + k_2 e^{-3x} \qquad (\text{since } e^0 = 1)$$

EXAMPLE 5

Solve the differential equation $\dfrac{d^2 y}{dx^2} - 5\dfrac{dy}{dx} + 4y = 0$ subject to the initial conditions that

$y = -2$ and $\dfrac{dy}{dx} = 1$ when $x = 0$.

Step 1: $\qquad D^2 y - 5\,Dy + 4y = 0$
$$m^2 - 5m + 4 = 0$$

Step 2: $\qquad (m - 4)(m - 1) = 0$
$$m_1 = 4 \quad \text{and} \quad m_2 = 1$$

Step 3: The general solution is

$$y = k_1 e^{m_1 x} + k_2 e^{m_2 x}$$
$$y = k_1 e^{4x} + k_2 e^{x} \tag{1}$$

Now find the particular solution subject to the conditions that $y = -2$ and $\dfrac{dy}{dx} = 1$ when

$x = 0$. First find $\dfrac{dy}{dx}$ from Equation (1).

$$\frac{dy}{dx} = 4k_1 e^{4x} + k_2 e^{x}$$

Next substitute $y = -2$, $\dfrac{dy}{dx} = 1$, and $x = 0$ into the equations for y and $\dfrac{dy}{dx}$:

$$y = k_1 e^{4x} + k_2 e^{x}$$
$$-2 = k_1 e^{4(0)} + k_2 e^0$$
$$-2 = k_1 + k_2 \qquad (\text{since } e^0 = 1)$$

and

$$\frac{dy}{dx} = 4k_1 e^{4x} + k_2 e^{x}$$
$$1 = 4k_1 e^{4(0)} + k_2 e^0$$
$$1 = 4k_1 + k_2$$

Now solve the resulting system of equations for k_1 and k_2:

$$-2 = k_1 + k_2$$
$$1 = 4k_1 + k_2$$

The solution of this system is $k_1 = 1$ and $k_2 = -3$. The particular solution is then

$$y = e^{4x} - 3e^{x}$$

Using a calculator,

F3 alpha C y 2nd = 2nd = -5y 2nd = +4y=0 2nd MATH 8 8 y(0)=-2 2nd MATH 8 8 y 2nd = (0)=1,x,y) ENTER

Exercises 12.1

State whether each differential equation is homogeneous or nonhomogeneous. Also state the order of each equation.

1. $3y^{(4)} - 7y''' + 2y = 0$　　**2.** $y'' - 7y' + 3y = 2$　　**3.** $y''' - 5y' + 2y + 6 = 0$

4. $y^{(5)} - 6y''' - y = 0$　　**5.** $4y'' - y' + 3y = 0$　　**6.** $y''' - y' + 2 = 0$

7. $y''' - y'' - 5y = 3$　　**8.** $y'' - 2y' = 3y$

Solve each differential equation.

9. $\dfrac{d^2y}{dx^2} - 5\dfrac{dy}{dx} - 14y = 0$　　**10.** $\dfrac{d^2y}{dx^2} + 4\dfrac{dy}{dx} - 5y = 0$　　**11.** $y'' - 2y' - 8y = 0$

12. $y'' - y' - 6y = 0$　　**13.** $y'' - y = 0$　　**14.** $y'' - 9y = 0$

15. $\dfrac{d^2y}{dx^2} - 3\dfrac{dy}{dx} = 0$　　**16.** $\dfrac{d^2y}{dx^2} + 2\dfrac{dy}{dx} = 0$　　**17.** $2D^2y - 13Dy + 15y = 0$

18. $3D^2y + 2Dy - 5y = 0$　　**19.** $3\dfrac{d^2y}{dx^2} - 7\dfrac{dy}{dx} + 2y = 0$　　**20.** $2y'' + y' - 15y = 0$

Find the particular solution of each differential equation subject to the given conditions.

21. $y'' - 4y' = 0$; $y = 3$ and $y' = 4$ when $x = 0$

22. $y'' + 3y' = 0$; $y = -1$ and $y' = 6$ when $x = 0$

23. $y'' - y' - 2y = 0$; $y = 2$ and $y' = 1$ when $x = 0$

24. $y'' + 3y' - 10y = 0$; $y = 7$ and $y' = 0$ when $x = 0$

25. $y'' - 8y' + 15y = 0$; $y = 4$ and $y' = 2$ when $x = 0$

26. $y'' + y' = 0$; $y = 2$ and $y' = 1$ when $x = 1$

12.2 REPEATED ROOTS AND COMPLEX ROOTS

When the auxiliary equation of a differential equation results in a repeated root $(m_1 = m_2 = m)$, then the general solution is in the form

> **REPEATED ROOTS**
>
> $$y = k_1e^{mx} + k_2xe^{mx}$$

Note carefully that the second term includes the factor x.

EXAMPLE 1

Solve the differential equation

$$y'' - 2y' + y = 0$$

Since $D^2y - 2\,Dy + y = 0$, the auxiliary equation is

$$m^2 - 2m + 1 = 0$$
$$(m - 1)^2 = 0$$
$$m_1 = m_2 = 1$$

The general solution is

$$y = k_1 e^{mx} + k_2 x e^{mx}$$
$$y = k_1 e^x + k_2 x e^x$$
$$y = e^x(k_1 + k_2 x)$$

EXAMPLE 2

Solve the differential equation $y'' + 4y' + 4y = 0$ subject to the conditions that $y = 2$ and $y' = 1$ when $x = 0$.

$$D^2y + 4\,Dy + 4y = 0$$
$$m^2 + 4m + 4 = 0$$
$$(m + 2)^2 = 0$$
$$m = -2$$

The general solution is

$$y = k_1 e^{-2x} + k_2 x e^{-2x}$$

Next,

$$y' = -2k_1 e^{-2x} - 2k_2 x e^{-2x} + k_2 e^{-2x}$$

Substitute $y = 2$, $y' = 1$, and $x = 0$ in the equations for y and y':

$$y = k_1 e^{-2x} + k_2 x e^{-2x}$$
$$2 = k_1 e^{-2(0)} + k_2(0)e^{-2(0)}$$
$$2 = k_1$$
$$y' = -2k_1 e^{-2x} - 2k_2 x e^{-2x} + k_2 e^{-2x}$$
$$1 = -2k_1 e^{-2(0)} - 2k_2(0)e^{-2(0)} + k_2 e^{-2(0)}$$
$$1 = -2k_1 + k_2$$

But since $k_1 = 2$, we have

$$1 = -4 + k_2 \quad \text{so} \quad k_2 = 5$$

The desired particular solution is then

$$y = 2e^{-2x} + 5xe^{-2x} \quad \text{so} \quad y = e^{-2x}(2 + 5x)$$

When the roots of the auxiliary equation of a differential equation are complex numbers, the method used in Section 12.1 still applies.

EXAMPLE 3

Solve the differential equation $y'' - y' + 2y = 0$.

Solve the auxiliary equation $m^2 - m + 2 = 0$ using the quadratic formula,

$$m = \frac{-(-1) \pm \sqrt{(-1)^2 - 4(1)(2)}}{2(1)}$$

$$= \frac{1 \pm \sqrt{1 - 8}}{2} = \frac{1 \pm \sqrt{-7}}{2} = \frac{1}{2} \pm \frac{j\sqrt{7}}{2}$$

So the general solution is

$$y = k_1 e^{[(1/2) + (j\sqrt{7}/2)]x} + k_2 e^{[(1/2) - (j\sqrt{7}/2)]x}$$

Simplify this expression

$$y = k_1 e^{x/2} e^{j(\sqrt{7}/2)x} + k_2 e^{x/2} e^{-j(\sqrt{7}/2)x}$$
$$y = e^{x/2}(k_1 e^{j(\sqrt{7}/2)x} + k_2 e^{-j(\sqrt{7}/2)x})$$

Recall that $z = re^{j\theta}$ represents the trigonometric form of a complex number. That is,

$$z = re^{j\theta} = r(\cos \theta + j \sin \theta) \qquad \text{(see Section 10.6.)}$$

or

$$e^{j\theta} = \cos \theta + j \sin \theta$$

For $\theta = \dfrac{\sqrt{7}}{2}x$,

$$e^{j(\sqrt{7}/2)x} = \cos \frac{\sqrt{7}}{2}x + j \sin \frac{\sqrt{7}}{2}x$$

and for $\theta = -\dfrac{\sqrt{7}}{2}x$,

$$e^{-j(\sqrt{7}/2)x} = \cos\left(-\frac{\sqrt{7}}{2}x\right) + j \sin\left(-\frac{\sqrt{7}}{2}x\right)$$
$$= \cos \frac{\sqrt{7}}{2}x - j \sin \frac{\sqrt{7}}{2}x$$

But then

$$k_1 e^{j(\sqrt{7}/2)x} = k_1 \cos \frac{\sqrt{7}}{2}x + jk_1 \sin \frac{\sqrt{7}}{2}x \qquad \textbf{(1)}$$

$$k_2 e^{-j(\sqrt{7}/2)x} = k_2 \cos \frac{\sqrt{7}}{2}x - jk_2 \sin \frac{\sqrt{7}}{2}x \qquad \textbf{(2)}$$

Adding Equations (1) and (2), we have

$$k_1 e^{j(\sqrt{7}/2)x} + k_2 e^{-j(\sqrt{7}/2)x} = (k_1 + k_2) \cos \frac{\sqrt{7}}{2}x + j(k_1 - k_2) \sin \frac{\sqrt{7}}{2}x$$

$$= C_1 \cos \frac{\sqrt{7}}{2}x + C_2 \sin \frac{\sqrt{7}}{2}x$$

where $C_1 = k_1 + k_2$ and $C_2 = j(k_1 - k_2)$.
The general solution can now be written as

$$y = e^{x/2}\left[C_1 \cos\left(\frac{\sqrt{7}}{2}x\right) + C_2 \sin\left(\frac{\sqrt{7}}{2}x\right)\right]$$

In general, if the auxiliary equation of a differential equation has complex roots $m = a \pm bj$, the solution of the differential equation is:

COMPLEX ROOTS

$$y = e^{ax}(k_1 \sin bx + k_2 \cos bx)$$

where k_1 and k_2 are arbitrary constants.

EXAMPLE 4

Solve the differential equation $y'' - 2y' + 3y = 0$.

The auxiliary equation is $m^2 - 2m + 3 = 0$. So

$$m = \frac{2 \pm \sqrt{-8}}{2} = 1 \pm j\sqrt{2}$$

Then $a = 1$ and $b = \sqrt{2}$, and the general solution is

$$y = e^x(k_1 \sin \sqrt{2}x + k_2 \cos \sqrt{2}x)$$

Using a calculator,

F3 alpha C y 2nd = 2nd = -2y 2nd = +3y=0,x,y) ENTER up arrow **2nd** right arrow

 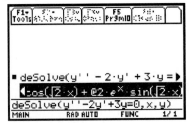

EXAMPLE 5

Solve the differential equation $y'' + 9y = 0$.

The auxiliary equation is $m^2 + 9 = 0$. So $m = \pm 3j$. Since $a = 0$ and $b = 3$, the general solution is

$$y = e^{0x}(k_1 \sin 3x + k_2 \cos 3x)$$
$$y = k_1 \sin 3x + k_2 \cos 3x$$

Exercises 12.2

Solve each differential equation.

1. $\dfrac{d^2y}{dx^2} - 4\dfrac{dy}{dx} + 4y = 0$

2. $\dfrac{d^2y}{dx^2} + 6\dfrac{dy}{dx} + 9y = 0$

3. $y'' - 4y' + 5y = 0$

4. $y'' - 2y' + 4y = 0$

5. $4y'' - 4y' + y = 0$

6. $y'' - 8y' + 16y = 0$

7. $\dfrac{d^2y}{dx^2} - 4\dfrac{dy}{dx} + 13y = 0$

8. $\dfrac{d^2y}{dx^2} + 2\dfrac{dy}{dx} + y = 0$

9. $D^2y - 10\,Dy + 25y = 0$

10. $D^2y + 12\,Dy + 36y = 0$

11. $D^2y + 9y = 0$

12. $D^2y + 16y = 0$

13. $\dfrac{d^2y}{dx^2} = 0$

14. $9\dfrac{d^2y}{dx^2} + 16y = 0$

Find the particular solution of each differential equation satisfying the given conditions.

15. $y'' - 6y' + 9y = 0$; $y = 2$ and $y' = 4$ when $x = 0$

16. $D^2y + 10\,Dy + 25y = 0$; $y = 0$ and $y' = 3$ when $x = 0$

17. $y'' + 25y = 0$; $y = 2$ and $y' = 0$ when $x = 0$

18. $y'' + 16y = 0$; $y = 1$ and $y' = -4$ when $x = 0$

19. $D^2y - 12\,Dy + 36y = 0$; $y = 1$ and $y' = 0$ when $x = 0$

20. $y'' - 6y' + 25y = 0$; $y = 0$ and $y' = 8$ when $x = 0$

Now consider the solutions of nonhomogeneous second-order linear differential equations with constant coefficients. We represent such an equation in the form

$$a_0\,D^2y + a_1\,Dy + a_2y = b$$

where $b = g(x)$ is a function of x (b may be a constant).

Obtaining a solution of a nonhomogeneous differential equation involves two basic steps. First, obtain a solution, called the *complementary solution* (denoted y_c). This is the solution of the homogeneous equation obtained by substituting 0 for b; that is,

$$a_0\,D^2y + a_1\,Dy + a_2y = 0$$

Next obtain a *particular solution* (denoted y_p) for the given nonhomogeneous equation. The general solution of the nonhomogeneous equation will then be

$$y = y_c + y_p$$

Since we have shown methods of obtaining the complementary solution y_c in Sections 12.1 and 12.2, our only problem is to determine a particular solution of a given nonhomogeneous equation. We find y_p using the *method of undetermined coefficients.*

The method relies on inspecting the expression $b = g(x)$ and determining a combination of all possible functions that after differentiating twice would result in at least one of the terms in the expression $g(x)$. In most applications the functions that most often occur are where $g(x)$ involves polynomials, exponentials, sines, and cosines. For these functions, see Table 12.1 for finding y_p.

To show how to use Table 12.1, six examples of a differential equation and its corresponding trial solution for y_p are shown in Example 1.

TABLE 12.1   Finding y_p

If $g(x) =$	Try $y_p =$
$b_nx^n + b_{n-1}x^{n-1} + \cdots + b_2x^2 +$ $b_1x + b_0$ (polynomial of degree n)	$A_nx^n + A_{n-1}x^{n-1} + \cdots + A_2x^2 + A_1x + A_0$
be^{nx}	Ae^{nx}
$b \sin nx$	
$b \cos nx$	$A \sin nx + B \cos nx$
$b \sin nx + c \cos nx$	

Note 1: If $g(x)$ is the sum of two or more types in the left column, try the corresponding sum in the right column.

Note 2: If a term of $g(x)$ is a solution of the homogeneous equation, try multiplying y_p by x or some higher power of x.

EXAMPLE 1

Find the trial solution y_p using Table 12.1 for each differential equation.

Differential equation	$g(x) =$	Try $y_p =$
(a) $y'' - 4y' - 5y = 6x^2 - 1$	$6x^2 - 1$	$Ax^2 + Bx + C$
(b) $y'' - y = 3e^{-4x}$	$3e^{-4x}$	Ae^{-4x}
(c) $y'' + 16y = 3 \sin 5x$	$3 \sin 5x$	$A \sin 5x + B \cos 5x$

Differential equation	$g(x) =$	Try $y_p =$
(d) $y'' - 9y = 2e^{3x}$	$2e^{3x}$	Axe^{3x}

Note: $2e^{3x}$ is a solution of the homogeneous equation.

(e) $y'' + y' = 6x + 4$	$6x + 4$	$Ax^2 + Bx$

Note: 4 is a solution of the homogeneous equation.

(f) $y'' + 9y = \cos 3x$	$\cos 3x$	$Ax \sin 3x + Bx \cos 3x$

Note: $\cos 3x$ is a solution of the homogeneous equation.

The method of undetermined coefficients is illustrated in the following examples.

EXAMPLE 2

Find a particular solution of the differential equation

$$y'' - 2y' - 3y = e^x$$

Try $y_p = Ae^x$. We need to find the value of A. First, find $y_p' = Ae^x$ and $y_p'' = Ae^x$. If y_p is a solution, it must satisfy the given differential equation.

$$y_p'' - 2y_p' - 3y_p = e^x$$
$$Ae^x - 2Ae^x - 3Ae^x = e^x$$
$$-4Ae^x = e^x$$
$$-4A = 1$$
$$A = -\frac{1}{4}$$

Then

$$y_p = -\frac{1}{4}e^x$$

EXAMPLE 3

Find a particular solution of the differential equation

$$y'' - 2y' - 3y = x^2 + e^{-2x}$$

Try $y_p = A + Bx + Cx^2 + De^{-2x}$, which is the sum of a second-degree polynomial and an exponential that appears in $g(x)$. To find the values of A, B, C, and D, we first find

$$y_p' = B + 2Cx - 2De^{-2x}$$

and

$$y_p'' = 2C + 4De^{-2x}$$

Then substitute into the given differential equation.

$$y_p'' - 2y_p' - 3y_p = x^2 + e^{-2x}$$

Then

$$(2C + 4De^{-2x}) - 2(B + 2Cx - 2De^{-2x}) - 3(A + Bx + Cx^2 + De^{-2x}) = x^2 + e^{-2x}$$
$$(2C - 2B - 3A) + (-4C - 3B)x + (-3C)x^2 + (4D + 4D - 3D)e^{-2x} = x^2 + e^{-2x}$$

From this last equation, we equate the coefficients on each side of the equation.

$$2C - 2B - 3A = 0 \quad \text{(constants)} \tag{1}$$
$$-4C - 3B \quad\quad = 0 \quad \text{(coefficients of } x\text{)} \tag{2}$$
$$-3C \quad\quad\quad = 1 \quad \text{(coefficients of } x^2\text{)} \tag{3}$$
$$5D \quad\quad\quad = 1 \quad \text{(coefficients of } e^{-2x}\text{)} \tag{4}$$

Solving this system of four equations, we find that

$$D = \frac{1}{5} \quad C = -\frac{1}{3} \quad B = \frac{4}{9} \quad \text{and} \quad A = -\frac{14}{27}$$

The particular solution is then

$$y_p = -\frac{14}{27} + \frac{4}{9}x - \frac{1}{3}x^2 + \frac{1}{5}e^{-2x}$$

GENERAL SOLUTION OF NONHOMOGENEOUS DIFFERENTIAL EQUATION

To find the general solution of the nonhomogeneous equation

$$a_0 D^2 y + a_1 Dy + a_2 y = b:$$

1. Find the complementary solution y_c by solving the homogeneous equation $a_0 D^2 y + a_1 Dy + a_2 y = 0$ using the methods developed in Sections 12.1 and 12.2.

2. Find a particular solution y_p by using the method of undetermined coefficients described in Examples 2 and 3.

3. Find the general solution y by adding the complementary solution y_c from step 1 and the particular solution y_p from Step 2:

$$y = y_c + y_p$$

EXAMPLE 4

Find the general solution of

$$y'' - 2y' - 3y = e^x$$

Step 1: Find the complementary solution y_c which is the solution of the homogeneous equation $y'' - 2y' - 3y = 0$. We solved this equation in Example 2, Section 12.1:

$$y_c = k_1 e^{3x} + k_2 e^{-x}$$

Step 2: Find a particular solution y_p, which we obtained in Example 2 of this section:

$$y_p = -\frac{1}{4}e^x$$

Step 3: The desired general solution is

$$y = y_c + y_p$$
$$y = k_1 e^{3x} + k_2 e^{-x} - \frac{1}{4}e^x$$

Using a calculator,

F3 alpha C y 2nd = 2nd = -2y 2nd = -3y= green diamond e^x x),x,y) ENTER

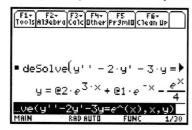

EXAMPLE 5

Find the general solution of $y'' - 2y' - 3y = x^2 + e^{-2x}$.

Step 1: From Example 4, the complementary solution is

$$y_c = k_1 e^{3x} + k_2 e^{-x}$$

Step 2: From Example 3, a particular solution is

$$y_p = -\frac{14}{27} + \frac{4}{9}x - \frac{1}{3}x^2 + \frac{1}{5}e^{-2x}$$

Step 3: The general solution is

$$y = y_c + y_p$$

$$y = k_1 e^{3x} + k_2 e^{-x} - \frac{14}{27} + \frac{4}{9}x - \frac{1}{3}x^2 + \frac{1}{5}e^{-2x}$$

EXAMPLE 6

Solve the differential equation

$$y'' + 6y' + 9y = x + \sin x$$

Step 1: Find y_c:

$$y'' + 6y' + 9y = 0$$
$$m^2 + 6m + 9 = 0$$
$$(m + 3)^2 = 0$$
$$m = -3 \qquad \text{(a repeated root)}$$
$$y_c = e^{mx}(k_1 + k_2 x)$$
$$= e^{-3x}(k_1 + k_2 x)$$

Step 2: Find y_p. Try $y_p = A + Bx + C \sin x + D \cos x$. Differentiating y_p, we have

$$y_p' = B + C \cos x - D \sin x$$
$$y_p'' = -C \sin x - D \cos x$$

Substitute y_p, y_p', and y_p'' in the given differential equation

$$y'' + 6y' + 9y = x + \sin x$$

$$(-C \sin x - D \cos x) + 6(B + C \cos x - D \sin x)$$
$$+ 9(A + Bx + C \sin x + D \cos x) = x + \sin x$$

$$(6B + 9A) + 9Bx + (-C - 6D + 9C) \sin x + (-D + 6C + 9D) \cos x = x + \sin x$$

Equating coefficients, we have

$$6B + 9A = 0$$
$$9B = 1$$
$$8C - 6D = 1$$
$$6C + 8D = 0$$

Solving for A, B, C, and D, we have

$$A = -\frac{2}{27} \qquad B = \frac{1}{9} \qquad C = \frac{2}{25} \quad \text{and} \quad D = -\frac{3}{50}$$

The particular solution is then

$$y_p = -\frac{2}{27} + \frac{1}{9}x + \frac{2}{25} \sin x - \frac{3}{50} \cos x$$

Step 3: The general solution is then

$$y = y_c + y_p$$

$$y = e^{-3x}(k_1 + k_2x) - \frac{2}{27} + \frac{1}{9}x + \frac{2}{25}\sin x - \frac{3}{50}\cos x$$

EXAMPLE 7

Solve the differential equation

$$y'' - 3y' - 4y = e^{4x}$$

Step 1: Find y_c:

$$m^2 - 3m - 4 = 0$$
$$(m - 4)(m + 1) = 0$$
$$m_1 = 4 \quad \text{and} \quad m_2 = -1$$
$$y_c = k_1e^{4x} + k_2e^{-x}$$

Step 2: Find y_p: First, note that $g(x) = e^{4x}$ is a solution of the homogeneous equation

$$y'' - 3y' - 4y = 0$$
$$16e^{4x} - 3(4e^{4x}) - 4(e^{4x}) = 0$$

Then, following Note 2 after Table 12.1 for finding y_p, we try

$$y_p = Axe^{4x}$$
$$y_p' = 4Axe^{4x} + Ae^{4x}$$
$$y_p'' = 16Axe^{4x} + 4Ae^{4x} + 4Ae^{4x} = 16Axe^{4x} + 8Ae^{4x}$$

Substituting into the differential equation, we have

$$y'' - 3y' - 4y = e^{4x}$$
$$(16Axe^{4x} + 8Ae^{4x}) - 3(4Axe^{4x} + Ae^{4x}) - 4(Axe^{4x}) = e^{4x}$$
$$16Axe^{4x} + 8Ae^{4x} - 12Axe^{4x} - 3Ae^{4x} - 4Axe^{4x} = e^{4x}$$
$$5Ae^{4x} = e^{4x}$$
$$5A = 1$$
$$A = \frac{1}{5}$$

Step 3: The general solution is

$$y = y_c + y_p$$

$$y = k_1e^{4x} + k_2e^{-x} + \frac{1}{5}xe^{4x}$$

Exercises 12.3

Solve each differential equation.

1. $y'' + y' = \sin x$
2. $y'' + 4y = 3e^{-x}$
3. $y'' - y' - 2y = 4x$
4. $y'' - 4y' + 4y = e^x$
5. $y'' - 10y' + 25y = x$
6. $y'' - 2y' - 3y = 3x^2$
7. $y'' - y = x^2$
8. $y'' + y = e^x$
9. $y'' + 4y = e^x - 2$
10. $y'' + 4y = e^x - 2x$
11. $y'' - 3y' - 4y = 6e^x$
12. $y'' - 4y' + 3y = 20\cos x$
13. $y'' + y = 5 + \sin 3x$
14. $y'' - y' - 6y = x + \cos x$
15. $y'' - y = e^x$
16. $y'' + 6y' = 18x^2 - 6x + 3$
17. $y'' + 4y = \cos 2x$
18. $y'' - 4y = e^{-2x}$

Find the particular solution of each differential equation satisfying the given conditions.

19. $y'' + y = 10e^{2x}$; $y = 0$ and $y' = 0$ when $x = 0$

20. $y'' + y' = \sin x$; $y = 0$ and $y' = 0$ when $x = 0$

21. $y'' + y = e^x$; $y = 0$ and $y' = 3$ when $x = 0$

22. $y'' - 4y = 2 - 8x$; $y = 0$ and $y' = 5$ when $x = 0$

12.4 APPLICATIONS OF SECOND-ORDER DIFFERENTIAL EQUATIONS

Mathematical models of numerous technical applications result in second-order linear differential equations. We will focus our attention on three basic applications: mechanical vibrations, buoyancy, and electric circuits.

We saw in Section 6.7 that the force required to stretch or compress a spring obeys Hooke's law. That is, the required force $F = f(x)$ is directly proportional to the distance x that the spring is stretched or compressed:

$$f(x) = kx$$

where k is called the *spring constant.*

Another basic mechanical principle, Newton's law, states that the force F acting on a mass m at any time t is equal to the product of the mass m and the acceleration a of the mass:

$$F = ma$$

We now relate these two mechanical principles and investigate the motion of a vibrating spring. Consider a spring hanging downward with a natural length l. If a weight W is attached to the spring, the force of gravity acting on the weight will stretch the spring downward a distance s until coming to rest (see Fig. 12.1).

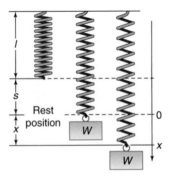

Figure 12.1

By Hooke's law the force used to stretch the spring a distance s to this rest position is

$$F = ks$$

and by Newton's law this same force is

$$F = mg$$

where m is the mass of the weight and g is the acceleration of the weight as a result of gravity (approximately 32 ft/s²). Then,

$$mg = ks$$

If the spring is now stretched beyond the rest length $l + s$ and then released, a vibrating motion will result. When in motion, at any time t, the force F acting on the weight is

$$F = mg - k(s + x)$$

where x is the distance measured from the rest position at time t. The term mg represents the force of gravity acting downward on the weight. The term $-k(s + x)$ represents the spring tension, which is a force acting to restore the system to the rest position.

Again, by Newton's law this same force acting on the weight is given by

$$F = ma$$

$$F = m\frac{d^2x}{dt^2}$$

since $\dfrac{d^2x}{dt^2}$ is the acceleration of the weight at any time t. Then

$$m\frac{d^2x}{dt^2} = mg - k(s + x)$$

$$= mg - ks - kx$$

$$= mg - mg - kx \qquad (\text{since } mg = ks)$$

$$= -kx$$

Then

$$m\frac{d^2x}{dt^2} + kx = 0$$

$$\frac{d^2x}{dt^2} + \frac{k}{m}x = 0$$

This is a second-order linear differential equation whose solution $x = f(t)$ describes the vibrating motion of the weight and spring. The motion described by this differential equation is called *simple harmonic motion.* Another visual example of simple harmonic motion is the motion of a simple pendulum like the pendulum of a clock.

The solution of this differential equation is obtained by using the methods of Section 12.2. The solution is

$$x = C_1 \sin \sqrt{\frac{k}{m}}\, t + C_2 \cos \sqrt{\frac{k}{m}}\, t$$

The period (time for one complete oscillation) of this simple harmonic motion is given by

$$P = \frac{2\pi}{\sqrt{\dfrac{k}{m}}}$$

since the period for $y = A \sin Bt$ and $y = A \cos Bt$ is $\dfrac{2\pi}{|B|}$.

EXAMPLE 1

Find the equation expressing the simple harmonic motion of a weight attached to a spring. A 32-lb weight stretches the spring 6 in. to a rest position. The motion is started by stretching the spring an additional 3 in. and then releasing the spring.

We have

$$m = \frac{W}{g} = \frac{32 \text{ lb}}{32 \text{ ft/s}^2} = 1 \text{ lb-s}^2/\text{ft}$$

$$k = \frac{F}{s} = \frac{32 \text{ lb}}{6 \text{ in.}} = \frac{32 \text{ lb}}{\frac{1}{2} \text{ ft}} = 64 \text{ lb/ft}$$

Then

$$x = C_1 \sin \sqrt{\frac{k}{m}}\, t + C_2 \cos \sqrt{\frac{k}{m}}\, t$$

$$x = C_1 \sin \sqrt{64}\, t + C_2 \cos \sqrt{64}\, t$$

$$x = C_1 \sin 8t + C_2 \cos 8t$$

To find constants C_1 and C_2, differentiate the preceding equation.

$$\frac{dx}{dt} = 8C_1 \cos 8t - 8C_2 \sin 8t$$

Substitute $x = 3$ in. $= 0.25$ ft and $\frac{dx}{dt} = 0$ when $t = 0$.

$$0.25 = C_1 \sin 0 + C_2 \cos 0$$

$$0 = 8C_1 \cos 0 - 8C_2 \sin 0$$

We find $C_1 = 0$ and $C_2 = 0.25$. The solution is

$$x = 0.25 \cos 8t \qquad \text{(See Fig. 12.2.)}$$

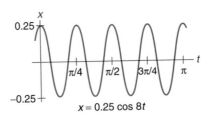

$$x = 0.25 \cos 8t$$

Figure 12.2

However, in reality, the vibrations do not continue forever. Resistant forces to the motion of vibration exist. The effect of these forces is called the *damping effect*. These resistant forces, such as friction, will cause the vibration to decrease. In many situations this force of resistance is directly proportional to the velocity of the vibrating system. That is, the damping force equals $p\dfrac{dx}{dt}$, where p is a constant.

The differential equation expressing the total forces acting on the system becomes

$$m\frac{d^2x}{dt^2} = -kx - p\frac{dx}{dt}$$

$$\frac{d^2x}{dt^2} + \frac{p}{m}\frac{dx}{dt} + \frac{k}{m}x = 0$$

The auxiliary equation is

$$\bar{m}^2 + \frac{p}{m}\bar{m} + \frac{k}{m}x = 0$$

(We use $\bar{m}$ in place of m in the auxiliary equation so it will not be confused with the mass m.) Solving for $\bar{m}$, we have

$$\bar{m} = \frac{-\dfrac{p}{m} \pm \sqrt{\left(\dfrac{p}{m}\right)^2 - \dfrac{4k}{m}}}{2} = -\frac{p}{2m} \pm \sqrt{\left(\frac{p}{2m}\right)^2 - \frac{k}{m}}$$

Now let

$$d = \left(\frac{p}{2m}\right)^2 - \frac{k}{m}$$

Case 1, $d > 0$: There are two real roots of the auxiliary equation:

$$\bar{m}_1 = -\frac{p}{2m} + \sqrt{d}$$

$$\bar{m}_2 = -\frac{p}{2m} - \sqrt{d}$$

The solution of the differential equation is

$$x = C_1 e^{\bar{m}_1 t} + C_2 e^{\bar{m}_2 t}$$

This case is called *overdamped* (see Fig. 12.3).

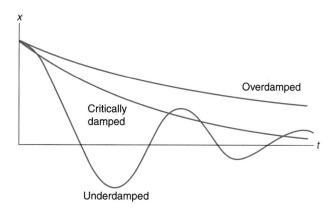

Figure 12.3 Damping effect.

Case 2, $d = 0$: There is a repeated real root of the auxiliary equation. The solution is

$$x = (C_1 + C_2 t)e^{\bar{m}t} \qquad \text{where } \bar{m} = -\frac{p}{2m}$$

This case is called *critically damped*.

Case 3, $d < 0$: There are two complex roots of the auxiliary equation. The solution is

$$x = e^{-[p/(2m)]t}(C_1 \sin \omega t + C_2 \cos \omega t) \qquad \text{where } \omega = \sqrt{|d|}$$

This case is called *underdamped* (see Fig. 12.3).

EXAMPLE 2

Find the general equation for the motion of the vibrating spring of Example 1 if there is a resistant force that exerts a force of 0.04 lb with a velocity of 2 in./s.

Since

$$F_{\text{resistant}} = p\frac{dx}{dt}$$

we have

$$p = \frac{F_{\text{resistant}}}{\dfrac{dx}{dt}} = \frac{0.04\ \text{lb}}{2\ \text{in./s}} = \frac{0.04\ \text{lb-s}}{\frac{1}{6}\ \text{ft}} = 0.24\ \text{lb-s/ft}$$

$$d = \left(\frac{p}{2m}\right)^2 - \frac{k}{m}$$

$$d = \left(\frac{0.24}{2(1)}\right)^2 - 64$$

$$d = -63.99 \quad \text{and} \quad \omega = \sqrt{|d|} = \sqrt{63.99} = 8$$

Case 3 applies and the general solution is

$$x = e^{-[p/(2m)]t}\,(C_1 \sin \omega t + C_2 \cos \omega t)$$
$$x = e^{-0.12t}\,(C_1 \sin 8t + C_2 \cos 8t) \qquad (\text{since } m = 1)$$

EXAMPLE 3

A cylindrical buoy with radius 1 ft is floating in water with the axis of the buoy vertical as in Fig. 12.4. If the buoy is pushed a small distance into the water and released, the period of vibration is found to be 4 s. Find the weight of the buoy given that the density of water is 62.4 lb/ft^3.

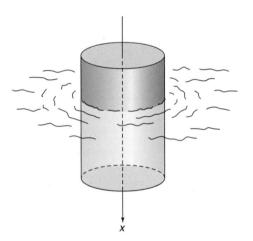

Figure 12.4

Let the origin be at the intersection of the axis of the buoy and the water when the buoy is in equilibrium. Let's choose the downward direction to be positive.

A body partly or wholly submerged in a fluid is buoyed up by a force equal to the weight of the fluid it displaces (Archimedes' principle). If the downward displacement is x, the change in submerged volume is $\pi(1)^2 x$ and the change in the buoyant force is $62.4\,\pi(1)^2 x$. If we let W (in pounds) be the weight of the buoy, then its mass is W/g, where $g = 32.2$ ft/s^2. The equation $F = ma$ thus becomes

$$-62.4\pi x = \frac{W}{g}\frac{d^2x}{dt^2}$$

$$\frac{d^2x}{dt^2} + \frac{2009\pi}{W}x = 0$$

By the methods of Section 12.2, the solution of this differential equation is

$$x = C_1 \sin\sqrt{\frac{2009\pi}{W}}t + C_2\cos\sqrt{\frac{2009\pi}{W}}t$$

Recall that the period is 4 s, so

$$P = \frac{2\pi}{\sqrt{\dfrac{2009\pi}{W}}} = 4$$

or

$$\frac{\pi}{2} = \sqrt{\frac{2009\pi}{W}}$$

$$\frac{\pi^2}{4} = \frac{2009\pi}{W} \qquad \text{(Square each side.)}$$

$$W = 2560 \text{ lb}$$

Electric circuits provide another example of the use of second-order linear differential equations. Kirchhoff's law states that the sum of the voltage drops across the elements of an electric circuit equals the voltage source V. If a circuit (as in Fig. 12.5) contains a capacitor C (in farads), an inductance L (in henrys), and a resistance R (in ohms), then Kirchhoff's law results in the differential equation:

$$L\frac{di}{dt} + Ri + \frac{1}{C}\int_0^t i\,dt = V$$

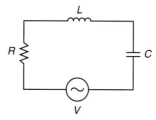

Figure 12.5   Circuit with capacitance, inductance, and resistance.

If the voltage source V is constant, then differentiating we obtain

$$L\frac{d^2i}{dt^2} + R\frac{di}{dt} + \frac{1}{C}i = \frac{dV}{dt} = 0$$

$$\frac{d^2i}{dt^2} + \frac{R}{L}\frac{di}{dt} + \frac{1}{CL}i = 0$$

The auxiliary equation is $m^2 + \dfrac{R}{L}m + \dfrac{1}{CL} = 0$, whose roots are

$$m = \frac{-\dfrac{R}{L} \pm \sqrt{\left(\dfrac{R}{L}\right)^2 - \dfrac{4}{CL}}}{2} = -\frac{R}{2L} \pm \sqrt{\frac{R^2}{4L^2} - \frac{1}{CL}}$$

The form of the solution depends on the value of $d = \dfrac{R^2}{4L^2} - \dfrac{1}{CL}$

Case 1, d > 0 (overdamped):

$$i = k_1 e^{m_1 t} + k_2 e^{m_2 t}$$

where

$$m_1 = -\frac{R}{2L} + \sqrt{d} \quad \text{and} \quad m_2 = -\frac{R}{2L} - \sqrt{d}$$

Case 2, d = 0 (critically damped):

$$i = (k_1 + k_2 t)e^{-[R/(2L)]t}$$

Case 3, d < 0 (underdamped):

$$i = e^{-[R/(2L)]t}(k_1 \sin \omega t + k_2 \cos \omega t) \qquad \text{where } \omega = \sqrt{|d|}$$

EXAMPLE 4

Find an expression for the current in an electric circuit containing an inductor of 0.2 H, a capacitor of 10^{-5} F, and a resistor of 300 Ω. The voltage source is 12 V. Assume that $i = 0$ and $q = 0$ at $t = 0$.

$$d = \frac{R^2}{4L^2} - \frac{1}{CL}$$

$$d = \frac{(300)^2}{4(0.2)^2} - \frac{1}{0.2 \times 10^{-5}} = 62{,}500 > 0$$

Then by Case 1, we have

$$-\frac{R}{2L} = -\frac{300}{2(0.2)} = -750$$

$$m_1 = -750 + \sqrt{62{,}500} = -500$$

$$m_2 = -750 - \sqrt{62{,}500} = -1000$$

The solution is then

$$i = k_1 e^{-500t} + k_2 e^{-1000t}$$

Differentiating, we have

$$\frac{di}{dt} = -500 k_1 e^{-500t} - 1000 k_2 e^{-1000t}$$

Using

$$L\frac{di}{dt} + Ri + \frac{1}{C}q = V$$

and the conditions $i = 0$ and $q = 0$ at $t = 0$, we have

$$(0.2)\frac{di}{dt} + 300(0) + \left(\frac{1}{10^5}\right)(0) = 12$$

$$\frac{di}{dt} = 60$$

Then, substituting $t = 0$ in the preceding equations for i and $\dfrac{di}{dt}$, we have

$$0 = k_1 + k_2$$
$$60 = -500k_1 - 1000k_2$$

Then

$$k_1 = 0.12$$
$$k_2 = -0.12$$

The particular solution is

$$i = 0.12e^{-500t} - 0.12e^{-1000t}$$

If the voltage source V in the electric circuit is not constant, then $\dfrac{dV}{dt}$ is not zero, and we obtain a nonhomogeneous differential equation:

$$\frac{d^2i}{dt^2} + \frac{R}{L}\frac{di}{dt} + \frac{1}{CL}i = \frac{1}{L}\frac{dV}{dt}$$

Similarly, if a force $f(t)$ external to a vibrating system of simple harmonic motion is applied (such as a periodic push or pull on the weight), then we also obtain a nonhomogeneous equation:

$$\frac{d^2x}{dt^2} + \frac{p}{m}\frac{dx}{dt} + \frac{k}{m}x = \frac{1}{m}f(t)$$

Exercises 12.4

1. A spring is stretched $\frac{1}{4}$ ft by a weight of 4 lb. If the weight is displaced a distance of 2 in. from the rest position and then released, find the equation of motion.

2. A spring is stretched 6 in. by a weight of 10 lb. If the weight is displaced a distance of 6 in. from the rest position and then released, find the equation of motion.

3. A spring is stretched 6 in. by a weight of 20 lb. A damping force exerts a force of 5 lb for a velocity of 4 in./s. If the weight is displaced from the rest position and then released, find the general equation of motion.

4. A spring is stretched 2 in. by a weight of 8 lb. A damping force exerts a force of 4 lb for a velocity of 6 in./s. If the weight is displaced from the rest position and then released, find the general equation of motion.

5. A cylindrical buoy with a radius of 1.5 ft is floating in water with the axis of the buoy vertical. When the buoy is pushed a small distance into the water and released, the period of vibration is 6 s. Find the weight of the buoy.

6. A cylindrical buoy with a radius of 2 ft is floating in water with the axis of the buoy vertical. When the buoy is pushed a small distance into the water and released, the period of vibration is 1.6 s. Find the weight of the buoy.

7. A cylindrical buoy weighing 1000 lb is floating in water with the axis of the buoy vertical. When the buoy is pushed a small distance into the water and released, the period of vibration is 2 s. Find the radius of the buoy.

8. A cylindrical buoy weighing 1600 lb is floating in water with the axis of the buoy vertical. When the buoy is pushed a small distance into the water and released, the period of vibration is 1.5 s. Find the radius of the buoy.

9. An electric circuit has an inductance $L = 0.1$ H, a resistance $R = 50\ \Omega$, and a capacitance $C = 2 \times 10^{-4}$ F. Find the equation for the current i.

10. An electric circuit has an inductance $L = 0.2$ H, a resistance $R = 400\ \Omega$, and a capacitance $C = 5 \times 10^{-4}$ F. Find the equation for the current i.

11. An electric circuit has an inductance $L = 0.4$ H, a resistance $R = 200\ \Omega$, and a capacitance $C = 5 \times 10^{-5}$ F. Find the equation for the current i if the voltage source is 12 V.

12. An electric circuit has an inductance $L = 0.5$ H, a resistance $R = 1000\ \Omega$, and a capacitance $C = 5.6 \times 10^{-6}$ F. Find the equation for the current i if the voltage source is 12 V.

12.5 THE LAPLACE TRANSFORM

The linear operator $\mathcal{L}$, called the *Laplace transform*, is very useful in solving differential equations with given initial conditions. The Laplace transform $\mathcal{L}$, like the differential operator D, operates on functions. The Laplace transform operates on a function $f(t)$ and transforms it into another function $F(s)$. That is, $\mathcal{L}[f(t)] = F(s)$. The transform is defined by means of integration:

LAPLACE TRANSFORM

$$F(s) = \mathcal{L}[f(t)] = \int_0^\infty e^{-st} f(t)\, dt$$

Note that this equation involves an integral with an infinite upper limit (∞). Such an integral is called an *improper integral,* whose value is found by a special limit process. In general, an integral of the form

$$\int_0^\infty g(t)\, dt$$

is evaluated by considering the limit of the integral $\int_0^b g(t)\, dt$ as $b \to \infty$ (as b takes on positive values without bound). That is,

$$\int_0^\infty g(t)\, dt = \lim_{b \to \infty} \int_0^b g(t)\, dt$$

And if $G(t)$ is an antiderivative of $g(t)$, then

$$\int_0^\infty g(t)\, dt = \lim_{b \to \infty} \int_0^b g(t)\, dt$$
$$= \lim_{b \to \infty} [G(t)]\Big|_0^b$$
$$= \lim_{b \to \infty} [G(b) - G(0)]$$

We illustrate this method of integration by using examples of Laplace transforms.

EXAMPLE 1

Find the Laplace transform of the function $f(t) = e^{at}$,

$$\mathcal{L}(e^{at}) = \int_0^\infty e^{-st}(e^{at})\, dt \qquad \text{(assume that } s > a\text{)}$$
$$= \int_0^\infty e^{-st+at}\, dt$$

$$= \int_0^\infty e^{-(s-a)t} \, dt$$

$$= \lim_{b \to \infty} \int_0^b e^{-(s-a)t} \, dt$$

$$= \lim_{b \to \infty} \left[\frac{-e^{-(s-a)t}}{s-a} \Big|_0^b \right]$$

$$= \lim_{b \to \infty} \left[\frac{-e^{-(s-a)b}}{s-a} - \frac{-e^0}{s-a} \right]$$

$$= \lim_{b \to \infty} \left[\frac{1}{s-a} \left(\frac{-1}{e^{(s-a)b}} + 1 \right) \right]$$

$$= \frac{1}{s-a}(0+1) \qquad \left(\frac{-1}{e^{(s-a)b}} \to 0 \text{ as } b \to \infty \right)$$

$$\mathscr{L}(e^{at}) = \frac{1}{s-a} \qquad (s > a)$$

EXAMPLE 2

Find the Laplace transform of the function $f(t) = 1$.

This is a special case of Example 1. If $a = 0$, $f(t) = e^{(0)t} = e^0 = 1$. Then

$$\mathscr{L}(1) = \mathscr{L}(e^{0t})$$

$$\mathscr{L}(1) = \frac{1}{s-0}$$

$$\mathscr{L}(1) = \frac{1}{s} \qquad (s > 0)$$

EXAMPLE 3

Find $\mathscr{L}(\sin kt)$.

$$\mathscr{L}(\sin kt) = \int_0^\infty e^{-st} \sin kt \, dt$$

$$= \lim_{b \to \infty} \int_0^b e^{-st} \sin kt \, dt$$

$$= \lim_{b \to \infty} \left[\frac{e^{-st}(-s \sin kt - k \cos kt)}{s^2 + k^2} \Big|_0^b \right] \qquad \text{(use Formula 78, Appendix B)}$$

$$= 0 - \frac{1(0-k)}{s^2 + k^2} \qquad (\text{for } s > 0, \, e^{-sb} \to 0 \text{ as } b \to \infty)$$

$$\mathscr{L}(\sin kt) = \frac{k}{s^2 + k^2}$$

Using a calculator,

2nd 7 green diamond e^x -s*t) 2nd SIN k*t),t,0, **green diamond** ∞)|s **2nd > 0 ENTER**

We now develop the expression for the Laplace transform of the derivative $f'(t)$ of a function $f(t)$:

$$\mathcal{L}[f'(t)] = \int_0^\infty e^{-st} f'(t) \, dt$$

Integrating by parts, we let $u = e^{-st}$ and $dv = f'(t) \, dt$. Then

$$\int u \, dv \qquad\qquad \int v \, du$$

$$\boxed{\begin{array}{l} u = e^{-st} \\ dv = f'(t) \, dt \end{array}} \longrightarrow \boxed{\begin{array}{l} du = -se^{-st} \, dt \\ v = f(t) \end{array}}$$

$$\int u \, dv = uv - \int v \, du$$

$$\begin{aligned}
\mathcal{L}[f'(t)] &= \int_0^\infty e^{-st} f'(t) \, dt = f(t)e^{-st} \Big|_0^\infty - \int_0^\infty -se^{-st} f(t) \, dt \\
&= \lim_{b \to \infty} \left[\frac{f(t)}{e^{st}} \Big|_0^b \right] + s \int_0^\infty e^{-st} f(t) \, dt \\
&= \lim_{b \to \infty} \left[\frac{f(b)}{e^{sb}} - \frac{f(0)}{e^0} \right] + s \, \mathcal{L}[f(t)] \\
&= s \, \mathcal{L}[f(t)] - f(0) \quad \left(\lim_{b \to \infty} \frac{f(b)}{e^{sb}} = 0 \right)
\end{aligned}$$

One can show that the Laplace transform of the second derivative is given as follows:

$$\mathcal{L}[f''(t)] = s^2 \, \mathcal{L}[f(t)] - s f(0) - f'(0)$$

One can also show that the Laplace transform is a linear operator. That is,

$$\mathcal{L}[a f(t) + b \, g(t)] = a \, \mathcal{L}[f(t)] + b \, \mathcal{L}[g(t)]$$

where $a f(t)$ represents a constant multiple of the function $f(t)$ and $b \, g(t)$ represents a constant multiple of the function $g(t)$.

Using this linearity property and the Laplace transforms for derivatives, we can obtain a Laplace transform for any linear combination of a function $y = f(t)$ and its first and second derivatives. Table 12.2 lists some commonly used Laplace transforms. A more complete table can be found in standard books of mathematical tables.

EXAMPLE 4

Express the Laplace transform for $y'' - 3y' + 2y$ in terms of $\mathcal{L}(y)$ and s, where $y = 2$ and $y' = 1$ when $t = 0$. That is, $y(0) = 2$ and $y'(0) = 1$.

$$\begin{aligned}
\mathcal{L}(y'' - 3y' + 2y) &= \mathcal{L}(y'') - 3 \, \mathcal{L}(y') + 2 \, \mathcal{L}(y) \\
&= [s^2 \, \mathcal{L}(y) - s \, y(0) - y'(0)] - 3[s \, \mathcal{L}(y) - y(0)] + 2 \, \mathcal{L}(y) \\
&= [s^2 \, \mathcal{L}(y) - s(2) - 1] - 3[s \, \mathcal{L}(y) - 2] + 2 \, \mathcal{L}(y) \\
&= (s^2 - 3s + 2) \, \mathcal{L}(y) - 2s + 5
\end{aligned}$$

EXAMPLE 5

Find the Laplace transform of $f(t) = te^{-3t}$

Using Formula 15 in Table 12.2, we have $a = -3$ and

$$\mathcal{L}(te^{-3t}) = \frac{1}{(s+3)^2}$$

TABLE 12.2   Laplace Transforms

$f(t) = \mathcal{L}^{-1}[F(s)]$	$\mathcal{L}[f(t)] = F(s)$
1. $a f(t) + b g(t)$	$a \mathcal{L}[f(t)] + b \mathcal{L}[g(t)]$
2. $f'(t)$	$s \mathcal{L}[f(t)] - f(0)$
3. $f''(t)$	$s^2 \mathcal{L}[f(t)] - s f(0) - f'(0)$
4. 1	$\dfrac{1}{s}$
5. t	$\dfrac{1}{s^2}$
6. $\dfrac{t^{n-1}}{(n-1)!}$	$\dfrac{1}{s^n}$ $\quad n = 1, 2, 3, \ldots$
7. e^{at}	$\dfrac{1}{s-a}$
8. $\sin kt$	$\dfrac{k}{s^2 + k^2}$
9. $e^{at} \sin kt$	$\dfrac{k}{(s-a)^2 + k^2}$
10. $\cos kt$	$\dfrac{s}{s^2 + k^2}$
11. $e^{at} \cos kt$	$\dfrac{s-a}{(s-a)^2 + k^2}$
12. $e^{at} - e^{bt}$	$\dfrac{a-b}{(s-a)(s-b)}$
13. $1 - e^{at}$	$\dfrac{-a}{s(s-a)}$
14. $ae^{at} - be^{bt}$	$\dfrac{s(a-b)}{(s-a)(s-b)}$
15. te^{at}	$\dfrac{1}{(s-a)^2}$
16. $\dfrac{t^{n-1}e^{at}}{(n-1)!}$	$\dfrac{1}{(s-a)^n}$ $\quad n = 1, 2, 3, \ldots$
17. $e^{at}(1 + at)$	$\dfrac{s}{(s-a)^2}$
18. $t \sin kt$	$\dfrac{2ks}{(s^2 + k^2)^2}$
19. $t \cos kt$	$\dfrac{s^2 - k^2}{(s^2 + k^2)^2}$
20. $\sin kt - kt \cos kt$	$\dfrac{2k^3}{(s^2 + k^2)^2}$
21. $\sin kt + kt \cos kt$	$\dfrac{2ks^2}{(s^2 + k^2)^2}$
22. $1 - \cos kt$	$\dfrac{k^2}{s(s^2 + k^2)}$
23. $kt - \sin kt$	$\dfrac{k^3}{s^2(s^2 + k^2)}$

When using the method of Laplace transforms to solve differential equations, we usually need to invert a transform at the end of the process. That is, we need to find the *inverse transform*

$$\mathscr{L}^{-1}[F(s)] = f(t)$$

where the symbol $\mathscr{L}^{-1}$ denotes an inverse transform.

If $F(s)$ is a function of s and $f(t)$ is a function of t such that

$$\mathscr{L}[f(t)] = F(s)$$

then

$$\mathscr{L}^{-1}[F(s)] = f(t) \text{ is the inverse transform.}$$

EXAMPLE 6

Find the inverse transform of $F(s) = \dfrac{5}{s(s + 5)}$.

Using Formula 13 in Table 12.2, we have $a = -5$ and

$$-\left\{ \mathscr{L}^{-1}\left[\frac{-5}{s(s + 5)}\right]\right\} = -(1 - e^{-5t}) = -1 + e^{-5t}$$

EXAMPLE 7

Find the inverse transform of $\mathscr{L}[f(t)]$, where

$$s(s^2 + 9)\,\mathscr{L}[f(t)] - 9 = 0.$$

Solving for $\mathscr{L}[f(t)]$, we have

$$\mathscr{L}[f(t)] = \frac{9}{s(s^2 + 9)} = F(s)$$

Applying Formula 22 in Table 12.2 with $k = 3$, we obtain

$$\mathscr{L}^{-1}[F(s)] = \mathscr{L}^{-1}\left[\frac{9}{s(s^2 + 9)}\right]$$

$$= 1 - \cos 3t$$

An inverse transform cannot always be found directly by using Table 12.2. Sometimes the method of completing the square or partial fractions can be used to rewrite a function $F(s)$ in a form applicable to the table.

EXAMPLE 8

Find the inverse transform of $F(s) = \dfrac{s + 1}{s^2 - 4s + 5}$.

Completing the square, we have

$$s^2 - 4s + 5 = (s^2 - 4s + 4) + 1 = (s - 2)^2 + 1$$

Then we can write

$$F(s) = \frac{s + 1}{s^2 - 4s + 5} = \frac{(s - 2) + 3}{(s - 2)^2 + 1}$$

$$= \frac{s - 2}{(s - 2)^2 + 1} + \frac{3}{(s - 2)^2 + 1}$$

and

$$\mathcal{L}^{-1}[F(s)] = \mathcal{L}^{-1}\left[\frac{s-2}{(s-2)^2+1} + \frac{3}{(s-2)^2+1}\right]$$

$$= \mathcal{L}^{-1}\left[\frac{s-2}{(s-2)^2+1}\right] + \mathcal{L}^{-1}\left[\frac{3}{(s-2)^2+1}\right]$$

$$= e^{2t}\cos t + 3(e^{2t}\sin t)$$

where the first term is found from Formula 11 using $a = 2$ and $k = 1$ and the second term is found from Formula 9 using $a = 2$ and $k = 1$.

Exercises 12.5

1. Derive the Laplace transform for $f(t) = t$.

2. Derive the Laplace transform for $f(t) = \cos kt$.

Using Table 12.2, Laplace Transforms, find the Laplace transform of each function f(t).

3. $f(t) = \sin 3t$

4. $f(t) = e^{5t}$

5. $f(t) = 1 - e^{4t}$

6. $f(t) = 1 - \cos 7t$

7. $f(t) = t^2$

8. $f(t) = te^{3t}$

9. $f(t) = \sin 2t - 2t\cos t$

10. $f(t) = e^{3t}\sin 5t$

11. $f(t) = t - e^{2t}$

12. $f(t) = t + \sin 3t$

13. $f(t) = t\sin 4t$

14. $f(t) = 6t - \sin 6t$

15. $f(t) = e^{-3t}\cos 5t$

16. $f(t) = t^5$

17. $f(t) = 8t + 4t^3$

18. $f(t) = t\sin 4t + t\cos 4t$

Express the Laplace transform of each expression with the given conditions in terms of $\mathcal{L}(y)$ and s.

19. $y'' - 3y'$; $y(0) = 0$ and $y'(0) = 0$

20. $y'' - y' + 2y$; $y(0) = 0$ and $y'(0) = 0$

21. $y'' + y' + y$; $y(0) = 0$ and $y'(0) = 1$

22. $y'' - 2y$; $y(0) = -1$ and $y'(0) = 2$

23. $y'' - 3y' + y$; $y(0) = 1$ and $y'(0) = 0$

24. $y'' + 4y$; $y(0) = 2$ and $y'(0) = -3$

25. $y'' + 8y' + 2y$; $y(0) = 4$ and $y'(0) = 6$

26. $y'' - 3y' + 6y$; $y(0) = 7$ and $y'(0) = -3$

27. $y'' - 6y'$; $y(0) = 3$ and $y'(0) = 7$

28. $y'' + 2y' + 3y$; $y(0) = -4$ and $y'(0) = 5$

29. $y'' + 8y' - 3y$; $y(0) = -6$ and $y'(0) = 2$

30. $y'' + 3y' - 2y$; $y(0) = -5$ and $y'(0) = -3$

Using Table 12.2, find the inverse transform of each function F(s).

31. $F(s) = \dfrac{1}{s}$

32. $F(s) = \dfrac{4}{s^2+16}$

33. $F(s) = \dfrac{1}{(s-5)^2}$

34. $F(s) = \dfrac{2}{s(s+2)}$

35. $F(s) = \dfrac{s}{s^2+64}$

36. $F(s) = \dfrac{1}{s^4}$

37. $F(s) = \dfrac{4}{(s-6)(s-2)}$

38. $F(s) = \dfrac{1}{s^2} - \dfrac{3}{(s-2)^2+9}$

39. $F(s) = \dfrac{2}{s^2-6s+13}$

40. $F(s) = \dfrac{1}{s^2+4s+13}$

41. $F(s) = \dfrac{2}{(s^2+1)^2}$

42. $F(s) = \dfrac{9s^2+3s+62}{(s^2+9)(s+5)}$

43. $F(s) = \dfrac{2}{(s-1)(s^2+1)}$

44. $F(s) = \dfrac{11s-3}{s^2-s-6}$

45. $F(s) = \dfrac{6s^2+42s+54}{s^3+9s^2+18s}$

46. $F(s) = \dfrac{4s^2-22s+38}{(s+1)(s-3)^2}$

47. $F(s) = \dfrac{s+11}{s^2+10s+29}$

48. $F(s) = \dfrac{3s-4}{(s+2)(s^2+s+3)}$

49. Show: $\mathcal{L}[f''(t)] = s^2\,\mathcal{L}[f(t)] - sf(0) - f'(0)$

12.6 SOLUTIONS BY THE METHOD OF LAPLACE TRANSFORMS

We find particular solutions y_p of linear differential equations using Laplace transforms as follows:

SOLVING LINEAR DIFFERENTIAL EQUATIONS USING LAPLACE TRANSFORMS

To solve $a_0 D^2 y + a_1 Dy + a_2 y = b$ subject to initial conditions $y(0) = c$ and $y'(0) = d$:

1. Take the Laplace transform of each side of the linear differential equation $a_0 D^2 y + a_1 Dy + a_2 y = b$. Substitute the initial values c for $y(0)$ and d for $y'(0)$ in this new equation.

2. Solve the resulting equation in Step 1 for $\mathcal{L}(y)$ to obtain an equation in the form $\mathcal{L}(y) = F(s)$.

3. Find the inverse transform of $\mathcal{L}(y) = F(s)$ from Step 2 to obtain the particular solution $y = \mathcal{L}^{-1}[F(s)]$.

EXAMPLE 1

Solve the homogeneous differential equation $y'' + 4y' + 4y = 0$ subject to the initial conditions that $y = 2$ and $y' = 1$ when $x = 0$. [Write this as $y(0) = 2$ and $y'(0) = 1$.]

Step 1: Take the Laplace transform of each side of the equation.

$$\mathcal{L}(y'' + 4y' + 4y) = \mathcal{L}(0)$$
$$[s^2 \mathcal{L}(y) - s\,y(0) - y'(0)] + 4[s\,\mathcal{L}(y) - y(0)] + 4\,\mathcal{L}(y) = 0$$
$$(s^2 + 4s + 4)\,\mathcal{L}(y) + (-s - 4)y(0) - y'(0) = 0$$
$$(s^2 + 4s + 4)\,\mathcal{L}(y) - (s + 4)(2) - 1 = 0$$
$$(s^2 + 4s + 4)\,\mathcal{L}(y) - 2s - 9 = 0$$

Step 2: Solve for $\mathcal{L}(y)$:

$$\mathcal{L}(y) = \frac{2s + 9}{s^2 + 4s + 4}$$

Step 3: Find the inverse transform of $\mathcal{L}(y) = F(s)$.

Although $F(s) = \dfrac{2s + 9}{s^2 + 4s + 4}$ does not appear in Table 12.2, we can rewrite $F(s)$ as follows:

$$F(s) = \frac{2s + 9}{s^2 + 4s + 4} = \frac{2s}{(s + 2)^2} + \frac{9}{(s + 2)^2}$$

then

$$y = \mathcal{L}^{-1}[F(s)]$$
$$= \mathcal{L}^{-1}\left[\frac{2s}{(s + 2)^2} + \frac{9}{(s + 2)^2}\right]$$
$$= 2\,\mathcal{L}^{-1}\left[\frac{s}{(s + 2)^2}\right] + 9\,\mathcal{L}^{-1}\left[\frac{1}{(s + 2)^2}\right]$$
$$= 2\,[e^{-2t}(1 - 2t)] + 9(te^{-2t})$$
$$= 2e^{-2t} - 4te^{-2t} + 9te^{-2t}$$

or

$$y = (2 + 5t)e^{-2t}$$

We used Formulas 15 and 17 with $a = -2$. Compare this result with the solution obtained in Example 2, Section 12.2.

EXAMPLE 2

Solve the nonhomogeneous differential equation

$$y'' + y = \sin t \quad \text{if} \quad y(0) = 1 \quad \text{and} \quad y'(0) = -1$$

Step 1: Take the Laplace transform of each side of the equation.

$$\mathcal{L}(y'' + y) = \mathcal{L}(\sin t)$$

$$\mathcal{L}(y'') + \mathcal{L}(y) = \mathcal{L}(\sin t)$$

$$s^2 \mathcal{L}(y) - s\,y(0) - y'(0) + \mathcal{L}(y) = \frac{1}{s^2 + 1}$$

$$s^2 \mathcal{L}(y) - s(1) - (-1) + \mathcal{L}(y) = \frac{1}{s^2 + 1}$$

$$(s^2 + 1)\,\mathcal{L}(y) - s + 1 = \frac{1}{s^2 + 1}$$

Step 2: Solve for $\mathcal{L}(y)$.

$$(s^2 + 1)\,\mathcal{L}(y) = \frac{1}{s^2 + 1} + s - 1$$

$$\mathcal{L}(y) = \frac{1}{(s^2 + 1)^2} + \frac{s - 1}{s^2 + 1}$$

Step 3: Find the inverse Laplace transform of $\mathcal{L}(y) = F(s)$.

$$y = \mathcal{L}^{-1}[F(s)]$$

$$= \mathcal{L}^{-1}\left[\frac{1}{(s^2 + 1)^2} + \left(\frac{s - 1}{s^2 + 1}\right)\right]$$

$$= \mathcal{L}^{-1}\left[\frac{1}{(s^2 + 1)^2}\right] + \mathcal{L}^{-1}\left(\frac{s - 1}{s^2 + 1}\right)$$

$$= \mathcal{L}^{-1}\left[\frac{1}{(s^2 + 1)^2}\right] + \mathcal{L}^{-1}\left(\frac{s}{s^2 + 1} - \frac{1}{s^2 + 1}\right)$$

$$= \mathcal{L}^{-1}\left[\frac{1}{(s^2 + 1)^2}\right] + \mathcal{L}^{-1}\left(\frac{s}{s^2 + 1}\right) - \mathcal{L}^{-1}\left(\frac{1}{s^2 + 1}\right)$$

Since

$$\mathcal{L}^{-1}\left[\frac{1}{(s^2 + 1)^2}\right] = \frac{1}{2}(\sin t - t\cos t) \qquad \text{(Formula 20 with } k = 1\text{)}$$

$$\mathcal{L}^{-1}\left(\frac{s}{s^2 + 1}\right) = \cos t \qquad \text{(Formula 10 with } k = 1\text{)}$$

$$\mathcal{L}^{-1}\left(\frac{s}{s^2 + 1}\right) = \sin t \qquad \text{(Formula 8 with } k = 1\text{)}$$

we have

$$y = \frac{1}{2}(\sin t - t\cos t) + \cos t - \sin t$$

$$y = -\frac{1}{2}\sin t + \left(1 - \frac{t}{2}\right)\cos t$$

EXAMPLE 3

Solve the differential equation $y'' + 4y' - 5y = 0$ with $y(0) = 1$ and $y'(0) = -4$.

Step 1: Take the Laplace transform of each side of the equation.

$$\mathcal{L}(y'' + 4y' - 5y) = \mathcal{L}(0)$$

$$[s^2 \mathcal{L}(y) - s\,y(0) - y'(0)] + 4[s\,\mathcal{L}(y) - y(0)] - 5\,\mathcal{L}(y) = 0$$

$$(s^2 + 4s - 5)\,\mathcal{L}(y) + (-s - 4)y(0) - y'(0) = 0$$

$$(s^2 + 4s - 5)\,\mathcal{L}(y) - (s + 4)(1) - (-4) = 0$$

$$(s^2 + 4s - 5)\,\mathcal{L}(y) - s = 0$$

Step 2: Solve for $\mathcal{L}(y)$.

$$(s^2 + 4s - 5)\,\mathcal{L}(y) = s$$

$$\mathcal{L}(y) = \frac{s}{s^2 + 4s - 5}$$

Step 3: Find the inverse Laplace transform of $\mathcal{L}(y) = F(s)$. $F(s) = \dfrac{s}{s^2 + 4s - 5}$ does not appear in Table 12.2. However, using partial fractions, we express $F(s)$ as a sum of two fractions:

$$\frac{s}{s^2 + 4s - 5} = \frac{s}{(s + 5)(s - 1)}$$

$$= \frac{A}{s + 5} + \frac{B}{s - 1}$$

$$= \frac{A(s - 1)}{(s + 5)(s - 1)} + \frac{B(s + 5)}{(s + 5)(s - 1)}$$

$$= \frac{As - A + Bs + 5B}{(s + 5)(s - 1)}$$

$$= \frac{(A + B)s + (-A + 5B)}{(s + 5)(s - 1)}$$

Setting

$$A + B = 1$$

$$-A + 5B = 0$$

we solve for A and B and find $A = \frac{5}{6}$ and $B = \frac{1}{6}$.

Then

$$y = \mathcal{L}^{-1}[F(s)]$$

$$= \mathcal{L}^{-1}\left(\frac{s}{s^2 + 4s - 5}\right)$$

$$= \mathcal{L}^{-1}\left[\left(\frac{\frac{5}{6}}{s + 5}\right) + \left(\frac{\frac{1}{6}}{s - 1}\right)\right]$$

$$= \frac{5}{6}\,\mathcal{L}^{-1}\left(\frac{1}{s + 5}\right) + \frac{1}{6}\,\mathcal{L}^{-1}\left(\frac{1}{s - 1}\right)$$

or

$$y = \frac{5}{6}e^{-5t} + \frac{1}{6}e^{t} \qquad \text{(Formula 7 with } a = -5 \text{ and } a = 1)$$

Exercises 12.6

Solve each differential equation subject to the given conditions by using Laplace transforms.

1. $y' - y = 0$; $y(0) = 2$

2. $y' + 3y = 0$; $y(0) = -1$

3. $4y' + 3y = 0$; $y(0) = 1$

4. $y' - 2y = 0$; $y(0) = 5$

5. $y' - 7y = e^t$; $y(0) = 5$

6. $y' + 2y = 3$; $y(0) = 0$

7. $y'' + y = 0$; $y(0) = 1$ and $y'(0) = 0$

8. $y'' + 3y = 0$; $y(0) = 2$ and $y'(0) = 5$

9. $y'' - 2y' = 0$; $y(0) = 1$ and $y'(0) = -1$

10. $y'' + 3y' = 0$; $y(0) = 0$ and $y'(0) = 2$

11. $y'' + 2y' + y = 0$; $y(0) = 1$ and $y'(0) = 0$

12. $y'' - 6y' + 9y = 0$; $y(0) = 1$ and $y'(0) = 2$

13. $y'' - 4y' + 4y = te^{2t}$; $y(0) = 0$ and $y'(0) = 0$

14. $y'' + 10y' + 25y = e^{-5t}$; $y(0) = 0$ and $y'(0) = 0$

15. $y'' + 2y' + y = 3te^{-t}$; $y(0) = 4$ and $y'(0) = 2$

16. $y'' - 6y' + 9y = e^{3t}$; $y(0) = 1$ and $y'(0) = 2$

17. $y'' + 3y' - 4y = 0$; $y(0) = 1$ and $y'(0) = -2$

18. $y'' - y' - 2y = 0$; $y(0) = 2$ and $y'(0) = 3$

CHAPTER SUMMARY

1. A linear differential equation of order n is represented by the form

$$a_0 \frac{d^n y}{dx^n} + a_1 \frac{d^{n-1} y}{dx^{n-1}} + \cdots + a_{n-1} \frac{dy}{dx} + a_n y = b$$

where $a_0, a_1, a_2, \ldots, a_{n-1}, a_n$, and b can be either functions of x or constants. If $b = 0$, the linear differential equation is called *homogeneous*. If $b \neq 0$, the equation is called *nonhomogeneous*.

2. *Differential operator D* operates on a function $y = f(x)$ as follows:

$$Dy = \frac{dy}{dx} \qquad D^2 y = \frac{d^2 y}{dx^2} \qquad D^3 y = \frac{d^3 y}{dx^3} \cdots$$

Operator D is an example of a linear operator. An algebraic expression in the form $ax + by$ where a and b are constants is called a *linear combination*. A *linear operator* is any process which after operating on a linear combination results in another linear combination. For the linear operator D, we have

$$D[a f(x) + b g(x)] = a D[f(x)] + b D[g(x)]$$

We are able to apply algebraic operations to linear operators.

3. The equation $a_0 m^2 + a_1 m + a_2 = 0$ is called the *auxiliary equation (characteristic equation)* for the differential equation

$$a_0 D^2 y + a_1 Dy + a_2 y = 0$$

Observe that $y = ke^{mx}$ will be a solution if m satisfies the auxiliary equation $a_0 m^2 + a_1 m + a_2 = 0$. The problem of determining a solution of the differential equation is now a problem of finding the two roots, m_1 and m_2, of the auxiliary equation.

The method may be summarized as follows:

(a) Write the auxiliary equation for a given differential equation:

$$a_0 D^2 y + a_1 Dy + a_2 y = 0$$
$$a_0 m^2 + a_1 m + a_2 = 0$$

(b) Determine the two roots, m_1 and m_2, of the auxiliary equation.

(c) Write the general solution as

$$y = k_1 e^{m_1 x} + k_2 e^{m_2 x}$$

4. When the auxiliary equation of a differential equation results in a repeated root $(m_1 = m_2 = m)$, then the general solution is in the form

$$y = k_1 e^{mx} + k_2 x e^{mx}$$

Note carefully that the second term includes the factor x.

5. If the auxiliary equation of a differential equation has complex roots in the form $m = a \pm bj$, then the solution of the differential equation is

$$y = e^{ax}(k_1 \sin bx + k_2 \cos bx)$$

where k_1 and k_2 are arbitrary constants.

6. To find the general solution of the nonhomogeneous equation

$$a_0 D^2 y + a_1 Dy + a_2 y = b,$$

(a) Find the complementary solution y_c by solving the homogeneous equation $a_0 D^2 y + a_1 Dy + a_2 y = 0$ using the methods summarized in 3–5 above.

(b) Find a particular solution y_p by using the method of undetermined coefficients.

(c) Find the general solution y by adding the complementary solution y_c from Step (a) and the particular solution y_p from Step (b):

$$y = y_c + y_p$$

7. The Laplace transform $\mathcal{L}$, like the differential operator D, operates on functions. The Laplace transform operates on a function $f(t)$ and transforms it into another function $F(s)$. That is, $\mathcal{L}[f(t)] = F(s)$. The transform is defined by means of integration:

$$F(s) = \mathcal{L}[f(t)] = \int_0^\infty e^{-st} f(t)\, dt$$

8. The particular solution y_p of a linear differential equation $a_0 D^2 y + a_1 Dy + a_2 y = b$ subject to given initial conditions $y(0) = c$ and $y'(0) = d$ can be found by using Laplace transforms as follows:

(a) Take the Laplace transform of each side of the linear differential equation $a_0 D^2 y + a_1 Dy + a_2 y = b$. Substitute the initial values c for $y(0)$ and d for $y'(0)$ in this new equation.

(b) Solve the resulting equation in Step (a) for $\mathcal{L}(y)$ to obtain an equation in the form $\mathcal{L}(y) = F(s)$.

(c) Find the inverse transform of $\mathcal{L}(y) = F(s)$ from Step (b) to obtain the particular solution $y = \mathcal{L}^{-1}[F(s)]$.

CHAPTER 12 REVIEW

State whether each differential equation is homogeneous or nonhomogeneous. Also state the order of each equation.

1. $2y'' + y' - 7y = 0$

2. $\dfrac{d^2 y}{dx^2} - 7y = 8$

3. $\dfrac{dy}{dx} - 3y - 5 = 0$

4. $y''' - y'' + y = 0$

Solve each differential equation.

5. $\dfrac{d^2y}{dx^2} + 4\dfrac{dy}{dx} - 5y = 0$ **6.** $\dfrac{d^2y}{dx^2} - 5\dfrac{dy}{dx} + 6y = 0$ **7.** $y'' - 6y' = 0$

8. $2y'' - y' - 3y = 0$ **9.** $y'' - 6y' + 9y = 0$ **10.** $D^2y + 10\,Dy + 25y = 0$

11. $D^2y - 2\,Dy + y = 0$ **12.** $9\,D^2y - 6\,Dy + y = 0$ **13.** $y'' + 16y = 0$

14. $4\,D^2y + 25y = 0$ **15.** $D^2y - 2\,Dy + 3y = 0$ **16.** $y'' - 3y' + 8y = 0$

17. $D^2y + Dy - 2y = x$ **18.** $y'' - 6y' + 9y = e^x$ **19.** $y'' + 4y = \cos x$

20. $y'' - 2y' + 3y = 6e^{2x}$

Find the particular solution of each differential equation satisfying the given conditions.

21. $\dfrac{d^2y}{dx^2} + 2\dfrac{dy}{dx} - 8y = 0;\ y = 6$ and $y' = 0$ when $x = 0$

22. $D^2y - 3\,Dy = 0;\ y = 3$ and $y' = 6$ when $x = 0$

23. $D^2y - 4\,Dy + 4y = 0;\ y = 0$ and $y' = 3$ when $x = 0$

24. $y'' + 6y' + 9y = 0;\ y = 8$ and $y' = 0$ when $x = 0$

25. $y'' + 4y = 0;\ y = 1$ and $y' = -4$ when $x = 0$

26. $D^2y - 8\,Dy + 25y = 0;\ y = 2$ and $y' = 11$ when $x = 0$

27. $y'' - y' = e^{2x};\ y = 1$ and $y' = 0$ when $x = 0$

28. $D^2y + 4y = \sin x;\ y = 2$ and $y' = \frac{7}{3}$ when $x = 0$

29. A spring is stretched 4 in. by a weight of 16 lb. If the weight is displaced a distance of 6 in. from the rest position and then released, determine the equation of motion.

30. A spring is stretched 2 in. by a weight of 12 lb. A damping force exerts a force of 8 lb for a velocity of 4 in./s. If the weight is displaced from the rest position and then released, determine the equation of motion.

31. An electric circuit has an inductance $L = 2$ H, a resistance $R = 400\ \Omega$, and a capacitance $C = 10^{-5}$ F. Find the equation for the current i.

32. An electric circuit has an inductance $L = 1$ H, a resistance $R = 2000\ \Omega$, and a capacitance $C = 4 \times 10^{-6}$ F. Find the equation for the current i.

Using Table 12.2, Laplace Transforms, find the Laplace transform for each function f(t).

33. $f(t) = e^{6t}$ **34.** $f(t) = e^{-2t}\cos 3t$ **35.** $f(t) = t^3 + \cos t$ **36.** $f(t) = 3t - e^{5t}$

Using Table 12.2, find the inverse transform for each function F(s).

37. $F(s) = \dfrac{1}{s^2}$ **38.** $F(s) = \dfrac{3}{s(s-2)}$ **39.** $F(s) = \dfrac{2}{(s-3)(s-4)}$ **40.** $F(s) = \dfrac{3}{s^2 + 9}$

Solve each differential equation subject to the given conditions by using Laplace transforms.

41. $4y' - 5y = 0;\ y(0) = 2$

42. $y'' + 9y = 0;\ y(0) = 3$ and $y'(0) = 0$

43. $y'' + 5y' = 0;\ y(0) = 0$ and $y'(0) = 2$

44. $y'' + 4y' + 4y = e^{-2t};\ y(0) = 0$ and $y'(0) = 0$

Weights and Measures

TABLE 1 English weights and measures

Units of length	Units of weight
Standard unit–inch (in. or ″)	Standard unit–pound (lb)
12 inches = 1 foot (ft or ′)	16 ounces (oz) = 1 pound
3 feet = 1 yard (yd)	2000 pounds = 1 ton (T)
$5\frac{1}{2}$ yards or $16\frac{1}{2}$ feet = 1 rod (rd)	
5280 feet = 1 mile (mi)	

Volume measure
Liquid
16 ounces (fl oz) = 1 pint (pt)
2 pints = 1 quart (qt)
4 quarts = 1 gallon (gal)
Dry
2 pints (pt) = 1 quart (qt)
8 quarts = 1 peck (pk)
4 pecks = 1 bushel (bu)

TABLE 2 Conversion tables

Length

	cm	m	km	in.	ft	mi
1 centimetre	1	10^{-2}	10^{-5}	0.394	3.28×10^{-2}	6.21×10^{-6}
1 metre	100	1	10^{-3}	39.4	3.28	6.21×10^{-4}
1 kilometre	10^5	1000	1	3.94×10^4	3280	0.621
1 inch	2.54	2.54×10^{-2}	2.54×10^{-5}	1	8.33×10^{-2}	1.58×10^{-5}
1 foot	30.5	0.305	3.05×10^{-4}	12	1	1.89×10^{-4}
1 mile	1.61×10^5	1610	1.61	6.34×10^4	5280	1

Area

Metric

$1 \text{ m}^2 = 10{,}000 \text{ cm}^2$
$\qquad = 1{,}000{,}000 \text{ mm}^2$
$1 \text{ cm}^2 = 100 \text{ mm}^2$
$\qquad = 0.0001 \text{ m}^2$
$1 \text{ km}^2 = 1{,}000{,}000 \text{ m}^2$
$1 \text{ ha} = 10{,}000 \text{ m}^2$

English

$1 \text{ ft}^2 = 144 \text{ in}^2$
$1 \text{ yd}^2 = 9 \text{ ft}^2$
$1 \text{ rd}^2 = 30.25 \text{ yd}^2$
$1 \text{ acre} = 160 \text{ rd}^2$
$\qquad = 4840 \text{ yd}^2$
$\qquad = 43{,}560 \text{ ft}^2$
$1 \text{ mi}^2 = 640 \text{ acres}$

	m^2	cm^2	ft^2	in^2
1 m^2	1	10^4	10.8	1550
1 cm^2	10^{-4}	1	1.08×10^{-3}	0.155
1 ft^2	9.29×10^{-2}	929	1	144
1 in^2	6.45×10^{-4}	6.45	6.94×10^{-3}	1

$1 \text{ mi}^2 = 2.79 \times 10^7 \text{ ft}^2 = 640 \text{ acres}$

$1 \text{ circular mil} = 5.07 \times 10^{-6} \text{ cm}^2 = 7.85 \times 10^{-7} \text{ in}^2$

$1 \text{ hectare} = 2.47 \text{ acres}$

Volume

	Metric	*English*
	$1 \text{ m}^3 = 10^6 \text{ cm}^3$	$1 \text{ ft}^3 = 1728 \text{ in}^3$
	$1 \text{ cm}^3 = 10^{-6} \text{ m}^3$	$1 \text{ yd}^3 = 27 \text{ ft}^3$
	$= 10^3 \text{ mm}^3$	

	m^3	cm^3	L	ft^3	in^3
1 m^3	1	10^6	1000	35.3	6.10×10^4
1 cm^3	10^{-6}	1	1.00×10^{-3}	3.53×10^{-5}	6.10×10^{-2}
1 L	1.00×10^{-3}	1000	1	3.53×10^{-2}	61.0
1 ft^3	2.83×10^{-2}	2.83×10^4	28.3	1	1728
1 in^3	1.64×10^{-5}	16.4	1.64×10^{-2}	5.79×10^{-4}	1

1 U.S. fluid gallon = 4 U.S. fluid quarts = 8 U.S. pints = 128 U.S. fluid ounces = 231 in^3 = 0.134 ft^3 = 3.79 litres

1 L = 1000 cm^3 = 1.06 qt

Other useful conversion factors

1 newton (N) = 0.225 lb
1 pound (lb) = 4.45 N
1 slug = 14.6 kg
1 joule (J) = 0.738 ft-lb
$\qquad = 2.39 \times 10^{-4}$ kcal
1 calorie (cal) = 4.185 J
1 kilocalorie (kcal) = 4185 J
1 foot-pound (ft-lb) = 1.36 J
1 watt (W) = 1 J/s = 0.738 ft-lb/s
1 kilowatt (kW) = 1000 W
$\qquad = 1.34$ hp
1 hp = 550 ft-lb/s = 746 W

1 atm = 101.32 kpa
$\qquad = 14.7$ lb/in^2
1 Btu = 0.252 kcal
1 kcal = 3.97 Btu
$F = \frac{9}{5}C + 32°$
$C = \frac{5}{9}(F - 32°)$
1 kg = 2.20 lb (on the
$\quad$ earth's surface)
1 lb = 454 g
$\qquad = 16$ oz
1 metric ton = 1000 kg
$\qquad\qquad = 2200$ lb

Table of Integrals

1. $\displaystyle\int u^n\, du = \frac{u^{n+1}}{n+1} + C \qquad (n \neq -1)$

2. $\displaystyle\int \frac{du}{a+bu} = \frac{1}{b}\ln|a+bu| + C$

3. $\displaystyle\int \frac{u}{a+bu}\, du = \frac{1}{b^2}\big[(a+bu) - a\ln|a+bu|\big] + C$

4. $\displaystyle\int \frac{u^2\, du}{a+bu} = \frac{1}{b^3}\left[\frac{1}{2}(a+bu)^2 - 2a(a+bu) + a^2\ln|a+bu|\right] + C$

5. $\displaystyle\int \frac{du}{u(a+bu)} = \frac{1}{a}\ln\left|\frac{u}{a+bu}\right| + C$

6. $\displaystyle\int \frac{du}{u^2(a+bu)} = -\frac{1}{au} + \frac{b}{a^2}\ln\left|\frac{a+bu}{u}\right| + C$

7. $\displaystyle\int \frac{u\, du}{(a+bu)^2} = \frac{1}{b^2}\left(\ln|a+bu| + \frac{a}{a+bu}\right) + C$

8. $\displaystyle\int \frac{u^2\, du}{(a+bu)^2} = \frac{1}{b^3}\left[a+bu - \frac{a^2}{a+bu} - 2a\ln|a+bu|\right] + C$

9. $\displaystyle\int \frac{du}{u(a+bu)^2} = \frac{1}{a(a+bu)} + \frac{1}{a^2}\ln\left|\frac{u}{a+bu}\right| + C$

10. $\displaystyle\int \frac{du}{u^2(a+bu)^2} = -\frac{a+2bu}{a^2u(a+bu)} + \frac{2b}{a^3}\ln\left|\frac{a+bu}{u}\right| + C$

Forms containing $\sqrt{a+bu}$

---

11. $\displaystyle\int u\sqrt{a+bu}\, du = -\frac{2(2a-3bu)(a+bu)^{3/2}}{15b^2} + C$

12. $\displaystyle\int u^2\sqrt{a+bu}\, du = \frac{2(8a^2 - 12abu + 15b^2u^2)(a+bu)^{3/2}}{105b^3} + C$

13. $\displaystyle\int \frac{u\, du}{\sqrt{a+bu}} = -\frac{2(2a-bu)\sqrt{a+bu}}{3b^2} + C$

14. $\displaystyle\int \frac{u^2\, du}{\sqrt{a+bu}} = \frac{2(3b^2u^2 - 4abu + 8a^2)\sqrt{a+bu}}{15b^3} + C$

15. $\displaystyle\int \frac{du}{u\sqrt{a+bu}} = \frac{1}{\sqrt{a}}\ln\left|\frac{\sqrt{a+bu}-\sqrt{a}}{\sqrt{a+bu}+\sqrt{a}}\right| + C \qquad (a>0)$

16. $\displaystyle\int \frac{du}{u\sqrt{a + bu}} = \frac{2}{\sqrt{-a}} \operatorname{Arctan} \sqrt{\frac{a + bu}{-a}} + C \qquad (a < 0)$

17. $\displaystyle\int \frac{\sqrt{a + bu}\, du}{u} = 2\sqrt{a + bu} + a \int \frac{du}{u\sqrt{a + bu}} + C$

Rational forms containing $a^2 \pm u^2$ and $u^2 \pm a^2$

---

18. $\displaystyle\int \frac{du}{a^2 + u^2} = \frac{1}{a} \operatorname{Arctan} \frac{u}{a} + C$

19. $\displaystyle\int \frac{du}{a^2 - u^2} = \frac{1}{2a} \ln \left| \frac{a + u}{a - u} \right| + C \qquad (a^2 > u^2)$

20. $\displaystyle\int \frac{du}{u^2 - a^2} = \frac{1}{2a} \ln \left| \frac{u - a}{u + a} \right| + C \qquad (a^2 < u^2)$

Irrational forms containing $\sqrt{a^2 - u^2}$

---

21. $\displaystyle\int (a^2 - u^2)^{1/2}\, du = \frac{u}{2} \sqrt{a^2 - u^2} + \frac{a^2}{2} \operatorname{Arcsin} \frac{u}{a} + C$

22. $\displaystyle\int \frac{du}{(a^2 - u^2)^{1/2}} = \operatorname{Arcsin} \frac{u}{a} + C \qquad (a > 0)$

23. $\displaystyle\int \frac{du}{(a^2 - u^2)^{3/2}} = \frac{u}{a^2\sqrt{a^2 - u^2}} + C$

24. $\displaystyle\int \frac{u^2\, du}{(a^2 - u^2)^{1/2}} = -\frac{u}{2} \sqrt{a^2 - u^2} + \frac{a^2}{2} \operatorname{Arcsin} \frac{u}{a} + C$

25. $\displaystyle\int \frac{u^2\, du}{(a^2 - u^2)^{3/2}} = \frac{u}{\sqrt{a^2 - u^2}} - \operatorname{Arcsin} \frac{u}{a} + C$

26. $\displaystyle\int \frac{du}{u(a^2 - u^2)^{1/2}} = -\frac{1}{a} \ln \left| \frac{a + \sqrt{a^2 - u^2}}{u} \right| + C$

27. $\displaystyle\int \frac{du}{u^2(a^2 - u^2)^{1/2}} = -\frac{\sqrt{a^2 - u^2}}{a^2 u} + C$

28. $\displaystyle\int \frac{(a^2 - u^2)^{1/2}\, du}{u} = \sqrt{a^2 - u^2} - a \ln \left| \frac{a + \sqrt{a^2 - u^2}}{u} \right| + C$

29. $\displaystyle\int \frac{(a^2 - u^2)^{1/2}\, du}{u^2} = -\frac{\sqrt{a^2 - u^2}}{u} - \operatorname{Arcsin} \frac{u}{a} + C$

Irrational forms containing $\sqrt{u^2 \pm a^2}$

---

30. $\displaystyle\int \sqrt{u^2 \pm a^2}\, du = \tfrac{1}{2}\left[u\sqrt{u^2 \pm a^2} \pm a^2 \ln|u + \sqrt{u^2 \pm a^2}| \right] + C$

31. $\displaystyle\int u^2\sqrt{u^2 \pm a^2}\, du = \tfrac{1}{8}u(2u^2 \pm a^2)\sqrt{u^2 \pm a^2} - \tfrac{1}{8}a^4 \ln|u + \sqrt{u^2 \pm a^2}| + C$

32. $\displaystyle\int \frac{\sqrt{u^2 + a^2}}{u}\, du = \sqrt{u^2 + a^2} - a \ln \left| \frac{a + \sqrt{u^2 + a^2}}{u} \right| + C$

33. $\displaystyle\int \frac{\sqrt{u^2 - a^2}}{u}\, du = \sqrt{u^2 - a^2} - a \operatorname{Arccos} \frac{a}{u} + C$

34. $\displaystyle\int \frac{\sqrt{u^2 \pm a^2}}{u^2}\, du = -\frac{\sqrt{u^2 \pm a^2}}{u} + \ln|u + \sqrt{u^2 \pm a^2}| + C$

35. $\displaystyle\int \frac{du}{\sqrt{u^2 \pm a^2}} = \ln|u + \sqrt{u^2 \pm a^2}| + C$

36. $\displaystyle\int \frac{du}{u\sqrt{u^2 - a^2}} = \frac{1}{a}\,\text{Arccos}\,\frac{a}{u} + C$

37. $\displaystyle\int \frac{du}{u\sqrt{u^2 + a^2}} = \frac{1}{a}\ln\left|\frac{u}{a + \sqrt{u^2 + a^2}}\right| + C$

38. $\displaystyle\int \frac{u^2\,du}{\sqrt{u^2 \pm a^2}} = \frac{1}{2}\left(u\sqrt{u^2 \pm a^2} \pm a^2\ln|u + \sqrt{u^2 \pm a^2}|\right) + C$

39. $\displaystyle\int \frac{du}{u^2\sqrt{u^2 \pm a^2}} = -\frac{(\pm\sqrt{u^2 \pm a^2})}{a^2 u} + C$

40. $\displaystyle\int \frac{du}{(u^2 \pm a^2)^{3/2}} = \frac{\pm u}{a^2\sqrt{u^2 \pm a^2}} + C$

41. $\displaystyle\int \frac{u^2\,du}{(u^2 \pm a^2)^{3/2}} = \frac{-u}{\sqrt{u^2 \pm a^2}} + \ln|u + \sqrt{u^2 \pm a^2}| + C$

Forms containing $a + bu \pm cu^2$ $(c > 0)$

42. $\displaystyle\int \frac{du}{a + bu + cu^2} = \frac{2}{\sqrt{4ac - b^2}}\,\text{Arctan}\,\frac{2cu + b}{\sqrt{4ac - b^2}} + C$ $(b^2 < 4ac)$

43. $\displaystyle\int \frac{du}{a + bu + cu^2} = \frac{1}{\sqrt{b^2 - 4ac}}\ln\left|\frac{2cu + b - \sqrt{b^2 - 4ac}}{2cu + b + \sqrt{b^2 - 4ac}}\right| + C$ $(b^2 > 4ac)$

44. $\displaystyle\int \frac{du}{a + bu - cu^2} = \frac{1}{\sqrt{b^2 + 4ac}}\ln\left|\frac{\sqrt{b^2 + 4ac} + 2cu - b}{\sqrt{b^2 + 4ac} - 2cu + b}\right| + C$

45. $\displaystyle\int \sqrt{a + bu + cu^2}\,du = \frac{2cu + b}{4c}\sqrt{a + bu + cu^2} - \frac{b^2 - 4ac}{8c^{3/2}}\ln|2cu + b + 2\sqrt{c}\sqrt{a + bu + cu^2}| + C$

46. $\displaystyle\int \sqrt{a + bu - cu^2}\,du = \frac{2cu - b}{4c}\sqrt{a + bu - cu^2} + \frac{b^2 + 4ac}{8c^{3/2}}\,\text{Arcsin}\left(\frac{2cu - b}{\sqrt{b^2 + 4ac}}\right) + C$

47. $\displaystyle\int \frac{du}{\sqrt{a + bu + cu^2}} = \frac{1}{\sqrt{c}}\ln|2cu + b + 2\sqrt{c}\sqrt{a + bu + cu^2}| + C$

48. $\displaystyle\int \frac{du}{\sqrt{a + bu - cu^2}} = \frac{1}{\sqrt{c}}\,\text{Arcsin}\left(\frac{2cu - b}{\sqrt{b^2 + 4ac}}\right) + C$

49. $\displaystyle\int \frac{u\,du}{\sqrt{a + bu + cu^2}} = \frac{\sqrt{a + bu + cu^2}}{c} - \frac{b}{2c^{3/2}}\ln|2cu + b + 2\sqrt{c}\sqrt{a + bu + cu^2}| + C$

50. $\displaystyle\int \frac{u\,du}{\sqrt{a + bu - cu^2}} = -\frac{\sqrt{a + bu - cu^2}}{c} + \frac{b}{2c^{3/2}}\,\text{Arcsin}\left(\frac{2cu - b}{\sqrt{b^2 + 4ac}}\right) + C$

Exponential and logarithmic forms

51. $\displaystyle\int e^u\,du = e^u + C$

52. $\displaystyle\int a^u\,du = \frac{a^u}{\ln a} + C$ $(a > 0, a \neq 1)$

53. $\displaystyle\int ue^{au}\,du = \frac{e^{au}}{a^2}(au - 1) + C$

54. $\displaystyle\int u^n e^{au}\,du = \frac{u^n e^{au}}{a} - \frac{n}{a}\int u^{n-1}e^{au}\,du$

55. $\displaystyle\int \frac{e^{au}}{u^n}\, du = -\frac{e^{au}}{(n-1)u^{n-1}} + \frac{a}{n-1} \int \frac{e^{au}}{u^{n-1}}\, du$

56. $\displaystyle\int \ln u\, du = u \ln u - u + C$

57. $\displaystyle\int u^n \ln u\, du = \frac{u^{n+1}\ln u}{n+1} - \frac{u^{n+1}}{(n+1)^2} + C$

58. $\displaystyle\int \frac{du}{u \ln u} = \ln|\ln u| + C$

59. $\displaystyle\int e^{au}\sin nu\, du = \frac{e^{au}(a\sin nu - n\cos nu)}{a^2 + n^2} + C$

60. $\displaystyle\int e^{au}\cos nu\, du = \frac{e^{au}(n\sin nu + a\cos nu)}{a^2 + n^2} + C$

Trigonometric forms

---

61. $\displaystyle\int \sin u\, du = -\cos u + C$

62. $\displaystyle\int \cos u\, du = \sin u + C$

63. $\displaystyle\int \tan u\, du = -\ln|\cos u| + C = \ln|\sec u| + C$

64. $\displaystyle\int \cot u\, du = \ln|\sin u| + C$

65. $\displaystyle\int \sec u\, du = \ln|\sec u + \tan u| + C$

66. $\displaystyle\int \csc u\, du = \ln|\csc u - \cot u| + C$

67. $\displaystyle\int \sec^2 u\, du = \tan u + C$

68. $\displaystyle\int \csc^2 u\, du = -\cot u + C$

69. $\displaystyle\int \sec u \tan u\, du = \sec u + C$

70. $\displaystyle\int \csc u \cot u\, du = -\csc u + C$

71. $\displaystyle\int \sin^2 u\, du = \tfrac{1}{2}u - \tfrac{1}{4}\sin 2u + C$

72. $\displaystyle\int \cos^2 u\, du = \tfrac{1}{2}u + \tfrac{1}{4}\sin 2u + C$

73. $\displaystyle\int \cos^n u \sin u\, du = -\frac{\cos^{n+1} u}{n+1} + C$

74. $\displaystyle\int \sin^n u \cos u\, du = \frac{\sin^{n+1} u}{n+1} + C$

75. $\displaystyle\int \sin mu \sin nu\, du = -\frac{\sin(m+n)u}{2(m+n)} + \frac{\sin(m-n)u}{2(m-n)} + C$

76. $\displaystyle\int \cos mu \cos nu\, du = \frac{\sin(m+n)u}{2(m+n)} + \frac{\sin(m-n)u}{2(m-n)} + C$

77. $\int \sin mu \, \cos nu \, du = -\dfrac{\cos(m+n)u}{2(m+n)} - \dfrac{\cos(m-n)u}{2(m-n)} + C$

78. $\int e^{au} \sin nu \, du = \dfrac{e^{au}(a \sin nu - n \cos nu)}{a^2 + n^2} + C$

79. $\int e^{au} \cos nu \, du = \dfrac{e^{au}(n \sin nu + a \cos nu)}{a^2 + n^2} + C$

80. $\int u \sin u \, du = \sin u - u \cos u + C$

81. $\int u \cos u \, du = \cos u + u \sin u + C$

82. $\int \sin^m u \cos^n u \, du = \dfrac{\sin^{m+1} u \cos^{n-1} u}{m+n} + \dfrac{n-1}{m+n} \int \sin^m u \cos^{n-2} u \, du$

83. $\int \sin^n u \, du = -\dfrac{1}{n} \sin^{n-1} u \cos u + \dfrac{n-1}{n} \int \sin^{n-2} u \, du$

84. $\int \cos^n u \, du = \dfrac{1}{n} \cos^{n-1} u \sin u + \dfrac{n-1}{n} \int \cos^{n-2} u \, du$

85. $\int \tan^n u \, du = \dfrac{\tan^{n-1} u}{n-1} - \int \tan^{n-2} u \, du$

86. $\int \cot^n u \, du = -\dfrac{\cot^{n-1} u}{n-1} - \int \cot^{n-2} u \, du$

87. $\int \sec^n u \, du = \dfrac{\sec^{n-2} u \tan u}{n-1} + \dfrac{n-2}{n+1} \int \sec^{n-2} u \, du$

88. $\int \csc^n u \, du = -\dfrac{\csc^{n-2} u \cot u}{n-1} + \dfrac{n-2}{n-1} \int \csc^{n-2} u \, du$

APPENDIX **C**

Using a Graphing Calculator

This appendix is included to provide faculty with the flexibility of integrating graphing calculators in their classes. Each section explains and illustrates important features of the Texas Instruments TI-83 and TI-83 Plus. Though this appendix was specifically designed to supplement the graphing calculator examples found throughout the text, the material is organized so that an interested student could also study it as a separate chapter.

C.1 INTRODUCTION TO THE TI-83 KEYBOARD

This section provides a guided tour of the keyboard of the TI-83 and TI-83 Plus graphing calculators. In this and the following sections, please have your calculator in front of you and be sure to try out the features as they are discussed.

First notice that the keys forming the bottom six rows of the keyboard perform the standard functions of a scientific calculator. The thin blue keys that form the very top row allow functions to be defined and their graphs to be drawn (see Section C.3 for details). The second, third, and fourth rows of keys provide access to menus full of advanced features and perform special tasks such as **INS**ert, **DEL**ete, **CLEAR**, and **QUIT** (to leave a menu, an editor, or a graph, and return to the home screen). Also found in these rows are the **2nd** and **ALPHA** shift keys, which give additional, color-coded meanings to almost every key on the calculator.

The **ON** key is in the lower left-hand corner. Note that pressing the (golden yellow) **2nd** key followed by the **ON** key will turn the calculator **OFF**. If the calculator is left unattended (or no buttons are pressed for a couple of minutes), the calculator will shut itself off. No work is lost when the unit is turned off. Just turn the calculator back **ON** and the display will be exactly as you left it. Due to different lighting conditions and battery strengths, the screen contrast needs adjustment from time to time. Press the **2nd** key, then *press and hold* the up (or down) arrow key to darken (or lighten) the screen contrast.

The **ENTER** key in the lower right-hand corner is like the = key on many scientific calculators; it signals the calculator to perform the calculation that you've been typing. Its (shifted) **2nd** meaning, **ENTRY**, gives you access to previously entered formulas, starting with the most recent one. If you continue to press **2nd ENTRY** you can access previous entries up to an overall memory limit of 128 characters. Depending on the length of your formulas, this means that about 10 to 15 of your most recent entries can be retrieved from the calculator's memory to be reused or modified.

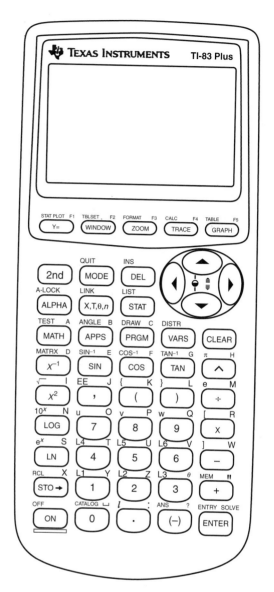

Courtesy of Texas Instruments

Just above **ENTER** is a column of four other blue keys that perform the standard operations of arithmetic. Please note, though, that the multiplication key, indicated by an ×, prints an asterisk on the screen and the division key prints a slash on the screen. Just above these four is the ^ key, which indicates that you're raising something to the power that follows; for example, 2^5 would mean 2^5. Moving to the left across that row, you will see the keys for the trigonometric functions: **SIN**, **COS**, and **TAN** (note that their standard setting is radians, but you can specify degrees by using the degree symbol, which is option 1 in the **ANGLE** menu, or the calculator can be set to always think in degrees by specifying that option in the **MODE** menu). Always press the trig key before typing the angle, as in $\cos(\pi)$ or $\sin(30°)$. Notice that the left-hand parenthesis is automatically included when you press any of the trig keys. To the left of these three is a key labeled $\mathbf{x^{-1}}$, which acts as a reciprocal key for ordinary arithmetic. It will also invert a matrix, as in $[A]^{-1}$, which explains why the key isn't labeled $\mathbf{1/x}$, as it would be on many scientific calculators. Beneath that key is $\mathbf{x^2}$

(the squaring key), whose shifted **2nd** meaning is square root. Below in that column are keys for logs, whose shifted **2nd** versions give exponential functions. Like the trig keys, the square root, **LOG**, **LN**, and exponential keys also precede their arguments. For example, log(2) will find the common logarithm of 2.

Between **LN** and **ON** is the **STO>** key, which is used to store a number (possibly the result of a calculation) into any of the 27 memory locations whose names are A, B, C, ..., Z, and θ. First indicate the number or calculation, then press **STO>** (which just prints an arrow on the screen) followed by the (green) **ALPHA** key, then the (green) letter name you want the stored result to have, and finally press **ENTER**. The computation will be performed and the result will be stored in the desired memory location as well as being displayed on the screen. If you have just performed a calculation and now wish that you had stored it, don't worry. Just press **STO>** on the next line followed by **ALPHA** and the letter name you want to give this quantity, then press **ENTER**.

Here are some examples:

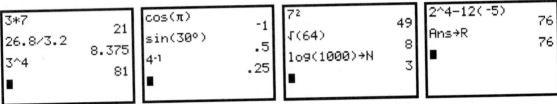

If you watched the last **STO>** example closely, you may have noticed that the calculator prints **Ans** (which stands for "the previous answer") on the screen whenever you don't indicate the first operand on a given line. For example, if you begin a formula with a plus sign, the calculator assumes that you want to add something to the previous result, so it displays "Ans+" instead of just "+". At times, you'll want to refer to the previous result somewhere other than at the beginning of your formula. In that case, press **2nd ANS** (the shifted version of the key to the left of **ENTER**) wherever you want the previous answer to appear in the computation.

The shifted **2nd** meaning of the **STO>** key is **RCL** (recall), as in **RCL Z**, which would display the *contents* of memory location Z at the current cursor position in your formula. It is usually easier to write the letter Z itself (press **ALPHA** followed by **Z**) in formulas instead of the current value that's stored there, so this recall feature isn't the best choice in most computations. However, the **RCL** feature is very useful in creating instant copies of functions (Rcl Y₁) and programs (Rcl prgmSIMPSON) so that newly modified versions don't have to destroy the old ones.

The key that changes the sign of a number is labeled **(-)** and is located just to the left of the **ENTER** key. Don't confuse this white (or gray) key with the dark blue subtraction key! Note also that the calculator consistently views the lack of an indicated operation between two quantities as an intended multiplication.

The parentheses keys are just above the 8 and 9 keys. These are used for all levels of parentheses. Do not be confused by symbols such as { } and [ ], which are the shifted **2nd** versions of these and other nearby keys. Braces { } are used *only* to indicate lists, and brackets [ ] are used *only* for matrices. Once again, these special symbols *cannot* be used to indicate higher levels of parentheses; just nest ordinary parentheses to show several levels of quantification. Also note that the comma key is used only with matrices, lists, multiple-argument functions, and certain commands in the calculator's programming language. Never use commas to separate digits within a number. The number three thousand should always be typed 3000 (not 3,000). The shifted **2nd** meaning of the comma key is **EE** (enter exponent), which is used to enter data in scientific notation; for example, **1.3** followed by **2nd EE (-)8** would be the keystrokes needed to enter 1.3×10^{-8} in a formula. It would be displayed on the screen as 1.3E-8.

The shifted **2nd** versions of the numbers 1 through 9 provide keyboard access to lists and sequences. The shifted **ALPHA** version of the zero key prints a blank space on the display. The shifted **2nd** version of the zero key is **CATALOG**, which provides alphabetical access to every feature of the calculator. Just press the first letter of the desired feature (without pressing **ALPHA**), then scroll from there using the down arrow key. Press **ENTER** when the desired feature is marked by the small arrow. The shifted **2nd** version of the decimal point is *i*, the imaginary unit (which is often called *j* in electronics applications). This symbol can be used in computations involving imaginary and complex numbers even when **MODE Real** has been selected.

The shifted **2nd** version of the plus sign is **MEM** (the memory management menu), which gives you a chance to erase programs, lists, and anything else stored in memory. Use this menu sparingly (remember, your calculator has a fairly large memory, so you don't usually need to be in a hurry to dispose of things which might prove useful later). If you get into **MEM** by accident, just press **2nd QUIT** to get back to the home screen. **2nd QUIT** always takes you back to the home screen from any menu, editor, or graph, but it will not terminate a running program on a TI-83 or TI-83 Plus. To interrupt a running program, just press the **ON** button, then choose "Quit" in the menu you'll see.

If you're looking for keys that will compute cube roots, absolute values, complex conjugates, permutations, combinations, or factorials, press the **MATH** key, and you'll see four submenus (selectable by using the right or left arrow key) which give you these options and many more. Especially interesting is >**Frac** (convert to fraction), which will convert a decimal to its simplified fractional form, provided that the denominator would be less than 10,000 (otherwise, it just writes the decimal form of the number). Other examples are also included below to give you a better idea of just how many options are available in the **MATH** menu.

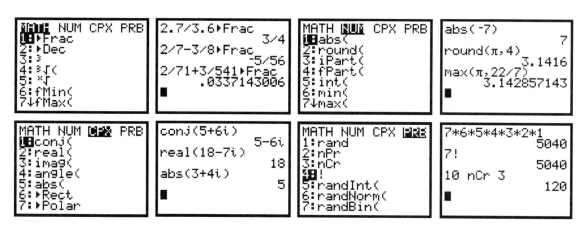

C.2 COMPUTATIONAL EXAMPLES

EXAMPLE 1

Compute the following:
(a) 7×6

(b) $3 \times 7 + 6(3 - 5)$

EXAMPLE 2

Compute $8\{3 + 5[2 - 7(8 - 9)]\}$

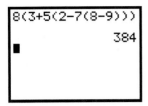

Note: The calculator uses only ordinary parentheses.

EXAMPLE 3

Express the following as a decimal and as a simplified fraction:

(a) $\dfrac{105}{100}$

(b) $\dfrac{3}{8} + \dfrac{21}{10} - \dfrac{17}{25}$

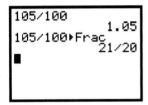
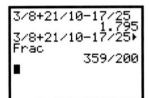

Note that a fraction is an indicated division operation and that the division key always prints a diagonal fraction bar line on the screen. The convert to fraction feature is the first item in the **MATH** menu and is accessed by pressing **MATH** then **1** (or **MATH** then **ENTER**) at the end of a formula. Note also that simplified improper fractions are the intended result. *Mixed numbers are not supported.* A decimal result would mean that the answer cannot be written as a simplified fraction with a denominator less than 10,000.

EXAMPLE 4

Compute the following, expressing the answer as a simplified fraction:

(a) $\dfrac{2^5}{6^2}$

(b) $\dfrac{5 - (-7)}{-2 - 12}$

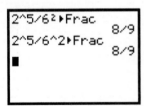
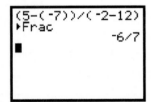

Squares can be computed by pressing the $\mathbf{x^2}$ key; similarly, a third power can be indicated by pressing **MATH** then **3**. Most other exponents require the use of the ^ key (found between **CLEAR** and the division key). In Part (b), notice the calculator's need for additional parentheses which enclose the numerator and denominator of the fraction. Also notice the difference between the calculator's negative sign (the key below **3**) and its subtraction symbol (the key to the right of **6**).

EXAMPLE 5

Compute the following complex numbers:

(a) $(3 + 4i)(-2 - 5i)$

(b) $\dfrac{7 + 29i}{30 + 10i}$

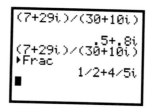

Note that the calculator key for the imaginary unit, i, is the shifted **2nd** version of the decimal point. This imaginary number is often called j in electronics applications.

EXAMPLE 6

Evaluate these expressions:

(a) $x^2 + 5x - 8$ when $x = 4$

(b) $x^2 y^3 + 4x - y$ when $x = 2$ and $y = 5$

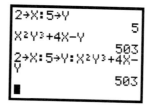

The **STO>** key (just above **ON**) is used to print the arrow symbol on the screen. The letter x can be typed on the screen by pressing the key next to **ALPHA**, labeled **X,T,θ,n**, or by pressing **ALPHA**, then **X**. Note that several steps can be performed on one line if the steps are separated by a colon (the shifted **ALPHA** version of the decimal point key). In such cases, all steps are performed in sequence, but only the result of the very last step is displayed on the screen.

EXAMPLE 7

Given that $f(x) = x^4 - 7x + 11$, find $f(3), f(5)$, and $f(-1)$.

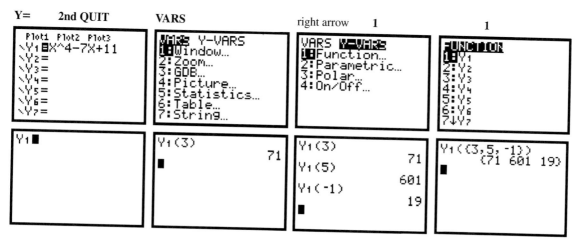

Note that using a stored function requires entering the function in the **Y=** menu, pressing **2nd QUIT** to return to the home screen, then finding the name of that function in the **FUNCTION** submenu under **Y-VARS**. Function evaluation requires the use of parentheses; without parentheses around the argument, multiplication would be assumed. To type the second and third uses of Y_1, press **2nd ENTRY**, then left arrow twice, modifying only the argument from the previous formula. The last screen shows how a list of arguments can be used to calculate a list of function values. Your list entries must be enclosed by braces { } and separated by commas (list entries output by the calculator are separated by spaces instead of commas).

C.3   GRAPHING FEATURES

The thin blue buttons along the top row of the calculator do most of its graphical work. The **Y=** key provides access to the calculator's list of ten functions (assuming that the calculator is set in **MODE Func**). Pressing the **Y=** key will reveal functions Y_1 through Y_7; the other three functions, Y_8, Y_9, and Y_0, can be seen by pressing the (blue) down arrow key nine times (or just press and hold the down arrow key). These functions are part of the calculator's memory, but the information stored on this screen can be easily edited, overwritten, or **CLEAR**ed. Functions are selected for graphing (turned "on") by highlighting their equal sign. This is done automatically when you type in a new function or modify an old one. In other cases, to change the status of a function (from "off" to "on" or vice versa), you will need to use the arrow keys to position the cursor over the equal sign (making it blink); then press **ENTER**. Functions marked with an ordinary equal sign are stored in memory but will *not* be graphed. To the left of each function name is a symbol indicating how it will be graphed. The normal setting looks like a backslash, \, and simply indicates that the graph will be drawn with a thin line. Other settings include a thicker line, shading above the graph, shading below the graph, two animated settings (one marks the path of motion on the screen; the other just shows the motion without marking its path), and finally an option that graphs with a dotted line. To switch from one option to another, just press the left-arrow key until the cursor is over the option marking (at the far left of that function's name), then press **ENTER** repeatedly until the desired option appears. One warning about the **Y=** menu is that the names Plot1, Plot2, and Plot3 at the top of the function list refer only to the calculator's **STAT**istical **PLOT**s. They have nothing to do with ordinary graphing and should *not* be highlighted if you are just trying to graph some functions.

The **WINDOW** key allows you to *manually* specify the extents of the *x*- and *y*-values that will be visible on the calculator's graphing screen (see **ZOOM** for *automatic* ways of doing this). The Xscl and Yscl options specify the meaning of a mark on the *x*- or *y*-axis. For example Xscl=5 means that each mark shown on the *x*-axis will mean an increment of 5 units (Xscl=1 is a common setting for algebraic functions; Xscl=$\pi/2$ is commonly used when graphing trigonometric functions). The last option, Xres, allows you to control how many points will actually be calculated when a graph is drawn. Xres=1 means that an accurate point will be calculated for each pixel on the *x*-axis (somewhat slow, but very accurate). Xres=2 will calculate only at every other pixel, etc.; Xres=8 only calculates a point for every 8th pixel (this is the fastest setting, but also the least accurate). In the examples that follow, all graphs are shown with Xres=1.

Pressing **2nd FORMAT** (the shifted version of the **ZOOM** key), reveals additional graphing options that allow you to change the way coordinates are displayed (polar instead of rectangular), turn coordinates off completely (inhibiting some **TRACE** features), provide a coordinate grid, hide the axes, label the axes, or inhibit printing expressions which describe the graphs. If you find your graphs looking cluttered or notice that axes, coordinates, or algebraic expressions are missing, the "standard" settings are all in the left-hand column. Like other menus where the options aren't numbered (**MODE** is similar), use your arrow keys to make a new option blink, then select it by pressing **ENTER**.

ZOOM accesses a menu full of *automatic* ways to set the graphical viewing window. **ZStandard** (option 6) is usually a good place to start, but you should consider option 7, **ZTrig**, if you're graphing trigonometric functions. **ZStandard** shows the origin in the exact center of the screen with x- and y-values both ranging from -10 to 10. From here you can **Zoom In** or **Zoom Out** (options 2 and 3), or draw a box around a portion of the graph that you would like magnified to fit the entire screen (option 1, **ZBox**). There is also an option to "square up" your graph so that units along the x-axis are equal in length to units along the y-axis (option 5, **ZSquare**); the *smaller* unit length from the axes of the previous graph will now be used on both axes. This option makes the graph look more like it would on regular graph paper; for example, circles really look like circles. **ZDecimal** and **ZInteger** (options 4 and 8) prepare the screen for **TRACE**s, which will utilize x-coordinates at exact tenths or integer values, respectively. Option 9, **ZoomStat**, makes sure that all of the data in a statistical plot will fit in the viewing window. Option 0, **ZoomFit**, calculates a viewing window using the present x-axis, but adjusting the y-axis so that the function fits neatly within the viewing window. All of these options work by making automatic changes to the **WINDOW** settings. Want to go back to the view you had before? The **MEMORY** submenu (press the right arrow key after pressing **ZOOM**) contains options to go back to your immediately previous view (option 1, **ZPrevious**) or to a window setting you saved a while ago (option 3, **ZoomRcl**). **ZoomSto** (option 2) is the way to save the current window setting for later (note that it can only retain one window setting, so the new information replaces whatever setting you had saved before). Option 4, **SetFactors...**, gives you the chance to control how dramatically your calculator will **Zoom In** or **Zoom Out**. These zoom factors are set by Texas Instruments for a magnification ratio of 4 on each axis. Many people prefer smaller factors, such as 2 on each axis. It is possible to set either factor to any number greater than or equal to 1; they don't need to be whole numbers, and they don't necessarily have to be equal.

The **TRACE** key takes you from any screen or menu to the current graph, displaying the x- and y-coordinates of specific points as you trace along a curve using the left and right arrow keys. Note that in **TRACE**, the up and down arrow keys are used to jump from one curve to another when several curves have been drawn on the same screen. The expression (formula) for the function you are presently tracing is shown in the upper left-hand corner of the screen (or its subscript number is shown in the upper right-hand corner if you have selected the **ExprOff** option from the **FORMAT** menu). If you press **ENTER** while in **TRACE**, the graph will be redrawn with the currently selected point in the exact center of the screen, even if that point is presently outside the current viewing window. This feature is a convenient way to pan up or down to see higher or lower portions of the graph. It is also the easiest way to locate a "lost" graph that doesn't appear anywhere in the current viewing window (just press **TRACE**, then **ENTER**). To pan left or right, just press and hold the left or right arrow key until new portions of the graph come into view. These useful features change the way the graph is centered on the screen without changing its magnification. Note also that these recentering features work *only* in **TRACE**. To exit **TRACE** without disturbing your view of the graph, just press **GRAPH** (or **CLEAR**). To return to the home screen, abandoning both **TRACE** and the graph, just press **2nd QUIT** (or press **CLEAR** twice). Note that using any **ZOOM** feature also causes an exit from **TRACE** (to resume tracing on the new zoomed version, you must press **TRACE** again).

The **GRAPH** key takes the calculator from any screen or menu to the current graph. Note that the calculator is smart enough that it will redraw the curves only if changes have been made to the function list (**Y=**). As previously mentioned, the **GRAPH** key can be used to turn off **TRACE**. You can also hide an unwanted free cursor by pressing **GRAPH**. When you're finished with viewing a graph, press **CLEAR** or **2nd QUIT** to return to the home screen.

C.4 EXAMPLES OF GRAPHING

EXAMPLE 1

To graph $y = x^2 - 5x$, first press **Y=**, then press **CLEAR** to erase the current formula in Y₁ (or use the down arrow key to find a blank function), then press the **X,T,θ,n** key (**ALPHA** then **X** will also work), followed by the **x²** button; now press the (blue) minus sign key, then **5**, followed immediately by the **X,T,θ,n** key (a multiplication sign is not needed). Your screen should look very much like the first one shown below. There is no need to press **ENTER** when you have finished typing a function's formula. To set up a good graphing window, we now press **ZOOM** and then **6** to choose **ZStandard**. This causes the graph to be immediately drawn on axes that range from −10 to 10. Notice that you did not have to press the **GRAPH** key; the **ZOOM** menu items and the **TRACE** key also activate the graphing screen. *Note:* If you have one or more unwanted graphs drawn on top of this one, go back to your function list (**Y=**) and turn "off" the unwanted functions by placing the cursor over their highlighted equal signs and pressing **ENTER**. After you have turned off the unwanted functions, just press **GRAPH** and you will finally see the last screen below.

 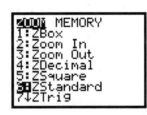 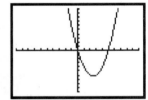

EXAMPLE 2

To modify this function to be $y = -x^2 + 4$, press **Y=**, then insert the negative sign by pressing **2nd INS** followed by the white (or gray) sign change key (**-**); now press the right arrow key twice to skip over the parts of the formula that are to be preserved. Note that the arrow keys also take you out of insert mode. Now type the plus sign and the 4 (replacing the −5), and finally press **DEL** to delete the extra X at the end of the formula. Press **TRACE** to plot this function. **TRACE** gives the added bonus of a highlighted point, with its coordinates shown at the bottom of the screen. Press the right or left arrow keys to highlight other points on the curve.

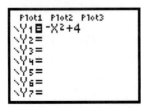 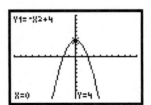 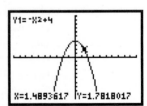

Perhaps $x = 1.4893617$, $y = 1.7818017$ was not a coordinate pair you had expected to investigate. Two special **ZOOM** features (options 4 and 8) can be used to make the **TRACE** option more predictable. Press **ZOOM**, then **4** to select **ZDecimal**; now press **TRACE**.

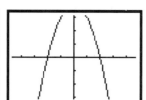

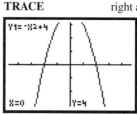

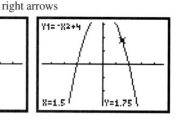

Try pressing the right or left arrow key about 15 times while watching the values at the bottom of the screen. You'll quickly notice that the *x*-values are now all *exact tenths* (**ZOOM** option 8, **ZInteger**, produces **TRACE**able *x*-values which are all integers). Another nice thing about **ZDecimal** is that the graph is "square" in the sense that units on the *x*- and *y*-axes have the same length. The main disadvantage to **ZDecimal** is that the graphing window is "small," displaying only points with *x*-values between -4.7 and 4.7 and *y*-values between -3.1 and 3.1. This disadvantage is apparent on the current graph, which runs off the top of the screen. To demonstrate how this problem can be overcome, **TRACE** the graph to the point $x = 1.5$, $y = 1.75$, and then press **ENTER**. This special feature of **TRACE** causes the graph to be redrawn with the highlighted point in the exact center of the screen (with no change in the magnification of the graph). This is the way to pan up or down from the current viewing window (to pan left or right, see Example 6).

ENTER

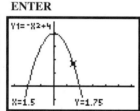

EXAMPLE 3

Another way to deal with the preceding problem is to **Zoom Out**, but first you'll want to set your ZOOM FACTORS to 2 (the factory setting is 4). Press **ZOOM**, then the right arrow key (**MEMORY**), then press **4** (**SetFactors**). To change the factors to 2, just type **2**, press **ENTER**, and then type another **2**.

ZOOM   right arrow   **4**   2 **ENTER**   2

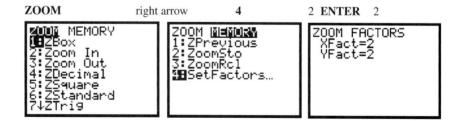

Now to reproduce our problem, press **ZOOM**, then **4** (**ZDecimal**). However, this time correct it by pressing **ZOOM** followed by **3** (**Zoom Out**). At first glance, it looks like nothing has happened, except that X=0 Y=0 is displayed at the bottom of the screen. The calculator is waiting for you to use your arrow keys to locate the point in the current window where you would like the exact center of the new graph to be (then press **ENTER**). Of course, if you like the way the graph is already centered, you will still have to press **ENTER** (you'll just skip pressing the arrow keys).

ZOOM   4

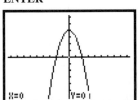 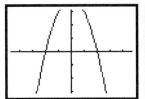

ZOOM   3

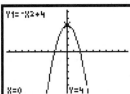 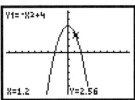

ENTER   **TRACE**   right arrows

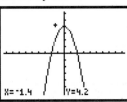

This extra keystroke has proven to be a bit confusing to beginners who think that **Zoom In**, **Zoom Out**, and **ZInteger** should work like the six zoom options (4, 5, 6, 7, 9, and 0) that do their job without pressing **ENTER**. Perhaps more interesting is the fact that you can continue to **Zoom Out** just by pressing **ENTER** again and again (of course, you can also press some arrow keys to recenter between zooms if you wish). Before you experiment with that feature, press **TRACE** and notice, by pressing the left or right arrow key a few times, that the *x*-values are now changing by .2 (instead of .1) and the graph is still "square." Other popular square window settings can be obtained by repeating this example with both zoom factors set to 2.5 or 5. These give "larger" windows where the *x*-values change by .25 or .5, respectively, during a **TRACE**.

EXAMPLE 4

The only other zoom option that needs extra keystrokes is **ZBox** (option 1). This is a very powerful option that lets you draw a box around a part of the graph which you would like enlarged to fit the entire screen. After selecting this option, use the arrow keys to locate the position of one corner of the box and press **ENTER**. Now use the arrow keys to locate the *opposite* corner, and press **ENTER** again.

ZOOM   1   left and up arrows   **ENTER**   right and down arrows

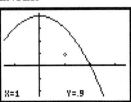 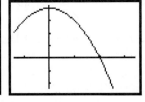 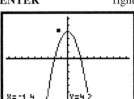 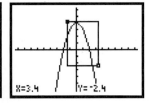

ENTER   **GRAPH**

Note that the resulting graph has a free cursor identifying the point in the exact center of the screen. What may not be apparent is that your calculator is ready for you to draw another box if you wish to zoom in closer. To get rid of this free cursor, just press **GRAPH** (or **CLEAR**).

EXAMPLE 5

Sometimes, you will know the precise interval on the *x*-axis (the domain) that you want for a graph, but a corresponding interval for the *y*-values (the range) may not be obvious. **ZoomFit** (the 10th **ZOOM** option) is designed for this circumstance. For example, to graph $f(x) = 2x^3 - 8x + 9$ on the interval $[-3, 2]$, manually set the **WINDOW** so that Xmin=–3 and Xmax=2 (the other values shown in the second frame are just leftovers from **ZStandard**). Now press **ZOOM**, then **0** to select **ZoomFit**. There is a noticeable pause while appropriate values of Ymin and Ymax are calculated, then the graph is drawn. To view the values calculated for the range, just press **WINDOW**. The minimum value of this function on the interval $[-3, 2]$ is -21 and its maximum value is approximately 15.16.

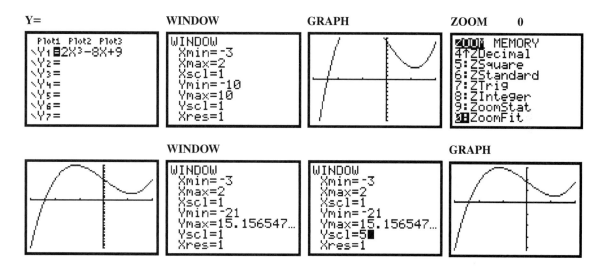

Note that **ZoomFit** changes only Ymin and Ymax. You may also wish to change Yscl.

EXAMPLE 6

Panning to the right is done by tracing a curve off the right-hand edge of the screen:

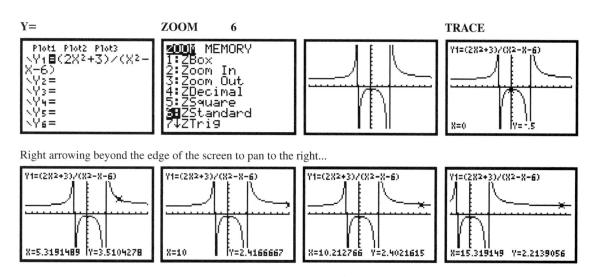

Right arrowing beyond the edge of the screen to pan to the right...

Panning to the left is done similarly. To pan *up or down*, see the last part of Example 2.

EXAMPLE 7

Creating, storing, and retrieving viewing windows:

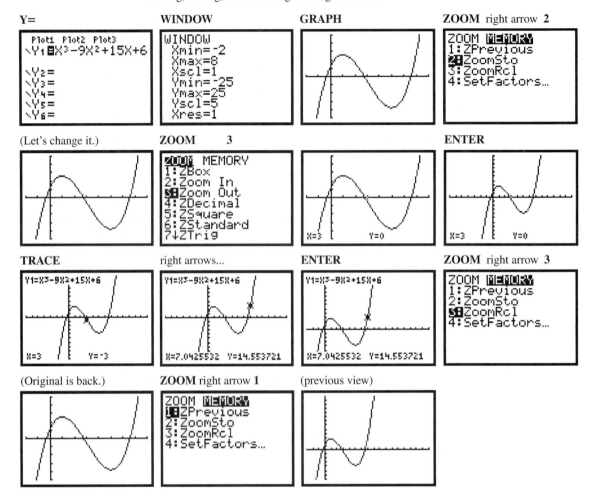

Y=

```
Plot1 Plot2 Plot3
\Y1■X³-9X²+15X+6
\Y2=
\Y3=
\Y4=
\Y5=
\Y6=
```

WINDOW

```
WINDOW
 Xmin=-2
 Xmax=8
 Xscl=1
 Ymin=-25
 Ymax=25
 Yscl=5
 Xres=1
```

GRAPH

ZOOM right arrow **2**

```
ZOOM MEMORY
1:ZPrevious
2■ZoomSto
3:ZoomRcl
4:SetFactors...
```

(Let's change it.)

ZOOM   3

```
ZOOM MEMORY
1:ZBox
2:Zoom In
3■Zoom Out
4:ZDecimal
5:ZSquare
6:ZStandard
7↓ZTrig
```

ENTER

X=3   Y=0   X=3   Y=0

TRACE

```
Y1=X³-9X²+15X+6

X=3     Y=-3
```

right arrows...

```
Y1=X³-9X²+15X+6

X=7.0425532 Y=14.553721
```

ENTER

```
Y1=X³-9X²+15X+6

X=7.0425532 Y=14.553721
```

ZOOM right arrow **3**

```
ZOOM MEMORY
1:ZPrevious
2:ZoomSto
3■ZoomRcl
4:SetFactors...
```

(Original is back.)

ZOOM right arrow **1**

```
ZOOM MEMORY
1■ZPrevious
2:ZoomSto
3:ZoomRcl
4:SetFactors...
```

(previous view)

EXAMPLE 8

Graphing and tracing more than one function:

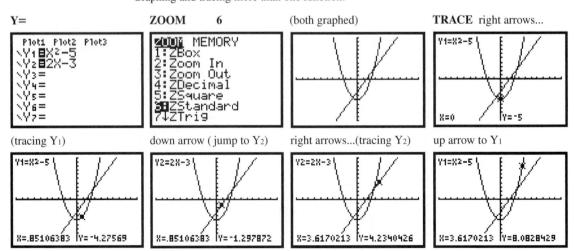

Y=

```
Plot1 Plot2 Plot3
\Y1■X²-5
\Y2■2X-3
\Y3=
\Y4=
\Y5=
\Y6=
\Y7=
```

ZOOM   6

```
ZOOM MEMORY
1:ZBox
2:Zoom In
3:Zoom Out
4:ZDecimal
5:ZSquare
6■ZStandard
7↓ZTrig
```

(both graphed)

TRACE right arrows...

```
Y1=X²-5

X=0      Y=-5
```

(tracing Y₁)

```
Y1=X²-5

X=.85106383 Y=-4.27569
```

down arrow (jump to Y₂)

```
Y2=2X-3

X=.85106383 Y=-1.297872
```

right arrows...(tracing Y₂)

```
Y2=2X-3

X=3.6170213 Y=4.2340426
```

up arrow to Y₁

```
Y1=X²-5

X=3.6170213 Y=8.0828429
```

Note that the down arrow increases the subscript of the function being traced and the up arrow decreases the subscript. The result has nothing to do with which graph is above or below the other.

EXAMPLE 9

Finding a point of intersection:

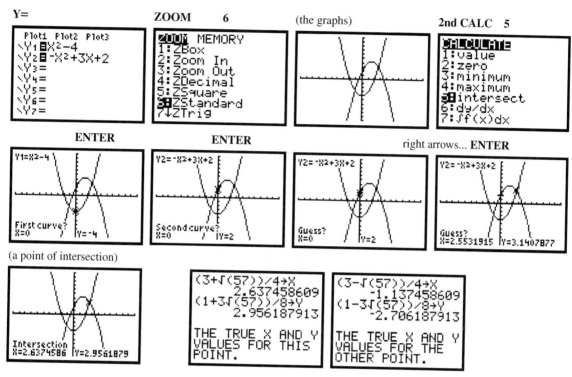

Note that the *Guess* was very important in determining which point of intersection would be calculated. Now find the other intersection point using **intersect**.

EXAMPLE 10

Calculating *y*-values and locating the resulting points on your graphs:
If you have just completed Example 9, please skip to the fourth frame.

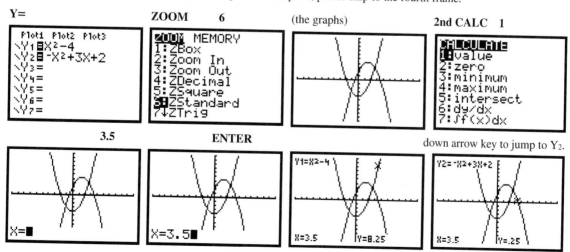

The *x*-value entered must be in the current viewing **WINDOW** (but the resulting *y*-value need not be). This *x*-value can be investigated for any of the functions that are presently graphed by pressing the up or down arrow keys. Note again that the subscript of a function increases as you jump from one curve to another by pressing the down arrow key. The up arrow key decreases this number, and either key can be used to wrap around and start over.

EXAMPLE 11

Another way to calculate a specific value of a function is to press **TRACE**, then type the *x*-value and press **ENTER**. The main difference between the **CALCULATE value** feature (see Example 10) and this special **TRACE** feature is that **TRACE** does *not* preserve the entered *x*-value if you jump from one curve to another (you would usually need to retype that *x*-value). The following frames assume **ZStandard (ZOOM 6) WINDOW** settings.

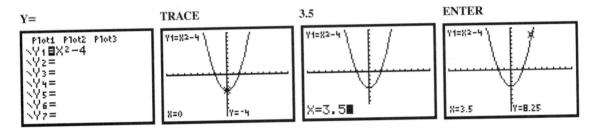

EXAMPLE 12

Using **TRACE** to quickly locate a "lost" graph that doesn't appear anywhere in the current viewing window:

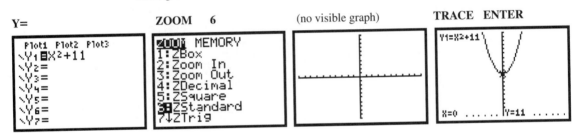

C.5 TRIGONOMETRIC FUNCTIONS AND POLAR COORDINATES

For calculating and graphing trigonometric functions on the TI-83 and TI-83 Plus, the standard default setting is **Radian** mode. To set the calculator to **Degree** mode, press the **MODE** key, arrow down and over to **Degree**, then press **ENTER**.

MODE down arrow *twice*, right arrow **ENTER**   **CLEAR** (or **2nd QUIT**) to exit

EXAMPLE 1

Evaluating trigonometric functions using degrees, minutes, and seconds:

  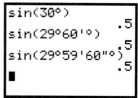

The symbols for degrees and minutes are the first two options in the menu found by pressing **2nd ANGLE** (just to the right of the **MATH** key). The symbol for seconds is the shifted **ALPHA** version of the addition key. Note that in **Radian MODE**, degrees can still be used, but an *additional* degree symbol must follow the angle's measure.

EXAMPLE 2

To graph $y = \sin x$, first press **Y=**, then press **CLEAR** to erase the current formula in Y_1 (or use the arrow keys to find a blank function), then press the **SIN** key followed by the **X,T,θ,n** key (**ALPHA X** will also work), and finally press the right parenthesis key. Your screen should look very much like the first one in the following figure. To set up a good graphing window, press **ZOOM** and then press **7** to choose **ZTrig**. This causes the graph to be immediately drawn on axes that range from roughly -2π to 2π (actually from $-352.5°$ to $352.5°$) in the x-direction and from -4 to 4 in the y-direction. Each mark along the x-axis represents a multiple of $\pi/2$ radians (90°). Notice also that you did not need to press the **GRAPH** key; the **ZOOM** menu items and the **TRACE** key also activate the graphing screen. *Note*: If you have one or more unwanted graphs drawn on top of this one, go back to your function list (**Y=**) and turn "off" the unwanted functions by placing the cursor over their highlighted equal signs and pressing **ENTER**. After you have turned off the unwanted functions, press **GRAPH** and you will finally see the last frame below.

Y=   **ZOOM 7**

  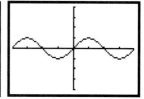

EXAMPLE 3

To modify this function to be $y = -3\sin 2x$, press **Y=**, then insert the -3 by pressing **2nd INS** followed by the white (or gray) sign change key (-) then the number **3**. Now press the right arrow key to skip over the part of the formula that's OK. Note that the arrow keys take you out of insert mode. Now type **2nd INS** then the **2**. Press **ZOOM 7**, then **TRACE** to plot this function. **TRACE** gives the added bonus of a highlighted point with its coordinates shown at the bottom of the screen. Press the right or left arrow keys to highlight other points on the curve. The highlighted coordinate pair in the 4th frame that follows is $x = 5\pi/12$, $y = -1.5$. **ZTrig** allows **TRACE** to display all points whose x-values are multiples of $\pi/24$ (of course, this includes such special values as 0, $\pi/6$, $\pi/4$, $\pi/3$, $\pi/2$, etc.), written in their decimal forms. In degrees, follow the same directions. The only difference is that the traced x-values are now multiples of 7.5° (which is the equivalent of $\pi/24$ radians). As this example illustrates, **ZTrig** has been carefully designed to produce the same graph for both radians and degrees. Other automatic ways of establishing a viewing window, such as **ZStandard** and **ZDecimal**, ignore the **MODE** setting and are not recommended for graphing trigonometric functions in degrees.

MODE	Y=	ZOOM 7 TRACE	right arrows... $[(5\pi/12,-1.5)]$

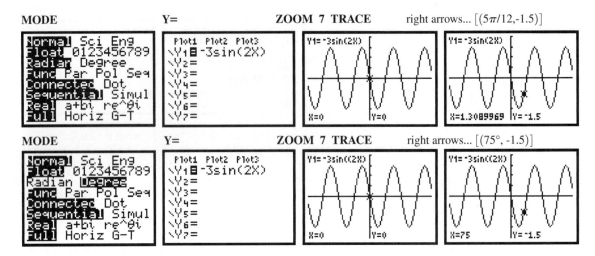

MODE	Y=	ZOOM 7 TRACE	right arrows... $[(75°, -1.5)]$

EXAMPLE 4

Several related trig functions can be drawn on the same screen, either by typing them separately in the function list **Y=** or by using a list of coefficients as shown in the following figure. A list consists of numbers separated by commas which are enclosed by braces { }. The braces are the shifted **2nd** versions of the parentheses keys. The first two frames indicate how to efficiently graph $y = \sin(x)$, $y = 2\sin(x)$, and $y = 4\sin(x)$ on the same screen. The last two frames graph $y = 2\sin(x)$ and $y = 2\sin(3x)$.

Y=	GRAPH	Y=	GRAPH

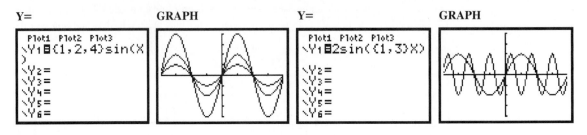

EXAMPLE 5

Multiple lists are allowed, but are not highly recommended. For example, to graph the two functions $y = 2\sin 3x$ and $y = 4\sin x$, you could do what's shown in the first frame or type them separately as shown in the third frame.

Y=	GRAPH	Y=	GRAPH

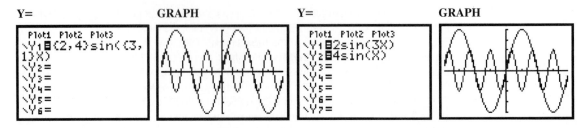

EXAMPLE 6

When graphing trig functions that have vertical asymptotes, remember that your calculator just evaluates individual points and arbitrarily assumes that it should connect those points if it is set in **Connected MODE**. The effect is shown in the following graph of $y = \sec x$. Some people like these "vertical asymptotes" being shown on the graph (they are really just nearly vertical lines which are trying to connect two points on the curve). The last frame shows the same graph in **Dot MODE**.

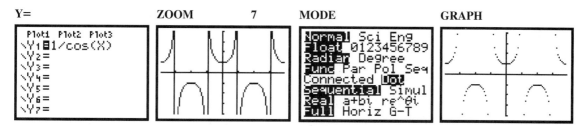

Y= ZOOM 7 MODE GRAPH

EXAMPLE 7

To graph in polar coordinates, the calculator's **MODE** must be changed from **Func** to **Pol**. You may also wish to change your **2nd FORMAT** options from **RectGC** to **PolarGC**, which will show values of r and θ (instead of x and y) when you **TRACE** your polar graphs. Note that the calculator treats these as two completely separate issues (it is possible to graph in one coordinate system and trace in the other). Pressing **Y=** will reveal the calculator's six polar functions $r_1, r_2, \ldots, r_6$. The **X,T,θ,n** key now prints θ on the screen.

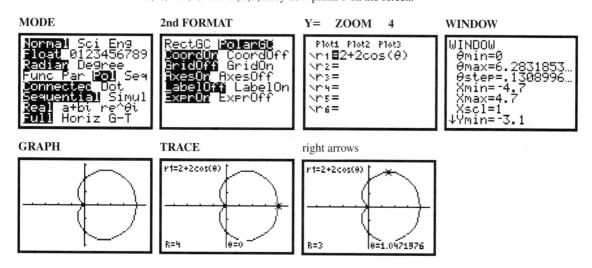

MODE 2nd FORMAT Y= ZOOM 4 WINDOW

GRAPH TRACE right arrows

Note in the preceding frames that the standard radian values of θmin, θmax, and θstep are 0, 2π, and $\pi/24$, respectively. A more accurate, smoother graph can be obtained by using θstep$=\pi/48$ or $\pi/96$. Also note that the *right* arrow key is used to **TRACE** in the standard, counterclockwise direction.

EXAMPLE 8

Graphing polar equations in **Degree MODE**:

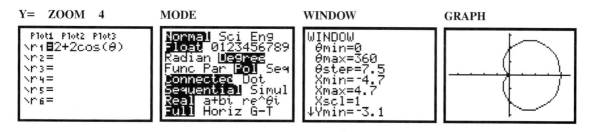

Y= ZOOM 4 MODE WINDOW GRAPH

WINDOW (change θstep)

WINDOW
θmin=0
θmax=360
θstep=2
Xmin=-4.7
Xmax=4.7
Xscl=1
↓Ymin=-3.1

GRAPH

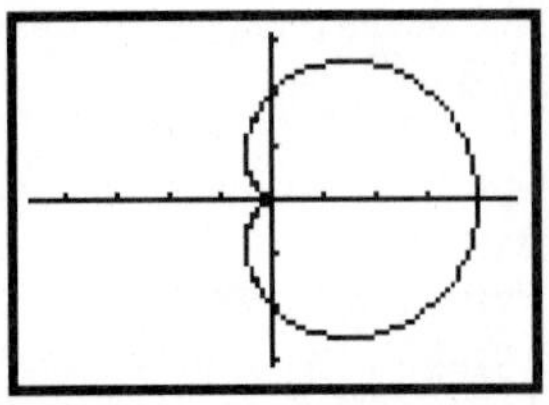

2nd FORMAT

RectGC PolarGC
CoordOn CoordOff
GridOff GridOn
AxesOn AxesOff
LabelOff LabelOn
ExprOn ExprOff

TRACE right arrows

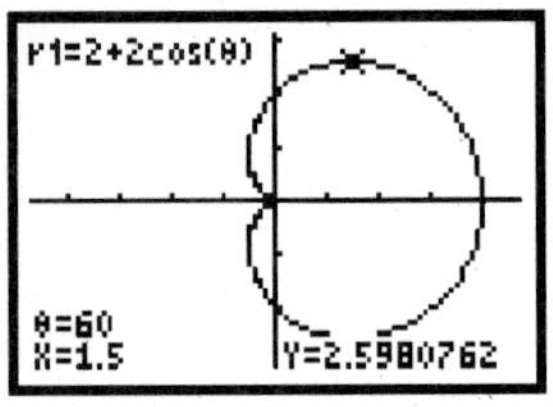
r1=2+2cos(θ)
θ=60
X=1.5
Y=2.5980762

The standard degree values of θmin, θmax, and θstep are 0, 360, and 7.5, respectively. Smoother (but slower) graphs can be obtained by using smaller values of θstep. The last two frames show a polar graph traced in **RectGC FORMAT**.

C.6 EQUATION-SOLVING AND TABLE FEATURES

EXAMPLE 1

Solving an equation on the home screen:

(a) Rewrite the equation on paper in the form $f(x) = 0$; for example, rewrite

$$x^3 + 15x = 9x^2 - 6$$

as

$$x^3 - 9x^2 + 15x + 6 = 0$$

(b) From the home screen, press **2nd CATALOG**, then press the letter **T** (the **4** key), next press the up arrow repeatedly until **solve(** comes into view; then press **ENTER**. You should now see **solve(** on the home screen.

(c) Finish the statement so that it looks like one of the following: **solve(X^3 − 9X^2 + 15X + 6, X, 3)** or **solve (Y$_1$, X, 3)**, presuming you've entered the function in Y_1 (to type the symbol Y_1 in a formula, press **VARS** then the right arrow key, then **1**, then **1** again).

Y=

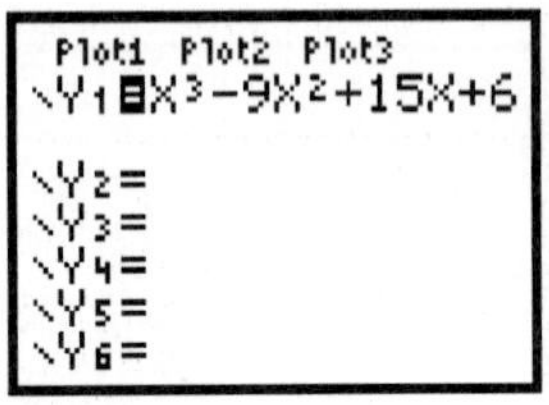
Plot1 Plot2 Plot3
\Y1=X³-9X²+15X+6
\Y2=
\Y3=
\Y4=
\Y5=
\Y6=

WINDOW

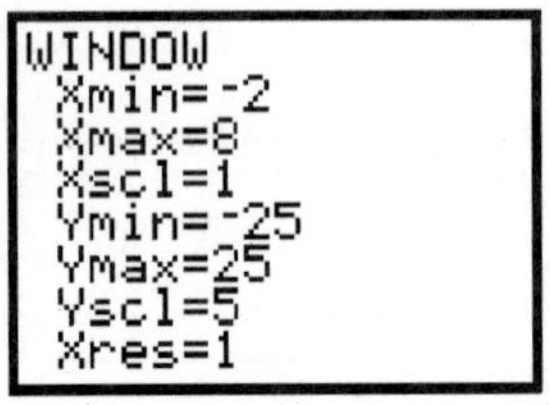
WINDOW
Xmin=-2
Xmax=8
Xscl=1
Ymin=-25
Ymax=25
Yscl=5
Xres=1

GRAPH

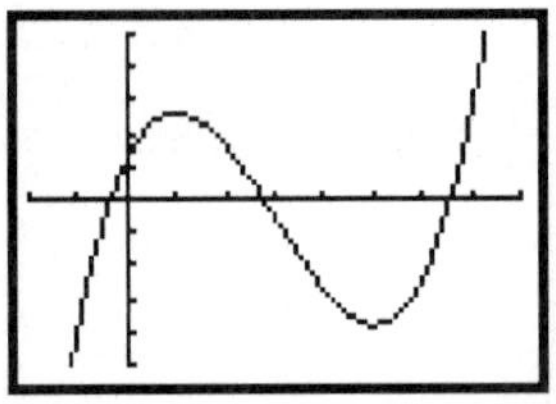

2nd CATALOG T up arrows

CATALOG
▸solve(
SortA(
SortD(
stdDev(
Stop
StoreGDB
StorePic

ENTER (home screen)

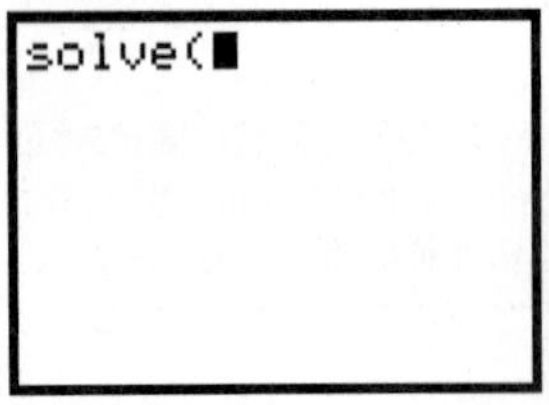
solve(

VARS right arrow **1**

VARS Y-VARS
1:Function...
2:Parametric...
3:Polar...
4:On/Off...

1

FUNCTION
1:Y1
2:Y2
3:Y3
4:Y4
5:Y5
6:Y6
7↓Y7

etc.

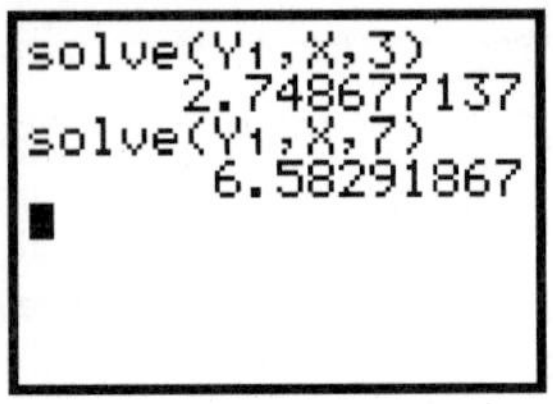
solve(Y1,X,3)
2.748677137
solve(Y1,X,7)
6.58291867

(d) Press **ENTER** and the calculator will try to find a zero of this function near 3 (answer: 2.748677137).

(e) Press **2nd ENTRY** to bring back your formula, then arrow left and change the 3 to a 7.

(f) Press **ENTER** and it will now find the zero near 7 (answer: 6.58291867).

(g) See if you can use the **solve** feature to find the other zero (answer: −.3315958073).

Notice that the **solve** feature finds solutions of an equation *one at a time*, with each new solution requiring its own estimate. Graphing the function and noticing where it crosses the x-axis is usually the easiest way to discover good estimates. If you prefer the **TABLE** feature (see Example 3), look for sign changes in the list of y-values; the corresponding x-values should be good estimates. Random guesses, although not recommended, can be effective when the equation has very few solutions.

EXAMPLE 2

Solving an equation on the graphics screen:

(a) As in Example 1(a), be sure to rewrite the equation as a function set equal to zero.

(b) Enter this function in your (**Y=**) list of functions and make sure that it is the only one selected for graphing.

(c) Press **2nd CALC** (the shifted **TRACE** key), and choose option 2, **zero**.

(d) The prompt "Left Bound?" is asking you to trace the curve using the arrow keys until you are just to the *left* of the desired zero (then press **ENTER**). Again, "Left Bound" just refers to an *x*-value that's too small to be the solution; do not consider whether the curve is above the axis or below the axis at that point. Similarly, the prompt "Right Bound?" is asking you to trace the curve until you are just to the *right* of the desired zero (then press **ENTER**). You'll notice in each case that a bracketing arrow is displayed near the top of the screen to graphically document the interval which will be searched for a solution.

(e) The prompt "Guess?" is asking you to trace the curve to a point as close as possible to where it crosses the axis (then press **ENTER**). This Guess is just an approximate solution like the **solve** feature uses (see Example 1).

(f) The solution (the "Zero") is displayed at the bottom of the screen (using 7 or 8 significant digits rather than the 10 digits you get on the home screen). An added bonus is that the *y*-value is also included (it should be exactly zero or extremely close to zero like "1E-12," which means 10^{-12}). Two of the solutions are found below. Try the third one on your own. The **WINDOW** from the previous example is assumed.

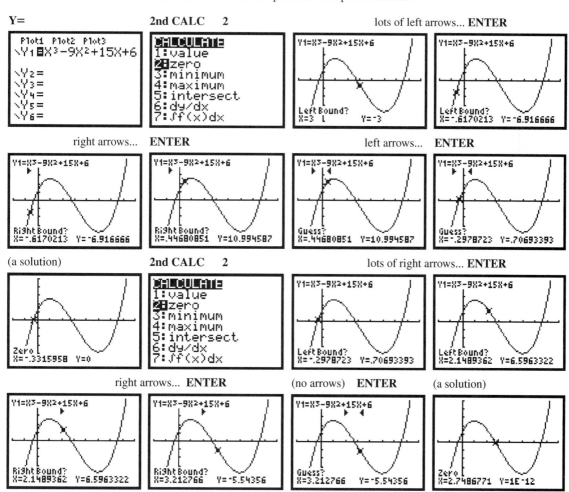

Note that a Right Bound can also be used as the Guess (see the last three frames).

EXAMPLE 3

Basic **TABLE** features:

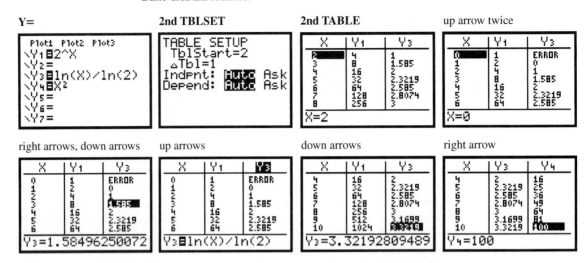

| **Y=** | **2nd TBLSET** | **2nd TABLE** | up arrow twice |

right arrows, down arrows · up arrows · down arrows · right arrow

Note that functions to be investigated using a **TABLE** need to be entered and turned on in the same sense as those you want to graph. To get to the TABLE SETUP screen, press **2nd TBLSET** (the shifted version of the **WINDOW** key). TblStart is just a beginning x-value for the table; you can scroll up or down using the arrow keys. ΔTbl is the incremental change in x. You can use ΔTbl$=1$ (as in the preceding example) to calculate the function at consecutive integers; in calculus, you could use 0.001 to investigate what is happening to a function as it approaches a limit; or you could use a larger number like 10 or 100 to study the function's numerical behavior as x goes to ∞.

EXAMPLE 4

Split-screen graphing with a table (**MODE G-T**):

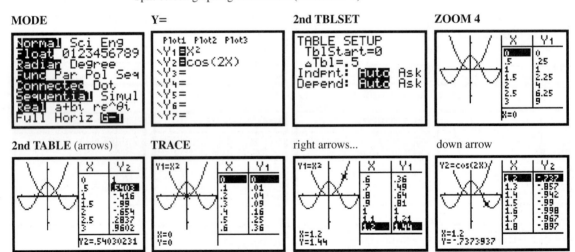

| **MODE** | **Y=** | **2nd TBLSET** | **ZOOM 4** |

2nd TABLE (arrows) · **TRACE** · right arrows... · down arrow

In **MODE G-T**, the graph and a corresponding table share the screen, but only one of them is "active" at any given moment. The **ZOOM** commands and the **GRAPH** key give control to the graphical side of the screen. This just means that the arrow keys refer to the graph rather than the table. Pressing **2nd TABLE** enables the arrow keys to be used to scroll through its values. The **TRACE** key links the table to the graph, with the graph in control (all previous TABLE SETUP specifications are replaced with values related to the **TRACE**).

Note in the last frame that jumping to a different function will display a different column of the table (the same value of x is highlighted).

C.7 THE NUMERIC SOLVER

The TI-83 and TI-83 Plus are equipped with a numeric **Solver** feature (press **MATH**, then **0**), composed of two specialized screens. One is an equation editor that shows **eqn:0=** on the screen, expecting you to fill in the right-hand side. To use the formula for the total surface area of a cylinder, $A = 2\pi R(R + H)$, you must first set one side equal to zero, entering the formula as $0 = 2\pi R(R + H) - A$ or $0 = A - 2\pi R(R + H)$. Pressing **ENTER** takes you to the second screen, where you can enter values and solve for a variable. Enter a value for each letter name except the variable you want to solve for. If there is more than one possible solution for that variable, you can control which one will be found by typing an estimate of it. Use the up and down arrow keys to place the cursor on the line which contains the variable you want to solve for, and press **ALPHA SOLVE (ALPHA**, then the **ENTER** key). The solution is marked by a small square to its left. Also marked (at the bottom of the screen) is a "check" of this solution, indicating the difference between the left- and right-hand sides of the equation using all of the values shown (this should be zero or something very close to zero, such as $1\text{E}-12$). Fourteen significant digits are calculated and displayed for the solution variable. Press **2nd**, then the right arrow key to view the last several digits or to perform arithmetic on the solution (for example, you could divide a solution by π to see *what multiple* of π it is).

The **bound=** option near the bottom of the screen allows you to specify an interval to be searched for the solution. The default interval $\{-1\text{E}99, 1\text{E}99\}$ essentially considers any number that the calculator is capable of representing. When solving trigonometric equations, a limited **bound** like the interval $\{0, 2\pi\}$ might be better, but in most cases, the default interval works well. If you accidentally erase this line and can't remember the syntax, just exit the **Solver** by pressing **2nd QUIT** and reenter it by pressing **MATH 0**. The default **bound** is restored whenever you enter the **Solver** (any **bound** interval you specify is valid only for that **Solver** session). The equation, however, stays in memory until you **CLEAR** it or replace it with another. Press the up arrow until the top line is reached. The calculator will switch immediately to the equation editor page, where you can modify or **CLEAR** the old equation. The following example assumes that you are starting with a blank equation (as if **Solver** has never been used before); you may **CLEAR** the present equation to achieve the same effect (if an equation is stored, you will always start on the solving screen).

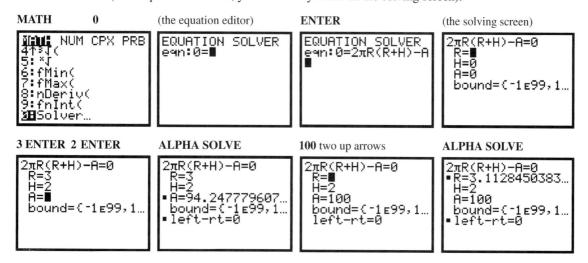

The first two frames in the bottom row of the previous example found the total surface area of a cylinder whose radius is 3 and whose height is 2. The last two frames found the radius of a cylinder whose height is 2 and whose total surface area is 100 square units. Use the **Solver** to find the height of a cylinder whose radius is 5 and whose total surface area is 440 square units (the answer is about 9).

The **Solver** can also be used on equations which contain only one variable, but think about the previous example (which had several variables) to understand why the **Solver** works as it does. In particular, it would seem natural to press **ENTER** after typing an estimate for the solution variable; instead, you must press **ALPHA SOLVE** (press **ALPHA**, *then* the **ENTER** key). If you accidentally press **ENTER** first, you will have to press the up arrow key to get back to the solution variable's line (if the cursor is on **bound=** rather than a variable's line, pressing **ALPHA SOLVE** does *nothing*). For direct comparison, the next example solves the same equation that was solved on both the home screen and the graphics screen in Section C.6. All three methods need one side to be zero and require each solution to be estimated. To solve

$$x^3 + 15x = 9x^2 - 6$$

rewrite it as
$$x^3 - 9x^2 + 15x + 6 = 0$$

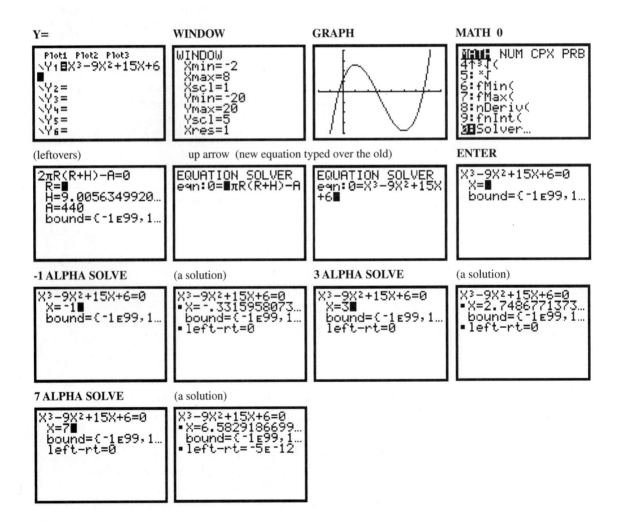

C.8 MATRIX FEATURES

Start by pressing the **MATRX** key on the TI-83 or **2nd MATRX** on the TI-83 Plus. This is the main keyboard difference between these two models. In the remainder of this section, directions will just refer to **MATRX**, which should be interpreted as **2nd MATRX** if you are using a TI-83 Plus. **MATRX** gives you access to three submenus: NAMES, MATH, and EDIT (which can be selected by pressing the right or left arrow keys). The MATH submenu has 16 options, but only 7 of them will fit on the screen at any one time (the arrows next to the 7 in the second frame and the 8 and D in the third frame indicate that there are additional options in those directions).

MATRX   right arrow   down arrows...

NAMES gives you the ability to insert the name of any one of the 10 user-addressable matrices into a formula on the home screen (or into a program statement in the **PRGM Edit**or). Choosing an item from this submenu is the *only* way to type the name of a matrix on a TI-83 or TI-83 Plus. In particular, typing a left bracket, followed by the letter A, followed by a right bracket, *looks* like the name of the matrix [A], but it will *not* be interpreted as a matrix by the calculator. Either dimension of a matrix can be as large as 99 (note, however, that there is not enough memory to handle a full 99 by 99 matrix). The EDIT submenu allows you to specify the dimensions and entries of the selected matrix. This example shows how to define matrix [C]:

MATRX left arrow **3**   **2 ENTER 3 ENTER**   **1 ENTER -4 ENTER**, etc.

2nd Quit (home screen)   **MATRX**   **3**   (home screen)   **ENTER**

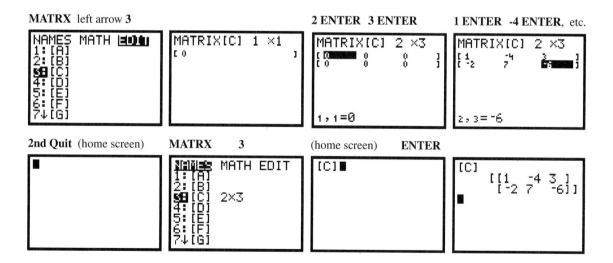

Note that the calculator left-justifies each column, which can make some matrices look a little bit ragged (as in the last frame). Note also that once a matrix has been defined, its dimensions appear in the NAMES and EDIT submenus (see the sixth frame).

The EDIT submenu can also be used to make changes to an existing matrix.

MATRX left arrow **3**   arrow to the incorrect entry, type the new value...**ENTER**

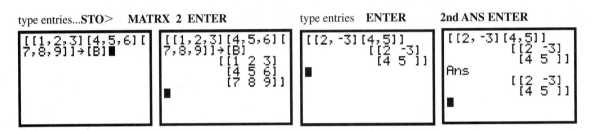

Press **2nd QUIT** to return to the home screen after you finish editing a matrix.

You can also create a new matrix or overwrite an existing one on the home screen (for most purposes, the matrix editor is much more convenient; see the previous example). Type each *row* within brackets, with the entries separated by commas, and enclose the entire matrix within an outer set of brackets. Typically, you'll want to store it in one of the ten matrix variables [A], . . . , [J]; but as the third and fourth frames show, there is also **Ans**, which stores the most recent computational result, even if it's a matrix result:

type entries...**STO>** **MATRX 2 ENTER**   type entries **ENTER**   **2nd ANS ENTER**

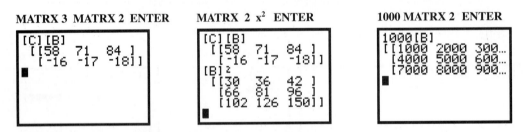

To multiply matrices, just type their names in the proper order (with or without a multiplication sign in between). You can square a matrix using the x^2 button. Multiplication by a scalar is shown in the final frame below:

MATRX 3 MATRX 2 ENTER   **MATRX 2 x^2 ENTER**   **1000 MATRX 2 ENTER**

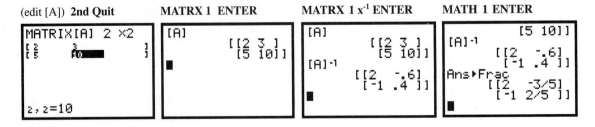

Notice that the three dots at the right of each row of the last frame indicate that there is more of the matrix in that direction; just use the right arrow key to reveal the hidden columns. To find the inverse of a (square) matrix, use the x^{-1} key following the name of the matrix. Fractional forms can be obtained by pressing **MATH 1**, then **ENTER**.

(edit [A]) **2nd Quit**   **MATRX 1 ENTER**   **MATRX 1 x^{-1} ENTER**   **MATH 1 ENTER**

The determinant of a (square) matrix is available as a feature in the MATH submenu.

MATRX right arrow **1** (home screen) **MATRX 1** **) ENTER**

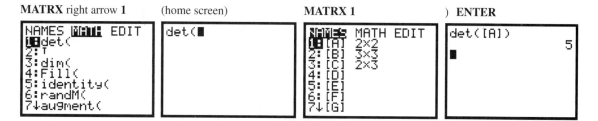

To row-reduce a matrix which might represent a system of equations, the option **rref(** is available in the MATH submenu.

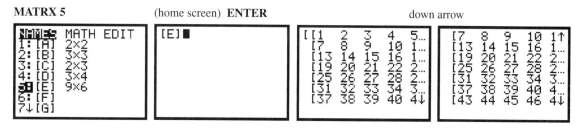

If a matrix ever has more rows than will fit on the viewing screen, an arrow appears to indicate that there are hidden rows in that direction. Just use the down (or up) arrow key to scroll to these hidden rows.

MATRX 5 (home screen) **ENTER** down arrow

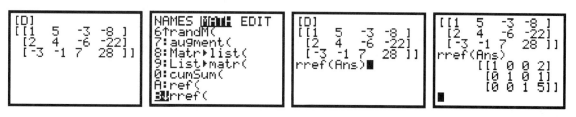

C.9 LIST FEATURES AND DESCRIPTIVE STATISTICS

The TI-83 has six built-in lists, named L_1, L_2, . . . , L_6 that can be accessed from the keyboard by pressing **2nd** followed by the subscript number of the list. You can create and name other lists (starting with a letter and using no more than five characters), and access them by pressing **2nd LIST** (**2nd**, then the **STAT** key). The TI-83 Plus will show L_1, L_2, . . . , L_6 in the **LIST** NAMES submenu, but an ordinary TI-83 will not. New list names can be created most easily in the **STAT Edit**or (press **STAT** then **1**). In this editor, go to the very top of any column and press **2nd INS** to create a new list. The example below creates a list named "RBI" and computes its descriptive statistics.

2nd LIST **STAT 1** up arrow

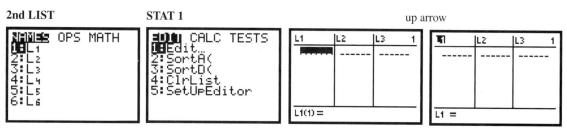

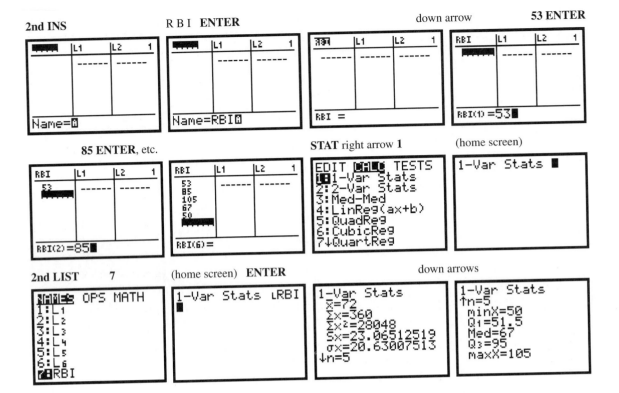

Note that when a list name is shown on the home screen, it is preceded by a small L. This character is available in the **LIST OPS** menu (press **2nd LIST**, right arrow, up arrow, **ENTER**), and is also in the **CATALOG** (press **2nd CATALOG L ENTER**). You can create a list from the home screen by enclosing your data in braces { }, separating the items with commas, and **STO**ring it in a new list name (preceded by the small L). However, a list created in this manner will *not* appear in the **STAT Edit**or until you include its name in a **SetUpEditor** command or **INS**ert its name along the top row of the **STAT Edit**or. To restore the standard setup, which shows just L1, L2, . . . , L6, execute **SetUpEditor** without specifying any list names. Deleting a list from the editor in this way (or by arrowing up to a list name at the top of the editor and pressing **DEL**) does not erase any data or delete the list from the **LIST NAMES** submenu. The list simply doesn't appear in the *editor* for the time being. To erase all data from a list (leaving it empty), arrow up to its name in the top row of the editor and press **CLEAR**, then **ENTER** (or **CLEAR**, then down arrow). To completely dispose of a list (name and all) on a TI-83 Plus, press **2nd MEM 2 4** and down arrow until the marker points at the name of the list you wish to get rid of, then press **DEL** (*not* **ENTER**). On an ordinary TI-83, press **ENTER** instead of **DEL**.

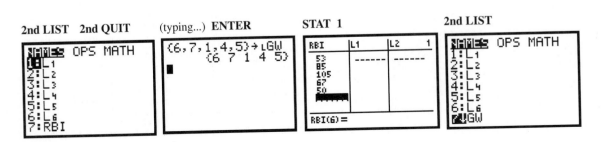

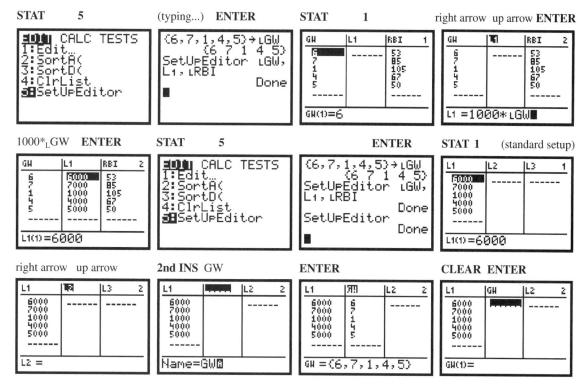

The first three frames in the bottom row show how to insert a preexisting list into the **STAT Edit**or. The last frame illustrates how to **CLEAR** a list, leaving it empty, but still named. If you highlight a list's name (in the very top row of the editor), you will be able to wrap around using the right or left arrow keys. If you wrap around using the *right* arrow, you will also notice a new blank list in the editor, ready to be named and filled with data. Just press **ENTER** or down arrow when it is highlighted, and you will be given a chance to name it. The **STAT Edit**or can hold up to 20 lists at once. Each list stored in a TI-83 or TI-83 Plus can have as many as 999 elements.

C.10 THE LINE OF BEST FIT (LINEAR REGRESSION)

EXAMPLE

Find and graph the equation of the line of best fit for the following data:

x	5	7	9	12	14
y	40	58	62	74	80

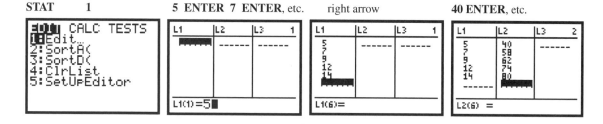

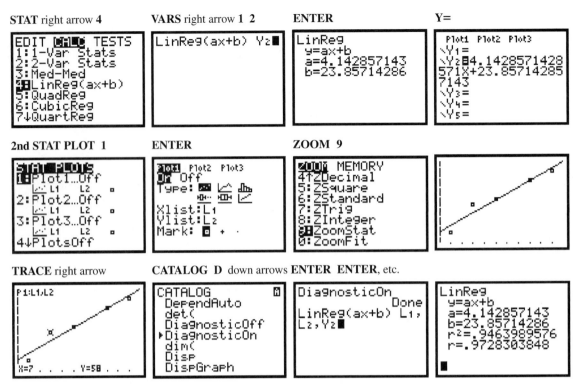

STAT right arrow **4**

```
EDIT CALC TESTS
1:1-Var Stats
2:2-Var Stats
3:Med-Med
4:LinReg(ax+b)
5:QuadReg
6:CubicReg
7↓QuartReg
```

VARS right arrow **1 2**

```
LinReg(ax+b) Y₂█
```

ENTER

```
LinReg
y=ax+b
a=4.142857143
b=23.85714286
```

Y=

```
Plot1 Plot2 Plot3
\Y₁=
\Y₂█4.1428571428
571X+23.85714285
7143
\Y₃=
\Y₄=
\Y₅=
```

2nd STAT PLOT 1

```
STAT PLOTS
1:Plot1...Off
   L₁  L₂  ▫
2:Plot2...Off
   L₁  L₂  ▫
3:Plot3...Off
   L₁  L₂  ▫
4↓PlotsOff
```

ENTER

```
Plot1 Plot2 Plot3
On Off
Type: ▦ ⤶ ⬛
      ⊶ ⊞ ⬉
Xlist:L₁
Ylist:L₂
Mark: ▫ + ·
```

ZOOM 9

```
ZOOM MEMORY
4↑ZDecimal
5:ZSquare
6:ZStandard
7:ZTrig
8:ZInteger
9:ZoomStat
0:ZoomFit
```

TRACE right arrow

```
P1:L₁,L₂




X=7 . . . . Y=58 . . .
```

CATALOG D down arrows **ENTER ENTER**, etc.

```
CATALOG        █
 DependAuto
 det(
 DiagnosticOff
▸DiagnosticOn
 dim(
 Disp
 DispGraph
```

```
DiagnosticOn
          Done
LinReg(ax+b) L₁,
L₂,Y₂█
```

```
LinReg
y=ax+b
a=4.142857143
b=23.85714286
r²=.9463989576
r=.9728303848
█
```

Please turn **STAT PLOT 1 Off** *now* (before trying to graph anything else).

In the last three frames, this linear regression is shown in more detail. **DiagnosticOn** enables the calculator to compute and print correlation coefficients. **DiagnosticOff** is the default setting. These options appear only in the **CATALOG**. The last linear regression command shows the full syntax of the **LinReg(ax+b)** statement. All three parameters are optional. The name of the first list provides the *x*-values, the second list provides the *y*-values, and the third parameter is the name of the function where the regression equation will be stored. If you omit the two list names, the calculator will use list L_1 for the *x*-values and list L_2 for the *y*-values (see the 6th frame of the example). If you don't plan to use or graph the regression equation, you may omit the function name. To turn **PLOT 1 Off**, press **Y=**, up arrow, then **ENTER**.

C.11 CALCULUS FEATURES

EXAMPLE 1

Finding local minimum and maximum values on the graphics screen:

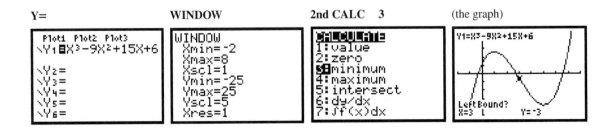

Y=

```
Plot1 Plot2 Plot3
\Y₁█X³-9X²+15X+6
\Y₂=
\Y₃=
\Y₄=
\Y₅=
\Y₆=
```

WINDOW

```
WINDOW
 Xmin=-2
 Xmax=8
 Xscl=1
 Ymin=-25
 Ymax=25
 Yscl=5
 Xres=1
```

2nd CALC 3

```
CALCULATE
1:value
2:zero
3:minimum
4:maximum
5:intersect
6:dy/dx
7:∫f(x)dx
```

(the graph)

```
Y1=X3-9X2+15X+6



LeftBound?
X=3  L      Y=-3
```

right arrows... **ENTER**　　　　　　　　　　　　right arrows... **ENTER**

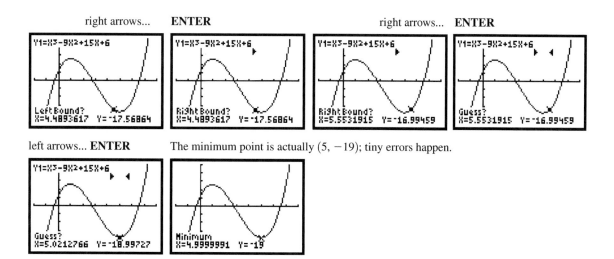

left arrows... **ENTER**　　　The minimum point is actually $(5, -19)$; tiny errors happen.

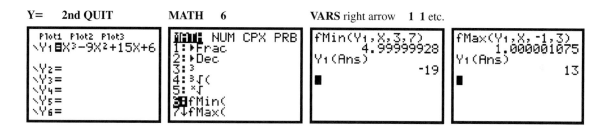

Now use **maximum** to find the local maximum point [answer: $(1, 13)$].

EXAMPLE 2

Finding local minimum and maximum values on the home screen: (Review Example 1 in Section C.6 if you don't remember how to type the symbol for Y_1 into a formula.)

Y=　　**2nd QUIT**　　　　**MATH**　　**6**　　　　**VARS** right arrow　**1 1** etc.

Plot1 Plot2 Plot3	MATH NUM CPX PRB	fMin(Y₁,X,3,7)	fMax(Y₁,X,-1,3)
\Y₁⊟X³−9X²+15X+6	1:▶Frac	4.99999928	1.000001075
	2:▶Dec	Y₁(Ans)	Y₁(Ans)
\Y₂=	3:³	−19	13
\Y₃=	4:³√(	■	■
\Y₄=	5:ˣ√		
\Y₅=	6⬛fMin(		
\Y₆=	7↓fMax(		

The only intended use of these maximum and minimum routines is to find *local* extreme values using a fairly short bracketing interval. These routines can give unpredictable results if you try to use them to compare local extrema with endpoint extrema on a given interval. Example 5 in Section C.4 shows how **ZoomFit** can be used to estimate the maximum and minimum of a function on a closed interval.

EXAMPLE 3

Finding values of the derivative using the graphics screen:

Y=　　　　　　**2nd CALC**　**6**　　　　　**7**　　**ENTER**

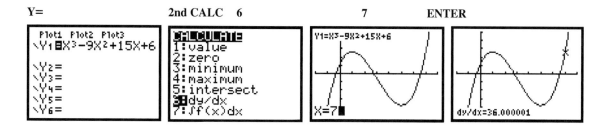

EXAMPLE 4

Finding values of the derivative using the home screen:

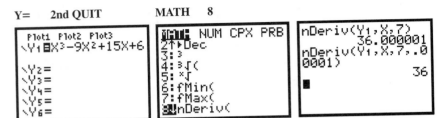

Note that **nDeriv** (numerical derivative) requires three arguments and has an optional fourth. The required arguments are the name of the function you wish to differentiate, the name of the variable to differentiate with respect to, and the value of that variable at the point where the derivative is to be calculated. This routine actually calculates the slope of the line connecting the two points $(x - h, f(x - h))$ and $(x + h, f(x + h))$, which are just a small increment on either side of the point $(x, f(x))$. The optional last argument gives you a chance to say just how small this increment h should be. The default value is .001. An h-value of .00001 often gives better accuracy for commonly studied functions. But don't go overboard on this or any other optional accuracy parameters. You risk severe loss-of-significance errors if you use tiny h-values such as .00000000000001. Note the last frame above. With $h = .00001$, there are nine significant digits in our answer (all but the last digit is correct); but with $h = .0000000001$, we get only 4 significant digits. In fact, **nDeriv(e^(X), X, 1, 1E-14)** calculates 0, a total loss of significant digits! This phenomenon is typical of computations that involve *subtracting* numbers that are *very* close together on a machine which can only store a limited number of digits.

EXAMPLE 5

Evaluating a function and its first and second derivatives (at $x = 4$):

Y= 2nd QUIT

EXAMPLE 6

Graphing a function and its first and second derivatives:

Y= ZOOM 7

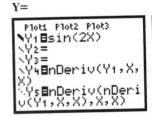

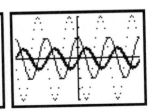

| Y= | | WINDOW | GRAPH |

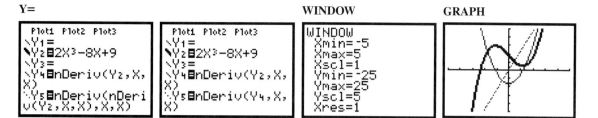

Note that the second derivative, Y5, is shown coded in two different, but equivalent, ways (choose whichever one you like). The first coding style is similar to the home-screen version of the second derivative (see Example 5). The other style exploits the formula for the first derivative, stored in Y4. It requires less typing, both in the initial example and also when all references to Y1 are changed to Y2. Many calculus students keep these derivative-graphing formulas permanently stored in Y4 and Y5, just turning them on or off at appropriate times. Third and higher derivatives are *not* supported in either manner of coding.

EXAMPLE 7

Numerical integration on the home screen:

MATH   9

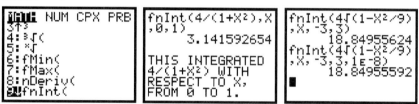

Note that **fnInt** (function integral) needs the name of the function, the name of the variable you're integrating with respect to, the lower limit of integration, and the upper limit of integration (in that order). The optional 5th argument gives you a chance to specify how little error you are willing to tolerate in the numerical result. The standard (default) value is .00001, which usually yields considerably less error than that. In the second frame in the preceding figure, the exact answer is π, which was computed to 14 significant digits (storing your answer in a **LIST** allows you to view all 14 digits that are stored in the calculator). In the last frame, the top calculation took 23 seconds and produced 8 significant digits; the second version produced 11 significant digits, but took 44 seconds to calculate. This is an example of a difficult numerical integration (finding the area enclosed by the ellipse $x^2/9 + y^2/4 = 1$); the exact answer is 6π. Most problems are like the middle frame, where you will get impressive accuracy, very quickly, without needing to specify the optional fifth argument. The **fnInt** feature is based on a powerful Gaussian method that will consistently outperform Simpson's rule and other elementary numerical procedures.

EXAMPLE 8

Numerical integration on the graphics screen:

| Y= | WINDOW | 2nd CALC   7 | 0 |

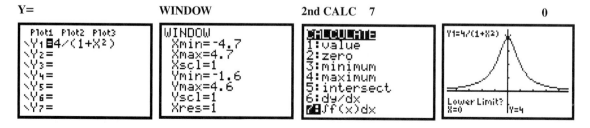

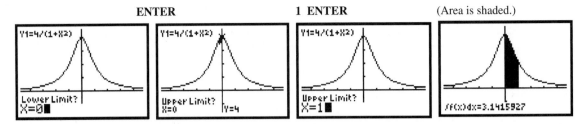

ENTER   **1 ENTER**   (Area is shaded.)

You can enter the lower and upper limits from the keyboard, as shown above or by arrowing over to the proper values using the left or right arrow keys (then pressing **ENTER**). The latter method has the disadvantage that the limits of integration you intended to use often aren't traceable *x*-values in the present viewing window (you can get close, but cannot get the exact limits desired). The result of the numerical integration is shown at the bottom of the screen and can be interpreted as the area of the shaded region. The accuracy of the calculation is the same as the default accuracy for **fnInt** (five decimal places are guaranteed).

C.12 SEQUENCES AND SERIES

On the TI-83 or TI-83 Plus, a sequence (**seq**) can be created and stored as a list. Series can be treated as the **sum** of such a list. Both features are found in the submenus under **2nd LIST** (the shifted version of the **STAT** key).

EXAMPLE 1

Enumerate the sequence of the first six squares ($a_n = n^2$ for $n = 1$ to 6), then evaluate its sum:

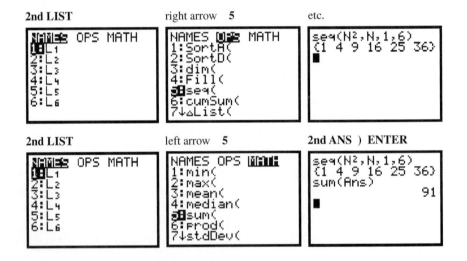

EXAMPLE 2

Evaluate the sum: $7 + 9 + 11 + 13 + 15 + \cdots + 121$.

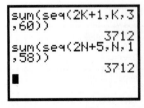

Note that there are many different coding possibilities.

EXAMPLE 3

Estimate the infinite geometric series $\sum\limits_{n=0}^{\infty} (2/3)^n$

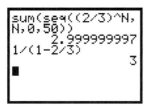

The top calculation shows the sum of the first 51 terms. The bottom calculation shows that the formula $a_0/(1 - r)$ gives an exact answer of 3.

EXAMPLE 4

Estimate the value of the infinite series $\sum\limits_{n=0}^{\infty} 1/n!$

Remember that the factorial symbol can be found by pressing **MATH**, left arrow, **4**, as shown in the first frame below.

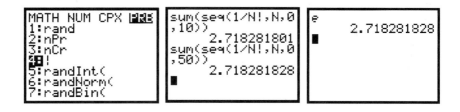

The exact answer is the number e, the shifted **2nd** version of the division key.

EXAMPLE 5

Use the 7th-degree and 21st-degree Taylor polynomials of $\sin(x)$ to estimate $\sin(\pi/6)$.

7th-degree estimate   21st-degree estimate

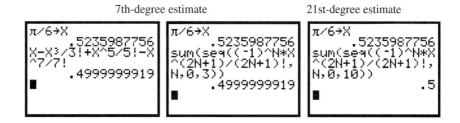

Note that the degree of the estimate was increased simply by changing the last **seq** index from 3 to 10.

EXAMPLE 6

Graph the 2nd-, 4th-, 6th-, and 8th-degree Taylor polynomials of $\cos(x)$ along with its exact graph. On a separate screen, graph the 20th-degree Taylor polynomial of $\cos(x)$. Be sure that your calculator is in **Radian MODE**.

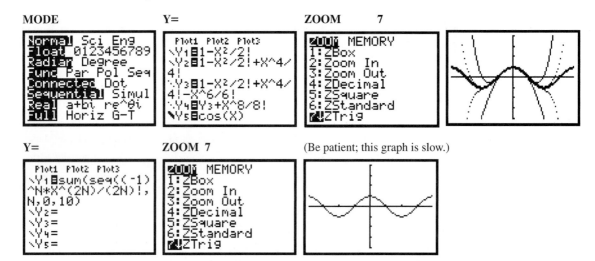

MODE   **Y=**   **ZOOM**   **7**

Y=   **ZOOM 7**   (Be patient; this graph is slow.)

You should try a similar exercise for $\sin(x)$, graphing its 1st-, 3rd-, 5th-, 7th-, and 9th-degree Taylor polynomials. See Example 5 for sample formulas.

Using an Advanced Graphing Calculator

This appendix is included to provide faculty with the flexibility of integrating advanced graphing calculators in their classes. Each section explains and illustrates important features of the Texas Instruments TI-89. Though this appendix was specifically designed to supplement the graphing calculator examples found throughout the text, the material is organized so that an interested student could also study it as a separate chapter.

D.1 INTRODUCTION TO THE TI-89 KEYBOARD

The Texas Instruments TI-89 has a 50-key layout, similar to earlier TI graphing calculators. The bottom six rows of keys perform the standard functions of a scientific calculator. The thin blue keys that form the top row are used to choose menus and to give quick access to specialized screens for graphing functions and calculating tables of function values. The 2nd, 3rd, and 4th rows of keys contain the calculator's editing and navigation features. Of special interest are the calculator's **MODE** key, which can be used to change any of its settings, and the **APPS** key, which gives access to all nine of the specialized editing and application screens. Also notice the (golden yellow) **2nd** key, the (purple) **alpha** key, and the **green diamond** key, which give additional, color-coded meanings to almost every key on the calculator. Start by pressing the **ON** key in the lower left-hand corner. There are two ways to turn the calculator **OFF**. If you want to be able to return to your work precisely as you left it, press the **green diamond** key, then **ON/OFF** (or just leave the calculator unattended for a couple of minutes). Pressing the **2nd** key, then **OFF**, exits any menus, editors, or application screens and puts the calculator back on the home screen as it turns off the power.

Look again at the scientific calculator features found in the bottom six rows of keys. The **LN** (natural logarithm), **SIN, COS,** and **TAN** functions are shown in gold as **2nd** shifted versions of the **X, Y, Z,** and **T** keys, respectively. The **green** shifted items in that row include e^x, the inverse trig functions, and the variable θ. The square root is the **2nd** version of the multiplication key. Unlike previous models, the TI-89 does *not* include specialized keys for squaring or taking reciprocals. Another change is that the common (base 10) logarithm, **log(**, is found only in the **CATALOG**. To access any item shown in green on the keyboard, press the **green diamond** key and then press the key which has the green writing above it; to increase (or decrease) screen contrast, *hold down* the **green diamond** key while pressing the addition (or subtraction) sign repeatedly. An interesting surprise is that many of the keys have **green** shifted meanings that are not marked on the

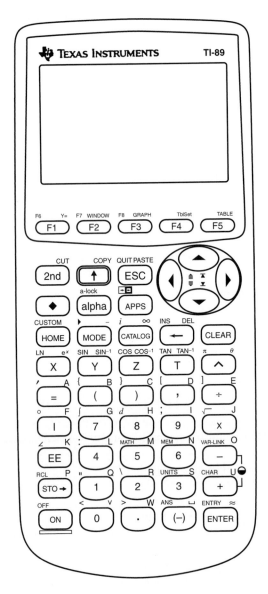

Courtesy of Texas Instruments.

face of the calculator. Among the most useful are $\le$, $\ge$, $\ne$, and ! (the factorial symbol). They are the **green** shifted versions of zero, the decimal point, the equal sign, and the division sign, respectively.

For a map of the hidden **green** features, press **green diamond** then **EE** (just to the left of **4**):

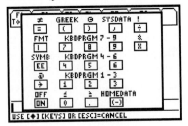

To reset the standard defaults without erasing programs or data:
(Newer TI-89s will use **2nd MEM** **F1** **1** **2**)

2nd MEM   **F1**   **3**   **ENTER**   **ENTER**

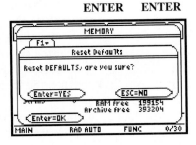

To reset the standard defaults on the calculator (without erasing programs or data), press **2nd MEM** (the **2nd** version of the **6** key), then press **F1** (RESET), then **3** (Default). Newer versions of the TI-89 operating system will use **2nd MEM F1** (RESET), then **1** (RAM), then **2** (Default). Now press **ENTER** twice. Doing this now will assure that your calculator's settings will be the same as the settings used in the examples that follow. The (blue) **ENTER** key in the lower right-hand corner is pressed after typing a command on the entry line. It signals the calculator to perform the indicated operation (don't confuse this with the = key, which is only used for writing equations and making comparisons on the TI-89). If the calculator shows **AUTO** or **EXACT** at the very bottom of the screen, pressing **ENTER** will usually produce an exact algebraic answer (a simplified fraction, a simplified radical, etc.). To obtain a decimal approximation, press **green diamond** then **ENTER** ($\approx$). In **AUTO** mode, another way to produce a decimal approximation is to include a decimal point when typing any of the numeric values in the expression. All decimal approximations are carried out to 14 significant digits, but the default is to display only 6 of them (up to 12 significant digits can be displayed by adjusting **MODE Display Digits**). Any subsequent calculations that refer to the approximation will use all 14 digits, regardless of the display setting. Exact integers up to 614 digits long can be calculated, displayed, and stored by the TI-89! The numerators and denominators of rational numbers also can be up to 614 digits long.

Some Numeric Examples

$12 \times 5 + 3 \div 7$ **ENTER**
green diamond ENTER

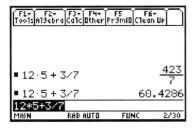

$312 \div 816$ **ENTER**
green diamond ENTER

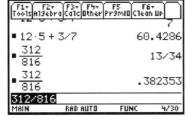

2nd multiplication sign 32)
ENTER green diamond ENTER

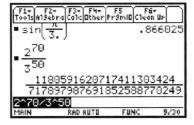

Note in the first frame that the multiplication key always prints an asterisk on the entry line and is shown as a raised dot in the history area. Likewise, the division key always prints a slanted fraction bar line on the entry line, but it is sometimes shown as a horizontal fraction bar line in the history area.

SIN is the **2nd** version of **Y**.
π is the **2nd** version of the **^** key.

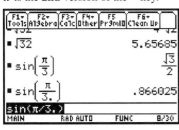

An exponent is indicated just after the **^** key is pressed.

Complex numbers are available;
i is the **2nd** version of **CATALOG**.

Note in the last frame that complex numbers on the TI-89 are shown with i representing the imaginary unit. This number is usually called j in electronics applications.

D.2 VARIABLES AND EDITING

The commonly used variables **X, Y, Z,** and **T** were each given their own key on the TI-89. All other letters of the alphabet are obtained by pressing the purple **alpha** key followed by the letter of the alphabet (written in purple above keys in the bottom 5 rows); a blank space can be obtained by pressing **alpha** and then the sign change key **(-)**. If you need to type several letters of the alphabet in succession, press **2nd a-lock** (the gold **2nd** key followed by the **alpha** key) or press **alpha** twice. To release alpha-lock, just press **alpha** one more time. The **backspace** key (a black key with a left arrow on it) is located just to the left of the **CLEAR** key. This key erases the character to the *left* of the cursor each time it is pressed. The **green diamond** version of this key (**DEL**) erases a character to the *right* of an insert cursor, and erases the character *highlighted* by an overstrike cursor. Both **backspace** and **DEL** will erase a block of highlighted characters on the entry line or in the history area. The right and left arrow keys are used to move the cursor one space at a time when editing the entry line. Pressing the **2nd** key just before pressing one of the (blue) arrow keys causes the cursor to travel as far as possible in that direction. Quite often, the screen is not wide enough to display the entire result of a calculation. A small arrow at the right means that there is more to view in that direction. Just arrow up to highlight that line of history and scroll using the right arrow key (or press **2nd** then right arrow to jump to the end of that line). There is also a **shift** key marked as a white up arrow on a black key (just to the right of the **2nd** key). The **shift** key enables you to type uppercase letters, but it should be noted that the symbolic algebra features of the TI-89 are *not* case sensitive. The symbols z and Z represent the *same* variable on the TI-89. To type several uppercase letters in a row, press **2nd shift alpha** or press **alpha shift alpha**. To release shift-lock, just press alpha one more time. This shift key is also used in conjunction with the arrow keys to highlight text which you may want to **cut, copy, paste,** or **delete.** *Hold down* the **shift** key while pressing the (blue) right or left arrow keys to highlight the text you want to manage, then press **green diamond CUT, COPY, PASTE,** or **DEL** (the **green** versions of the **2nd, shift, ESC,** or **backspace** keys, respectively). These text-managing options can also be found in the **F1 (Tools)** menu. Editing text or other input can be done with either an "insert" or "overstrike" cursor. The insert cursor is indicated by a blinking vertical line, which fits between the characters where the insertion will occur, while the overstrike cursor highlights a single character, which will be replaced by the one you are typing. To switch between the two cursors, just press **2nd INS** (the **2nd** version of the **backspace** key). All keystroke examples in this book will assume that the calculator currently shows an insert cursor.

Variable names for the TI-89 can be from one to eight characters in length and must start with a letter (the other characters may be alphabetic or numeric). The one-letter variables a–z can be deleted easily (press **2nd F6**, then press **ENTER** twice) and are popular for symbolic manipulation and other short-term use. Variables that store valuable, long-term information should probably be named using more than one letter. Don't forget, though, that the TI-89 expression **4ac** does *not* mean **4** times **a** times **c**. Instead, it means **4** times a variable named **ac**. Note also that parentheses immediately following a variable name always indicate function notation (*not* multiplication) on the TI-89. For example $q(x/z)$ indicates a function q of the expression x/z (not q times x/z). The times sign (which prints an asterisk on the screen) is used much more often than proximity to indicate multiplication on this calculator. If multiplication is intended in the preceding examples, you could enter $4a*c$, $4*a*c$, or $(4a)c$ in the first example and $q*x/z$, $q*(x/z)$, or $(q*x)/z$ in the second example.

To erase everything to the right of the cursor on the entry line, press **CLEAR** once; pressing **CLEAR** again (with nothing to the right of the cursor) will erase the entire entry line. The **CLEAR** key (or the **backspace** key) can also be used to erase unwanted items in the history area. To erase an individual history item, just press the (blue) up arrow key until the item is highlighted, then press **CLEAR** (or **backspace**). Both the entry and its answer are erased at the same time. To scroll to the very top of the history area on the home screen, enter the history area by pressing the (blue) up arrow key, then press **green diamond**, then up arrow. Pressing **green diamond** then down arrow scrolls to the bottom of the history area. Press the **ESC** key to escape the history area and return to the entry line. Pressing **ESC** can also escape any unwanted menu without leaving the current application. The **HOME** key escapes from any menu or application and returns you to the home screen. To erase all history on the home screen, be sure that you have the standard home screen menu, then press **F1** (the blue key in the upper left-hand corner) then **8 (Clear Home)**.

D.3   THE HOME SCREEN MENUS

The home screen is used for computation and symbolic manipulation (there are specialized screens for graphing, program input/output, tables, and data editing). Two different menu systems are available for the home screen. The standard menu system (shown in the first and second frames below) allows quick access to the algebra and calculus features of the TI-89. The default custom menu system (shown in the third frame) contains useful templates, common units of measurement, and special symbols, which can save many keystrokes. As their name implies, the custom menus can also be modified to reflect an individual user's needs and preferences (see "Creating a Custom Menu" in the owner's manual that came with your calculator). The custom menu system is activated or exited by pressing **2nd CUSTOM** (the **2nd** version of the **HOME** key). It is also exited (and the standard menu system is shown) whenever you go to another application screen and later return to the home screen.

Press **F1** then **8** to erase all history   To switch between these two, press **2nd CUSTOM.**

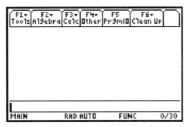

Options **1 Open** and **3 New** are   The standard menu system.   The (default) custom menu system.
dimmed, which indicates that they
are not active on the home screen.

The Standard Home Screen Menus

To select an individual menu, press the (thin blue) key in the top row that has the same marking as the menu (to access the **Algebra** menu, press the **F2** key). Note that **F6, F7,** and **F8** are the **2nd** versions of the **F1, F2,** and **F3** keys, respectively. The top of the **Tools** menu was shown in the previous example.

The bottom of the **F1 Tools** menu.

Keyboard shortcuts are shown at the right of options 7 and 9. Options 4, 5, and 6 are also on the keyboard.

The **F2 Algebra** menu (top).

The algebra features include a wide variety of solving and simplifying commands (see Section D.14).

The **F2 Algebra** menu (bottom).

To get to the bottom of a menu quickly, press **2nd**, then the down arrow key.

The **F3 Calculus** menu (top).

Options 1 and 2 are available on the keyboard as **2nd** versions of the **8** and **7** keys, respectively.

The **F3 Calculus** menu (bottom).

The **F4 Other** menu (top).

Of particular importance is option 4, which is used to free up a specific variable.

The **F4 Other** menu (bottom).

Option 9 allows reference to previous results; ans(1) means the most recent answer, etc. This feature is **2nd (-)** on the keyboard.

F5 Program Input/Output screen.

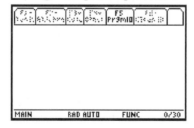

The **F5** key accesses a specialized screen for user-written programs (rather than a menu). Press **F5**, **ESC**, or **HOME** to exit.

The **F6 Clean Up** menu.

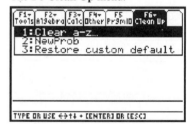

Clear a-z frees up only these variables. **NewProb** clears a–z, turns off all functions, and clears all screens.

The (Default) CUSTOM Menus for the Home Screen

Press **2nd CUSTOM** (the **2nd** version of the **HOME** key) to switch between the standard home screen menus and the custom menus for the home screen.

The **F1 Variables** menu (top).

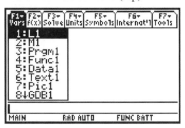

The **F1 Variables** menu (bottom).

The **F2 f(x)** menu.

The **F1 Variables** menu assumes that many people will name their lists L1, L2, . . . , and their matrices M1, M2, . . . , etc. These are *not* predefined variable names on the TI-89.

This menu saves many keystrokes. Option 6 is particularly useful in setting up function notation.

The **F3 Solve** menu.

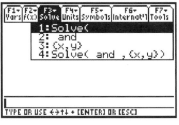

The **F4 Units** menu (top).

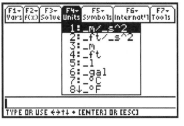

The **F4 Units** menu (bottom).

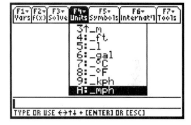

Option 4 is a template for solving systems of equations in two unknowns. Equations go on both sides of **and**.

A more complete list of units can be found by pressing **2nd UNITS** on the keyboard.

The **F5 Symbols** menu.

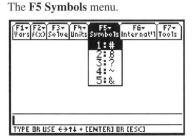

The **F6 International** menu (top).

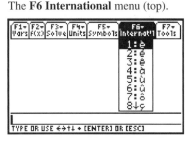

The **F7 Tools** menu.

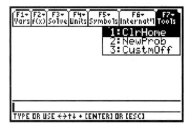

These symbols are seldom used.

For a more complete list of special characters, press **2nd CHAR**.

These tools require **ENTER** to be pressed after the option is selected.

D.4   THE KEYBOARD MENUS

Eight other menus are shown only when you specify them from the keyboard. Three of the keyboard menus have their own dedicated key (**APPS, MODE,** and **CATALOG**); the five others (**MATH, MEM, VAR-LINK, UNITS,** and **CHAR**) are accessed as **2nd** versions of keys in the lower right-hand portion of the keyboard. All of the keyboard menus can be accessed from any screen on the calculator (not just the home screen).

The most important navigation menu is accessed by pressing the (blue) **APPS** key (see the following two screens). Each of the options activates one of the calculator's specialized application screens. The first five options are also readily available from the keyboard (the **HOME** key and the **green diamond** versions of **F1, F2, F3,** and **F5,** respectively). Newer versions of the TI-89 will show option 1 as **FlashApps** instead of **Home**.

The **APPS** menu (top).   The **APPS** menu (bottom).

The MODE Options

Pressing the **MODE** key reveals page 1 of the three-page **MODE** menu (use **F2** and **F3** to change pages).

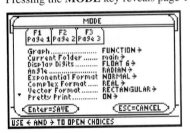      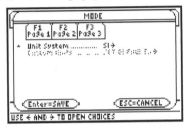

Page 1   Page 2   Page 3

The **Graph** options.   The **Display Digits** options.   The **Angle** options.

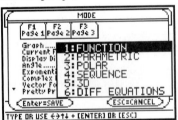      

The **Exponential Format** options.   The **Complex Format** options.   The **Vector Format** options.

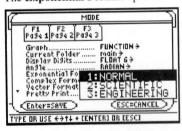      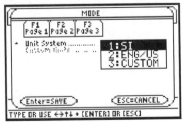

F2 The **Split Screen** options.   The **Exact /Approx.** options.   **F3** The **Unit System** options.

   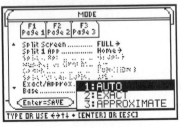   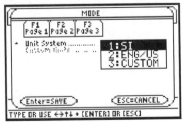

Page 2   Page 2   Page 3

On page 2, many of the options are dimmed (not available) unless a split screen mode other than **FULL** is in effect. On page 3, the second option is dimmed unless a custom unit system has been selected. To change an option, first arrow down to the category you wish to change, then press the right arrow key to open the options list. Next select the number of the option (or arrow down and press **ENTER** when the option is highlighted). This closes the option box and gives you a chance to make further changes in the **MODE** menu if you wish. Unfortunately, this closing of the option box has led many to the wrong conclusion that their changes were finalized. To SAVE your changes, you *must* press **ENTER** to exit the **MODE** menu. Exiting this menu by pressing **ESC, 2nd QUIT,** or an application screen key (like **green diamond Y=,** etc.) will *not* make any of the changes that you had specified! The following example shows how to set the calculator's Angle status to **DEG**rees.

MODE   three down arrows   right arrow   down arrow   **ENTER**   **ENTER**

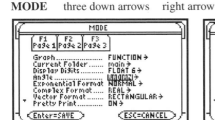  

Note that even after **DEGREE** has been selected in the last frame, the calculator is still in **RAD**ian mode (as indicated in the status line at the bottom of the screen)! None of your changes are put into effect until you press **ENTER** to exit the **MODE** menu. Exiting this menu in any other manner (**ESC, 2nd Quit, green diamond Y=,** etc.) discards all of the changes that you had specified.

The CATALOG

Pressing the **CATALOG** key reveals an alphabetized listing of almost every TI-89 feature. Press the first letter of the desired command (without pressing **alpha**), then scroll from there using the down arrow key. Press **ENTER** to select the command marked by the small arrow on the left. As you browse, the lower left-hand corner of the screen shows the necessary operands (in the required order) for each command. In fact, you may want to access the **CATALOG** simply to obtain this information about an unfamiliar command. The **CATALOG** can also be searched a page at a time by pressing **2nd** down arrow (or **2nd** up arrow).

CATALOG   C   **2nd** down arrow (twice)   down arrows   **ENTER**

 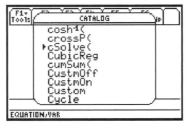 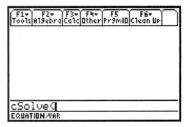

Notice that the operands "EQUATION,VAR" of the **cSolve(** command are shown in the lower left-hand corner when it is highlighted (and even after it is selected). This means that **cSolve(** should be followed by an equation, then a comma, then the name of a variable. Of course, be sure to include a closing parenthesis.

The MATH Submenus

Pressing **2nd MATH** (**2nd**, then the **5** key) gives access to more than 100 different commands. Because there are so many options available, they have been divided into 13 submenus. On the TI-89, a submenu is marked by a small arrow to the right of its name. You can open a submenu just by pressing its number or letter, or you can arrow down to the submenu you want, then press the right arrow key (or **ENTER**) to open it. Press the left arrow key (or **ESC**) if you want to close a submenu without selecting one of its options. Some of the most important **MATH** submenus are shown in the frames that follow.

2nd MATH (top of **MATH** menu)

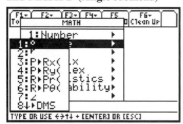

1 (top of the **Number** submenu)

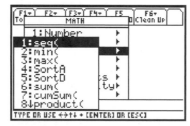

(bottom of the **Number** submenu)

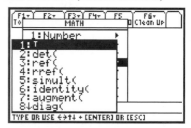

2nd MATH 2 (**Angle** submenu)

2nd MATH 3 (**List** submenu)

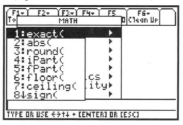

2nd MATH 4 (**Matrix** submenu)

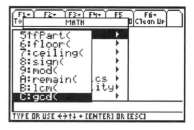

2nd MATH 5 (**Complex** submenu)

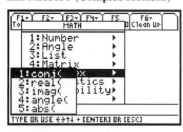

2nd MATH 6 (**Statistics** submenu)

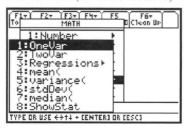

3 (**Regressions** sub-submenu)

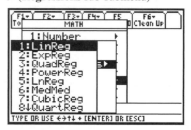

2nd MATH 7 (**Probability** submenu)

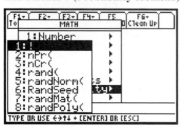

2nd MATH 8 (**Test** submenu)

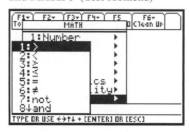

(bottom of the **MATH** menu)

The bottom five submenus of **MATH** (shown in the last frame) include exact duplicates of the standard home screen **Algebra** (**F2**) and **Calculus** (**F3**) menus, as well as access to **Hyperbolic** trig functions, **String** manipulation features, and other number **Bases**.

The MEMory Menu

Press **2nd MEM** (**2nd**, then the **6** key) to see how much memory is free and how much is allocated to the various data types in storage. The only memory management tool available in this menu is **F1 RESET**. Look for less drastic memory management options under the **VAR-LINK** menu. Of particular interest is option **3 Default**, which you used earlier to set your calculator to its original factory settings, without disturbing any other data or programs.

Newer versions of the TI-89 will have a more elaborate **RESET** menu.

2nd MEM   **F1**

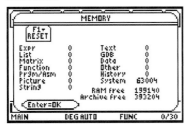 

The VAR-LINK Menu

Pressing **2nd VAR-LINK** (the **2nd** version of the subtraction sign) takes you to a menu where all of your variables, programs, and other data items are listed and can be managed individually (see **2nd MEM** for ways to reset the entire calculator). Put a check mark (press **F4**) by any variable, program, or other data item that you wish to manage; then select the proper option from the **F1 Manage** menu. This menu is commonly used to **Delete** variable names, **Rename** programs, etc. Pressing **F3** shows the menu options required to transfer data or programs from one calculator to another using the TI-89's link cable.

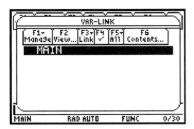  

Option 9 is **Unarchive Variable.**

The UNITS Menu

A menu of physical constants and units of measure can be found by pressing **2nd UNITS** (press **2nd** then the **3** key). Press the down arrow key to select a submenu, then press the right arrow key to open it. These submenus can be viewed a page at a time (as shown in the next figure) by pressing **2nd**, then down arrow.

2nd UNITS   **2nd** down arrow   **2nd** down arrow

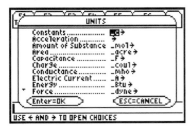  

The CHARacter Menu

Press **2nd CHAR** (**2nd,** then the plus sign) to reveal a menu that contains more than 90 special characters and mathematical symbols. They are organized into five submenus.

D.5   GRAPHING FUNCTIONS

The **green diamond** versions of the thin blue buttons along the top row of the calculator access most of its graphical features. The **green diamond Y=** key provides access to the calculator's list of 99 functions (assuming that the calculator's graph **MODE** is **Function**). Pressing **green diamond Y=** will most likely reveal functions y1 through y7. Press **2nd**, then the down or up arrow key to view the functions a page at a time. These functions are part of the calculator's memory, but the information stored on this screen can be easily edited, overwritten, or **CLEAR**ed. Functions are selected for graphing (turned "on") if they have a check mark immediately to the left of the function's name. The check mark appears automatically when you type in a new function or modify an old one. In other cases, to change the status of a function (from "off" to "on" or vice versa), you will need to use the up and down arrow keys to highlight the function, then press **F4** (the check mark). Unmarked functions are stored in memory, but will *not* be graphed. The default graphing style is **Line**, which means that the individually calculated points will be connected with thin line segments (giving an appearance of continuity). The optional graphing styles can be found by pressing **2nd F6**. These include options to plot only isolated **Dot**s or (larger) **Square**s for each of the calculated points, an option to plot a **Thick** continuous graph, an option that will **Animate** the curve as a path of motion and another that will mark that **Path** during the animation. There are also options to shade **Above** or **Below** the curve. To view the style setting of a particular function, highlight that function, then press **2nd F6** (**2nd**, then the **F1** key). Its current style setting is indicated by a check mark. Of course, you can also change the style setting in this menu by pressing the number of the appropriate option.

Pressing **green diamond WINDOW** allows you to *manually* specify the extents of the *x*- and *y*-values that will be visible on the calculator's graphing screen (see **ZOOM** for *automatic* ways of doing this). The **xscl** and **yscl** options specify the meaning of a mark on the *x*- or *y*-axis. For example **xscl=5** means that each mark shown on the *x*-axis will mean an increment of 5 units (**xscl=1** is a common setting for algebraic functions; **xscl=π/2** is commonly used when graphing trigonometric functions). The last option, **xres**, allows you to control how many points will actually be calculated when a graph is drawn. Setting **xres=1** means that an accurate point will be calculated for each pixel on the *x*-axis (slow, but very accurate). The default setting is **xres=2**, which will calculate a point at every other pixel. The fastest (but least accurate) setting, **xres=10**, calculates a point only at every tenth pixel on the *x*-axis. In the examples that follow, all graphs are shown with **xres=2** unless otherwise indicated.

To see how the graphing screen is presently formatted, press **green diamond Y=** or **green diamond GRAPH**, then press **F1** followed by **9**. This menu contains options that allow you to change the way coordinates are displayed (polar instead of rectangular, or off completely, inhibiting some **Trace** features), change the way that two or more graphs will be drawn (**Seq**uentially or **Simult**aneously), provide a coordinate grid, hide the axes, provide a graphing cursor, or label the axes. If you find your graphs looking cluttered or notice that axes or coordinates are missing, the default settings are shown in the next figure.

green diamond Y=   **F1   9**   (These are the default graph formats.)

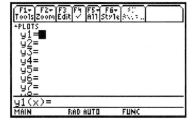  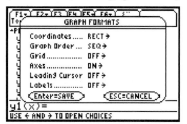

The **F2 Zoom** menu can be found on both the **Y=** editor and the graphing screen. It provides access to *automatic* ways to set the graphical viewing window. **ZoomStd** (option 6) is usually a good place to start, but you should consider option 7, **ZoomTrig,** if you're graphing trigonometric functions. **ZoomStd** shows the origin in the exact center of the screen with *x*- and *y*-values both ranging from -10 to 10. **ZoomStd** also sets **xscl=1, yscl=1,** and **xres=2.** From here you can **ZoomIn** or **ZoomOut** (options 2 and 3), or draw a box around a portion of the graph that you would like magnified to fit the entire screen (option 1, **ZoomBox**). There is also an option to "square up" your graph so that units along the *x*-axis are equal in length to units along the *y*-axis (option 5, **ZoomSqr**); the *smaller* unit length from the axes of the previous graph will now be used on both axes. This option makes the graph look more like it would on regular graph paper; for example, circles really look like circles. **ZoomDec** and **ZoomInt** (options 4 and 8) prepare the screen for **Traces,** which will utilize *x*-coordinates at exact tenths or integer values, respectively. Option 9, **ZoomData,** makes sure that all of the data in a statistical plot will fit in the viewing window. Option A, **ZoomFit,** calculates a viewing window using the present *x*-axis, but adjusting the *y*-axis so that the function fits neatly within the viewing window. All of these options work by making automatic changes to the **WINDOW** settings. Want to go back to the view you had before? The **Memory** submenu (**Zoom** option B) includes features to go back to your immediately previous view (**Memory** option 1, **ZoomPrev**) or to a window setting you saved a while ago (**Memory** option 3, **ZoomRcl**). ZoomSto (**Memory** option 2) is the way to save the current window setting for later (note that it can only retain one window setting, so the new information replaces whatever setting you had saved before). **Zoom** option C, **SetFactors...,** gives you the chance to control how dramatically your calculator will **ZoomIn** or **ZoomOut.** These zoom factors are set by Texas Instruments for a magnification ratio of 4 on each axis. Many people prefer smaller factors, such as 2 on each axis. It is possible to set either factor to any number greater than or equal to 1; they don't need to be whole numbers, and they don't necessarily have to be equal.

The **F3 Trace** feature is found only on the graphing screen. Its purpose is to display the *x*- and *y*-coordinates of specific points as you trace along a curve using the left and right arrow keys. Note that in **Trace,** the up and down arrow keys are used to jump from one curve to another when several curves have been drawn on the same screen. The subscript of the function you are presently tracing is shown in the upper right-hand corner of the screen. If you press **ENTER** while in **Trace,** the graph will be redrawn with the currently selected point in the exact center of the screen, even if that point is presently outside the current viewing window. This feature is a convenient way to pan up or down to see higher or lower portions of the graph. It is also the easiest way to locate a "lost" graph that doesn't appear anywhere in the current viewing window (just press **F3 Trace**, then **ENTER**). To pan left or right, just press and hold the left or right arrow key until new portions of the graph come into view. These useful features change the way the graph is centered on the screen without changing its magnification. Note also that these recentering features work *only* in **Trace**. To exit **Trace** without disturbing your view of the graph, just press **CLEAR** (or **green diamond GRAPH**). To return to the home screen, abandoning both **Trace** and the graph, just press **HOME** or **2nd QUIT** (or press **CLEAR** twice). Note that using any **F2 Zoom** feature also causes an exit from **Trace** (to resume tracing on the new zoomed version, you must press **F3 Trace** again).

Pressing **green diamond GRAPH** takes the calculator from any screen or menu to the current graph. Note that the calculator is smart enough that it will redraw the curves only if changes have been made to the function list (**green diamond Y=**). As previously mentioned, **green diamond GRAPH** can be used to turn off **Trace**. You can also hide an unwanted free cursor by pressing **green diamond GRAPH**. When you're finished with viewing a graph, press **HOME** or **CLEAR** or **2nd QUIT** to return to the home screen.

D.6   EXAMPLES OF GRAPHING

EXAMPLE 1

To graph $y = x^2 - 5x$, first press **green diamond Y=**, then highlight y1 (using the up arrow key, if necessary) and press **CLEAR** to erase its current formula (or use the down arrow key to find a blank function), then press **ENTER** or **F3** (**Edit**), which activates the editor line near the bottom of the screen. Type the function by pressing the **X** key, followed by the ^ button and **2**; now press the (black) minus sign key, then **5**, followed immediately by the **X** key (a multiplication sign is not needed). Your screen should look very much like the first frame shown in the following figure. Note that you are not required to press **ENTER** when you have finished typing a function's formula. To set up a good graphing window, press **F2** (**Zoom**) then **6** to choose **ZoomStd**. This causes the graph to be immediately drawn on axes that range from -10 to 10. Notice that you did not have to press **green diamond GRAPH**; the **F2 Zoom** menu items also activate the graphing screen. If you have one or more unwanted graphs drawn on top of this one, go back to your function list (**Y=**) and turn "off" the unwanted functions by highlighting them one at a time and pressing **F4** to remove their check mark. After you have turned off the unwanted functions, just press **green diamond GRAPH** and you will finally see the last screen below.

green diamond Y=   (typing...)   **F2   6**

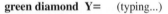
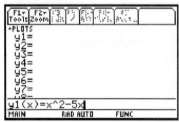      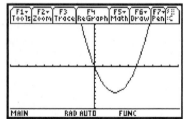

EXAMPLE 2

To modify this function to be $y = -x^2 + 6$, press **green diamond Y=**, then **ENTER** or **F3** (**Edit**). Insert the negative sign by pressing the left arrow key (which takes you to the left-hand edge of the formula), then the gray sign change key (-); now press the right arrow key three times to skip over the parts of the formula that are to be preserved. Type the plus sign and the **6**, then press **CLEAR** (only once) to delete the unwanted right-hand end of the formula. Press **green**

diamond GRAPH to plot this function, then press **F3** (**Trace**). **Trace** gives the added bonus of a highlighted point, with its coordinates shown at the bottom of the screen. Press the right or left arrow keys to highlight other points on the curve.

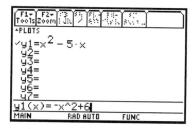 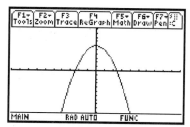 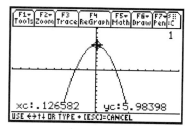

Perhaps $x = .126582$, $y = 5.98398$ was not a coordinate pair that you had expected to investigate. Two special **Zoom** features (options 4 and 8) can be used to make the **Trace** option more predictable. Press **F2 Zoom**, then **4** to select **ZoomDec**; now press **F3 Trace**.

F2   4   **F3**

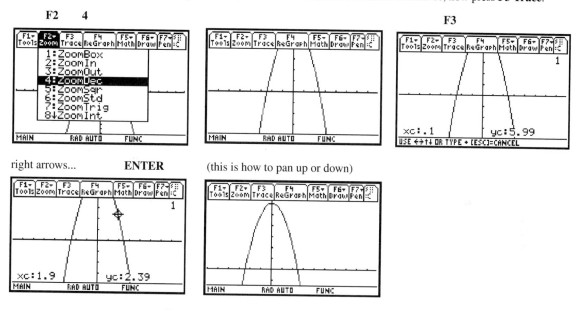

right arrows...   **ENTER**   (this is how to pan up or down)

Try pressing the right or left arrow key about 10 times watching the values at the bottom of the screen. You'll quickly notice that the x-values are now all *exact tenths*, .1, .3, .5, etc. (**Zoom** option 8, **ZoomInt**, produces **Trace**able x-values which are all integers). If you want to see each and every decimal x-value (0, .1, .2, .3, etc.), press **green diamond WINDOW** and set **xres=1**. However, the default value, **xres=2**, will be restored whenever you use **ZoomStd**. Another nice thing about **ZoomDec** is that the graph is "square" in the sense that units on the x- and y-axes have the same length. The main disadvantage to **ZoomDec** is that the graphing window is "small," displaying only points with x-values between -7.9 and 7.9 and y-values between -3.8 and 3.8. This disadvantage is apparent on the current graph, which runs off the top of the screen. To demonstrate how this problem can be overcome, **Trace** the graph to the point $x = 1.9$, $y = 2.39$ and then press **ENTER**. This special feature of **Trace** causes the graph to be redrawn with the highlighted point in the exact center of the screen (with no change in the magnification of the graph). This is the way to pan up or down from the current viewing window (to pan left or right, see Example 6).

EXAMPLE 3

Another way to deal with the preceding problem is to **ZoomOut**, but first you'll want to set your ZOOM FACTORS to 2 (the factory setting is 4). Press **F2** (**Zoom**), then **alpha C**. Now press **2**, down arrow, **2**, down arrow, **2**, then press **ENTER** twice.

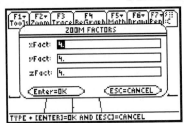
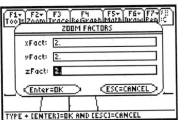

To reproduce our problem, press **F2**, then **4** (**ZoomDec**). However, this time correct it by pressing **F2** followed by **3** (**ZoomOut**). At first glance, it looks like nothing has happened, except that $xc:0$, $yc:0$, and the question "New Center?" are displayed near the bottom of the screen. The calculator is waiting for you to use your arrow keys to locate the point in the current window where you would like the exact center of the new graph to be (then press **ENTER**). Of course, if you like the way the graph is already centered, you will still have to press **ENTER** (you'll just skip pressing the arrow keys). The resulting graph is drawn on a window whose x- and y-axes represent lengths twice as long (remember the ZOOM FACTORS are set to 2) as the original axes. The scaling marks on each axis also represent twice what they originally did (**xscl** and **yscl** are now both equal to 2 instead of 1). Press **F3** (**Trace**) and notice, by pressing the left or right arrow key a few times, that the x-values are now changing by .4 (instead of .2) and the graph is still "square." Other popular square window settings can be obtained by repeating this example with both zoom factors set to 2.5 or 5. These give "larger" windows where the x-values can be changed by .25 or .5, respectively, during a **Trace** (if you also change the **WINDOW** options to **xres=1**).

F2　　4　　　　　　　(the problem graph)　　　　　　**F2　　3**

ENTER　　　　　　　　(the new graph)　　　　　　**F3**　　right arrows...

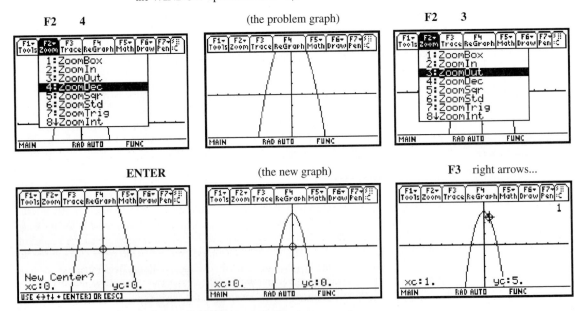

EXAMPLE 4

The extra keystroke needed to activate **ZoomOut** in the previous example has proven to be a bit confusing to beginners who think that **ZoomIn**, **ZoomOut**, and **ZoomInt** should work like the six zoom options (4, 5, 6, 7, 9, and A) that do their job without pressing **ENTER**. The only other zoom option that needs extra keystrokes is **ZoomBox** (option 1). This is a very powerful option that lets you draw a box around a part of the graph that you would like enlarged to fit the entire screen. After selecting this option, use the arrow keys to locate the position of one corner of the box and press **ENTER**. Now use the arrow keys to locate the *opposite* corner, and press **ENTER** again.

F2   1   left arrows and up arrows (to position one corner of the box)   **ENTER**

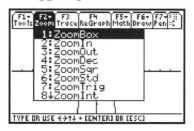

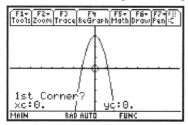

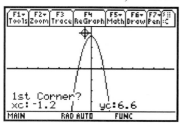

(the placement of one corner)   right and down arrows (opposite corner)   **ENTER**

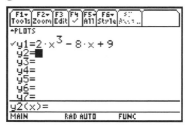

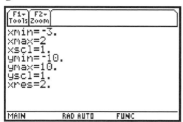

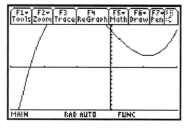

EXAMPLE 5

Sometimes, you will know the precise interval on the *x*-axis (the domain) that you want for a graph, but a corresponding interval for the *y*-values (the range) may not be obvious. **ZoomFit** (**Zoom** option A) is designed for this circumstance. For example, to graph $f(x) = 2x^3 - 8x + 9$ on the interval $[-3, 2]$, manually set the **WINDOW** so that **xmin=−3** and **xmax=2** (the other values shown in the second frame are just leftovers from **ZoomStd**). Now press **F2** (**Zoom**), then **alpha A** to select **ZoomFit**. There is a noticeable pause while appropriate values of **ymin** and **ymax** are calculated, then the graph is drawn. To view the values calculated for the range, just press **green diamond WINDOW**.

green diamond Y=   **green diamond WINDOW**   **green diamond GRAPH**

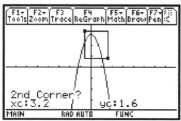

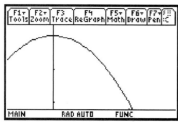

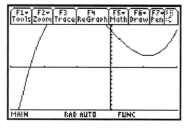

F2   alpha A   (the new graph)   **green diamond WINDOW**

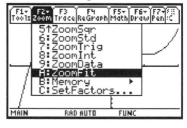

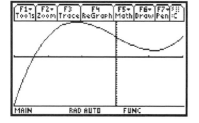

Note that **ZoomFit** changes only **ymin** and **ymax**. You may wish to change **yscl**. Note also that the maximum value of this function on $[-3, 2]$ is approximately 15.158 and that its minimum value appears to be -21. By accessing this information from the final **WINDOW** settings, **ZoomFit** can be used to find the range of a function on an interval.

EXAMPLE 6

Panning to the right is done by tracing a curve off the right-hand edge of the screen.

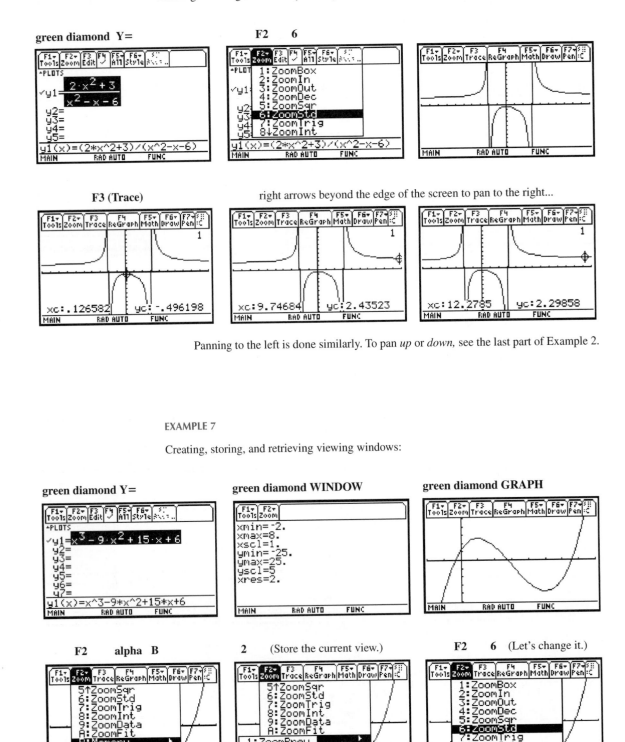

green diamond Y=

F2   6

F3 (Trace)

right arrows beyond the edge of the screen to pan to the right...

Panning to the left is done similarly. To pan *up* or *down,* see the last part of Example 2.

EXAMPLE 7

Creating, storing, and retrieving viewing windows:

green diamond Y=

green diamond WINDOW

green diamond GRAPH

F2   alpha B

2   (Store the current view.)

F2   6   (Let's change it.)

F3 (Trace)   ENTER   (We've now made several changes.)   **F2   alpha   B   3**

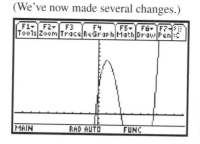
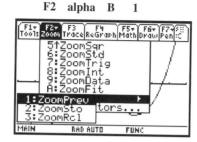
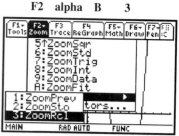

(The stored view has been recalled.)   **F2   alpha   B   1**   (the immediately previous view)

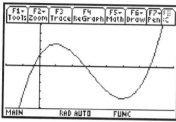
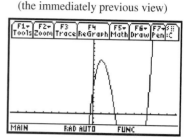
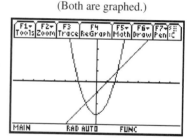

EXAMPLE 8

Graphing and tracing more than one function:

green diamond Y=   **F2   4**   (Both are graphed.)

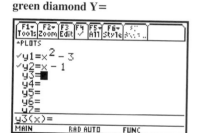
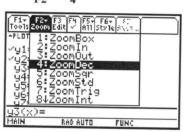
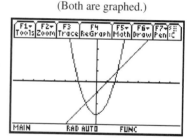

F3 (Trace)   right arrows... (**Trac**ing y1)   down arrow (to jump to y2)

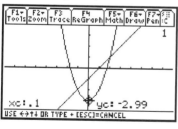
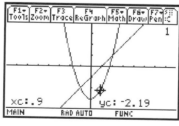
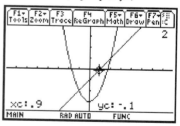

left arrows... (**Trac**ing y2)   up arrow (to jump to y1)

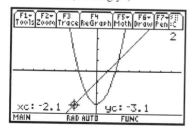
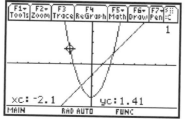

Note that the subscript of the function being investigated is shown in the upper right-hand corner of the **Trace** screens. Pressing the down arrow increases this subscript; pressing the up arrow decreases it. Both of these arrow keys will wrap around and start over. The result has nothing to do with which graph lies above or below the other.

EXAMPLE 9

Finding a point of intersection:

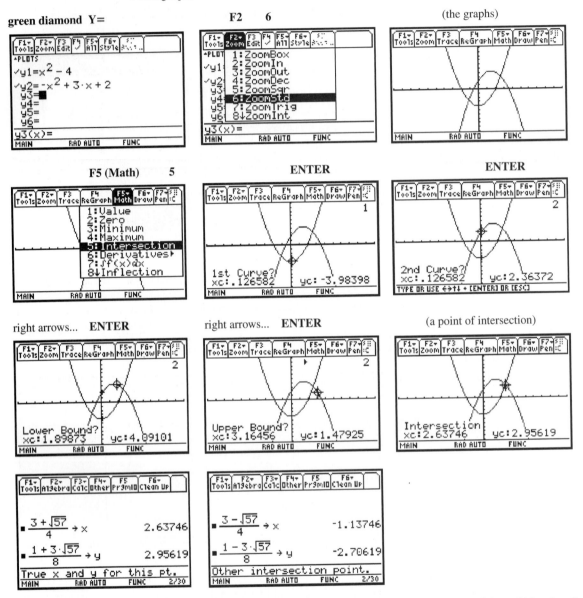

green diamond Y=

F2   6

(the graphs)

F5 (Math)   5

ENTER

ENTER

right arrows... **ENTER**

right arrows... **ENTER**

(a point of intersection)

Note that the upper and lower bounds were very important in specifying *which* point of intersection would be calculated. You should now find the other point using **Intersection**.

EXAMPLE 10

Calculating *y*-values and locating the resulting points on your graphs:
If you have just completed Example 9, please skip to the third frame.

green diamond Y= F2 6   (the graphs)   F5 (Math) 1

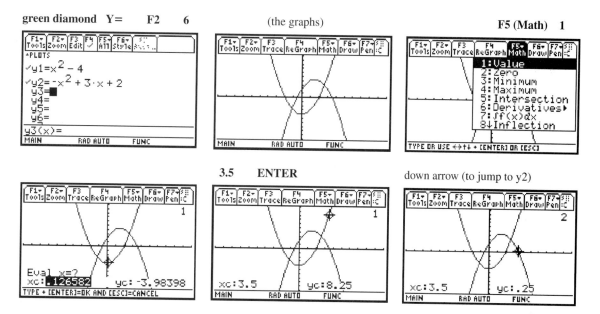

3.5 ENTER   down arrow (to jump to y2)

Note again that the subscript of a function (shown in the upper right-hand corner) increases as you jump from one curve to another by pressing the down arrow key. The up arrow key decreases this number; and either key can be used to wrap around and start over.

EXAMPLE 11

Another way to calculate a specific value of a function is to press **F3** (**Trace**), then type the *x*-value and press **ENTER**. The main difference between the **Value** feature (see Example 10) and this special **Trace** feature is that **Trace** does *not* preserve the entered *x*-value if you jump from one curve to another (you would usually need to retype that *x*-value). The following frames assume that Example 10 has just been completed.

F3 (Trace)   1.6   ENTER

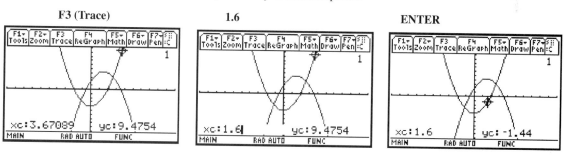

EXAMPLE 12

Using **Trace** to quickly locate a "lost" graph that doesn't appear anywhere in the current viewing window:

green diamond Y= x^2 + 11 **ENTER** F2 6 (**ZoomStd**) *no visible graph*   F3 (Trace)   ENTER   F3 (Trace)

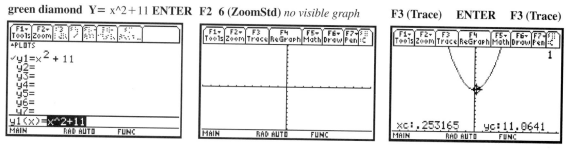

D.7 TRIG FUNCTIONS AND POLAR COORDINATES

EXAMPLE 1

Evaluating trigonometric functions using degrees, minutes, and seconds:

MODE three down arrows right arrow **2 ENTER**

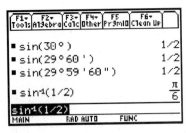  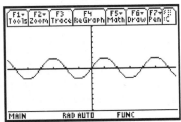

Note that degrees (**2nd**, then the I key), minutes (**2nd** then the = key), and seconds (**2nd** then the 1 key) can easily be specified, even when the calculator is in **RADIAN MODE** (indicated by **RAD** in the status line). Comparing the first and third frames shows that the main difference is in the evaluation of an *inverse* trig function, as shown in the fourth calculations. Note also that a degree symbol is optional if the calculator is in **DEGREE MODE** (indicated by **DEG** in the status line) and no minutes or seconds are shown.

EXAMPLE 2

To graph $y = \sin x$, first press **green diamond Y=**, then press **CLEAR** to erase the current formula for y1 (or use the arrow keys to find a blank function), then press **2nd SIN** (**2nd** then the **Y** key) followed by **X**, and finally press the right parenthesis key. Your screen should look very much like the first one in the following figure. To set up a good graphing window, press **F2** and then press **7** to choose **ZoomTrig**. This causes the graph to be immediately drawn on axes that range from $-79\pi/24$ to $79\pi/24$ ($-592.5°$ to $592.5°$) in the x-direction and from -4 to 4 in the y-direction. Each mark along the x-axis represents a multiple of $\pi/2$ radians ($90°$), and each mark along the y-axis represents 0.5. The **ZoomTrig** feature is especially designed to show the same graph whether the calculator is set for **RADIAN**s or **DEGREE**s. Other automatic ways to set the viewing window (such as **ZoomStd** and **ZoomDec**) ignore the **MODE Angle** setting and should *not* be used to graph trigonometric functions in **DEGREE**s. If you have one or more unwanted graphs drawn on top of this one, go back to your function list (**green diamond Y=**) and turn "off" the unwanted functions by placing the cursor over their formulas and pressing **F4** (to erase their check mark). After you have turned off the unwanted functions, just press **green diamond GRAPH** and you will finally see the last frame in the following figure.

green diamond Y=   **F2**   **7**   (the graph)

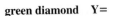
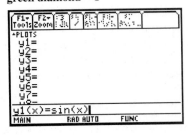 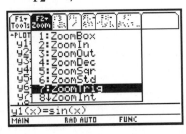 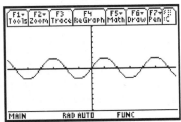

EXAMPLE 3

To modify this function to be $y = -3 \sin 2x$, press **green diamond Y=**, then **ENTER** or **F3** (to bring the current y1 to the edit line). Press the left arrow key (to start editing from the left) and type -3; now press the right arrow key four times (to skip over the part of the formula that's

OK), then press **2**. Press **F2**, then **7** (**ZoomTrig**), and **F3** (**Trace**) to graph and begin tracing this function. **Trace** gives the added bonus of a highlighted point with its coordinates shown at the bottom of the screen. Press the right or left arrow keys to highlight other points on the curve. For best results when using **Trace** with **ZoomTrig**, set **xres=1** on the **green diamond WINDOW** screen. This allows you to see a calculated point for all multiples of $\pi/24$ (7.5°), shown in decimal form. Of course, this includes all special angles values such as 0, $\pi/6$, $\pi/4$, $\pi/3$, $\pi/2$, etc. (in degrees: 0, 30, 45, 60, 90, etc.). Be careful to remember that the default, **xres=2**, is restored whenever **ZoomStd** is used. Therefore, switching back and forth between **ZoomTrig** and **ZoomStd** is not highly recommended.

green diamond Y=

Note that the old function **sin(x)** is still shown as y1 while it is being modified.

green diamond WINDOW

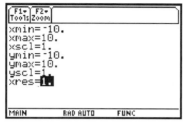

Set **xres=1** for best results with **Trace**. All other settings will be changed by **ZoomTrig**.

F2   7

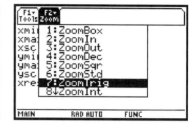

(the graph)

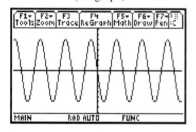

F3 (Trace)

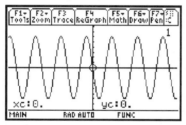

right arrows... $[(5\pi/12, -1.5)]$

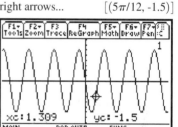

MODE three down arrows right arrow **2 ENTER   F2   7**

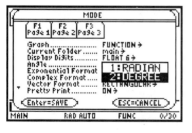

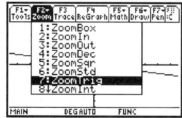

F3 right arrows... $[(75°, -1.5)]$

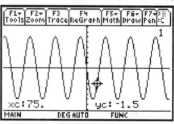

Switching to **DEGREE MODE** initially distorts the graph (note the graph in the background of the middle frame), but **ZoomTrig** restores its appearance. The last frame shows a **DEGREE MODE Trace**.

EXAMPLE 4

Several related trig functions can be drawn on the same screen, either by typing them separately in the function list **green diamond Y=** or by using a list of coefficients as shown in the following figure. A list consists of numbers separated by commas, which are enclosed by braces { }. The braces are the shifted **2nd** versions of the parentheses keys. The first row of frames indicates how to efficiently graph $y = \sin(x)$, $y = 2\sin(x)$, and $y = 4\sin(x)$ on the same screen. The last row graphs $y = 2\sin(x)$ and $y = 2\sin(3x)$.

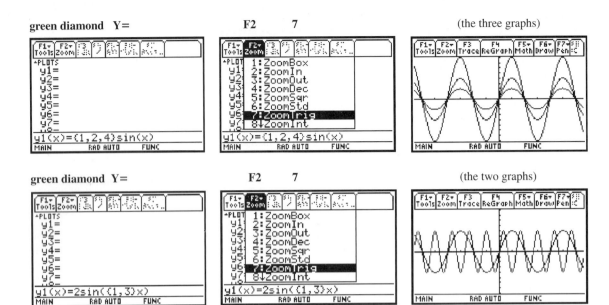

green diamond Y=   **F2   7**   (the three graphs)

green diamond Y=   **F2   7**   (the two graphs)

EXAMPLE 5

Multiple lists are allowed but are not highly recommended. To graph the two functions $y = 2 \sin 3x$ and $y = 4 \sin x$, you could do what's shown in the first frame or type them separately as shown in the third frame.

green diamond Y=   **green diamond GRAPH**   **green diamond Y=**

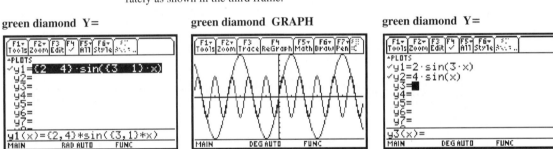

EXAMPLE 6

When graphing trig functions that have vertical asymptotes, remember that your calculator just evaluates individual points and arbitrarily assumes that it should connect those points if the function's graph style is **Line, Thick,** or **Path**. The effect is shown in the following graph of $y = \sec x$. Some people like these "vertical asymptotes" being shown on the graph (they are really just nearly vertical lines which are trying to connect two points on the curve). The last frames show the same graph in **Dot** style.

green diamond Y= 2nd F6 (Style)   **green diamond WINDOW**   **green diamond GRAPH**

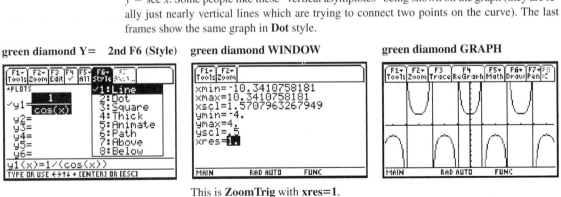

This is **ZoomTrig** with **xres=1**.

green diamond Y= 2nd F6 2

green diamond GRAPH

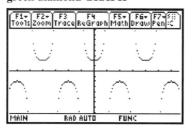

EXAMPLE 7

To graph in polar coordinates, the calculator's **Graph MODE** must be changed from **FUNCTION** to **POLAR**. You may also wish to change your **Coordinates GRAPH FORMAT** from **RECT**angular to **POLAR**, which will show values of r and θ (instead of x and y) when you **Trace** your polar graphs. Note that the calculator treats these as two completely separate issues (it is possible to graph in one coordinate system and trace in the other). Pressing **green diamond Y=** will reveal the calculator's 99 polar functions r1, r2, ..., r99. It is important to note that θ is the **green diamond** version of the ^ key and that θ must be used as the independent variable for all **POLAR** graphs.

MODE   right arrow **3**   **ENTER**   **green diamond Y= F1 9** right arrow **2 ENTER**   etc.

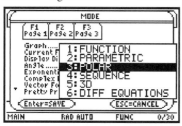

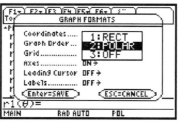

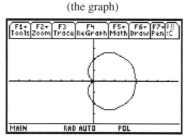

F2   4   (the graph)   **F3 (Trace)**   right arrows

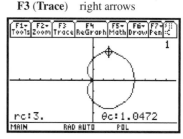

Pressing **green diamond WINDOW** reveals that the standard radian values of θmin, θmax, and θstep are 0, 2π, and $\pi/24$, respectively. A more accurate, smoother graph can be obtained by using θstep $= \pi/48$ or $\pi/96$. Also note that the *right* arrow key is used to **Trace** in the standard, counterclockwise direction.

EXAMPLE 8

Graphing polar equations in **DEGREE MODE**:

MODE down arrows right arrow **2 ENTER**   **green diamond Y=**   **F2   6**

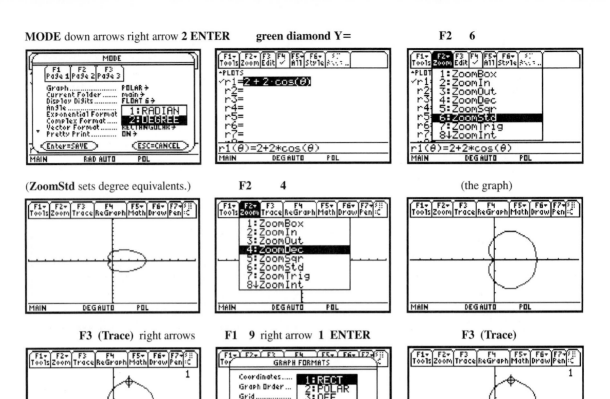

(**ZoomStd** sets degree equivalents.)   **F2   4**   (the graph)

F3 (Trace) right arrows   **F1   9** right arrow **1 ENTER**   **F3 (Trace)**

ZoomStd sets the standard degree values of θmin, θmax, and θstep, which are 0, 360, and 7.5, respectively. Smoother (but slower) graphs can be obtained by using smaller values of θstep (such as θstep = 2). The last frame shows a polar graph traced in **RECT**angular **Coordinates** GRAPH FORMAT.

D.8   NUMERICAL GRAPH AND TABLE FEATURES

EXAMPLE 1

To numerically estimate solutions of equations on the graphing screen:

(a) Rewrite the equation on paper in the form $f(x) = 0$; for example, rewrite

$$x^3 + 15x = 9x^2 - 6$$

as

$$x^3 - 9x^2 + 15x + 6 = 0$$

(b) Enter this function in the **green diamond Y=** menu, and make sure that it is the only one selected for graphing.

(c) Graph it using an appropriate viewing **WINDOW** or **Zoom** option (**F2**). The zeros being estimated must appear in the viewing window.

(d) Press **F5 (Math)** and choose option 2, **Zero**.

(e) The prompt "Lower Bound?" is asking for an x-value to the *left* of the desired zero. Either type a value from the keyboard or arrow over to it, then press **ENTER**. Likewise, "Upper Bound?" is asking for an x-value to the *right* of the desired solution. Pressing **ENTER** then produces an estimated zero in the specified interval and also shows the y-value of the point as a check (it should be zero or extremely close to zero, such as 1.E-12, which means 10^{-12}).

green diamond Y=

green diamond WINDOW

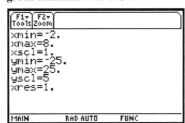

green diamond GRAPH   F5   2

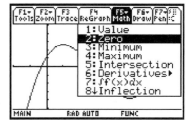

left arrows...   ENTER

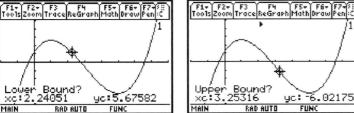

right arrows...   ENTER

(a solution)

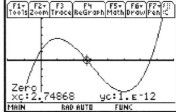

EXAMPLE 2

The functions you expect to investigate using a **TABLE** must be turned on (indicated by a check mark) in the same sense as those you want to graph. To get to the TABLE SETUP screen, press **green diamond TblSet** (**green diamond**, then the **F4** key). The specification for **tblStart** is just a beginning *x*-value for the table; you can scroll up or down using the arrow keys. Δ**tbl** is the incremental change in *x*. You can give Δ**tbl** the value 1 (as in the following frames) to calculate the function at consecutive integers; in calculus, you could use .001 to investigate what is happening as a function approaches a limit; or you could use a larger number like 10 or 100 to study the function's numerical behavior as *x* goes to ∞.

green diamond Y=

Even y5 is active for this **TABLE**, since it is defined on the edit line.

green diamond TblSet

These standard values start the **TABLE** at 0 and change by 1.

green diamond TABLE

up arrow

down arrows   right arrows

two more right arrows (to reveal y5)

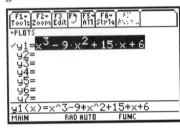

EXAMPLE 3

Finding local maximum (and minimum) values on the graphing screen:

green diamond   Y=   **green diamond   WINDOW**   **green diamond GRAPH   F5   4**

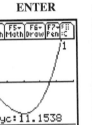
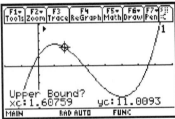
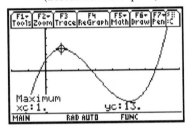

left arrows…   **ENTER**   right arrows…   **ENTER**   (a local maximum point)

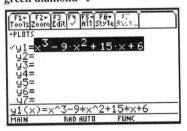
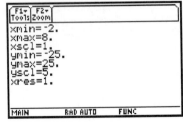
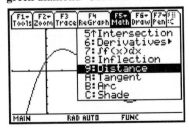

The only intended use of these graphical maximum and minimum features is to find *local* extreme values using a fairly short bracketing interval. These routines can give unpredictable results if you try to use them to compare local extrema with endpoint extrema on a given interval. **ZoomFit** and the home screen features **f Min** and **f Max** (options 6 and 7 in the **F3 Calculus** menu) are more suitable features if your problem is determining a function's overall maximum and minimum values on a specified interval.

EXAMPLE 4

Find the (straight line) distance between two points on the same curve:

green diamond   Y=   **green diamond   WINDOW**   **green diamond GRAPH   F5   9**

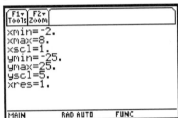
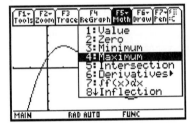

arrows... **ENTER** (the first point)

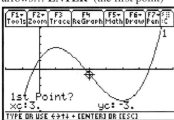

arrows... **ENTER** (the second point)

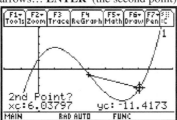

(the distance)

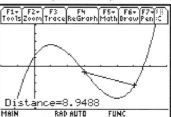

D.9   SEQUENCES AND SERIES

EXAMPLE 1

On the TI-89, a sequence can be created and stored as a list. To find **seq(** in the menus, press **2nd MATH 3** (**List**) then **1** or use the **CATALOG**. You may find it even easier just to type **seq(** as separate characters on the entry line (this works for any feature whose name you can remember).

2nd MATH   3

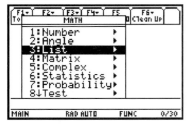

1

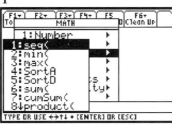

(the squares from 1 to 5)

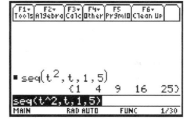

EXAMPLE 2

Evaluate the sum $7 + 9 + 11 + 13 + 15 + \cdots + 121$.

F3   4

(There are many different ways in which this sum can be expressed.)

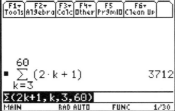

EXAMPLE 3

Estimate the infinite geometric series $1 + \dfrac{2}{3} + \dfrac{4}{9} + \dfrac{8}{27} + \cdots$

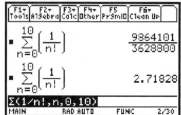

The sum of the first 21 terms is given as both a fraction and a decimal.

The exact infinite sum is given by the formula $a_0/(1 - r)$, where $a_0 = 1$ and $r = \frac{2}{3}$.

EXAMPLE 4

Find the value of the infinite series $1/0! + 1/1! + 1/2! + 1/3! + 1/4! + \cdots$

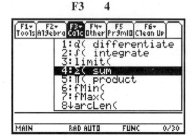

The factorial symbol is **green diamond**, then the ÷ key.

D.10 THE NUMERIC SOLVER

The **Numeric Solver** feature (press **APPS**, then **9**) is composed of two specialized screens. One is an equation editor, which shows **eqn:** on the screen, waiting for you to write the equation or formula you want to investigate. To use the formula for the total surface area of a cylinder, $A = 2\pi r(r + h)$, remember to place a times sign before the parentheses (otherwise, the TI-89 will see function notation instead of multiplication). Whether or not you capitalize the A makes no difference to the TI-89 (it's *not* case sensitive). Pressing **ENTER** takes you to the second screen, where you can enter values and solve for a variable. Enter a value for each variable name except the unknown variable you intend to solve for. If there is more than one possible solution for that variable, you can control which one will be found by typing an estimate of it. Use the up and down arrow keys to place the cursor on the line that contains the unknown variable and press **F2 (Solve)**. The solution is marked by a small square to its left. Also marked (at the bottom of the screen) is a "check" of this solution, indicating the difference between the left- and right-hand sides of the equation using all of the values shown (this should be zero or something very close to zero, such as 1.E-12). Fourteen significant digits are calculated and displayed for the solution variable.

The **bound=** option near the bottom of the screen allows you to specify an interval to be searched for the solution. The default interval {−1.E14,1.E14} is restored whenever you enter the **Numeric Solver** (any **bound** interval you specify is valid only for that session). When solving trigonometric equations, a limited bound like the interval {0,2π} might be better; but in most cases, the default interval works well. If you accidentally erase this line and can't remember the syntax, just exit the **Numeric Solver** by pressing **HOME**; then press **APPS 9** to reenter it (restoring the default **bound** in the process). The equation stays in memory until you **CLEAR** it or replace it with another. Press the up arrow until the top line is reached. The calculator will switch immediately to the equation editor page, where you can modify or **CLEAR** the old equation, or even replace it with a previous equation from the **F5 Eqns** list. The following example assumes that you are starting with a blank equation (as if **Solver** has never been used before); you may **CLEAR** the present equation to achieve the same effect.

APPS   9   (the equation editor)   **ENTER**   (the solving screen)

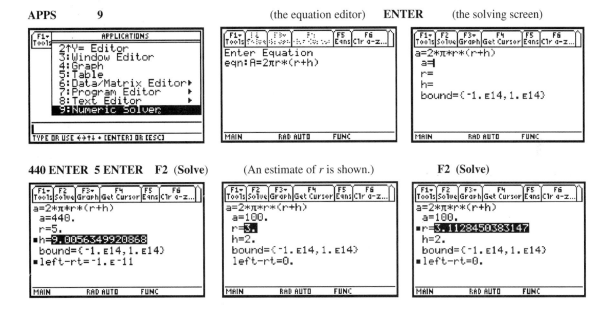

440 ENTER 5 ENTER   F2 (Solve)   (An estimate of *r* is shown.)   **F2 (Solve)**

The first frame in the bottom row of the previous example found the height of a cylinder whose total surface area is 440 square units and whose radius is 5 (the answer is about 9). The last frame found the radius of a cylinder whose total surface area is 100 square units and whose height is 2. Note that the estimate $r = 3$ determined which of the two solutions would be found (the other is negative and not physically important).

The **Numeric Solver** can also be used on equations which contain only one variable, but think about the previous example (which had several variables) to understand why the **Numeric Solver** works as it does. In particular, it would seem natural to press **ENTER** after typing an estimate for the solution variable; instead, you must press **F2** (**Solve**). If you accidentally press **ENTER** first, you will have to press the up arrow key to get back to the solution variable's line (if the cursor is on **bound=** rather than a variable's line, pressing **F2** (**Solve**) does *nothing*). For direct comparison, the next example solves the same equation that was solved on the graphics screen in Section D.8. Estimates of the solutions are critical in determining which of the three solutions will be found. Solve: $x^3 + 15x = 9x^2 - 6$.

APPS   9   (typing)   **ENTER**   **3   F2**   (a solution)

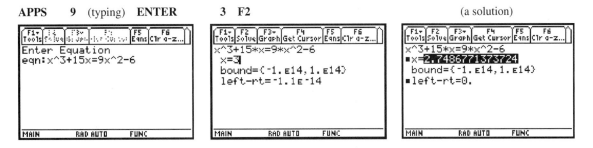

7   F2   (another solution)   **-1   F2** (the third solution)

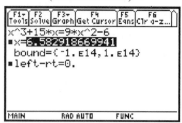

D.11   MATRIX FEATURES

Matrices on the TI-89 can easily be entered on the home screen. The entire matrix should be enclosed in brackets [ ], with rows separated by semicolons, and entries within a row separated by commas. The brackets are the **2nd** versions of the comma and ÷ keys. The semicolon is the **2nd** version of the **9** key. The effect is shown in the following frames.

Matrix multiplication example.   **2nd MATH   4**   **2**

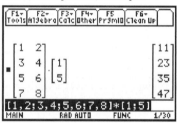
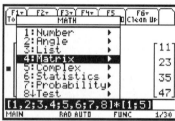
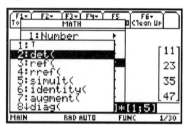

Determinant example.   **2nd MATH   4   4**   Row reduction example.

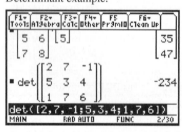
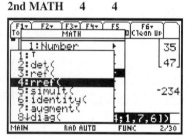
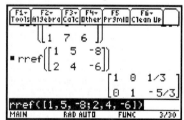

Naming and storing a matrix.   **APPS   6**   (the Matrix Editor)   **2**   (Open the Matrix m1.)

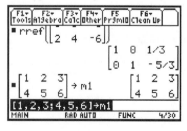

right arrow **2**

(Select m1.) **ENTER ENTER**

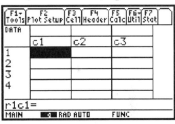

Once in the Matrix Editor, use the arrow keys to highlight any entries you may want to change and type over them. Pressing **ENTER** takes you to the next entry on the right. When the end of a row is reached, **ENTER** puts you at the beginning of the next row. The effect is very nice when you are entering a new matrix, which can thus be entered row by row by pressing **ENTER** after each matrix entry. Press **2nd QUIT** or **HOME** to return to the home screen after you finish editing a matrix.

D.12 THE DATA EDITOR AND DESCRIPTIVE STATISTICS

APPS 6 (Data/Matrix Editor) **3**

(Name the Data Variable "**rbi**.")

(the Data Editor) **alpha**

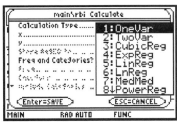

(Press **ENTER** after each entry.)

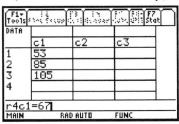

(The bottom of the data.)

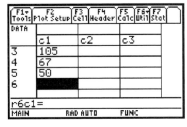

F5 (**Calc**) right arrow **1**

(Enter **c1** as the *x*-variable, *not* **rbi**.) **ENTER**

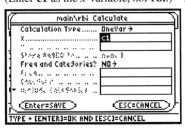

down arrows

Note that the *x*-variable required in the **F5** (**Calc**) menu was the column heading, **c1** (not the data variable name **rbi**). A more descriptive title can be given to this column by pressing the up arrow key to the blank space above **c1**; but even if you enter such a title, use **c1** as the variable name in the **F5 (Calc)** menu. Data variables can have as many as 999 rows.

D.13 THE LINE OF BEST FIT (LINEAR REGRESSION)

EXAMPLE

Find and graph the equation of the line of best fit for the following data:

x	5	7	9	12	14
y	40	58	62	74	80

APPS   6 (Data Editor) **3**

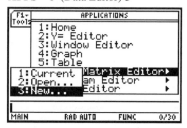

(Create the Data Variable **lin**.)

Top of the data entered.

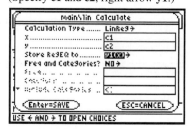

Bottom of the data entered.

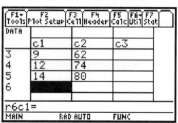

F5 (Calc) right arrow **5**

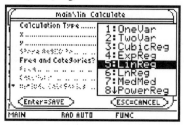

(Specify **c1** and **c2**, right arrow **y1**.)

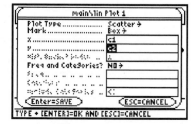

ENTER

green diamond Y= up arrow **ENTER**

(Specify **c1** and **c2**.)   **ENTER**

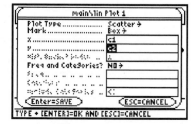

(Statistical Plot 1 is now defined.)

F2   9

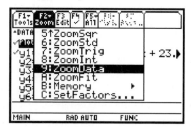

The data and the regression line.

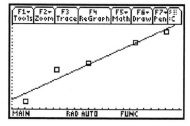

D.14 SYMBOLIC ALGEBRA FEATURES

The symbolic algebra features of the TI-89 are all found in the **F2 Algebra** menu of the home screen. If the menu **F2 f(x)** appears instead, just press **2nd CUSTOM** to restore the standard home screen menus.

Factoring Whole Numbers, Polynomials, and Fractions

The **factor** feature can factor whole numbers, fractions, polynomials, and rational expressions. If you just use **factor** followed by an expression in parentheses, the feature tries to factor using only integer or rational coefficients. In the case of fractions or rational expressions, the *simplified* versions of the numerator and denominator are shown in factored form.

Press **F2**, then **2**; finish by typing **392200)** and pressing **ENTER**.

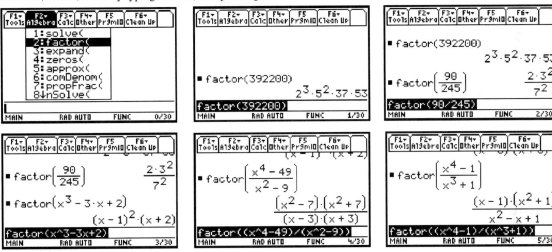

By typing an additional comma and the variable name at the end of the expression, factorizations involving irrational numbers are obtained. Complex factors can be found using **cFactor** instead of **factor** (just type a **c** in front of **factor** or press **F2** then **alpha A** then **2**). This means that there are actually four levels of factoring on a TI-89: **factor** and **cFactor**, with and without the variable name specified (real factorizations and complex factorizations with and without irrational numbers used in the coefficients).

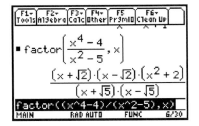

Solving Equations

Equations can be solved exactly for their real solutions using the **solve** or **zeros** features. **Zeros** assumes that the expression is set equal to zero and returns a list of solutions. **Solve** requires you to type the entire equation and returns the solutions in an algebraic format.

Both features require a comma and the variable's name near the end of the command. The commands **cSolve** and **cZeros** will find all real and complex solutions (press **F2**, then **alpha A** to see these options).

Press **F2**, then **4**; finish by typing **x^3−3x+2,x**) and pressing **ENTER**.

 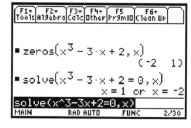 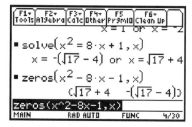

To restrict the domain of an equation, use the "with" operator (|) located just to the left of the **7** key. In the first frame below, the restriction that y is negative eliminates the well-known solutions 2 and 4. Note that decimal estimates of solutions are computed if no exact form can be found. In the second frame, the trig equation has infinitely many solutions, which are all odd multiples of π. Arbitrary integer constants are shown as @n1, @n2, etc. In the last frame, **nSolve** is illustrated. It approximates solutions one at a time, using an estimated x-value to control which solution will be approximated.

(Restricting an equation's domain.)   (Infinitely many periodic solutions.)   **F2 alpha A 1**

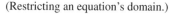
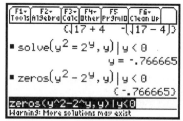

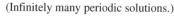
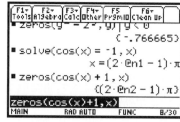

(Complex-valued solutions.)   **F2   8**   (Numerically estimated solutions.)

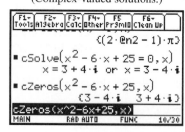 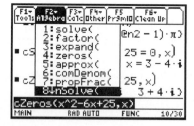 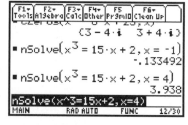

Systems of Equations

To solve systems of equations, use the solve feature with the **and** operator placed between the equations and a list of variables at the end of the command. There is a template for this purpose in the custom menu (press **2nd CUSTOM F3 4**). When you are finished with this section press **2nd CUSTOM** again to return to the standard menu. Returning to the **HOME** screen from a specialized screen also restores the standard menu.

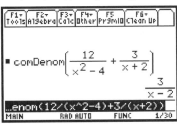
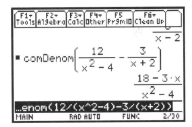

Combining Expressions into a Single Fraction

The feature, which adds, subtracts, and otherwise combines several rational expressions into a single, simplified fraction, is called **comDenom**. This feature is *not* for computing a common denominator or least common multiple. The result is always a simplified fraction in expanded form. If you prefer a factored form, just use the **factor** feature in conjunction with **comDenom: factor(comDenom(***expression***))**.

Press **F2**, then **6**; finish by typing **12/(x^2−4)+3/(x+2))** and pressing **ENTER**.

Products and Partial Fraction Expansions

The **expand** feature multiplies out indicated products and powers. For rational expressions, **expand** produces a partial fraction expansion.

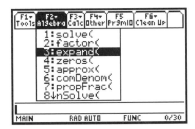
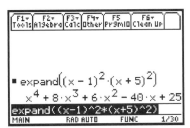

Polynomial Division

The feature that performs polynomial division is called **propFrac**. It can also be used to find a mixed number representation for improper numeric fractions.

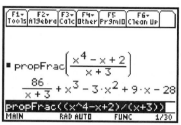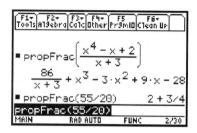

Expanding or Simplifying Trigonometric Expressions

To expand a multiple-angle trig formula in powers of sin (*x*) and cos (x), use the **tExpand** feature. The **tCollect** feature is the opposite of **tExpand**, taking an expression involving powers of trig functions and returning a multiple-angle equivalent.

Press **F2**, then **9**, then **1**; finish by typing **sin(3x))** and pressing **ENTER**.

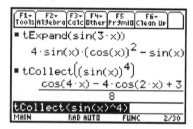

D.15 BASIC CALCULUS FEATURES

The calculus features of the TI-89 can be found in the **F3 (Calc)** menu. Note that the differentiation operator and the integral sign are both accessible from the TI-89 keyboard as the **2nd** shifted versions of the **8** and **7** keys, respectively.

Limits

The **limit** feature requires the name of the function, the name of the variable, and the value that the variable is approaching (in that order). An optional fourth parameter allows you to specify one-sided limits. Just use any *positive* number to indicate a right-hand limit or any *negative* number to indicate a left-hand limit. Entering *zero* or no fourth parameter at all means an ordinary (two-sided) limit.

Press **F3**, then **3**; finish by typing **sin(h)/h,h,0)** and pressing **ENTER**.   (∞ is **green diamond CATALOG**)

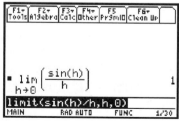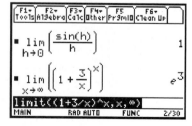

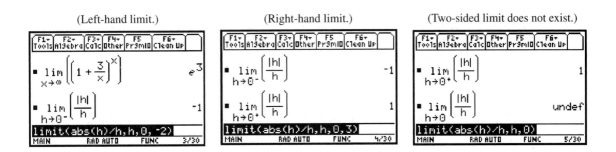

Differentiation

To find the derivative of a function, just press **2nd**, then the **8** key (or **F3** then **1**) followed by the function and the name of the variable you are differentiating with respect to. An optional third parameter allows you to specify that you want a 2nd- or higher-order derivative. Simply indicate the number of the higher-order derivative you want calculated; for example, $d(sin(4x),x,3)$ finds the third derivative of $\sin(4x)$ with respect to x. If you want a specific value of a derivative, you can use the "with" operator (|) located just to the left of the **7** key (see the second frame below). It is also possible to differentiate both sides of an equation (for implicit differentiation or related rate problems). Be sure to write any implicit functions in function notation (write $v(t)$, not just v, if v varies with respect to t).

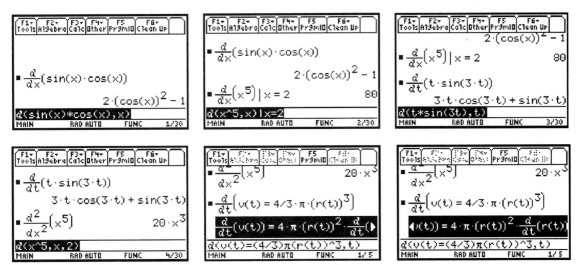

Up arrow, right arrows, to see the rest of the answer.

Integration

Indefinite, definite, and multiple integrals are all easy to specify on a TI-89. To find an indefinite integral (antiderivative), just press **2nd** then **7** (or **F3** then **2**) followed by the integrand and the name of the variable you are integrating with respect to. For a definite integral, you will simply include the lower and upper limits of integration as the (optional) third and fourth parameters, respectively. A double or triple integral can be specified by "nesting" the integration command (see the third frame that follows).

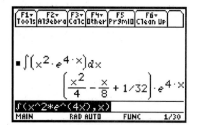  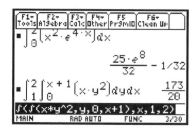

Finding Maximum and Minimum Values on an Interval

The **fMax** and **fMin** features in the **F3** (**Calc**) menu are powerful symbolic features of the TI-89 and should not be confused with the (less capable) **maximum** and **minimum** features available in the graphing screen's **F5** (**MATH**) menu. Specifying **fMax**, the function, and its independent variable calculates the value(s) of the independent variable where the function achieves its greatest height (this might be at ∞ or $-\infty$). To narrow the search down to a specific interval, use the "with" operator (|), located just to the left of the **7** key.

(Infinitely many max points all tied.)   (The absolute min is a local min.)   (A tie: local min and an end point.)

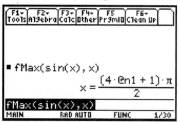  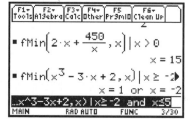

D.16   GRAPHING IN 3D

To graph functions of two variables, set the **Graph MODE** to **3D**. The **green diamond Y=** menu then shows the functions z1, z2, . . . , z99. When a 3D graph is displayed, the **X**, **Y**, or **Z** key can be pressed to see the view looking down the x-, y-, or z-axis, respectively. To rotate the graph in a given direction, press and hold the appropriate arrow key for a second or two. The + and − keys control the speed. To stop the rotation, press any arrow key or try the multiplication sign if you want a closer look.

MODE right arrow **5 ENTER**   **green diamond   Y=**   **F2   6**

 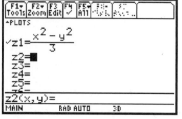 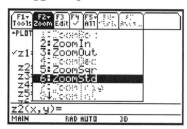

(Default is **WIRE FRAME** Style.)   **F1 9** three down arrows, right arrow **2 ENTER** (**HIDDEN SURFACE** Style)

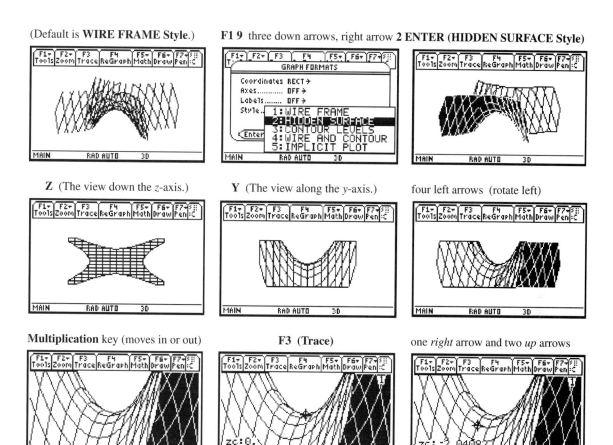

Z (The view down the *z*-axis.)   **Y** (The view along the *y*-axis.)   four left arrows (rotate left)

Multiplication key (moves in or out)   **F3** (**Trace**)   one *right* arrow and two *up* arrows

As illustrated in the previous example, the default **3D GRAPH FORMAT Style** is **WIRE FRAME**. You can choose **HIDDEN SURFACE** or other **Style** options by pressing **F1** (**Tools**), then **9** (**Format**). In **Trace**, press the right arrow key to increase the *x*-coordinate (press left arrow to decrease *x*). Similarly, the up arrow key always increases the *y*-coordinate in **Trace** (pressing the down arrow decreases *y*). Just remember that, in many views of a 3D graph, the direction that increases *x* will not be "to the right" on the calculator's screen (see the last two frames in the previous example). In **Trace**, you are also allowed to enter any *x*- and *y*-values that are in the current viewing window. The corresponding *z*-value is computed and the resulting point on the surface is marked by the **Trace** cursor. The next frame shows the point (5, 4, 3) obtained in this way.

F3 (**Trace**) **5 ENTER 4 ENTER**

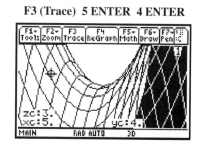

D.17 ADVANCED CALCULUS FEATURES

The TI-89 can automatically compute Taylor polynomials of a specified degree, partial derivatives, and symbolic solutions to first and second order differential equations (with or without initial conditions).

Taylor Polynomials

To find a Taylor polynomial for a function, press **F3** (**Calc**), then **9** (**taylor**), followed by the name of the function, the name of the variable, and the degree of the polynomial approximation you want. The optional fourth parameter allows you to specify the point about which the Taylor series will be expanded. For instance, specifying 1 would yield powers of $x - 1$ in the **taylor** expansion (see the third frame that follows). The default is zero, yielding powers of x itself. (A Maclaurin series.)

F3   9   (7th degree, using powers of x.)   (3rd degree, using powers of $x - 1$.)

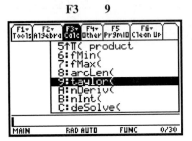   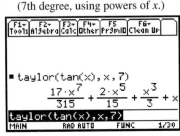   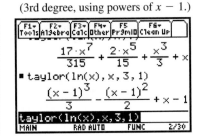

Partial Derivatives

To compute a partial derivative, simply use the differentiation operator (**2nd**, then **8**) followed by the function of several variables, then the name of the variable you want to differentiate with respect to (all other variable names will be treated like *constants*).

(The partial with respect to x.)   (The partial with respect to y.)

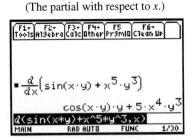   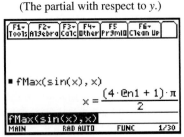

Differential Equations

To solve differential equations and initial value problems, use the **deSolve** feature found at the very bottom of the Calculus menu. Press **F3 2nd** down arrow **ENTER** (or **F3 alpha C**) from the standard home screen menus. The "prime" sign is the shifted **2nd** version of the = key (it doubles as the symbol for "minutes" in trigonometry). This symbol is used in *describing* differential equations on the TI-89; it is *not* used for calculating derivatives. Any initial conditions must be attached to the differential equation by using **and** (**2nd MATH 8 8**). The second and third parameters are the names of the independent and dependent variables, respectively. Arbitrary constants in a general solution are shown as @1, @2, etc.

F3   alpha   C

(The general solution.)

(The particular solution if $y(0) = 3$.)

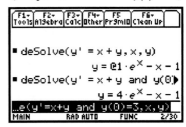

(Homogeneous second-order DE.)

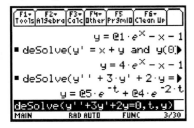

(Second-order initial value problem.)

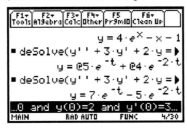

(Nonhomogeneous second order.)

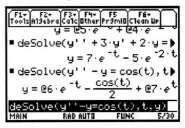

Answers to Odd-Numbered Exercises and Chapter Reviews

CHAPTER 1

Exercises 1.1, Page 7

	Function	Domain	Range
1.	Yes	$\{2, 3, 9\}$	$\{2, 4, 7\}$
3.	No	$\{1, 2, 7\}$	$\{1, 3, 5\}$
5.	Yes	$\{-2, 2, 3, 5\}$	$\{2\}$
7.	Yes	Real numbers	Real numbers
9.	Yes	Real numbers	Real numbers where $y \geq 1$
11.	No	Real numbers where $x \geq -2$	Real numbers
13.	Yes	Real numbers where $x \geq -3$	Real numbers where $y \geq 0$
15.	Yes	Real numbers where $x \geq 4$	Real numbers where $y \geq 6$

17. (a) 20 **(b)** -12 **(c)** -28 **19. (a)** 35 **(b)** 15 **(c)** -25 **21. (a)** 95 **(b)** 0 **(c)** 4
23. (a) 2 **(b)** $\frac{2}{3}$ **(c)** 0 is not in the domain of $f(t)$. **25. (a)** $6a + 8$ **(b)** $24a + 8$ **(c)** $6c^2 + 8$
27. (a) $4x^2 + 4x - 8$ **(b)** $4x^2 - 36x + 72$ **(c)** $16x^2 - 8x - 8$
29. (a) $x^2 - 3x$ **(b)** $-x^2 + 9x - 2$ **(c)** $3x^3 - 19x^2 + 9x - 1$ **(d)** $3x + 3h - 1$ **31.** All real numbers $x \neq 2$
33. All real numbers $t \neq 6$ or $t \neq -3$ **35.** All real numbers $x < 5$

Exercises 1.2, Pages 14–16

1.

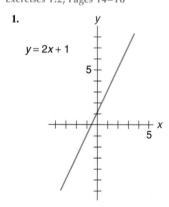

$y = 2x + 1$

3.

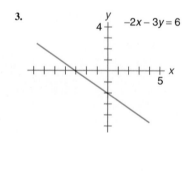

$-2x - 3y = 6$

5.

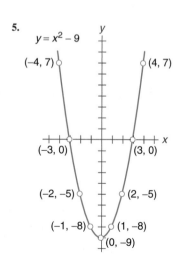

$y = x^2 - 9$

(−4, 7)   (4, 7)

(−3, 0)   (3, 0)

(−2, −5)   (2, −5)

(−1, −8)   (1, −8)

(0, −9)

7.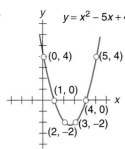
$y = x^2 - 5x + 4$

9.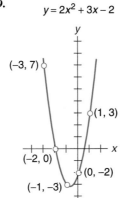
$y = 2x^2 + 3x - 2$

11.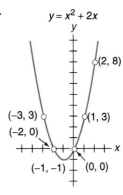
$y = x^2 + 2x$

13.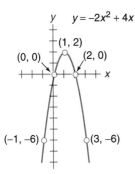
$y = -2x^2 + 4x$

15.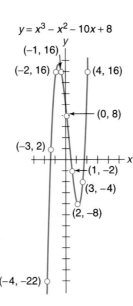
$y = x^3 - x^2 - 10x + 8$

17.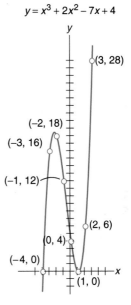
$y = x^3 + 2x^2 - 7x + 4$

19.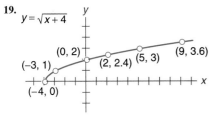
$y = \sqrt{x + 4}$

21.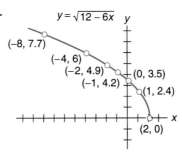
$y = \sqrt{12 - 6x}$

23. 3 and -3, 2 and -2, 3.3 and -3.3 **25.** 1 and 4, 4.5 and 0.5, no solution

27. -2 and $\frac{1}{2}$, 1 and -2.5, 1.3 and -2.8 **29.** 0 and -2, 1 and -3, 1.7 and -3.7

31. 0 and 2, no solution, 2.7 and -0.7, 2.3 and -0.3 **33.** 3.3, -3.1, 0.8; 3.4, -3.0, 0.6; 3.2, -3.2, 1.0

35. -4, 1; -3.8, 0, 1.8; -3.6, -0.5, 2.1

37. -4 and 1, -5 and 2, 0.5 and -3.5 **39.** 2 and -2, no solution, 3.5 and -3.5 **41.** 2.9, -0.5, 0.7; 2.5, -0.9, 1.3; 2.8, -0.6, 0.8

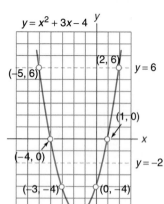

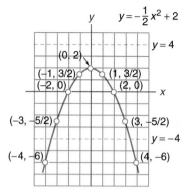

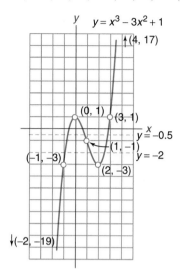

43. 2.6 ms, 4 ms, 5.5 ms   **45.** 0.27 ms, 0.48 ms, 0.94 ms
47. 2.7 s, 3.1 s   **49.** $A(1.18, -1.62)$, $B(2.35, -3.24)$, $C(3.53, -4.85)$

Exercises 1.3, Pages 22–23

1. 1 **3.** -4 **5.** 0 **7.** $\frac{5}{8}$ **9.**

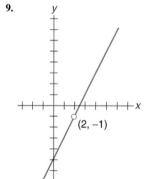

11.

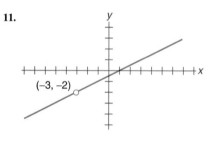

13.

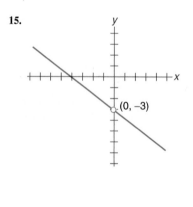

15.

17. $3x + y - 2 = 0$ **19.** $x - 2y - 5 = 0$ **21.** $x + y - 5 = 0$ **23.** $x + 2y + 10 = 0$
25. $y = -5x - 2$ **27.** $y = 2x + 7$ **29.** $y = 5$ **31.** $x = -2$ **33.** $y = -3$ **35.** $x = -7$
37. $m = -\frac{1}{4}$, $b = 3$ **39.** $m = 2$, $b = 7$ **41.** $m = 0$, $b = 6$

43.

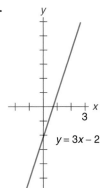

$y = 3x - 2$

45.

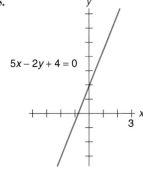

$5x - 2y + 4 = 0$

47.

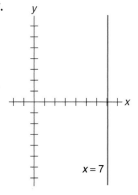

$x = 7$

49.

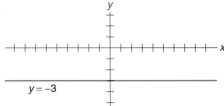

$y = -3$

51.

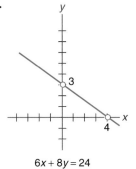

$6x + 8y = 24$

53.

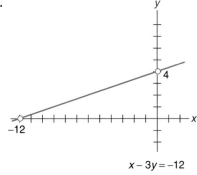

$x - 3y = -12$

55. $\frac{1}{350}$

Exercises 1.4, Page 27

1. Perpendicular   **3.** Neither   **5.** Parallel   **7.** $2x - y + 7 = 0$ **9.** $5x + y + 31 = 0$ **11.** $3x - 4y = 0$
13. $3x - 2y = 18$ **15.** $y = 8$ **17.** $x = 7$ **19. (a)** Yes, slopes of opposite sides are equal.   **(b)** No, slopes of adjacent sides are not negative reciprocals.

Exercises 1.5, Pages 29–30

1. 15   **3.** 7   **5.** $4\sqrt{2}$ **7.** 7   **9.** $(3\frac{1}{2}, 5)$ **11.** $(1\frac{1}{2}, -1)$ **13.** $(0, -2\frac{1}{2})$
15. (a) 24   **(b)** Yes   **(c)** No   **(d)** 24   **17. (a)** $10 + \sqrt{82} + \sqrt{58}$ or 26.7   **(b)** No   **(c)** No   **19.** $4\sqrt{2}$
21. $x - 2y = 10$ **23.** $x + 2y = 21$

1.
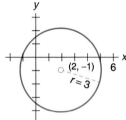

3.
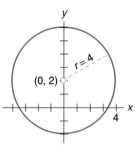

5. $(x - 1)^2 + (y + 1)^2 = 16$ **7.** $(x + 2)^2 + (y + 4)^2 = 34$ **9.** $x^2 + y^2 = 36$ **11.** $(0, 0); r = 4$
13. $(-3, 4); r = 8$ **15.** $(4, -6); r = 2\sqrt{15}$ **17.** $(6, 1); r = 7$ **19.** $(-\frac{7}{2}, -\frac{3}{2}); r = \sqrt{94}/2$
21. $x^2 + y^2 - 2y - 9 = 0; (0, 1); r = \sqrt{10}$ **23.** $x^2 + y^2 + 10x - 40y = 0; (-5, 20); r = 5\sqrt{17}$

1.
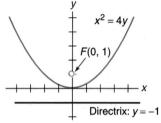

3.
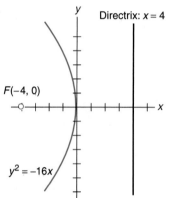

5.
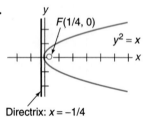

7.
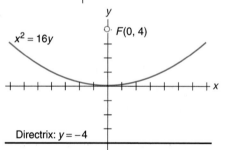

9.
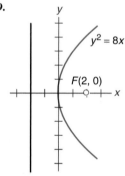

11. $y^2 = 8x$ **13.** $y^2 = -32x$ **15.** $x^2 = 24y$ **17.** $y^2 = -16x$ **19.** $y^2 - 6y + 8x + 1 = 0$ **21.** 15 m, 7 m
23. $x^2 = 32y$ **25.**
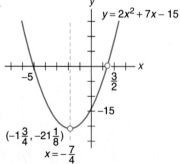

27.
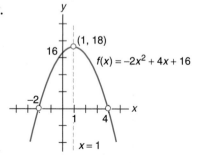

29. (a) 1024 m;   **(b)** 1024 m   **31.** 3600 m²

Vertices	Foci	Major axis	Minor axis
1. $(5, 0)(-5, 0)$	$(3, 0)(-3, 0)$	10	8
3. $(4, 0)\,(-4, 0)$	$(\sqrt{7}, 0)\,(-\sqrt{7}, 0)$	8	6
5. $(0, 6)(0, -6)$	$(0, \sqrt{35})\,(0, -\sqrt{35})$	12	2
7. $(0, 4)\,(0, -4)$	$(0, \sqrt{7})\,(0, -\sqrt{7})$	8	6

1.

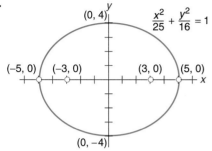

3.

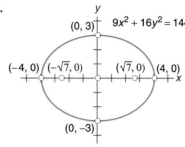

5.

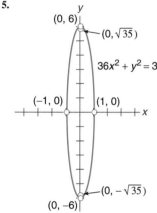

7.

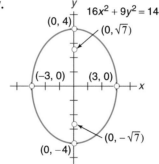

9. $\dfrac{x^2}{16} + \dfrac{y^2}{12} = 1$ or $3x^2 + 4y^2 = 48$ **11.** $\dfrac{x^2}{45} + \dfrac{y^2}{81} = 1$ or $9x^2 + 5y^2 = 405$

13. $\dfrac{x^2}{36} + \dfrac{y^2}{25} = 1$ or $25x^2 + 36y^2 = 900$ **15.** $\dfrac{x^2}{39} + \dfrac{y^2}{64} = 1$ or $64x^2 + 39y^2 = 2496$

17. $\dfrac{x^2}{5600^2} + \dfrac{y^2}{5000^2} = 1$ or $625x^2 + 784y^2 = 1.96 \times 10^{10}$

	Vertices	Foci	Transverse axis	Conjugate axis	Asymptotes
1.	$(5, 0)\,(-5, 0)$	$(13, 0)\,(-13, 0)$	10	24	$y = \pm\dfrac{12}{5}x$
3.	$(0, 3)\,(0, -3)$	$(0, 5)\,(0, -5)$	6	8	$y = \pm\dfrac{3}{4}x$
5.	$(\sqrt{2}, 0)\,(-\sqrt{2}, 0)$	$(\sqrt{7}, 0)\,(-\sqrt{7}, 0)$	$2\sqrt{2}$	$2\sqrt{5}$	$y = \pm\sqrt{\dfrac{5}{2}}x$
7.	$(0, 1)\,(0, -1)$	$(0, \sqrt{5})\,(0, -\sqrt{5})$	2	4	$y = \pm\dfrac{1}{2}x$

1.

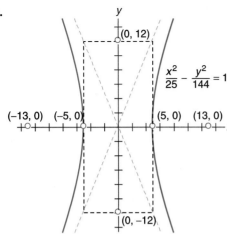

$$\frac{x^2}{25} - \frac{y^2}{144} = 1$$

3.

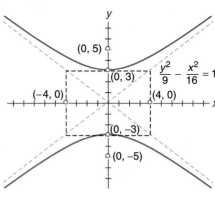

$$\frac{y^2}{9} - \frac{x^2}{16} = 1$$

5.

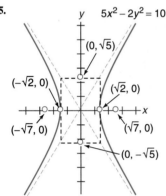

$$5x^2 - 2y^2 = 10$$

7.

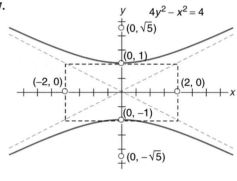

$$4y^2 - x^2 = 4$$

9. $\dfrac{x^2}{16} - \dfrac{y^2}{20} = 1$ or $5x^2 - 4y^2 = 80$ **11.** $\dfrac{y^2}{36} - \dfrac{x^2}{28} = 1$ or $7y^2 - 9x^2 = 252$

13. $\dfrac{x^2}{9} - \dfrac{y^2}{25} = 1$ or $25x^2 - 9y^2 = 225$ **15.** $\dfrac{x^2}{25} - \dfrac{y^2}{11} = 1$ or $11x^2 - 25y^2 = 275$

17.

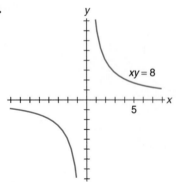

$$xy = 8$$

Exercises 1.10, Pages 57–58

1. $\dfrac{(x-1)^2}{16} + \dfrac{(y+1)^2}{12} = 1$ **3.** $\dfrac{(y-1)^2}{36} - \dfrac{(x-1)^2}{28} = 1$ **5.** $(y+1)^2 = 8(x-3)$

7. Parabola; vertex: $(2, -3)$

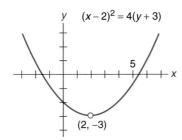

$(x - 2)^2 = 4(y + 3)$

9. Hyperbola; center: $(-2, 0)$

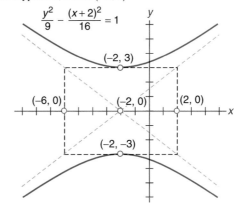

$$\frac{y^2}{9} - \frac{(x + 2)^2}{16} = 1$$

11. Ellipse; center: $(2, 0)$

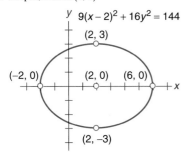

$9(x - 2)^2 + 16y^2 = 144$

13. Ellipse; center: $(3, 1)$

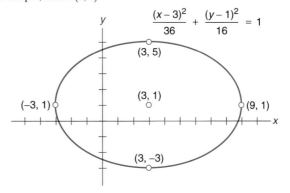

$$\frac{(x - 3)^2}{36} + \frac{(y - 1)^2}{16} = 1$$

15. Parabola; vertex: $(1, -3)$

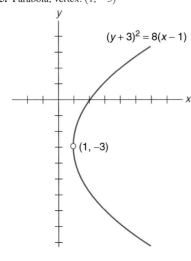

$(y + 3)^2 = 8(x - 1)$

17. Hyperbola; center: $(-1, -1)$

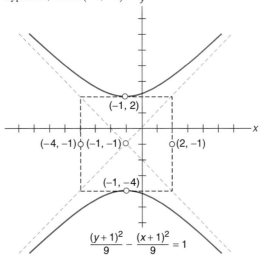

$$\frac{(y + 1)^2}{9} - \frac{(x + 1)^2}{9} = 1$$

19. Parabola; vertex: $(2, -1)$

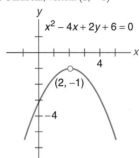

$x^2 - 4x + 2y + 6 = 0$
$(2, -1)$

21. Ellipse; center: $(-2, 1)$

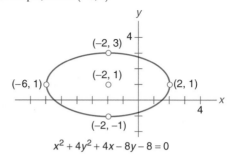

$(-2, 3)$
$(-6, 1)$ $(-2, 1)$ $(2, 1)$
$(-2, -1)$
$x^2 + 4y^2 + 4x - 8y - 8 = 0$

23. Hyperbola; center: $(1, 1)$

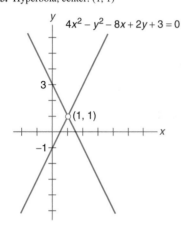

$4x^2 - y^2 - 8x + 2y + 3 = 0$
$(1, 1)$

25. Hyperbola; center: $(-3, 3)$

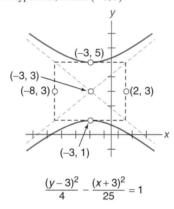

$(-3, 5)$
$(-3, 3)$
$(-8, 3)$ $(2, 3)$
$(-3, 1)$

$$\frac{(y-3)^2}{4} - \frac{(x+3)^2}{25} = 1$$

27. Parabola; vertex: $(-8, -2)$

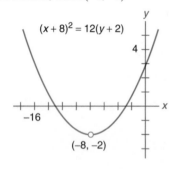

$(x + 8)^2 = 12(y + 2)$
-16
$(-8, -2)$

29. Ellipse; center: $(-6, -2)$

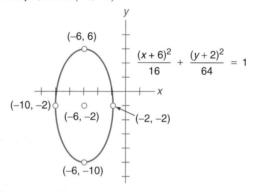

$(-6, 6)$
$$\frac{(x+6)^2}{16} + \frac{(y+2)^2}{64} = 1$$
$(-10, -2)$
$(-6, -2)$
$(-2, -2)$
$(-6, -10)$

Exercises 1.11, Page 59

1. Ellipse **3.** Parabola **5.** Hyperbola **7.** Circle **9.** Circle **11.** Ellipse **13.** Hyperbola
15. Parabola

Exercises 1.12, Page 63

1. $(3, 3)$ **3.** $(1, \sqrt{3}), (1, -\sqrt{3})$ **5.** $(2\sqrt{6}, 6), (-2\sqrt{6}, 6)$ **7.** $(-2, 0), (2, 0)$ **9.** $(-6, 6), (6, 6)$
11. $(-2, 2), (-2, -2)$ **13.** $(2.3, 5.5), (-2.3, 5.5)$ **15.** $(5, 4), (5, -4), (-5, 4), (-5, -4)$
17. $(1, 4), (-1, -4), (4, 1), (-4, -1)$

Exercises 1.13, Pages 69–70

1.

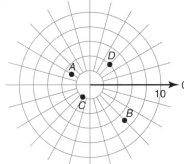

3.

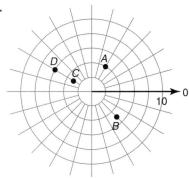

5. $(-3, 240°), (-3, -120°), (3, -300°)$ **7.** $(5, 135°), (-5, -45°), (5, -225°)$

9. $(-4, -315°), (-4, 45°), (4, 225°)$ **11.** $(-3, 7\pi/6), (-3, -5\pi/6), (3, -11\pi/6)$

13. $(9, 5\pi/3), (9, -\pi/3), (-9, -4\pi/3)$ **15.** $(4, -3\pi/4), (-4, \pi/4), (4, 5\pi/4)$

17.

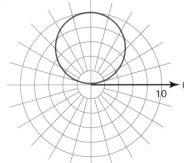

$r = 10 \sin \theta$

19.

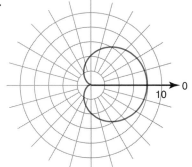

$r = 4 + 4 \cos \theta$

21.

$r \cos \theta = 4$

23.

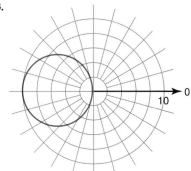

$r = -10 \cos \theta$

25.

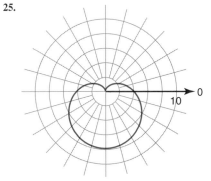

$r = 4 - 4 \sin \theta$

27.

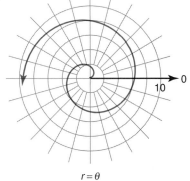

$r = \theta$

29. $(3\sqrt{3}/2, 3/2)$ **31.** $(1, \sqrt{3})$ **33.** $(2\sqrt{3}, -2)$ **35.** $(0, 6)$ **37.** $(2.5, -4.33)$ **39.** $(1.4, 1.4)$
41. $(7.1, 45°)$ **43.** $(4, 90°)$ **45.** $(4, 240°)$ **47.** $(4\sqrt{2}, 3\pi/4)$ **49.** $(2\sqrt{2}, 5\pi/6)$ **51.** $(4, 3\pi/2)$
53. $r\cos\theta = 3$ **55.** $r = 6$ **57.** $r + 2\cos\theta + 5\sin\theta = 0$ **59.** $r = 12/(4\cos\theta - 3\sin\theta)$
61. $r^2 = 36/(9 - 5\sin^2\theta)$ **63.** $r = 4\sec\theta\tan^2\theta$ **65.** $y = -3$ **67.** $x^2 + y^2 = 25$ **69.** $y = x$
71. $x^2 + y^2 - 5x = 0$ **73.** $x^2 + y^2 - 3x + 3\sqrt{3}y = 0$ **75.** $y^2 = 3x$ **77.** $xy = 1$
79. $x^4 + 2x^2y^2 + y^4 - 2xy = 0$ **81.** $y^2 = x^2(x^2 + y^2)$ **83.** $x^2 + 6y - 9 = 0$
85. $x^4 + 2x^2y^2 + y^4 + 4y^3 - 12x^2y = 0$ **87.** $x^4 + 2x^2y^2 + y^4 - 8x^2y - 8y^3 - 4x^2 + 12y^2 = 0$ **89.** $\sqrt{13}$
91. $d = \sqrt{r_1^2 + r_2^2 - 2r_1r_2\cos(\theta_1 - \theta_2)}$

Exercises 1.14, Pages 76–77

1.

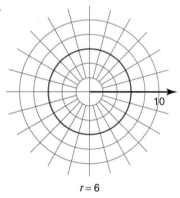

$r = 6$

3.

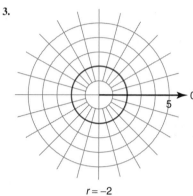

$r = -2$

5.

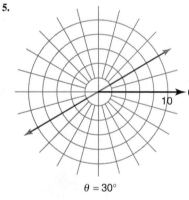

$\theta = 30°$

7.

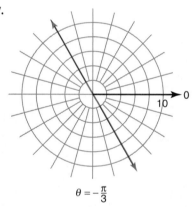

$\theta = -\dfrac{\pi}{3}$

9.

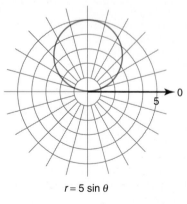

$r = 5\sin\theta$

11.

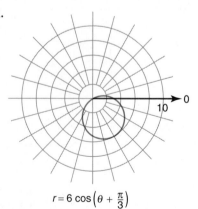

$r = 6\cos\left(\theta + \dfrac{\pi}{3}\right)$

13.

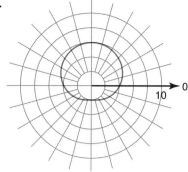

$r = 4 + 2 \sin \theta$

15.

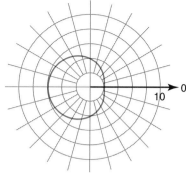

$r = 4 - 2 \cos \theta$

17.

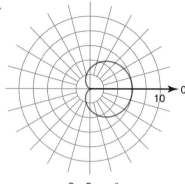

$r = 3 + 3 \cos \theta$

19.

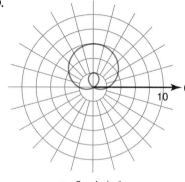

$r = 2 + 4 \sin \theta$

21.

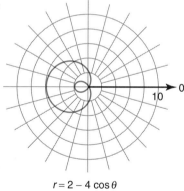

$r = 2 - 4 \cos \theta$

23.

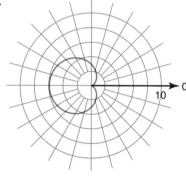

$r = 3 - 3 \cos \theta$

25.

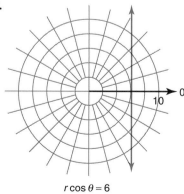

$r \cos \theta = 6$

27.

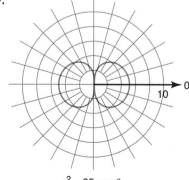

$r^2 = 25 \cos \theta$

Answers to Odd-Numbered Exercises and Chapter Reviews   **597**

29.

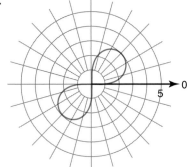

$r^2 = 9 \sin 2\theta$

31.

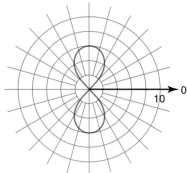

$r^2 = -36 \cos 2\theta$

33.

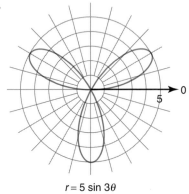

$r = 5 \sin 3\theta$

35.

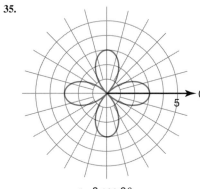

$r = 3 \cos 2\theta$

37.

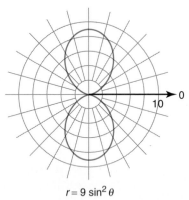

$r = 9 \sin^2 \theta$

39.

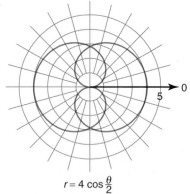

$r = 4 \cos \frac{\theta}{2}$

41.

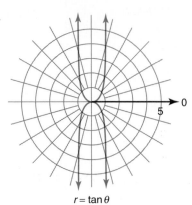

$r = \tan \theta$

43.

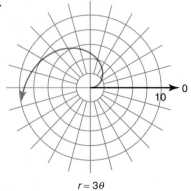

$r = 3\theta$

45.

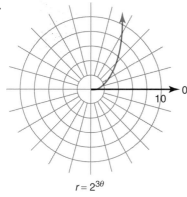

$r = 2^{3\theta}$

47.

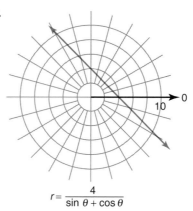

$r = \dfrac{4}{\sin \theta + \cos \theta}$

49.

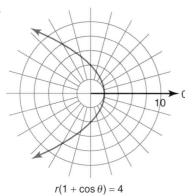

$r(1 + \cos \theta) = 4$

Chapter 1 Review, Pages 81–83

	Function	Domain	Range
1.	Yes	$\{2, 3, 4, 5\}$	$\{3, 4, 5, 6\}$
2.	No	$\{2, 4, 6\}$	$\{1, 3, 4, 6\}$
3.	Yes	Real numbers	Real numbers
4.	Yes	Real numbers	Real numbers where $y \geq -5$
5.	No	Real numbers where $x \geq 4$	Real numbers
6.	Yes	Real numbers where $x \leq \frac{1}{2}$	Real numbers where $y \geq 0$

7. (a) 24 **(b)** 14 **(c)** -6 **8. (a)** 10 **(b)** -12 **(c)** 38 **9. (a)** 5 **(b)** $\frac{85}{4}$ **(c)** -15 is not in the domain of $h(x)$.
(d) 1 is not in the domain of $h(x)$. **10. (a)** $a^2 - 6a + 4$ **(b)** $4x^2 - 12x + 4$ **(c)** $z^2 - 10z + 20$

11.

$y = 4x + 5$

$(0, 5)$
$(-1, 1)$
$(-2, -3)$

12. $y = x^2 + 4$

$(-3, 13)$ $(3, 13)$
$(-2, 8)$ $(2, 8)$
$(-1, 5)$ $(1, 5)$
$(0, 4)$

13. $y = x^2 + 2x - 8$

$(-5, 7)$ $(3, 7)$
$(-4, 0)$
$(2, 0)$
$(-3, -5)$ $(1, -5)$
$(-2, -8)$ $(0, -8)$
$(-1, -9)$

14. $y = 2x^2 + x - 6$

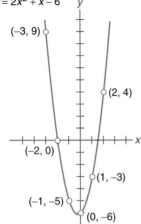

15. $y = -x^2 - x + 4$

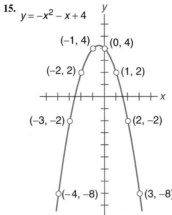

16. $y = \sqrt{2x}$

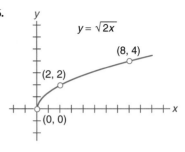

17. $y = \sqrt{-2 - 4x}$

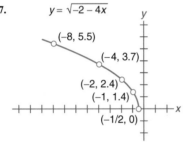

18.

$y = x^3 - 6x$

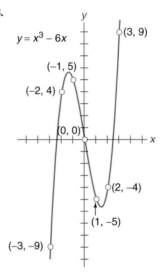

19. 1 and -1, 1.7 and -1.7, no solution   **20.** -4 and 2, 1.6 and -3.6, -4.5 and 2.5

21. -2 and 1, -2.6, and 1.6, -3 and 2   **22.** 0, 2.4, -2.4; 2.6, -2.3, -0.3; 2.1, -2.7, 0.5   **23.** 1, 1.7, 2

24. 2.1, 2.4   **25.** $-\frac{2}{9}$ **26.** $\sqrt{85}$ **27.** $\left(-\frac{3}{2}, -3\right)$ **28.** $11x + 2y - 58 = 0$ **29.** $2x - 3y + 9 = 0$

30. $x + 3y + 9 = 0$ **31.** $x = -3$ **32.** $m = \frac{3}{2}; b = -3$ **33.**

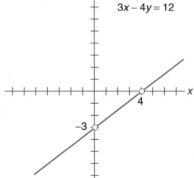

$3x - 4y = 12$

34. Perpendicular   **35.** Parallel   **36.** Neither   **37.** Perpendicular   **38.** Parallel   **39.** $2x - y - 8 = 0$

40. $5x - 3y + 20 = 0$ **41.** $(x - 5)^2 + (y + 7)^2 = 36$ or $x^2 + y^2 - 10x + 14y + 38 = 0$ **42.** $(4, -3); 7$

43. $(0, \frac{3}{2})$; $y = -\frac{3}{2}$

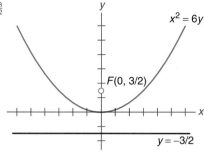

$x^2 = 6y$

$F(0, 3/2)$

$y = -3/2$

44. $y^2 = -16x$ **45.** $(y-3)^2 = 8(x-2)$ or $y^2 - 6y - 8x + 25 = 0$

46. $V(7, 0), (-7, 0)$; $F(3\sqrt{5}, 0), (-3\sqrt{5}, 0)$

47. $\dfrac{x^2}{4} + \dfrac{y^2}{16} = 1$ or $4x^2 + y^2 = 16$

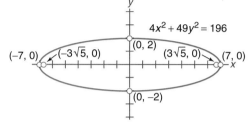

$4x^2 + 49y^2 = 196$

$(-7, 0)$ $(-3\sqrt{5}, 0)$ $(0, 2)$ $(3\sqrt{5}, 0)$ $(7, 0)$

$(0, -2)$

48. $V(6, 0), (-6, 0)$; $F(2\sqrt{13}, 0), (-2\sqrt{13}, 0)$

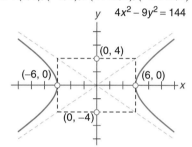

$4x^2 - 9y^2 = 144$

$(-6, 0)$ $(0, 4)$ $(6, 0)$

$(0, -4)$

49. $\dfrac{y^2}{25} - \dfrac{x^2}{16} = 1$ or $16y^2 - 25x^2 = 400$ **50.** $\dfrac{(x-3)^2}{9} + \dfrac{(y+4)^2}{25} = 1$ **51.** $\dfrac{(x+7)^2}{81} - \dfrac{(y-4)^2}{9} = 1$

52. Hyperbola

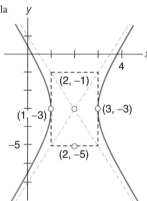

4

$(2, -1)$

$(1, -3)$ $(3, -3)$

-5

$(2, -5)$

$16x^2 - 4y^2 - 64x - 24y + 12 = 0$

53. $(0, 0), (-12, -6)$ **54.** $(4, \sqrt{3}), (4, -\sqrt{3}), (-4, \sqrt{3}), (-4, -\sqrt{3})$

55.

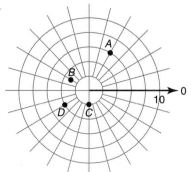

56.

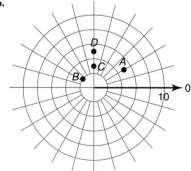

57. $(5, -225°), (-5, -45°), (-5, 315°)$ **58.** $(2, -11\pi/6), (-2, -5\pi/6), (2, \pi/6)$
59. (a) $(-2.6, -1.5)$ (b) $(-1, -1.7)$ (c) $(-4.3, 2.5)$ (d) $(0, 6)$ **60.** (a) $(4.2, 135°)$ (b) $(6, 270°)$ (c) $(2, 120°)$
61. (a) $(5, \pi)$ (b) $(12, 5\pi/6)$ (c) $(\sqrt{2}, 7\pi/4)$ **62.** $r = 7$ **63.** $r \sin^2 \theta = 9 \cos \theta$ **64.** $r = 8/(5 \cos \theta + 2 \sin \theta)$
65. $r^2 = 12/(1 - 5 \sin^2 \theta)$ **66.** $r = 6 \csc \theta \cot^2 \theta$ **67.** $r = \cos \theta \cot \theta$ **68.** $x = 12$ **69.** $x^2 + y^2 = 81$
70. $y = -\sqrt{3}x$ **71.** $x^2 + y^2 - 8x = 0$ **72.** $y^2 = 5x$ **73.** $xy = 4$ **74.** $x^4 + 2x^2y^2 + y^4 + 4y^2 - 4x^2 = 0$
75. $y = 1$ **76.** $x^4 + y^4 + 2x^2y^2 - 2x^2y - 2y^3 - x^2 = 0$ **77.** $x^2 = 4(y + 1)$

78.

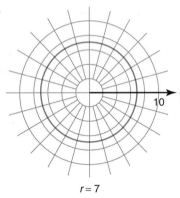

$r = 7$

79.

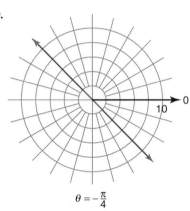

$\theta = -\dfrac{\pi}{4}$

80.

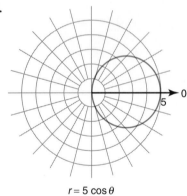

$r = 5 \cos \theta$

81.

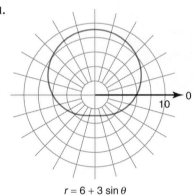

$r = 6 + 3 \sin \theta$

82.

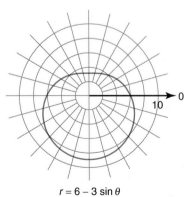

$r = 6 - 3 \sin \theta$

83.

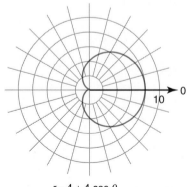

$r = 4 + 4 \cos \theta$

84.

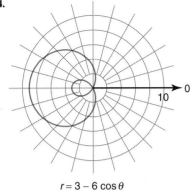

$r = 3 - 6 \cos \theta$

85.

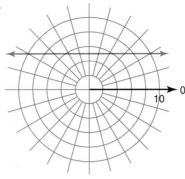

$r \sin \theta = 5$

86.

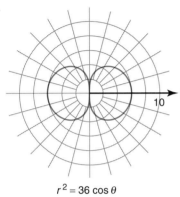

$r^2 = 36 \cos \theta$

87.

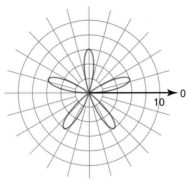

$r = 6 \sin 5\theta$

88.

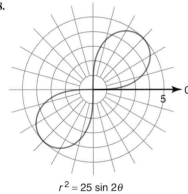

$r^2 = 25 \sin 2\theta$

89.

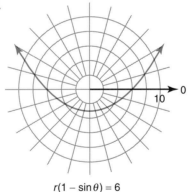

$r(1 - \sin \theta) = 6$

CHAPTER 2

Exercises 2.1, Pages 89–90

1. 9   **3.** -16 **5.** $\frac{1}{7}$ **7.** $2\sqrt{7}$ **9.** $3h + 7$ **11.** $3(\Delta t)^2 + 14(\Delta t) + 11$ **13.** $3(\Delta t)$; 3
15. $2(\Delta t)^2 + 4t(\Delta t)$; $2(\Delta t) + 4t$ **17.** $(\Delta t)^2 + 2t(\Delta t) - 2(\Delta t)$; $\Delta t + 2t - 2$ **19.** 50 m/s   **21.** 17 m/s
23. 6×10^5 m/s   **25.** 100 μA   **27.** 6 m/s   **29.** 10 m/s   **31.** $-\frac{1}{18}$ m/s   **33.** $-\frac{1}{4}$ m/s   **35.** 64 ft/s

Exercises 2.2, Pages 97–98

1. 4   **3.** 3   **5.** 1   **7.** -6 **9.** 1   **11.** 2   **13.** -6 **15.** 1   **17.** No limit   **19.** 0   **21.** $\frac{3}{4}$
23. 13   **25.** 11   **27.** 6   **29.** 102   **31.** -12 **33.** -11 **35.** $\frac{12}{5}$ **37.** -14 **39.** 10   **41.** 152
43. $2x$ **45.** $-1/x^2$ **47.** $1/(2\sqrt{a})$ **49.** Does not exist   **51.** b **53.** Does not exist   **55.** b **57.** No
59. No   **61.** No   **63.** No   **65.** Does not exist   **67.** 0   **69.** b **71.** Does not exist

Exercises 2.3, Pages 100–101

1. 6   **3.** −6   **5.** 9   **7.** −5   **9.** 12   **11.** $y = -4x - 4$ **13.** $y = -8x - 11$ **15.** $y = -13x - 3$
17. $y = -13x + 29$ **19.** $y = 4x - 3$ **21.** $\left(-\frac{1}{6}, \frac{1}{36}\right)$ **23.** $(1, 2), (-1, -2)$

Exercises 2.4, Pages 104–105

1. 3   **3.** −2   **5.** $6x$ **7.** $2x - 2$ **9.** $6x - 4$ **11.** $-12x$ **13.** $3x^2 + 4$ **15.** $-1/x^2$ **17.** $-2/(x - 3)^2$
19. $-2/x^3$ **21.** $2x/(4 - x^2)^2$ **23.** $1/(2\sqrt{x} + 1)$ **25.** $-1/\sqrt{1 - 2x}$ **27.** $-1/(2(x - 1)^{3/2})$ **29.** 6
31. 3/2   **33.** $y = x - 4$ **35.** $x - 4y = 3$ **37.** $(4, 1), (2, -1)$ **39.** $(-3, 1)$

Exercises 2.5, Pages 110–111

1. 0   **3.** $5x^4$ **5.** 4   **7.** −3   **9.** $10x$ **11.** $2x - 3$ **13.** $8x - 3$ **15.** $-16x$ **17.** $9x^2 + 4x - 6$
19. $20x^4 - 6x^2 + 1$ **21.** $20x^7 - 6x^4 + 30x^3 - 3x^2$ **23.** $4\sqrt{7}x^3 - 3\sqrt{5}x^2 - \sqrt{3}$ **25.** −4   **27.** 92   **29.** 26
31. 2   **33.** 7107   **35.** $y = -5x + 2$ **37.** 120 W/A   **39.** 0.4 V/Ω   **41.** $\frac{3}{2}x^{1/2}$ or $\frac{3\sqrt{x}}{2}$ **43.** $-4x^{-5}$ or $\frac{-4}{x^5}$
45. $120x^{19}$ **47.** $-112x^{-9}$ or $\frac{-112}{x^9}$ **49.** $-\frac{5}{3}x^{-4/3}$ or $\frac{-5}{3x\sqrt[3]{x}}$ **51.** 0.5 V/Ω

Exercises 2.6, Pages 113–114

1. $6x^2 + 2x$ **3.** $24x^2 + 12x - 10$ **5.** $20x + 7$ **7.** $12x^2 + 14x - 4$
9. $(x^2 + 3x + 4)(3x^2 - 4) + (x^3 - 4x)(2x + 3)$ or $5x^4 + 12x^3 - 24x - 16$

11. $(x^4 - 3x^2 - x)(6x^2 - 4) + (2x^3 - 4x)(4x^3 - 6x - 1)$ or $14x^6 - 50x^4 - 8x^3 + 36x^2 + 8x$ **13.** $\dfrac{5}{(2x + 5)^2}$

15. $\dfrac{-2x - 1}{(x^2 + x)^2}$ **17.** $\dfrac{7}{2(x + 2)^2}$ **19.** $\dfrac{2x^2 + 2x}{(2x + 1)^2}$ **21.** $\dfrac{-x^2 + 2x + 2}{(x^2 + x + 1)^2}$ **23.** $\dfrac{-12x^3 - 81x + 72}{x^3(3x - 4)^2}$ **25.** −7
27. 10   **29.** −13/64   **31.** $y = -5x + 21$ **33.** 10.4 V/s

Exercises 2.7, Page 118

1. $160(4x + 3)^{39}$ **3.** $5(3x^2 - 7x + 4)^4(6x - 7)$ **5.** $\dfrac{-12x^2}{(x^3 + 3)^5}$ **7.** $\dfrac{10x - 7}{2\sqrt{5x^2 - 7x + 2}}$ **9.** $\dfrac{16x^2 + 2}{\sqrt[3]{8x^3 + 3x}}$
11. $-\frac{3}{2}(2x + 3)^{-7/4}$ **13.** $15(4x + 1)(4x + 5)^3$ **15.** $6x^3(2x^2 - 1)(x^3 - x)^2$ **17.** $4(2x + 1)(x^2 + 1)(3x^2 + x + 1)$
19. $\dfrac{27x^3 - 5x^2 + 9x - 1}{\sqrt{9x^2 - 2x}}$ **21.** $\dfrac{33x^2 + 32x + 18}{(3x + 4)^{1/4}}$ **23.** $-\dfrac{4}{x^5} - 8(2x + 1)^3$ **25.** $\dfrac{-10x}{(3x - 1)^3}$
27. $\dfrac{(x^3 + 2)^3(40x^4 - 33x^3 - 16x + 6)}{(4x^2 - 3x)^2}$ **29.** $\dfrac{3(3x + 2)^4(4x - 9)}{(2x - 1)^4}$ **31.** $\dfrac{2x + 8}{(4x + 3)^{3/2}(3x - 1)^{1/3}}$
33. $\dfrac{8(1 + x)^3}{(1 - x)^5}$ **35.** 0.177 m/s

Exercises 2.8, Page 121

1. −4/3   **3.** x/y **5.** $\dfrac{-x}{y + 2}$ **7.** $\dfrac{2x - y}{y^2 + x}$ **9.** $\dfrac{y^2 - 2x}{4y^3 - 2xy}$ **11.** $\dfrac{2xy^2 - 3x}{2y^3 - 2x^2y}$ **13.** $\dfrac{(9x^2 - 12)(x^3 - 4x)^2}{4y^3 + 8y}$
15. $\dfrac{2x - 2y + 8 - 3x^2 - 6xy - 3y^2}{3x^2 + 6xy + 3y^2 + 2x - 2y + 8}$ **17.** $\dfrac{y}{x - y(x - y)^2}$ **19.** $\frac{4}{5}$ **21.** $\frac{1}{3}$ **23.** $y = 4x + 9$
25. $12x + 5y = -8$ **27.** −1   **29.** $\frac{1}{6}$

Exercises 2.10, Page 127

1. $y' = 5x^4 + 6x, y'' = 20x^3 + 6, y''' = 60x^2, y^{(4)} = 120x$
3. $y' = 25x^4 + 6x^2 - 8, y'' = 100x^3 + 12x, y''' = 300x^2 + 12, y^{(4)} = 600x$ **5.** $-6/x^4$ **7.** 162
9. $-\frac{1215}{16}(3x + 2)^{-7/2}$ **11.** $\dfrac{6x^2 - 2}{(x^2 + 1)^3}$ **13.** $\dfrac{4}{(x - 1)^3}$ **15.** $y' = -x/y; y'' = -1/y^3$
17. $y' = \dfrac{2x - y}{x - 2y}; y'' = \dfrac{6}{(x - 2y)^3}$ **19.** $y' = -y^{1/2}/x^{1/2}; y'' = \dfrac{1}{2x^{3/2}}$ **21.** $y' = y^2/x^2; y'' = \dfrac{2y^3 - 2xy^2}{x^4}$ or $\dfrac{2y^3}{x^3}$
23. $y' = \dfrac{1}{2(1 + y)}; y'' = \dfrac{-1}{4(1 + y)^3}$ **25.** $a = 6t^2 - 36t - 8$ **27.** $a = \dfrac{-9}{(6t - 4)^{3/2}}$ **29.** $x - 5y = -14; 4x - 5y = 14$

1. $6t(\Delta t) + 3(\Delta t)^2$; $6t + 3(\Delta t)$ **2.** $10t(\Delta t) + 5(\Delta t)^2$; $10t + 5(\Delta t)$ **3.** $2t(\Delta t) + (\Delta t)^2 - 3(\Delta t)$; $2t + \Delta t - 3$
4. $6t(\Delta t) + 3(\Delta t)^2 - 6(\Delta t)$; $6t + 3(\Delta t) - 6$ **5.** 27 m/s   **6.** 20 m/s   **7.** 10 m/s   **8.** 33 m/s   **9.** 12 m/s
10. 10 m/s   **11.** 4 m/s   **12.** 17 m/s   **13.** 4   **14.** -11 **15.** -4 **16.** 10   **17.** No limit   **18.** $\sqrt{15}$
19. $\frac{22}{9}$ **20.** $-\frac{37}{24}$ **21.** 168   **22.** 112   **23.** $-\frac{1}{4}$ **24.** $\frac{1}{2}$ **25.** $\frac{5}{2}$ **26.** $\frac{7}{10}$ **27.** Does not exist   **28.** c
29. c **30.** Does not exist   **31.** No   **32.** Yes   **33.** -10; $y = -10x + 2$ **34.** -1; $y = -x - 16$
35. 14; $y = 14x - 11$ **36.** -24; $y = -24x - 13$ **37.** -0.048 cm/s   **38.** -64 ft/s   **39.** $20x^3 - 9x^2 + 4x + 5$
40. $100x^{99} + 400x^4$ **41.** $6x^5 - 4x^3 + 15x^2 - 4$ **42.** $21x^6 - 25x^4 + 12x^3 - 36x^2 - 10x + 20$
43. $\dfrac{3x^2 - 8x - 3}{(3x - 4)^2}$ **44.** $\dfrac{6x^5 - 18x^4 - 4x + 4}{(3x^4 + 2)^2}$ **45.** $30x(3x^2 - 8)^4$ **46.** $\frac{3}{4}(x^4 + 2x^3 + 7)^{-1/4}(4x^3 + 6x^2)$
47. $\dfrac{-12}{(3x + 5)^5}$ **48.** $\dfrac{(7x^2 + 21x)(7x^2 - 5)^{-1/2} - 2\sqrt{7x^2 - 5}}{(x + 3)^3}$ or $\dfrac{-7x^2 + 21x + 10}{(x + 3)^3\sqrt{7x^2 - 5}}$ **49.** $\dfrac{-3x^2 - 45x + 20}{2(x + 5)^2\sqrt{2 - 3x}}$
50. $\dfrac{2y^3 - x}{y - 6xy^2}$ **51.** $\dfrac{y}{2y^3 - y - x}$ **52.** $\dfrac{4x}{3y(y^2 + 1)^2}$ **53.** $\dfrac{9x^2(2x^3 - 3)^2}{2(y + 2)^3}$ **54.** 648   **55.** 0   **56.** -7
57. $\dfrac{4}{9\sqrt{2}}$ or $\dfrac{2\sqrt{2}}{9}$ **58.** $11x + y + 14 = 0$ **59.** $26x - y + 62 = 0$ **60.** $x - y + 10 = 0$
61. $13x + 4y - 55 = 0$ **62.** 4 m/s   **63.** 48 m/s   **64.** 2.75 m/s   **65.** 0.0788 m/s   **66.** $2x + y + 18 = 0$
67. 50.3 A   **68.** $\dfrac{dc}{dT} = 0.5 + 0.000012T$ **69.** $-\dfrac{1}{4\pi C\sqrt{LC}}$
70. $y' = 24x^5 - 32x^3 + 27x^2 - 6$, $y'' = 120x^4 - 96x^2 + 54x$, $y''' = 480x^3 - 192x + 54$, $y^{(4)} = 1440x^2 - 192$
71. $\dfrac{-1}{(2x - 3)^{3/2}}$ **72.** $\dfrac{72x^2 - 12}{(2x^2 + 1)^3}$ **73.** $y' = \dfrac{-y}{x + y}$, $y'' = \dfrac{4}{(x + y)^3}$
74. $y' = \dfrac{y^{3/2}}{x^{3/2}}$, $y'' = \dfrac{3y^2 - 3y^{3/2}x^{1/2}}{2x^3}$ or $\dfrac{3y^2}{2x^{5/2}}$ **75.** $a = -\frac{3}{4}(2t + 3)^{-7/4}$

CHAPTER 3

Exercises 3.1, Pages 140–141

1.
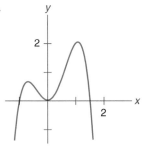
$y = 2x(x + 1)(x - 4)$

3.
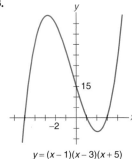
$y = (x - 1)(x - 3)(x + 5)$

5.
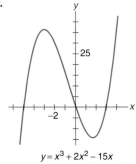
$y = x^3 + 2x^2 - 15x$

7.
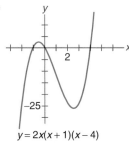
$y = x^2(x + 1)(3 - 2x)$

9.
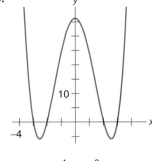
$y = x^4 - 13x^2 + 36$

11.
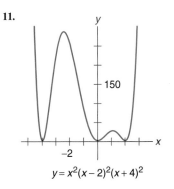
$y = x^2(x - 2)^2(x + 4)^2$

13.

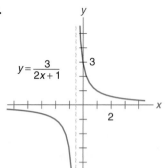

$y = \dfrac{3}{2x+1}$

15.

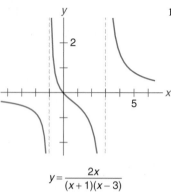

$y = \dfrac{2x}{(x+1)(x-3)}$

17.

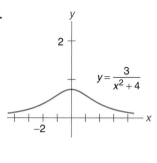

$y = \dfrac{3}{x^2+4}$

19.

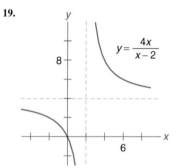

$y = \dfrac{4x}{x-2}$

21.

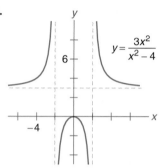

$y = \dfrac{3x^2}{x^2-4}$

23.

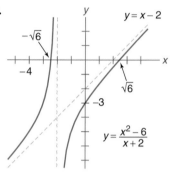

$y = x - 2$

$-\sqrt{6}$

$\sqrt{6}$

$y = \dfrac{x^2-6}{x+2}$

25.

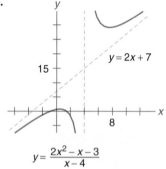

$y = 2x + 7$

$y = \dfrac{2x^2 - x - 3}{x - 4}$

27.

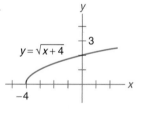

$y = \sqrt{x+4}$

29.

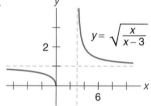

$y = \sqrt{\dfrac{x}{x-3}}$

31.

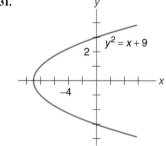

$y^2 = x + 9$

33.

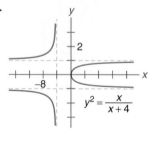

$y^2 = \dfrac{x}{x+4}$

35.

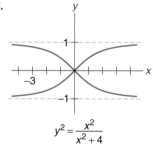

$y^2 = \dfrac{x^2}{x^2+4}$

1.

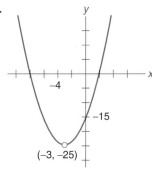

$$y = x^2 + 6x - 16$$

3.

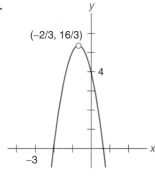

$$y = 4 - 4x - 3x^2$$

5.

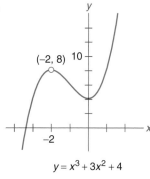

$$y = x^3 + 3x^2 + 4$$

7.

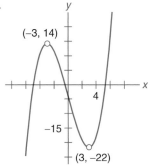

$$y = \frac{1}{3}x^3 - 9x - 4$$

9.

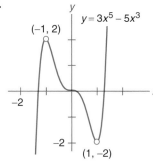

11.

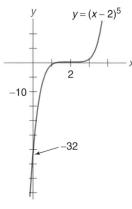

13.

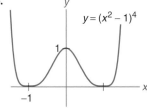

15.

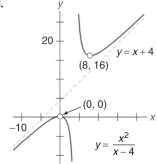

17.

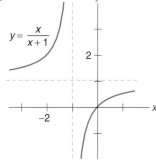

19.

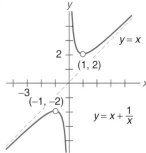

21.

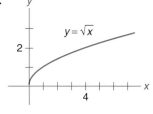

23.

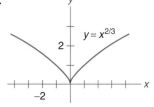

1. (a) Increasing for $x > 2$, decreasing for $x < 2$ **(b)** Relative minimum at $(2, -4)$ **(c)** Concave upward for all values of x
(d) No points of inflection   **(e)** See graph.
3. (a) Increasing for $x > 0$, decreasing for $x < 0$ **(b)** Relative minimum at $(0, 0)$ **(c)** Concave upward for all values of x
(d) No points of inflection   **(e)** See graph.

1e.

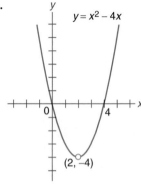

3e.

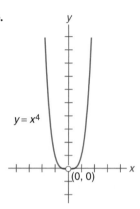

5. (a) Increasing for $-1 < x < 1$, decreasing for $x < -1$ and $x > 1$ **(b)** Relative minimum at $(-1, -2)$, relative maximum at
$(1, 2)$ **(c)** Concave upward for $x < 0$, concave downward for $x > 0$ **(d)** Point of inflection at $(0, 0)$ **(e)** See graph.
7. (a) Increasing for $x < 2$, decreasing for $x > 2$ **(b)** Relative maximum at $(2, 3)$ **(c)** Concave downward for all values of x
(d) No points of inflection   **(e)** See graph.

5e.

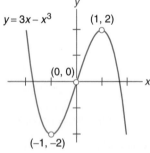

7e.

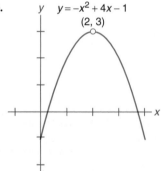

9. (a) Increasing for $-2 < x < 0$ and $x > 2$, decreasing for $x < -2$ and $0 < x < 2$ **(b)** Relative maximum at $(0, 5)$, relative
minimums at $(2, -11)$ and $(-2, -11)$

(c) Concave upward for $x < -\dfrac{2}{\sqrt{3}}$ and $x > \dfrac{2}{\sqrt{3}}$, concave downward for $-\dfrac{2}{\sqrt{3}} < x < \dfrac{2}{\sqrt{3}}$

(d) Points of inflection at $\left(\dfrac{-2}{\sqrt{3}}, \dfrac{-35}{9}\right)$ and $\left(\dfrac{2}{\sqrt{3}}, \dfrac{-35}{9}\right)$ **(e)** See graph.

11. (a) Increasing for $x < 0$, decreasing for $x > 0$ **(b)** No relative maximum or minimum   **(c)** Concave upward for $x > 0$ and
$x < 0$ **(d)** No points of inflection   **(e)** See graph.

9e.

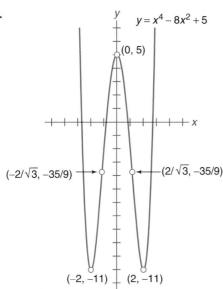

$y = x^4 - 8x^2 + 5$

$(0, 5)$

$(-2/\sqrt{3}, -35/9)$ → $(2/\sqrt{3}, -35/9)$

$(-2, -11)$ $(2, -11)$

11e.

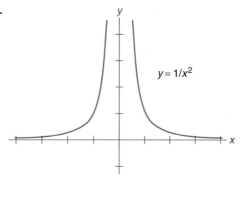

$y = 1/x^2$

13. **(a)** Increasing for $x < -2$ and $x > -2$ **(b)** No relative maximum or minimum   **(c)** Concave upward for $x < -2$, concave downward for $x > -2$ **(d)** No points of inflection   **(e)** See graph.

15. **(a)** Increasing for $x < -1$ and $x > -1$ **(b)** No relative maximum or minimum   **(c)** Concave upward for $x < -1$, concave downward for $x > -1$ **(d)** No points of inflection   **(e)** See graph.

13e.

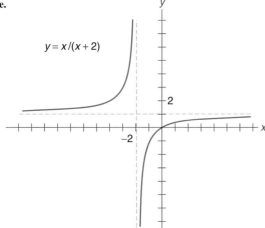

$y = x/(x + 2)$

2

-2

15e.

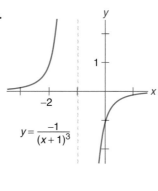

1

-2

$y = \dfrac{-1}{(x+1)^3}$

17. **(a)** Increasing for $-1 < x < 1$, decreasing for $x < -1, x > 1$ **(b)** Relative maximum at $\left(1, \frac{1}{2}\right)$, relative minimum at $\left(-1, -\frac{1}{2}\right)$ **(c)** Concave upward for $-\sqrt{3} < x < 0, x > \sqrt{3}$, concave downward for $x < -\sqrt{3}, 0 < x < \sqrt{3}$ **(d)** Points of inflection at $(\sqrt{3}, \sqrt{3}/4), (0, 0), (-\sqrt{3}, -\sqrt{3}/4)$ **(e)** See graph.

19. **(a)** Increasing for $x < -3, x > -3$ **(b)** No relative maximum or minimum   **(c)** Concave upward for $x < -3$, concave downward for $x > -3$ **(d)** No points of inflection   **(e)** See graph.

17e.

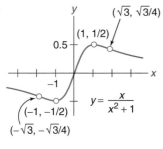

$(\sqrt{3}, \sqrt{3}/4)$

$(1, 1/2)$

0.5

-1

$(-1, -1/2)$

$(-\sqrt{3}, -\sqrt{3}/4)$

$y = \dfrac{x}{x^2 + 1}$

19e.

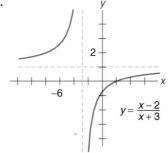

2

-6

$y = \dfrac{x - 2}{x + 3}$

21. (a) Increasing for $x < 0$, decreasing for $x > 0$ **(b)** Relative maximum at $(0, 1)$ **(c)** Concave upward for $x < -2/\sqrt{3}$, $x > 2/\sqrt{3}$, concave downward for $-2/\sqrt{3} < x < 2/\sqrt{3}$ **(d)** Inflection points at $(2/\sqrt{3}, \frac{3}{4})$ and $(-2/\sqrt{3}, \frac{3}{4})$ **(e)** See graph.
23. (a) Increasing for $0 < x < 2$, decreasing for $x < 0, x > 2$ **(b)** Relative maximum at $(2, \frac{1}{4})$ **(c)** Concave upward for $x > 3$, concave downward for $x < 0, 0 < x < 3$ **(d)** Point of inflection $(3, \frac{2}{9})$ **(e)** See graph.

21e.

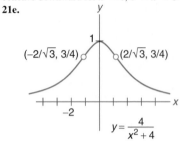

23e.

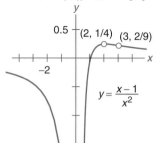

Exercises 3.4, Pages 157–159

1. 28, 28 **3.** 2 cm × 2 cm × 0.5 cm **5.** 200 m × 400 m **7.** 81 cm² **9.** 1600 m **11.** $96\sqrt{3}$ **13.** $\frac{32}{3}$
15. 4 Ω **19.** $m = \frac{3}{2}$ at $x = \frac{1}{2}$ **21.** $i = -16$ A at $t = 2$ s **23.** 20 cm × 20 cm × 10 cm **25.** 20
27. $r = 2\sqrt{2}$ cm, $h = 4\sqrt{2}$ cm **29.** $w = 2r/\sqrt{3}, d = 2r\sqrt{6}/3$ **31.** 4 km

Exercises 3.5, Pages 161–163

1. 36 **3.** −6 **5.** 0.06 Ω/s **7.** 1.92 cm²/min **9.** 43.2 cm³/min **11.** −0.277 cm²/min **13.** 3/(80π) m/s
15. −160 W/s **17.** 0.509 cm/s **19.** $\frac{4}{3}$ m/s **21.** 12 lb/in²/min **23.** −2.4 Ω/s **25.** 240 ΩA **27.** 8 V/s
29. (a) 8 m/min **(b)** $\frac{8}{9}$ m/min

Exercises 3.6, Page 166

1. $dy = (10x - 24x^2)\,dx$ **3.** $dy = \dfrac{-7\,dx}{(2x - 1)^2}$ **5.** $dy = 16t(2t^2 + 1)^3\,dt$

7. $ds = -2(t^4 - t^{-2})^{-3}(4t^3 + 2t^{-3})\,dt$ or $ds = \dfrac{-4t^3(2t^6 + 1)\,dt}{(t^6 - 1)^3}$ **9.** $dy = \dfrac{-x\,dx}{4y}$

11. $dy = \dfrac{x^{-1/2} - 6(x + y)^2}{6(x + y)^2 - y^{-1/2}}\,dx$ or $dy = \dfrac{y^{1/2} - 6x^{1/2}y^{1/2}(x + y)^2}{6x^{1/2}y^{1/2}(x + y)^2 - x^{1/2}}\,dx$ **13.** 43.2 **15.** 1.2 **17.** 282.7

19. (a) 1.20 cm² **(b)** 1.2025 cm² **(c)** 0.833% **21. (a)** 84.9 m³ **(b)** 662,000 kg **23.** 0.3 hp **25.** 4 V

Chapter 3 Review, Pages 169–170

1.

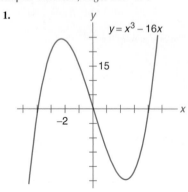

$y = x^3 - 16x$

2.

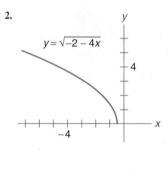

$y = \sqrt{-2 - 4x}$

3.

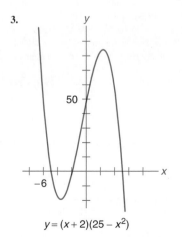

$y = (x + 2)(25 - x^2)$

4.

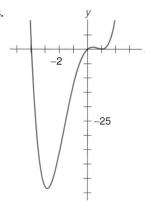

$$y = (x^2 + 4x)(x-1)^2$$

5.

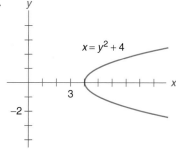

$x = y^2 + 4$

6.

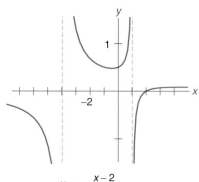

$$y = \frac{x-2}{(x+4)(x-1)}$$

7.

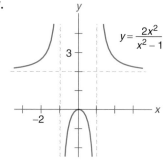

$y = \frac{2x^2}{x^2 - 1}$

8.

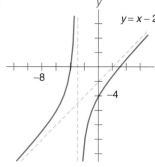

$y = x - 2$

$$y = \frac{x^2 + x - 12}{x+3}$$

9.

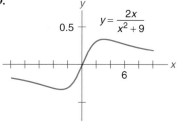

$y = \frac{2x}{x^2 + 9}$

10.

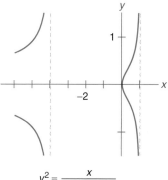

$$y^2 = \frac{x}{(1-x)(x+4)}$$

11. (a) Increasing for $-\sqrt{2} < x < \sqrt{2}$, decreasing for $x < -\sqrt{2}$ and $x > \sqrt{2}$ **(b)** Relative maximum at $(\sqrt{2}, 4\sqrt{2})$, relative minimum at $(-\sqrt{2}, -4\sqrt{2})$ **(c)** Concave upward for $x < 0$, concave downward for $x > 0$ **(d)** Point of inflection at $(0, 0)$ **(e)** See graph.

12. (a) Increasing for $x > \frac{3}{2}$, decreasing for $x < \frac{3}{2}$ **(b)** Relative minimum at $(\frac{3}{2}, -\frac{25}{4})$ **(c)** Concave upward for all values of x **(d)** No points of inflection **(e)** See graph.

13. (a) Increasing for all values of x **(b)** No relative maximum or minimum **(c)** Concave upward for $x > 0$, concave downward for $x < 0$ **(d)** Inflection point at $(0, -7)$ **(e)** See graph.

11e.

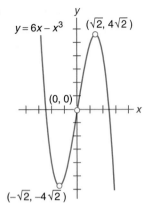

$y = 6x - x^3$

$(\sqrt{2}, 4\sqrt{2})$

$(0, 0)$

$(-\sqrt{2}, -4\sqrt{2})$

12e.

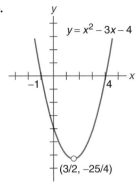

$y = x^2 - 3x - 4$

$(3/2, -25/4)$

13e.

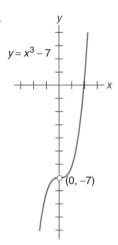

$y = x^3 - 7$

$(0, -7)$

14. (a) Increasing for $x < -1$, $x > 4$, decreasing for $-1 < x < 4$ (b) Relative maximum at $(-1, 11)$, relative minimum at $(4, -114)$ (c) Concave upward for $x > \frac{3}{2}$, concave downward for $x < \frac{3}{2}$ (d) Point of inflection $(1.5, -51.5)$ (e) See graph.
15. (a) Increasing for $x < -1$, decreasing for $x > -1$ (b) No relative maximum or minimum (c) Concave upward for $x < -1$, $x > -1$ (d) No points of inflection (e) See graph.
16. (a) Increasing for $x > 0$, decreasing for $x < 0$ (b) Relative minimum at $(0, -\frac{1}{4})$ (c) Concave upward for $-2/\sqrt{3} < x < 2/\sqrt{3}$, concave downward for $x < -2/\sqrt{3}$, $x > 2/\sqrt{3}$ (d) Points of inflection at $(2/\sqrt{3}, \frac{1}{16})$, $(-2/\sqrt{3}, \frac{1}{16})$
(e) See graph.

14e.

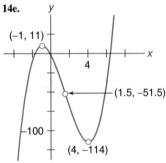

$(-1, 11)$

4

$(1.5, -51.5)$

-100

$(4, -114)$

$y = 2x^3 - 9x^2 - 24x - 2$

15e.

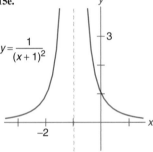

$y = \dfrac{1}{(x+1)^2}$

3

-2

16e.

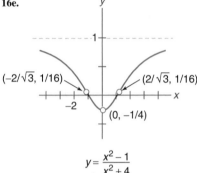

1

$(-2/\sqrt{3}, 1/16)$

$(2/\sqrt{3}, 1/16)$

-2

$(0, -1/4)$

$y = \dfrac{x^2 - 1}{x^2 + 4}$

17. (a) Increasing for $x < 0$, decreasing for $x > 0$ (b) Relative maximum at $(0, 10)$ (c) Concave upward for $x < -1/\sqrt{3}$, $x > 1/\sqrt{3}$, concave downward for $-1/\sqrt{3} < x < 1/\sqrt{3}$ (d) Points of inflection at $(-1/\sqrt{3}, \frac{15}{2})$, $(1/\sqrt{3}, \frac{15}{2})$
(e) See graph.
18. (a) Increasing for $-2 < x < 0$, decreasing for $x < -2$, $x > 0$ (b) Relative minimum at $(-2, -\frac{1}{4})$ (c) Concave upward for $-3 < x < 0$, $x > 0$, concave downward for $x < -3$ (d) Point of inflection $(-3, -\frac{2}{9})$ (e) See graph.

17e.

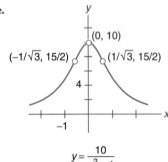

$(0, 10)$

$(-1/\sqrt{3}, 15/2)$

$(1/\sqrt{3}, 15/2)$

4

-1

$y = \dfrac{10}{x^2 + 1}$

18e.

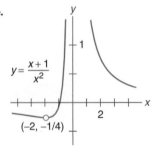

1

$y = \dfrac{x+1}{x^2}$

2

$(-2, -1/4)$

19. 900 ft   **20.** $b = h = 10\sqrt{2}$ **21.** 1 Ω   **22.** 12 m²   **23.** $3\sqrt{2}$ cm   **24.** $(\frac{1}{2}, \sqrt{2}/2)$ **25.** $\dfrac{1}{288\pi}$ ft/s

26. 18 V/s   **27.** -0.0087 A/s   **28.** -0.8π cm²/min or -2.51 cm²/min   **29.** 0.255 km/day

30. 1.92 km/s   **31.** $dy = (20x^4 - 18x^2 + 2)\,dx$ **32.** $dy = -2(3x - 5)^{-5/3}\,dx$ **33.** $ds = \dfrac{(15t^2 + 6t + 20)\,dt}{(5t + 1)^2}$

34. $dy = \dfrac{2 - 4x(x^2 + y^2)}{4y(x^2 + y^2) - 1}\,dx$ **35.** 2.6   **36.** -0.000158 **37.** 45.2 in^3 **38.** 0.3 m   **39.** 9.58 gal

40. 8.04 cm^3 **41.** -74.1 kN

CHAPTER 4

Exercises 4.1, Pages 178–180

37. $\sin 4\theta$ **39.** $\cos \theta$ **41.** $\tan 5\theta$ **43.** $2 \sin \theta \cos \phi$ **45.** $\sin \dfrac{x}{2}$ **47.** $\cos 6x$ **49.** $\cos \dfrac{\theta}{8}$ **51.** $\cos \dfrac{x}{3}$

53. $10 \sin 8\theta$ **55.** $4 \cos 2\theta$

57.

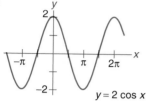
$y = 2 \cos x$

59.

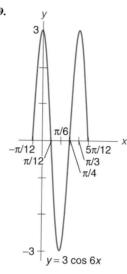
$y = 3 \cos 6x$

61.

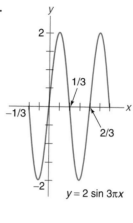
$y = 2 \sin 3\pi x$

63.

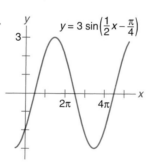
$y = -\sin\left(4x - \dfrac{2\pi}{3}\right)$

65.

$y = 3 \sin\left(\dfrac{1}{2}x - \dfrac{\pi}{4}\right)$

67.

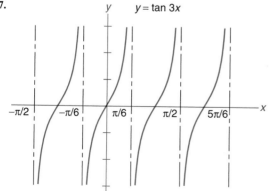
$y = \tan 3x$

Exercises 4.2, Pages 184–185

1. $7\cos 7x$ **3.** $-10\sin 5x$ **5.** $6x^2\cos x^3$ **7.** $-24x\sin 4x^2$ **9.** $-4\cos(1-x)$ **11.** $6x\cos(x^2+4)$
13. $-4(10x+1)\sin(5x^2+x)$ **15.** $-4(x^3-x)\sin(x^4-2x^2+3)$ **17.** $-6\cos(3x-1)\sin(3x-1)$
19. $6\sin^2(2x+3)\cos(2x+3)$ **21.** $4(2x-5)\cos(2x-5)^2$ **23.** $-12x^2(x^3-4)^3\sin(x^3-4)^4$
25. $\cos 3x\cos x - 3\sin x\sin 3x$ **27.** $-6\sin 5x\sin 6x + 5\cos 6x\cos 5x$ **29.** $-7\cos 4x\sin 7x - 4\cos 7x\sin 4x$
31. $2(x+1)\cos x^3\cos(x^2+2x) - 3x^2\sin(x^2+2x)\sin x^3$ **33.** $5(x^2+3x)\cos(5x-2) + (2x+3)\sin(5x-2)$
35. $\dfrac{5x\cos 5x - \sin 5x}{x^2}$ **37.** $\dfrac{2x\cos 3x + 3(x^2-1)\sin 3x}{\cos^2 3x}$ **39.** $5\cos 5x - 6\sin 6x$
41. $(2x-3)\cos(x^2-3x) - 4\sin 4x$ **43.** $\sec^2 x$ **45.** $-\cos x$ **47.** $-\cos x$ **49.** $-25\sin 5x - 36\cos 6x$
51. $6\sqrt{3}$ **53.** $y = -10x + \pi$

Exercises 4.3, Page 189

1. $3\sec^2 x$ **3.** $7\sec 7x\tan 7x$ **5.** $-6x\csc^2(3x^2-7)$ **7.** $-9\csc(3x-4)\cot(3x-4)$
9. $10\tan(5x-2)\sec^2(5x-2)$ **11.** $-24\cot^2 2x\csc^2 2x$ **13.** $\dfrac{2x+1}{2\sqrt{x^2+x}}\sec\sqrt{x^2+x}\tan\sqrt{x^2+x}$
15. $-\dfrac{\csc x(x\cot x + 1)}{3x^2}$ **17.** $3\sec^2 3x - 2x\sec(x^2+1)\tan(x^2+1)$ **19.** $\sec x(2\sec^2 x - 1)$ **21.** $\cos 2x$
23. $\sec x(x\tan x + 1)$ **25.** $2x\sec^2 x(x\tan x + 1)$ **27.** $-3\csc 3x(2\csc^2 3x - 1)$ **29.** $-3\csc 3x\cot 3x$
31. $-2\cos 2x$ **33.** $4(x+\sec^2 3x)^3(1 + 6\sec^2 3x\tan 3x)$ **35.** $3\sec x(\sec x + \tan x)^3$ **37.** $\cos(\tan x)\sec^2 x$
39. $-\sin x\sec^2(\cos x)$ **41.** $-2\sin x\sin(\cos x)\cos(\cos x)$ or $-\sin x\sin(2\cos x)$ **43.** $\dfrac{1 + \sin^2 x}{\cos^3 x}$
45. $-2\sin x\cos x$ or $-\sin 2x$ **47.** $\dfrac{\cos x + \sin x - \sin x\sec^2 x}{(1 + \tan x)^2}$ **49.** $18\sec^2 3x\tan 3x$
51. $2\csc^2 x(x\cot x - 1)$ **53.** 2

Exercises 4.4, Pages 198–199

1. $\dfrac{\pi}{3}, \dfrac{2\pi}{3}$ **3.** $\dfrac{\pi}{4}, \dfrac{5\pi}{4}$ **5.** $\dfrac{5\pi}{6}, \dfrac{7\pi}{6}$ **7.** $\dfrac{2\pi}{3}, \dfrac{4\pi}{3}$ **9.** $\dfrac{3\pi}{4}, \dfrac{5\pi}{4}$ **11.** $\dfrac{\pi}{3}, \dfrac{4\pi}{3}$ **13.** 0 **15.** $1.10, 4.24$
17. $\dfrac{2\pi}{3} + 2n\pi, \dfrac{5\pi}{3} + 2n\pi$, for every integer n **19.** $\dfrac{5\pi}{6} + 2n\pi, \dfrac{7\pi}{6} + 2n\pi$, for every integer n
21. $\dfrac{\pi}{6} + 2n\pi, \dfrac{5\pi}{6} + 2n\pi$, for every integer n **23.** $\pi + 2n\pi$ or $(2n+1)\pi$, for every integer n
25. $\dfrac{\pi}{6} + 2n\pi, \dfrac{7\pi}{6} + 2n\pi$, for every integer n **27.** $\dfrac{\pi}{6} + 2n\pi, \dfrac{11\pi}{6} + 2n\pi$, for every integer n
29. $2n\pi$, for every integer n **31.** $4.17 + 2n\pi, 5.25 + 2n\pi$, for every integer n **33.** $x = \frac{1}{3}\arcsin y$
35. $x = \arccos\dfrac{y}{4}$ **37.** $x = 2\arctan\dfrac{y}{5}$ **39.** $x = 4\text{ arccot }\dfrac{2y}{3}$ **41.** $x = 1 + \arcsin\dfrac{y}{3}$ **43.** $x = -\frac{1}{3} + \frac{1}{3}\arccos 2y$
45. $\dfrac{\pi}{3}$ **47.** $-\dfrac{\pi}{6}$ **49.** $\dfrac{5\pi}{6}$ **51.** $\dfrac{2\pi}{3}$ **53.** $\dfrac{\pi}{4}$ **55.** $\dfrac{\pi}{3}$ **57.** $\dfrac{\pi}{4}$ **59.** $-\dfrac{\pi}{3}$ **61.** $\dfrac{1}{2}$ **63.** $\dfrac{1}{\sqrt{2}}$
65. 0 **67.** $\dfrac{\sqrt{3}}{2}$ **69.** 0.8 **71.** -0.1579 **73.** $\sqrt{1-x^2}$ **75.** $\dfrac{\sqrt{x^2-1}}{x}$ **77.** $\dfrac{1}{x}$ **79.** x
81. $\sqrt{1-4x^2}$ **83.** $2x\sqrt{1-x^2}$

Exercises 4.5, Pages 201–202

1. $\dfrac{5}{\sqrt{1-25x^2}}$ **3.** $\dfrac{3}{1+9x^2}$ **5.** $\dfrac{1}{|1-x|\sqrt{x^2-2x}}$ **7.** $\dfrac{-3}{\sqrt{2x-x^2}}$ **9.** $\dfrac{-12x}{1+9x^4}$ **11.** $\dfrac{15}{|x|\sqrt{x^6-1}}$
13. $\dfrac{3\,\text{Arcsin}^2 x}{\sqrt{1-x^2}}$ **15.** $\dfrac{-12\,\text{Arccos}\,3x}{\sqrt{1-9x^2}}$ **17.** $\dfrac{6\,\text{Arctan}^3\sqrt{x}}{\sqrt{x}(1+x)}$ **19.** 0 **21.** $\dfrac{1-x}{\sqrt{1-x^2}}$
23. $\dfrac{3x}{\sqrt{1-9x^2}} + \text{Arcsin}\,3x$ **25.** $\dfrac{x}{1+x^2} + \text{Arctan}\,x$ **27.** $\text{Arcsin}\,x$ **29.** $\dfrac{\sqrt{1-x^2}\,\text{Arcsin}\,x - x}{\sqrt{1-x^2}\,\text{Arcsin}^2 x}$
31. $2/\sqrt{3}$ **33.** $\dfrac{-\pi - 2}{4}$

1.
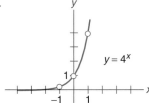
$y = 4^x$

3.
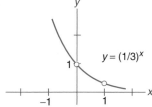
$y = (1/3)^x$

5.
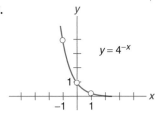
$y = 4^{-x}$

7.
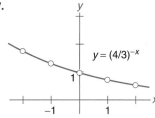
$y = (4/3)^{-x}$

9. $\log_3 9 = 2$ **11.** $\log_5 125 = 3$ **13.** $\log_9 3 = \frac{1}{2}$ **15.** $\log_{10} 0.00001 = -5$ **17.** $5^2 = 25$ **19.** $25^{1/2} = 5$
21. $2^{-2} = \frac{1}{4}$ **23.** $10^{-2} = 0.01$

25.
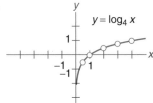
$y = \log_4 x$

27.
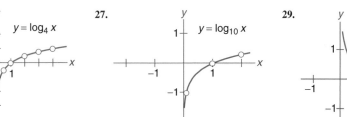
$y = \log_{10} x$

29.
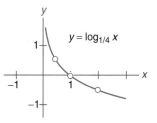
$y = \log_{1/4} x$

31. 64 **33.** $\frac{1}{2}$ **35.** 3 **37.** $\frac{1}{2}$ **39.** 5 **41.** 3 **43.** 144 **45.** 27 **47.** 4
49. $\log_2 5 + 3 \log_2 x + \log_2 y$ **51.** $3 \log_b y + \frac{1}{2} \log_b x - 2 \log_b z$ **53.** $\frac{2}{3} \log_b x - \frac{1}{3} \log_b y$
55. $\frac{1}{2} \log_2 y - \log_2 x - \frac{1}{2} \log_2 z$ **57.** $3 \log_b z + \frac{1}{2} \log_b x - \frac{1}{3} \log_b y$ **59.** $2 \log_b x + \log_b (x + 1) - \frac{1}{2} \log_b (x + 2)$

61. $\log_b xy^2$ **63.** $\log_b \dfrac{xy^2}{z^3}$ **65.** $\log_3 \dfrac{x\sqrt[3]{y}}{\sqrt{z}}$ **67.** $\log_{10} \dfrac{x^2}{(x + 1)\sqrt{x - 3}}$ **69.** $\log_b \dfrac{x^5 \sqrt[3]{x - 1}}{x + 2}$

71. $\log_{10} \dfrac{x(x - 1)^2}{\sqrt[3]{(x + 2)(x - 5)}}$ **73.** 3 **75.** 2 **77.** 3 **79.** -2 **81.** -3 **83.** 0 **85.** 5 **87.** 36 **89.** $\frac{1}{25}$

1. $\dfrac{4 \log e}{4x - 3}$ **3.** $\dfrac{\log_2 e}{x}$ **5.** $\dfrac{6x^2}{2x^3 - 3}$ **7.** $\dfrac{3 \sec^2 3x}{\tan 3x}$ or $3 \sec 3x \csc 3x$ **9.** $\dfrac{x \cos x + \sin x}{x \sin x}$ **11.** $\dfrac{3}{2(3x - 2)}$

13. $\dfrac{x^2 + 3}{x(x^2 + 1)}$ **15.** $\dfrac{\sec^2 (\ln x)}{x}$ **17.** $\dfrac{1}{x \ln x}$ **19.** $\dfrac{2}{x(1 + \ln^2 x^2)}$ **21.** $\dfrac{-2}{(\text{Arccos } x)\sqrt{1 - x^2}}$

23. $(3x + 2)(6x - 1)^2(x - 4)\left(\dfrac{3}{3x + 2} + \dfrac{12}{6x - 1} + \dfrac{1}{x - 4}\right)$

25. $\dfrac{(x + 1)(2x + 1)}{(3x - 4)(1 - 8x)}\left(\dfrac{1}{x + 1} + \dfrac{2}{2x + 1} - \dfrac{3}{3x - 4} + \dfrac{8}{1 - 8x}\right)$

27. $x^x(1 + \ln x)$ **29.** $2x^{2/x}\left(\dfrac{1 - \ln x}{x^2}\right)$ **31.** $(\sin x)^x[x \cot x + \ln (\sin x)]$ **33.** $(1 + x)\left[\dfrac{x^2}{1 + x} + 2x \ln (1 + x)\right]$
35. $y = x - 1$ **37.** $y = \sqrt{3}(x - \pi/6) - \ln 2$

1. $5e^{5x}$ **3.** $12x^2 e^{x^3}$ **5.** $\dfrac{3(10^{3x})}{\log_{10} e}$ **7.** $\dfrac{-6}{e^{6x}}$ **9.** $\dfrac{e^{\sqrt{x}}}{2\sqrt{x}}$ **11.** $(\cos x)e^{\sin x}$ **13.** $6e^{x^2 - 1}(2x^2 + 1)$

15. $e^{3x^2}(6x \cos x - \sin x)$ **17.** $-5e^{5x} \tan e^{5x}$ **19.** $e^x + e^{-x}$

21. $\dfrac{6xe^x(2-x) - 2x^2}{(3e^x - x)^2}$ or $\dfrac{12xe^x - 6x^2e^x - 2x^2}{(3e^x - x)^2}$ **23.** $\dfrac{6e^{3x}}{1 + e^{6x}}$ **25.** $\dfrac{6e^{-2x} \operatorname{Arccos}^2 e^{-2x}}{\sqrt{1 - e^{-4x}}}$ **27.** xe^x **29.** $2x$

Exercises 4.9, Pages 217–219

1.

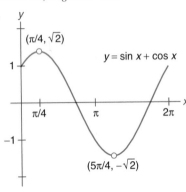

3.

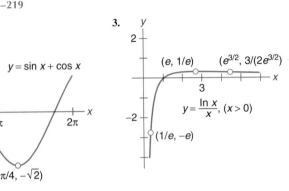

5.

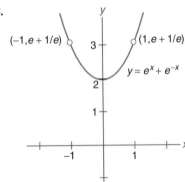

7.

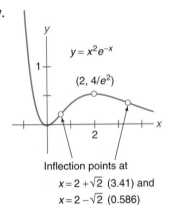

9.

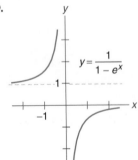

11. $\left(\dfrac{1}{2}, \dfrac{1}{2e}\right)$ maximum   **13.** $(1, e^{-2})$ maximum; $(0, 0)$ minimum   **15.** $\left(e^{1/2}, \dfrac{1}{2e}\right)$ maximum   **17.** $y = -2x + \pi$

19. $y = 3x - 2$ **21.** $-4e^x \sin x$ **23.** $V_L = -300 \sin 2t$ **25.** 33.2 V   **27.** 2.12 W/s

29. $-0.1e^{-0.02t}(\sin 2t + 0.01 \cos 2t)$ **31.** $v = 6e^{3t} - 15e^{-3t}, a = 18e^{3t} + 45e^{-3t}$ **33.** $v = \dfrac{\pi}{2e^2}, a = -\dfrac{\pi}{2e^2}$

35. -0.0672

11. $\frac{1}{2}\sin 2\theta$ **12.** $\cos 6\theta$ **13.** $\cos^2 2\theta$ **14.** $\cos\dfrac{2\theta}{3}$ **15.** $\cos 5x$ **16.** $\sin x$

17.
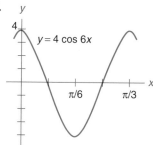

18.
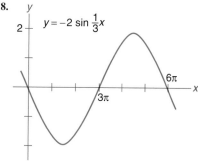

19.
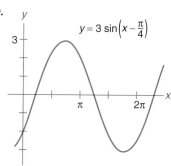

20.
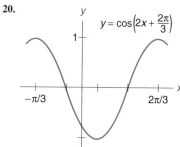

21.
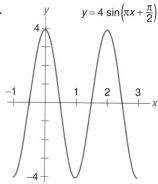

22.
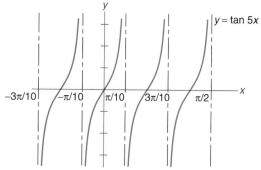

23.

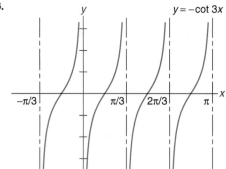

$y = -\cot 3x$

24.

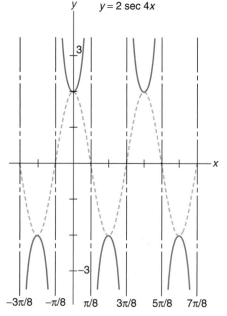

$y = 2 \sec 4x$

25. $2x \cos(x^2 + 3)$ **26.** $-8 \sin 8x$ **27.** $-15 \cos^2(5x - 1) \sin(5x - 1)$

28. $-2 \sin 2x \sin 3x + 3 \cos 2x \cos 3x$ **29.** $3 \sec^2(3x - 2)$ **30.** $4 \sec(4x + 3) \tan(4x + 3)$

31. $-12x \csc^2 6x^2$ **32.** $-2(16x + 1) \csc^2(8x^2 + x) \cot(8x^2 + x)$ **33.** $2 \sec^3 x - \sec x$ **34.** $2x + 2 \csc^2 x \cot x$

35. $\sec x \tan x \sec^2(\sec x)$ **36.** $\dfrac{-1}{1 + \sin x}$ **37.** $-3 \cos x(1 - \sin x)^2$ **38.** $8(1 + \sec 4x) \sec 4x \tan 4x$

39. $\dfrac{\pi}{6}, \dfrac{5\pi}{6}$ **40.** $\dfrac{5\pi}{6}, \dfrac{11\pi}{6}$ **41.** $\dfrac{\pi}{4}, \dfrac{7\pi}{4}$ **42.** $\dfrac{2\pi}{3}, \dfrac{4\pi}{3}$ **43.** $\dfrac{2\pi}{3} + 2n\pi, \dfrac{5\pi}{3} + 2n\pi$, for every integer n

44. $\dfrac{3\pi}{2} + 2n\pi$, for every integer n **45.** $x = \tfrac{4}{3} \arcsin 2y$ **46.** $x = \dfrac{1}{2}\left(1 - \arctan \dfrac{y}{5}\right)$ **47.** $\dfrac{\pi}{4}$ **48.** $-\dfrac{\pi}{6}$

49. π **50.** $\dfrac{2\pi}{3}$ **51.** $\dfrac{\sqrt{3}}{2}$ **52.** $\sqrt{3}$ **53.** $\dfrac{\sqrt{x^2 + 1}}{x^2 + 1}$ **54.** $\dfrac{3x^2}{\sqrt{1 - x^6}}$ **55.** $\dfrac{3}{1 + 9x^2}$

56. $\dfrac{3}{|x|\sqrt{4x^2 - 1}}$ **57.** $\dfrac{2}{|x|\sqrt{16x^2 - 1}}$ **58.** $\dfrac{3 \operatorname{Arcsin} 3\sqrt{x}}{\sqrt{x - 9x^2}}$ **59.** $\dfrac{x}{\sqrt{1 - x^2}} + \operatorname{Arcsin} x$

60.

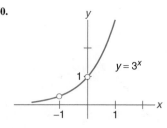

$y = 3^x$

61.

$y = \log_3 x$

62. 81 **63.** 2 **64.** 5 **65.** $\log_4 6 + 2\log_4 x + \log_4 y$ **66.** $\log_3 5 + \log_3 x + \tfrac{1}{2}\log_3 y - 3\log_3 z$

67. $2\log x + 3\log(x + 1) - \tfrac{1}{2}\log(x - 4)$ **68.** $3\ln x + 3\ln(x - 1) - \tfrac{1}{2}\ln(x + 1)$ **69.** $\log_2 \dfrac{xy^3}{z^2}$

70. $\log \dfrac{\sqrt{x + 1}}{(x - 2)^3}$ **71.** $\ln \dfrac{x^4}{(x + 1)^5(x + 2)}$ **72.** $\ln \dfrac{\sqrt{x(x + 2)}}{(x - 5)^2}$ **73.** 3 **74.** x^2 **75.** 2 **76.** x

77. $\dfrac{3x^2}{x^3 - 2}$ **78.** $\dfrac{4\log_3 e}{4x + 1}$ **79.** $\dfrac{6}{x^3 + 3x}$ **80.** $-\dfrac{1}{x}\sin(\ln x)$

81. $\dfrac{\sqrt{x + 1}(3x - 4)}{x^2(x + 2)}\left[\dfrac{1}{2(x + 1)} + \dfrac{3}{3x - 4} - \dfrac{2}{x} - \dfrac{1}{x + 2}\right]$ **82.** $x^{1-x}\left[\dfrac{1 - x}{x} - \ln x\right]$ **83.** $2xe^{x^2 + 5}$

84. $\dfrac{3(8^{3x})}{\log_8 e}$ **85.** $-2e^{2x}\csc^2 e^{2x}$ **86.** $2x\, e^{\sin x^2} \cos x^2$ **87.** $\dfrac{-4e^{-4x}}{\sqrt{1 - e^{-8x}}}$ **88.** $x^2 e^{-4x}(3 - 4x)$

89.

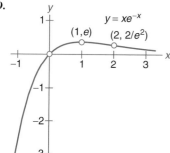

$y = xe^{-x}$
$(1, e)$
$(2, 2/e^2)$

90.

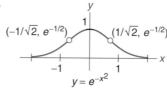

$(-1/\sqrt{2}, e^{-1/2})$
$(1/\sqrt{2}, e^{-1/2})$
$y = e^{-x^2}$

91. $-1000 \sin 5t$ **92.** $p = -360 \cos 3t \sin 3t$ or $-180 \sin 6t$ **93.** $y = 2x - 2$ **94.** $v = (\cos t)e^{\sin t}$
95. $-4e^{-1/2}$ or -2.43

CHAPTER 5

Exercises 5.1, Pages 230–231

1. $\frac{1}{8}x^8 + C$ **3.** $\frac{1}{3}x^9 + C$ **5.** $4x + C$ **7.** $\frac{54}{11}x^{11/6} + C$ **9.** $-\dfrac{3}{x^2} + C$ **11.** $\dfrac{5x^3}{3} - 6x^2 + 8x + C$

13. $x^3 - \frac{1}{2}x^2 - \dfrac{5}{2x^2} + C$ **15.** $\frac{4}{5}x^5 - 4x^3 + 9x + C$ **17.** $\frac{1}{9}(6x + 2)^{3/2} + C$ **19.** $(x^2 + 3)^4 + C$

21. $\frac{3}{40}(5x^2 - 1)^{4/3} + C$ **23.** $\frac{1}{10}(x^2 - 1)^5 + C$ **25.** $2\sqrt{x^2 + 1} + C$ **27.** $\frac{1}{4}(x^3 + 2x)^4 + C$ **29.** $\dfrac{-1}{3(x^3 - 4)} + C$

31. $\frac{2}{3}(5x^2 - x)^{3/2} + C$ **33.** $2\sqrt{x^2 + x} + C$ **35.** $\frac{1}{6}(2x + 3)^3 + C$ **37.** $\frac{1}{10}(2x - 1)^5 + C$

39. $\dfrac{x^7}{7} + \dfrac{3x^5}{5} + x^3 + x + C$ **41.** $\frac{1}{2}(x^2 + 1)^4 + C$ **43.** $\frac{2}{5}(5x^3 + 1)^5 + C$ **45.** $4\sqrt{x^3 + 1} + C$

47. $3(x^3 + 3x)^{2/3} + C$ **49.** $-1/x + 1/(2x^2) + C$

Exercises 5.2, Pages 234–235

1. $y = \frac{3}{2}x^2 + 1$ **3.** $y = x^3 + 3x + 6$ **5.** $y = \frac{1}{6}(x^2 - 3)^3 + 1$ **7.** $s = \frac{1}{2}t^3 + 16t + 50$
9. 64 ft (36 ft from the ground at $t = 2$); -80 ft/s   **11.** 240 ft/s   **13. (a)** 31.9 m **(b)** 5.10 s **(c)** 25 m/s
15. (a) $s = -16t^2 + 30t + 200$ **(b)** 4.59 s   **17.** 213 rev   **19.** 633 V   **21.** $q = \dfrac{(t^2 + 1)^{3/2} - 1}{3}; \dfrac{2\sqrt{2} - 1}{3}$

Exercises 5.3, Pages 241–242

1. 2   **3.** $\frac{52}{3}$ **5.** 4   **7.** $\frac{3}{2}$ **9.** $\frac{14}{9}$ **11.** 4   **13.** $\frac{8}{5}$ **15.** $\frac{98}{3}$ **17.** $\frac{3}{4}$ **19.** $\frac{4}{5}$ **21.** $\frac{3}{7}$ **23.** 36
25. $\frac{4}{3}$ **27.** $\frac{1}{12}$ **29.** $\frac{4}{15}$

Exercises 5.4, Pages 245–246

1. $\frac{5}{2}$ **3.** $\frac{16}{3}$ **5.** -6 **7.** $\frac{14}{3}$ **9.** 40   **11.** $\frac{88}{3}$ **13.** 5546.2   **15.** $\frac{208}{3}$ **17.** $-\frac{15}{2}$ **19.** $-\frac{3}{10}$
21. $\dfrac{2\sqrt{2} - 1}{3}$ **23.** 12   **25.** $2\sqrt{10} - 2\sqrt{2}$

Chapter 5 Review, Page 247

1. $\frac{5}{3}x^3 - \frac{1}{2}x^2 + C$ **2.** $\frac{3}{8}x^8 + x^2 + 4x + C$ **3.** $\frac{4}{3}x^{9/2} + C$ **4.** $\frac{12}{5}x^{5/3} + C$ **5.** $\dfrac{-3}{4x^4} + C$ **6.** $\dfrac{-2}{\sqrt{x}} + C$
7. $\frac{1}{8}(3x^4 + 2x - 1)^4 + C$ **8.** $\frac{5}{16}(7x^2 + 8x + 2)^{8/5} + C$ **9.** $2\sqrt{x^2 + 5x} + C$ **10.** $3(5x^3 + 4x)^{1/3} + C$
11. $y = x^3 - 4$ **12.** $s = -16t^2 + 25t + 100$ **13.** 69.2 Ω   **14.** 2.97   **15.** $\frac{14}{3}$ **16.** 6   **17.** $\frac{15}{64}$ **18.** 1
19. $\dfrac{144 - 4\sqrt{6}}{5}$ **20.** $\dfrac{16\sqrt{2} - 8}{3}$ **21.** $\frac{17}{12}$ **22.** $\frac{108}{5}$ **23.** $\frac{62}{3}$ **24.** $\frac{7}{18}$ **25.** $\dfrac{12\sqrt{6} - 4\sqrt{2}}{3}$
26. $\dfrac{4\sqrt{7} - 2}{3}$ **27.** -6 **28.** $\dfrac{6 - 3\sqrt{2}}{4}$

CHAPTER 6

Exercises 6.1, Pages 254–255

1. $\frac{1}{3}$ **3.** $\frac{1}{2}$ **5.** $\frac{9}{2}$ **7.** $\frac{32}{3}$ **9.** $\frac{1}{6}$ **11.** $\frac{9}{2}$ **13.** $\frac{1}{2}$ **15.** $\frac{9}{2}$ **17.** 4 **19.** $\frac{44}{15}$ **21.** $\frac{343}{6}$ **23.** $\frac{22}{5}$ **25.** $\frac{37}{12}$

Exercises 6.2, Page 262

1. $\frac{26\pi}{3}$ **3.** $\frac{178\pi}{15}$ **5.** $\frac{8\pi}{3}$ **7.** 2π **9.** $\frac{\pi}{3}$ **11.** $\frac{\pi}{3}$ **13.** 2π **15.** 8π **17.** $\frac{4\pi}{15}$ **19.** $\frac{5\pi}{6}$
21. $\frac{8\pi}{3}$ **23.** $\frac{32\pi}{3}$ **25.** 60π **27.** $\frac{100\pi\sqrt{5}}{9}$

Exercises 6.3, Page 266

1. 2π **3.** 2π **5.** 4π **7.** $\frac{128\pi}{7}$ **9.** $\frac{8\pi}{3}$ **11.** $\frac{\pi}{3}$ **13.** $\frac{5\pi}{6}$ **15.** $\frac{4\pi}{21}$ **17.** $\frac{\pi}{2}$ **19.** $\frac{16\pi}{5}$

Exercises 6.4, Pages 269–270

1. 15/7 **3.** −6.75 **5.** −6 **7.** $-\frac{11}{3}$ **9.** 15 **11.** 40 **13.** 11 mi north of Flatville **15.** (20/7, 22/7)
17. (−3.8, −7.2) **19.** (2.1, −8.4) **21.** 100 **23.** 1.50 mi east and 2.56 mi south of A

Exercises 6.5, Pages 281–282

1. 10 cm **3.** $\frac{20}{3}$ cm **5.** 8.77 cm **7.** 4 cm from given end **9.** $(4\frac{2}{3}, 4\frac{2}{3})$ **11.** (8, 2) **13.** (0, 2.95)
15. $(\frac{27}{5}, \frac{9}{8})$ **17.** (1, −0.4) **19.** (0, 1.6) **21.** $(\frac{8}{15}, \frac{8}{21})$ **23.** $\left(0, \frac{4}{3\pi}\right)$ **25.** $(\frac{7}{8}, 0)$ **27.** $(\frac{3}{4}, 0)$ **29.** $(0, \frac{2}{3})$

Exercises 6.6, Pages 288–289

1. 516; 3.79 **3.** 1965; 5.98 **5.** 963; 5.17 **7.** 759; 4.87 **9.** $\frac{64}{3}$; 0.894 **11.** $\frac{1}{7}$; 0.463 **13.** $\frac{64}{5}$; 0.894
15. 2; 1.41 **17.** 576π; 1.55 **19.** 2.29×10^4; 8.43 **21.** 1458π; 1.73 **23.** 4096π; 2.53

Exercises 6.7, Pages 298–299

1. $\frac{63}{4}$ **3.** 25 in.-lb **5.** 675 N cm or 6.75 J **7.** 2.896×10^{-14} J **9.** (a) 900 ft-lb (b) 1875 ft-lb (c) 2500 ft-lb
11. 225,800 ft-lb **13.** 602,200 ft-lb **15.** 58,810 ft-lb **17.** 25,000 lb **19.** 352,800 N **21.** 710,500 N
23. 166.4 lb **25.** 54,660 lb **27.** 21,902 lb **29.** $\frac{13}{3}$ **31.** $\frac{2}{5}$ **33.** 1.70 A

Chapter 6 Review, Pages 303–304

1. $\frac{16}{3}$ **2.** $\frac{2}{3}$ **3.** $\frac{1}{12}$ **4.** 8 **5.** $\frac{128}{15}$ **6.** 36 **7.** 8π **8.** 8π **9.** $\frac{\pi}{30}$ **10.** $\frac{8\pi}{3}$ **11.** $\frac{243\pi}{10}$
12. $\frac{99\pi}{2}$ **13.** $\frac{256\pi}{5}$ **14.** $\frac{63\pi}{2}$ **15.** 7.5 **16.** (−1, −6.8) **17.** (9.6, 5.3) **18.** $(2\frac{2}{3}, 6\frac{2}{3})$ **19.** $(\frac{3}{2}, \frac{27}{5})$
20. (3, −3.6) **21.** $(0, \frac{3}{2})$ **22.** $(\frac{5}{6}, 0)$ **23.** (0, 2) **24.** 462; 4.39 **25.** 192; 2.83 **26.** 576; 4.90
27. $\frac{8}{15}$; 0.447 **28.** $\frac{2\pi}{13}$; 0.519 **29.** $\frac{8\pi}{7}$; 0.845 **30.** 126π; 2.65 **31.** 200 in.-lb **32.** 2.62×10^{-16} J
33. 130,000 ft-lb **34.** 49,920 lb **35.** 816,700 N **36.** 9.5 V **37.** 67.52 A **38.** 20 W

CHAPTER 7

Exercises 7.1, Page 308

1. $\frac{2}{9}(3x + 2)^{3/2} + C$ **3.** $2\sqrt{4 + x} + C$ **5.** $\frac{2}{7}(x^2 + 4x)^{7/4} + C$ **7.** $-\frac{1}{4}\cos^4 x + C$ **9.** $\frac{1}{16}\tan^4 4x + C$
11. $-\frac{1}{8}(\cos 4x + 1)^2 + C$ **13.** $\frac{2}{3}(9 + \sec x)^{3/2} + C$ **15.** $\frac{1}{3}(1 + e^{2x})^{3/2} + C$ **17.** $\sqrt{1 + e^{x^2}} + C$
19. $\frac{1}{6}\ln^2|3x - 5| + C$ **21.** $-\dfrac{1}{\ln|x|} + C$ **23.** $\frac{1}{6}\text{Arcsin}^2 3x + C$ **25.** $\frac{1}{4}\sin^4 x + C$ **27.** $\frac{1}{3}\text{Arctan}^3 x + C$
29. $\frac{64}{3}$ **31.** $\frac{2}{3}(\sqrt{1 + e^3} - \sqrt{2})$ **33.** $\frac{1}{4}\ln^2 3$ **35.** $\frac{1}{3}$

Exercises 7.2, Page 312

1. $\frac{1}{3}\ln|3x + 2| + C$ **3.** $-\frac{1}{4}\ln|1 - 4x| + C$ **5.** $-2\ln|1 - x^2| + C$ **7.** $\frac{1}{4}\ln|x^4 - 1| + C$
9. $-\ln|\cot x| + C$ or $\ln|\tan x| + C$ **11.** $\frac{1}{3}\ln|1 + \tan 3x| + C$ **13.** $-\ln|1 + \csc x| + C$ **15.** $\ln|1 + \sin x| + C$

17. $\ln|\ln|x|| + C$ **19.** $\frac{1}{2}e^{2x} + C$ **21.** $\frac{-1}{4e^{4x}} + C$ **23.** $\frac{1}{2}e^{x^2} + C$ **25.** $-\frac{1}{2}e^{-x^2-9} + C$ **27.** $-e^{\cos x} + C$

29. $\frac{e^2}{2}(e^4 - 1)$ **31.** $\frac{4^x}{\ln 4} + C$ **33.** $2\ln|e^x + 4| + C$ **35.** $\frac{1}{2}\ln 2$ or 0.347 **37.** $2\ln 9$ or $\ln 81$ or 4.39

39. $2(e - 1)$ or 3.44 **41.** $\frac{1}{3}(e - 1)$ or 0.573 **43.** $\ln 2$ or 0.693 **45.** $\frac{1}{2}\ln 3$ or 0.549 **47.** $\frac{1}{2}(e^8 - 1)$ or 1490

Exercises 7.3, Pages 317–318

1. $-\frac{1}{5}\cos 5x + C$ **3.** $\frac{1}{3}\sin(3x - 1) + C$ **5.** $-\frac{1}{2}\cos(x^2 + 5) + C$ **7.** $\sin(x^3 - x^2) + C$ **9.** $-\frac{1}{5}\cot 5x + C$

11. $\frac{1}{3}\sec 3x + C$ **13.** $\frac{1}{4}\tan(4x + 3) + C$ **15.** $-\frac{1}{2}\csc(2x - 3) + C$ **17.** $\frac{1}{2}\tan(x^2 + 3) + C$

19. $-\frac{1}{3}\csc(x^3 - 1) + C$ **21.** $-\frac{1}{4}\ln|\cos 4x| + C$ **23.** $\frac{1}{5}\ln|\sec 5x + \tan 5x| + C$ **25.** $\ln|\sin e^x| + C$

27. $x + 2\ln|\sec x + \tan x| + \tan x + C$ **29.** $5\ln|\sec x + \tan x| - \ln|\cos x| + C$ **31.** $\frac{1}{2}$ **33.** 3 **35.** 1

37. $\frac{\sqrt{2} - 1}{2}$ **39.** 2 **41.** 1 **43.** $\frac{1}{2}\ln 2$ **45.** π

Exercises 7.4, Page 321

1. $\frac{1}{3}\cos^3 x - \cos x + C$ **3.** $\sin x - \frac{2}{3}\sin^3 x + \frac{1}{5}\sin^5 x + C$ **5.** $\frac{1}{3}\sin^3 x + C$ **7.** $\frac{-1}{2\cos^2 x} + C$

9. $-\frac{1}{3}\cos^3 x + \frac{1}{5}\cos^5 x + C$ **11.** $\frac{x}{2} - \frac{1}{4}\sin 2x + C$ **13.** $\frac{3x}{8} + \frac{1}{12}\sin 6x + \frac{1}{96}\sin 12x + C$

15. $\frac{x}{8} - \frac{1}{32}\sin 4x + C$ **17.** $\frac{x}{16} - \frac{1}{64}\sin 4x + \frac{1}{48}\sin^3 2x + C$ **19.** $\frac{1}{2}\tan^2 x + \ln|\cos x| + C$

21. $x - \frac{1}{6}\cot^3 2x + \frac{1}{2}\cot 2x + C$ **23.** $\tan x + \frac{2}{3}\tan^3 x + \frac{1}{5}\tan^5 x + C$

25. $\frac{1}{6}\tan^3 2x - \frac{1}{2}\tan 2x + x + C$ **27.** $\frac{\pi}{2}$

Exercises 7.5, Page 324

1. $\frac{1}{3}\text{Arcsin }3x + C$ **3.** $\text{Arcsin }\frac{x}{3} + C$ **5.** $\frac{1}{5}\text{Arctan }\frac{x}{5} + C$ **7.** $\frac{1}{6}\text{Arctan }\frac{3x}{2} + C$ **9.** $\frac{1}{5}\text{Arcsin }\frac{5x}{6} + C$

11. $\frac{1}{2\sqrt{3}}\text{Arcsin }2x + C$ **13.** $\frac{1}{2}\text{Arctan}\left(\frac{x-1}{2}\right) + C$ **15.** $\frac{1}{4}\text{Arctan}\left(\frac{x+3}{4}\right) + C$ **17.** $\text{Arcsin }e^x + C$

19. $-\text{Arctan}(\cos x) + C$ **21.** $\frac{\pi}{4}$ **23.** 0.215 **25.** 10.9 N

Exercises 7.6, Page 331

1. $\frac{5}{x+2} + \frac{3}{x-7}$ **3.** $\frac{3}{2x+3} - \frac{2}{x-4}$ **5.** $\frac{7}{x} + \frac{2}{3x-4} + \frac{5}{2x+1}$ **7.** $\frac{1}{x+1} + \frac{1}{(x+3)^2}$

9. $\frac{3}{4x-1} + \frac{1}{(4x-1)^2} - \frac{7}{(4x-1)^3}$ **11.** $\frac{3}{x} + \frac{8}{x-1} - \frac{4}{(x-1)^2}$ **13.** $\frac{x-1}{x^2+1} - \frac{x}{x^2-3}$

15. $\frac{4x+1}{x^2+x+1} - \frac{1}{x^2-5}$ **17.** $\frac{5x-2}{x^2+5x+3} + \frac{1}{x+3} - \frac{2}{x-3}$ **19.** $\frac{5}{x} + \frac{3x-1}{x^2+1} - \frac{5}{(x^2+1)^2}$

21. $\frac{1}{x} - \frac{2}{x^2} - \frac{4x}{(x^2+2)^2}$ **23.** $\frac{2}{x+3} + \frac{4}{x-3} - \frac{6x}{x^2+9}$ **25.** $x + \frac{\frac{1}{2}}{x+1} + \frac{\frac{1}{2}}{x-1}$ **27.** $x - 1 + \frac{3}{x-2} + \frac{1}{x+2}$

29. $3x - 2 + \frac{5}{x} - \frac{8x}{x^2+1}$

Exercises 7.7, Pages 333–334

1. $\frac{1}{2}\ln\left|\frac{x+1}{x-1}\right| + C$ **3.** $\frac{1}{6}\ln\left|\frac{x-2}{x+4}\right| + C$ **5.** $2\ln|x-2| - \ln|x-1| + C$ or $\ln\left|\frac{(x-2)^2}{x-1}\right| + C$

7. $\frac{2}{3}\ln|x+5| + \frac{1}{3}\ln|x-1| + C$ or $\frac{1}{3}\ln|(x+5)^2(x-1)| + C$ **9.** $\ln\left|\frac{x}{x+1}\right| + \frac{1}{x+1} + C$

11. $2\ln|x+3| - \frac{1}{x} + C$ **13.** $\frac{x^2}{2} - 3x + \ln\left|\frac{(x+2)^8}{x+1}\right| + C$ **15.** $\frac{3}{2}\ln|x^2+1| - 2\ln|x| + C$

17. $\frac{1}{2}\ln|x^2+9| + \text{Arctan }\frac{x}{3} + \frac{1}{x} + C$ **19.** $\frac{1}{2(x^2+1)} + \frac{1}{2}\ln|x^2+1| + C$ **21.** $\frac{3}{2}\ln\frac{2}{3}$ **23.** $\frac{11}{6}\ln 3 - \frac{5}{6}\ln 7$

25. $\ln 3 + 3\ln\frac{7}{5}$

1. $x \ln|x| - x + C$ **3.** $xe^x - e^x + C$ **5.** $\frac{2}{3}x^{3/2}\ln|x| - \frac{4}{9}x^{3/2} + C$ **7.** $x \ln x^2 - 2x + C$

9. $x \operatorname{Arccos} x - \sqrt{1 - x^2} + C$ **11.** $\frac{1}{2}e^x(\sin x + \cos x) + C$ **13.** $x^2 \sin x + 2x \cos x - 2 \sin x + C$

15. $x \tan x + \ln|\cos x| + C$ **17.** $x(\ln|x|)^2 - 2x \ln|x| + 2x + C$ **19.** $x \sec x - \ln|\sec x + \tan x| + C$

21. $\frac{1}{9}(2e^3 + 1)$ **23.** $\frac{16}{15}$ **25.** $3 \ln 3 - 2 \ln 2 - 1$ or $\ln \frac{27}{4} - 1$ **27.** $\ln 2 - \frac{1}{2}$ **29.** 2π

1. $\frac{1}{2}\ln|2x + \sqrt{9 + 4x^2}| + C$ **3.** $-\frac{x}{18}\sqrt{4 - 9x^2} + \frac{2}{27}\operatorname{Arcsin}\frac{3x}{2} + C$ **5.** $\operatorname{Arcsin}\frac{1}{3}$ **7.** $\frac{1}{4\sqrt{3}}$

9. $\frac{1}{2}\ln\left|\frac{\sqrt{x^2 + 4} - 2}{x}\right| + C$ **11.** $\ln|x + \sqrt{x^2 - 9}| - \frac{\sqrt{x^2 - 9}}{x} + C$ **13.** $\frac{-\sqrt{16 - x^2}}{16x} + C$

15. $\sqrt{9 + x^2} - 3\ln\left|\frac{3 + \sqrt{9 + x^2}}{x}\right| + C$ **17.** $\frac{x}{25\sqrt{25 - x^2}} + C$ **19.** $\ln|x + \sqrt{x^2 + 9}| + C$

21. $\frac{9x^2 - 8}{243}\sqrt{9x^2 + 4} + C$ **23.** $\ln|x - 3 + \sqrt{x^2 - 6x + 8}| + C$ **25.** $-\frac{x + 4}{\sqrt{x^2 + 8x + 15}} + C$ **27.** $\ln(1 + \sqrt{2})$

1. $\frac{1}{\sqrt{5}}\ln\left|\frac{\sqrt{x + 5} - \sqrt{5}}{\sqrt{x + 5} + \sqrt{5}}\right| + C; 15$ **3.** $\ln|x + \sqrt{x^2 - 4}| + C; 35$ **5.** $-\frac{(3 - x)\sqrt{2x + 3}}{3}; 13$

7. $\frac{1}{4}\left(\frac{\sin 4x}{2} - \frac{\sin 10x}{5}\right) + C; 75$ **9.** $-\frac{x}{2}\sqrt{9 - x^2} + \frac{9}{2}\operatorname{Arcsin}\frac{x}{3} + C; 24$ **11.** $\frac{1}{1 + 9x} + \ln\left|\frac{x}{1 + 9x}\right| + C; 9$

13. $\frac{1}{10}\ln\left|\frac{x - 5}{x + 5}\right| + C; 20$ **15.** $\frac{x}{2}\sqrt{x^2 + 4} + 2\ln|x + \sqrt{x^2 + 4}| + C; 30$ **17.** $\frac{1}{16}\left(\frac{3}{3 + 4x} + \ln|3 + 4x|\right) + C; 7$

19. $\frac{1}{4}\operatorname{Arccos}\frac{4}{3x} + C; 36$ **21.** $\frac{e^{3x}}{25}(3 \sin 4x - 4 \cos 4x) + C; 59$ **23.** $\frac{1}{2}\sin(2x - 3) - \frac{(2x - 3)}{2}\cos(2x - 3) + C; 80$

25. $-\frac{1}{4}\sin^3 x \cos x + \frac{3x}{8} - \frac{3}{8}\sin x \cos x + C; 83$ **27.** $\sqrt{9x^2 - 16} - 4\operatorname{Arccos}\frac{4}{3x} + C; 33$

1. 1.117 **3.** 0.783 **5.** 1.913 **7.** 7.395 **9.** 0.925 **11.** 219 ft-lb **13.** 270 **15.** 1.443 **17.** 0.105

19. 1.464 **21.** 0.186 **23.** 1.910 **25. (a)** 6.889 **(b)** 6.998 **27. (a)** 1.023 **(b)** 1.000

29. (a) 1.006 **(b)** 1.006 **31. (a)** 373,650 ft^2 **(b)** 378,200 ft^2

1. $\frac{4\pi}{3}$ **3.** $\frac{9\pi}{4}$ **5.** 9 **7.** $\frac{3\pi}{2}$ **9.** 8π **11.** 8π **13.** $\frac{41\pi}{2}$ **15.** $\frac{1}{4}(e^{2\pi} - 1)$ **17.** $\pi - \frac{3\sqrt{3}}{2}$

19. 2π **21.** $\frac{\pi}{4} - \frac{3\sqrt{3}}{16}$ **23.** $\frac{2\pi}{3} + \sqrt{3}$ **25.** $\frac{11\pi}{12} + \sqrt{3}$ **27.** $9\sqrt{2} + \frac{27\pi}{8} + \frac{9}{4}$

Chapter 7 Review, Pages 362–364

1. $\frac{2}{3}\sqrt{2 + \sin 3x} + C$ **2.** $\frac{(5 + \tan 2x)^4}{8} + C$ **3.** $\frac{1}{3}\sin 3x + C$ **4.** $\frac{1}{2}\ln|x^2 - 5| + C$ **5.** $\frac{1}{6}e^{3x^2} + C$

6. $\frac{1}{3}\operatorname{Arctan}\frac{2x}{3} + C$ **7.** $\operatorname{Arcsin}\frac{x}{4} + C$ **8.** $\frac{1}{7}\tan(7x + 2) + C$ **9.** $\frac{1}{5}\ln|3 + 5\tan x| + C$

10. $-\frac{1}{3}\cos(x^3 + 4) + C$ **11.** $\frac{1}{12}\operatorname{Arctan}\frac{3}{4}$ **12.** $\frac{1}{\pi}$ **13.** $-\ln\frac{\sqrt{2}}{2}$ or $\ln\sqrt{2}$ **14.** $\frac{1}{2}\operatorname{Arcsin}\frac{2}{3}$

15. $\frac{1}{3}\operatorname{Arccos}\frac{3}{4x} + C$ **16.** $-\frac{1}{12}\cos^4 3x + C$ **17.** $\frac{1}{6}\operatorname{Arctan}^2 3x + C$ **18.** $\frac{1}{5}\ln|\sec 5x + \tan 5x| + C$

19. $-\frac{1}{2}\ln|\cos x^2| + C$ **20.** $-\frac{\cos^3 2x}{6} + \frac{1}{5}\cos^5 2x - \frac{\cos^7 2x}{14} + C$ **21.** $\frac{x}{16} - \frac{1}{192}\sin 12x + \frac{1}{144}\sin^3 6x + C$

22. $\frac{1}{4}\ln\left|\frac{x - 1}{x + 3}\right| + C$ **23.** $\frac{e^{3x}}{25}(3 \cos 4x + 4 \sin 4x)$ **24.** $\cos x(1 - \ln|\cos x|) + C$ **25.** $\frac{\tan^3 x}{3} - \tan x + x + C$

26. $\frac{x}{2} + \frac{\sin 10x}{20} + C$ **27.** $\frac{2}{3}\ln|3x + 1| - \frac{1}{2}\ln|2x - 1| + C$ **28.** $\frac{2}{3}x^{3/2}(\ln|x| - \frac{2}{3}) + C$ **29.** $\frac{\cos^3(e^{-x})}{3} + C$

30. $\ln\left|\frac{x - 2}{x + 2}\right| + C$ **31.** $-x^2 \cos x + 2x \sin x + 2 \cos x + C$ **32.** $\frac{1}{10}(\operatorname{Arcsin} 5x)^2 + C$ **33.** $\operatorname{Arcsec} 3 - \operatorname{Arcsec} 2$

34. $\dfrac{3e^4 + 1}{16}$ **35.** $\ln 2$ **36.** $2 \ln 3$ or $\ln 9$ **37.** $\dfrac{-1}{(x - 2)} + 3 \ln |x| + C$ **38.** $\dfrac{1}{4} \ln \left| \dfrac{4 - \sqrt{16 - 9x^2}}{3x} \right| + C$

39. $\ln 3$ **40.** $e^5 - e^3$ or $e^3(e^2 - 1)$ **41.** $\frac{1}{2}(e^2 - 1)$ **42.** $\dfrac{\pi}{4}$ **43.** $\dfrac{\pi}{6}$ **44.** $\frac{1}{2} + \ln 2$ **45.** 1 **46.** $\dfrac{\pi}{8}$

47. $\frac{1}{2} \ln 2$ **48.** $\pi(2 \ln^2 2 - 4 \ln 2 + 2)$ **49.** $\dfrac{\pi}{2}(e^2 - 1)$ **50.** $-\dfrac{4t}{3} \cos 3t + \dfrac{4}{9} \sin 3t + C$

51. $3 \ln |t + 2| + 2 \ln |t - 1| + C$ **52.** $\dfrac{e}{2}(e^3 - 1)$ **53.** $\frac{1}{5} \ln \left| \dfrac{x}{5 + 3x} \right| + C$; 5

54. $-\dfrac{\sqrt{16 - x^2}}{x} - \arcsin \dfrac{x}{4} + C$; 29 **55.** $\ln |2x + 6 + 2\sqrt{3 + 6x + x^2}| + C$; 47 **56.** $\frac{1}{5} e^x (\sin 2x - 2 \cos 2x) + C$; 78

57. $\sqrt{4x^2 - 9} - 3 \operatorname{arcsec} \dfrac{2x}{3} + C$; 33 **58.** $-\frac{1}{15} \cos^5 3x + C$; 73 **59.** $2\sqrt{9 + 4x} + 3 \ln \left| \dfrac{\sqrt{9 + 4x} - 3}{\sqrt{9 + 4x} + 3} \right| + C$; 17, 15

60. $\frac{1}{5} \tan^5 x - \frac{1}{3} \tan^3 x + \tan x - x + C$; 85 **61.** 1.011 **62.** 0.155 **63.** 12.359 **64.** 6.489 **65.** 22.2

66. 0.96 **67.** 18.365 **68.** 5.608 **69.** 0.605 **70.** 71.632 **71.** 4 **72.** $\dfrac{3\pi}{2}$ **73.** 11π **74.** $\dfrac{\pi}{16}$

75. $\dfrac{\pi^3}{8}$ **76.** $\dfrac{\sqrt{3}}{2} + \dfrac{\pi}{3}$ **77.** $4\pi - 6\sqrt{3}$ **78.** $2 - \pi/4$ **79.** $4\sqrt{3} - \dfrac{4\pi}{3}$ **80.** $1 - \sqrt{2}/2$

CHAPTER 8

Exercises 8.1, Page 373

1.

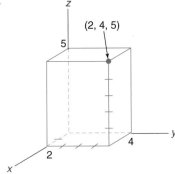

3.

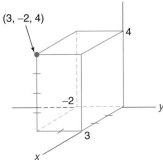

5.

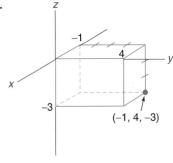

7. Plane

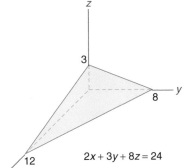

$2x + 3y + 8z = 24$

9. Plane

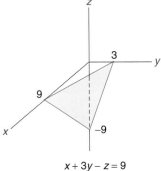

$x + 3y - z = 9$

11. Plane

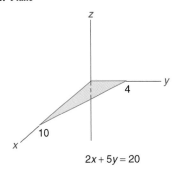

$2x + 5y = 20$

13. Plane

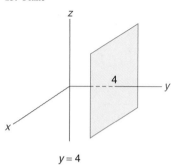

$y = 4$

15. Sphere

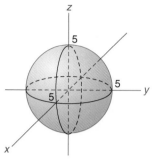

$x^2 + y^2 + z^2 = 25$

17. Sphere

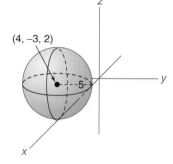

$(4, -3, 2)$

$x^2 + y^2 + z^2 - 8x + 6y - 4z + 4 = 0$

19. Cylindrical surface

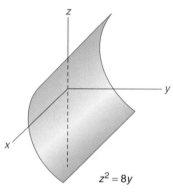

$z^2 = 8y$

21. Cylindrical surface

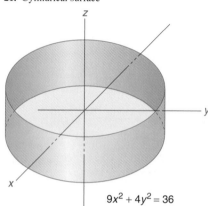

$9x^2 + 4y^2 = 36$

23. Ellipsoid

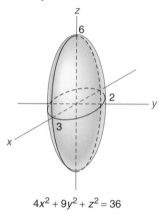

$4x^2 + 9y^2 + z^2 = 36$

25. Elliptic paraboloid

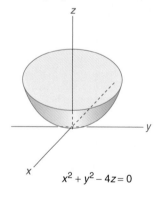

$x^2 + y^2 - 4z = 0$

27. Hyperboloid of two sheets

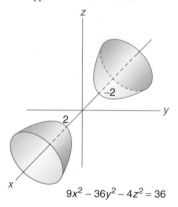

$9x^2 - 36y^2 - 4z^2 = 36$

29. Cylindrical surface

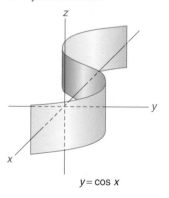

$y = \cos x$

31. Hyperbolic paraboloid

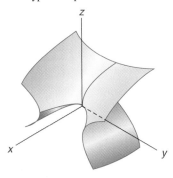

$$y^2 - z^2 = 8x$$

33. Elliptic cone

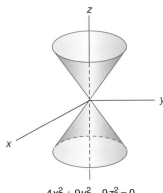

$$4x^2 + 9y^2 - 9z^2 = 0$$

35. Hyperboloid of one sheet

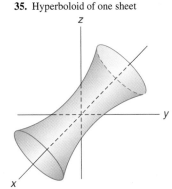

$$81y^2 + 36z^2 - 4x^2 = 324$$

37. $(x - 3)^2 + (y + 2)^2 + (z - 4)^2 = 36$ **39.** $(x - 3)^2 + (y + 2)^2 + (z - 4)^2 = 16$ **43.** 6

Exercises 8.2, Pages 376–377

1. (a) $12x^2y^2$ **(b)** $8x^3y$ **3. (a)** $12xy^4 + 2y^2$ **(b)** $24x^2y^3 + 4xy$ **5. (a)** $\dfrac{x}{\sqrt{x^2 + y^2}}$ **(b)** $\dfrac{y}{\sqrt{x^2 + y^2}}$

7. (a) $\dfrac{x^2 + y^2}{2x^2y}$ **(b)** $\dfrac{-x^2 - y^2}{2xy^2}$ **9. (a)** ay **(b)** ax **11. (a)** $1/x$ **(b)** $-1/y$ **13. (a)** $\sec^2(x - y)$ **(b)** $-\sec^2(x - y)$

15. (a) $e^{3x}(y \cos xy + 3 \sin xy)$ **(b)** $xe^{3x} \cos xy$ **17. (a)** $\dfrac{e^{xy}(xy - 1)}{x^2 \sin y}$ **(b)** $\dfrac{e^{xy}(x \sin y - \cos y)}{x \sin^2 y}$

19. (a) $-2 \sin x \sin y$ **(b)** $2 \cos x \cos y$ **21. (a)** $xy^2 \sec^2 xy + y \tan xy$ **(b)** $x^2y \sec^2 xy + x \tan xy$

23. $2IR$ **25.** $\dfrac{-E}{(R + r)^2}$ **27.** $\dfrac{R}{\sqrt{R^2 + X_L^2}}$ **29.** $2\pi Ef \cos 2\pi ft$ **31.** $R_2 + R_3$ **33.** $\dfrac{-E}{R^2C}e^{-t/(RC)}$

35. $Ee^{-t/(RC)}\left(\dfrac{t}{RC} + 1\right)$ **37.** $\dfrac{-X_L}{R^2 + X_L^2}$ **39. (a)** 18 **(b)** -16 **41. (a)** $\frac{5}{3}$ **(b)** $-\frac{12}{5}$ **45.** 96π cm^3

Exercises 8.3, Pages 384–385

1. $(6x + 4y)\,dx + (4x + 3y^2)\,dy$ **3.** $2x \cos y\,dx - x^2 \sin y\,dy$ **5.** $\dfrac{1}{x^2}\,dx - \dfrac{1}{y^2}\,dy$

7. $\dfrac{y}{2(1 + xy)}\,dx + \dfrac{x}{2(1 + xy)}\,dy$ **9.** 195 cm^3 **11.** 17 Ω **13.** -64.1 cm^3 **15.** Maximum, $(1, -2, 20)$

17. Minimum, $(1, 1, 3)$ **19.** Saddle point, $(1, -2, -1)$ **21.** Saddle point, $(3, 2, -5)$

23. Minimum, $(3, 1, -109)$; saddle point, $(-1, 1, 19)$ **25.** 10 cm $\times$ 10 cm $\times$ 5 cm   **27.** 10, 10, 10

Exercises 8.4, Page 390

1. 20   **3.** $\frac{17}{8}$ **5.** $\frac{14}{5}$ **7.** $\frac{2}{3}$ **9.** $\frac{1}{2}e^2 - e + \frac{1}{2}$ **11.** $\frac{244}{3}$ **13.** $1 - \pi/4$ **15.** 16   **17.** 2   **19.** 16

Chapter 8 Review, Page 392

1. Plane

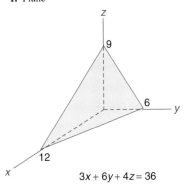

$$3x + 6y + 4z = 36$$

2. Elliptic paraboloid

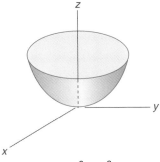

$$9x^2 + 9y^2 = 3z$$

3. Hyperboloid of one sheet

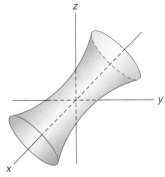

$$36y^2 + 9z^2 - 16x^2 = 144$$

4. Ellipsoid

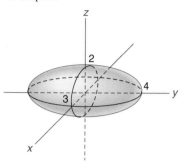

$$16x^2 + 9y^2 + 36z^2 = 144$$

5. Cylindrical surface

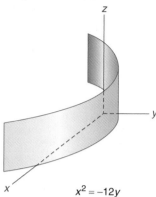

$$x^2 = -12y$$

6. Hyperbolic paraboloid

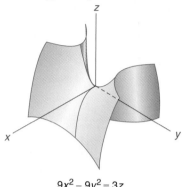

$$9x^2 - 9y^2 = 3z$$

7. Hyperboloid of two sheets

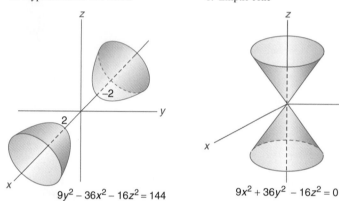

$$9y^2 - 36x^2 - 16z^2 = 144$$

8. Elliptic cone

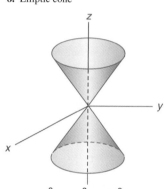

$$9x^2 + 36y^2 - 16z^2 = 0$$

9. $(x + 3)^2 + (y - 2)^2 + (z - 1)^2 = 9$ **10.** $5\sqrt{2}$ **11.** (a) $3x^2 + 6xy$ (b) $3x^2 + 4y$

12. (a) $6xe^{2y}$ (b) $6x^2e^{2y}$ **13.** (a) $2/x$ (b) $1/y$ **14.** (a) $3\cos 3x \sin 3y$ (b) $3\sin 3x \cos 3y$

15. (a) $\dfrac{ye^{x^2}(2x^2 - 1)}{x^2 \ln y}$ (b) $\dfrac{e^{x^2}(\ln y - 1)}{x \ln^2 y}$ **16.** (a) $\dfrac{e^x \sin y(\sin x - \cos x)}{y \sin^2 x}$ (b) $\dfrac{e^x(y\cos y - \sin y)}{y^2 \sin x}$

17. $\dfrac{-p^{1/2}}{2w^{3/2}}$ **18.** LI **19.** $\dfrac{X_C}{\sqrt{R^2 + X_C^2}}$ **20.** $\dfrac{rE}{(R + r)^2}$ **21.** (a) 8 (b) 2   **22.** $\dfrac{2x + y}{2(x^2 + xy)}\,dx + \dfrac{x}{2(x^2 + xy)}\,dy$

23. $\dfrac{2y}{(x + y)^2}\,dx - \dfrac{2x}{(x + y)^2}\,dy$ **24.** 452 L   **25.** 0.000666 A or 0.666 mA   **26.** Saddle point (5, 2, −33)

27. Saddle point, (1, 2, −7)   **28.** Maximum, (0, −4, 25)   **29.** 2 m × 2 m × 2 m   **30.** $\frac{1}{20}$ **31.** 516

32. $\frac{14}{3}$ **33.** $\ln \sqrt{2}$ or $\frac{1}{2}\ln 2$ **34.** $abc/6$ **35.** $\frac{2}{3}$ **36.** $\frac{4}{15}$

CHAPTER 9

Exercises 9.1, Pages 395–396

1. 17   **3.** 24   **5.** −95   **7.** 57   **9.** 202.5   **11.** −546   **13.** 2, −1, −4, −7, −10   **15.** 5, $5\frac{2}{3}$, $6\frac{1}{3}$, 7, $7\frac{2}{3}$

17. 4   **19.** 1,000,000   **21.** $31,200

Exercises 9.2, Pages 399–400

1. $\frac{20}{2187}$ **3.** 8   **5.** $-\frac{1}{64}$ **7.** $\dfrac{65,600}{2187}$ **9.** $7\sqrt{2} + 14$ **11.** $\frac{341}{64}$ **13.** 3, $\frac{3}{2}$, $\frac{3}{4}$, $\frac{3}{8}$, $\frac{3}{16}$ **15.** 5, $-\frac{5}{4}$, $\frac{5}{16}$, $-\frac{5}{64}$, $\frac{5}{256}$

17. −4, −12, −36, −108, −324   **19.** $\frac{1}{2}$ **21.** $17,531   **23.** $\frac{3}{8}$ ft   **25.** 15.1°C   **27.** $\frac{14}{3}$ **29.** $\frac{8}{3}$

31. No sum   **33.** No sum   **35.** $\frac{1}{3}$ **37.** $\frac{2}{165}$ **39.** $\frac{13}{15}$

Exercises 9.3, Page 402

1. $27x^3 + 27x^2y + 9xy^2 + y^3$ **3.** $a^5 - 10a^4 + 40a^3 - 80a^2 + 80a - 32$ **5.** $16x^4 - 32x^3 + 24x^2 - 8x + 1$

7. $64a^6 + 576a^5b + 2160a^4b^2 + 4320a^3b^3 + 4860a^2b^4 + 2916ab^5 + 729b^6$

9. $\frac{32}{243}x^5 - \frac{160}{81}x^4 + \frac{320}{27}x^3 - \frac{320}{9}x^2 + \frac{160}{3}x - 32$ **11.** $a^2 + 12a^{3/2}b^2 + 54ab^4 + 108a^{1/2}b^6 + 81b^8$

13. $\dfrac{x^4}{y^4} - \dfrac{8x^3}{y^3z} + \dfrac{24x^2}{y^2z^2} - \dfrac{32x}{yz^3} + \dfrac{16}{z^4}$ **15.** $-126x^4y^5$ **17.** $41{,}184a^5b^8$ **19.** $280x^3y^8$ **21.** $4320x^3y^3$

23. $-8064x^5$

Chapter 9 Review, Page 403

1. 47   **2.** $\frac{1}{16}$ **3.** -81 **4.** -62 **5.** $\frac{2}{81}$ **6.** 95   **7.** 300   **8.** $\frac{127}{16}$ **9.** $\dfrac{-80\sqrt{3}}{1+\sqrt{3}}$ **10.** -348

11. $\frac{728}{81}$ **12.** 500   **13.** 1,001,000   **14.** \$2988 (approx.)   **15.** No sum   **16.** $\frac{35}{6}$ **17.** $\frac{3}{2}$ **18.** No sum

19. $\frac{5}{11}$ **20.** $\frac{152}{165}$ **21.** $a^6 - 6a^5b + 15a^4b^2 - 20a^3b^3 + 15a^2b^4 - 6ab^5 + b^6$

22. $32x^{10} - 80x^8 + 80x^6 - 40x^4 + 10x^2 - 1$ **23.** $16x^4 + 96x^3y + 216x^2y^2 + 216xy^3 + 81y^4$

24. $1 + 8x + 28x^2 + 56x^3 + 70x^4 + 56x^5 + 28x^6 + 8x^7 + x^8$ **25.** $90x^2$ **26.** $1280a^3b^3$ **27.** $8064x^5b^{10}$

28. $3{,}247{,}695x^{16}$

CHAPTER 10

Exercises 10.1, Pages 410–411

1. $5 + 9 + 13 + 17 + 21 + 25$ **3.** $10 + 17 + 26 + 37 + 50 + 65$

5. $\dfrac{1}{2} + \dfrac{4}{3} + \dfrac{9}{4} + \dfrac{16}{5} + \cdots + \dfrac{n^2}{n+1}$ **7.** $-1 + \dfrac{1}{4} - \dfrac{1}{9} + \dfrac{1}{16} - \dfrac{1}{25} + \cdots$ **9.** $\displaystyle\sum_{n=1}^{12} n$ **11.** $\displaystyle\sum_{n=1}^{50} (2n)$

13. $\displaystyle\sum_{k=1}^{n} (2k-1)$ **15.** $\displaystyle\sum_{k=3}^{n} (k^2+1)$ **17.** Diverges   **19.** Diverges   **21.** Diverges   **23.** Converges

25. Converges   **27.** Converges   **29.** Diverges   **31.** Converges   **33.** Diverges   **35.** Converges

37. Converges   **39.** Diverges   **41.** Converges   **43.** Diverges

Exercises 10.2, Page 414

1. Converges   **3.** Converges   **5.** Converges   **7.** Diverges   **9.** Converges   **11.** Diverges   **13.** Diverges

15. Diverges   **17.** Diverges   **19.** Converges

Exercises 10.3, Page 416

1. Converges conditionally   **3.** Converges absolutely   **5.** Diverges   **7.** Converges conditionally

9. Converges absolutely   **11.** Diverges   **13.** Diverges   **15.** Converges conditionally   **17.** Diverges

19. Converges absolutely

Exercises 10.4, Page 420

1. $-2 < x < 2$ **3.** $x = 0$ **5.** $-\infty < x < \infty$ **7.** $-1 \le x \le 1$ **9.** $-1 \le x < 1$ **11.** $-\frac{3}{2} < x < \frac{3}{2}$

13. $-2 < x \le 2$ **15.** $1 < x \le 3$ **17.** $-1 \le x \le 1$ **19.** $-\infty < x < \infty$ **21.** $-\frac{3}{2} \le x < \frac{3}{2}$

23. $2 \le x \le 3$

Exercises 10.5, Page 423

1. $x - \dfrac{x^3}{3!} + \dfrac{x^5}{5!} - \cdots$ **3.** $1 - x + \dfrac{x^2}{2!} - \dfrac{x^3}{3!} + \cdots$ **5.** $x - \dfrac{x^2}{2} + \dfrac{x^3}{3} - \dfrac{x^4}{4} + \cdots$ **7.** $1 - \dfrac{4x^2}{2!} + \dfrac{16x^4}{4!} - \cdots$

9. $x + x^2 + \dfrac{x^3}{2!} + \dfrac{x^4}{3!} + \cdots$ **11.** $2 - \dfrac{x}{4} - \dfrac{x^2}{(32)(2!)} - \dfrac{3x^3}{256(3!)} - \cdots$ **13.** $-1 + \dfrac{x^2}{2!} - \dfrac{x^4}{4!} + \cdots$

15. $1 + 2x + 3x^2 + 4x^3 + \cdots$ **17.** $1 + 5x + 10x^2 + 10x^3 + 5x^4 + x^5$ (Sum is finite.)

19. $x - x^2 + \frac{1}{3}x^3 - \frac{1}{30}x^5 + \cdots$

Exercises 10.6, Pages 426–427

1. $1 - x + \dfrac{x^2}{2!} - \dfrac{x^3}{3!} + \dfrac{x^4}{4!} - \cdots$ **3.** $1 + x^2 + \dfrac{x^4}{2!} + \dfrac{x^6}{3!} + \cdots$ **5.** $-x - \dfrac{x^2}{2} - \dfrac{x^3}{3} - \dfrac{x^4}{4} - \cdots$

7. $1 - \dfrac{25x^4}{2!} + \dfrac{625x^8}{3!} - \cdots$ **9.** $x^3 - \dfrac{x^9}{3!} + \dfrac{x^{15}}{5!} - \cdots$ **11.** $x + x^2 + \dfrac{x^3}{2!} + \dfrac{x^4}{3!} + \dfrac{x^5}{4!} + \cdots$

13. $-\dfrac{x}{2!} + \dfrac{x^3}{4!} - \dfrac{x^5}{6!} + \cdots$ **15.** 0.743 **17.** $\ln 2 + \frac{7}{4}$ **19.** 0.602 **21.** $x + \dfrac{x^3}{3!} + \dfrac{x^5}{5!} + \cdots$

23. 0.041481 coulomb

1. $-\left(x - \dfrac{\pi}{2}\right) + \dfrac{\left(x - \dfrac{\pi}{2}\right)^3}{3!} - \dfrac{\left(x - \dfrac{\pi}{2}\right)^5}{5!} + \cdots$ **3.** $e^2\left[1 + (x - 2) + \dfrac{(x - 2)^2}{2!} + \dfrac{(x - 2)^3}{3!} + \cdots\right]$

5. $3 + \dfrac{(x - 9)}{6} - \dfrac{(x - 9)^2}{108(2!)} + \dfrac{(x - 9)^3}{648(3!)} + \cdots$ **7.** $\dfrac{1}{2} - \dfrac{(x - 2)}{4} + \dfrac{(x - 2)^2}{4(2!)} - \dfrac{3(x - 2)^3}{8(3!)} + \cdots$

9. $(x - 1) - \dfrac{(x - 1)^2}{2!} + \dfrac{2(x - 1)^3}{3!} - \dfrac{6(x - 1)^4}{4!} + \cdots$ **11.** $1 - \tfrac{1}{2}(x - 1) + \tfrac{3}{8}(x - 1)^2 - \tfrac{5}{16}(x - 1)^3 + \cdots$

13. $1 - 2(x - 1) + 3(x - 1)^2 - 4(x - 1)^3 + \cdots$ **15.** $-1 + \dfrac{(x - \pi)^2}{2!} - \dfrac{(x - \pi)^4}{4!} + \cdots$

1. 1.10517   **3.** 0.99985   **5.** -0.68229 **7.** 1.0488   **9.** 3.66832   **11.** 0.48481   **13.** 0.029996

1. $-\pi + 2 \sin x + \sin 2x + \tfrac{2}{3} \sin 3x + \cdots$ **3.** $\dfrac{\pi}{3} - \dfrac{2}{3} \sin x - \dfrac{1}{3} \sin 2x - \dfrac{2}{9} \sin 3x - \cdots$

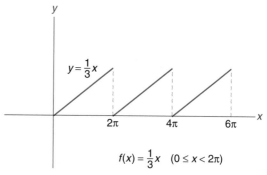

5. $\dfrac{1}{2} - \dfrac{2}{\pi} \sin x - \dfrac{2}{3\pi} \sin 3x - \dfrac{2}{5\pi} \sin 5x - \cdots$

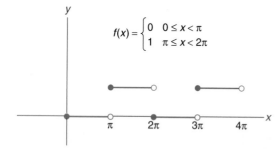

7. $\dfrac{4}{\pi} \sin x + \dfrac{4}{3\pi} \sin 3x + \dfrac{4}{5\pi} \sin 5x + \cdots$

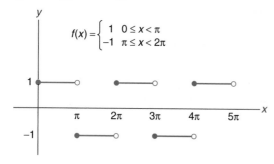

9. $3 + \dfrac{12}{\pi} \sin \dfrac{\pi x}{5} + \dfrac{4}{\pi} \sin \dfrac{3\pi x}{5} + \dfrac{12}{5\pi} \sin \pi x + \cdots$

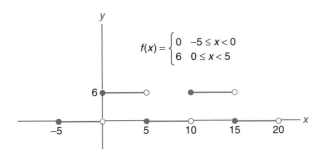

$$f(x) = \begin{cases} 0 & -5 \le x < 0 \\ 6 & 0 \le x < 5 \end{cases}$$

11. $\dfrac{\pi}{2} - \dfrac{4}{\pi} \cos x - \dfrac{4}{9\pi} \cos 3x - \dfrac{4}{25\pi} \cos 5x - \cdots$

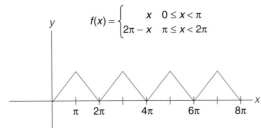

$$f(x) = \begin{cases} x & 0 \le x < \pi \\ 2\pi - x & \pi \le x < 2\pi \end{cases}$$

13. $\dfrac{e^{2\pi} - 1}{2\pi} + \dfrac{1}{\pi} \cdot \dfrac{e^{2\pi} - 1}{2} \cos x + \dfrac{1}{\pi} \cdot \dfrac{e^{2\pi} - 1}{5} \cos 2x + \dfrac{1}{\pi} \cdot \dfrac{e^{2\pi} - 1}{10} \cos 3x + \cdots + \dfrac{1}{\pi} \cdot \dfrac{1 - e^{2\pi}}{2} \sin x +$ $\dfrac{1}{\pi} \cdot \dfrac{2 - e^{2\pi}}{5} \sin 2x + \cdots$

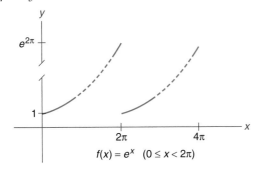

$$f(x) = e^x \quad (0 \le x < 2\pi)$$

15. $\dfrac{1}{\pi} + \dfrac{1}{2} \sin x - \dfrac{2}{3\pi} \cos 2x - \dfrac{2}{15\pi} \cos 4x - \cdots$

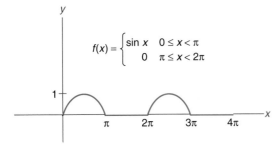

$$f(x) = \begin{cases} \sin x & 0 \le x < \pi \\ 0 & \pi \le x < 2\pi \end{cases}$$

1. $-2 - 5 - 8 - 11 - 14 - 17$ **2.** $2 + \dfrac{3}{2} + \dfrac{4}{3} + \dfrac{5}{4} + \cdots + \dfrac{n+1}{n}$ **3.** $\displaystyle\sum_{n=1}^{7} \dfrac{1}{3^n}$ **4.** $\displaystyle\sum_{n=1}^{10} \dfrac{n}{n+3}$ **5.** Converges

6. Diverges **7.** Converges **8.** Converges **9.** Diverges **10.** Converges **11.** Converges **12.** Diverges

13. Diverges **14.** Converges **15.** Converges absolutely **16.** Converges absolutely **17.** Diverges

18. Converges conditionally **19.** $1 < x < 3$ **20.** $2 < x < 4$ **21.** $-\infty < x < \infty$ **22.** $\frac{11}{3} \le x \le \frac{13}{3}$

23. $1 + x + x^2 + x^3 + \cdots$ **24.** $1 + \dfrac{x}{2} - \dfrac{x^2}{8} + \dfrac{x^3}{16} - \cdots$ **25.** $1 + x - \dfrac{x^2}{2!} - \dfrac{x^3}{3!} + \dfrac{x^4}{4!} + \cdots$ **26.** $x + x^2 + \frac{1}{3}x^3 + \cdots$

27. $-1 - \dfrac{x}{2!} - \dfrac{x^2}{3!} - \cdots$ **28.** $1 - \dfrac{x^4}{2!} + \dfrac{x^8}{4!} - \dfrac{x^{12}}{6!} + \cdots$ **29.** $3x - \dfrac{9x^3}{2} + \dfrac{81x^5}{40} - \cdots$

30. $1 + \sin x + \dfrac{\sin^2 x}{2!} + \dfrac{\sin^3 x}{3!} + \cdots$ **31.** 0.09772 **32.** 0.09994

33. $\dfrac{1}{2} - \sqrt{3}\left(x - \dfrac{\pi}{6}\right) - 2\left(x - \dfrac{\pi}{6}\right)^2 + 4\sqrt{3}\left(x - \dfrac{\pi}{6}\right)^3 + 8\left(x - \dfrac{\pi}{6}\right)^4 + \cdots$

34. $\ln 4 + \dfrac{x-4}{4} - \dfrac{(x-4)^2}{32} + \dfrac{(x-4)^3}{192} - \cdots$ **35.** $e\left[1 + 2(x-1) + \dfrac{6(x-1)^2}{2!} + \dfrac{20(x-1)^3}{3!} + \cdots\right]$

36. $-1 + \dfrac{\left(x - \dfrac{3\pi}{2}\right)^2}{2!} - \dfrac{\left(x - \dfrac{3\pi}{2}\right)^4}{4!} + \cdots$ **37.** 0.5150 **38.** 3.3199 **39.** 0.18227 **40.** 2.024846

41. $-\dfrac{1}{2} + \dfrac{2}{\pi}\sin x + \dfrac{2}{3\pi}\sin 3x + \dfrac{2}{5\pi}\sin 5x + \cdots$

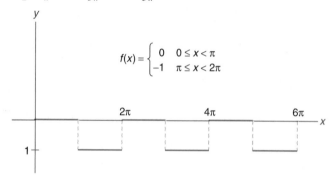

42. $\dfrac{\pi^2}{6} - 2\cos x + \dfrac{1}{2}\cos 2x - \dfrac{2}{9}\cos 3x + \cdots + \dfrac{\pi^2 - 4}{\pi}\sin x - \dfrac{\pi}{2}\sin 2x + \dfrac{9\pi^2 - 4}{27\pi}\sin 3x - \cdots$

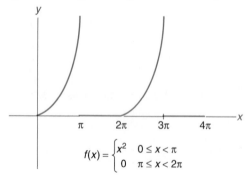

CHAPTER 11

Exercises 11.1, Pages 445–446

1. Order 1; degree 1 **3.** Order 2; degree 1 **5.** Order 3; degree 1 **7.** Order 2; degree 3

Exercises 11.2, Page 451

1. $\dfrac{-1}{y} = \ln x - C$ or $y \ln x + 1 = Cy$ **3.** $y = \dfrac{C}{x}$ or $xy = C$ **5.** $-2y^{-1/2} = x + C$ **7.** Arctan $y = \dfrac{x^3}{3} + C$

9. $-\ln(3 - y) = \ln x + C$ or $x(3 - y) = C$ **11.** $3y^2 = 2x^3 + C$ **13.** $1 = 2y^2(\sin x + C)$ **15.** $2y + e^{-2x} = C$

17. $y = \text{Arctan } x + C$ **19.** $\text{Arctan } y = x + \dfrac{x^3}{3} + C$ **21.** $e^y = e^x + C$ **23.** $\text{Arctan } \dfrac{y}{2} = 2\ln(x + 1) + C$

25. $(x^2 + 1)^2 = C(y + 3)^5$ **27.** $1 = (2 - x^3)y^3$ **29.** $y^2 = 2\ln(x^2 + 1) + 16$ **31.** $y^2 = 2e^x + 34$

33. $x^{3/2} + y^{3/2} = 9$ **35.** $y^2 = \ln^2 x$

Exercises 11.3, Page 454

1. $3xy = y^3 + C$ **3.** $y = 5xy + Cx$ **5.** $x + xy + Cy + 3 = 0$ **7.** $x^2 = 4\sqrt{x^2 + y^2} + C$

9. $\ln\sqrt{x^2 + y^2} = xy + C$ **11.** $xy = x^2 + y^2 - 1$ **13.** $\text{Arctan } \dfrac{y}{x} = xy + \dfrac{\pi - 16}{4}$

Exercises 11.4, Page 456

1. $2y + e^{3x} = Ce^{5x}$ **3.** $14x^3y = 2x^7 - 7x^4 + C$ **5.** $y = e^{3x} + Ce^{-x^2}$ **7.** $3y = x^3e^{4x} + Ce^{4x}$

9. $y = x^6 - x + Cx^5$ **11.** $(1 + x^2)y = x^3 + C$ **13.** $5x^2y = x^5 - 35x + C$ **15.** $y = e^{-x} + Ce^{-2x}$

17. $y = 3x^2 + Cx$ **19.** $(y - 1)e^{\sin x} = C$ **21.** $y = e^{2x}(3e^x - 1)$ **23.** $y \sin x = x + \pi$ **25.** $y = \dfrac{e^x}{2} + e^{-x}$

27. $y = 3 - \dfrac{5}{x}$ **29.** $xy = x^4 - 10$

Exercises 11.5, Pages 462–463

1. $v = 5t + 10$; $v = 25$ m/s   **3.** $3xy = x^3 + 2$ **5.** $i = \frac{1}{2}(3 + e^{-800t})$ **7.** $i = 1 - e^{-5t^2}$ **9.** $Q = e^{-1.54 \times 10^{-10}t}$

11. $Q = 5e^{-0.00293t}$; 237 yr   **13.** 72.9 lb   **15.** 24.3°C   **17.** 3.2×10^7 **19.** $P = CV^k$

Chapter 11 Review, Pages 464–465

1. Order 2; degree 1   **2.** Order 1; degree 2   **3.** Order 2; degree 3   **4.** Order 1; degree 2

9. $2x^2 \ln y = -1 + Cx^2$ **10.** $2\ln y + 3e^{-2x} = C$ **11.** $2y^3 + 3\ln(9 + x^2) = C$ **12.** $\tan x + e^{-y} = C$

13. $y = x^4 + Cx$ **14.** $(y - 2)e^{2x^3} = C$ **15.** $(y + 5)e^{1/x} = C$ **16.** $\ln(x^2 + y^2) = y^2 + C$

17. $3\text{ Arctan } \dfrac{y}{x} = x^3 + C$ **18.** $3xy = 7x^6 + C$ **19.** $ye^{3x} = e^x + C$ **20.** $4y = -4x^4 - 7x + Cx^5$

21. $y = 4x^3 \ln x - x^2 + Cx^3$ **22.** $2y \sin x = \sin^2 x + C$ **23.** $3x^2y - 10y + 2 = 0$ **24.** $y^2 = 3e^{-2x} + 1$

25. $4y = x^5 + 3x$ **26.** $xy = \dfrac{x^2}{4}(2\ln x - 1) + 1$ **27.** $4y = e^{5x} - 13e^x$ **28.** $2y = x^4 - 10x^2 \ln x + 5x^2$

29. $i = 2 - e^{-300t}$ **30.** $Q = e^{-7.88 \times 10^{-10}t}$ **31.** 4.524 kg   **32.** 20.6°C   **33.** $y^3 = 8e^{x^3}$ **34.** $v = t^4$; 81 m/s²

CHAPTER 12

Exercises 12.1, Page 471

1. Homogeneous, 4   **3.** Nonhomogeneous, 3   **5.** Homogeneous, 2   **7.** Nonhomogeneous, 3

9. $y = k_1e^{7x} + k_2e^{-2x}$ **11.** $y = k_1e^{4x} + k_2e^{-2x}$ **13.** $y = k_1e^x + k_2e^{-x}$ **15.** $y = k_1 + k_2e^{3x}$

17. $y = k_1e^{3x/2} + k_2e^{5x}$ **19.** $y = k_1e^{2x} + k_2e^{x/3}$ **21.** $y = 2 + e^{4x}$ **23.** $y = e^{2x} + e^{-x}$ **25.** $y = 9e^{3x} - 5e^{5x}$

Exercises 12.2, Page 474

1. $y = e^{2x}(k_1 + k_2x)$ **3.** $y = e^{2x}(k_1 \sin x + k_2 \cos x)$ **5.** $y = e^{x/2}(k_1 + k_2x)$ **7.** $y = e^{2x}(k_1 \sin 3x + k_2 \cos 3x)$

9. $y = e^{5x}(k_1 + k_2x)$ **11.** $y = k_1 \sin 3x + k_2 \cos 3x$ **13.** $y = k_1 + k_2x$ **15.** $y = 2e^{3x}(1 - x)$

17. $y = 2 \cos 5x$ **19.** $y = e^{6x}(1 - 6x)$

Exercises 12.3, Pages 479–480

1. $y = k_1 + k_2e^{-x} - \frac{1}{2}\sin x - \frac{1}{2}\cos x$ **3.** $y = k_1e^{2x} + k_2e^{-x} - 2x + 1$ **5.** $y = e^{5x}(k_1 + k_2x) + \dfrac{x}{25} + \dfrac{2}{125}$

7. $y = k_1e^x + k_2e^{-x} - 2 - x^2$ **9.** $y = k_1 \sin 2x + k_2 \cos 2x + \dfrac{e^x}{5} - \dfrac{1}{2}$ **11.** $y = k_1e^{4x} + k_2e^{-x} - e^x$

13. $y = k_1 \sin x + k_2 \cos x + 5 - \frac{1}{8}\sin 3x$ **15.** $y = k_1e^x + k_2e^{-x} + \frac{1}{2}xe^x$ **17.** $y = k_1 \sin 2x + k_2 \cos 2x + \frac{1}{4}x \sin 2x$

19. $y = 2(e^{2x} - \cos x - 2 \sin x)$ **21.** $y = \frac{1}{2}(5 \sin x - \cos x + e^x)$

1. $x = 0.167 \cos 8\sqrt{2}\, t$ **3.** $x = c_1 e^{-3.06t} + c_2 e^{-20.9t}$ **5.** 13,000 lb   **7.** 1.25 ft   **9.** $i = k_1 e^{-138t} + k_2 e^{-362t}$
11. $i = 0.136 e^{-140t} - 0.136 e^{-360t}$

3. $\dfrac{3}{s^2 + 9}$ **5.** $\dfrac{-4}{s(s-4)}$ **7.** $\dfrac{2}{s^3}$ **9.** $\dfrac{16}{(s^2+4)^2}$ **11.** $\dfrac{-s^2 + s - 2}{(s-2)s^2}$ **13.** $\dfrac{8s}{(s^2+16)^2}$ **15.** $\dfrac{s+3}{s^2 + 6s + 34}$
17. $\dfrac{8s^2 + 24}{s^4}$ **19.** $(s^2 - 3s)\mathcal{L}(y)$ **21.** $(s^2 + s + 1)\mathcal{L}(y) - 1$ **23.** $(s^2 - 3s + 1)\mathcal{L}(y) - s + 3$
25. $(s^2 + 8s + 2)\mathcal{L}(y) - 4s - 38$ **27.** $(s^2 - 6s)\mathcal{L}(y) - 3s + 11$ **29.** $(s^2 + 8s - 3)\mathcal{L}(y) + 6s + 46$
31. 1   **33.** te^{5t} **35.** $\cos 8t$ **37.** $e^{6t} - e^{2t}$ **39.** $e^{3t} \sin 2t$ **41.** $\sin t - t \cos t$ **43.** $e^t - \sin t - \cos t$
45. $2e^{-3t} + e^{-6t} + 3$ **47.** $e^{-5t}(3 \sin 2t + \cos 2t)$

1. $y = 2e^t$ **3.** $y = e^{-3t/4}$ **5.** $6y = 31e^{7t} - e^t$ **7.** $y = \cos t$ **9.** $2y = 3 - e^{2t}$
11. $y = e^{-t}(1 + t)$ **13.** $6y = t^3 e^{2t}$ **15.** $y = (4 + 6t + \tfrac{1}{2} t^3)e^{-t}$ **17.** $5y = 3e^{-4t} + 2e^t$

1. Homogeneous; order 2   **2.** Nonhomogeneous; order 2   **3.** Nonhomogeneous; order 1   **4.** Homogeneous; order 3
5. $y = k_1 e^{-5x} + k_2 e^x$ **6.** $y = k_1 e^{3x} + k_2 e^{2x}$ **7.** $y = k_1 + k_2 e^{6x}$ **8.** $y = k_1 e^{(3/2)x} + k_2 e^{-x}$ **9.** $y = k_1 e^{3x} + k_2 x e^{3x}$
10. $y = k_1 e^{-5x} + k_2 x e^{-5x}$ **11.** $y = k_1 e^x + k_2 x e^x$ **12.** $y = k_1 e^{(1/3)x} + k_2 x e^{(1/3)x}$ **13.** $y = k_1 \sin 4x + k_2 \cos 4x$
14. $y = k_1 \sin \tfrac{5}{2} x + k_2 \cos \tfrac{5}{2} x$ **15.** $y = e^x (k_1 \sin \sqrt{2}\, x + k_2 \cos \sqrt{2}\, x)$ **16.** $y = e^{3x/2} \left(k_1 \sin \dfrac{\sqrt{23}}{2} x + k_2 \cos \dfrac{\sqrt{23}}{2} x \right)$
17. $y = k_1 e^x + k_2 e^{-2x} - \tfrac{1}{4} - \tfrac{1}{2}x$ **18.** $y = k_1 e^{3x} + k_2 x e^{3x} + \tfrac{1}{4} e^x$ **19.** $y = k_1 \sin 2x + k_2 \cos 2x + \tfrac{1}{3} \cos x$
20. $y = e^x (k_1 \sin \sqrt{2}\, x + k_2 \cos \sqrt{2}\, x) + 2e^{2x}$ **21.** $y = 4e^{2x} + 2e^{-4x}$ **22.** $y = 1 + 2e^{3x}$ **23.** $y = 3x e^{2x}$
24. $y = 8e^{-3x}(1 + 3x)$ **25.** $y = -2 \sin 2x + \cos 2x$ **26.** $y = e^{4x}(\sin 3x + 2 \cos 3x)$ **27.** $2y = e^{2x} - 2e^x + 3$
28. $y = \sin 2x + 2 \cos 2x + \tfrac{1}{3} \sin x$ **29.** $x = \tfrac{1}{2} \cos 4\sqrt{6}\, t$ **30.** $x = k_1 e^{(-32 + 8\sqrt{13})t} + k_2 e^{(-32 - 8\sqrt{13})t}$
31. $i = e^{-100t}(k_1 \sin 200t + k_2 \cos 200t)$ **32.** $i = k_1 e^{(-1000 + 500\sqrt{3})t} + k_2 e^{(-1000 - 500\sqrt{3})t}$ **33.** $\dfrac{1}{s - 6}$
34. $\dfrac{s + 2}{(s + 2)^2 + 9}$ **35.** $\dfrac{6}{s^4} + \dfrac{s}{s^2 + 1}$ **36.** $\dfrac{3}{s^2} - \dfrac{1}{s - 5}$ **37.** t **38.** $\tfrac{3}{2}(e^{2t} - 1)$ **39.** $2(e^{4t} - e^{3t})$
40. $\sin 3t$ **41.** $y = 2e^{(5/4)t}$ **42.** $y = 3 \cos 3t$ **43.** $y = \tfrac{2}{5}(1 - e^{-5t})$ **44.** $y = \dfrac{t^2 e^{-2t}}{2}$

Subject Index

633

Common Trigonometric Identities

1. $\sin \theta = \dfrac{1}{\csc \theta}$

2. $\cos \theta = \dfrac{1}{\sec \theta}$

3. $\tan \theta = \dfrac{1}{\cot \theta}$

4. $\cot \theta = \dfrac{1}{\tan \theta}$

5. $\sec \theta = \dfrac{1}{\cos \theta}$

6. $\csc \theta = \dfrac{1}{\sin \theta}$

7. $\tan \theta = \dfrac{\sin \theta}{\cos \theta}$

8. $\cot \theta = \dfrac{\cos \theta}{\sin \theta}$

9. $\sin^2 \theta + \cos^2 \theta = 1$

10. $1 + \tan^2 \theta = \sec^2 \theta$

11. $\cot^2 \theta + 1 = \csc^2 \theta$

12. $\sin (-\theta) = -\sin \theta$

13. $\cos (-\theta) = \cos \theta$

14. $\tan (-\theta) = -\tan \theta$

15. $\cot (-\theta) = -\cot \theta$

16. $\sec (-\theta) = \sec \theta$

17. $\csc (-\theta) = -\csc \theta$

18. $\sin (\theta + \phi) = \sin \theta \cos \phi + \cos \theta \sin \phi$

20. $\cos (\theta + \phi) = \cos \theta \cos \phi - \sin \theta \sin \phi$

22. $\tan (\theta + \phi) = \dfrac{\tan \theta + \tan \phi}{1 - \tan \theta \tan \phi}$

24. $\sin 2\theta = 2 \sin \theta \cos \theta$

19. $\sin (\theta - \phi) = \sin \theta \cos \phi - \cos \theta \sin \phi$

21. $\cos (\theta - \phi) = \cos \theta \cos \phi + \sin \theta \sin \phi$

23. $\tan (\theta - \phi) = \dfrac{\tan \theta - \tan \phi}{1 + \tan \theta \tan \phi}$

25. **(a)** $\cos 2\theta = \cos^2 \theta - \sin^2 \theta$
     (b) $\cos 2\theta = 2 \cos^2 \theta - 1$
     (c) $\cos 2\theta = 1 - 2 \sin^2 \theta$

26. $\tan 2\theta = \dfrac{2 \tan \theta}{1 - \tan^2 \theta}$

27. $\sin \dfrac{\theta}{2} = \pm \sqrt{\dfrac{1 - \cos \theta}{2}}$

28. $\cos \dfrac{\theta}{2} = \pm \sqrt{\dfrac{1 + \cos \theta}{2}}$

29. $\tan \dfrac{\theta}{2} = \dfrac{1 - \cos \theta}{\sin \theta}$

30. $\sin^2 \theta = \dfrac{1}{2}(1 - \cos 2\theta)$

31. $\cos^2 \theta = \dfrac{1}{2}(1 + \cos 2\theta)$

32. $\sin \theta + \sin \phi = 2 \sin \left(\dfrac{\theta + \phi}{2}\right) \cos \left(\dfrac{\theta - \phi}{2}\right)$

33. $\sin \theta - \sin \phi = 2 \cos \left(\dfrac{\theta + \phi}{2}\right) \sin \left(\dfrac{\theta - \phi}{2}\right)$

34. $\cos \theta + \cos \phi = 2 \cos \left(\dfrac{\theta + \phi}{2}\right) \cos \left(\dfrac{\theta - \phi}{2}\right)$

35. $\cos \theta - \cos \phi = -2 \sin \left(\dfrac{\theta + \phi}{2}\right) \sin \left(\dfrac{\theta - \phi}{2}\right)$

36. $\sin (\theta + \phi) + \sin (\theta - \phi) = 2 \sin \theta \cos \phi$

38. $\cos (\theta + \phi) + \cos (\theta - \phi) = 2 \cos \theta \cos \phi$

37. $\sin (\theta + \phi) - \sin (\theta - \phi) = 2 \cos \theta \sin \phi$

39. $\cos (\theta + \phi) - \cos (\theta - \phi) = -2 \sin \theta \sin \phi$